FUNGAL PATHOGENESIS IN PLANTS AND CROPS

BOOKS IN SOILS, PLANTS, AND THE ENVIRONMENT

Soil Biochemistry, Volume 1, edited by A. D. McLaren and G. H. Peterson
Soil Biochemistry, Volume 2, edited by A. D. McLaren and J. Skujiņš
Soil Biochemistry, Volume 3, edited by E. A. Paul and A. D. McLaren
Soil Biochemistry, Volume 4, edited by E. A. Paul and A. D. McLaren
Soil Biochemistry, Volume 5, edited by E. A. Paul and J. N. Ladd
Soil Biochemistry, Volume 6, edited by Jean-Marc Bollag and G. Stotzky
Soil Biochemistry, Volume 7, edited by G. Stotzky and Jean-Marc Bollag
Soil Biochemistry, Volume 8, edited by Jean-Marc Bollag and G. Stotzky
Soil Biochemistry, Volume 9, edited by G. Stotzky and Jean-Marc Bollag

Organic Chemicals in the Soil Environment, Volumes 1 and 2, edited by C. A. I. Goring and J. W. Hamaker
Humic Substances in the Environment, M. Schnitzer and S. U. Khan
Microbial Life in the Soil: An Introduction, T. Hattori
Principles of Soil Chemistry, Kim H. Tan
Soil Analysis: Instrumental Techniques and Related Procedures, edited by Keith A. Smith
Soil Reclamation Processes: Microbiological Analyses and Applications, edited by Robert L. Tate III and Donald A. Klein
Symbiotic Nitrogen Fixation Technology, edited by Gerald H. Elkan
SoilWater Interactions: Mechanisms and Applications, Shingo Iwata and Toshio Tabuchi with Benno P. Warkentin
Soil Analysis: Modern Instrumental Techniques, Second Edition, edited by Keith A. Smith
Soil Analysis: Physical Methods, edited by Keith A. Smith and Chris E. Mullins
Growth and Mineral Nutrition of Field Crops, N. K. Fageria, V. C. Baligar, and Charles Allan Jones
Semiarid Lands and Deserts: Soil Resource and Reclamation, edited by J. Skujiņš
Plant Roots: The Hidden Half, edited by Yoav Waisel, Amram Eshel, and Uzi Kafkafi
Plant Biochemical Regulators, edited by Harold W. Gausman
Maximizing Crop Yields, N. K. Fageria

Transgenic Plants: Fundamentals and Applications, edited by Andrew Hiatt
Soil Microbial Ecology: Applications in Agricultural and Environmental Management, edited by F. Blaine Metting, Jr.
Principles of Soil Chemistry: Second Edition, Kim H. Tan
Water Flow in Soils, edited by Tsuyoshi Miyazaki
Handbook of Plant and Crop Stress, edited by Mohammad Pessarakli
Genetic Improvement of Field Crops, edited by Gustavo A. Slafer
Agricultural Field Experiments: Design and Analysis, Roger G. Petersen
Environmental Soil Science, Kim H. Tan
Mechanisms of Plant Growth and Improved Productivity: Modern Approaches, edited by Amarjit S. Basra
Selenium in the Environment, edited by W. T. Frankenberger, Jr., and Sally Benson
Plant–Environment Interactions, edited by Robert E. Wilkinson
Handbook of Plant and Crop Physiology, edited by Mohammad Pessarakli
Handbook of Phytoalexin Metabolism and Action, edited by M. Daniel and R. P. Purkayastha
Soil–Water Interactions: Mechanisms and Applications, Second Edition, Revised and Expanded, Shingo Iwata, Toshio Tabuchi, and Benno P. Warkentin
Stored-Grain Ecosystems, edited by Digvir S. Jayas, Noel D. G. White, and William E. Muir
Agrochemicals from Natural Products, edited by C. R. A. Godfrey
Seed Development and Germination, edited by Jaime Kigel and Gad Galili
Nitrogen Fertilization in the Environment, edited by Peter Edward Bacon
Phytohormones in Soils: Microbial Production and Function, W. T. Frankenberger, Jr., and Muhammad Arshad
Handbook of Weed Management Systems, edited by Albert E. Smith
Soil Sampling, Preparation, and Analysis, Kim H. Tan
Soil Erosion, Conservation, and Rehabilitation, edited by Menachem Agassi
Plant Roots: The Hidden Half, Second Edition, Revised and Expanded, edited by Yoav Waisel, Amram Eshel, and Uzi Kafkafi
Photoassimilate Distribution in Plants and Crops: Source–Sink Relationships, edited by Eli Zamski and Arthur A. Schaffer
Mass Spectrometry of Soils, edited by Thomas W. Boutton and Shinichi Yamasaki
Handbook of Photosynthesis, edited by Mohammad Pessarakli
Chemical and Isotopic Groundwater Hydrology: The Applied Approach, Second Edition, Revised and Expanded, Emanuel Mazor
Fauna in Soil Ecosystems: Recycling Processes, Nutrient Fluxes, and Agricultural Production, edited by Gero Benckiser
Soil and Plant Analysis in Sustainable Agriculture and Environment, edited by Teresa Hood and J. Benton Jones, Jr.
Seeds Handbook: Biology, Production, Processing, and Storage: B. B. Desai, P. M. Kotecha, and D. K. Salunkhe

Modern Soil Microbiology, edited by J. D. van Elsas, J. T. Trevors, and E. M. H. Wellington

Growth and Mineral Nutrition of Field Crops: Second Edition, N. K. Fageria, V. C. Baligar, and Charles Allan Jones

Fungal Pathogenesis in Plants and Crops: Molecular Biology and Host Defense Mechanisms, P. Vidhyasekaran

Plant Pathogen Detection and Disease Diagnosis, P. Narayanasamy

Agricultural Systems Modeling and Simulation, edited by Robert M. Peart and R. Bruce Curry

Additional Volumes in Preparation

Plant–Microbe Interactions and Biological Control, edited by Gregory J. Boland and L. David Kuykendall

Handbook of Soil Conditioners, edited by Arthur Wallace and Richard E. Terry

Sulfur in the Environment, edited by Doug Maynard

Agricultural Biotechnology, edited by Arie Altman

FUNGAL PATHOGENESIS IN PLANTS AND CROPS

MOLECULAR BIOLOGY AND HOST DEFENSE MECHANISMS

P. VIDHYASEKARAN
Centre for Plant Protection Studies
Tamil Nadu Agricultural University
Coimbatore, India

MARCEL DEKKER, INC. NEW YORK • BASEL • HONG KONG

ISBN: 0-8247-0039-2

The publisher offers discounts on this book when ordered in bulk quantities. For more information, write to Special Sales/Professional Marketing at the address below.

This book is printed on acid-free paper.

MARCEL DEKKER, INC.
270 Madison Avenue, New York, New York 10016
http://www.dekker.com

Current printing (last digit):
10 9 8 7 6 5 4 3 2 1

PRINTED IN THE UNITED STATES OF AMERICA

Preface

Fungal pathogens cause heavy crop losses amounting to several billion dollars. Molecular biology of pathogenesis is an important field that has changed disease management strategy from chemical control to development of transgenic disease-resistant plants or "induced systemic resistant (ISR)" plants. Host plants are endowed with several potential defense mechanisms, while pathogens produce an array of pathogenic weapons to evade or suppress these defense mechanisms. The nature, timing, and spatial coordination of the actions taken by either host or pathogen are crucial in defining the result of any given interaction and thus the overall outcome of the infection. A vast literature has accumulated in this new field during this decade. In studying the available literature, I have created an entirely new concept that explains the molecular basis of fungal evasion of the host plant's defense mechanisms. This book critically discusses in great depth each of the fungal infection processes from initial contact and penetration to subsequent evasion of postpenetration defense mechanisms. This book is unique in providing research results and theories on the molecular events of fungal pathogenesis.

Molecular plant pathology has attracted the interest of molecular biologists who are using the modern tools of genetic engineering to develop transgenic plants with built-in resistance against fungal diseases. In many host–pathogen interactions, pathogens do not simply avoid the host's defense mechanisms but actually suppress them. It would be a futile exercise for genetic engineers to try to clone defense genes and develop transgenic plants to express defense genes for management of disease if these defense genes are suppressed by the potential pathogens. This book provides detailed information on molecular events leading to suppression of defense mechanisms by fungal pathogens. Hence it will be a valuable resource for researchers and students studying plant molecular biology, plant biotechnology, plant cell biology, genetic engineering, plant biochemistry, molecular plant pathology, plant physiology, applied biology, experimental botany, and other branches of biological sciences. This book will also be highly useful to students of botany and plant pathology taking courses on molecular biology of parasitism and susceptibility.

P. Vidhyasekaran

Contents

Part I

Molecular Events During Early Recognition Process

1

Pathogen Recognizes Host, Host Recognizes Pathogen

I. FUNGAL PATHOGENS RECOGNIZE THE HOST WHEN THEY COME INTO CONTACT WITH HOST SURFACE

A. Adhesion of Fungal Spores to Plant Surface

Vidhyasekaran (1988a, 1993a,b) described early molecular events during fungal pathogenesis and it appears that pathogenesis starts when fungal spore lands on surface of the plant. Molecular events arising from first fungal contact to attachment to host surface seem to play a critical role in the recognition process. Successful pathogens adhere to the host surface (Lippincott and Lippincott, 1984; Epstein et al., 1985; Beckett et al., 1990; Nicholson and Epstein, 1991; Ding et al., 1994; Mercure et al., 1994a). Conidia of several species of plant pathogenic fungi adhere to host surface prior to germination (Young and Kauss, 1984; Hamer et al., 1988; Jones and Epstein, 1989; Sela-Buurlage et al., 1991). Mutants with adhesion-deficient macroconidia of *Nectria haematococca* were found to be avirulent and spore adhesion appears to be a virulence factor (Jones and Epstein, 1990). Adhesion has been shown to be important in the pathogenesis of *Colletotrichum graminicola* (Mercure et al., 1994a,b) and *Phytophthora megasperma* f.sp. *glycinea* (Ding et al., 1994). Adhesion of the pathogen to the host surface has been shown to be important for infection, recognition, and signal transduction (Odermatt et al., 1988; Nicholson and Epstein, 1991).

B. Molecules Favoring Adhesion of Ungerminated Spores

Spores adhere to the host surface when they come into contact with it. Adhesion of ungerminated conidia has been reported for *Nectria haematococca* (Jones and Epstein, 1989; Kwon and Epstein, 1993), *Uromyces appendiculatus* (Epstein et al., 1987; Terhume and Hoch, 1993), *Uromyces viciae-fabae* (Beckett et al., 1990; Deising et al., 1992), *Botrytis cinerea* (Doss et al., 1993), *Colletotrichum musae*

(Sela-Buurlage et al., 1991), *C. lindemuthianum* (Young and Kauss, 1984), *C. graminicola* (Mercure et al., 1994a), and *Magnaporthe grisea* (Hamer et al., 1988). Adhesion of ungerminated conidia of *C. graminicola* begins immediately after contact of the conidia with a leaf (Mercure et al., 1994a).

The spores adhere to the leaf cuticle, which is a hydrophobic layer. Surface hydrophobicity appears to be important for adhesion of the conidia of several plant pathogens (Jones and Epstein, 1989; Beckett et al., 1990). Conidia of *Colletotrichum graminicola* adhered to a greater extent to the hydrophobic surface of polystyrene and they did not adhere to glass, which is a hydrophobic surface (Mercure et al., 1994b). The ungerminated conidia of *Colletotrichum musae* ((Sela-Buurlage et al., 1991), *Botrytis cinerea* (Doss et al., 1993), and *Uromyces appendiculatus* (Terhume and Hoch, 1993) require a hydrophobic surface for adhesion.

The influence of physical properties of the surface on adhesion is not well understood. Doss et al. (1993) showed that the oxidation of a polyethylene hydrophobic surface lowered the water contact angle and decreased adhesion of ungerminated *Botrytis cinerea* conidia.

The capacity of conidia of some fungi to adhere is affected by the exposure to respiration inhibitors (Young and Kauss, 1984; Jones and Epstein, 1989). Temperatures also affect conidial adhesion (Jones and Epstein, 1989; Sela-Buurlage et al., 1991). Temperature is known to affect mitochondrial respiration. Hence it is possible that adhesion depends upon active metabolism of the pathogen, but requirement of active metabolism for adhesion has not been found in some fungi. Treatment of conidia with low (4°C) or high (50°C) temperatures did not affect adhesion of *Colletotrichum graminicola* conidia. Sodium azide, a mitochondrial respiration inhibitor, also had no effect on the adhesion of *C. graminicola* conidia (Mercure et al., 1994b). Conidia that were heat-killed by autoclaving still retained some ability to adhere (Mercure et al., 1994b). Thus a preformed adhesive may exist in *C. graminicola* conidia. Even in *C. graminicola,* however, treatment of the conidia with either cycloheximide (a protein synthesis inhibitor) or brefeldin A (a glycoprotein synthesis and transport inhibitor) significantly reduced adhesion (Mercure et al., 1994b). It suggests that blocking of synthesis of a new material (protein or glycoprotein) may lead to loss of adhesion of conidia. Scanning electron microscopy revealed the presence of a film of material in the contact surface of *C. graminicola* conidia that adhered to corn (*Zea mays*) leaves (Mercure et al., 1994a). It indicates that a material is released from conidia as a result of contact with a substrate (Mercure et al., 1994b). All these observations suggest that a preformed adhesive exists on *C. graminicola* conidial surface and additional adhesive is produced upon contact of the conidium with the substrate. This conclusion is strengthened by the observations that the number of *C. graminicola* conidia that adhere to a substrate increases over time, about 10% of the population of heat-killed conidia adhered by 10 min, and this level of adhesion did not increase when conidia were incubated for longer period (Mercure et al., 1994b). Two kinds of adhesive materials have been detected in *C. graminicola* conidia. The preformed adhesive material is a protein while the secreted material is a glycoprotein (Mercure et al., 1994b). Ramadoss et al. (1985) showed that conidia of *C. graminicola* are produced in a mucilage composed primarily of a complex mixture of glycoproteins. The glycoprotein adhesive material is released from ungerminated *C. graminicola* conidia at an extremely early time in the infection process (Mercure et al., 1995).

Release of adhesive materials from spores of different pathogens has been demonstrated using electron microscopy (Sing and Bartnicki Garcia, 1975; Hamer et al., 1988) or by fluorescein-labeled lectins or antibodies (Hardham, 1985; Hardham and Suzaki, 1986; Gubler and Hardham, 1988; Freytag and Mendgen, 1991; O'Connell, 1991; Kwon and Epstein, 1993). A mucilage is released from the apex of *Magnaporthe grisea* conidia during adhesion (Hamer et al., 1988). Macroconidia of *Fusarium solani* release an extracellular material from the tip of the conidium and this material was involved in adhesion (Schuerger and Mitchell, 1993). The involvement of conidial tip mucilage in the initial adhesion of ungerminated conidia to substrate has been demonstrated for *Nectria haematococca* (Jones and Epstein, 1989; 1990).

Zoospores of *Phytophthora palmivora* secrete an adhesive material that is labeled by concanavalin A (ConA) (Sing and Bartnicki-Garcia, 1975). Secretion of ConA-positive adhesive materials from the small peripheral vesicles of encysting zoospores of *Phytophthora cinnamomi* has been reported (Hardham and Suzaki, 1986; Gubler and Hardham, 1988). The lectin-binding activity of these adhesive materials indicates the glycoprotein nature of these adhesives. The material associated with adhesion of *Uromyces viciae-fabae* urediniospores has also been identified as a glycoprotein (Clement et al., 1993a,b).

Macroconidia of the cucurbit pathogen *Nectria haematococca* did not attach to a polystyrene substratum. Within minutes after exposure to its plant host (zucchini) extract, however, the macroconidia became adhesion competent (Jones and Epstein, 1989). The conidia produced fluorescein isothiocyanate-conjugated ConA-labeled mucilage at the spore apex and adhered at the macroconidial apex. The macroconidia produced a 90 kDa glycopeptide when incubated in zucchini plant extract. Similar glycopeptide was not produced when the macroconidia was incubated in Czapek-Dox medium or water and the macroconidia did not adhere to the substratum (Table 1; Kwon and Epstein, 1993). Thus production of the 90 kDa glycopeptide was associated with the induction of adhesion competence.

The materials associated with adhesion have been identified as proteins or glycoproteins in many ungerminated spores of fungal pathogens (Sing and Bartnicki-Garcia, 1975; Hinch and Clark, 1980; Hamer et al., 1988; Gubler and Hardham, 1988; Gubler et al., 1989; Estrada-Garcia et al., 1990a; Sela-Buurlage

Table 1 Relationship between macroconidial adhesion and production of 90 kDa glycoprotein by *Nectria haematococca*

Incubation medium	Macroconidial adhesion (%)	90 kDa band present
Zucchini fruit extract	65	++
Czapek Dox	17	–
Water	1	–

++, Strong labeling; –, no labeling.
Source: Kwon and Epstein, 1993.

et al., 1991). A protein has been shown to be involved in attaching *Phytophthora cinnamomi* zoospores to the host surface and its molecular weight was over 200 kDa (Gubler and Hardham, 1988).

Some of the adhesive materials have properties of lectins in binding specific sugar moieties. Attachment of zoospores of *Pythium aphanidermatum* to root surfaces of cress, *Lepidium sativum* (Longman and Callow, 1987), attachment of *Phytophthora* to root surface slime (Hinch and Clark, 1980), and adhesion of *Phytophthora* to cells of soybean (*Glycine max*) and other plants (Hohl and Balsiger, 1986; Guggenbuhl and Hohl, 1992) have been reported due to lectins.

Cutinases and other esterases have been shown to be involved in adhesion of some fungi. Conidia of *Erysiphe graminis* f.sp. *hordei,* the barley (*Hordeum vulgare*) powdery mildew pathogen, rapidly released a liquid upon their contact with a barley leaf. The liquid contained esterase activity resembling fungal cutinase (Nicholson et al., 1988). Esterase activity was released in two stages. The first stage of release began within 2 min of the contact stimulus and the second began between 10 and 15 min after contact. The release of the liquid was completed within 30 min of the contact stimulus. The liquid film flowed off the conidium and onto the contact surface again forming a film on the contact surface (Kunoh et al., 1988, 1990; Nicholson et al., 1988). The release of exudate onto surface of barley leaves resulted in an apparent loss of structural integrity of the underlying cuticular surface (Kunoh et al., 1988; 1990). The exudate containing cutinase activity was involved in the erosion of the leaf cuticle (Pascholati et al., 1992). Cuticular erosion presents a surface that triggers recognition that initiates the infection process (Nicholson and Epstein, 1991). Such erosion may be necessary either for recognition of the host surface by the spores or for adhesion (Pascholati et al., 1992).

A possible role of cutinases in adhesion has been reported for many fungi (Purdy and Kolattukudy, 1975; Nicholson and Epstein, 1991). Cutinase and nonspecific esterases have been shown to be present on the surface of the wall of urediniospores of the rust fungus *Uromyces viciae-fabae* (Deising et al. 1992). Upon hydration, the enzymes are released from the spore wall and are located at the interface of the spore and the underlying substratum. These enzymes have been shown to function in the adhesion of the spore to the leaf surface (Deising et al., 1992). It has been suggested that fungal spore senses contact with the plant surface via the unique monomers of plant cuticle generated by the small amount of cutinase carried by the spore (Kolattukudy, 1980, Woloshuk and Kolattukudy, 1986).

Not all components found in conidial mucilage may be involved in adhesion. The conidial mucilage that has been shown to be essential for conidium survival had no effect on adhesion (Nicholson and Moraes, 1980; Nicholson et al., 1986; Leite and Nicholson, 1992). An extracellular conidial mucilage of *Fusarium solani* f.sp. *phaseoli* is not involved in adhesion (Schuerger and Mitchell, 1993; Schuerger et al., 1993). The water-soluble mucilage that surrounds *C. graminicola* conidia was removed by repetitive washing. The removal of water-soluble mucilage from the ungerminated conidia did not affect adhesion of those conidia, indicating that the conidial mucilage may not be involved in adhesion (Mercure et al., 1994b).

C. Molecules Triggering Spore Eclosion

The fungal spores undergo major structural changes after adhesion to the substrate (Caesar-TonThat and Epstein, 1991). The complex sequence of events prior to any germ tube formation is called spore eclosion (Chapela et al., 1990). Within 5 min of contact the reticulate surface of *Erysiphe graminis* f.sp. *hordei* conidium began to disappear. By 10 min, only the spinelike surface protrusions of the conidium were visible, and by 30 min globose bodies appeared on the conidium surface. These events took place prior to germination of the conidium (Kunoh et al., 1988). The release of liquid containing esterase activity from the conidium within 5 min of contact with the barley leaf has been suggested to be responsible for this eclosion (Nicholson et al., 1988; Kunoh et al., 1990). The surface of macroconidia of *Nectria haematococca* undergoes major structural changes hours before germ tube emergence (Caesar TonThat and Epstein, 1991). Ascospores of *Hypoxylon fragiforme,* a pathogen of hazel (*Coryllus avellana*), are irreversibly activated, prior to germination, within minutes of contact with a potential host (Chapela et al., 1991). Some molecules found in the host would have activated the spore eclosion. Chapela et al. (1991) suspended the ascospores of *H. fuscum,* a pathogen of beech (*Fagus sylvatica*), and *H. fragiforme* in extracts obtained from beech or hazel. *H. fragiforme* ascospores were more sensitive to beech than to hazel extracts, while *H. fuscum* showed the inverse sensitivity. Each fungus reached a higher maximal response in the extract of the corresponding host than in that from heterologous tree.

The eclosion-inducing factors in the beech extract were identified as compounds closely related to monolignol glucosides z-isoconiferin (glucosylated form of coniferyl alcohol) and z-syringin. These compounds are cell-wall-related compounds. The fungal spores ecloded only in close vicinity to even small wounds, strongly suggesting that these eclosion-inducing compounds are accumulated in inner bark layers and released in significant amounts only after wounding. In solution, *cis-trans* photoisomerization of monolignols can occur. In nature, however, E-(*trans*)-isomers are fixed as polymerized lignin in the plant cell walls, while their z-(*cis*)-counterparts accumulate unpolymerized in the bark of beech and other angiospermous trees (Morelli et al., 1986; Lewis et al., 1988). These low-molecular-weight compounds may play an important role in recognition. Host wounds may signal their presence to the potential colonizer, and initiate infection by the fungus (Chapela et al., 1991).

D. Molecules Favoring Adhesion of Germ Tubes and Appressorium

Spore germlings also adhere to the substrate (Epstein et al. 1987; Chaubal et al., 1991; Clement et al., 1993a). They may also require hydrophobic surface for adhesion. Germlings of *Uromyces appendiculatus* required a hydrophobic surface for adhesion. By using interference reflection microscopy, Terhume and Hoch (1993) demonstrated that the area of contact of germ tubes was greater on a hydrophobic than on a hydrophilic surface. Spore germlings of *Uromyces viciae-fabae* adhered in association with the release of an extracellular matrix or mucilage (Clement et al., 1993a). Mucilages produced by fungi during germination have been implicated in adhesion (Nicholson and Epstein, 1991).

Urediniospore germlings of *Uromyces appendiculatus,* the bean (*Phaseolus vulgaris*) rust fungus, grow in a specific direction that apparently increases the probability that they will encounter a stomate, their site of penetration (Dickinson, 1969; Pring, 1980). After contact with a stomatal guard cell, the germlings form a series of early infection structures such as appressorium, infection peg, and vesicle (Wynn, 1976; 1981). The germlings adhered to the substrate during this process (Epstein et al., 1985). Pronase E treatment reduced appressorium formation process and cell–substratum adhesion without affecting germination or growth (Epstein et al., 1985). The results suggest that extracellular protein may be involved in adhesion of the bean rust fungus to an inducive surface.

In a further detailed study Epstein et al. (1987) showed that pronase E treatment effectively reduced adhesion of bean rust germlings. It also reduced the germling directional growth and nuclear division (appressorium formation). Germlings treated with pronase E appeared to grow as vigorously as the controls, but were rounded, lacked visible extracellular material between the fungus and the substratum, and were not tightly adhered to the surface. Pronase E was also effective in removing germlings previously adhered to bean leaf discs. Washing with water or heat-denatured pronase E did not significantly affect germling adhesion. Bean rust germlings adhered to the sides of glass vials when incubated in stirred solutions of heat-denatured pronase E. In contrast, germlings with pronase E remained in suspension. All these observations strongly suggest the involvement of an extracellular protein in adhesion (Epstein et al., 1987). The composition of the bean rust pathogen's extracellular material was partially characterized. Six predominant extracellular peptides were detected (Epstein et al. 1987).

When a drop of suspension of conidia of *Magnaporthe grisea,* a rice (*Oryza sativa*) pathogen, was dispersed on the surface of a polycarbonate film, the conidia settled to the contact surface and some of the settled conidia were observed to adhere to the contact surface within 30 min prior to germ tube emergence. Germinated conidia with or without forming appressoria adhered firmly to the contact surface and were completely resistant to removal by rotating in water for 2 min or even overnight (Xiao et al., 1994a). Scanning electron microscopy showed abundant mucilaginous substances around germ tubes and appressoria (Xiao et al., 1994a, b). α-Glucosidase, α-mannosidase, and protease strongly inhibited appressorium formation as well as conidial adhesion. In the presence of these enzymes, germinated conidia failed to adhere to the contact surface and floated in the water droplet. The mucilage substance may be a glycoprotein since the proteolytic and glycolytic enzymes degaded the mucilage. The mucilaginous substances were observed around germ tubes and appressoria on plant also. The results suggest that the adhesion of germinated conidia on plant surfaces may be due to a glycoprotein molecule (Xiao et al., 1994a).

Adhesion of appressoria has been suggested to be due to the action of cutinases also. When conidia of *Colletotrichum graminicola,* the maize pathogen, are produced in acervuli on infected plant tissues or in culture, they are surrounded by a mucilage. The mucilage contains four cutinases (Pascholati et al., 1993). The enzyme activity in the mucilage remained in association with conidia even after they were diluted into water. Diisopropyl fluorophosphate (DIPF) inhibits the cutinase activity. DIPF-treated conidia developed normal appressoria but failed to cause disease. DIPF prevented firm adhesion of appressoria and firm adhesion was necessary for penetration. The results suggest that cutinases play a role in adhesion (Pascholati et al., 1993).

E. Molecules Favoring Adhesion of Germ Tubes and Hyphae with Mesophyll Cells of the Host

The outermost cuticle layer of the host is a hydrophobic one. After penetration of the cuticle, the fungal pathogen comes into contact with mesophyll cells. Adhesion of the hyphae to mesophyll cell walls has also been reported (Ding et al., 1994). In contrast to cuticular adhesion, the interactions between host cell wall

Table 2 Influence of different monosaccharides and Concanavalin A on colonization of soybean leaf discs by *Phytophthora megasperma* f.sp. *glycinea*

Treatment	Infection[a]
Water control	100
D-mannose	45
L-mannose	118
D-glucose	126
D-galactose	104
L-fucose	97
Concanavalin A	9

[a]Values are given as the percentage of radial growth of the pathogen into the leaf disc. The water control was set to 100%.
Source: Ding et al., 1994.

and fungal hyphae involve primarily hydrophilic surfaces (Ding et al., 1994). Lectin–ligand types of binding appear to be involved in adhesion between *Phytophthora megasperma* f.sp. *glycinea* germ tube and mesophyll cells of its soybean host (Guggenbuhl and Hohl, 1992). Mannosyl/glycosyl residues were detected on the surface of germinated cysts of the pathogen and also on the host cell wall (Hohl, 1991; Guggenbuhl and Hohl, 1992). A mannose-binding receptor has been detected on the surface of mesophyll cells (Hohl, 1991). D (+)-mannose greatly reduced lesion formation in *Phytophthora megasperma* f.sp. *glycinea*-inoculated soybean leaf disc (Table 2; Ding et al., 1994). ConA also strongly inhibited colonization (Table 2; Ding et al., 1994). With ConA there was almost no attachment of the fungal germ tubes to the wounded and exposed tissues of the leaf discs (Ding et al., 1994). D-mannose inhibited colonization by inhibiting adhesion (Guggenbuhl and Hohl, 1992; Ding et al., 1994). Without adhesion, even colonization of the exposed intercellular spaces of the host tissue did not occur (Ding et al., 1994). The putative adhesins have been identified as (manno)-glycoproteins and a 65 kDa glycoprotein appears to be instrumental in adhesion (Ding et al., 1994).

F. Possible Role of Hydrophobins in Adhesion

Hydrophobins are components of fungal cell walls that contribute to cell surface hydrophobicity. They are required for the formation of the hydrophobic rodlet layer of spores (Stringer et al., 1991; Bell-Pedersen et al., 1992; Lauter et al., 1992). The hydrophobins are also secreted (Wessels, 1993).

Hydrophobins are relatively small cysteine-rich proteins. The hydrophobins are between 96 and 157 amino acids in length, contain 8 cysteine residues, and are strongly hydrophobic (Wessels et al., 1991; Stringer and Timberlake, 1993). Another common feature is that the second and third cysteines form a doublet and are usually followed by an asparagine residue (Templeton et al., 1994).

The hydrophobic nature of the surfaces of fungal pathogens is important for adhesion of pathogens to host structures (Beever and Dempsey, 1978; Stringer et al., 1991). The external location of the rodlets means that they could mediate the initial contact in fungal interactions (Templeton et al., 1994).

MPG1, a gene involved in pathogenicity of *Magnaporthe grisea,* the rice blast fungus, has been isolated and characterized (Talbot et al., 1993). MPG1 potentially encodes a small, secreted, cysteine-rich, moderately hydrophobic protein with the characteristics of a fungal hydrophobin. MPG1 mutants show an "easily wettable" phenotype due to lack of hydrophobin. MPG1 mutants show an impaired ability to undergo appressorium formation and they had a reduced ability to cause disease symptoms. Thus hydrophobins may have a role in the elaboration of infective structures by fungi (Talbot et al., 1993). Hydrophobins may also be involved in recognition (Templeton et al., 1994).

G. Possible Binding Molecules of Host Origin

In mammalian systems, substrate adhesion molecules such as fibronectin, vitronectin, laminin, and collagen, and their receptors have been shown to be important in cell–substratum adhesion processes. These molecules have been detected in plant cells also (Sanders et al., 1991; Tronchin et al., 1991; Calderone and Braun, 1991; Axelos et al., 1993; Zhu et al., 1993, 1994). Vitronectin is a glycoprotein. Vitronectin connects the extracellular matrix with the intracellular network through a subset of plasma membrane receptors known as integrins (Vogel et al., 1993). One of the vitronectin-like proteins, PVN1, has been isolated from tobacco (*Nicotiana tabacum*) cells and binds tobacco cells to glass surfaces. This plant adhesion protein has been found localized in the cell wall of tobacco cells (Zhu et al., 1994). The vitronectin-like proteins have been implicated in attachment of bacterial pathogens to plant cells (Wagner and Mathysse, 1992). Similar adhesion molecules may also be involved in fungal adhesion. Further studies are needed to assess the role of the host-binding molecules in binding fungal spores.

H. What is the Importance of Adhesion in Fungal Pathogenesis?

Adhesion of fungi to host surfaces is not a specific process. The fungi adhere to surface of their host tissues as well as to their nonhost. Even host substrates may not be needed for adhesion. They may adhere to any hard substrate. Host tissue extracts may not be needed for adhesion of many fungi. When the spores are suspended even in water, adhesion may occur. Fungal spores adhere to hydrophobic surfaces and the outermost layer of the host's surface is a hydrophobic cuticle layer. Some of the fungal spores, however, adhere to hydrophobic as well as hydrophilic surfaces (Bartnicki-Garcia and Sing, 1987; Hamer et al., 1988). Both preformed and induced adhesive may be present in the fungal spores.

All these observations suggest that adhesion may simply aid the spores in adhering to the host surface without any displacement from the leaf surface (Mercure et al., 1994a). Convincing evidence has now been obtained to show that adhesion may be important for chemical communication (signal transduction) between the pathogen and host (Xiao et al., 1994a). Glycoprotein molecule has been reported to be responsible for adhesion of *Magnaporthe grisea* germ tubes and appressoria (Xiao et al., 1994a). ConA and WGA, the lectins with sugar-binding specificity, suppressed appressorium formation, but they did not affect adhesion of germ tubes. In the scanning electron microscope, conidia suppressed in appressorium formation by ConA were observed to have extensive growth of germ tubes with abundant mucilage-like materials bound to the germ tubes. Treatment of germinating conidia with FITC-conjugated ConA yielded strong, specific labeling around germ tubes and appressoria. ConA inhibition of appressorium was blocked by the potential ConA competitors, methyl α-D-mannoside, methyl-α-D-glucoside, and D-mannose. These results suggest that ConA suppressed appressorium formation by binding to the adhesive glycoprotein, specifically blocking the sensing and transmission of information about appressorium induction (Xiao et al., 1994a).

Appressorial formation is an active process and is a prerequisite for invasion of host plants by several fungi (Emmett and Prabery, 1976; Ding et al., 1994). A relation between formation of infection structures and protein synthesis has been reported for pathogenic fungi (Miehle and Lukezic, 1972; Staples and Yaniv, 1976; Staples et al., 1976; Lovett, 1976; Furusawa et al., 1977; Wessels, 1993; Ding et al., 1994). Formation of infection structures in rust uredospores was blocked by inhibitors of protein and RNA synthesis (Dunkle, 1969; Ramakrishnan and Staples, 1970). The polypeptide with a molecular mass of 95 kDa was found associated with function of appressoria in *Colletotrichum lagenarium* (Suzuki et al. 1981), although the polypeptide was not associated with formation of appressoria.

Suppression of function of appressoria (and not formation of appressoria) has been reported due to suppression of adhesion of appressoria to the host surface by DIPF, an inhibitor of cutinase (Pascholati et al., 1993).

Without adhesion, there was no infection by *Phytophthora megasperma* f.sp. *glycinea* in soybean (Ding et al., 1994). Materials released from conidia of *Erysiphe graminis* upon their contact with a barley leaf are essential to the success of the infection process (Kunoh et al., 1988b; Nicholson et al., 1988,

1993). Thus adhesion may be necessary to elicit information to activate entire infection process.

Besides the adhesive molecules, several other molecules may also be released from the adhering spores. Proline-rich proteins found in the mucilage of *Colletotrichum graminicola* bind to and detoxify fungitoxic phenols produced by the plant within lesions (Nicholson et al., 1986). By detoxification of phenolic compounds, these proteins may ensure that conidia are not inhibited or killed as they are dispersed within the acervulus to potentially new infection sites (Nicholson et al., 1986). This is necessary since conidia must pass over necrotic tissue, which releases toxic phenols upon contact with water (Nicholson et al., 1989). High-molecular-weight glycoproteins found in the mucilage of *C. graminicola* conidia are responsible for the antidesiccant property of the mucilage, which maintains conidium viability over long periods of time even at low relative humidity (Nicholson and Moraes, 1980). The mucilage also contains several enzymes (Nicholson and Epstein, 1991).

Physical contact of the spores with host surface releases elicitors and suppressors from the fungal cell wall; these molecules have been shown to determine the ability of the pathogen to infect a particular host. These molecules activate or suppress the host's defense mechanisms.

II. PLANT RECOGNIZES PATHOGENS WHEN PHYSICAL CONTACT BETWEEN THEM OCCURS

A. Elicitors

Host cells are endowed with the ability to sense or recognize the presence of microorganisms, irrespective of whether they are compatible or incompatible pathogens, and produce defense chemicals to suppress them. Elicitors are the molecules that elicit defense mechanisms in plants (Scheel and Parker, 1990). Most of the fungal pathogens, whether compatible or not, are known to have elicitors particularly in their cell walls. It is believed that when fungal cell wall comes in contact with plant cell, elicitors are released and defense mechanism is activated.

B. Structure of Fungal Cell Wall

The fungal cell wall is made up of 80–90% polysaccharides with most of the remainder consisting of protein and lipid. The principal wall component of *Phytophthora* and *Pythium* is glucan(s). Glucose is the principal monomer and small amounts of mannose are also detected in oomycete cell walls. Oomycetes do not have any appreciable amount of chitin but a small amount of hexosamine (usually less than 1%) has been detected (Bartnicki-Garcia, 1968). Chitin and glucan are commonly present in mycelial cell walls of Ascomycetes, Basidiomycetes, and Deuteromycetes. The linkage most commonly reported in the glucans is β-1,3. Presence of minor sugars in the cell walls has been reported. Ascomycetes have galactose and galactosamine in their walls, while lacking xylose and fucose. Basidiomycetes have xylose and fucose, lacking galactose and galactosamine. Mannose occurs in both classes. The fungal cell walls have glycoprotein complexes. Lipid is found firmly bound to the cell wall of fungi in many cases (Bartnicki-Garcia, 1968).

C. Nature of Elicitors

1. Glycoproteins

Elicitors isolated from several pathogens have been identified as glycoproteins. Elicitors from *Cladosporium fulvum,* the leaf mold pathogen of tomato (*Lycopersicon esculentum*), have been identified as glycoproteins. The carbohydrate moiety of the glycoprotein has a mannose/galactose ratio of 1.21:1 and contains traces of glucose. The protein moiety is rich in alanine, aspartic acid/asparagine, glutamic acid/glutamine, proline, serine, and theronine (DeWit and Roseboom, 1980; DeWit and Kodde, 1981a, b). There was a strong positive correlation between the mannose and galactose content of the peptidogalactoglucomannan and its phytoalexin-eliciting activity since the glucoprotein, which contained mainly glucose, was not a very active elicitor. Mild acid or alkaline hydrolysis of the peptidogalactoglucomannan, removing its galacto-furanosyl residues and its mannose-containing oligomer side chains, respectively, also led to a significant inhibition of eliciting activity (DeWit and Kodde, 1981b).

Three elicitors have been purified from the cell wall and culture filtrate of *Colletotrichum lagenarium,* a pathogen of melon (*Cucumis melo*) (Toppan and Esquerre-Tugaye, 1984). All three elicitors were glycoproteins containing amino acid, sugar, and phosphate residues. Elicitor 1 was richer in aspartic and glutamic residues, elicitor 2 in the basic residues lysine and arginine, and elicitor 3 in serine, threonine, and glycine. The three elicitors also contained the same sugar residues: mannose, galactose, and glucose. Mannose was the most abundant in elicitor 1, rhamnose was detected in elicitor 2, and glucosamine in elicitor 3 (Table 3). The elicitor activity appears to remain in the carbohydrate moiety rather than in protein moiety (Toppan and Esquerre-Tugaye, 1984).

A crude extract from the germ tube wall of uredosporelings as well as mycelium of *Puccinia graminis* f.sp. *tritici* elicited lignification (Moreschbacher et al., 1986a, b; Kogel et al., 1988). The elicitor was purified and identified as a glycoprotein with a molecular weight higher than 130,000. The predominant sugars in the polysaccharide moieties of the fungal cell wall glycoproteins were mannose (50%), galactose (47%), and glucose (3%). The carbohydrate and protein contents were in the ratio of 9:1. Digestion with pronase or trypsin had no effect on elicitor activity, but complete loss of activity was observed after

Table 3 Sugar and amino acid composition of elicitors of *Colletotrichum lagenarium*

Composition	Elicitor 1	Elicitor 2	Elicitor 3
Major amino acids	% of total amino acids		
Aspartic acid	12.6	10.2	9.9
Glutamic acid	12.3	11.6	11.7
Lysine	2.8	7.0	5.2
Arginine	2.3	4.3	2.2
Serine	6.7	7.0	9.2
Threonine	5.9	6.2	7.3
Glycine	8.7	9.8	10.7
Sugars	% of total sugar		
Mannose	49.4	95.9	32.5
Galactose	39.5	Trace	35.6
Glucose	8.9	4.0	28.2
Rhamnose	2.2	0.0	0.0
Glucosamine	0.0	0.0	3.8
Phosphate (μg/g)	9.0	3.1	315.6

Source: Toppan and Esquerre-Tugaye, 1984.

periodate treatment (Table 4; Kogel et al., 1988). The results suggest that the carbohydrate moiety contains the active site of the elicitor.

In the glycoprotein elicitor of *P. graminis* f.sp. *tritici,* two types of glyco-moieties were linked to the protein core (Beissmann and Reisener, 1990). A mannan residue was linked N-glycosidically to asparagine and an α-galactan residue was O-glycosidically linked to serine or threonine (Beissmann and

Table 4 Induction of activities of phenylalanine ammonialyase on wheat leaves after treatment with chemically or enzymatically modified elicitor of *Puccinia graminis* f.sp. *tritici*

Treatment of elicitor	% Phenylalanine ammonialyase induction
Untreated	100
Pronase	93
Trypsin	86
Sodium periodate	0

Source: Kogel et al., 1988.

Reisener, 1990). The mannan-residue could be removed from the protein core by endo-β-N-acetylglucosaminidase H (Endo H) cleavage without loss of elicitor activity, but complete loss of Concanavalin A affinity suggested that mannan residue is not important for the elicitor activity (Kogel and Beissmann, 1992).

The intercellular washing fluid (IWF) obtained from wheat (*Triticum aestivum*) leaves inoculated with *Puccinia graminis* f.sp. *tritici* showed elicitor activity (Beissmann et al., 1992). Elicitor active glycoproteins were isolated from IWF by affinity chromatography using Concanavalin A–sepharose. Elicitor-active material was found in both the concanavalin-A binding (Fraction 2,3 and 4) and nonbinding fraction (fraction 5), indicating that elicitor activity does not rely on terminal α-mannose or α-glucose residues. The elicitor fraction that lacked the mannan residue (fraction 5) appears to contain the galacto-residues of the glycoprotein elicitor. The elicitor activity in fraction 5 may reflect cleavage of the intact 67 kDa IWF glycoprotein during pathogenesis, resulting in a separation of elicitor-active and concanavalin A-binding structures (Beissmann et al., 1992).

An elicitor has been isolated from *Pyricularia oryzae* (Schaffrath et al., 1995). The elicitor is a glycoprotein with a molecular weight of 15.6 kDa. The elicitor activity was not lost when the elicitor was treated with pronase, but the activity was completely abolished when it was treated with sodium periodate, indicating that carbohydrate is the active moiety (Schaffrath et al., 1995). The carbohydrate consists mainly of mannose with some glucose and galactose (Schaffrath et al., 1995).

The elicitor from *Uromyces phaseoli*, the bean rust fungus, has been found to contain carbohydrates and proteins. The carbohydrate components of the elicitor were composed of glucose. The protein moiety of the elicitor was composed of 16 amino acids of which serine, glycine, and alanine dominated. The molecular weight of the protein moiety was about 50,000 (Humme et al., 1981). Both high- and low-molecular-weight elicitors have been isolated from *Rhizopus stolonifer*, the pathogen of castor bean (*Ricinus communis*). The high-molecular-weight elicitor was a glycoprotein and the molecular weight was about 30,000 (Stekoll and West, 1978).

Many elicitor fractions have been identified in culture filtrates of *Colletotrichum lindemuthianum* (Hamdan and Dixon, 1987). All fractions were predominantly polysaccharides, but they contained about 5% protein also. Galactose and mannose were the major monosaccharide components after acid hydrolysis (Table 5; Hamdan and Dixon, 1987).

Several studies have indicated that elicitor preparations from *C. lindemuthianum* cell walls and culture filtrates contain high proportions of mannose and galactose (Anderson-Prouty and Albersheim, 1975; Anderson, 1978; Dixon et al., 1981; Whitehead et al., 1982; Dixon, 1986). Although these molecules were related structurally, they were heterogeneous with different molecular weights

Table 5 Composition of *Colletotrichum lindemuthianum* elicitor fractions

Elicitor fraction	Monosaccharide composition (% composition based on total sugars) Galactose	Glucose	Mannose	Rhamnose	Ribose	Protein (% w/w)
1	49	7	35	9	+	4.5
2	39	5	50	6	+	2.8
3	20	6	70	4	+	6.5
4	43	8	42	7	+	0.0
5	47	7	39	7	+	0.0

+, Trace.
Source: Hamdan and Dixon, 1987.

(Whitehead et al., 1982). Protein has been shown not to be necessary for elicitor activity (Hamdan and Dixon, 1986), although elicitor-active molecules may be attached to protein (Anderson, 1978).

In some glycoprotein elicitors, only protein moiety has been found to be the active fraction. Glycoproteins extracted from isolted cell walls and culture filtrates of *Phytophthora megasperma* f.sp. *glycina* (Pmg) elicited phytoalexin in soybean hypocotyls (Keen, 1971; Keen et al. 1972; Keen and Legrand, 1980). The glycoprotein was purified and its molecular weight was 42 kDa. Proteinase digestion and deglycosylation treatments of the pure glycoprotein were performed to determine which portion was responsible for elicitor activity. Activity of the 42 kDa glycoprotein was unaffected by autoclaving the elicitor. Treatment with proteinases, pronase, and trypsin destroyed the activity in parsley (*Petroselinum crispum*) cells, whereas deglycosylation had no effect. The results suggest that the elicitor activity is conferred by the protein portion (Parker et al., 1991).

Two endoglycanases were used to test if the Pmg elicitor activity is due to N-linked glycopeptides (Basse and Boller, 1992). Endo-β-N-acetylglucosaminidase H (endo H) hydrolyzes N-linked glycans between the two 2-acetamido-2-deoxyglucopyranoside (GlcNAc) residues that are linked to the peptide, releasing oligosaccharides with one GlcNAc residue at its reducing end. N-glycanase hydrolyzes various classes of N-linked oligosaccharides between asparagine and GlcNAc and produces oligosaccharide fragments with two GlcNAc residues at the reducing end. Both endoH and N-glycanase did not affect the Pmg-elicitor (Basse and Boller, 1992) indicating that the union of carbohydrate and the peptide part is not important for elicitor activity. Periodate is known to degrade glycosyl residues with adjacent unsubstituted hydroxyl groups and, therefore is expected to attack terminal or 1,6-or 1,4-linked glycosyl residues. Periodate treatment did not affect Pmg-elicitor activity (Basse and Boller, 1992). Pmg-elicitor was completely inactivated by boiling in 20 mM β-mercaptoethanol, a treatment known to reduce S-S bridges in peptides. It indicates that S-S bridges are involved in the activity of Pmg-elicitor (Basse and Boller, 1992). These results show that the elicitor-active component of crude Pmg-elicitor consists of polypeptides without essential carbohydrate parts.

2. Carbohydrates

Two different elicitor fractions have been identified in *Colletotrichum graminicola,* the pathogen of corn. One of them is a carbohydrate (Yamaoka et al., 1990). One of the elicitors of *Phytophthora megasperma* f.sp. *glycinea* has been reported to be β-1,3-glucan (Ayers et al., 1976a, b, c; Ebel et al., 1976; Valent and Albersheim, 1977; Albersheim and Valent, 1978; Ebel, 1981; Yoshikawa et al., 1983; Schmidt and Ebel, 1987; Mayer and Ziegler, 1988; Graham and Graham, 1991). β-1,3-Glucan fragments isolated from cell walls of the fungus elicited phytoalexins in soybean (Sharp et al., 1984a, b, c). Keen et al. (1983) purified various elicitors from cell walls of *P. megasperma* f.sp. *glycinea.* They belong to three unique classes of carbohydrates: β-1,3-glucans, glucomannans, and glycopeptides. Glucomannans were the most active elicitors and were approximately 10 times more active than β-1,3-glucan elicitor (Table 6; Keen et al., 1983).

A branched β-1,3-, β-1,6-linked heptaglucan of defined structure from *P. megasperma* f.sp. *glycinea* has been shown to be the smallest fragment with high elicitor activity in soybean (Sharp et al., 1984a, b c: Parker et al., 1988). The chemical structure of this elicitor is hexa-β-glucosyl glucitol (Sharp et al., 1984a). A related branched β-1,3-glucan with 30 glucose residues, mycolaminaran, has also been shown to have elicitor activity, but only at higher concentrations than reported for the glucan (Keen et al., 1983; Yoshikawa et al., 1983).

Chitin, a linear β(1-4)-linked polymer of N-acetyl-D-glucosamine, and chitosan, a polymer of β-1,4-glucosamine, are common in the cell walls of many fungi (Bartnicki-Garcia, 1970). Chitosan and soluble derivatives of chitin are elicitors of β-1,3-glucanase and chitinase in pea (*Pisum sativum*) (Mauch et al.,

Table 6 Elicitor activity of various fractions from cell walls of *Phytophthora megasperma*

Elicitor fraction	Concentration for half-maximal activity (μg/ml)
Glucomannan	0.3
β-1,3-glucan	2.0
Glycopeptides	5.2
Mycolaminaran	43.0

Source: Keen et al., 1983.

1984), phytoalexin in castor bean (West, 1981; Walker-Simmons et al., 1984), callose synthesis in soybean (Kauss, 1984), terpenoids in lodgepole pine (*Pinus contorti* var.*latifolia*) (Miller et al., 1986), and proteinase inhibitors in tomato (Walker-Simmons et al., 1984; Walker-Simmons and Ryan, 1984, 1985; and alfalfa (*Medicago sativa*) (Brown and Ryan, 1984). Chitin preparation from *Fusarium moniliforme* elicited chitinase activity in rice cells (Ren and West, 1992). The chitosan heptamer is highly active in inducing pisatin synthesis in peas, whereas monomers and dimers are inactive (Kendra and Hadwiger, 1984).

3. Fatty Acids

Early reports on the elicitors from *Phytophthora infestans,* the potato (*Solanum tuberosum*) late blight pathogen, indicated that the elicitors may be carbohydrates (Varns et al., 1971). However, Lisker and Kuc (1977) showed that several β-glucans from *P. infestans* were relatively poor elicitors. An ether-extractable mycelial elicitor was insensitive to various glucanase treatments, suggesting that factors other than carbohydrate were involved in elicitor activity (Kuc and Lisker, 1978). Lipid was an active component of the elicitor (Kurantz and Zacharius, 1981). The highly active elicitors were identified as arachidonic and eicosapentaenoic acids (Bostock et al., 1982). The eliciting activity of these fatty acids was greatly enhanced by a carbohydrate from *P. infestans* (Kurantz and Osman, 1983). Kurantz and Zacharius (1981) also observed that lipid fraction alone lacked the elicitor activity and glucans alone were inactive as elicitors (Preisig and Kuc, 1985; Bostock et al., 1986).

Poly and oligosaccharides isolated from mycelium of *P. infestans* enhanced the elicitation of phytoalexins by arachidonic acid in potato tubers. The saccharides were partially characterized as branched chain β-linked glucans. Of various acids and sugars tested, only arachidonic acid elicited accumulation of the phytoalexins (Maniara et al., 1984). Bryan et al. (1985) observed that the most active fraction was composed of carbohydrate, protein, and lipid. A heat-released preparation containing very low levels of eicosapentaenoic, arachidonic, and dihomo-y-linolenic acids, was more active than a lipid extract of the mycelium. The bulk of the elicitor activity of the mycelium remained in the wall residue after extraction, which contained low level of lipid. The results suggested that the most active elicitor in the mycelium may be a lipoglycoprotein. Chalova et al. (1977a, b) also reported that the elicitor may be a lipoglycoprotein. Another report suggests that there may be elicitors of phytoalexins devoid of arachidonic acid and eicosapentaenoic acid (Keenan et al., 1985). Significance of the fatty acids in elicitation of phytoalexins in potato has also been questioned. The concentration of arachidonic or eicosapentaenoic acid required for phytoalexin accumulation in potato tubers is approximately 330 nmol or more, which can be obtained from approximately 5 mg lypophilized mycelium (Bostock et al., 1982). It is unlikely that such a high concentration of elicitors will be available at the infection site. The contribution of β-glucans appears to be important in this regard for their marked enhancement of fatty acid elicitor activity (Maniara et al., 1984; Preisig and Kuc, 1985). The role of the fatty acids is further complicated by the diversity of fractions from *P. infestans* mycelium capable of eliciting the response (Bostock et al., 1981).

However, Creamer and Bostock (1988) provided convincing evidence that eicosapolyenoic fatty acids may contribute more to the phytoalexin elicitor activities. They showed that eicosapentaenoic acid and arachidonic acid were abundant in sporangia, zoospores, and cystospores of *P. infestans* (Table 7; Creamer and Bostock, 1988).

Bromine treatment completely eliminated the unsaturated fatty acids in lyophilized sporangia, zoospore, and cystospore tissue and abolished their sesquiterpenoid phytoalexin elicitor activity. Addition of brominated spore tissue to pure arachidonic acid significantly enhanced the phytoalexins levels (Table 8; Creamer and Bostock, 1988).

The results indicate that the eicosapolyenoic acid lipids play an important role in elicitation of defense chemicals. Besides arachidonic and eicosapentaenoic acids, another fatty acid, homo-y-linolenic acid (*cis*-8,11,14-eicosatrienoic acid) is also present in small amounts in lipids of *P. infestans* and has significant elicitor activity in the presence of mycelial glucans (Kurantz and Osman, 1983; Bryan et al., 1985; Preisig and Kuc, 1985; Creamer and Bostock, 1986).

Nonanoic and linoleic acids are also present in mycelial cell walls of *P. infestans*. Nonanoic acid elicited phytoalexin only to a certain extent, but it caused necrosis almost similar to arachidonic acid. Linoleic acid did not induce phytoalexins or necrosis (Table 9; Maina et al., 1984).

4. Peptides

Some of the elicitor molecules have been identified as peptides. Yamaoka et al. (1990) showed that elicitor activity was present in a peptide as well as in a carbohydrate component extracted from conidia of *Colletorichum graminicola.*

Table 7 Arachidonic and eicosapentaenoic acid concentrations (μg/mg dry weight) in different spore types of *Phytophthora infestans*

Spore type	Arachidonic acid	Eicosapentaenoic acid
Sporangia	2.7	31.4
Zoospores	2.0	18.5
Cystospores	1.4	14.4

Source: Creamer and Bostock, 1988.

Table 8 Phytoalexin elicitor activities of killed spores of *Phytophthora infestans* before and after treatment with bromine

Treatment	Rishitin + lubimin (μg/g fresh weight)
Sporangia	520
Sporangia + bromine	0
Cystospores	258
Cystopores + bromine	0
Zoospores	188
Zoospores + bromine	0
Arachidonic acid	57
Cystospores + bromine + arachidonic acid	161
Zoospores + bromine + arachidonic acid	141
Water	0

Source: Creamer and Bostock, 1988.

Fractions with elicitor activity were separated by Sephadex G50 column choromatography. Elicitor-active fractions eluted in two peaks that coincided with the elution of protein and carbohydrate components of the crude preparation. The elicitor activity in the protein- and carbohydrate-containing fractions elicited almost similar response from mesocotyls relative to their accumulation of phytoalexins. Both elicitor preparations caused accumulation of the same phytoalexins in sorghum (*Sorghum vulgare*), but the quality of each phytoalexin that accumulated in the tissue varied somewhat among the treatments.

Several *Phytophthora* spp. produce a novel family of protein elicitors called elicitins (Huet and Pernollet, 1989; Ricci et al., 1992). The elicitins are holoproteins with molecular masses of 10 kDa (Huet et al., 1992; Nespoulous et al., 1992). The holoproteins are devoid of side chain modification (Terce-Laforgue et al., 1992). When applied to tobacco plants, elicitins induce systemic remote leaf

Table 9 Visible necrosis and phytoalexin content of potato tuber discs incubated with various fatty acids

Fatty acid	Necrosis	Phytoalexin (μg/g fresh weight)
Arachidonic acid	++++	55.1
Linoleic acid	+	0.0
Nonanoic acid	+++	5.4
Buffer (check)	+	0.0

+ to ++++ indicate increased intensity of necrosis.
Source: Maina et al., 1984.

necrosis after migration toward the leaf tissue from the inoculation site (Zanetti et al., 1992). They induce the accumulation of pathogenesis-related proteins (Bonnet et al., 1986).

Elicitins from six *Phytophthora* species have been purified. The complete 98 amino acid sequences of these elicitins are known (Huet and Pernollet, 1989; Ricci et al., 1989; Huet et al., 1992; Nespoulous et al., 1992). Based on the complete sequence, amino acid composition, isoelectric point, and hydropathy index data, the elicitins have been classified into two classes. The α class is with a valyl residue at position 13 and an acidic isoelectric point; the β class has a hydrophilic residue at position 13 and a basic isoelectric point (Table 10; Pernollet et al., 1993; Huet et al., 1994). *P. crytopgea, P. cinnamomi, P. drechsleri,* and *P. megasperma* var. *megasperma* produce several elicitin isoforms of both α and β classes (Huet et al., 1992, Pernollet et al., 1993).

Cryptogein, the elicitor from *P. cryptogea,* is synthesized as a preprotein with a signal peptide removed cotranslationally before the secretion. It accumulates in the mycelium in its mature form (Terce-Laforgue et al., 1992). It elicits the production of ethylene and phytoalexins in tobacco cell suspension cultures (Blein et al., 1991). The protein elicitor migrates for long distances in tobacco plants (Devergne et al., 1992).

A necrosis-inducing peptide called NIP1 was purified from the culture filtrate of *Rhynchosporium secalis,* the barley pathogen. The peptide strongly induced accumulation of PRHv-1, a thaumatin-like pathogenesis-related protein

Table 10 Properties of elicitins produced by *Phytophthora* spp.

Phytophthora spp.	Elicitin	PI	13th residue
Alpha class			
P. cactorum	Cacto	4.0	Valine
P. capsici	Capsicein	3.5	Valine
P. cryptogea	Cryptogein α	3.6	Valine
P. citrophthora	Citro	3.5	Valine
P. megasperma var. *megasperma*	MgM α	3.5	Valine
P. drechsleri	Dre α	4.6	Valine
P. cinnamomi	Cinnamomin α	4.4	Valine
P. infestans	Infestin	3.9	Valine
Beta class			
P. megasperma var. *megasperma*	MgM β	8.9	Lysine
P. drechsleri	Dre β	9.6	Threonine
P. cryptogea	Cryptogein β	9.8	Lysine
P. cinnamomi	Cinnamomin β	8.9	Lysine

Source: Pernollet et al., 1993; Huet et al., 1994.

in resistant barley varieties. In susceptible varieties, similar induction was not observed. Protease treatment of NIP1 completely destroyed elicitor activity, suggesting that the elicitor activity resides in the peptide (Hahn et al., 1993).

A peptide elicitor has been isolated from apoplastic fluids of compatible interactions between *Cladosporium fulvum* and tomato, which induced necrosis. The necrosis-inducing peptide was detected in apoplastic fluids from uninoculated plants or from incompatible interactions. The necrosis inducing peptide was detected 8–10 days after inoculation of the tomato plants, coinciding with the time at which significant fungal growth was observed. The peptide was purified and contained 27 amino acids. The first amino acid in the sequence was tyrosine and contained 6 cysteine residues (Schottens-Toma and DeWit, 1988).

A proteinaceous elicitor has been isolated from *Fusarium oxysporum* f.sp. *pisi* and *Macrophomina phaseolina* (Dean et al., 1984). Similar elicitor has been isolated from a saprophytic fungus, *Trichoderma viride* (Dean et al., 1989; Dean and Anderson, 1990). The elicitor has been identified as endoxylanase (Dean and Anderson, 1990). The elicitor isolated from *T. viride* has been purified and identified as a 22 kDa protein. The amino acid composition of the 22 kDa polypeptide is enriched by glycine, serine, threonine, tyrosine, and tryptophan, but depleted in alanine, leucine, glutamine, and lysine. The protein lacks sulfur-containing amino acids. The protein is glycosylated and synthesized as a 25 kDa precursor protein that is processed to 22 kDa during secretion (Dean and Anderson, 1990). The elicitor induces ethylene (Dean and Anderson, 1990), PR proteins (Lotan and Fluhr, 1990), phytoalexin production (Farmer and Helgeson, 1987), tissue necrosis (Bailey et al., 1990), lipid peroxidation (Ishii, 1988), electrolyte leakage (Bailey et al., 1990), and cell death (Bucheli et al., 1990).

5. Other Types of Elicitors

An elicitor has been isolated from *Uromyces appendiculatus* that, when treated with protease or periodate, did not lose activity on bean leaves. It suggests that the elicitor-active components are not proteins or carbohydrates (Ryerson and Heath, 1992). Elicitation also was apparently not due to the presence of chitin or deacetylated chitin (chitosan), since particles of these compounds were found not to be associated with a significant number of plant cell wall responses, and no responses were elicited by soluble chitin tetramer. The elicitor may not be β-1,3-glucan: injection of laminarin prior to fungal inoculation did not alter the normal development of the fungus. The treatment with laminarinase (β-1,3-glucanase) and the activity of the endoglucanase on laminarin also failed to produce an elicitor. Prior treatment of wall fragments of either fungus with laminarinase had no effect of their subsequent elicitation of cell wall autofluorescence and wall refractivity. All these data suggest that these wall responses are not elicited by components released by plant β-glucanases from the fungal cell wall. Thus the actual nature of the elicitor is unknown, but it may be different from already known elicitors (Ryerson and Heath, 1992).

Victorin is the host-specific toxin produced by *Helminthosporium victoriae*. Treatment of oats (*Avena sativa*) susceptible to *H. victoriae* with low concentrations of victorin elicits production of the oat phytoalexin avenalumin, whereas treatment with higher concentrations greatly reduces the accumulation of avenalumin (Mayama et al., 1986). Victorin induced the phytoalexin avenalumin in oat varieties resistant to the crown rust pathogen *Puccinia coronata* f.sp. *avenae* (Mayama et al., 1995). It suggests that a toxin may also act as elicitor.

D. Several Kinds of Elicitors are Produced by a Single Pathogen

It appears that almost all kinds of components of fungal cell walls act as elicitors. Several kinds of elicitors have been isolated and characterized from a single pathogen. *Phytophthora infestans,* the potato late blight pathogen, has been shown to contain arachidonic acid, eicosapentaenoic acid, homo-y-linolenic acid, nonanoic acid, linoleic acid, β-glucan, and infestin (an elicitin) (Varns et al., 1971; Kuc and Lisker, 1978; Maina et al. 1984; Bryan et al., 1985; Bostock et al., 1986; Ricci et al., 1989). *Phytophthora megasperma* f.sp. *glycinea,* the soybean pathogen, contains glycoprotein, β-1,3-glucan, glucomannan, mycolaminarin, hexa-β-glucosyl glucitol, and alpha and beta classes of elicitins as elicitors (Keen, 1971; Ayers et al., 1976a; Ricci et al., 1989; Basse and Boller, 1972). *Cladosporium fulvum,* the tomato pathogen, contains a glycoprotein and a peptide as elicitors (DeWit and Roseboom, 1980; Schottens-Toma and DeWit, 1988). *Colletotrichum graminicola,* the corn pathogen, is known to contain two kinds of elicitors: one is a carbohydrate and another is a peptide (Yamaoka et al., 1990). Several elicitor fractions of glycoprotein in nature with different molecular weights have been isolated from cell walls of *Colletotrichum lindemuthianum,* the bean pathogen (Dixon, 1986; Hamdan and Dixon, 1987). Several chitin, chitosan, and β-1,3-glucan molecules have been detected in cell walls of several fungal pathogens and they also act as elicitors (Bartnicki-Garcia, 1970; Kauss, 1984, Ren and West, 1992). Some of the toxins produced by the pathogens are also known to act as elicitors (Mayama et al., 1986; 1995). In general, elicitor molecules are commonly present in fungal cell walls. These elicitor molecules may activate several kinds of defense mechanisms of the host.

III. ELICITOR MOLECULE SIGNALS INDUCTION OF VARIOUS DEFENSE MECHANISMS OF PLANTS

Elicitors from fungal cell wall signal induction of defense mechanisms in different ways in plants. Application of chitosan, a common elicitor detected in cell walls of many fungal pathogens, induced cell wall appositions such as papillae (Table 11; Benhamou and Theriault, 1992) and occlusion of xylem with a coating material in tomato roots inoculated with *Fusarium oxysporum* f.sp. *radicis-lycopersici* (Benhamou and Theriault, 1992). In addition, the accumulation of amorphous deposits infused with phenolics was observed in most intercellular spaces and some host cells. These deposits intefered with the walls of invading hyphae causing severe alterations. These observations suggest that the plant has been signalled to produce walling-off structures for halting fungal invasion by the elicitor molecule (Benhamou and Theriault, 1992). The newly formed barriers are important determinants of resistance (Beckman, 1987; Kuc, 1987).

Elicitors isolated from bean rust fungus (*Uromyces appendiculatus*) elicited autofluorescence (due to phenolic compounds) and an increase in cell wall refractivity (due to silica deposition) in plant cells when they were injected into bean leaves. These two responses are the important plant defense mechanisms in bean (Ryerson and Heath, 1992).

Arachidonic acid, the elicitor isolated from *Phytophthora infestans,* suppresses the synthesis of sterol derivatives while increasing the synthesis of highly fungitoxic phytoalexins in potato (Bostock et al., 1982, 1986). These changes are associated with changes in the activities of 3-hydroxy-3-methylglutaryl coenzyme A reductase (HMGR), squalene synthetase, and sesquiterpene cyclase (Oba et al., 1985; Stermer and Bostock, 1987; Vogeli and Chappell, 1988; Zook and Kuc, 1991a, b). HMGR catalyzes the first step in the biosynthetic pathway of the phytoalexins by converting HMG coenzyme A to mevalonic acid (Bach, 1987; Kleinig, 1989). Many HMGR isoforms have been reported. There are four genes for HMGR in tomato (Choi et al., 1992), two genes in *Arabidopsis* (Caelles et al., 1989), two genes in *Hevea brasilieus* (Chye et al., 1992), and three to four genes in potato (Stermer et al., 1991; Yang et al., 1991; Choi et al., 1992).

Treatment of potato tuber discs with arachidonic acid strongly enhanced the accumulation of total HMGR transcripts (Choi et al., 1992). Three HMGR genes (hmg 1, hmg 2, and hmg 3) were cloned. Levels of hmg 1 mRNA were strongly

Table 11 Effect of chitosan treatment on the number of papillae formed in tomato root tissues 4 days after inoculation with *Fusarium oxysporum* f.sp. *radicis-lycopersici*

	Number of papillae in root tissue	
Organ treated with chitosan	Control	Chitosan-treated
Root	1	8
Leaf	1	11

Source: Benhamou and Theriault, 1992.

suppressed by the elicitor treatment. In contrast, transcript levels for hmg 2 and hmg 3 were strongly enhanced (two- to threefold) by elicitor treatment (Choi et al., 1992). The differential expression of HMGR isoforms to the elicitor treatment leads to suppression of steroid synthesis and induction of phytoalexin synthesis. HMGR 1 appears to have coordinate regulation with squalene synthetase (Zook and Kuc, 1991a), leading to sterol/steroid glucoalkaloid synthesis. Inhibition of HMGR 1 by the elicitor leads to inhibition of steroid synthesis. HMGR 2 and HMGR 3 alone are responsible for induction of sesquiterpenoid phytoalexin synthesis (Choi et al., 1992). These studies suggest that the elicitor can have such a high selective and specific action selecting even among the isoforms of the same enzyme.

Disease resistance in potato against *P. infestans* is characterized by a rapid accumulation of sesquiterpenoid phytoalexins (mainly rishitin and lubimin) as well as browning due to phenolic oxidation (Friend, 1981; Maniara et al., 1984). 5-Lipoxygenase from potato converts arachidonic acid to 5-S-hydroperoxyeicosatetraenoic acid (5-S-HPETE), which, at a concentration of 1 μg per potato tuber slice induces phytoalexin accumulation to higher levels than those induced by arachidonic acid (AA) at 20 μg per slice. Tissue browning elicited by the 5-S-HPETE was lower than that elicited by AA (Table 12; Castoria et al., 1992). The results indicated that the two elicitor components may induce the two different mechanisms in a different manner.

The elicitor isolated from *Colletotrichum graminicola* induced phytoalexins in roots of sorghum seedlings (Table 13; Ransom et al., 1992). The elicitor isolated from *Magnaporthe grisea* induces HMGR, the key enzyme in synthesis of the diterpene phytoalexins oryzalexins and monilactones in rice (Nelson et al., 1994).

The elicitor isolated from *Uromyces appendiculatus* induces production of ethylene in French bean (*Phaseolus vulgaris*) plants (Paradies et al., 1979). The chitin isolated from fungal cell walls induces lignification in wheat leaves (Pearce and Ride, 1982). Chitosan isolated from fungal cell wall elicited proteinase inhibitors in tomato (Walker-Simmons and Ryan, 1984). Elicitor from mycelial

Table 12 Phytoalexin accumulation and degree of browning induced by elicitors of *Phytophthora infestans*

Compound	Concentration (μg/disc)	Phytoalexins (μg/g) Rishitin	Phytoalexins (μg/g) Lubimin	Intensity of browning (OD at 440 nm)
Arachidonic acid	20	19.6	13.4	0.140
5-S-HPETE	1	38.6	40.2	0.080

Source: Castoria et al., 1992.

Table 13 Accumulation of 3-deoxyanthocyanidin in elicitor-treated roots of sorghum seedlings

Treatment	Time (h)	Apigeninidin (μg)	Luteolinidin (μg)	Caffeoyl–arabinosyl–apigeninidin (μg)
Elicitor	0	n.d.	n.d.	n.d.
	8	0.9	0.2	n.d.
	24	5.2	5.1	2.6
	32	12.2	14.2	6.8
Water	32	n.d.	n.d.	n.d.

n.d., Not detected.
Source: Ransom et al., 1992.

walls of *Chaetomium globosum* stimulated phenylalanine ammonialyase activity and the accumulation of phenolic acids in carrot (*Daucus carota*) cells (Kurosaki et al., 1986). The elicitor from *Colletotrichum lindemuthianum* induced accumulation of phenolics and phytoalexins. It was preceded by accumulation of phenylalanine ammonia–lyase and chalcone synthase mRNAs (Tepper et al., 1989). The elicitor from *Cladosporium fulvum* induces callose deposition in tomato (Peever and Higgins, 1989). The elicitor from *Phytophthora megasperma* f.sp. *glycinea* induced a rapid insolubilization of preexisting hydroxyproline-rich structural proteins in the cell wall (Brady et al., 1992).

The elicitor isolated from hyphal wall of *Phytophthora infestans* activated a O_2^- generating NADPH oxidase system in potato tubers (Doke and Miura, 1995). Activation of the O_2^- generating system has been reported in tomato, tobacco, and sweet pepper (*Capsicum annuum*) (Sanchez et al., 1992), soybeans (Lindner et al., 1988), and rice (Sekizawa et al., 1990). Thus elicitors are known to activate several types of defense mechanisms in the host under controlled experimental conditions. It is not known whether all these elicitors are released from the fungal cell wall when the fungal pathogen comes into contact with the host surface. The release of elicitors may be important event in fungal pathogenesis.

IV. HOST ENZYMES RELEASE ELICITORS FROM FUNGAL CELL SURFACE

A. β-1-3-Glucanase

It is now well established that elicitors are present in fungal cell surface. Physical contact of the fungal pathogens on plant cell surface results in activation of host defense mechanisms (Esnault et al., 1987; Cuypers et al., 1988). Elicitors are signal molecules and elicit synthesis of phytoalexins, lignins, callose, phenolics, pathogenesis-related (PR) proteins, and HRGP and the syntheses occur inside the cell. The signals should be received by some receptors in host cell surface and transducted, since the cell wall receptors cannot interact intracellulary. The receptors may be present in plasma membrane (Keen and Bruegger, 1977; Yoshikawa et al., 1983). If so, how could pathogen-surface elicitors that exist innately as insoluble components in fungal cell walls contact receptors on the plasma membrane. The possibility is that the host enzymes may rapidly solubilize the elicitors and release them from fungal cell walls. Yoshikawa et al. (1981) showed the release of a soluble phytoalexin elicitor from mycelial walls of *Phytophthora megasperma* f.sp. *glycinea* by a factor contained in soybean tissues. Two enzymes from soybean cotyledons released elicitor-active carbohydrates from cell walls of *P. megasperma* f.sp. *glycinea.* They were identified as isoenzymes of β-1-3-endoglucanase. The purified enzymes hydrolyzed several β-1-3-glucans in a strictly random manner (Keen and Yoshikawa, 1983).

The amount of carbohydrate solubilized from the *P. megasperma* f.sp. *glycinea* mycelial walls by β-1-3-glucanase correlated with the amount of elicitor activity solubilized from the *P. megasperma* f.sp. *glycinea* mycelial walls (Fig. 1; Ham et al., 1991). Heat-inactivated β-1-3 glucanase did not solubilize elicitor activity and did not solubilize carbohydrates (Ham et al., 1991). The purified β-1-3-glucanase hydrolyzed glycosidic linkages of fungal cell wall solubilizing oligo- and/or polysaccharides. The elicitor-active molecules solubilized were β-1-3-glucans (Ham et al., 1991).

The purified soybean β-1-3-glucanase at 0–500 μg/ml did not show any direct fungitoxic effect on zoospores motility, cystospore germination, or mycelial growth of *P. megasperma* f.sp. *glycinea.* Rather the purified glucanase possessed the activity to release soluble elicitors from insoluble fungal cell walls (Yoshikawa et al., 1990).

Exogenous application of purified glucanase to inoculation sites resulted in higher levels of phytoalexin (glyceollin) accumulation in soybean. Conversely laminarin, which has been shown to inhibit the release of elicitors from fungal cell walls in vitro, partially suppressed glyceollin synthesis (Table 14; Yoshikawa et al., 1990). Ethylene increased 50–100 fold the level of β-1-3-endoglucanase mRNA (Takeuchi et al., 1990) and exogenous ethylene increased the level of β-1-3-endoglucanase and accumulation of the phytoalexin glyceollin (Yoshikawa et al., 1990). All these results suggest that the process of elicitor release mediated by host β-1-3-endoglucanase is an important process leading to phytoalexin synthesis.

Cline and Albersheim (1981a, b) isolated an exoglucanase enzyme from soybean cell walls and it solubilized fungal glucans. Nichols et al., (1980) showed that β-1,3-glucanase activity increased within 6 h in pea pod tissue inoculated with *Fusarium solani* f.sp. *pisi*. If endoglucanase should be considered important to release the elicitors from fungal cell wall, it should be present near the plant

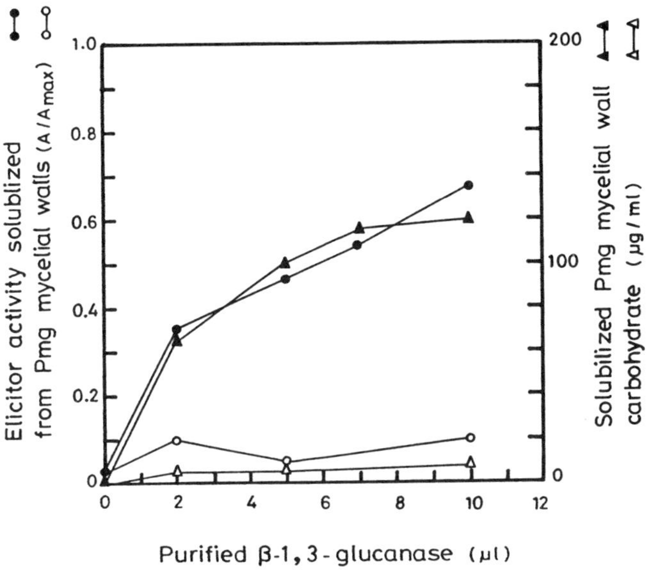

Figure 1 Elicitor-releasing activity of purified β-1,3-glucanase from *Phytophthora megasperma* f.sp. *glycinea* cell wall. The figure shows the results obtained when *P. megasperma* f.sp. *glycinea* cell wall suspension was incubated with various amounts of the purified β-1,3-glucanase (open symbols) or with heat-inactivated (closed symbols) purified β-1,3-glucanase. (From Ham et al., 1991.)

Table 14 Effect of exogenous application of glucanase and laminarin on phytoalexin production in soybean

Treatment	Concentration (µg/ml)	Glyceollin (µg/g hypocotyls)
Glucanase	0	120
	50	578
	100	564
	500	625
Laminarin	0	1112
	2	816
	5	909
	10	670
	20	330

Source: Yoshikawa et al., 1990.

cell wall. β-1,3-Glucanase has been reported to accumulate extracellularly based on its occurrence in intercellular washing fluids collected from infected leaves (Kauffman et al., 1987; Legrand et al., 1987; Kombrink et al., 1988). Mauch and Staehelin (1989) used immunocytochemical methods to locate the glucanase. Antiglucanase antibodies labeled the expanded middle lamella region of the cell wall, indicating that the glucanase is present extracellularly. When fungal pathogens initially grow in the intercellular space of their host plants, they may make contact with the β-1,3-glucanase localized in middle lamella. Upon contact, the β-1,3-glucanase may release oligosaccharide fragments from the β-1,3-glucan-containing fungal cell wall. These oligosaccharides released by β-1,3-glucanase from fungal cell walls may act as elicitors of phytoalexin production.

B. Chitinase

Chitin is another important insoluble elicitor in fungal cell wall. The enzyme chitinase hydrolyzes chitin into soluble oligosaccharide fragments (Molano et al., 1979; Pegg and Young, 1981; Boller et al., 1983) and chitinase is commonly present in plants (Prowning and Irzykiewiz, 1965; Abeles et al., 1970; Pegg and Vessey, 1973).

The role of soluble chitin oligomers in eliciting lignification in wheat leaves has been studied (Barber et al., 1989). The chitin monomer and dimer were inactive, while the trimer possessed very slight elicitor activity. In contrast, the chitin tetramer, pentamer, and hexamer all possessed significant elicitor activity. Digestion of the tetramer with purified wheat leaf chitinase completely abolished elicitor activity (Table 15; Barber et al., 1989). The identification of a soluble chitin oligomer with elicitor activity suggests that wheat leaves possess an enzyme, presumably a chitinase, capable of releasing chitin oligomers from

Table 15 Ability of chitin oligomers to elicit lignification in wheat leaves

Chitin oligomer	Lignification (%)
Water	0
Monomer	4
Dimer	5
Trimer	13
Tetramer	77
Pentamer	80
Hexamer	75

Source: Barber et al., 1989.

chitin. Chitinase activity has been reported in wheat germ (Molano et al., 1979) and leaves (Boller et al., 1983).

Nichols et al., (1980) observed that chitinase activity increased within 30 min in pea pod tissue following contact with macroconidia of *F. solani* f.sp. *pisi*. Incubation of the fungal cell walls with the enzyme mixture released solubilized components.

Chitinases purified from carrot cells released sugar components from insoluble mycelial walls of *Chaetomium globosum*. These partial hydrolysates stimulated phenylalanine ammonia–lyase activity and accumulation of phenolic acids in carrot cells (Kurosaki et al., 1986). The substances responsible for elicitation lost their activity after prolonged digestion by chitinase, but were released from insoluble mycelial walls by mild hydrolysis with this enzyme (Kurosaki et al., 1987a, b). Chitinase activity was induced in cultured carrot cells when they were treated with insoluble mycelial walls of *C. globosum*. Most of the induced chitinase activity was found in the soluble fraction of carrot cells (Kurosaki et al., 1987a). Kurosaki et al. (1987b) showed that some of chitinase was secreted into the culture medium immediately after the addition of the mycelial walls. Monensin, an inhibitor of transport from the Golgi complex to the apoplastic space, inhibited secretion of chitinase but did not affect the total activity of intra- and extracellular chitinases (Kurosaki et al., 1987b). These results suggest that the chitinase induced in infected carrot cells is transported into the apoplastic space of the cells by an active transport system, and not as a result of the destruction of cellular structure. The fungal contact on the host cell may increase chitinase activity of the host and the chitinase will be available extracellularly at the infection site to make the insoluble chitin of the fungal cell wall soluble.

C. Chitosan-Degrading Enzymes

Chitosan, a partially soluble polymer of β-1,4-glucosamine, is also a component of cell walls of many fungi. Water extracts of the cell wall of *F. solani* f.sp. *pisi* do not induce the phytoalexin pisatin in pea pods (Hadwiger and Beckman, 1980). However, the fungal cell wall extract, when incubated with crude pea plant enzyme preparation, induced high amounts of the phytoalexin. The results suggest that phytoalexin inducers can be released from fungal cell walls by crude plant enzyme preparation. Chitosan was detected in the fungal cell wall and its content increased up to 43% within 2 h after contacting plant tissue. Chitosan effectively induced phytoalexin pisatin (Hadwiger and Beckman, 1980). Effectiveness of chitosan increased as the particle size decreased. The cleaved chitosan was highly active (Hadwiger and Loschke, 1981). Plant cell wall is regarded as a barrier to higher-molecular-weight compounds such as chitosan (Carpita et al., 1979). Pea tissues possess enzymes capable of cleaving chitosan molecules to molecules with lower molecular weight that could conceivably pass through the cell wall barrier (Nichols et al., 1980). When [^{3}H]chitosan was applied to pea tissue, the large crystals remained outside the cell. Only lower-molecular-weight cleaved chitosan entered the cell within 15 min after application and the label was readily detectable within the plant cytoplasm and nucleus (Hadwiger et al., 1981).

Both chitinase and β-glucanase were applied to pregerminated macroconidia of *F. solani* f.sp. *pisi* to determine their effect on chitosan oligomer accumulation (Kendra et al., 1989). [^{3}H] Label was applied for 45 min followed by 2 h exposure to the enzyme mixture. The label was detected in monomer, dimer, trimer, pentamer, and heptamer (Kendra et al., 1989). The chitosan heptamer was highly effective in eliciting phytoalexin, while the pentamer was less active and the smaller oligomers were totally inactive (Kendra and Hadwiger, 1984). The minimum oligomer size necessary to elicit the host response had a degree of polymerization (DP) of approximately seven or greater and was essentially nonacetylated. The measurable biological activity observed for heptamers accumulating from the digestion with chitinase and β-glucanase-rich proteins derived from *F. solani* f.sp. *phaseoli*-challenged pea tissues suggests that these enzymes may be important in the release of chitosan (Kendra et al., 1989).

Chitosan oligomers have been shown to elicit proteinase inhibitory factors in tomato. The dimer possessed weak activity, and the elicitor activity increased with oligomer size with the pentamer and hexamer showing maximal activity (Walker-Simmons and Ryan, 1984).

Barber et al. (1989) prepared deacetylated chitosan oligomers with degree of polymerization 1–4. None of the four oligomers (monomer, dimer, trimer, and tetramer) possessed any elicitor activity in wheat leaves, but partially or fully deacetylated polymeric chitosans were very active. Probably the acetyl groups may be required for elicitor activity. Totally deacetylated chitosan possessed considerably less elicitor activity than either chitosan or chitin (Barber and Ride, 1988).

D. Lipoxygenase

Fatty acids have been reported to be elicitors in cell walls of *Phytophthora infestans* (Bostock et al., 1982). Arachidonic acid and eicosapentaenoic acid are present primarily in esterified form in the fungus (Creamer and Bostock, 1986). However, free arachidonic acid elicits a faster response than do esterified forms (Preisig and Kuc, 1987a). Free carboxyl group is required for optimal elicitor activity (Preisig and Kuc, 1985). Potato lipoxygenase enhances the activity of arachidonic acid. Salicylhydroxamic acid is an inhibitor of lipoxygenase and it effectively inhibited arachidonic acid-elicited hypersensitivity (HR) (Preisig and Kuc, 1987b). When potato tissue was treated with radiolabeled free arachidonic acid, most of the fatty acid was incorporated without further modification into potato acyl lipids. A low level of radiolabel from arachidonic acid was recovered continually in the region of hydroxy fatty acids suggesting that oxidation of the fatty acid molecule was occurring. Arachidonic acid incorporated into acyl lipids was recovered primarily from phospholipids. More than 90% of the radiolabel in polar lipids was in two lipid classes: phosphatidyl choline (PC) and phosphatidyl ethanolamine (PE). Arachidonic acid was released from PC by potato tuber by potato lipid acyl hydrolase and lipase (Preisig and Kuc, 1985; 1987b). Thus release of arachidonic acid from esterified form, the form in which it occurs in the fungus, appears to be important for the elicitation activity (Preisig and Kuc, 1988).

Glucans from *P. infestans* enhanced metabolism of arachidonic acid to compounds comigrating with oxidized fatty acids. Calcium chloride also enhanced the elicitor activity and it also enhanced oxidation of arachidonic acid. Salicylhydroxamic acid inhibited oxidative metabolism of arachidonic acid and it inhibited HR (Preisig and Kuc, 1988). All these results revealed that arachidonic acid has to be oxidized by lipoxygenase (Corey and Lansbury, 1983; Shimizu et al., 1984; Preisig and Kuc, 1985; 1988). The importance of lipoxygenase in arachidonic acid-elicited defense mechanisms was demonstrated by using lipoxygenase-deficient potato cell lines. These cell lines showed diminished response to arachidonic acid (Vaughan and Lulai, 1992).

However, Ricker and Bostock (1994) questioned the importance of lipoxygenase in release of elicitor activity from arachidonic acid. The purified 5-lipoxygenase from potato tuber produced all positional hydroperoxyeicosatetraenoic acid (HpETE) isomers such as 5-, 8-, 11-, 12- and 15-HpETE following incubation with arachidonic acid. HpETEs were formed in tuber discs within 10 min after addition of arachidonic acid (Ricker and Bostock, 1994). The β-glucan preparation, which enhances arachidonic acid elicitor activity, suppressed HpETE formation. Abscisic acid treatment, which suppresses phytoalexin accumulation in potato, had no effect on type or amount of HpETEs formed in tuber discs. None of the purified lipoxygenase products of arachidonic acid (HpETEs) possessed elicitor activity (Ricker and Bostock, 1994). These observations do not support the importance of lipoxygenase in elicitor activity of arachidonic acid. However, metabolites other than HpETEs may be important for arachidonic acid elicitor activity.

Several of these studies have clearly indicated that fungal cell wall elicitors can be released only by host enzymes. Most of these host enzymes are in a low level in healthy plants. Their enzyme activity increases only when the pathogen invades host tissues. Hence contact of the fungal cell surface with host cell surface should have activated these enzymes.

However, the mechanism of induction of these host enzymes is not known. Probably the fungal cell wall chitin, β-1,3-glucan, and arachidonic acid would have served as substrate for chitinase, β-1,3-glucanase, and lipoxygenase, respectively, but in such case higher induction of these enzymes should occur in susceptible interactions in which more fungal growth (with more chitin, β-1,3-glucan and arachidonic acid) occurs. In contrast, more induction of these enzymes is seen in incompatible interactions. It suggests that additional signal molecules may be involved in these plant–microbe interactions. This is discussed in Chapter 3.

V. ENZYMES OF PATHOGENS RELEASE ELICITORS OF HOST ORIGIN

Although host enzymes release elicitors from fungal cell wall, there are reports that fungal enzymes can release elicitors from host cell walls. Elicitors of host origin are called constitutive or endogenous elicitors while elicitors of pathogen origin are called exogenous elicitors (Hargreaves and Bailey, 1978; Hargreaves and Selby, 1978, Hahn et al., 1981). Pectic fragments induce accumulation of proteinase inhibitor proteins in tomato (Ryan, 1974; 1978; Bishop et al., 1981; 1984; Ryan et al., 1981; 1987; Doherty et al., 1983; Baydoun and Fry, 1985; Farmer et al., 1989). The elicitor has been found to be a host cell wall-derived polygalacturonide with a degree of polymerization (DP) of about 20 (Bishop et al., 1984). The pectic fragment that showed proteinase inhibitor-inducing activity in sycamore (*Acer pseudoplatanus*) has been identified as rhamnogalacturonan (Ryan et al., 1981). Elicitor of phytoalexin accumulation in soybean has been identified as dodeca-α-1,4-D galacturonide (Nothnagel et al., 1983). Elicitor-active components from french bean hypocotyls were identified as oligogalacturonides with nine galacturonosyl units (Dixon et al., 1989). A host elicitor that induces synthesis of hydroxyproline-rich glycoprotein in cucumber (*Cucumis sativus*) has been reported (Roby et al., 1985). Oligogalacturonide elicitor has been detected in parsley cells (Davis and Hahlbrock, 1987). The elicitor active substances found in carrot cells are a mixture of heterogeneous molecules in which uronide and peptide are essential components (Kurosaki et al., 1985). The elicitor in castor bean has been identified as trideca-α-1,4-D-galacturonide (Jin and West, 1984). A noncarbohydrate elicitor has also been identified in tomato (Farmer et al., 1989). These constitutive elicitors in the host are released by the fungal enzymes.

Casbene is a phytoalexin of castor bean (Sitton and West, 1975). Elicitors of casbene synthetase activity were detected in culture filtrates from *Rhizopus stolonifer,* a common fungus found in castor bean seeds (Stekole and West, 1978). Polygalacturonase activity was readily detected in the partially purified elicitor preparation (Lee and West, 1981a,b). Heat treatment of the enzyme preparations led to equivalent losses of both enzymic activity and elicitor ability, suggesting that the elicitor abilities of these enzymes may be dependent on their catalytic activities (Lee and West, 1981b). The catalytic action could cause the breakdown of a normal plant constituent to one or more degradation products that themselves could function as elicitors. Incubation of castor bean homogenate with the pure enzyme led to the isolation of soluble heat-stable elicitor components lacking enzyme. Analysis of the crude heat-stable elicitor fraction indicated that it was a pectic fragment containing galacturonic acid or its methyl ester. More complete digestion of the heat-stable elicitor by endopolygalacturonase at optimum pH conditions for prolonged period resulted in loss of activity (Bruce and West, 1982). Only partial digestion of polygalacturonic acid with polygalacturonase produced the heat-stable elicitor. The partial digestion produced a mixture of α-1,4-D-galacturonic oligomers. The individual oligomers were assayed for casbene synthetase elicitor activity. The fractions corresponding to octomers and smaller oligomers showed no significant elicitor activity, while fractions corresponding to nonamers through pentadecamers were active, with the tridecamer eliciting the greatest response (Jin and West, 1984). Methyl esterification of the carboxylate groups greatly diminished the elicitor activity of the oligomers, suggesting a requirement for the polyanionic character of the oligomers for full activity (Jin and West, 1984).

Endopolygalacturonase (endo PG) from *Cladosporium cucumerinum,* the cucumber pathogen, elicited lignification in cucumber hypocotyls (Robertson, 1986). The enzyme released elicitors of lignification from polygalacturonic acid and cucumber cell wall. When the endo-PG was incubated with polygalacturonic acid, a series of oligosaccharides were produced starting with galacturonic acid. The mixture of oligomers with 9-12 galacturonosyl units showed elicitor activity at concentration down to about 3 μg uronic acid units/ml, which confirms that the endopolygalacturonase is able to release oligogalacturonide elicitors from polygalacturonic acid (Robertson, 1986).

Endopolygalacturonase from *Aspergillus niger* elicited necrosis in *Vigna unguiculata* (Cervone et al., 1987). The enzyme elicited synthesis of proteinase inhibitors (Walker-Simmons et al., 1984), and lignification (Robertson, 1986). The eliciting activity of the enzyme was by releasing oligogalacturonide elicitors from the pectic polysaccharide of plant cell walls (Hahn et al., 1981; Nothnagel et al., 1983; Walker-Simmons et al., 1984). The ability of oligogalacturonides to induce phytoalexin and lignin synthesis was strictly dependent upon their degree of polymerization (DP). Oligomers of galacturonic acid with a DP between 10 and 13 were effective, while shorter oligomers had little or no activity (Hahn et al., 1981; 1989; Nothnagel et al., 1983; Davis et al., 1984, 1986; Robertson 1986; Ryan, 1987, 1988; Lorenzo et al., 1990).

Endopolygalacturonase from different races of *Colletotrichum lindemuthianum* elicited phytoalexins in bean by forming elicitor-active oligogalacturonides (Lorenzo et al., 1990; Tepper and Anderson, 1990). Endo-α-1,4-polygalacturonase from *C. lindemuthianum* solubilized purified walls of sycamore cells producing a large pectic polysaccharide, called rhamnogalacturonan I. It elicited proteinase inhibitor. Galacturonic acid was the major glycosyl residue in rhamnogalacturonan I. Arabinosyl, galactosyl, and rhamnosyl residues were also present (Ryan et al., 1981).

Endo-pectin lyase (PL) produced by *Botrytis cinerea,* a pathogen of carrot, induced phytoalexin synthesis in carrot. Both lower and higher concentrations of the enzyme decreased phytoalexin accumulation (Table 16; Movahedi and Heale, 1990). Induction of carrot cell walls with 10 units of endo-PL resulted in high elicitor activity. The cell wall preparations alone did not elicit the phytoalexin (Movahedi and Heale, 1990). Kurosaki et al. (1984, 1985) also reported that carrot cells treated with pectinase released host wall fragments that elicited the production of the phytoalexin.

A pectic lyase purified from the fungus *Fusarium solani* f.sp. *pisi* produced fragments that, when supplied to leaves of tomato plants, induced accumulation of proteinase inhibitor I. Isolation and analysis of oligomeric fragments of a partial digest of polygalacturonic acid revealed that α-1,4-digalacturonic acid with 4,5-unsaturated nonreducing terminal galacturonosyl residue and the unsaturated trimer were effective inducers of the synthesis of proteinase inhibitor I when supplied to tomato plants through their cut petioles (Ryan, 1987). The galacturonosyl oligomers that elicit phytoalexin responses are larger than the oligomers that induce proteinase inhibitors. Although accumulation of phytoalexins and lignin (Davis et al., 1984; Robertson, 1986) is elicited by oligomers of DP-10-13, proteinase inhibitor synthesis is even induced by dimers (Bishop et al., 1984; Ryan, 1987). Movahedi and Heale (1990) reported that *Botrytis cinerea* produced an aspartic proteinase both in vitro and in infected carrot tissue. When the enzyme was applied at low concentrations, it induced the phytoalexin 6-methoxymellein. The enzyme-released endogenous-elicitors (mixture of uronide and peptide) induced the phytoalexin.

Table 16 Elicitor activity of *Botrytis cinerea* enzyme

Concentration of enzyme applied as pretreatment units/ml	6 methoxymellein content (μg/g fresh weight)
Control	43
0.0001	46
0.001	474
0.01	360
0.1	46
1.0	23
10.0	8

Source: Movahedi and Heale, 1990.

VI. SYNERGISTIC ACTION OF FUNGAL CELL WALL ELICITORS AND HOST CELL WALL ELICITORS

The relative importance of endogenous and exogenous elicitors in induction of host defense mechanisms has been studied. Hexa-β-glucosyl glucitol is the elicitor isolated from *Phytophthora megasperma* f.sp. *glycinea,* while decagalacturonic is a pectic fragment released from polygalacturonic acid by action of endopolygalacturonase produced by the pathogen of soybean. The elicitor activity of combinations of various amounts of the decagalacturonide and 150 ng/ml hexa-β-glucosyl glucitol was studied. An approximately linear increase in the elicitor activity of combinations of the decagalacturonide and the hexa-β-glucosyl glucitol was observed over a concentration of 3–22 μg/ml decagalacturonide. A 35-fold stimulation above the calculated additive response was obtained at 22 μg/ml decagalacturonide. Similarly a significant increase in the elicitor activity of combinations of hexa-β-glucosyl glucitol and 11 μg/ml decagalacturonide was observed at concentrations of 50 to 150 ng/ml hexa-β-glucosyl glucitol, with approximately a 15-fold stimulation above the calculated response at 150 ng/ml hexa-β-glucosyl glucitol (Davis et al., 1986a, b).

Pectic fragments were active as elicitor in dark red kidney bean cotyledons. Preparation from both α and β races of *Colletotrichum lindemuthianum* also acted as elicitors, although those from α races were more active. The presence of pectic fragments enhanced the elicitor activity of both the α and β race preparations, particularly when low concentrations, at which the elicitor activity of each factor was very much negligible, were used for the study (Table 17; Tepper and Anderson, 1990).

To investigate nature of the fungal components that could act synergistically with the pectic fragments, the fractionated culture filtrate products obtained by DEAE-Sephadex chromatography were examined (Tepper and Anderson, 1990). Adsorbed fraction 1 from α race and pectic fragments acted synergistically

Table 17 Elicitor activity in dark red kidney bean cotyledons following treatment with elicitors of *Colletotrichum lindemuthianum* α and β races and pectic fragments

Treatment	Elicitor units	Phaseollin (μg/cotyledon)
Pectic fragments	15	2
α Race elicitor	46	0
α Race elicitor	82	50
β Race elicitor	0	0
β Race elicitor + pectic fragments	20	17

Source: Tepper and Anderson, 1990.

for the production of soluble phenolics, phytoalexins and condensed phenolics. Adding fractions 3, 4, and 5 with pectic fragments enhanced phytoalexin production while fractions 2 and 3 interacted with pectin fragments to increase the production of condensed phenolics. Pectic fragments with β race fractions 2, 3, and 4 increased the production of soluble phenolics and phytoalexins, but did not affect the accumulation of condensed phenolics. Galactoglucomannan exhibited a complex pattern of stimulation depending on the concentration of both fungal product and pectic fragments. At the higher concentrations of galactoglucomannan there was little stimulation in the production of soluble phenolics and phytoalexins. At lower galactoglucomannan concentrations, pectic fragments had a greater effect on the accumulation of soluble and condensed phenolics including phytoalexins (Tepper and Anderson, 1990). These data suggest that several fungal components interact with pectic fragments to stimulate the accumulation of defense chemicals in bean.

Davis and Hahlbrock (1987) reported that the plant cell wall elicitor acted synergistically with the glucan elicitor from *Phytophthora megasperma* f.sp. *glycinea* in the induction of coumarin phytoalexins in parsley cells (Fig. 2). About 10-fold stimulation in coumarin accumulation above the calculated additive response was observed in cell cultures treated with combinations

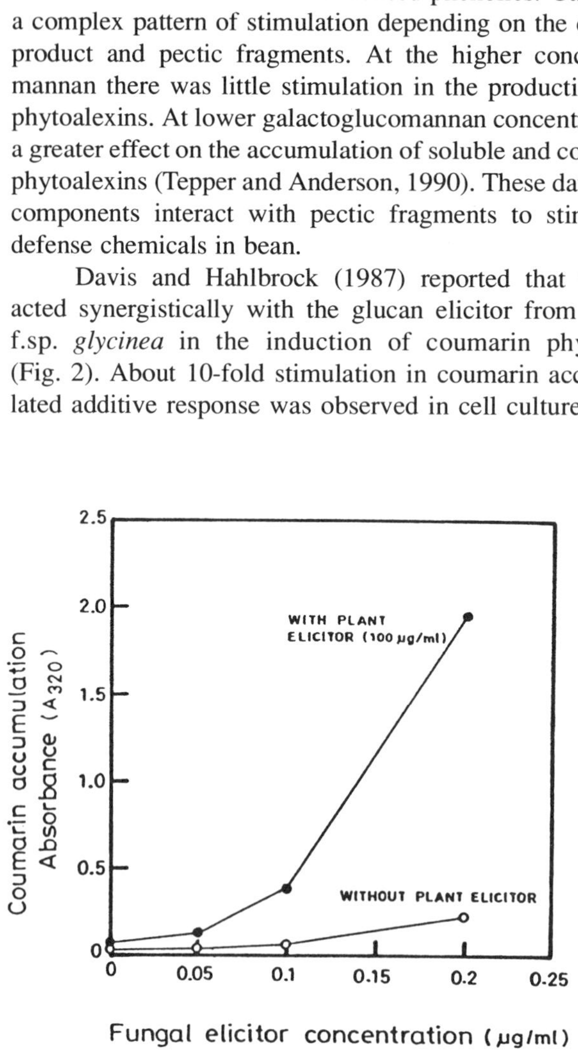

Figure 2 Induction of coumarin accumulation in parsley cell cultures by different concentrations of fungal elicitor in the absence (○—○) or presence (●—●) of plant elicitor. (From Davis and Hahlbrock, 1987.)

of plant and fungal elicitors. The synergistic effect was also observed for the induction of phenylalanine ammonialyase, 4-coumarate-CoA ligase and S-adenosyl-L-methionine-xanthotoxol O-methyl transferase (XMT), the key enymes involved in phenyl propanoid metabolism and furanocoumarin biosynthetic pathway.

Both elicitors of pathogen and host origins are involved in signaling defense mechanisms of the host. These signal molecules require receptor sites for their action.

VII. RECEPTOR SITES FOR ELICITORS MAY EXIST IN HOST CELL MEMBRANE

Binding sites for the elicitors may exist in host cell membranes (Dixon, 1986). Garas and Kuc (1981) showed that elicitors from *Phytophthora infestans* were precipitated by potato lectin. The lectin–potato elicitor complexes nonetheless retained most of the elicitor activity when applied to potato discs. These results suggested that lectins in potato may serve as binding or receptor sites for elicitor. Several indirect lines of evidence have been presented to show that the fungal elicitors have to be bound with the receptor sites of the host for their action. The elicitors from *P. infestans* agglutinated rapidly with potato protoplasts (Peter et al., 1978). The activity of elictor from *Phytophthora megasperma* var. *sojae* was shown to be inhibited by certain methyl sugar derivatives that were presumed to act by competing for elicitor binding sites (Ayers et al., 1976b). Methyl glycosides inhibited the activity of elicitor from *P. infestans* presumably by competing for the binding site (Marcan et al., 1979). Dextran-bound-p-chloromercuribenzoate, which is unable to enter the cells, inhibited the effects of *P. infestans* elicitors on potato protoplasts (Doke and Furuichi, 1982). Yoshikawa et al. (1983) used ^{14}C-labeled mycolaminarin isolated from *Phytophthora* spp. as elicitor. Mycolaminarin bound with membrane preparations from soybean cotyledons. The binding was inhibited by pretreatment of the membranes with heat or pronase, indicating the presence of a proteinaceous binding site. The binding was characterized by a dissociation constant of 11.5 μM with respect to the one class of binding site identified and a total of 16500 binding sites per cell was calculated. Maximum specific binding per mg protein was associated with a fraction containing plasma membranes.

Schmidt and Ebel (1987) used β-1,3-[^3H]glucan elicitor fraction from *P. megasperma* f.sp. *glycinea,* the soybean pathogen to identify putative receptor sites in soybean tissues. The studies successfully demonstrated the presence of saturable, high-affinity, and specific binding site(s) for the glucan elicitor in a microsomal fraction of soybean roots. Highest binding activity was associated with a plasma membrane-enriched fraction. The binding was abolished by pronase treatment of the microsomal fraction and stabilized in the presence of dithiothreitol, indicating proteinaceous nature of the binding site. The maximum number of binding sites was 0.5 pmol/mg protein. Competition studies with the [^{3}H]glucan elicitor and a number of polysaccharides demonstrated that only polysaccharides of a branched β-glucan type effectively displaced the radiolabeled ligand from membrane binding. The differential effects of a number of glucans on radiolabeled ligand binding corresponded well with their phytoalexin elicitor activity. For example, mycolaminaran showed similar weak displacing and phytolaexin elicitor activities and laminaran exhibited intermediate activities.

The binding of the glucan elicitor with soybean cell membrane was reversible, and labeled glucan elicitor could be displaced by unlabeled elicitative derivatives but not by inactive glucans. High-specific-activity [^{125}I]glucan with elicitor activity was used to demonstrate high-affinity binding to soybean protoplasts in vivo. The most binding was observed in a plasma membrane-enriched fraction (Cosio et al., 1988; Cosio and Ebel, 1988).

The presence of elicitor-binding sites with a high affinity for the protein elicitor cryptogein has been demonstrated in tobacco cells (Blein et al., 1991). The binding was saturable and susceptible to displacement by unlabeled ligand (Blein et al., 1991). Except for these few studies, receptor molecules have not been isolated, purified, and characterized. Further studies are needed to characterize receptor molecules.

VIII. SIGNAL TRANSDUCTION

Elicitors are signal molecules and when bound with binding sites in plasma membrane three kinds of signals are produced for activation of defense genes: intracellular signals, short-distance intercellular signals, and systemic signals. Phenylpropanoid and chitinase genes are activated directly following intracellular transduction of the initial external signal (Lamb et al., 1989). In elicitor-treated bean cells, activation of genes encoding hydroxyproline-rich cell wall protein (HRGP) appears at a distance with a time lag of about 1 h indicating transmission of intercellular signals (Lamb et al., 1989). Short distance intercellular signals may account for the transcriptional activation of chalcone synthase observed in tissue adjacent to hypersensitively responding cells in bean infected with *Colletotrichum lindemuthianum* (Bell et al., 1986). Transcripts encoding proteinase inhibitors and certain PR proteins accumulate to high levels in tissue distant from the site of infection. Sometimes the induction occurs throughout the plant (Ryan, 1988). In these cases systemic signals may be involved. The microbial elicitors cannot themselves enter the host cells and hence some second messenger or response coupler to transmit the elicitation signal intracellularly may exist (Dixon, 1986).

IX. INTRACELLULAR SIGNAL TRANSDUCTION

A. Calcium Ion May Act as Second Messenger

Calcium ion acts as a second messenger in eukaryotic cells by transducting extracellular primary stimuli into intracellular events (Pitt and Kaile, 1990). The chemical properties of Ca^{2+} render it more suitable for the role of second messenger than other more abundant cellular cations (Carafoli and Penniston, 1985). Elicitation of defense chemicals was more effective in the presence of Ca^{2+} in plants (Ebel, 1984; Pelissier and Esquerre-Tugaye, 1984). Ebel (1984) showed that chalcone synthase activity was induced more in elicitor-treated soybean cells after calcium treatment. Calcium ion requirement for 1,3-β-glucan synthase in soybean cells has been demonstrated (Kauss et al., 1983). This Ca^{2+}-dependent enzyme is directly activated by the influx of Ca^{2+} occurring concomitantly with the leakage of cell constituents (Kohle et al., 1985).

Phytoalexin synthesis in soybean cell suspension cultures was stimulated by the Ca^{2+} ionophore A23187. Elicitation by *P. megasperma* f.sp. *glycinea* elicitor was inhibited by external Ca^{2+} channel depletion. Verapamil is a blocker of Ca^{2+} channel and it inhibited the elicitation (Stab and Ebel, 1987). The Ca^{2+} ionophore A23187 induced the accumulation of 6-methoxymellein, the phytoalexin in carrot cell suspension cultures, and the induction by a crude elicitor from carrot tissues was inhibited by addition of verapamil within 30 min of elicitation (Kurosaki et al., 1987d).

B. Mode of Action of Calcium as Second Messenger

Regulation of metabolism via altered calcium levels is mediated through calcium-binding protein, in particular calmodulin (Means and Dedman, 1980; Bino et al., 1984; Dieter, 1984; Lukas et al., 1984; Collinge and Trewavas, 1989; Trewavas and Gilbroy, 1991). Eukaryotic cells possess a system of proteins that interact with the calcium ion and govern the transmission of the intercellular message (Carafoli and Penniston, 1985). Similar mechanisms exist in plants (Hepler and Wayne, 1985; Poovaiah, 1988). The important calcium-dependent enzymes are protein kinases, Ca^{2+} and H^+ transport ATPases, phospholipases, NAD kinase, and quinate/NAD^+ oxidoreductase. Protein kinases have been purified from plant cells (Elliott and Skinner, 1987; Bogre et al., 1988). cDNA sequences encoding for protein kinases have been cloned (Breviario et al., 1995).

Unlike the microbial elicitors, calcium can pass through plasma membrane. Calcium passively crosses plasma membranes, usually via an electrochemical gradient (Zocchi and Rabotti, 1993). The structures through which the ions permeate are "pore"- or "channel"-forming proteins embedded in the lipid bilayer of the membrane (Reuter, 1983). Calcium regulates ion channels in the plasma membrane (Schroeder and Hagiwara, 1989).

Cytoplasmic calcium homeostasis is important for cellular functions. The second messenger role of calcium requires that its concentration inside the cell should be very low, ranging between 10^{-6} and 10^{-8} M (Hanson, 1984; Poovaiah and Reddy, 1987). In plants this is achieved principally through the activity of two different mechanisms: Ca^{2+} transport ATPases operating at the plasmalemma and endoplasmic reticulum (ER) level (Rasi-Caldogno et al., 1982; Zocchi and Hanson, 1983; Schumaker and Sze, 1985; Evans, 1988; Zocchi, 1988); and an H^+/Ca^{2+} antiport mechanism dependent on the proton electrochemical gradient operating at the tonoplast (Schumaker and Sze, 1985; Zocchi, 1988).

In plant cell, Ca^{2+} is maintained far from its electrochemical equilibrium, and its passive influx is strongly favored (Zocchi and Rabotti, 1993). Calcium channels are present in plant cells (Johannes et al. 1991) and may be opened by IAA (McAinsh et al., 1990). As an alternative, Ca^{2+} can be released from internal stores by the action of inositol 1,4,5-triphosphate (Boss and Massel, 1985; Drobak and Ferguson, 1985). A calcium-induced Ca^{2+} release at the plasmalemma may operate at this point (Zocchi and Rabotti, 1993).

C. Calcium-Regulated Protein Phosphorylation as a Component of Signal Transduction

Phosphorylation of proteins is an important component in the integration of external and internal stimuli in animal system (Cohen, 1982). The coupling of the stimulus to the response involves perception of the stimulus by membrane-bound receptors, opening of channels in the membrane, an increase in the concentration of second messengers like calcium, activation of second-messenger dependent enzymes like protein kinases, phosphorylation of proteins and amplification of stimulus and dephosphorylation, and return to a resting situation for turning off the stimulus (Nishizuka, 1984).

In plants a similar system may exist. The plants may use protein phosphorylation as an effective device for responding to endogenous stimuli (Ranjeva and Boudet, 1987). Phosphorylation of cell membrane-bound proteins may be crucial in regulating the structure and function of membrane components. These reversible posttranslational changes play an important role in signal amplification and are able to sense the fluctuations in the concentration of substrates and regulatory molecules (Ranjeva and Boudet, 1987). Phosphorylation of membrane-located proteins has been reported in many plants (Salimath and Marme, 1983; Morre et al., 1984a, b, Polya et al., 1984; Veluthambi and Poovaiah, 1984a, b, c; Paliyath and Poovaiah, 1985).

The phosphorylation of membrane proteins is dependent on Ca^{2+} and calmodulin in many cases (Polya et al., 1984; Veluthambi and Pooviah, 1984b,c, Teulieres et al., 1985). Protein phosphorylation reaction is catalyzed by protein kinases. Calcium-regulated protein kinases have been reported in plants

(Hetherington and Trewavas, 1982; Ranjeva et al., 1983; Marme et al., 1986). The addition of micromolar concentrations of Ca^{2+} to the nuclear preparation from pea plumules increased the level of phosphorylation in several nuclear proteins (Ranjeva and Boudet, 1987). The protein kinase from pea plasma membranes is activated 5–15-fold by micromolar levels of Ca^{2+} ions (Hetherington and Trewavas, 1984). Calmodulin enhanced the effect of Ca^{2+} ions (Salimath and Marme, 1983). Low concentrations of calmodulin antagonists inhibited the calcium-induced repsonse (Raghothama et al., 1985).

Protein kinase C occurs in many plants (Schafer et al., 1985). In animals, extracellular signals that activate cellular functions and proliferation through interaction with membrane receptors result in the breakdown of inositol phospholipids and formation of inositol triphosphate and diacylglycerol. Inositol triphosphate may be responsible for the transient release of Ca^{2+} from internal stores and diacyl glycerol stimulates protein kinase C (Berridge, 1984). Protein kinase C occurs as a soluble or a membrane-bound enzyme. A shuttle system between cytosol and membrane would allow the phosphorylation of either soluble or membrane-bound proteins and therefore would allow protein kinase C to participate in transmembrane signaling (Ranjeva and Bouder, 1987).

Elicitor treatment increased inositol triphosphate and this increase preceded phytoalexin accumulation in cultured carrot cells. It suggests the involvement of a phosphatidyl inositol turnover-mediated signal pathway in carrot (Kurosaki et al., 1987a). Glucan elicitors induced changes in the phosphorylation of specific proteins in soybean cells (Ebel, 1989). Oligogalacturonide and noncarbohydrate elicitors of proteinase inhibitors cause the phosphorylation of specific plasma membrane proteins in vitro (Farmer et al., 1989).

D. Phospholipases in Signal Transduction Process

Phospholipases have been shown to play important role in signal response transduction mechanisms (Michell, 1982; Berridge and Irvine, 1984, Nishizuka, 1984). Phospholipase C activity has been reported in plants (Irvine, 1982; Helsper et al., 1986a, b) and phosphoinositol phosphates occur in plants (Boss and Massel, 1985; Heim and Wagner, 1986; Morse et al., 1986; Sandeihus and Sommarin, 1986). Pfaffmann et al. (1987) reported the distribution of phosphatidylinositol-specific phospholipase C in plasma membranes of soybean and bushbean (*Phaseolus lunatus*). The soluble and membrane-bound activity of the enzyme was stimulated by calcium and not by calmodulin. Enough evidence is available at present to show that both phosphoinositides and phosphatidylinositol-specific phospholipases are associated with plant membranes (Pfaffmann et al., 1987). Inositol 1,4,5-triphosphate (IP_3) has been reported to stimulate calcium efflux (Drobak and Ferguson, 1985; Rincon and Boss, 1987).

E. H_2^+-Transport ATPases in Signal Transduction Process

Membrane potential of plant cells consists of two components, respiration-dependent electrogenic potential and energy-independent potential (Higginbotham, 1973; Poole, 1978; Spanswick, 1981). Energy-dependent electrogenic potential pump depends upon plasmalemma-H^+-ATPase, a proton pump. The transport of solutes across the plasma membrane is driven by the proton pump (H^+-ATPase) that produces an electric potential and pH gradient. Cellular metabolism is coupled to solute transport by means of the pH and electrical gradient generated by the proton pump embedded within the plasma membrane (Spanswick, 1981).

On the basis of polypeptide composition and sensitivity to inhibitors the plant plasma membrane H^+-ATPase is readily differentiated from proton pumps found in membranes derived from the chloroplast, mitochondria, and vacuole. The purified enzyme contains a single polypeptide of about 100,000 Da that show similarities in a reaction mechanism and structure to group of cation pumping ATPases: H^+-ATPase, Na^+, K^+, and Ca^{2+}-ATPase (Harper et al., 1989).

The membrane potential difference (Pd) of soybean cells is in the range of –180 mV. Addition of potassium cyanide reduced it to –80 mV, suggesting that the energy-dependent component of the membrane potential of soybean is –100 mV (Mayer and Ziegler, 1988). The energy-dependent component is a proton translocating ATPase. Treatment of soybean cotyledonary tissue with glucan elicitors isolated from *Phytophthora megasperma* f.sp. *glycinea* cell walls causes rapid Pd changes. The elicitor induced a transient depolarization within 2 min of contact with the cells, which was followed by a sustained hyperpolarization (Mayer and Ziegler, 1988). Hyperpolarization was always approximately 10 mV and independent of the applied elicitor concentration (1–5 μg/ml). The hyperpolarization induced by the elicitor was maximal at pH 6 and at the same pH the elicitor induced the phytoalexin synthesis to the maximum extent (Mayer and Ziegler, 1988). Fusicoccin hyperpolarizes the membranes of soybean cotyledonary tissue by stimulating a plasma membrane proton pump and also induces phytoalexin synthesis in soybean (Mayer and Ziegler, 1988). These results suggest that the proton pump may be involved in signaling induction of phytoalexins.

Vanadate, an abiotic elicitor, activated the genes that code for a set of enzymes synthesizing the phytoalexins (stibenes) in peanut. The enzyme activities of phenylalanine ammonia-lyase, stilbene synthase, and cinnamate 4-hydroxylase increased 10–100-fold. The dose–response of vanadate as an elicitor of gene expression in intact cells matched precisely its inhibitory effect on the ATPase activity of isolated plasma membrane (Steffens et al., 1989). It was hypothesized that membrane potentials created or modulated by ATPases may be intermediates in the signal chain, starting with the recognition process at the plasma membrane and eventually leading to the production of phytoalexins (Steffens et al., 1989).

Vera-Estrella et al. (1994) studied the possible involvement of plasma membrane H^+-ATPase in elicitation of defense responses in tomato using a cell line of tomato near isogenic for the resistance gene Cf5 and race-cultivar specific elicitors from *Cladosporium fulvum*. A fourfold increase in H^+-ATPase activity was detected immediately after elicitor treatment. Okadaic acid is a specific inhibitor of protein phosphatase 1 and 2A, while staurosporine is a general inhibitor of protein kinases. Okadaic acid almost completely inhibited the increase in H^+-ATPase activity, while the presence of staurosporine did not affect the increase in ATPase activity induced by the elicitor. These results indicated that a phosphatase and not a kinase is involved in the elicitor-induced increase in H^+-ATPase. Phosphokinase is involved in phosphorylation, while phosphatase is involved in dephosphorylation. Hence the protein dephosphorylation may be required for increased H^+-ATPase activity (Vera-Estrella et al. 1994).

F. Cyclic AMP as a Second Messenger

Cyclic AMP-binding proteins and a cAMP-dependent protein kinase have been detected in plant cells (Trewavas and Gilroy, 1990). Kurosaki et al. (1987d) have shown that dibutyryl cAMP induced phytoalexin accumulation in carrot cultures, and the elicitor induced a rapid but transient increase in cAMP levels. However, further evidence is necessary to show the involvement of cAMP in signal transduction mechanism in plants.

G. G Proteins That Control Second Messenger Systems

Guanosine triphosphte-binding (GTP-binding) regulatory proteins or G proteins are found in plasma membranes of several plant cells. G proteins function as mediators in the transduction of signals that interact with receptors on cell surfaces. cDNAs for G protein have been isolated from different crops (Ma et al., 1991; Ishikawa et al., 1995; Seo et al., 1995). G proteins control the second messenger gel systems such as calcium, phospholipase, and cyclic AMP (Freissmuth and Gilman, 1991). There is evidence that GTP can cause Ca^{2+} release from membrane vesicles. GTP-binding proteins may therefore regulate Ca^{2+} movement (Trewavas and Gilroy, 1991). Stimulation of H^+-ATPase in tomato cells treated with elicitor preparations containing the avr5 gene products from race 4 of *Cladosporium fulvum* has been shown as a result of the protein being dephosphorylated by a membrane-bound phosphatase activated through a G-protein (Vera-Estrella et al., 1994).

H. H_2O_2 as Second Messenger

A sudden burst in H_2O_2 production has been observed in cells treated with elicitors. When fluorescent dyes such as pyranine are equilibrated with soybean cells, the cells absorb the dyes and display a bright fluorescence. This fluorescence suddenly changes when an elicitor is added to the cell suspension (Low and Heinstein, 1986; Apostol et al., 1987). The decrease in fluorescence density is seen after about 8 min of addition of elicitor. Since the bleached fluorescent probes contained readily oxidizable functional groups, an elicitor-active oxidative process might be involved in probe modification. Cell suspensions were treated with superoxide dismutase, mannitol, or catalase prior to elicitation in order to destroy any elicitation-generated O_2^-, OH, or H_2O_2, respectively. Superoxide dismutase and mannitol had no effect on the elicitor-induced fluorescence transition of pyranine. Catalase nearly completely obliterated the quenching reaction (Apostol et al., 1987). These observations indicate that H_2O_2 is required in the fluorescence bleaching process.

To evaluate whether elicitation triggers the oxidative burst by activating the peroxidase or by stimulating production of hydrogen peroxide, H_2O_2 was added to the labeled cells in the absence of elicitor (Apostol et al., 1989). The dye was instantaneously bleached, indicating that the oxidative enzymes are constitutively present and simply await the elicitor-stimulated production of H_2O_2 to destroy susceptible compounds. When H_2O_2 was added to the suspension cell culture, a significant production of phytoalexins was observed. When catalase was introduced prior to elicitation, the phytoalexin production was reduced by nearly 80%. When addition of catalase was delayed until 1 h after elicitor addition, no significant inhibition of phytoalexin biosynthesis was observed. This strict dependence of catalase inhibition on its time introduction demonstrates that catalase must interfere with one of the initial events of elicitation, a step or process that is complete within 1 h of the initial extracellular stimulus but considerably before any phytoalexins are produced (Apostol et al., 1989). Normally, phytoalexins are produced after 2–4 h after stimulation with elicitor (Keen and Bruegger, 1977; Ebel et al., 1984; Ryan et al., 1985). All of the inhibitors of the oxidative burst examined blocked glyceollin production (Table 18; Apostol et al., 1989). These data suggest the participation of H_2O_2 in the signal transduction.

An increase in chemiluminescence in the presence of luminol occurred rapidly after treatment of bean cells with elicitor from α-race of *Colletotrichum lindemuthianum* (Anderson et al., 1991). Increase in chemiluminescence occurred immediately and was maximal about 30 min after treatment. The increase in chemiluminescence indicates production of activated oxygen species, and the increase in chemiluminescence was inhibited totally by the presence of catalase, suggesting that the response involves production of H_2O_2. Following the production of H_2O_2, proteins appeared in elicitor-treated cells by 6 h. Phenolic accumulation in cells treated with the elicitor was above those of control cells by 6 h. The phytoalexins were detected in the medium of elicitor-treated cells by 8 h. These observations suggest that the activated oxygen species may act as second

Table 18 Effect of peroxidase inhibitors on elicitor-stimulated pyranine oxidation and glyceollin production in cultured soybean cells

	Inhibition (%)	
Inhibitor	Pyranine oxidation	Glyceollin formation
None	0	0
Catechol	60	43
Diethyldithiocarbamate	75	56
$Fe(CN)_6^{-4}$	90	74
$FeSo_4$	100	50
Salicylylhydroxamic acid	75	41

Source: Apostol et al., 1989.

messengers for the synthesis of phenolics, phytoalexins and new proteins (Anderson et al., 1991).

The elicitor-stimulated production of H_2O_2 may represent another class of receptor-linked transplasma membrane redox components, since the interaction of several hormones with their receptors has been found to regulate transplasma membrane redox enzymes capable of oxidizing NADH to generate H_2O_2 (Ramasarma, 1982; Crane et al., 1985) and H_2O_2 mimicks many of the effects of hormones like insulin (Koshio et al., 1988).

I. Ethylene as Second Messenger

The increased production of ethylene is one of the earliest chemically detectable events in pathogen-infected plants or treated with elicitors (Toppan and Esquerre-Tugaye, 1984). Ethylene stimulates defense mechanisms against pathogens, in some instances. Ethylene production and cell wall hydroxyproline-rich glycoprotein (HRGP) biosynthesis are greatly enhanced in melon seedlings infected with *Colletotrichum lagenarium* (Toppan et al., 1982). Treatment of melon petioles with an elicitor from *Colletotrichum lindemuthianum* induced ethylene and HRGP. The ethylene induction was observed even at 6 h after treatment while induction of HRGP occurred after 24 h (Table 19; Roby et al., 1985). Aminoethoxyvinylglycine (AVG) is an inhibitor of biosynthesis of ethylene. In the presence of AVG, the elicitor-induced ethylene was decreased, and this decrease was paralleled by a quantitatively similar inhibition of HRGP synthesis. 1-Aminocyclopropane-1-carboxylic acid (ACC), a precursor of ethylene, stimulated HRGP biosynthesis within 4–6 h after the addition of the compound and ACC triggers the synthesis of HRGP to the same extent as 0.1 and 0.3 mg of

Table 19 Time course elicitation of ethylene and cell wall HRGP in melon petioles treated with elicitor

		Ethylene production		HRGP	
Incubation time (h)	Assay	nl/g petioles	% of control	n Bq/g petioles	% of control
6	Control	0.18	—	0.085	—
	+ Elicitor	0.28	155	0.076	89
18	Control	0.35	—	0.501	—
	+ Elicitor	0.53	151	0.485	96
24	Control	0.38	—	0.363	—
	+ Elicitor	0.83	218	0.504	139
30	Control	0.85	—	0.619	—
	+ Elicitor	1.25	147	0.907	146

Source: Roby et al., 1985.

elicitor (Roby et al., 1985). These results suggest that ethylene may be a signal in induction of HRGP.

ACC synthase is the first enzyme in the biosynthetic pathway of ethylene (Boller and Kende, 1980; Yu and Yang, 1980; De Laat and Van Loon, 1982; Fuhrer, 1982). The elicitor isolated from cell walls of *Phytophthora megasperma* induced ACC synthase activity about 10-fold within 1 h after treatment in parsley cell cultures. PAL activity started to increase after 1 h and reached its peak after 8 h of elicitor treatment. Chitinase activity started to increase in parsley cells after 3 h of elicitor treatment and reached its peak activity at about 12 h after treatment (Chappell et al., 1984). This induction of ethylene is the first event, which is followed by induction of defense mechanisms.

AVG, a specific inhibitor of ACC synthase, inhibited the induction of PAL activity by elicitor. AVG almost completely inhibited the enhancement of ethylene biosynthesis in elicitor-treated cells. The results suggest that ethylene may be a signal for PAL induction (Chappell et al., 1984).

AVG partially inhibited the induction of PAL in *Phaseolus vulgaris* by elicitors of *C. lindemuthianum* and this inhibition could be reversed by ACC and ethylene. Unless introduced within 2 h of elicitor application, ethylene was without effect on PAL production. Likewise, the inhibitory effect of AVG was minimal if not injected within the same time period following elicitor treatment. Elicitor rapidly induced the synthesis of ACC during this 2 h period leading to the production of ethylene. This was completely inhibited by AVG and completely reversed by ACC. It has been concluded that ethylene is involved in the activation of the induced response leading to the synthesis of PAL, and thereby phytoalexins, particularly when the response to elicitor is suboptimal (Hughes and Dickerson, 1989).

J. Other Messengers

Polyamines are considered as messengers that stimulate phosphorylation of various membrane proteins. Most of these proteins are different from those phosphorylated in the presence of calcium (Veluthambi and Poovaiah, 1984a–c; Datta et al., 1986).

A phospholipid that stimulates protein kinase activity has been suggested to be a plant regulatory molecule (Schafer et al., 1985). Glutathione, a tripeptide found throughout the plant kingdom, has been suggested to be a signal molecule in bean. Gluthathione induces high levels of expression of defense genes in bean cell suspension cultures (Wingate et al., 1988). It activates the bean chalcone synthase (CHS) promoter linked to a reporter gene in electroporated soybean or alfalfa protoplasts (Dron et al. 1988).

X. SYSTEMIC SIGNAL TRANSDUCTION

A. Oligogalacturonides as Systemic Signal Molecules

Oligogalacturonides have been reported to be one of the important systemic signal transduction molecules. In tomato plants injury of one leaf induces in neighboring uninjured leaves of the same plant a "defense response" by synthesizing proteinase inhibitor (PI) proteins. It suggests that messenger released at the site of injury is capable of moving to other leaves where it induces the defense response. The hypothetical messenger (hormone) was named "proteinase inhibitor inducing factor" (PIIF) (Green and Ryan, 1972). PIIF was translocated away from the injury at approximately 5 cm/h (Green and Ryan, 1973). It appeared to move via the phloem (Zuroske et al., 1980). The PIIF was identified as pectic polysaccharides (Ryan, 1974, 1987; McNeill et al., 1984; Ryan et al., 1981, 1986; Bishop et al., 1984; Lau et al., 1985). Bishop et al. (1981) reported that pectic polysaccharides, and oligosaccharides derived by partial enzymic hydrolysis of them, evoked the synthesis of PI when imported into tomato leaves via the transpiration stream. Hence it was speculated that such oligosaccharides are mobile wound-hormones that move from injured leaves to neighboring uninjured leaves, where they induce PI synthesis. Active pectic fragments may arise from degradation of the plant cell wall by hydrolytic enzymes that are activated during wounding (Bishop et al., 1981). Very small injuries were made more effective at PI induction by application of exogenous endopolygalacturonase (Ryan et al., 1985). However, Baydoun and Fry (1985) showed that pectic oligosaccharides with DP 6-9 and 10-14 did not move acropetally or basipetally (Table 20). They suggested that pectic substances are not themselves long-distance wound hor-

Table 20 Possible systemic movement of pectic oligosaccharide

Carbohydrate applied	Distribution of radioactivity after 20 h		
	Cpm in treated leaf	Cpm moving acropetally	Cpm moving basipetally
[^{3}H]pectic oligosaccharide			
DP_{6-9}	23,900	9.9	3.5
DP_{10-14}	19,500	14.3	3.1
[^{14}C] Sucrose[a]	41,300	3148	2010

[a]Sucrose alone moved systemically.
Source: Baydoun and Fry, 1985.

mones. Probably they may function as the alarm signal or cause its release. Calcium polypectates are resistant to hydrolytic enzymes. Sequestering of Ca^{2+} from host tissues may make the pectic substances more susceptible to hydrolytic enzymes. Solutions of K_3PO4, K_2H PO4, Na_2PO4, and NA_2HPO4 sprayed on the undersides of the first and second true leaves of cucumber-induced systemic resistance in leaves 3 and 4 to anthracnose caused by *Colletotrichum lagenarium* (Gottstein and Kuc, 1989; Descalzo et al., 1990). A pH above 7 markedly enhanced the activity of phosphates, and these phosphates form highly water-insoluble complexes with Ca^{2+} at alkaline pH values. The sequestering of Ca^{2+} from host tissues by the phosphates may affect membranes, destroy cell compartmentalization, and cause the release or synthesis of hydrolytic enzymes. It may lead to breakdown of pectic substances and the oligogalacturonides formed may function as the alarm signal (Gottstein and Kuc, 1989). Spraying of extracts from spinach (*Spinacia oleracea*) or rhubarb (*Rheum rhaponticum*) on cucumber leaves also induced systemic resistance against *C. lagenarium*. The active component of both extract was identified as oxalate (Doubrava et al., 1988). Oxalic acid may precipitate calcium from the middle lamellae to form calcium crystals, leaving pectic materials more susceptible to enzymatic degradation resulting in formation of oligogalacturonide elicitors.

The mechanism by which oligosaccharide molecules activate the proteinase inhibitor genes is not known. The oligosaccharide molecules have been associated with membrane receptors (Schmidt and Ebel, 1987) and with changes in protein phosphorylation patterns of membranes (Farmer et al., 1989; 1991) and cellular proteins during the induction process (Farmer et al., 1991). The induction of proteinase inhibitors in tomato leaves by pectic fragments is sensitive to a range of chemicals that affects proton transport (Doherty and Bowles, 1990; Thain et al., 1990). The oligosaccharide fragments cause rapid and reversible fluctuations in cell membrane potential (Thain et al., 1990). Ca^{2+}-uronide complexes form aggregated "egg-box" structures with polygalacturonic acid and larger uronides of polygalacturonic acid (Kohn, 1985). In these structures, the uronides are the "boxes," complexed with Ca^{2+} as the eggs (Ryan, 1992). Egg-box structures do not appear to form between oligouronides with DPs below about 10 uronide units and Ca^{2+} (Grant et al., 1973). It suggests that active species of polygalacturonides that induce proteinase inhibitors in vivo and that induce protein phosphorylation in vivo may be large Ca^{2+}–uronide aggregates (Farmer and Ryan, 1990; Ryan, 1992). Induction of proteinase inhibitors due to fungal invasion and elicitor treatment has been reported in plants (Walker-Simmons and Ryan, 1984; Graham et al., 1985a,b; 1986; Lee et al., 1986; Thornburg et al., 1987) and it is possible that polygalacturonides may also be involved in these interactions.

B. Salicylic Acid as Systemic Signal Molecule

Salicylic acid or ortho-hydroxy benzoic acid belongs to a diverse group of plant phenolics. It has been reported in all the 34 plant species tested for its content (Raskin et al., 1990). The compound is an aromatic ring bearing a hydroxy group.

Salicylic acid (O-hydroxybenzoic acid) is synthesized from *trans*-cinnamic acid. PAL is the key enzyme in the synthesis of salicylic acid. Two pathways of biosynthesis of salicylic acid have been suggested (Fig. 3; Funk and Brodelius, 1990a,b; Schnitzler et al., 1992).

Radioactive salicylic acid was formed via O-coumaric acid by leaf segments of *Primula acaulis* and *Gaultheria procumbens* after they were fed ^{14}C-labeled phenylalanine or cinnamic acid (El-Basyouni et al., 1964). In the same species, labeled salicylic acid was also formed after treatment with [^{14}C] benzoic acid (El-Basyouni et al., 1964). In tomato salicylic acid has likewise been reported to be formed through both the pathways (Chadha and Brown, 1974), but benzoic acid alone serves as a precursor for the formation of salicylic acid in potato, pea, and sunflower (*Helianthus annuus*) (Klambt, 1962). In tobacco and

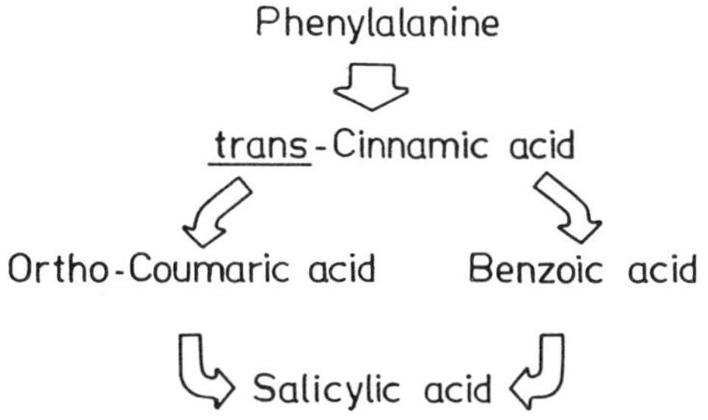

Figure 3 Biosynthesis of salicylic acid.

rice salicylic acid is formed from cinnamic acid via benzoic acid (Yalpani et al., 1993; Leon et al., 1993; Silverman et al., 1995).

Plants inoculated with necrotophic pathogens such as fungi, bacteria, or viruses react by inducing a transient resistance against subsequent fungal, bacterial, or viral infection (Ross, 1966; Vidhyasekaran et al., 1971; Sequeira, 1983). This induced resistance becomes systemic and it is termed systemic acquired resistance (SAR). SAR is mediated by an endogenous signal that is produced in the infected leaf and translocated in the phloem to other plant parts, where it activates resistance mechanisms. Salicylic acid has been shown to increase at the onset of systemic acquired resistance in cucumber (Metraux et al., 1990). When cucumber leaves were inoculated with *Colletotrichum lagenarium,* a distinct increase of a fluorescing metabolite was detected in the phloem after inoculation. This increase appeared before necrotization had taken place on the infected leaf and preceded the induction of resistance observed in the upper uninfected leaves. The concentration of salicylic acid in the phloem sap before resistance occurs ranged typically between 0.2 and 7 μM in infected plants, compared to 0 and 0.7 μM in control plants (Metraux et al., 1990). The ability to synthesize salicylic acid by the plants has been shown as requirement to induce resistance against pathogens (Gaffney et al., 1993).

Salicylic acid induced resistance against *C. lagenarium* when exogenously applied to cucumber cotyledons at a concentration of 14.5 mM (Metraux et al., 1990). It has no direct antifungal activity against *C. lagenarium* and no fungitoxic metabolites were detected in salicylate-treated cucumber tissue. Exogenously applied salicylic acid has been shown to be translocated in cucumber. When ^{14}C-labeled salicylic acid was infiltrated into a lower leaf, 1.3% of the initial salicylic acid could be found in the treated leaf 1 day after application, whereas in the upper leaf 0.4% was recovered (Metraux et al., 1990). Along with SAR, induction of PR proteins was common. In cucumber the major PR protein is an extracellular endochitinase. This protein and its mRNA accumulated in leaves treated with either pathogens or salicylic acid. The induction of the chitinase gene by salicylic acid shows that salicylic acid is a signal compound. In tobacco, exogenously applied salicyclic acid induces PR proteins and resistance (White, 1979; Van Loon and Antoniw, 1982). The SAR is not limited to one pathogen and it is shown against fungal pathogens including *Cercospora nicotianae, Peronospora tabacina,* and *Phytopthora parasitica* var. *nicotianae,* bacterial pathogens such as *Pseudomonas syringae* pv. *tabaci,* and viruses including tobacco mosaic virus (Ward et al., 1991).

Salicylic acid application induced nine gene families that encode PR proteins in tobacco (Ward et al., 1991). They are PR-1 (acidic, extracellular, function unknown); PR-2 (acidic, extracellular β-1,3-glucanase); PR-3 (acidic, extracellular chitinase), PR-4 (acidic, extracellular, unknown function), PR-5 (acidic, extracellular, thaumatin-like protein and bifunctional amylase/proteinase inhibitor of maize), PR-1 basic (basic form of PR-1), basic class III chitinase (structurally unrelated to PR-3) acidic class III chitinase (extracellular; approximately 60% identical to basic form), and PR-Q′ (acidic, extracellular β-1,3-glucanase). These gene families include the PR proteins PR-1 (PR-1a, PR-1b, and PR-1c), PR-2 (PR-2a, PR-2b, and PR-2c), PR-3 (PR-3a and PR-3b), PR-4 (PR-4a and PR-4b), and PR-5 (PR-5a and PR-5b), as well as the basic form of PR-1, the acidic and basic forms of class III chitinase, and PR-Q′ (Ward et al., 1991). Salicylic acid treatment induced expression of the same set of genes as TMV infection induces and the same kind of accumulation of PR proteins. High level of expression of these genes correlated well with the onset of a resistant state (Ward et al., 1991). Since salicylic acid induced the genes comparable with biologically induced resistance, it is probable that salicylic acid acts as a signal molecule. Salicylic acid may play a role in coordinating plant defense gene expression (Enyedi et al., 1992).

The fact that salicylic acid acts as a systemic signal has been proved further by the studies of Yalpani et al. (1991) in tobacco. Excised healthy leaves were fed salicylic acid for 72 h through the cut petiole and PR-1 protein levels were analyzed in opposite half-leaves. The level of salicylic acid in a leaf was proportional to the concentration of salicylic acid in the solution in which the petiole was immersed. Induction of PR-1 proteins was positively correlated with leaf salicylic acid. The average basal level of salicylic acid in control leaves was 34 ng salicylic acid/g fresh weight. A 59% increase in tissue salicylic acid levels, to 54 ng/g fresh weight, caused detectable induction of PR-1a in the extracellular fluid.

A fungal cell wall preparation from *Chaetomium globosum* induced an accumulation of salicylic acid and induced the production of chitinase in carrot suspension culture (Schneider-Muller et al., 1994). Induction of salicylic acid has been shown to require Ca^{2+}. A specific calcium chelator (EGTA) and a calcium blocker (verapamil) inhibited the production of salicyclic acid and, in turn, inhibited accumulation of chitinase (Schneider-Muller et al., 1994). The results suggest that salicylic acid may play a key role in transferring intracellular signal transmitted by calcium ion. A soluble binding site for salicylic acid that appears as an aggregate of approximately 650 kDa has been reported in tobacco (Chen and Klessig, 1991).

The measured physical properties of salicylic acid, such as pKa = 2.98 and log Kow (octanol/water partitioning coefficient) = 2.26, are nearly ideal for long distance transport in the phloem (Yalpani et al., 1991). All these observations suggest that salicyclic acid acts as an endogenous signal in induction of PR proteins and some other components of defense mechanisms. This conclusion is based on the fact that salicyclic acid meets the essential criteria of a signal molecule: salicylic acid induces resistance to pathogens; salicylic acid induces PR proteins; salicylic acid levels increase locally and systemically following pathogen attack; and salicylic acid moves throughout the plant via phloem.

When *Pseudomonas syringae* pv. *syringae* was inoculated, salicylic acid was detected in the phloem exudate from cucumber leaves. The earliest detectable increase in salicyclic acid was observed 8 h after inoculation with the bacterial pathogen. The systemic accumulation of salicyclic acid was observed even when the inoculated leaf remained attached to the plant for only 4 h (Rasmussen et al., 1991). It suggests that another chemical signal may be required for the systemic accumulation of salicylic acid in cucumber (Raskin, 1992).

C. Systemin as Systemic Signal Molecule

A polypeptide free of carbohydrates was detected in tomato leaf extracts and induced proteinase inhibitor activity when supplied to young tomato plants. Slightly more than 1 μg of the active factor was isolated from approximately 60 lbs. tomato leaves. The compound was purified and sequenced (Fig. 4; Pearce et al., 1991).

No significant similarities were found in known protein sequences. This sequence was, however, a palindrome: X × Q × BPP × BB × PPB × QXX (X, any residue: B, lysine or arginine; Q, glutamine; and P, proline). A polypeptide corresponding to residues 2 through 18 was synthesized. The synthetic polypeptide was as effective as the native polypeptide for inducing the synthesis and accumulation of both proteinase inhibitor I and II when supplied to the cut stems of young plants. About 40 fmol of the polypeptide per plant was required to produce half maximal accumulation of inhibitors I and II, which represents about 10^5 times more activity on a molar basis than the PIIF oligogalacturonide inducers from plant cell walls. The polypeptide is transported to distal tissues. ^{14}C-labeled polypeptide was synthesized and placed on fresh wounds of tomato plants. Within 30 min the radioactivity had moved throughout the leaf, and within 1–2 h it was identified in the phloem exudate. The polypeptide was named "systemin" (Pearce

$+H_3N$ - AVQSKPPSKRDPPKMQTD - COO^-

A - Alanine;	**D - Aspartic acid;**	**K - Lysine;**
M - Methionine;	**P - Proline;**	**Q - Glutamine;**
R - Arginine;	**S - Serine;**	**T - Threonine;**
V - Valine;		

Figure 4 The amino acid sequence of systemin.

et al., 1991). The role of systemin in induction of proteinase inhibitors is further discussed in chapter 4.

D. Methyl Jasmonate and Jasmonic Acid as Systemic as Well as Interplant Signal Molecules

Jasmonic acid and methyl jasmonate are naturally occurring compounds in plants from at least nine families (Farmer and Ryan, 1990). About 10 ng to as much as 3 μg/g fresh weight of them have been reported in different tissues (Falkenstein et al., 1991; Staswick, 1992). The acid seems to be the more prevalent form (Farmer et al., 1992). Methyl jasmonate is a constituent of the fragrance from several flowers including jasmine (Staswick, 1992). Jasmonate has been reported to induce proteins in 26 plant species (Herman et al., 1989). When methyl jasmonate was sprayed on leaves of tomato plants, it powerfully induced the synthesis and accumulation of proteinase inhibitor I protein. The chemical induced the accumulation of inhibitor I to levels higher than that could be induced by wounding. Control plants that had not been sprayed with methyl jasmonate, but incubated in the same chambers with the sprayed plants, accumulated low levels of proteinase inhibitor I protein (Table 21; Farmer and Ryan, 1990).

The results suggest that volatile methyl jasmonate was inducing synthesis of proteinase inhibitors in the nearly untreated control plants. When tomato plants were placed in air-tight chambers together with cotton-tipped wooden dowels onto which various dilutions of methyl jasmonate in ethanol has been applied, synthesis and accumulation of proteinase inhibitors I and II were observed in tomato leaves in a dose-dependent manner (Farmer and Ryan, 1990). Tomato plants began to accumulate proteinase inhibitors I and II about 5 h after the initial exposure to the volatile compound (Farmer and Ryan, 1990).

Table 21 Accumulation of proteinase inhibitor I in tomato leaves induced by methyl jasmonate

Plants	Proteinase inhibitor I (μg/g of tissue)
Chamber A	
Methyl jasmonate-sprayed	210
Control	28
Chamber B	
Wounded	86
Control	0

Control plants incubated in separate chambers did not accumulate proteinase inhibitor I at all.
Source: Farmer and Ryan, 1990.

When tobacco and alfalfa plants were exposed to methyl jasmonate, trypsin inhibitor content increased several-fold (Table 22; Farmer and Ryan, 1990).

When tomato plants were incubated in air-tight chambers along with leafy branches of *Artemisia tridentata,* the leaves of tomato plants exhibited elevated levels of proteinase inhibitors I and II. *A. tridentata* contains methyl jasmonate in its leaves (Farmer and Ryan, 1990). These results strongly suggest a role for methyl jasmonate in interplant and systemic signal transduction system.

Methyl jasmonate and/or jasmonic acid induce many other defense genes encoding phenylalanine ammonia-lyase (Gundlach et al., 1992), proline-rich cell wall protein (Creelman et al., 1992), leaf polypeptides (Wiedhase et al., 1987), and lipoxygenase (Bell and Mullet, 1991; Grimes et al., 1992; Melan et al., 1993). The role of methyl jasmonate in induction of proteinase inhibitors is discussed further in Chapter 4.

E. Fatty Acids as Systemic Signal Molecules

Plant lipoxygenases play an important role in defense mechanisms against pathogens (Preisig and Kuc, 1987a, b; Hildebrand et al., 1988; Bostock and Stermer, 1989; Siedow, 1991). Normal substrates for plant lipoxygenases are linoleic and linolenic acids. Lipoxygenase products of these fatty acids are metabolized to compounds that may function as signal molecules. Methyl jasmonate, the important systemic signal molecule, is formed due to the action of lipoxygenase (Farmer and Ryan, 1992). Another signal molecule, traumatin, is also synthesized by action of lipoxygenase (Vick and Zimmerman, 1987).

Arachidonic and eicosapentaenoic acids are the important elicitors of *Phytophthora infestans* and they are also substrates for lipoxygenase (Bostock et al., 1981; Preisig and Kuc, 1985). Lipoxygenase has been suggested to be an intermediate enzyme in the pathway between treatment with arachidonic acid and

Table 22 Increase in trypsin inhibitor in tobacco and alfalfa plants exposed to methyl jasmonate

Plants	Trypsin inhibitor (μg/g leaf tissue)
Tobacco	
Exposed to methyl jasmonate	116
Control	7
Alfalfa	
Exposed to methyl jasmonate	385
Control	33

Source: Farmer and Ryan, 1990.

induced defense mechanism. Inhibitors of lipoxygenase abolished the elicitor activity of arachidonic acid (Preisig and Kuc, 1987b).

The role of lipoxygenase in signaling defense mechanisms of the host will be discussed in detail in Chapter 3. Cohen et al. (1991) demonstrated that both arachidonic acid and eicosapentaenoic acid induced systemic resistance to *Phytophthora infestans* in foliage. Ricker and Bostock (1992) showed that arachidonic acid diffuses within developing lesions and may therefore provide a signal directly to the surrounding cells.

F. Abscisic Acid as Systemic Signal Molecule

Abscisic acid may also be a key factor in the systemic induction of proteinase inhibitor genes (Pena-Cortes et al., 1989). Spraying abscisic acid on potato plants induces proteinase inhibitor mRNA synthesis in leaves (Pena-Cortes et al., 1991).

G. Ethylene as Systemic Signal Molecule

Ethylene also induces systemic resistance and modulates gene expression at the posttranscriptional level, suggesting a role as a downstream modulator of response initiated by elicitor-induced transcriptional activation (Lincoln and Fischer, 1988).

XI. HOW DO PATHOGENS AVOID OR OVERCOME ELICITOR-INDUCED HOST DEFENSE MECHANISMS?

A. Delayed Release of Elicitors into Host Tissues May Provide Time for the Pathogen to Establish Itself in the Host

It is now well established that elicitor molecules are present in fungal cell surface. Virulent pathogens or avirulent mutants or even saprophytes do have elicitors in their cell wall and they induce defense mechanisms in all kinds of plants, whether they are susceptible or resistant to the pathogen (Frank and Paxton, 1971; Beretta et al., 1977; Humme et al., 1978; Cline et al., 1978; Paradies et al., 1979; Mayama, 1983; Beardmore et al., 1983; Tietjen et al., 1983; Tietjen and Matern, 1983; Darvill and Albersheim, 1984; Kogel et al., 1985; Albersheim et al., 1986; Ebel, 1986; Moerschbacher et al., 1986 b; 1988; Dangel et al., 1987; and Sutherland et al., 1989). Even the amount of elicitors present in the cell wall does not differ much between pathogens and nonpathogens. Both compatible and incompatible races of *P. infestans* do not differ significantly in the amounts of the two elicitors, arachidonic acid (AA) and eicosapentaenoic acid (EPA) present in their cell walls (Creamer and Bostock, 1986; 1988). Upon contact with the host cell surface, the elicitors are released from the fungal cell wall due to action of host enzymes. Release of elicitors appears to be an important phenomenon during pathogenesis. During successful pathogenesis, the elicitor may not be released or the release may be delayed. A rise in lipoxygenase activity was observed during the incompatible interaction in cotyledonary leaves of coffee (*Coffea arabica*) infected with the rust fungus *Hemileia vastatrix* (Rojas et al., 1993). Similar increase was not observed during the compatible interaction, but an elicitor preparation from germ tubes of the rust fungus stimulated a rise in lipoxygenase activity in both resistant and susceptible cultivars (Rojas et al., 1993). The results suggest that the elicitor can induce defense activity in both the susceptible and resistant varieties, but the fungus could induce it only in the resistant variety. Hence the release of elicitor from the fungal cell wall may be important in determining the susceptibility or resistance of a host.

The release of elicitors from fungal wall into the host tissues was well documented in potato infected with *Phytophthora infestans* (Ricker and Bostock, 1992). The fungal sporangia were radiolabeled with [^{14}C]AA to trace the release of the elicitor into potato leaf tissues. Microautoradiography of these sproangia revealed that radioactivity was distributed throughout fungal cytoplasm and cell walls, and was often associated with lipid bodies. Potato leaves were inoculated with the radiolabeled sporangia of both compatible and incompatible races. At 3 and 6 h after inoculation radioactivity was not observed within the leaf tissue but was restricted to sporangia. By 9–10 h after inoculation most sporangia of both races had germinated and had formed appressoria, but there was little or no ramification of infection hyphae into leaf tissue. In the compatible interaction the label was detected only by 12 h after inoculation, while in the incompatible interaction the label was found consistently in plant tissue at 9 h after inoculation itself. By 12 and 14 h after inoculation the radioactivity was present throughout the tissue sections in both interactions and no difference between compatible and incompatible interactions could be seen (Ricker and Bostock, 1992).

Hence the only difference between susceptible and resistant reactions was a 3 h delay in release of elicitor in the former interaction. If this delay is compensated for by providing the elicitor at an early period of infection process, even the susceptible interaction can become a resistant one. It has been demonstrated that pretreatment of potato tuber slices with arachidonic acid or eicosapentaenoic acid suppressed colonization of the tubers by compatible races of *P. infestans* (Bostock et al., 1986). Similar observations have been reported in rice–*Pyricularia oryzae* interactions (Schaffrath et al., 1995). The elicitor isolated from *P. oryzae*, the blast pathogen, was infiltrated into the intercellular spaces of leaves of a susceptible rice cultivar. After 4 h, the leaves were inoculated with a virulent race of *P. oryzae*. There was a significant reduction of blast symptoms on elicitor-treated leaves to water-infiltrated leaves (Schaffrath et al., 1995). The growth of the pathogen was stopped by early induction of defense mechanisms by infiltrating the elicitor in the genetically susceptible rice cultivar. This response was similar to that observed in the pathogen-inoculated genetically resistant cultivar without any prior treatment with the elicitor (Schaffrath et al., 1995). The results suggest that only due to nonrelease of elicitor in the initial stages of infection, is the susceptible cultivar invaded by the pathogen.

What is the possible mechanism through which the release of elicitor is delayed during successful pathogenesis? Very few attempts have been made to answer this question. Bostock (1989) studied the method of release of elicitors from cell walls of *P. infestans*. The lipids containing the radioisotope-labeled elicitors arachidonic and eicosapentaenoic acids were extracted from *P. infestans* and applied to potato discs subsequently inoculated with spores of an incompatible or compatible isolate of *P. infestans*. Inoculation with either race resulted in a significant decline in the proportion of radioactivity recovered in triglycerides and concomitant increase in polar lipid and free fatty acid pools, indicating degradation of lipids containing esterified eicosapolyenoic acids, with a corresponding increase in the free fatty acid pool. Activities of the lipid-degrading enzymes, acyl hydrolase and lipoxygenase were similar in both compatible and incompatible interactions. During the first 12 h after inoculation, lipase activity was suppressed in compatible interaction but not in the incompatible interaction (Bostock, 1989). Triglycerides are the most abundant sources of arachidonic acid and eicosapentaenoic acid in *P. infestans* (Creamer and Bostock, 1986) and suppression of lipase activity may favor the development of the pathogen due to reduction in the availability of the elicitors in the initial period of infection (Bostock, 1989; Ricker and Bostock, 1992).

B. Elicitors May be Released During Successful Pathogenesis but May Not be Active in Susceptible Cultivars

In some host–pathogen interactions, elicitors may be released, but they may not be active during successful pathogenesis. Elicitors of resistance have been isolated from basidiospore-derived infection structures of *Uromyces vignae,* the cowpea (*Vigna unguiculata*) rust pathogen (Chen and Heath, 1990). Intercellular washing fluid from teliospore-inoculated leaves of a susceptible variety showed appreciable elicitor activity when tested on a resistant variety, but no appreciable elicitor activity was observed when it was tested on a susceptible variety (Table 23; Chen and Heath, 1992). The results show that the elicitor may be released in the susceptible variety during pathogenesis, but it may not be active on the susceptible cultivar (Chen and Heath, 1992). The elicitor is active only in the incompatible plant–fungus interaction.

An elicitor that is a glycoprotein has been isolated from germ tube walls of wheat stem rust uredo sporelings (Kogel et al., 1988; Beissmann and Reisener,

Table 23 Necrosis-inducing activity of intercellular washing fluid from *Uromyces vignae*-infected cowpea leaves

Source of intercellular washing fluid[a]	Number of necrotic cells	
	Resistant variety	Susceptible variety
Inoculated with teliospores	4923	831
Uninoculated control	364	328

[a]Susceptible variety was used to prepare the intercellular washing fluid.
Source: Chen and Heath, 1992.

1990). When the pathogen was inoculated on primary leaves of a susceptible wheat cultivar, elicitor activity was detected in intercellular washing fluid (IWF). The elicitor activity was tested by assessing phenylalanine ammonia-lyase (PAL) inducing acitivity. PAL-inducing activity of IWF from rust-infected leaves significantly increased at 5 days after inoculation, reaching a maximum at 6 days after inoculation (Beissmann et al., 1992). Elicitor-activity was detected in IWF from the compatible interaction of various wheat cultivars such as Prelude, Kanzler, Little Club, and Ares with *P. graminis* race 32 at 6 days after inoculation, but no activity was found in IWF from incompatiable interactions of race 32 and the resistant cultivars Feldkrone and Prelude. The results suggest that elicitors are released only in the susceptible interactions. Probably enough quantity of elicitor for extraction tissues is produced only when large amount of mycelium accumulates, which can occur only in susceptible interactions. In the resistant reactions, reduced fungal growth occurs and hence only trace amounts of elicitor may be present in the intercellular fluid, which the authors could not have extracted. Although the elicitor was not detected in the infected resistant tissues, the elicitor induced more PAL activity in resistant variety than in the susceptible variety (Beissmann et al., 1992).

Cladosporium fulvum produces elicitors in culture that can elicit a hypersensitive response in host tissue, but these elicitors are not race-specific (Dow and Callow, 1979; Lazarovitz et al., 1979; Lazarovitz and Higgins, 1979; Dewit and Roseboom, 1980; Dewit and Kodde, 1981a, b). However race-specific elicitors have been identified in apoplastic fluids from susceptible cultivars infected with a compatible race of the pathogen (DeWit and Spikman, 1982). A number of physiological races of the fungus and many tomato cultivars with different genes for resistance have been reported (Day, 1956; Kerr et al., 1971; Schotten-Toma and DeWit, 1988) (Table 24).

The elicitor accumulated only in tomato leaves of susceptible cultivars

Table 24 Differential interactions between races with different avirulence genes and tomato cultivars with different genes for resistance to *Cladosporium fulvum*

		Races of *Cladosporium fulvum*			
Tomato cultivar	Genes for resistance	4 $(A_2A_5A_9)^a$	5 $(A_2A_4A_9)^a$	2.4 $(A_5A_9)^a$	2.4.5.9 (None)
Moneymaker	None	C	C	C	C
Near-isogenic line Cf4 of Moneymaker	Cf4	C	I	C	C
Near-isogenic line Cf5 of Moneymaker	Cf5	I	C	I	C
Sonato	Cf2 Cf4	I	I	C	C
Sonatine	Cf2 Cf4 Cf9	I	I	I	C

[a]Genes for avirulence
C, Compatible interaction; I, Incompatible interaction.
Source: Schotten-Toma and Dewit, 1988.

infected with races containing avirulence gene A9. It was not detected in compatible interactions involving race 2.4.5.9, which lack avirulence gene A_9 (Schottens-Toma and DeWit, 1988). The elicitor was not detectable in apoplastic fluids from uninoculated plants or from incompatible interactions, as mentioned in Table 24. The elicitor was detected 8–10 days after inoculation of the tomato plants. Mannitol is a specific reserve sugar of *C. fulvum* and was used as a marker of fungal biomass. The appearance of the race-specific elicitor coincided with the time when mannitol started to accumulate (Table 25; Schottens-Toma and DeWit, 1988). The results indicate that the elicitor is of pathogen origin.

The race-specific elicitor was purified and found to be a peptide. The peptide contained 28 amino acids, the first one being tyrosine. It contained six cysteine residues and the molecular weight of the elicitor was 3049 Da (Schottens-Toma and DeWit, 1988). This peptide is processed in the host from a 63 amino acid precursor protein (Van Kan et al., 1991, Vanden Ackerveken et al., 1993). The amino acid sequences of the peptides isolated from various compatible interactions cf4/race, 4, cf2,cf4/race 2,4, and cf5/race 5 were identical and it is possible that the peptide is a product of avirulence gene A9 of *C. fulvum* (Schottens-Toma and DeWit, 1988).

These studies suggest that the elicitors are released during the compatible interaction without being able to elicit defense mechanisms and it is probable that the elicitation is suppressed by an unknown mechanism.

Table 25 Relationship between elicitor production and fungal growth in the apoplastic fluids of tomato cvr. Moneymaker (Cf5) inoculated with race 5 of *Cladosporium fulvum*

Time after inoculation (days)	Elicitor activity (Dilution end point)	Fungal growth (mannitol concentration: μg/ml)
3	—	0
8	1	24
10	16	216
14	64	500
17	128	648

—, No necrosis-inducing activity.
Necrosis-inducing activity was determined on cv. Sonatine (Cf2 Cf4 Cf9), an incompatible cultivar.
Source: Schottens-Toma and De Wit, 1988.

C. Some Elicitors Do Not Act or Show Little Activity on Susceptible Cultivars

Some of the elicitors do not induce defense mechanisms in the susceptible host of the pathogen, although they induce accumulation of defense chemicals in the incompatible hosts. Tepper and Anderson (1986) could isolate a galacto-gluco-mannan from *Colletotrichum lindemuthianum* that showed cultivar specificity. The elicitor induced accumulation of phytoalexins in bean cultivar resistant to the pathogen and did not induce any phytoalexin in the susceptible cultivar (Table 26; Tepper et al., 1989).

Different types of elicitors have been purified from *Phytophthora megasperma* f.sp. *glycinea*. Keen and Legrand (1980) could isolate an elicitor from

Table 26 Induction of phytoalexin accumulation in bean cultivars susceptible, moderately resistant, or resistant to *Colletotrichum lindemuthianum* by galactoglucomannan elicitor of the pathogen

	μg/Cotyledon		
Cultivar	Kievitone	Phaseollin–isoflavan	Phaseollin
Susceptible	0	0	0
Moderately resistant	0.96	0	0.78
Resistant	16.80	4.30	4.64

Source: Tepper et al., 1989.

Phytophthora megasperma f.sp. *glycinea* that was active more on a resistant soybean variety than on the susceptible variety. Keen et al. (1983) purified the elicitor and found it to be glucomannan. The glucomannan induced less phytolaexin in the compatible variety, while it induced substantial amount of phytolaexin in the incompatible variety.

Race-specific elicitor of resistance has been isolated from cowpea rust fungus (*Uromyces vignae*) (Chen and Heath, 1990). Aqueous extracts from appressorium-bearing basidiospore germlings of race 1 of the cowpea rust fungus elicited necrosis of mesophyll cells when injected into leaves of the resistant cowpea cv. Dixie Cream. Levels of necrosis were significantly higher than those elicited in a susceptible cultivar. In contrast, the exudates from race N2, which is compatible with both Dixie Cream and California Blackeye, induced much fewer necrotic cells in cv. Dixie Cream than did exudates from race 1 and there was no significant difference between the two cultivars tested and water controls (Table 27; Chen and Heath, 1990). Some of the elicitors act selectively on varieties containing specific resistance genes. Deverall and Deakin (1985) detected an elicitor in intercellular washing fluids from wheat leaves infected with the leaf rust fungus, *P. recondita* f.sp. *tritici*. In infected tissue, rust hyphae present in the intercellular spaces contribute the elicitor materials to these washing fluids (Holden and Rohringer, 1985). The presence of an elicitor in intercellular washing fluids from wheat leaves infected with the stem rust fungus, *P. graminis* f.sp. *tritici* has also been reported (Sutherland et al., 1989). These elicitors induced hypersensitive response in a group of wheat cultivars, although this group of cultivars was variable in disease reaction. The cultivar-specific responsiveness to the elicitor was controlled by a factor present on chromosome 5A in the wheat cultivars (Deverall and Deakin, 1987; Sutherland et al., 1989). Two sets of chromosome substitution lines in wheat were developed. These two sets provided each of 21 chromosome pairs from either cv. Cappelle-Desprez or a cv.Marquis selection substituted separately into cv.Chinese Spring. The source cultivars for these sets of lines showed a clear positive bioassay response, while the back-

Table 27 Number of necrotic cells elicited by exudates from rate 1 and race N2 in cowpea cvs Dixie Cream and California Blackeye

	Exudates from		
Cultivars	Race 1	Race N2	Water
Dixie Cream	296	40	27
California Blackeye	21	33	23

Source: Chen and Heath, 1990.

ground cultivar Chinese spring did not. Among the Chinese Spring/Cappelle-Desprez or Marquis lines only the stocks bearing chromosome 5A from either Cappelle-Desprez or Marquis showed a positive hypersensitive response. Both leaf rust and stem rust germ tube elicitors elicited this response. Only the source cultivars, Cappelle-Desprez and Marquis, and the two substitution lines containing 5A chromosomes in the nonreactive background of Chinese Spring showed a significant increase in PAL activity following elicitor injection. These two separate assays, based on the elicitation of macroscopically visible symptoms and of PAL activity, clearly demonstrated that a factor or factors controlled by chromosome 5A in these cultivars is involved in the recognition of and or response to the elicitors (Sutherland et al., 1989). The short arm of chromosome 5A is known to increase field resistance in adult plants to the stripe rust and powderly mildew (Pink et al., 1983).

Another kind of elicitor specificity has been reported by Yamaoka et al. (1990). *Colletotrichum graminicola* is a pathogen of maize. It is unable to infect sorghum. The extracts of conidia and the conidial mucilage of *C. graminicola* contained materials that elicited the accumulation of deoxyxanthocyanidin phytoalexins in sorghum mesocotyls and juvenile sorghum leaves. The elicitor preparations did not elicit a visible response from maize. Although the host-pathogen specificity was linked with the response to the elicitor in case of sorghum (nonhost) and maize (host), the elicitor failed to induce resistance in several other nonhost plants including soybean, cotton (*Gossypium hirsutum*), cucumber, tomato, radish (*Raphanus sativus*), pea, green bean, and cantaloupe (*Cucumis melo*) (Yamaoka et al., 1990).

Parker et al. (1988) isolated a new elicitor from *P. megasperma* f.sp. *glycinea,* the soybean pathogen. The elicitor consists of approximately 80 μg protein and 500 μg carbohydrate/1 mg freeze-dried material. Pronase treatment of the elicitor led to a complete inactivation of elicitor-mediated coumarin synthesis in parsley cells. The previously described active elicitors of *P. megasperma* f.sp. *glycinea* are glucans (Ayers et al., 1976b, c) and glucomannan (Keen et al., 1983) and both are resistant to pronase treatments. Hence the elicitor described by Parker et al. (1988) is different from the earlier ones. This elicitor elicited phytoalexin only in parsley (nonhost) cells and not in soybean (host) cells. The elicitor after pronase treatment elicited phytoalexins in soybean and not in parsley cells. Deglycosylation of the elicitor resulted in loss of elicitor activity on soybean cotyledons; but deglycosylation did not diminish the elicitor activity in parsley protoplasts (Table 28; Parker et al., 1988). The reciprocal responses of the parsley and soybean systems to deglycosylated and proteinase-digested *P. megasperma* f.sp. *glycinea* elicitor suggest that structurally different components of the fungal cell walls elicit phytoalexin production in parsley and soybean cells (Parker et al., 1988). Different elicitors may exist in the fungal cell wall and some of them may not elicit defense mechanisms in susceptible hosts (Parker et al., 1988).

Table 28 Effect of proteinase treatment and deglycosylation of *Phytophthora megasperma* f.sp. *glycinea* elicitor on parsley and soybean cells

	Phytoalexin accumulation (% of maximum)	
Treatment	Parsley cells	Soybean cells
Elicitor only	100	100
Elicitor + pronase	5	63
Deglycosylated elicitor	49	4

Source: Parker et al., 1988.

Thus some elicitors will be active only in incompatible interactions in inducing defense mechanisms of the host. In the compatible interactions, host defense mechanism is not activated and hence pathogens successfully invade the host tissue.

D. Amount of Elicitors Released May Determine Susceptibility

The quantity of elicitors available for induction of defense mechanisms may also determine susceptibility or resistance. Higher concentration of the elicitor isolated from *Pyricularia oryzae* induces higher chitinase activity in rice leaves (Schaffrath et al., 1995). Peroxidase and cinnamyl-alcohol dehydrogenase activities in rice leaves were also increased with increase in concentration of the elicitor (Schaffrath et al., 1995). If additional elicitor is infiltrated into rice leaves, even the susceptible rice variety becomes resistant to *P. oryzae*, probably due to higher amounts of elicitor present in the intercellular spaces of the host tissue during penetration of the pathogen (Schaffrath et al., 1995).

Activity of glucomannan elicitor isolated from *Phytophthorra megasperma* f.sp. *glycinea* increased with increase in its concentration (Keen et al., 1983). With increased concentration of glucan elicitor from *P. megasperma* f.sp. *glycinea*, increased induction of coumarin phytoalexins in parsley cells has been reported (Davis and Hahlbrock, 1987). Similar concentration-dependent induction of phytoalexins in bean has been reported with the elicitor isolated from *Colletotrichum lindemuthianum* (Tepper and Anderson, 1990). Hence amount of elicitor in the site of infection in planta may determine the disease reaction, but no effort has been made to assess the exact amount of accumulation of elicitors in the infection court. When chitin was administered to wheat leaves at an eliciting concentration, oligosaccharides up to the hexamer could be detected with

1 h of treatment. The elicitor-active oligosaccharides, β(1-4) linked N-acetyl-D-glucosamine $(GlcNAc)_{4-6}$, released in the leaves increased over the monitoring period (Barber and Ride, 1994). However, the relative amount of release during compatible and incompatible interactions has not been studied.

E. Pathogens May Have Suppressors to Suppress Action of Elicitors

Some plant pathogens have been shown to release suppressors that inhibit or delay active defense mechanisms of host plants. This delay provides the pathogen with sufficient time to enter host tissues and establish successful colonization (Shiraisihi et al., 1978; Kessmann and Barz, 1986; Arase et al. 1989). Suppressors have been detected in exudates of the bean rust fungus (*Uromyces phaseoli* var. *typica*) that suppressed the deposition of silicon-containing deposits in French bean leaves, which would otherwise prevent the development of the first haustorium (Heath, 1980; 1981).

Pycnospore germination fluid of *Mycosphaerella pinodes,* a pathogen of pea, contains both elicitors (high-molecular-mass glycoproteins of > 70,000 Da) and suppressors (low-molecular-mass glycopeptides of < 5000 Da) of the accumulation of the pea phytoalexin, pisatin (Oku et al. 1977, 1987; Shirasihi et al., 1978). The suppressor was purified into two active components, F_2 and F_5 (Shiraishi et al., 1978). The concomitant presence of F_5 with the elicitor negated the activity of the elicitor (Hiramatsu et al., 1986).

Pea epicotyl segments were treated with the elicitor alone or in the concomitant presence of the suppressor (elicitor+ suppressor) (Yamada et al., 1989). The apparent activation of pisatin biosynthesis was initiated approximately 6 h after treatment with the elicitor, and pisatin accumulation peaked at about 36 h, followed by a gradual decline. In the presence of the suppressor, however, the apparent activation of pisatin biosynthesis was delayed by 6–9 h compared to the treatment with elicitor alone, although the pattern of pisatin accumulation had very similar kinetics (Yamada et al., 1989). The activation of PAL activity had very similar kinetics in elicitor-treated and elicitor- plus suppressor-treated tissues, except for a delay of about 6 h as observed in the pattern of pisatin biosynthesis. Elicitor treatment caused a marked and prolonged accumulation of the pea PAL mRNA. Dramatic increases in 2.8 kb RNA species were first observed about 1 h after elicitor treatment, after which the mRNA remained at high levels in the elicitor-treated tissues. The accumulation of the 2.0 and 1.5 kb RNA species followed a similar pattern. In contrast, the dramatic increase in the 2.8-kb RNA species was first observed about 4 h after elicitor plus suppressor treatment, and accumulation of only a 2.0-kb RNA followed a similar pattern (Yamada et al. 1989).

A significant increase in the chalcone synthase (CHS) mRNA was observed about 1 h after elicitor treatment and about 4 h after elicitor plus suppressor treatment, which is a pattern very similar to that observed in PAL mRNA (Yamada et al. 1989). These results suggest that the suppressor may be acting at or before the transcriptional level for the genes coding for the key enzymes, PAL and CHS, leading to phytoalexin production. The limited time of the delay of the defense reactions in elicitor plus suppressor-treated tissues suggests that the suppressor would have been degraded by the host. This conclusion is supported by the fact that the subsequent addition of fresh suppressor 3 h after the initial treatment with elicitor plus suppressor treatment further delayed the activation of pisatin biosynthesis (Yamada et al., 1989). It also suggests that duration of suppressor activity is limited only to a few hours and factors limiting the duration of suppressor activity may reside in the host cells. Proteinase K-treated suppressor loses its activity dramatically (Yamada et al., 1989) and host cells may have the proteolytic enzymes to inactivate the suppressor.

The suppressor isolated from *M. pinodes* markedly inhibits the ATPase activity in pea plasma membranes in vitro in an uncompetitive manner, similarly to an inhibitor of P-type ATPase, orthovanadate (Yoshioka et al., 1990; Shiraishi et al., 1991). In pea epicotyl tissues, orthovanadate suppresses the accumulation of pisatin induced by the fungal elicitor (Yoshioka et al., 1990). It appears that the primray site of action of the suppressor may be the ATPase in the pea plasma membranes (Yoshioka et al. 1990; Shiraishi et al., 1991). Orthovanadate delayed accumulation of mRNAs encoding PAL and chalcone synthase in pea epicotyls induced by an elicitor from *M. pinodes* (Yoshioka et al. 1992). Orthovanadate acted in a manner similar to the fungal suppressor (Yamada et al., 1989; Yoshioka et al., 1990).

The activity of an ATPase in the pea plasma membranes, which is markedly suppressed by 6 h, recovers within 9 h after the start of the treatment with the suppressor from *M. pinodes* (Shiraishi et al., 1991). Accumulation of mRNA for a putative P-type ATPase was not affected by the treatment with orthovanadate. The inhibition of a P-type ATPase might cause a temporary suppression of expression of the PAL and CHS genes that are responsible for production of pisatin. Expression of these genes may gradually recover as a result of the biosynthesis of new ATPase molecules from the accumulated transcripts (Yoshioka et al., 1992).

Prior attack of barley coleoptile cells by the pathogen *Erysiphe graminis* induced accessibility in those cells to a challenge by a nonpathogen *E. pisi*, suggesting that *E. graminis* might have the ability to suppress the resistance mechanisms of host cells (Kunoh et al., 1985, 1986, 1988a,b; 1989, 1990a,b). To assess whether *E. graminis* may have a suppressor(s), an experiment was designed using a system that involved the coordinate inoculation of a single barley coleoptile cell with *E. graminis* and *E. pisi*. The coleoptiles were inoculated with *E. pisi* conidia. Germinating conidia of this fungus were removed from the host surface at or after maturation of the appressorium. Germinating conidia of *E. graminis* that had been inoculated onto another coleoptile were then transferred to the coleoptile inoculated initially with *E. pisi*. This was done so that the *E. graminis* germling would attempt to penetrate the same cell from which *E. pisi* germling had been removed. Induced inaccessibility of cells that suppresses the penetration by *E. graminis* was expressed when *E. graminis* was transferred to the cells at or after the maturation of the *E. pisi* appressorium. If *E. graminis* was transferred before the time of the maturation of the *E. pisi* appressorium, it suppressed the enhancement of inaccessibility caused by *E. pisi*. The results suggest that *E. pisi* may release an elicitor to enhance the inaccessibility either by or at the time of maturation of its appressoria. However, such elicitor from *E. pisi* becomes ineffective when a germling of *E. graminis* is present on the same cell before the *E. pisi* appressoria had matured. It suggests that germlings of *E. graminis* release a suppressor that blocks the effect of elicitor from *E. pisi* (Komura et al., 1990). However, the suppressor(s) has not been isolated from *E. graminis*.

A suppressor has been isolated from the culture filtrate of *Ascochyta rabiei*, the pathogen of chickpea (*Cicer arietinum*). The suppressor compound could be precipitated from the culture medium by trichloroacetic acid or fractionated with ammonium sulfate indicating that it is a protein. The suppressor had a molecular weight of less than 10,000 daltons and inhibited accumulation of constitutive phenolics including biochanin A, biochanin A 7-0-gluoside-6′-0-malonate (BGM), formononetin and formononetin 7-0-glucoside-6′-0-malonate (FGM), and elicitor-induced pterocarpan phytoalexins (medicarpin and maackiain) (Table 29; Kessmann and Barz, 1986).

Suppressor has been isolated from *Phytophthora infestans*. It has been characterized as glucans containing β-1-3 and β-1-6 linkages and 17–23 glucose units. The glucans from both mycelia and zoospores contained a nonanionic glucan and an anionic glucan. One of two residues of the latter was esterified with a phosphoryl monoester. The anionic glucan was more active than nonanionic glucan (Doke et al., 1979). Suppressor could be isolated from both compatible

Table 29 Suppression of accumulation of phenolics and phytoalexins in chickpea by the suppressor from Ascochyta rabiei

	Phenolics and phytoalexins (n mol/g fr. wt.)					
Treatment	Formonometin	Biochanin A	FGM	BGM	Medicarpin	Maackiain
Control	37	104	140	57	29	32
Suppressor	30	78	45	27	13	10

Source: Kessmann and Barz, 1986.

and incompatible races of the pathogen. There was no significant difference between compatible and incompatible races in the amount of suppressor in the germination fluid of cystospores of the fungus, but those suppressors were qualitatively different. Suppressor from race 4 (incompatible with the cultivar Kennebec) was less active thant suppressor from compatible race 1.2.3.4 in suppressing the defense action induced by the elicitor in the cvr. Kennebec (Table 30; Doke et al., 1979).

Pretreatment of protoplasts prepared from nine potato cultivars having different resistant genes (r, R_1, R_3, R_4, R_1R_2, R_1R_3, R_1R_4, R_2R_3 and R_2R_4) with suppressors from seven races of *P. infestans* (0, 1, 3, 4, 1.2, 1.4 and 1.2.3.4) suppressed the hypersensitive reaction of protoplasts elicited by elicitors isolated from the pathogen. Greater suppressive activity of the suppressor was characteristic of the compatible relationships between protoplasts and races used as a source of suppressor (Doke and Tomiyama, 1980a,b).

It is yet to be assessed whether suppressors act by competition for elicitor-binding sites or by the induction of a separate response that overcomes the consequences of elicitor action. The suppressor isolated from *M. pinodes* caused only a delay in the accumulation of PAL and CHS transcripts in response to the elicitors but did not affect their final levels (Yamada et al., 1989). It suggests that the suppressor does not compete for elicitor receptors but acts at a later stage in the signal pathway (Dixon and Lamb, 1990). *Ascochyta rabiei* suppressor inhibited the synthesis of elicitor-induced phytoalexins as well as the constitutive phenolics, suggesting the existence of sites of action at least partly different from those of the elicitor (Kessmann and Barz, 1986).

Suppressor may be present as part of elicitor itself (Basse and Boller, 1992). A suppressor has been isolated from the glycopeptide elicitors of yeast extract. When the elicitor was digested with endo β-N-acetyl-gluosaminidase H, the suppressor activity was detected in the released oligosaccharides. Periodate oxidation completely destroyed the suppressor activity. Carbohydrates released

Table 30 Relative efficacy of suppressors from compatible (race 1,2,3,4) and incompatible (race 4) races of *Phytophthora infestans* in suppressing action of elicitors

Treatment	Intensity of browning in potato cultivar Kennebec
Elicitor from race 4	+++
Suppressor from race 1,2,3,4 + elicitor from race 4	+
Suppressor from race 4 + elicitor from race 4	+++

Source: Doke et al., 1979.

from the yeast extract elicitor by N-glycanase also showed suppressor activity. These observations suggest that hydrolysis of the elicitor-active glycopeptides into the glyco- and peptide parts by endoglycanases yields oligosaccharides that act as suppressors of the activity (Basse and Boller, 1992). α-Mannosidase destroyed the suppressor activity, indicating that mannose residues are important for the suppressor activity. It has also been shown that the mannose residues are important for the elicitor activity. It has been well documented that N-linked glycans are essential for elicitor activity, but they act as suppressors of the same activity when released from glycopeptides (Basse and Boller, 1992).

The suppressor from the yeast extract-elicitor inhibited induction of ethylene biosynthesis and phenylalanine ammonia-lyase in tomato cells by the elicitor from the yeast extract. It suggests that the elicitor and suppressors may compete for a single recognition site (Basse and Boller, 1992). When elicitor from *Phytophthora megasperma* f.sp. *glycinea* (Pmg) was employed to induce ethylene biosynthesis, the suppressors from yeast extract-elicitor did not suppress the induction in tomato cells. The active compounds in Pmg-elicitor are proteinaceous in nature and differ from the yeast extract-derived elicitor. The results suggest that the suppressor activity is highly specific, and that tomato cells contain at least two distinct elicitor recognition sites with different specificities (Basse and Boller, 1992).

Enzymatic nature of suppressor produced by *Colletotrichum* spp. has been reported (Siegrist and Kauss, 1990). Chitin deacetylase is produced by several races of *Colletotrichum lindemuthianum* (Kauss et al., 1983) and *C. lagenarium* (Kauss and Bauch, 1988). The enzyme produces chitosan from chitin. The enzyme is relatively heat-stable. Fragments arising from the action of chitinase provided a better substrate than crystalline chitin for the deacetylase. Chitin is insoluble in water and its elicitor activity is mediated by a soluble chitin oligomer. Chitinase is capable of releasing chitin oligomers from chitin (Carratu et al., 1985). The elicitor activity of chinin oligomers arising from chitinase action may be destroyed by chitin deacetylase, since fragments arising from the action of chitinase provided a better substrate than crystalline chitin for the deacetylase. This enzyme was detected in cucumber leaves infected with *C. lagenarium* also (Siegrist and Kauss, 1990).

F. Host May Have Suppressor to Suppress Action of Fungal Elicitors

Suppressor(s) of host origin has been reported by Peever and Higgins (1989). *Cladosporium fulvum* produces elicitors in culture (Lazarovitz and Higgins, 1979; Lazarovitz et al., 1979). Coinjection of elicitor and intercellular fluids from *C. fulvum*-infected tomato consistently resulted in complete suppression of elicitor-induced necrosis and callose deposition in the injected panels. Elicitor plus distilled water controls injected into the opposite site of the leaflet always induced high levels of necrosis and callose deposition. Several preparations of intercellular fluids from uninfected plants also suppressed elicitor-induced necrosis. Intercellular fluids from rust-infected bean also suppressed the induction of necrosis by the elicitor when coinjected into tomato leaves (Peever and Higgins, 1989).

The presence of suppressor activity in some preparations of intercellular fluids from uninfected, healthy tomato indicates that it may be a host-produced factor. Heat treatment immediately following mixture of elicitor and intercellular fluids eliminated suppressor activity but not elicitor activity. Following a 20 h incubation between mixing and heat treatment, however, almost complete suppression of elicitor-induced necrosis occurred similar to that seen in the unheated treatment (Peever and Higgins, 1989). The experiments suggested an enzymatic type of interaction that does not depend on the presence of the host.

However, further studies revealed that the suppressor may not be an enzyme (Lu and Higgins, 1993). The suppression may be largely as a result of low-molecular-weight heat-stable components. The retention of suppressor activity following treatment of intercellular fluid (IF) with protease, mild base, or acid suggests the nonproteinaceous nature of the suppressor. A pectate digest, produced from sodium polypectate by pectinase, suppressed the elicitor-induced necrosis. Coinjection of the elicitor with active pectinase resulted in suppression. Pectinase digestion of host cell walls would have released the suppressor-active molecules (Lu and Higgins, 1993). The results suggest that the suppressor may be a carbohydrate.

G. Host May Degrade Fungal Elicitors

When the yeast extract-elicitor was incubated in the medium containing rice cells, the elicitor activity disappeared in time-dependent manner (Felix et al., 1991). Degradation of the elicitor results in suppression of the induced defense mechanisms in rice cells (Felix et al., 1991).

Constitutive wheat leaf endochitinases capable of releasing elicitor-active oligosaccharides from chitin (Ride and Barber, 1990) and N-acetylhexosaminidases (Barber and Ride, 1989) hydrolyze elicitor-active oligosaccharides to inactive derivatives such as smaller oligomers (Ride et al., 1989). When $(GlcNAc)_4$ was administered at an eliciting concentration to wheat leaves, its total level decreased from 48,100 to 29,300 pmol/wound over the 12 h period (a decrease of 39%). This decrease in $(GlcNAc)_4$ could be accounted for by the increases in the levels of the smaller oligosaccharides. The primary breakdown products detected were $(GlcNAc)_2$ and GlcNAc, with $(GlcNAc)_3$ being present at very much lower concentrations (Barber and Ride, 1994). When chitin was administered to wounded wheat leaves, GlcNAc started accumulating. The monomer became the predominant product at 12 h after treatment (Table 31; Barber and

Table 31 Levels of $(GlcNAc)_n$ in wheat leaves treated with chitin

Time (h)	Oligosaccharide level (p mol/wound)					
	GlcNac	$(GlcNAc)_2$	$(GlcNAc)_3$	$(GlcNAc)_4$	$(GlcNAc)_5$	$(GlcNAc)_6$
0	0	0	0	0	0	0
8	640	219	200	169	85	38
12	1109	323	315	249	107	44

Source: Barber and Ride, 1994.

Ride, 1994) probably due to the action of N-acetylhexosamionidase or exochitinase or chitobiase (Barber and Ride, 1994). At the end of the 12 h period approximately 2% by weight of the applied chitin had been solubilized in wheat leaves. Of this amount the elicitor-inactive oligomers GlcNAc, $(GlcNAc)_2$ $(GlcNAc)_2$ accounted for 84% on a molar basis. Extremely little of the applied chitin polymer (only 0.03% by weight) is actually present within the leaf in the form of the known elicitor-active oligomers $(GlcNAc)_{4\text{-}6}$ by the end of the 12 h period (Barber and Ride, 1994).

The degradation of the elicitor may be due to host enzymes. It has been demonstrated that elicitor activity of the chitin tetramer can be totally abolished by prolonged exposure to wheat leaf chitinases (Barber et al., 1989). The possibility of these enzymes in degrading the fungal elicitor during fungal pathogenesis has not been studied.

H. Fungal Pathogens May Degrade Host Elicitors

Endogenous elicitors of host origin are oligomers of pectic fragments. Polygalaturonides with a degree of polymerization (DP) of about 20 (Bishop et al. 1984), 9-12 galacturonosyl units (Robertson, 1986), or nonamers through pentadecamers (Jin and West, 1984), are considered as elicitors and are produced due to partial digestion of polygalacturonic acid (Bruce and West, 1982) by fungal enzymes. In susceptible tissues, pathogen would have produced more of these pectic enzymes and higher concentration of these enzymes would have degraded the endogenous elicitors into inactive galacturonic acids. Induction of pectic enzymes in susceptible reactions is much higher than that in resistant reactions (see Chapter 2).

XII. CONCLUSION

Fungal pathogenesis commences when the spore contacts the host surface. The pathogen recognizes host by producing proteinaceous adhesin and esterases including cutinase and interacting with cuticle monomers and monolignol glucosides of the host. Host recognizes pathogen and produces enzymes including glycohydrolases, chitinases, β-1,3-glucanases, lipoxygenases, acetyl hydrolases, and lipases that act on the fungal cell surface releasing fungal elicitors. Pathogen produces enzymes including endopolygalacturonase, endopectin-lyase, and aspartic proteinase that release elicitors from host. Elicitors from pathogen as well as from host synergistically act as signal molecules. These signals are carried intra- and intercellularly by second messengers. Calcium ion, protein phosphorylase, protein kinase, phospholipase, ATPases, H_2O_2, ethylene, abscisic acid, polyamines, phospholipids, and glutathione act as second messengers. Systemic signal molecules have also been detected. Oligogalacturonides, salicyclic acid, systemin, methyl jasmonate, and fatty acids act as systemic signal molecules. These signals induce defense mechanisms of plants such as cell wall appositions, reinforcement of cell walls by deposition of lignin, callose, and hydroxyproline-rich glycoproteins and induce synthesis of phytoalexins and pathogenesis-related proteins. During successful pathogenesis, delayed release of elicitors or inactivation of elicitors in the susceptible host occurs. Pathogens produce suppressors to suppress the action of elicitor action. Host also may have suppressors to suppress the defense mechanisms induced by the elicitor. Host may have enzymes to degrade the elicitor molecules.

Thus early molecular events during fungal pathogenesis appear to be highly complex. Presence of molecules that can induce resistance as well as susceptibility in fungal cell walls and also in host cell walls complicate the interaction. Involvement of host enzymes in release of these molecules from fungal cell wall and involvement of fungal enzymes in release of the molecules from host cell wall determine the type of disease reaction.

De Wit (1992) and Vidhyasekaran (1993c) suggested that the resistance or susceptibility of host plants to different races of a fungal pathogen is determined by the match of dominant resistance genes in the plant with dominant avirulence genes in the pathogen. The avirulence genes may code for elicitor molecules while resistance genes may code for receptor molecule. Although this theory appears to be attractive, convincing experimental evidence is lacking in many host–pathogen interactions. Elicitor as an avirulence product has been probably demonstrated well only in *Cladosporium fulvum*–tomato interactions. The peptide elicitor of *C. fulvum* has been shown to be a product of avirulence gene, avr 9 (Schottens-Toma and De Wit, 1988) and avr 4 (Joosten et al., 1994). A single base pair mutation in one of three different cysteine codons in avr 4 makes the avirulent race into a virulent one on hosts carrying genes for resistance to the race. The mutation was from TGT to tyrosine codon TAT and mutation from cysteine to tyrosine in avr gene product (peptide elicitor) results in loss of avirulence (Joosten et al., 1994).

Other host-specific elicitors from *Uromyces vignae* (Chen and Heath,

1992), *Colletotrichum lindemuthianum* (Tepper et al., 1989). *Phytophthora megasperma* f.sp. *glycinea* (Keen et al., 1983), and *Rhynchosporium secalis* (Hahn et al., 1993) may be considered as avirulence gene products, but there are several other non-specific elicitors. Elicitors have been isolated even from saprophytes. If so, what is the role of elicitors? A single pathogen contains several kinds of elicitors. Even *Cladosporium fulvum* produces a nonspecific elicitor. Do they have any function in fungal pathogenesis? There is an argument that every pathogen may have a host-specific elicitor; we have not developed techniques to isolate, purify, and identify them. Crude elicitor preparation from *Colletotrichum lindemuthianum* did not show any host specificity, but on purification one host-specific elicitor could be identified (Tepper and Anderson, 1986). However, in spite of several attempts host-specific elicitor could not be identified in most of the pathogens.

It is not the presence of elicitor but release of the elicitor from the fungal cell wall may be important in pathogenesis. Unfortunately not much work has been done in this direction. Although many elicitors have been detected in a single fungus, it is not known which elicitor is released. What is the amount of release of elicitor in susceptible and resistant interactions? It is particularly important since the elicitor action is concentration-dependent. What is the time of release of elicitor during the infection process? If it is a delayed release in the susceptible interaction, what is the mechanism that delays the release of elicitor?

When does the endogenous elicitor get released in the susceptible interaction? How much of these elicitors is released? What is the relative importance of endogenous and exogenous elicitors in susceptible and resistant interactions? Suppressors have been isolated along with elicitors from fungal cell walls. In some cases, suppressor has been detected as a part of elicitor molecule itself, but not much work has been done on suppressor molecule. It is not known whether suppressor is host-specific. Can suppressor be called a virulence gene product if elicitor is considered a avirulence gene product? Suppressor is degraded quickly in the host. The suppressor delays the action of elicitor only and it disappears long before the disappearance of elicitor in the host tissue. Both suppressor and elicitor are degraded in host tissues. What are the enzymes involved in their degradation? What is the speed with which they are expressed in susceptible and resistant interactions?

Many of the above questions are unanswered. The mechanism of release of elicitors should be a priority for study. If this release is increased and accelerated by genetic engineering techniques, disease incidence can be minimized.

REFERENCES

Abeles, F. B., Bosshart, R. P., Forrence, L. E., and Habig, W. H. (1970). Preparation and purification of glucanase and chitinase from bean leaves. Plant Physiol., 47:129–134.

Albersheim, P., and Valent, B. S. (1978). Host–pathogen interactions in plants. Plants when exposed to oligosaccharides of fungal origin defend themselves by accumulating antibiotics, J. Cell. Biol., 78:627–643.

Albersheim, P., Darvill, A. G., Sharp, J. K., Davis, K. R., and Doares, S. H. (1986). Studies on the role of carbohydrates in host–microbe interactions. Recognition in Microbe–Plant Symbiotic and Pathogenic Interactions (B. Lugtenberg, ed.), Springer, Berlin, pp. 297–309.

Anderson, A. (1978). Initation of resistant responses in bean by mycelial wall fractions from three races of the bean pathogen *Colletotrichum lindemuthianum,* Can. J. Bot., 56:2247–2251.

Anderson, A. J., Rogers, K., Tepper, C. S., Blee, K., and Cardon, J. (1991). Timing of molecular events following elicitor treatment of plant cells, Physiol. Mol. Plant Pathol., 38:1–13.

Anderson-Prouty, A. J., and Albersheim, P. (1975). Host–pathogen interactions. VIII. Isolation of a pathogen synthesized fraction rich in glucan that elicits a defence response in the pathogen's host, Plant Physiol., 56:286–291.

Apostol, I., Low, P. S., Heinstein, P. F., Stipanovic, R. D., and Altman, D. W. (1987). Inhibition of elicitor-induced phytoalexin formation in cotton and soybean cells by citrate, Plant Physiol., 84:1276–1280.

Apostol, I., Heinstein, P. F., and Low, P. S. (1989). Rapid stimulation of an oxidative burst during elicitation of cultured plant cells. Role in defense and signal transduction, Plant Physiol., 90:109–116.

Arase, S., Tanaka, E., and Nishimura, S. (1989). Production of susceptibility-inducing factor(s) in spore germination fluids of *Pyricularia oryzae.* Host-specific Toxins: Recognition and Specificity Factors in Plant Disease (K. Kohmoto and R. D. Durbin, eds.), Tottori University, Tottori, pp. 59–73.

Axelos, M., Bardet, C., and Lescure, B. (1993). An Arabidosis cDNA encoding a 33-kilodalton laminin receptor homolog. Plant Physiol., 103:299–300.

Ayers, A. R., Ebel, J., Finelli, F., Berger, N., and Albersheim, P. (1976a). Host–pathogen interactions. IX. Quantitative assays of elicitor activity and characterization of the elicitor present in the extracellular medium of cultures of *Phytophthora megasperma* var. *sojae,* Plant Physiol., 57:751–759.

Ayers, A. R., Ebel, J., Valent, B. S., and Albersheim, P. (1976b). Host–pathogen interactions. X. Fractionation and biological activity of an elicitor isolated from the mycelial walls of *Phytophthora megasperma* var. *sojae,* Plant Physiol., 57:760–765.

Ayers, A. R., Valent, B. S., Ebel, J., and Albersheim, P. (1976c). Host–pathogen interactions. XI. Composition and structure of wall-released elicitor fractions, Plant Physiol., 57:766–774.

Bach, T. J. (1987). Synthesis and metabolism of mevalonic acid in plants, Plant Physiol., 25:163–178.

Bailey, B. A., Dean, J. F. D., and Anderson, A. J. (1990). An ethylene biosynthesis-inducing endoxylanase elicts electrolyte leakage and necrosis in *Nicotiana tabacum* cv. Xanthi leaves, Plant Physiol., 94:1849–1854.

Barber, M. S., and Ride, J. P. (1988). A quantitative assay for induced lignification in wounded wheat and its use to serve potential elicitors of the response, Physiol. Mol. Plant Pathol., 32:185–197.

Barber, M. S., Bertram, R. E., and Ride, J. P. (1989). Chitin oligosaccharides elicit lignification in wounded wheat leaves, Physiol. Mol. Plant Pathol., 34:3–12.

Barber, M. S., and Ride, J. P. (1989). Purification and properties of a wheat leaf N-acetyl-β-D-hexosaminidase, Plant Sci., 60:163–172.

Barber, M. S., and Ride, J. P. (1994). Levels of elicitor-active β (1-4) linked N-acetyl-D-glucosamine oligosaccharides in the lignifying tissues of wheat, Physiol. Mol. Plant Pathol., 45:37–45.

Bartnicki-Garcia, S. (1968). Cell wall chemistry, morphogenesis and taxonomy of fungi, Annu. Rev., Microbiol., 22:87–108.

Bartnicki-Garcia, S. (1970). Cell wall composition and other biochemical markers in fungal phylogeny. Phytochemicals (J. B. Harbourne, ed.), Academic Press, London, pp. 81–101.

Bartnicki-Garcia, S., and Sing, V. O. (1987). Adhesion of zoospores of *Phytophthora* to solid surfaces. Zoosporic Fungi in Teaching and Research (M. S. Fuller, and A. Janorski, eds.), Southeastern Publishing Corporation, Athens, pp. 279–283.

Basse, C. W., and Boller, T. (1992). Glycopeptide elicitors of stress responses in tomato cells. N-linked glycans are essential for activity but act as suppressors of the same activity when released from the glycopeptides, Plant Physiol., 98:1239–1247.

Baydoun, E. A. H., and Fry, S. C. (1985). The mobility of pectic substances in injured tomato leaves and its bearing on the identity of the wound hormone, Planta, 165:269–276.

Beardmore, J., Ride, J. P., and Granger, J. W. (1983). Cellular lignification as a factor in the hypersensitive resistance of wheat to stem rust, Physiol. Plant Pathol., 22:209–220.

Beckett, A., Tatnell, J. A., and Taylor, N. (1990). Adhesion and pre-invasion behaviour of urediniospores of *Uromyces viciae-fabae* during germination on host and synthetic surfaces, Mycol. Res., 94:865–875.

Beckman, C. H. (1987). The Nature of Wilt Diseases of Plants. American Phytopathological Society Press, St. Paul, Minnesota, pp. 175.

Beissmann, B., Engels, W., Kogel, K., Marticke, K. H., and Reisener, H. J. (1992). Elicitor-active glycoproteins in apoplastic fluids of stem-rust infected wheat leaves, Physiol. Mol. Plant Pathol., 40:79–89.

Beissmann, B., and Reisener, H. J. (1990). Isolation and purity determination of a glycoprotein elicitor from wheat steam rust by medium pressure liquid chromatography, J. Chromatography, 521:187–197.

Bell, J. N., Ryder, T. B., Wingate, V. P. M., Bailey, J. A., and Lamb, C. J. (1986). Differential accumulation of plant defense gene transcripts in compatible and incompatible plant–pathogen interactions, Mol. Cell. Biol., 6:1615–1623.

Bell, E., and Mullet, J. E. (1991). Lipoxygenase gene expression is modulated in plants by water deficit, wounding, and methyl jasmonate, Mol. Gen. Genet., 230:456–462.

Bell-Pedersen, D., Dunlap, J. C., and Loros, J. J. (1992). The neurospra circadian clock-controlled gene *cog-2*, is allelic to *eas* and encodes a fungal hydrophobin required for formation of the conidial rodlet layer, Genes Dev., 6:2382–2394.

Benhamou, N., and Theriault, G. (1992). Treatment with chitosan enhances resistance of tomato plants to crown and root pathogen *Fusarium oxysporumd* f.sp. *radicis-lycopersici,* Physiol. Mol. Plant Pathol., 41:33–52.

Beretta, M. J. G., Martins, E. M. F., and Moraes, W. B. C. (1977). Induced protection to *Hemileia vastatrix* at a distance from the site of the inducing action in coffee plants. Summa Phytopathol., 3:66–70.

Berridge, M. J. (1984). Inositol triphosphate and diacylglycerol as second messenger, Biochem. J., 220:345–360.

Bino, R. L., Daye, S., Serlin, B. S., Terry, M. E., Datta, N., Sopory, S. K., and Roux, S. J. (1984). Characterization of oat calmodulin and radio-immunoassay of its subcellular distribution, Plant Physiol., 75:182–186.

Bishop, P. D., Pearce, G., Bryant, J. E., and Ryan, C. A. (1984). Isolation and characterization of the proteinase inhibitor-inducing factor from tomato leaves. Identity and activity of poly- and oligogalacturonide fragments. J. Biol. Chem., 259:13172–13177.

Bishop, P. D., Makus, D. J., Pearce, G., and Ryan, C. A. (1981). Proteinase inhibitor-inducing activity in tomato leaves resides in oligosaccharides enzymically released from cell walls, Proc. Natl. Acad. Sci. USA, 78:3536–3540.

Blein, J. P., Milat, M. L., and Ricci, P. (1991). Responses of cultured tobacco cells to cryptogein a proteinaceous elicitor from *Phytophthora cryptogea,* Plant Physiol., 95:486–491.

Bogre, L., Olah, Z., and Dudits, D. (1988). Ca^{2+}-dependent protein kinase from alfalfa (*Medicago varia*): partial purification and autophosphorylation, Plant Sci., 85:135–144.

Boller, T., Gehri, A., Mauch, F., and Vogeli, U. (1983). Chitinase in bean leaves: induction by ethylene, purification, properties and possible function, Planta, 157:22–31.

Boller, T., and Kende, H. (1980). Regulation of wound ethylene synthesis in plants. Plant Growth Substances, (P. F. Wareing, ed.), Academic Press, London, pp. 303–312.

Bonnet, P., Poupet, A., Abad, P., Venard, P., and Cardin, L. (1986). Induction de nécroses foliaires, de proteins b et de resistance dans les intéractions tobacco–*Phytophthora,* Agronomie, 6:829–837.

Boss, W. F., and Massel, M. O. (1985). Polyphosphoinositides are present in plant tissue culture cells, Biochem. Biophys. Res. Commun., 132:1018–1023.

Bostock, R. M. (1989). Metabolism of lipids containing arachidonic and eicosapentaenoic acids in race-specific interactions between *Phytophthora infestans* and potato, Phytopathology, 79:989–902.

Bostock, R M., Kuc, J. A., and Laine, R. A. (1981). Eicosapentaenoic and arachidonic acids from *Phytophthora infestans* elicit fungitoxic sesquiterpenes in potato, Science, 212:67–69.

Bostock, R M., Schaeffer, D. A., and Hammerschmidt, R. (1986). Comparison of elicitor activities of arachidonic acid, fatty acids and glucans from *Phytophthora infestans* in hypersensitivity expression in potato tuber, Physiol. Mol. Plant Pathol., 29:349–360.

Bostock, R. M., and Stermer, B. A. (1989). Perspectives on wound healing in resistance to pathogens, Annu. Rev. Phytopathol., 27:343–371.

Bostock, R. M., Yamamoto, H., Choi, D., Ricker, K. E., and Ward, B. L. (1992). Rapid stimulation of 5-lipoxygenase activity in potato by the fungal elicitor arachidonic acid, Plant Physiol., 100:1448–1456.

Bradley, P., Kjellbom, P., and Lamb, C. J. (1992). Elicitor and wound-inducible oxidative cross-linking of a proline rich plant cell wall protein: a novel, rapid defense response, Cell, 70:21–30.

Breviario, D., Morello, L., and Gianai, S. (1995). Molecular cloning of two novel rice cDNA sequences encoding putative calcium-dependent protein kinases, Plant Mol. Biol., 27:953–967.

Brown, W., and Ryan, C. A. (1984). Isolation and characterization of a wound induced trypsin inhibitor from alfalfa leaves, Biochemistry, 23:3418–3422.

Bruce, R J., and Ryan, C. A. (1982). Elicitation of casbene synthetase activity in castor bean. The role of pectic fragments of the plant cell wall in elicitation by a fungal endopolygalacturonease, Plant Physiol., 69:1181–1188.

Bryan, I. B., Rathmell, W. G., and Friend, J. (1985). The role of lipid and non-lipid components of *Phytophthora infestans* in the elicitation of the hypersensitive response in potato tuber tissue, Physiol. Plant Pathol., 26:331–335.

Bucheli, P., Doares, S. H., Albersheim, P., and Darvill, A. G. (1990). Host–pathogen interactions. XXXVI. Partial purification and characterization of heat-labile molecules secreted by the rice blast pathogen that solubilize plant cell wall fragments (Oligosaccharins) that kill plant cells, Physiol. Mol. Plant Pathol., 36:159–173.

Caelles, C., Ferrer, A., Balcells, L., Hegendt, F. G., and Boronat, A. (1989). Isolation and structural characterization of cDNA encoding *Arabidopsis thaliana* 3-hydroxy-3-methylglutaryl Coenzyme A reductase, Plant Mol. Biol., 13:627–638.

Caesar-TonThat, T. C., and Epstein, L. (1991). Adhesion reduced mutants and the wild type *Nectria haematococca:* an ultrastructural component of the macroconidial walls. Exp. Mycol., 15:193–205.

Calderone, R. A., and Braun, P. C. (1991). Adherence and receptor relationships of *Candida albicans,* Microbiol. Rev., 55:1–20.

Carafoli, E., and Penniston, J. T. (1985). The calcium signal, Sci. Am., 253:50–58.

Carratu, G., Colacino, C., Conti, S., and Giannattasio, M. (1985). β-N-acetylglucosaminidase and β-galactosidase from aleurone layers of resting wheat grains, Phytochemistry, 24:1465–1469.

Castoria, R., Fanelli, C., Fabbri, A. A., and Passi, S. (1992). Metabolism of arachidonic acid involved in its eliciting activity, Physiol. Mol. Plant Pathol., 41:127–137.

Cervone, F., De Lorenzo, G., Degra, L., Salvi, G., and Bergami, M. (1987). Purification and characterization of a polygalacturonase-inhibiting protein from *Phaseolus vulgaris* L., Plant Physiol., 85:631–637.

Cervone, F., Hahn, M. G., De Lorenzo, G., Darvill, A., and Albersheim, P. (1989). Host–pathogen interactions. XXXIII. A plant protein converts a fungal pathogenesis factor into an elicitor of plant defence responses, Plant Physiol., 90:542–548.

Chadha, K. C., and Brown, S. A. (1974). Biosynthesis of phenolic acids in tomato plants infected with *Agrobacterium tumefaciens,* Can. J. Bot., 52:2041–2047.

Chalova, L. I., Baramidze, V. G., Yurkanova, T. Y., Yakov, D., Lozeretskovskaya, O. L., and Metliski, L. V. (1977a). Isolation and characterization of the inducer of protective responses of the potato from the cytoplasmic contents of the phytophthorosis pathogen, Doklady Akademii Naauk SSR, 235:1215–1218.

Chalova, L. I., Ozeretskovskaya, O. L., Yurganova, L. A., Baramidze, V. G., Protsenko, M. A., D'Yakov, Y., and Metlizki, L. V. (1977b). Metabolites of phytopathogenic fungi as elicitors of defense reactions in plant (in the potato–*Phytophthora infestans* interaction), Doklady Biol. Sci., 230:433–436.

Chapela, I. H., Petrini, O., and Hagmann, L. (1991). Monolignol glucosides as specific recognition messengers in fungus–plant symbioses, Physiol. Mol. Plant Pathol., 39:289–298.

Chapela, I. H., Petrini, O., and Petrini, L. E. (1990). Unusual ascospore germination in *Hypoxylon fragiforme:* first steps in the establishment of an endophytic symbiosis, Can. J. Bot., 68:2571–2575.

Chappell, J., Hahlbrock, K., and Boller, T. (1984). Rapid induction of ethylene biosynthesis in cultured parsley cells by fungal elicitor and its relationship to the induction of phenylalanine ammonia-lyase, Planta, 161:475–480.

Chaubal, R., Wilmot, V. A., and Wynn, W. K. (1991). Visualization, adhesiveness and cytochemistry of the extracellular matrix produced by the urediniospore germtubes of *Puccinia sorghi,* Can. J. Bot., 69:2044–2054.

Chen, C. Y., and Heath, M. C. (1990 Cultivar-specific induction of necrosis by exudates from basidiospore germlings of the cowpea rust fungus, Physiol. Mol. Plant Pathol., 37:169–177.

Chen, C. Y., and Heath, M. C. (1992). Effect of stage of development of the cowpea rust fungus on the release of a cultivar-specific elicitor of necrosis, Physiol. Mol. Plant Pathol., 40:23–30.

Chen, Z., and Klessig, D. F. (1991). Identification of a soluble salicylic acid-binding protein that may function in signal transduction in the plant disease-resistance response, Proc. Natl. Acad. Sci. USA, 88:8179–8183.

Choi, D., Ward, B. L., and Bostock, R. M. (1992). Differential induction and suppression of potato 3-hydroxy-3-methyl glutaryl Coenzyme A reductase genes in response to *Phytophthora infestans* and to its elicitor, arachidonic acid, Plant Cell, 4:1333–1344.

Chye, M. L., Tan, C. T., and Chua, N. H. (1992). Three genes encode HMG-CoA reductase in *Hevea brasiliensis: hmg* 1 and *hmg* 3 are differentially expressed, Plant Mol. Biol., 19:473–484.

Clement, J. A., Butt, I. M., and Beckett, A. (1993a). Characterization of the extracellular matrix produced in vitro by urediniospores and sporelings of *Uromyces viciae-fabae,* Mycol. Res., 97:594–603.

Clement, J. A., Mastin, S. G., Porter, R., Butt, T. M., and Beckett, A. (1993b). Germination and the role of extracellular matrix in adhesion of urediniospores of *Uromyces viciae-fabae* to synthetic surfaces, Mycol. Res., 97:585–593.

Cline, K., and Albershim, P. (1981a). Host–pathogen interactions. XVI. Purification and characterization of a α-glucosyl hydrolase/transferase present in the walls of soybean cells, Plant Physiol., 68:207–220.

Cline, K., and Albersheim, P. (1981b). Host–pathogen interactions. XVII. Hydrolysis of biologically active fungal glucans by enzymes isolated from soybean cells, Plant Physiol., 68:221–228.

Cline, K., Wade, M., and Albersheim, P. (1978). Host–pathogen interactions. XV. Fungal glucans which elicit phytoalexin accumulation in soybean also elicit the accumulation of phytoalexins in other plants, Plant Physiol., 62:918–921.

Cohen, P. (1982). The role of protein phosphorylation in neutral and hormonal control of cellular activity, Nature, 296:613–620.

Cohen, Y., Gisi, U., and Mosinger, E. (1991). Systemic resistance of potato plants against *Phytophthora infestans* induced by unsaturated fatty acids, Physiol. Mol. Plant Pathol., 38:255–263.

Collinge, M., and Trewavas, A. J. (1989). The location of calmodulin in pea plasma membrane, J. Biol. Chem., 264:8865–8872.

Corey, E. J., and Lansbury, P. T. Jr. (1983). Stereochemical course of 5-lipoxygenation of arachidonate by rat basophil leukemic cells (RBL-1) and potato enzymes, J. Am. Chem. Society, 105:4093–4094.

Cosio, E. G., Popperl, H., Schmidt, W. E., and Ebel, J. (1988). High-affinity binding of fungal glucan fragments to soybean (*Glycine max* L.) microsomal fractions and protoplasts, Eur. J. Biochem., 175:309–315.

Cosio, E. G., and Ebel, J. (1988). Solubilization of fungal oligoglucan binding proteins from soybean cell membranes, Plant Physiol. Suppl., 86:22.

Crane, F. I., Sun, H., Clark, M. G., Grebing, C., and Low, H. (1985). Transplasma-membrane redox systems in growth and development, Biochem. Biophys. Acta, 811:233–264.

Creamer, P., and Bostock, R. M. (1986). Characterization and biological activity of phospholipids from *Phytophthora infestans* in the hypersensitive response of potato tuber, Physiol. Mol. Plant Pathol., 28:216–225.

Creamer, P., and Bostock, R. M. (1988). Contribution of eicosapolyenoic fatty acids to the sesquiterpenoid phytoalexin elicitor activities of *Phytophthora infestans* spores, Physiol. Mol. Plant Pathol., 32:49–59.

Creelman, R. A., Tierney, M. L., and Mullet, J. E. (1992). Jasmonic acid/methyl jasmonate accumulate in wounded soybean hypocotyls and modulate wound gene expression, Proc. Natl. Acad. Sci. USA, 89:4938–4941.

Cuypers, B., Schmelzer, E., and Hahlbrock, K. (1988). *In situ* localization of rapidly accumulated phenylalanine ammonialyase mRNA around penetration sites of *Phytophthora infestans* in potato leaves. Mol. Plant–Microbe Interact., 1:157–160.

Dangel, J. L., Hanfle, K. D., Lipphardt, S., Hahlbrock, K., and Scheel, D. (1987). Parsley protoplasts retain differential responsiveness to u.v. light and fungal elicitor, EMBO J., 6:2551–2556.

Darvill, A. G., and Albersheim, P. (1984). Phytoalexins and their elicitors—a defense against microbial infection in plants, Annu. Rev. Plant Physiol., 35:243–275.

Datta, N., Hardison, L. K., and Roux, S. J. (1986). Polyamine stimulation of protein phosphorylation in isolated pea nuclei, Plant Physiol. Suppl., 80:15.

Davis, K. R., Darvill, A. G., and Albersheim, P. (1984). Host pathogen interactions. XXV. Endopolygalacturonic acid lyase from *Erwinia carotovora* elicits phytoalexin accumulation by releasing plant cell wall fragments, Plant Physiol., 74:52–60.

Davis, K. R., Darvill, A. G., and Albersheim, P. (1986a). Host–pathogen interactions. XXX. Characteriztion of elicitors of phytoalexin accumulation in soybean released from soybean cell walls by endopolygalacturonic acid lyase, Z. Naturforschung, 41c:39–48.

Davis, K. R., Darvill, A. G., Albersheim, P., and Dell, A. (1986b). Host–pathogen interactions. XXIX. Oligogalacturonides released from sodium polypectate by endopolygalacturonic acid lyase are elicitors of phytoalexins in soybean, Plant Physiol., 80:568–577.

Davis, K. R., and Hahlbrock, K. (1987). Induction of defense responses in cultured parsley cells by plant cell wall fragments. Plant Physiol., 85:1286–1290.

Day, P. R. (1956). Race names of *Cladosporium fulvum,* Rep. Tomato Genet. Coop., 6:13–14.

De Laat, A. M. M., and Van Loon, L. C. (1982). Regulation of ethylene biosynthesis in virus-infected tobacco leaves. II. Time course of levels of intermediates and in vivo conversion rates, Plant Physiol., 69:240–245.

Dean, J. F. D., and Anderson, J. D. (1990). Ethylene biosynthesis-inducing xylanase. II. Purification and physical characterization of the enzyme produced by *Trichoderma viride,* Plant Physiol., 94:1849–1854.

Dean, J. F. D., Gamble, H. R. and Anderson, A. J. (1989). The ethylene biosynthesis-inducing xylanase: its induction in *Trichoderma viride* and certain plant pathogens, Phytopathology, 79:1071–1078.

Deising, H., Nicholson, R. L., Haug, M., Howard, R. J., and Mendgen, K., (1992). Adhesion pad formation and the involvement of cutinase and esterases in the attachment of uredospores to the host cuticle, Plant Cell, 4:1101–1111.

Descalzo, R. C., Rahe, J. E., and Mauza, B. (1990). Comparative efficacy of induced resistance for selected diseases of greenhouse cucumber, Can. J. Plant Pathol., 12:16–24.

Deverall, B. J., and Deakin, A. L. (1985). Assessment of Lr 20 gene-specificity of symptom elicitation by intercellular fluids from leaf rust-infected wheat leaves, Physiol. Plant Pathol., 27:99–107.

Deverall, B. J., and Deakin, A. L. (1987). Genetic tests of the basis of wheat cultivar selectivity in symptom elicitation by preparations from rust pathogens, Physiol. Mol. Plant Pathol., 30:225–232.

Devergne, J. C., Bonnet, P., Panabieres, F., Blein, J. P., and Ricci, P. (1992). Migration of the fungal protein cryptogein within tobacco plants, Plant Physiol., 99:843–847.

DeWit, P. J. G. M. (1992). Molecular characterization of gene-for-gene sytems in plant-fungus interactions and the application of avirulence genes in control of plant pathogens, Annu. Rev. Phytopathol., 30:391–418.

DeWit, P. J. G. M., and Kodde, E. (1981). Induction of polyacetylenic phytoalexins in *Lycopersicon esculentum* after inoculation with *Cladosporium fulvum,* Physiol. Plant Pathol., 18:143–148.

DeWit, P. J. G. M., and Kodde, E. (1981). Further characterization and cultivar-specificity of glycoprotein elicitors from culture filtrates and cell walls of *Cladosporium fulvum* (syn. *Fulvia fulva*), Physiol. Plant Pathol., 18:297–314.

DeWit, P. J. G. M., and Roseboom, P. H. M. (1980). Isolation, partial characterization and specificity of glycoprotein elicitors from culture filtrates, mycelium and cell walls of *Cladosporium fulvum* (syn. *Fulvia fulva*), Physiol. Plant Pathol., 16:391–408.

DeWit, P. J. G. M., and Spikman, G. (1980). Evidence for the occurrence of race and cultivar-specific elicitors of necrosis in intercellular fluids of compatible interactions of *Cladosporium fulvum* and tomato, Physiol. Plant Pathol., 21:1–11.

DeWit, P. J. G. M., and Spikman, G. (1982). Evidence for the occurrence of race and cultivar specific elicitors of necrosis in intercellular fluids from leaf rust-infected wheat leaves, Physiol. Plant Pathol., 27:99–107.

Dickinson, S. (1969). Studies in the physiology of obligate parasitism. VI. Directed growth, Phytopath. Z., 114:38–49.
Dieter, P. (1984). Calmodulin and calmodulin-mediated processes in plants, Plant Cell Environ., 7:371–380.
Ding, H., Balsiger, S., Guggenbuhl, C., and Hohl, H. R. (1994). A putative IgG-binding 65 kDa adhesin involved in adhesion and infection of soybeans by *Phytophthora megasperma* f.sp. *glycinea,* Physiol. Mol. Plant Pathol., 44:363–378.
Dixon, R. A. (1986). The phytoalexin response: elicitation, signalling and control of host gene expression, Biol. Rev., 61:239–291.
Dixon, R. A., Dey, P. M., Murphy, D. L., and Whitehead, I. M. (1981). Dose response of *Colletotrichum lindemuthianum* elicitor mediated enzyme induction in French bean cell suspension cultures, Planta, 151:272–280.
Dixon, R. A., and Harrison, M. J. (1990). Activation, strucutre, and organization of genes involved in microbial defense in plants, Adv. Genet., 28:165–234.
Dixon, R. A., and Lamb, C. J. (1990). Molecular communication in interactions between plants and microbial pathogens, Annu. Rev. Plant Physiol. Plant Mol. Biol., 41:339–367.
Dixon, R.A., Jennings, A. C., Davies, L. A., Gerrish, C., and Murphy, D. L. (1989). Elicitor-active components from French bean hypocotyls, Physiol. Mol. Plant Pathol., 34:99–115.
Doherty, H. M., and Bowles, D. J. (1990). The role of pH and ion transport in oligosaccharide-induced proteinase inhibitor accumulation in tomato plants, Plant Cell Environ., 13:851–855.
Doherty, H. M., Selvendran, R. R., and Bowles, D. J. (1988). The wound response of tomato plants can be inhibited by aspirin and related hydroxy-benzoic acids, Physiol. Mol. Plant Pathol., 33:377–384.
Doke, N., and Furuichi, N. (1982). Response of protoplast to hyphal wall components in relationship to resistance of potato to *Phytophthora infestans,* Physiol. Plant Pathol., 21:23–30.
Doke, N., Garas, N. A., and Kuc, J. (1979). Partial characterization and aspects of the mode of action of a hypersensitivity-inhibiting factor (HIF) isolated from *Phytophthora infestans,* Physiol. Plant Pathol., 15:127–140.
Doke, N., and Tomiyama, K. (1980a). Suppression of the hypersensitive response of potato tuber protoplast to hyphal wall components by water soluble glucans isolated from *Phytophthora infestans,* Physiol. Plant Pathol., 16:177–186.
Doke, N., and Tomiyama, K. (1980b). Effect of hyphal wall components from *Phytophthora infestans* on protoplasts of potato tuber tissues, Physiol. Plant Pathol., 16:169–176.
Doke, N., and Miura, Y. (1995). In vitro activation of NADPH-dependent O_2^- generating system in a plasma membrane-rich fraction of potato tuber tissues by treatment with an elicitor from *Phytophthora infestans* or with digitonin, Physiol. Mol. Plant Pathol., 46:17–28.
Doss, R. P., Potter, S. W., Chastagner, G. A., and Christian, J. K. (1993). Adhesion of nongerminted *Botrytis cinerea* conidia to several substrate, Appl. Environ. Microbiol., 59:1786–1791.
Doubrava, N. S., Dean, R. A., and Kuc, J. (1988). Induction of systemic resistance to

anthracnose caused by *Colletotrichum lagenarium* in cucumber by oxalate and extracts from spinach and rhubarb leaves, Physiol. Mol. Plant Pathol., 33:69–79.

Dow, J. M., and Callow, J. A. (1979). Leakage of electrolytes from isolated leaf mesophyll cells of tomato induced by glycoproteins from culture filtrates of *Fulvia fulva* (Cooke) Ciferri (syn. *Cladosporium fulvum*), Physiol. Plant Pathol., 15:27–34.

Drobak, B. K., and Ferguson, I. B. (1985). Release of Ca^{2+} from plant hypocotyl microsomes by inositol-1,4,5-trisphosphate, Biochem. Biophys. Res. Comm., 130:1241–1246.

Dron, M., Clouse, S. D., Lawton, M. A., Dixon, R. A., and Lamb, C. J. (1988). Glutathione and fungal elicitor regulation of a plant defense gene promoter in electroporated protoplasts, Proc. Natl. Acad. Sci. USA, 85:6738–6742.

Dunkle, L. D., Maheswari, R., and Allen, P. J. (1969). Infection structures from bean rust uredospores: effect of RNA and protein synthesis inhibitors, Science, 163:481–482.

Ebel, J. (1981). Elicitor-induced phytoalexin synthesis in soybean (*Glycine max*), Regulation of Secondary Product and Plant Hormone Metabolism (M. Luckner and K. Schreiber, eds.), Pergamon Press, Oxford, pp. 155–162.

Ebel, J. (1984). Induction of phytoalexin synthesis in soybean, 4th Congress of the Federation of European Societies of Plant Physiology, Strasbourg, p. 153.

Ebel, J. (1986). Phytoalexin synthesis: the biochemical analysis of the induction process, Annu. Rev. Phytopathol., 24:235–264.

Ebel, J. (1989). Elicitor-binding proteins and signal transduction in elicitor action, J. Cell Biochem., 13D(Suppl.):248.

Ebel, J, Ayers, A. R., and Albersheim, P. (1976). Host–parasite interactions. XII. Response of suspension-cultured soybean cells to the elicitor isolated from *Phytophthora megasperma* var. *sojae,* a fungal pathogen of soybeans, Plant Physiol., 57:775–779.

El-Basyouni, S. Z., Chen, D., Ibrahim, R. K., Neish, A. C., and Towers, G. H. N. (1964) The biosynthesis of hydroxybenzoic acids in higher plants, Phytochemistry, 3:485–492.

Elliott, D. C., and Skinner, J. D. (1986). Calcium-dependent, phospholipid-activated protein kinase in plants, Phytochemistry, 25:39–44.

Emmett, R. W., and Parbery, D. G. (1975). Appressoria, Annu. Rev. Phytopathol., 13:147–167.

Enyedi, A. J., Yalpani, N., Paul, S., and Raskin, I. (1992). Signal molecules in systemic plant resistance to pathogens and pests. Cell, 70:879–886.

Epstein, L., Laccetti, L., Staples, R. C., Hoch, H. C., and Hoose, W. A. (1985). Extracellular proteins associated with induction of differentiation in bean rust uredospore germlings, Phytopathology, 75:1073–1076.

Epstein, L., Laccetti, L., Staples, R. C., and Hoch, H. C. (1987). Cell substratum adhesive protein involved in surface contact responses of the bean rust fungus, Physiol. Mol. Plant Pathol., 30:373–378.

Esnault, R., Chibbar, R. N., Lee, D., Van Huystee, R. B., and Ward, E. W. B. (1987). Early differences in production of mRNAs for phenylalanine ammonia-lyase and chalcone synthase in resistant and susceptible cultivars of soybean inoculated with *Phytophthora megasperma* f.sp. *glycinea,* Physiol. Mol. Plant Pathol., 30:293–297.

Estrada-Garcia, M. T., Callow, J. A., and Green, J. R. (1990). Monoclonal antibodies to the adhesive cell coat secreted by *Pythium aphanidermatum* zoospores recognize 200 ×

10^3 Mr. Glycoproteins stored within large peripheral vesicles, J. Cell Sci., 95:199–206.

Evans, D. E. (1988). Regulation of cytosolic calcium by cell membranes, Cell Biol. Int. Rep., 12:383–396.

Falkenstein, E., Groth, B., Mithofer, A., and Weiler, E. W. (1991). Methyl jasmonate and α-linolenic acid are potent inducers of tendril coiling, Planta, 185:316–322.

Farmer, E. E., and Helgeson, J. P. (1987). An extracellular protein from *Phytophthora parasitica* var. *nicotianae* is associated with stress metabolite accumulation in tabacco callus, Plant Physiol., 85:733–740.

Farmer, E. E., Johnson, R. R., and Ryan, C. A. (1992). Regulation of expression of proteinase inhibitor genes by methyl jasmonate and jasmonic acid, Plant Physiol., 98:995–1002.

Farmer, E. E., Moloshok, T. D., Saxton, M. J., and Ryan, C. A. (1991). Oligosaccharide signalling in plants: Specificity of oligouronide-enhanced plasma membrane protein phosphorylation, J. Biol. Chem., 266:3140–3145.

Farmer, E. E., Pearce, G., and Ryan, C. A. (1989). In vitro phosphorylation of plant plasma membrane proteins in response to the proteinase inhibitor inducing factor, Proc. Natl. Acad. Sci. USA, 86:1539–1542.

Farmer, E. E. and Ryan, C. A. (1990). Interplant communication: airborne methyl jasmonate induces synthesis of proteinase inhibitors in plant leaves, Proc. Natl. Acad. Sci. USA, 87:7713–7716.

Farmer, E. E. and Ryan, C. A. (1992). Octadecanoid precursors of jasmonic acid activate synthesis of wound-inducible proteinase inhibitors, Plant Cell, 4:129–139.

Felix, G., Grosskopf, D. G., Regenass, M., Basse, C. W., and Boller, T. (1991). Elicitor-induced ethylene biosynthesis in tomato cells. Characterization and use as a bioassay for elicitor action, Plant Physiol., 97:19–25.

Frank, J. A., and Paxton, J. D. (1971). An inducer of soybean phytoalexin and its role in the resistance of soybeans to *Phytophthora* rot, Phytopathology, 61:954–958.

Freissmuth, M., and Gilman, A. G. (1991). G proteins and the regulation of second messenger systems. Principles of Internal Medicine (J. D. Wilson, E. Braunwald, K. J. Isselbacher, R. G. Petersdorf, J. B. Martin, A. S. Fauci and R. F. Root, eds.), McGraw Hill, New York, pp. 393–397.

Freytag, S., and Hardham, A. R. (1991). Surface carbohydrates and cell wall structure of *in vitro* induced uredospore infection structures of *Uromyces viciae-fabae* before and after treatment with enzymes and alkali, Protoplasma, 161:94–103.

Friend, J. (1981). Plant phenolics, lignification and disease resistance, Prog. Phytochem., 7:197–261.

Fuhrer, J. (1982). Ethylene biosynthesis and cadmium toxicity in leaf tissue of beans (*Phaseolus vulgaris* L.), Plant Physiol., 70:162–167.

Funk, C., and Brodelius, P. E. (1990a). Phenylpropanoid metabolism in suspension cultures of *Vanilla planifolia* Andr. II. Effects of precursor feeding and metabolic inhibitors, Plant Physiol., 94:102–108.

Funk, C., and Brodelius, P. E. (1990b). Phenyl-propanoid metabolism in suspension cultures of *Vanilla planifolia* Andr. III. Conversion of 4-methoxy-cinnamic acids into 4-hydrobenzoic acids, Plant Physiol., 94:102–108.

Furusawa, I., Nishiguchi, M., Tani, M., and Ishida, N. (1977). Evidence of early protein synthesis-essential to the spore germination of *Colletotrichum lagenarium,* J. Gen. Microbiol., 101:307–310.

Garas, N. A., and Kuc, J. (1981). Potato lectin lyses zoospores of *Phytophthora infestans* and precipitates elicitors of terpenoid accumulation produced by the fungus, Physiol. Plant Pathol., 18:227–237.

Gottstein, H. D., and Kuc, J. (1989). Induction of systemic resistance to anthracnose in cucumber by phosphates, Phytopathology, 79:176–179.

Graham, T. L., and Graham, M. Y. (1991). Glyceollin elicitors induce major but distinctly different shifts in isoflavonoid metabolism in proximal and distal soybean cell populations. Mol. Plant–Microbe Interact., 4:60–68.

Graham, J. S., Hall, G., Pearce, G., and Ryan, C. A. (1986). Regulation of synthesis of proteinase inhibitors I and II mRNAs in leaves of wounded tomato plants, Planta, 169:399–405.

Graham, J. S., Pearce, G., Merryweather, J., Titani, K., Ericsson, L. and Ryan, C. A. (1985a). Wound induced proteinase inhibitors from tomato leaves. (I). The cDNA deduced primary structure of pre-inhibitor I and its post-translational processing, J. Biol. Chem., 260:6555–6564.

Graham, J. S., Pearce, G., Merryweather, J., Titani, K., Ericsson, L. and Ryan, C. A. (1985b). Wound induced proteinase inhibitors from tomato leaves. II. The cDNA deduced primary structure of pre-inhibitor II, J. Biol. Chem., 260:6561–6564.

Grant, G. T., Morris, B. R., Rees, D. A., Smith, P. J. C., and Thorn, D. (1973). Biological interactions between polysaccharides and divalentcations: the eggbox model, FEBS Lett., 32:195–198.

Green, T. R., and Ryan, C. A. (1972). Wound-induced proteinase inhibitor in plant leaves: a possible defense mechanism against insects, Science, 175:776–777.

Green, T. R., and Ryan, C. A. (1973). Wound-induced protinase inhibitor in tomato leaves: some effects of light and temperature on the wound response, Plant Physiol., 51:19–21.

Grimes, H. D., Koetze, D. S., and Franscehi, V. R. (1992). Expression, activity, and cellular accumulation of methyl jasmonate-responsive lipoxygenase in soybean seedlings, Plant Physiol., 100:433–443.

Gubler, F., and Hardham, A. R. (1988). Secretion of adhesive material during encystment of *Phytophthora cinnamomi* zoospores, characterized by immunogold labelling with monoclonal antibodies to components of peripheral vesicles, J. Cell Sci., 90:225–235.

Gubler, F., Hardham, A. R., and Duniec, J. (1989). Characterizing adhesiveness of *Phytophthora cinnamomi* zoospores during encystment, Protoplasma, 149:24–30.

Guggenbuhl, C., and Hohl, H. R. (1992). Mutual adhesion and interactions of isolated soybean mesophyll cells and germtubes of *Phytophthora megasperma* f.sp. *glycinea,* Bot. Helv., 102:247–261.

Gundlach, H., Muller, M. J., Kutchan, T., and Zenk, M. H. (1992). Jasmonic acid is a signal inducer in elicitor-induced plant cell cultures, Proc. Natl. Acad. Sci. USA, 89:2389–2393.

Hadwiger, L. A., and Beckman, J. M. (1980). Chitosan as a component of pea–*Fusarium solani* interactions, Plant Physiol., 66:205–211.

Hadwiger, L. A., Beckman, J. M., and Adams, M. J. (1981). Localization of fungal components in the pea–*Fusarium* interaction detected immunochemically with anti-chitosan and anti-fungal cell wall antisera, Plant Physiol., 67:170–175.

Hadwiger, L. A., and Loschke, D. C. (1981). Molecular communication in host-parasite interactions: hexosamine polymers (chitosan) as regulatory compounds in race-specific and other interactions, Phytopathology, 71:756–763.

Hahn, M., Jungling, S., and Knogge, W. (1993). Cultivar-specific elicitation of barley defense reactions by the phytotoxic peptide NIP1 from *Rhynchosporium secalis,* Mol. Plant–Microbe Interact., 6:745–754.

Hahn, M. G., Bucheli, P., Cervone, F., Doares, S. H., O'Neill, R. A., Darvill, A. G., and Albersheim, P. (1989). The role of cell wall constituents in plant–pathogen interactions. Plant-Microbe Interactions (E. Nester and T. Kosuge, eds.), MacMillan Press, New York, pp. 131–181.

Hahn, M. G., Darvill, A. G., and Albersheim, P. (1981). Host–pathogen interactions. XIX. The endogenous elicitor, a fragment of a plant cell wall polysaccharide that elicits phytoalexin accumulation insoybean, Plant Physiol., 68:1161–1169.

Ham, K. S., Kauffmann, S., Albersheim, P., and Darvill, A. G. (1991). Host–pathogen interactions. XXXIX. A soybean pathogenesis-related protein with β-1,3-glucanase activity releases phytoalexin elicitor-active heat-stable fragments from fungal walls, Mol. Plant Microbe Interact., 4:545–552.

Hamdan, M. A. M. S., and Dixon, R. A. (1986). Differential biochemical effects of elicitor preparations from *Colletotrichum lindemuthianum,* Physiol. Mol. Plant Pathol., 28:329–344.

Hamdan, M. A. M. S., and Dixon, R. A. (1987). Fractionation and properties of elicitors of the phenyl propanoid pathway from culture filtrates of *Colletotrichum lindemuthianum,* Physiol. Mol. Plant Pathol., 31:91–103.

Hamer, J. E., Howard, R. J., Chumley, P. G., and Valent, B., (1988). A mechanism for surface attachment in spores of a pathogenic fungus, Science, 239:288–290.

Hanson, J. B. (1984). The functions of calcium in plant nutrition. Advances in Plant Nutrition (P. B. Tinker and A. Lauchli, eds.), Praeger Publishers, New York, pp. 149–208.

Hardham, A. R. (1985). Studies on the surface of zoospores and cysts of the fungus *Phytophthora cinnamomi:* the influence of fixation on patterns of lectin binding, J. Histochem. Cytochem., 33:110–118.

Hardham, A. R., and Suzaki, E. (1986). Encystment of zoospores of the fungus, *Phytophthora cinnamomi,* is induced by specific lectin and monoclonal antibody binding to the cell surface. Protoplasma, 133:165–173.

Hargreaves, J. A., and Bailey, J. A. (1978). Phytoalexin production by hypocotyls of *Phaseolus vulgaris* in response to constitutive metabolites released by damaged bean cells. Physiol. Plant Pathol., 13:89–100.

Hargreaves, J. A., and Selby, C. (1978). Phytoalexin formation in cell suspensions of *Phaseolus vulgaris* in response to an extract of bean hypocotyl. Phytochemistry, 17:1099–1102.

Harper, J. F., Surowy, T. K., and Sussman, M. R. (1989). Molecular cloning and sequence of cDNA encoding the plasma membrane proton pump (H^+-ATPase) of *Arabidopsis thaliana*. Proc. Natl. Acad. Sci. USA, 86:1234–1238.

Heath, M. C. (1980). Effect of infection by compatible species or injection of tissue extracts on the susceptibility of non-host plants to rust fungi, Phytopathology, 76:356–360.

Heath, M. C. (1981). The suppression of the development of silicon-containing deposits in French bean leaves by exudates of the bean rust fungus and extracts from bean rust-infected tissue, Physiol. Plant Pathol., 18:149–155.

Heim, S., and Wagner, K. G. (1986). Evidence of phosphorylated phosphatidylinositols in the growth cycle of suspension cultured plant cells. Biochem. Biophys. Res. Commun., 134:1175–1181.

Helsper, J. P. F. G., Groot, P. F. M. De., Linskens, H. F., and Jackson, J. F. (1986a) Phosphatidyl monophosphate in *Lilium* pollen and turnover of phospholipid during pollen tube extension, Phytochemistry, 25:2193–2199.

Helsper, J. P. F. G., Groot, P. F. M. De., Linskens, H. F., and Jackson, J. F. (1986b). Phosphatidylinositol phospholipase C activity in pollen of *Lilium longiflorum,* Phytochemistry, 25:2053–2055.

Hepler, P. K., and Wayne, R. O. (1985). Calcium and the cell wall, Plant Cell Environ., 7:441–448.

Hermann, G., Lehmann, J., Peterson, A., Sembder, G., Weidhase, R. A., and Parthier, B. (1989). Species and tissue specificity of jasmonate-induced abundant proteins, J. Plant Physiol., 134:703–709.

Hetherington, A., and Trewavas, A. (1982). Calcium-dependent protein kinase in pea shoot membranes, FEBS Lett., 145:67–71.

Hetherington, A., and Trewavas, A. (1984). Activation of pea membrane protein kinase by calcium ions, Planta, 161:409–417.

Higinbotham, N. (1973). Electropotentials of plant cells, Annu. Rev. Plant Physiol., 24:25–46.

Hildebrand, D. F., Hamilton-Kemp, T. R., Legg, C. S., and Bookjans, G. (1988). Plant lipoxygenases: occurrence, properties, and possible functions, Curr. Top. Plant Biochem. Physiol., 7:201–219.

Hinch, J. M. and Clarke, A. E. (1980). Adhesion of fungal zoospores to root surfaces is mediated by carbohydrate determinants of the root slime, Physiol. Plant Pathol., 16:303–307.

Hiramatsu, I., Ichinose, Y., Shiraishi, T., Oku, H., and Ouchi, S. (1986). Regulation of pisatin biosynthesis in pea leaves by elicitor and suppressor produced by *Mycosphaerella pinodes*, Ann. Phytopathol. Soc. Japan, 52:53–58.

Hohl, H. R. (1991). Surface-related host-pathogen interactions in *Phytophthora*. Phytophthora (J. A. Lucas, R. C. Shattock, D. S. Shaw and L. R. Cooke, eds.), Cambridge University Press, Cambridge, pp. 70–89.

Hohl, H. R., and Balsiger, S. (1986). Probing the surfaces of soybean protoplasts and germtubes of the soybean pathogen *Phytophthora megasperma* f.sp. *glycinea* with lectins, Bot. Helv., 96:289–297.

Holden, D. W., and Rohringer, R. (1985). Proteins in intercellular washing fluid from noninoculated and rust-affected leaves of wheat and barley, Plant Physiol., 78:715–723.

Huet, J. C., Nespoulos, C., and Pernollet, J. C. (1992). Structures of elicitin isoforms secreted by *Phytophthora drechsleri,* Phytochemistry, 31:1471–1476.

Huet, J. C., and Pernollet, J. C. (1989). Amino acid sequence of cinnamomin, a new member of the elicitin family, and its comparison to cryptogein and capsicein, FEBS Lett., 257:302–306.

Huet, J. C., Salle-Tourne, M., and Pernollet, J. C. (1994). Amino acid sequence and toxicity of the α elicitin secreted with ubiquitin by *Phytophthora infestans,* Mol. Plant-Microbe Interact., 7:302–304.

Hughes, R. K., and Dickerson, A. G. (1989). The effect of ethylene on phenylalanine ammonia lyase (PAL) induction by a fungal elicitor in *Phaseolus vulgaris,* Physiol. Mol. Plant Pathol., 34:361–378.

Humme, B., Hoppe, H. H., and Heitefuss, R. (1978). Resistenzin duzierende Faktoren isoliert aus Zellwanden der Uredos porenkeimschlauche des Bohnenerostes (*Uromyces phaseoli*), Phytopathol. Z., 92:281–284.

Humme, B., Hoppe, H. H., and Heitefuss, R. (1981). Glucans from cell walls of uredospore germtubes of the bean rust fungus (*Uromyces phaseoli*) as elicitors of phytoalexin accumulation in *Phaseolus vulgaris,* Phytopathol. Z., 101:51–64. p2.

Irvine, R. F. (1982). The enzymology of stimulated inositol lipid turnover, Cell Calcium, 3:295–309.

Ishii, S. (1988). Factors influencing protoplast viablity of suspension-cultured rice cells during isolation process. Plant Physiol., 88:26–29.

Ishikawa, A., Tsubouchi, H., Iwasaki, Y., and Asahi, T. (1995). Molecular cloning and characterization of cDNA for the α subunit of a G protein from rice, Plant Cell Physiol., 36:353–359.

Johannes, E., Brosnan, J. M., and Sanders, D. (1991). Calcium channels and signal transduction in cells, Bioessay, 13:331–336.

Jin, D. F., and West, C. A. (1984). Characteristics of galacturonic acid oligomers as elicitors of casbene synthetase activity in castor bean seedlings. Plant Physiol., 74:989–992.

Jones, M. J., and Epstein, L. (1989). Adhesion of *Nectria haematococca* macroconidia, Physiol. Plant Pathol., 35:453–461.

Jones, M. J., and Epstein, L. (1990). Adhesion of macroconidia to the plant surface and virulence of *Nectria haematococca,* Appl. Environ. Microbiol., 56:3772–3778.

Joosten, M. H. A. J., Cozijnsen, A. J., and DeWit, P. J. G., (1994). Host resistance to a fungal tomato pathogen lost by a single base pair change in an avirulence gene, Nature, 367:384–386.

Kauffman, S., Legrand, M., Geoffroy, P., and Fritig, B. (1987). Biological function of 'pathogenesis-related' proteins: four PR proteins of tobacco have β-1,3-glucanase activity, EMBO J., 6:3209–3212.

Kauss, H., and Bauch, B. (1984). Chitosan-elicited synthesis of callose in soybean cells: a rapid and Ca^{++}-dependent process. 6th John Innes Symp., Cell Surface in Plant Growth and Development, Norwich, p. 53.

Kauss, H., and Bauch, B. (1988). Chitin deacetylase from *Colletotrichum lindemuthianum,* Methods Enzymol., 161B:518–523.

Kauss, H., Jeblick, W., and Young, D. H. (1982). Chitin deacetylase from the plant pathogen *Colletotrichum lindemuthianum,* Plant Science Lett., 28:231–236.

Kauss, H., Hohle, H., and Jeblick, W. (1983). Proteolytic activation and stimulation of Ca^{2+} of glucan synthase from soybean cells, FEBS Lett., 158:84–88.

Keen, N. T. (1971). Hydroxyphaseollin production by soybeans resistant and susceptible to *Phytophthora megasperma* var. *sojae,* Physiol. Plant Pathol., 1:265–275.

Keen, N. T., and Bruegger, B. (1977). Phytoalexins and chemicals that elicit their production in plants, ACS Symp. Ser., 62:1–26.

Keen, N. T., and Legrand, M. (1980). Surface of glycoproteins: Evidence that may function as the race specific phytoalexin elicitors of *Phytophthora megasperma* f.sp. *glycinea,* Physiol. Plant Pathol., 17:175–192.

Keen, N. T., Partridge, J. E., and Zaki, A. I. (1972). Pathogen-produced elicitor of a chemical defense mechanism in soybeans monogenically resistant to *Phytophthora megasperma* var. *sojae,* Phytopathology, 62:768.

Keen, N. T., and Yoshikawa, M. (1983). β-1,3-Endoglucanase from soybean releases elicitor-active carbohydrates from fungus cell walls, Plant Physiol., 71:460–465.

Keen, N. T., Yoshikawa, M. L., and Wang, M. C. (1983). Phytoalexin elicitor activity of carbohydrates from *Phytophthora megasperma* f.sp. *glycinea* and other sources, Plant Physiol., 71:466–471.

Keenan, P., Bryan, I. B., and Friend, J. (1985). The elicitation of the hypersensitive response of potato tuber tissue by a component of the culture filtrate of *Phytophthora infestans,* Physiol. Plant Pathol., 26:343–355.

Kendra, D. F., and Hadwiger, L. A. (1984). Characterization of the smallest chitosan oligomer that is maximally antifungal to *Fusarium solani* and elicits pisatin formation in *Pisum sativum,* Exp. Mycol., 8:276–281.

Kendra, D. F., Christian, D., and Hadwiger, L. A. (1989). Chitosan oligomers from *Fusarium solani*/pea interations, chitinase/β-glucanase digestion of sporelings and from fungal wall chitin actively inhibit fungal growth and enhance disease resistance, Physiol. Mol. Plant Pathol., 35:215–230.

Kerr, E. A., Patrick, Z. A., and Bailey, D. L. (1971). Resistance in tomato species to new races of leaf mold (*Cladosporium fulvum* (Cke)), Hort. Res. 11:84–92.

Kessmann, H., and Barz, W. (1986). Elicitation and suppression of phytoalexin and isoflavone accumulation in cotyledons of *Cicer arietinum* L. as caused by wounding and polymeric components from the fungus *Ascochyta rabiei,* J. Phytopathol., 117:321–335.

Klambt, H. D. (1962). Conversion of plants of benzoic acid to salicylic acid and its βd-glucoside, Nature, 196:491.

Kleinig, H. (1989). The role of plastids in isoprenoid biosynthesis, Annu. Rev. Plant Physiol. Plant Mol. Biol., 49:39–59.

Kogel, G., Beissmann, B., Reisener, H. J., and Kogel, K. H. (1988). A single glycoprotein from *Pucinia graminis* f.sp. *tritici* cell walls elicits the hypersensitive lignification response in wheat, Physiol. Mol. Plant Pathol., 33:173–185.

Kogel, K. H. and Beissmann, B. (1992). Isolation and characterization of elicitors. Modern Methods of Plant Analysis: Plant Toxin Analysis, (J. K. Linskens, and J. F. Jackson, eds.), Springer Verlag, Berlin, pp. 239–257.

Kogel, K. H., Heck, B., Kogel, G., Moerschbacher, B., and Reisener, H. J. (1985). A fungal elicitor of the resistance response in wheat, Z. Naturforschung 40:743–744.

Kohle, H., Jeblick, W., Poten, F., Blaschek, W., and Kauss, H. (1985). Chitosan-elicited callose synthesis in soybean cells as a Ca^{2+}-dependent process, Plant Physiol., 77:544–551.

Kohn, R. (1985). Ion binding in polyuronates, alginate and pectin, Pure Appl. Chem., 42:371–397.

Kolattukudy, P. E. (1980). Cutin, suberin, and waxes, The Biochemistry of Plants. Vol. 4. Lipids: Structure and Function (P. K. Stumpf, ed.), New York, pp. 571–645.

Kombrink, E., Schroeder, M., and Hahlbrock, K. (1988). Several 'pathogenesis-related' proteins in potato are β-1,3-glucanases and chitinases, Proc. Natl. Acad. Sci. USA, 85:782–786.

Komura, T., Kobayashi, I., Yamaoka, N., and Kunoh, H. (1990). Induced accessibility and enhanced inaccessibility at the cellular level in barley coleoptiles. VIII. Cytological evidence for suppressor(s) of host inaccessibility from *Erysiphe graminis,* Physiol. Mol. Plant Pathol., 37:409–416.

Koshio, O., Akanuma, Y., and Kasuga, M. (1988). Hydrogen peroxide stimulates tyrosine phosphorylation of the insulin receptor and its tyrosine kinase activity in intact cells, Biochem. J., 250:95–101.

Kuc, J. (1987). Plant immunization and its applicability to disease control. Approaches to Plant Disease Control (I. Chet, ed.), John Wiley, New York, pp. 255–274.

Kuc, J., and Lisker, N. (1978). A toxin from *Phytophthora infestans* which regulates phenolic and terpenoid metabolism in potato. Toxin: Animal, Plant and Microbial, (P. Rosenberg, ed.), Pergamon Press, Oxford, pp. 829–835.

Kunoh, H., Hayashimoto, A., Harui, M., and Ishizaki, H. (1985). Induced susceptibility and enhanced resistance at cellular level in barley coleoptiles. I. The significance of timing of fungal invasion, Physiol. Plant Pathol., 27:43–54.

Kunoh, H., Katsuragawa, N., Yamaoka, N., and Hayashimoto, A. (1988a). Induced accessibility and enhanced inaccessibility at the cellular level in barley coleoptiles. III. Timing and localization of enhanced inaccessibility in a single coleoptile cell and its transfer to an adjacent cell, Physiol. Plant Pathol., 33:81–93.

Kunoh, H., Komura, T., Kobayashi, I., and Yamaoka, N. (1990a). Induced accessibility and enhanced inaccessibility at the cellular level in barley coleoptiles. VII. Enhancement of inaccessibility at the prepenetration stage of a nonpathogen, Physiol. Mol. Plant Pathol., 37:399–407.

Kunoh, H., Kuroda, K., Hayashimoto, A., and Ishizaki, H. (1986). Induced susceptibility and enhanced resistance at cellular level in barley coleoptile. II. Timing and localization of induced susceptibility in a single coleoptile cell and its transfer to an adjacent cell, Can. J. Bot., 64:889–895.

Kunoh, H., Nicholson, R. L., Yoshioka, H., Yamaoka, N., and Kobayashi, I. (1990b). Preparation of the infection court by *Erysiphe graminis:* degradation of the host cuticle, Physiol. Mol. Plant Pathol., 36:397–407.

Kunoh, H., Toyoda, K., Yamaoka, N., and Kobayashi, I. (1989). Induced accessibility and enhanced inaccessibility at the cellular level in barley coleoptiles. V. Duration of stimulus by a non-pathogen in relation to enhanced inaccessibility, Physiol. Mol. Plant Pathol., 35:507–518.

Kunoh, H., Yamaoka, N., Yoshioka, H., and Nicholson, R. L. (1988b). Preparation of the infection court by *Erysiphe graminis.* I. Contact-mediated changes in morphology of the conidium surface, Exp. Mycol., 12:325–335.

Kurantz, M. J., and Osman, S. F. (1983). Class distribution, fatty acid composition and elicitor activity of *Phytophthora infestans* mycelial lipids, Physiol. Plant Pathol., 22:363–370.

Kurantz, M. J., and Zacharius, R. M. (1981). Hypersensitive response in potato tuber:

elicitation by combination of non-eliciting components from *Phytophthora infestans,* Physiol. Plant Pathol., 18:67–77.

Kurosaki, F., Amin, M., and Nishi, A. (1986). Induction of phytoalexin production and accumulation of phenolic compounds in cultured carrot cells, Physiol. Mol. Plant Pathol., 28:359–370.

Kurosaki, F., Futamura, K., and Nishi, A. (1985). Factors affecting phytoalexin production in cultured carrot cells, Plant Cell Physiol., 26:693–700.

Kurosaki, F., Matsui, K., and Nishi, A. (1984). Production and metabolism of 6-methoxymellein in cultured carrot cells, Physiol. Plant Pathol., 25:313–322.

Kurosaki, F., Tashiro, N., and Nishi, A. (1986). Induction of chitinase and phenylalanine ammonia-lyase in cultured carrot cells treated with fungal walls, Plant Cell Physiol., 27:1587–1591.

Kurosaki, F., Tashiro, N., and Nishi, A. (1987a). Induction, purification and possible function of chitinase in cultured carrot cells, Physiol. Mol. Plant Pathol., 31:201–210.

Kurosaki, F., Tashiro, N., and Nishi, A. (1987b). Secretion of chitinase from cultured carrot cells treated with fungal mycelial walls. Physiol. Mol. Plant Pathol., 31:211–216.

Kurosaki, F., Tsurusawa, Y., and Nishi, A. (1985). Partial purification and characterization of elicitors for 6-methoxymellein production in cultured carrot cells, Physiol. Plant Pathol., 27:209–217.

Kurosaki, F., Tsurusawa, Y., and Nishi, A. (1987c). Breakdown of phosphatidylinositol during the elicitation of phytoalexin and production in cultured carrot cells, Plant Physiol., 85:601–604.

Kurosaki, F., Tsurusawa, Y., and Nishi, A. (1987d). The elicitation of phytoalexins by Ca^{2+} and cyclic AMP in carrot cells, Phytochemistry, 26:1919–1923.

Kwon, Y. H., and Epstein, L. (1993). A 90-kDa glycoprotein associated with adhesion of *Nectria haematococca* macroconidia to substrate, Mol. Plant–Microbe Interact., 6:481–487.

Lamb, C. J., Lawton, M. A., Dron, M., and Dixon, R. A. (1989). Signals and transduction mechanism for activation of plant defenses against microbial attack, Cell, 56:215n224.

Lau, J. M., McNeil, M., Darvill, A. G., and Albersheim, P. (1985). Structure of the backbone of rhamnogalacturonan I, a pectic polysaccharide in the primary cell walls of plants, Carbohydrate Res., 137:111–125.

Lauter, F. R., Russo, V. E. A., and Yanofsky, C. (1992). Development and light regulation of *eas,* the structural gene for the rodlet protein, Genes Dev., 6:2373–2381.

Lazarovitz, G., Bhullar, B. S., Sugiyama, H. J., and Higgins, V. J. (1979). Purification and partial characterization of a glycoprotein toxin produced by *Cladosporium fulvum,* Phytopathology, 69:1062–1068.

Lazarovitz, G., and Higgins, V. J. (1979). Biological activity and specificity of a toxin produced by *Cladosporium fulvum,* Phytopathology, 69:1051–1061.

Lee, S. C., and West, C. A. (1981a). Polygalacturonase from *Rhizopus stolonifer,* an elicitor of casbene synthetase activity in castor bean (*Ricinus communis* L.) seedlings, Plant Physiol., 67:633–639.

Lee, S. C., and West, C. A. (1981b). Properties of *Rhizopus stolonifer* polygalacturonase, an elicitor of casbene synthetase activity in castor bean (*Ricinus communis* L.) seedlings, Plant Physiol., 67:640–645.

Lee, J. S., Brown, W. E., Graham, J. S., Pearce, G., Fox, E. A., Dreher, T. W., Ahern, K. G., Pearson, G. D., and Ryan, C. A. (1986). Molecular characterization and phylogenetic inhibitor I gene in *Lycopersicon* species, Proc. Natl. Acad. Sci. USA, 83:7277–7281.

Legrand, M., Kauffman, S., Geoffroy, P., and Fritig, B. (1987). Biological function of pathogenesis-related proteins: four tobacco pathogenesis-related proteins are chitinases, Proc. Natl. Acad. Sci. USA, 84:6750–6754.

Leite, B., and Nicholson, R. L. (1992). Mycosporine-alanine: A self-inhibitor of germination from the conidial mucilage of *Colletotrichum graminicola,* Exp. Mycol., 16:76–86.

Leon, J., Yalpani, N., Raskin, I., and Lawton, M. A. (1993). Induction of benzoic acid 2-hydroxylase in virus-inoculated tobacco, Plant Physiol., 103:323–328.

Lewis, N. G., Ingiong, M. E., Ohashi, H., Towers, G. H. N., and Yamamoto, E. (1988). Exclusive accumulation of Z-isomers of monolignols and their glucosides in bark of *Fagus grandifolia,* Phytochemistry, 27:2119–2121.

Lincoln, J. E., and Fischer, R. L. (1988). Diverse mechanisms for the regulation of ethylene-inducible gene expression, Mol. Gen. Genet., 212:71–75.

Lindner, W., Hoffmann, C., and Grisebach, H. (1988). Rapid elicitor-induced chemiluminescence in soybean cell suspension cultures, Phytochemistry, 27:2501–2503.

Lippincott, J. A., and Lippincott, B. B. (1984). Concepts and experimental approaches in host-microbe recognition. Plant Microbe Interactions: Molecular and Genetic Perspectives (T. Kosuge and E.W. Nester, eds.) Macmillan, New York, pp. 195–214.

Lisker, N., and Kuc, J. (1977). Elicitors of terpenoid accumulation in potato tuber slices, Phytopathology, 67:1356–1359.

Longman, D., and Callow, J. A. (1987). Specific saccharide residues are involved in the recognition of plant root surfaces by zoospores of *Pythium aphanidermatum,* Physiol. Mol. Plant Pathol., 30:139–150.

Lorenzo, G. De., Ito, Y., D'Ovidio, R., Cervone, F., Albersheim, P., and Darvill, A. G. (1990). Host-pathogen interactions. XXXVII. Abilities of the polygalacturonase-inhibiting proteins from three races of *Colletotrichum lindemuthianum,* Physiol. Mol. Plant Pathol., 36:421–435.

Lotan, T., and Fluhr, R. (1990). Xylanase, a novel elicitor of pathogenesis-related proteins in tobacco, uses a nonethylene pathway for induction, Plant Physiol., 93:811–817.

Lovett, J. S. (1976). Regulation of protein metabolism during fungal spore germination. Fungal Spore (D. J. Weber, and W. H. Hess, eds.), John Wiley & Sons, New York, pp. 189–242.

Low, P. S., and Heinstein, P. F. (1986). Elicitor stimulation of the defense response in cultured plant cells monitored by fluorescent dyes, Arch. Biochem. Biophys., 249:472–479.

Lu, H. and Higgins, V. J. (1993). Partial characterization of a non-proteinaceous suppressor of non-specific elicitors from *Cladosporium fulvum* (syn. *Fulvia fulva*), Physiol. Mol. Plant Pathol., 42:427–439.

Lukas, T. J., Iverson, D. B., Schleicher, M., and Watterson, D. M. (1984). Structural characterization of a higher plant calmodulin, Plant Physiol., 75:788–795.

Ma, H., Yanofsky, M. F., and Meyerowitz, E. M. (1991). Isolation and sequence analysis of TGA1 cDNAs encoding a tomato G protein α subunit, Gene, 107:189–195.

Maina, G., Allen, R. D., Bhatia, S. K., and Stelzic, D. A. (1984). Phenol metabolism, phytoalexins, and respiration in potato tuber tissue treated with fatty acid, Plant Physiol., 76:735–738.

Maniara, G., Laine, R., and Kuc, J. (1984). Oligosaccharides from *Phytophthora infestans* enhance the elicitation of sesquiterpenoid stress metabolites by arachidonic acid in potato, Physiol. Plant Pathol., 24:177–186.

Marcan, H., Jarvis, M. C., and Friend, J. (1979). Effect of methyl glycosides and oligosaccharides on cell death and browning of potato tuber discs induced by mycelial components of *Phytophthora infestans*, Physiol. Plant Pathol., 14:1–9.

Marme, D. (1989). The role of calcium calmodulin in signal transduction. Second Messengers in Plant Growth and Development, (W. F. Boss, and D. J. Morre, eds.), Alan R. Liss, New York, pp. 57–80.

Mauch, F., Hadwiger, L. A., and Boller, T. (1984). Ethylene: symptom, not signal for the induction of chitinase and β-1,3-glucanase in pea pods by pathogens and elicitors, Plant Physiol., 76:607–611.

Mauch, F., and Staehelin, L. A. (1989). Functional implications of the subcellular localization of ethylene-induced chitinase and β-1,3-glucanase in bean leaves, Plant Cell, 1:447–457.

Mayama, S. (1983). The role of avenalumin in the resistance of oats to crown rust, Mem. Fac. Agric., Kagawa Univ., No. 42:36–44.

Mayama, S., Bordin, A. P. A., Morikawa, T., Tanpo, H., and Kato, H. (1995). Association of avenalumin concentration with co-segregation victorin sensitivity and crown rust resistance in oat lines carrying the Pc-2 gene, Physiol. Mol. Plant Pathol., 46:263–274.

Mayama, S., Tani, T., Ueno, T., Midland, S. L., Sims, J. J., and Keen, N. T. (1986). The purification of victorin and its phytoalexin elicitor activity in oat leaves, Physiol. Mol. Plant Pathol., 29:1–18.

Mayer, M. G., and Ziegler, E. (1988). An elicitor from *Phytophthora megasperma* f.sp. *glycinea* influences the membrane potential of soybean cotyledonary cells, Physiol. Mol. Plant Pathol., 33:397–407.

McAinsh, M. R., Brownlee, C., and Hetherington, A. M. (1991). Abscisic acid-induced elevation of guard cell cytosolic Ca^{2+} precedes stomatal closure, Nature, 343:186–188.

McNeill, M., Darvill, A. G., Fry, S. C., and Albersheim, P. (1984). Structure and function of the primary cell walls of plants, Annu. Rev. Biochem., 53:625–663.

Means, A. R., and Dedman, J. R. (1980). Calmodulin an intracellular receptor, Nature, 285:73–77.

Melan, M. A., Dong, X., Endara, M. E., Davis, K. R., Ausubel, F. M., and Peterman, T. K. (1993). An *Arabidopsis thaliana* lipoxygenase gene can be induced by pathogens, abscisic acid, and methyl jasmonate, Plant Physiol., 101:441–450.

Mercure, E. W., Kunoh, H., and Nicholson, R. L. (1994a). Adhesion of *Colletotrichum graminicola* to corn leaves: a requirement for disease development, Physiol. Mol. Plant Pathol., 45:407–420.

Mercure, E. W., Leite, B., and Nicholson, R. L. (1994b). Adhesion of ungerminated conidia of *Colletotrichum graminicola* to artificial hydrophobic surfaces, Physiol. Mol. Plant Pathol., 45:421–440.

Mercure, E. W., Kunoh, H., and Nicholson, R. L. (1995). Visualization of materials released from adhered, ungerminated conidia of *Colletotrichum graminicola,* Physiol. Mol. Plant Pathol., 46:121–135.

Metraux, J. P., Signer, H., Ryals, J., Ward, E., Wyss-Benz, M., Gaudin, J., Raschdorf, K., Schmid, E., Blum, W., and Inverardi, B. (1990). Increase in salicylic acid in the onset of systemic acquired resistance in cucumber, Science, 250:1004–1006.

Michell, R. H. (1982). Stimulated inositol lipid metabolites: an introduction, Cell Calcium, 3:285–294.

Miehle, B. R., and Lukezic, F. L. (1972). Studies on conidial germination and appressorium formation by *Colletotrichum trifolii* Bain and Essary, Can. J. Microbiol., 18:1263–1269.

Miller, R. H., Berryman, A. A., and Ryan, C. A. (1986). Biotic elicitors of defense reactions in lodgepole pine, Phytochemistry, 25:611–612.

Moerschbacher, B. M., Heck, B., Kogel, K. H., Obst, O., and Reisener, H. J. (1986). An elicitor of the hypersensitivity lignification response in wheat leaves isolated from the rust fungus, *Puccinia graminis* f.sp. *tritici,* Z. Naturforschung, 41:839–844.

Moerschbacher, B. M., Kogel, K. H., Noll, U., and Reisener, H. J. (1986). An elicitor of the hypersensitivity lignification response in wheat leaves isolated from the rust fungus *Puccinia graminis* f.sp. *tritici.* I. Partial purification and characterization, Z. Naturforschung, 41:830–838.

Moerschbacher, B. M., Noll, U., Flott, B. E., and Reisener, H. J. (1988). Lignin biosynthetic enzymes in stem rust infected, resistant and susceptible near-isogenic wheat lines, Physiol. Mol. Plant Pathol., 33:33–46.

Molano, J., Polacheck, I., Duran, A., and Cabib, E. (1979). An endochitinase from wheat germ, J. Biol. Chem., 254:4901–4907.

Morelli, E., Rej, N., Lewis, N. G., Just, G., and Towers, G. H. N. (1986). Cis-monolignols in *Fagus grandifolia* and their possible involvement in lignification, Phytochemistry, 25:1701–1704.

Morre, D. J., Gripshover, B., Monroe, A., and Morre, J. T. (1984a). Phosphatidylinositol turnover in isolated soybean membranes stimulated by the synthetic growth hormone 2,4-dichlorophenoxyacetic acid, J. Biol. Chem., 259:15364–15468.

Morre, D. J., Morre, J. T., and Varnold, R. L. (1984b). Phosphorylation of membrane located proteins of soybean *in vitro* and response to auxin, Plant Physiol., 75:265–268.

Morse, M. J., Crain, R. C., and Satter, R. L. (1986). Phosphatidylinositol turnover in *Samanca pulvini:* a mechanism of phosphotransduction, Plant Physiol., 80:S-92.

Movahedi, S., and Heale, J. B. (1990). The roles of aspartic proteinase and endo-pectin lyase enzymes in the primary stages of infection and pathogenesis of various host tissues by different isolates of *Botrytis cinerea* Pers ex. Pers., Physiol. Mol. Plant Pathol., 36:303–324.

Nelson, A. J., Doerner, P. W., Zhu, Q., and Lamb, C. J. (1994). Isolation of a monocot 3-hyroxy-3-methylglutaryl coenzyme A reductase gene that is elicitor-inducible, Plant Mol. Biol., 25:401–412.

Nespoulos, C., Huet, J. C., and Pernollet, J. C. (1992). Structure-function relationships of α and β elicitin signal proteins involved in the plant–*Phytophthora* interaction, Planta, 186:551–557.

Nichols, E. J., Beckman, J. M., and Hadwiger, L. A. (1980). Glycosidic enzyme activity in pea tissue and pea-*Fusarium solani* interactions, Plant Physiol., 66:199–204.

Nicholson, R. L., Butler, L. G., and Asquith, T. N. (1986). Glycoproteins from *Colletotrichum graminicola* that bind phenols: Implications for survival and virulence of phytopathogenic fungi, Phytopathology, 76:1315–1318.

Nicholson, R. L., and Epstein, L. (1991). Adhesion of fungi to plant surface: prerequisite for pathogenesis. The Fungal Spore and Disease Initiation in Plants and Animals, (G. T. Cole, and H. G. Hoch, eds.), Plenum Press, New York, pp. 3–23.

Nicholson, R. L., Hipskind, J., and Hanau, R. M. (1989). Protection against phenol toxicity by the spore mucilage of *Colletotrichum graminicola,* an aid to secondary spread, Physiol. Mol. Plant Pathol., 35:243–252.

Nicholson, R. L., Kunoh, H., Shiraishi, T., and Yamada, T. (1993). Initiation of the infection process by *Erisiphe graminis:* conversion of the conidial surface from hydrophobicity to hydrophilicity and influence of the conidial exudate on the hydrophobicity of the barley leaf surface, Physiol. Mol. Plant Pathol., 43:307–318.

Nicholson, R. L., and Moraes, W. B. C. (1980). Survival of *Colletotrichum graminicola:* importance of the spore matrix, Phytopathology, 70:255–261.

Nicholson, R. L., Yoshioka, H., Yamaoka, N. and Kunoh, H. (1988). Preparation of the infection court by *Erysiphe graminis:* II. Release of esterase enzyme from conidia in response to a contact stimulus, Exp. Mycol., 12:336–349.

Nishizuka, Y. (1984). The role of protein kinase C in cell surface signal transduction and tumour promotion, Nature, 308:693–698.

Nothnagel, E. A., McNeil, M., Albersheim, P., and Dell, A. (1983). Host-pathogen interactions. XXII. A galacturonic acid oligosaccharide from plant cell walls elicits phytoalexins, Plant Physiol., 71:916–926.

O'Connell, R. J. (1991). Cytochemical analysis of infection structures of *Colletotrichum lindemuthianum* using fluorochrome-labelled lectins, Physiol. Mol. Plant Pathol., 39:189–200.

O'Connell, R. J., Nash, C., and Bailey, J. A. (1992). Lectin cytochemistry: a new approach to understanding cell differentiation, pathogenesis, and taxonomy in *Colletotrichum,* Colletotrichum: Biology, Pathology and Control (J. A. Bailey, and M. J. Jeger, eds.), CAB Press, London, pp. 67–87.

Oba, K., Kondo, K., Doke, N., and Uritani, I. (1985). Induction of 3-hydroxy-3-methylglutaryl CoA reductase in potato tuber after slicing, fungal infection or chemical treatment, and some properties of the enzymes, Plant Cell Physiol., 26:873–880.

Odermatt, M., Rothlisberger, A., Werner, C., and Hohl, H. R. (1988). Interactions between agarose-embedded plant protoplasts and germtubes of *Phytophthora,* Physiol. Mol. Plant Pathol., 33:209–220.

Oku, H., Shiraishi, T., and Ouchi, S. (1977). Suppression of induction of phytoalexin, pisatin by low-molecular substances from spore germination fluid of a pea pathogen, *Mycosphaerella pinodes,* Z. Naturforschung, 64:643.

Oku, H., Shiraishi, T., and Ouchi, S. (1995). Role of specific suppressors in pathogenesis of *Mycosphaerella* species. Molecular Determinants of Plant Diseases (S. Nishimura,

C. P. Vance and N. Doke, eds.), Japan Scientific Society Press and Springer-Verlag, Tokyo and Berlin, pp. 144–156.

Paliyath, G., and Poovaiah, B. W. (1985). Calcium and calmodulin-promoted phosphorylation of membrane proteins during senescence in apples, Plant Cell Physiol., 26:977–986.

Paradies, I., Humme, B., Hoppe, H. H., Heitefuss, R., and Elstner, E. F. (1979). Induction of ethylene formation in bean (*Phaseolus vulgaris,* hypocotyl segments by preparations isolated from germtube cell walls of *Uromyces phaseoli,* Planta, 146:193–197.

Parker, J. E., Hahlbrock, K., and Scheel, D. (1988). Different cell wall components from *Phytophthora megasperma* f.sp. *glycinea* elicit phytoalexin production in soybean and parsley, Planta, 176:75–82.

Parker, J. E., Schulte, W., Hahlbrock, K., and Scheel, D. (1991). An extracellular glycoprotein from *Phytophthora megasperma* f.sp. *glycinea* elicits phytoalexin synthesis in cultured parsley cells and protoplasts, Mol. Plant-Microbe Interact., 4:19–27.

Pascholati, S. F., Deising, H., Leite, B., Anderson, D., and Nicholson, R. L. (1993). Cutinase and non-specific esterase activities in the conidial mucilage of *Colletotrichum graminicola,* Physiol. Mol. Plant Pathol., 42:37–51.

Pascholati, S. F., Yoshioka, H., Kunoh, H., and Nicholson, R. L. (1992). Preparation of the infection court by *Erysiphe graminis* f.sp. *hordei* cutinase is a component of the conidial exudate, Physiol. Mol. Plant Pathol., 41:53–59.

Pearce, G., Strydom, D., Johnson, S., and Ryan, C. A. (1991). A polypeptide from tomato leaves induces wound-inducible proteinase inhibitor proteins, Science, 253:895–898.

Pearce, R. B., and Ride, J. P. (1982). Chitin and related compounds as elicitors of the lignification response in wounded wheat leaves, Physiol. Plant Pathol., 20:119–123.

Peever, T. L., and Higgins, V. J. (1989). Suppression of the activity of non-specific elicitor from *Cladosporium fulvum* intercellular fluids from tomato leaves. Physiol. Mol. Plant Pathol., 34:471–482.

Pegg, G. F., and Young, D. H. (1981). Changes in glycosidase activity and their relationship to fungal colonisation during infection of tomato by *Verticillium albo-atrum,* Physiol. Plant Pathol., 19:371–382.

Pegg, G. F., and Vessey, J. C. (1973). Chitinase activity in *Lycopersicon esculentum* and its relationship to the in vitro lysis of *Verticillium albo-atrum* mycelium, Physiol. Plant Pathol., 3:207–222.

Pelissier, B., and Esquerre-Tugaye, M. T. (1984). Membrane effects of fungal elicitors in plants. 4th Congress of the Federation of European Societies of Plant Physiology, Strasbourg, p. 304.

Pena-Cortes, H., Willmitzer, L., and Sanchez-Serrano, J. (1991). Abscisic acid mediates wound-induction but not developmental expression of the proteinase inhibitor II gene family, Plant Cell, 3:963–972.

Pena-Cortes, H., Serrano, J. J., Mertens, R., and Willmitzer, L. (1989). Abscisic acid is involved in the wound-induced expression of the proteinase inhibitor II gene in potato and tomato, Proc. Natl. Acad. Sci. USA, 86:9851–9855.

Pernollet, J. C., Sallankin, M., Salle-Tourne, M., and Huet, J. C. (1993). Elicitin isoforms from *Phytophthora* species: comparison of their physico-chemical properties and toxicity to tobacco and other plant species, Physiol. Mol. Plant Pathol., 42:53–67.

Peters, B. M., Cribbs, D. H., and Stelzig, D. A. (1978). Agglutination of plant protoplasts by fungal cell wall glucans, Science, 201:364–365.

Pfaffmann, H., Hartmann, E., Brightman, A.O., and Morre, D. J. (1987). Phosphatidylinositol specific phospholipase C of plant stems. Membrane associated activity concentrated in plasma membranes, Plant Physiol., 85:1151–1155.

Pink, D. A. C., Bennet, F. G. A., Caten, C. E., and Law, C. N. (1983). Correlated effects of homoeologous group 5 chromosomes upon infection of wheat by yellow rust and powdery mildew, *Z. Naturforschung,* 91:278–294.

Pitt, D., and Kaile, A. (1990). Transduction of the calcium signal with special reference to Ca^{2+}-induced conidiation in *Penicillium notatum.* Biochemistry of Cell Walls and membranes in Fungi, (J. Kuhn, A. P. J. Trinci, M. J. Jung, M. W. Goosey and L. G. Copping, eds.), Springer-Verlag, Berlin, pp. 283–298.

Polya, G. M., Schibeci, A., and Miucci, V. (1984). Phosphorylation of membrane proteins from cultured *Lolium multiflorum* (rye grass) endosperm cells, Plant Sci. Lett., 36:51–57.

Poole, R. J. (1978). Energy coupling for membrane transport, Annu. Rev. Plant Physiol., 29:437–460.

Poovaiah, B. W. (1988). Molecular and cellular aspects of calcium action in plants, Hortscience, 23:267–271.

Poovaiah, B. W., and Reddy, A. S. N. (1987). Calcium messenger system in plants, CRC Crit. Rev. Plant Sci., 6:47–103.

Preisig, C. L., and Kuc, J. A. (1985). Arachidonic acid-related elicitors of the hypersensitive response in potato and enhancement of their activities by glucans from *Phytophthora infestans* (Mont.) de Bary, Arch. Biochem. Biophys., 236:379–389.

Preisig, C. L., and Kuc, J. A. (1987a). Phytoalexins, elicitors, enhancers, suppressors and other considerations in the regulation of R-gene resistance to *Phytophthora infestans* in potato. Biochemical and Molecular Determinants of Plant Diseases, (S. Nishimura, C. Vance and N. Doke, eds.), Japan Scientific Society, Tokyo, pp. 203–221.

Preisig, C. L., and Kuc, J. A. (1987b). Inhibition by salicylhydroxamic acid, BW 755C, eicosatetrynoic acid and disulfiram of the hypersensitive response elicited in potato by arachidonic acid or poly-L-lysine, Plant Physiol., 84:891–894.

Preisig, C. L., and Kuc, J. A. (1988). Metabolism by potato tuber of arachidonic acid, and elicitor of hypersensitive resistance, Physiol. Mol. Plant Pathol., 32:77–88.

Pring, R. J. (1980). A fine-structural study of the infection of leaves of *Phaseolus vulgrais* by uredospores of *Uromyces phaseoli,* Physiol. Plant Pathol., 17:269–276.

Prowning, R. F., and Irzykiewicz, H. (1965). Studies on the chitinase system in bean and other seeds, Comp. Biochem. Physics, 14:127–133.

Purdy, R. E., and Kolattukudy, P. E. (1975). Hydrolysis of plant cuticle by plant pathogens. Properties of cutinase I, cutinase II and a nonspecific esterase isolated from *Fusarium solani pisi,* Biochemistry, 14:2832–2840.

Ragothama, K. G., Veluthambi, K., and Poovaiah, B. W. (1985). Stage-specific changes in calcium-regulated protein phosphorylation in developing tomato fruits, Plant Cell Physiol., 26:1565–1572.

Ramadoss, C. S., Uhlig, J., Carlson, D. M., Butler, L. G., and Nicholson, R. L. (1985).

Composition of the mucilaginous spore matrix of *Colletrotrichum graminicola,* a pathogen of corn, sorghum and other grasses, J. Agric. Food Chem., 33:728–732.
Ramakrishnan, L., and Staples, R. C. (1970). Changes in ribonucleic acids during uredospore differentiation, Phytopathology, 60:1087–1091.
Ramasarma, T. (1982). Generation of H_2O_2 in biomembranes, Biochim. Biophys. Acta, 694:69–93.
Ranjeva, R., and Boudet, A. M. (1987). Phosphorylation of proteins in plants: regulatory effects and potential involvement in stimulus/response coupling, Annu. Rev. Plant Physiol., 38:73–93.
Ranjeva, R., Graziana, A., Ranty, B., Cavalie, G., and Boudet, A. M. (1984). Phosphorylation of proteins in plants: a step in the integration of extra and intracellular stimuli? Physiol. Veg., 22:365–376.
Ranjeva, R., Refeno, G., Boudet, A. M., and Marme, D. (1983). Activation of plant quinate: NAD^+ 3 oxidoreductase by calcium and calmodulin, Proc. Natl. Acad. Sci. USA, 80:5222–5224.
Ransom, R. F., Hipskind, J., Leite, B., Nicholson, R. L., and Dunkle, L. D. (1992). Effects of elicitor from *Colletotrichum graminicola* on the response of sorghum to *Periconia circinata* and its pathotoxin, Physiol. Mol. Plant Pathol., 41:75–84.
Rasi-Caldogno, F., De Michelis, M. I., and Puliarello, M. C. (1982). Active transport of Ca^{2+} in membrane vesicles from pea. Evidence for a H^+/Ca^{2+} antiport, Biochim. Biophys. Acta, 693:287–295.
Raskin, I. (1992). Salicylate, a new plant hormone, Plant Physiol., 99:799–803.
Raskin, I., Ehmann, A., Melander, W. R., and Meeuse, B. J. D. (1987). Salicylic acid—a natural inducer of heat production in Arun lilies, Science, 237:1545–1556.
Raskin, I., Skubatz, Z., Tang, W., and Meeuse, B. J. D. (1990). Salicylic acid levels in thermogenic and non-thermogenic plants, Ann. Bot., 66:376–373.
Raskin, I., Turner, I. M., and Melander, W. R. (1989). Regulation of heat production in the inflorescences of an Arum lily by endogenous salicylic acid, Proc. Natl. Acad. Sci. USA, 86:2214–2218.
Rasmussen, J. B., Hammerschmidt, R., and Zook, M. N. (1991). Systemic induction of salicylic acid accumulation in cucumber after inoculation with *Pseudomonas syringae* pv. *syringae,* Plant Physiol., 97:1342–1347.
Ren, Y. Y., and West, C. A. (1992). Elicitation of diterpene biosynthesis in rice (*Oryza sativa* L.) by chitin, Plant Physiol., 99:1169–1178.
Reuter, H. (1983). Calcium channel modulation by neurotransmitters, enzymes and drugs, Nature, 301:569–579.
Ricci, P., Bonnet, P., Huet, J. C., Sallantin, M., Beauvais-Cante, F., Bruneteau, M., Billard, V., Michel, G., and Pernollet, J. C. (1989). Structure and activity of proteins from pathogenic fungi *Phytophthora* eliciting necrosis and acquired resistance in tobacco, Eur. J. Biochem., 183:555–563.
Ricci, P., Trentin, F., Bonnet, P., Vernard, P. J., Mouton-Perronnet, F., and Bruneteau, M. (1992). Differential production of parasiticein, an elicitor of necrosis and resistance in tobacco, by isolates of *Phytophthora parasitica,* Plant Pathology, 41:298–307.
Ricker, K. E., and Bostock, R. M. (1992). Evidence for release of the elicitor arachidonic acid and its metabolites from sporangia of *Phytophthora infestans* during infection of potato, Physiol. Mol. Plant Pathol., 41:61–72.

Ricker, K. E., and Bostock, R. M., (1994). Eicosanoids in the *Phytophthora infestans*–potato interaction: lipoxygenase metabolism of arachidonic acid and biological activites of selected lipoxygenase products, Physiol. Mol. Plant Pathol., 44:65–80.

Ride, J. P., and Barber, M. S. (1990). Purification and characterization of multiple forms of endochitinase from wheat leaves. Plant Sci., 71:185–197.

Ride, J. P., Barber, M. S., and Bertram, R. E. (1989). Infection-reduced lignification in wheat. Plant Cell Wall Polymers: Biogenesis and Biodegradation, (N. G. Lewis, and M. G. Paide, eds.), American Chemical Society, Washington, pp. 361–369.

Rincon, M., and Boss, W. F. (1987). Myo-inositol triphosphate mobilizes calcium from fusogenic carrot (Daucus carota L.) protoplasts, Plant Physiol., 83:395–398.

Robertson, B. (1986). Elicitors of the production of lignin-like compounds in cucumber hypocotyls, Physiol. Mol. Plant Pathol., 28:137–148.

Roby, D., Toppan, A., and Esquerre-Tugaye, M. T. (1985). Cell surfaces in plant-microorganism interactions. V. Elicitors of fungal and of plant origin trigger the synthesis of ethylene and of cell wall hydroxyproline-rich glycoprotein in plants, Plant Physiol., 77:700–704.

Rojas, M. L., Montes De Gomez, V., and Ocampo, C. A. (1993). Stimulation of lipoxygenase activity in cotyledonary leaves of coffee reacting hypersensitively to the coffee leaf rust, Physiol. Mol. Plant Pathol., 43:209–219.

Ross, A. F. (1966). Systemic effects of local lesion formation. Viruses of Plants, (A. B. R. Beemster, and J. Dijkstra, eds.), North Holland, Amsterdam, pp. 127–150.

Ryan, C. A. (1974). Assay and biochemical properties of the proteinase inhibitor-inducing factor, a wound hormone, Plant Physiol., 54:328–332.

Ryan, C. A. (1978). Proteinase inhibitors in plant leaves: a biochemical model for pest-induced plant protection, Trends Biochem. Sci., 5:148–157.

Ryan, C. A. (1987). Oligosaccharide signalling in plants. Annu. Rev. Cell Biol., 3:295–317.

Ryan, C. A. (1988). Oligosaccharides as recognition signals for the expression of defensive genes in plants, Biochemistry, 27:8879–8883.

Ryan, C. A. (1992). The search for the proteinase inhibitor-inducing factor, PIIF, Plant Mol. Biol., 19:123–133.

Ryan, C. A., Bishop, P. D., Pearce, G., Darvill, A. G., McNeil, M., and Albersheim, P. (1981). A sycamore cell wall polysaccharide and a chemically related tomato leaf oligosaccharide possess similar proteinase inhibitor-inducing activities, Plant Physiol., 68:616–618.

Ryan, C. A., Bishop, P. D., Walker-Simmons, M., Brown, W. E., and Graham, J. S. (1985). Pectin fragments regulate the expression of proteinase inhibitor genes in plants. Cellular and Molecular Biology of Plant Stress (J. L. Key and T. Kosuge, eds.), Alan R. Liss, New York, pp. 319–334.

Ryerson, D. E., and Heath, M. C. (1992). Fungal elicitation of wall modifications in leaves of *Phaseolus vulgaris* L. cv. Pinto II. Effects of fungal wall components, Physiol. Mol. Plant Pathol., 40:283–298.

Salimath, B. P., and Marme, D. (1983). Protein phosphorylation and its regulation by calcium and calmodulin in membrane fractions from *Zucchini* hypocotyls, Planta, 158:560–568.

Sanchez, L. M., Doke, N., and Kawakita, K. (1992). Elicitor-induced chemiluminescence in cell suspension cultures of tomato, sweet pepper and tobacco plants and its inhibition by suppressors from *Phytophthora* spp., Plant Sci., 88:141–148.

Sandeihus, and Sommarin, M. (1986). Phosphorylation of phosphatidylinositols in isolated plant membranes, FEBS Lett., 201:282–286.

Sanders, L. C., Wang, C. S., Walling, L. L., and Lord, E. M. (1991). A homologue of the substrate adhesion molecule vitronectin occurs in 4 species of flowering plants, Plant Cell, 3:629–635.

Schafer, A., Bygrave, F., Matzenauer, S., and Marme, D. (1985). Identification of a calcium and phospholipid dependent protein kinase in plant tissue, FEBS Lett., 187:25–28.

Schaffrath, U., Scheinpflug, H., and Reisner, H. J. (1995). An elicitor from *Pyricularia oryzae* induces resistance responses in rice: isolation, characterization and physiological properties, Physiol. Mol. Plant Pathol., 46:293–307.

Scheel, D., and Parker, J. E. (1990). Elicitor recognition and signal transduction in plant defense gene activation, Z. Naturforschung, 45c:569–575.

Schmidt, W. E., and Ebel, J. (1987). Specific binding of a fungal glucan phytoalexin elicitor to membrane fractions from soybean *Glycine max,* Proc. Natl. Acad. Sci. USA, 84:4117–4121.

Schneider-Muller, S., Kurosaki, F., and Nishi, A. (1994). Role of salicylic acid and intracellular Ca^{2+} in the induction of chitinase activity in carrot suspension culture, Physiol. Mol. Plant Pathol., 45:101–109.

Schnitzler, J. P., Madlung, J., Rose, A., and Seitz, H. U. (1992). Biosynthesis of p-hydroxybenzoic acid in elicitor-treated carrot cell cultures, Planta, 188:594–600.

Schottens-Toma, I. M. J., and DeWitt, P. J. G. M. (1988). Purification and primary structure of a necrosis-inducing peptide from the apoplastic fluids of tomato infected with *Cladosporium fulvum* (syn. *Fulvia fulva*), Physiol. Mol. Plant Pathol., 33:59–67.

Schroeder, J. I., and Hagiwara, S. (1989). Cytosolic calcium regulates ion channels in the plasma membrane of *Vicia faba* guard cells. Nature, 338:427–430.

Schuerger, A. C., and Mitchell, D. J. (1993). Influence of mucilage secreted by macroconidia of *Fusarium solani* f.sp. *phaseoli* on spore attachment to roots of *Vigna radiata* in hydroponic nutrient solution, Phytopathology, 83:1162–1170.

Schuerger, A. C., Mitchell, D. J., and Kaplan, D. T. (1993). Influence of carbon source on attachment and germination of macroconidia of *Fusarium solani* f.sp. *phaseoli* on roots of *Vigna radiata* grown in hydroponic nutrient solutions, Phytopathology, 83:1171–1177.

Schumaker, K. S., and Sze, H., (1985). A Ca^{2+}/H^+ antiport driven by the proton electrochemical gradient of a tonoplast H^+-ATPase from oat roots, Plant Physiol., 79:1111–1117.

Sekizawa, Y., Haga, M., and Kantou, K. (1990). A superoxide forming enzyme, NADPH oxidase of rice blade tissue stimulated with a blast fungus elicitor, Ann. Phytopahtol. Soc., Japan, 56:565–567.

Sela-Buurlage, M. B., Epstein, L., and Rodriguez, R. J. (1991). Adhesion of ungerminated *Colletotrichum musae* conidia, Physiol. Mol. Plant Pathol., 39:345–352.

Seo, H. S., Kim, H. Y., Jeong, J. Y., Lee, S. Y., Cho, J. J., and Bahk, J. D. (1995). Molecular cloning and characterization of RGA1 encoding a G protein α subunit from rice (*Oryza sativa* L. IR-36), Plant Mol. Biol., 27:1119–1131.

Sequeira, L. (1983). Mechanisms of induced resistance in plants, Annu. Rev. Microbiol., 37:51–79.

Sharp, J. A., Albersheim, P., Ossowski, P., Pilotti, A., Garegg, P., and Lindberg, B. (1984a). Comparison of the structures and elicitor activities of a synthetic and a mycelial-wall derived hexa (β-D-glucopyranosyl)-D-glucitol, J. Biol. Chem., 259:11341–11345.

Sharp, J. A., McNeil, M. and Albersheim, P. (1984b). The primary structure of one elicitor active and seven elicitor-inactive hexa (β-D-glucopyranosyl)-D-glucitols isolated from the mycelial walls of *Phytophthora megasperma* f.sp. *glycinea,* J. Biol. Chem., 259:11321–11326.

Sharp, J. A., Valent, B., and Albersheim, P. (1984c). Purification and partial characterization of a β-glucan fragment that elicits phytoalexin accumulation in soybean, J. Biol. Chem., 259:11312–11320.

Shimizu, T., Radmark, O., and Samuelsson, B. (1984). Enzyme with dual lipoxygenase activities catalyzes leukotriene A_4 synthesis from arachidonic acid, Proc. Natl. Acad. Sci. USA, 81:689–693.

Shiraishi, T., Oku, H., Yamashita, M. and Ouchi, S. (1978). Elicitor and suppressor of pisatin induction in spore germination fluid of pea pathogen, *Mycosphaerella pinodes,* Ann. Phytopathol. Soc., Japan, 44:659–665.

Shiraishi, T., Araki, M., Yoshioka, H., Kobayashi, I., Yamada, T., Ichinose, Y., Kunoh, H., and Oku, H. (1991). Inhibition of ATPase activity in pea plasma membranes in situ by a suppressor from a pea pathogen, *Mycosphaerella pinodes,* Plant Cell Physiol., 32:1067–1075.

Siedow, J. N. (1991). Plant lipoxygenase: structure and function, Annu. Rev. Plant Physiol. Plant Mol. Biol., 42:145–188.

Siegrist, J., and Kauss, H. (1990). Chitin deacetylase in cucumber leaves infected by *Colletotrichum lagenarium,* Physiol. Mol. Plant Pathol., 36:267–275.

Silverman, P., Seskar, M., Kanter, D., Schweizer, P., Metraux, J. P., and Raskin, I. (1995). Salicylic acid in rice. Biosynthesis, conjugation, and possible role, Plant Physiol., 108:633–639.

Sing, V. O., and Bartnicki-Garcia, S. (1975). Adhesion of *Phytophthora palmivora* zoospores: detection and ultrastructural visualization of Concanavalin A-receptor sites appearing during encystment, J. Cell Sci., 19:11–20.

Sitton, D., and West, C. A. (1975). Casbene, an antifungal diterpene produced in cell-free extracts of *Ricinus communis* seedlings, Phytochemistry, 14:1921–1925.

Spanswick, R. M. (1981). Electrogenic ion pumps, Annu. Rev. Plant Physiol., 32:267–289.

Stab, M. R., and Ebel, J. (1987). Effects of Ca^{2+} on phytoalexin induction by fungal elicitor in soybean cells, Arch. Biochem. Biophys., 257:416–423.

Staples, R. C., Laccetti, L., and Yaniv, Z. (1976). Appressorium formation and nuclear division in *Colletotrichum truncatum*, Arch. Microbiol., 109:75–84.

Staples, R. C., and Yaniv, Z. (1976). Protein and nucleic acid metabolism during germination. Encyclopedia of Plant Physiology, (R. Heitefuss and R. H. Williams, eds.), Springer Verlag, Berlin, pp. 86–103.

Staswick, P. E. (1992). Jasmonate, genes, and fragrant signals, Plant Physiol., 99:804–807.

Steffens, M., Ettl, F., Kranz, D., and Kindl, H. (1989). Vanadate mimicks effects of fungal cell wall in eliciting gene activation in plant cell cultures, Planta, 177:160–168.

Stekoll, M. S., and West, C. A. (1978). Purification and properties of an elicitor of castor bean phytoalexin from culture filtrates of the fungus *Rhizopus stolonifer*, Plant Physiol., 61:38–45.

Stermer, B. A., and Bostock, R. M. (1987). Involvement of 3-hydroxy-3-methylglutaryl Coenzyme A reductase in the regulation of sesquiterpenoid phytoalexin synthesis in potato, Plant Physiol., 84:404–408.

Stermer, B. A., Edwards, L. A., Edgington, B. V., and Dixon, R. A. (1991). Analysis of elicitor-inducible transcripts encoding 3-hydroxy-3-methylglutaryl Coenzyme A reductase in potato, Physiol. Mol. Plant Pathol., 39:135–145.

Stringer, M. A., Dean, R. A., Sewall, T. C., and Timberlake, W. I. (1991). Rodletless, a new *Aspergillus* development mutant induced by directed gene inactivation, Genes Dev., 5:1161–1171.

Stringer, M. A., and Timberlake, W. E. (1993). Cerato-ulmin, a toxin involved in Dutch elm disease, is a fungal hydrophobin, Plant Cell, 5:145–146.

Sutherland, M. W., Deverall, B. J., Moerschbacher, M., and Reisener, H. J. (1989). Wheat cultivar and chromosomal selectivity of two types of eliciting preparations from rust pathogens, Physiol. Mol. Plant Pathol., 35:535–541.

Suzuki, K., Furusawa, I., Ishida, N., and Yamamoto, M. (1981). Protein synthesis during germination and appressorium formation of *Colletotrichum lagenarium* spores, J. Gen. Microbiol., 124:61–69.

Takeuchi, Y., Yoshikawa, M., Takeba, G., Tanaka, K., Shibata, D., and Horino, O. (1990). Molecular cloning and ethylene induction of mRNA encoding a phytoalexin elicitor-releasing factor, β-1,3-endoglucanase, in soybean, Plant Physiol., 93:673–682.

Talbot, N. J., Ebbole, D. J., and Hamer, J. E. (1993). Identification and characterization of MGP1, a gene involved in pathogenicity from rice blast fungus, *Magnoporthe grisea*, Plant Cell, 5:1575–1590.

Templeton, M. D., Rikkerink, E. H. A., and Beever, R. E. (1994). Small, cysteine-rich proteins and recognition in fungal-plant interactions, Mol. Plant–Microbe Interact., 7:320–325.

Tepper, C. S., and Anderson, A. J. (1986). Two cultivars of bean display a differential response to extracellular components from *Colletotrichum lindemuthianum*, Physiol. Mol. Plant Pathol., 29:411–420.

Tepper, C. S., and Anderson, A. J. (1990). Interactions between pectic fragments and extracellular components from the fungal pathogen *Colletotrichum lindemuthianum*, Physiol. Mol. Plant Pathol., 36:147–158.

Tepper, C. S., Albert, F. G., and Anderson, A. J. (1989). Differential mRNA accumulation in three cultivars of bean in response to elicitors from *Colletotrichum lindemuthianum*, Physiol. Mol. Plant Pathol., 34:85–98.

Terce-Laforgue, T., Huet, J. C., and Pernollet, J. C. (1992). Biosynthesis and secretion of cryptogein, a protein elicitor secreted by *Phytophthora cryptogea*, Plant Physiol., 98:936–941.

Terhume, B. T., and Hoch, H. C. (1993). Substrate hydrophobicity and adhesion of *Uromyces* urediospores and germlings, Exp. Mycol., 17:241–252.

Teulieres, C., Alibert, G., and Ranjeva, R. (1985). Reversible phosphorylation of tonoplast proteins involves tonoplast-bound calcium-calmodulin dependent protein phosphatase(s), Plant Cell Rep., 4:199–201.

Thain, J. F., Doherty, H. M., Bowles, K. J., and Wildon, D. C. (1990). Oligosaccharides

that induce proteinase inhibitor activity in tomato plants cause depolarization of tomato leaf cells, Plant Cell Environ., 13:569–574.
Thornburg, R. W., An, G., Cleveland, T. E., Johnson, R., and Ryan, C. A. (1987). Wound inducible expression of a potato inhibitor II. Chloramphenicol acetyl transferase gene fusion in transgenic tobacco plants, Proc. Natl. Acad. Sci. USA, 84:744–748.
Tietjen, K. G., Hunkler, D., and Matern, U. (1983). Differential response of cultured parsley cells to elicitors from two non-pathogenic strains of fungi. I. Identification of induced products as coumarin derivatives, Eur. J. Biochem., 131:401–407.
Tietjen, K. G., and Matern, U. (1983). Differential response of cultured parsley cells to elicitors from two non-pathogenic strains of fungi. II. Effects of enzyme activities, Eur. J. Biochem., 131:409–413.
Toppan, A., and Esquerre-Tugaye, M. T. (1984). Cell surfaces in plant-microorganism interactions. IV. Fungal glycopeptides which elicit the synthesis of ethylene in plants, Plant Physiol., 75:1133–1138.
Toppan, A., Roby, D., and Esquerre-Tugaye, M. T. (1982). Cell surfaces in plant microorganisms interactions. III. In vivo effect of ethylene on hydroxyproline rich glycoprotein accumulation in the cell wall of diseased plants, Plant Physiol., 70:82–86.
Trewavas, A., and Gilroy, S. (1991). Signal transduction in plant cells, Trend. Genet., 7:356–361.
Tronchin, G., Bouchara, J. P., Annaix, V., Robert, R., and Senet, J. M. (1991). Fungal cell adhesion molecules in *Candida albicans*, Eur. J. Epidemiol., 7:23–33.
Valent, B. S., and Albersheim, P. (1977). Role of elicitors of phytoalexin accumulation in plant disease resistance, ACS Symp. Ser., 62:27–34.
Van den Ackerveken, G. F. J. M., van Kan, J. A. L., Joosten, M. H. A. J., Muisers, J. M., Verbakel, H. M., and DeWit, P. J. G. M. (1993). Characterization of two putative pathogenicity genes of the fungal tomato pathogen, *Cladosporium fulvum*, Mol. Plant–Microbe Interact., 6:210–215.
van Kan, J. A., van den Ackerveken, G. F. J. M., and DeWit, P. J. G. M. (1991). Cloning and characterization of cDNA of a avirulence gene avr 9 of the fungal pathogen, *Cladosporium fulvum*, causal agent of tomato leaf mold, Mol. Plant–Microbe Interact., 4:52–59.
Van Loon, L. C., and Antoniw, J. F. (1982). Comparison of the effects of salicylic acid and ethephon with virus-induced hypersensitivity and acquired resistance in tobacco, Neth. J. Plant Pathol., 88:237–256.
Varns, J. L., Kuc, J., and Williams, W. B. (1971). Terpenoid accumulation as a biochemical response of the potato tuber to *Phytophthora infestans*, Phytopathology, 61:174–177.
Vaughn, S. F., and Lulai, E. C. (1992). Further evidence that lipoxygenase activity is required for arachidonic acid-elicited hypersensitivity in potato callus cultures, Plant Sci., 84:91–98.
Veluthambi, K., and Poovaiah, B. W. (1984a). Polyamine-stimulated phosphorylation of proteins from corn (*Zea mays* L.) coleoptiles, Biochem. Biophys. Res. Commun., 122:1374–1380.
Veluthambi, K., and Poovaiah, B. W. (1984b). Calcium promoted protein phosphorylation in plants, Science, 223:167–169.

Veluthambi, K., and Poovaiah, B. W. (1984c). Calcium- and calmodulin-regulated phosphorylation of soluble and membrane proteins from corn coleoptiles, Plant Physiol., 76:359–365.

Vera-Estrella, R., Barkla, B. J., Higgins, V. J., and Blumwald, E. (1994). Plant defense response to fungal pathogens. Activation of host-plasma membrane H^+-ATPase by elicitor-induced enzyme dephosphorylation, Plant Physiol., 104:209–215.

Vick, B. A., and Zimmerman, D. C. (1987). Oxidative systems of modification of fatty acids: the lipoxygenase pathway. The Biochemistry of Plants. A comprehensive Treatise, (P. K. Stumpf, ed.), Academic Press, Orlando, pp. 53–90.

Vidhyasekaran, P. (1988). Physiology of Disease Resistance in Plants. Vol. I. CRC Press, Boca Raton, FL, p. 149.

Vidhyasekaran, P. (1993). Defense genes for crop disease management. Genetic Engineering, Molecular Biology and Tissue Culture for Crop Pest and Disease Management, (P. Vidhyasekaran, ed.), Daya Publishing House, Delhi, pp. 17–30.

Vidhyasekaran, P. (1993). Chitinase gene and pathogenesis-related proteins for crop disease management. Genetic Engineering, Molecular Biology, and Tissue Culture for Crop Pest and Disease Management, (P. Vidhyasekaran, ed.), Daya Publishing House, Delhi, pp. 41–63.

Vidhyasekaran, P. (1993). Avirulence gene for crop disease management. Genetic Engineering, Molecular Biology, and Tissue Culture for Crop Pest and Disease Management, (P. Vidhyasekaran, ed.), Daya Publishing House, Delhi, pp. 65–74.

Vidhyasekaran, P., Kandasamy, D., Durairaj, P., and Kandaswamy, T. K. (1971). Protection of sorghum plants from helminthosporiose by immunization, Indian J. Agric. Sci., 41:284–288.

Vogel, B. E., Lee, S. J., Hildebrand, A., Craig, W., Pierschbacher, M. D., Wong-Stall, F., and Ruoslahti, E. (1993). A novel integrin specifically exemplified by binding of the α-vβ5 integrin to the basic domain of the HIV Tat protein and vitronectrin, J. Cell Biol., 121:461–468.

Vogeli, U., and Chappell, J. (1988). Induction of sesquiterpene cyclase and suppression of squalene synthase activity in plant cell culture treated with fungal elicitor, Plant Physiol., 88:1291–1296.

Wagner, V. I., and Matthysse, A. G. (1992). Involvement of a vitronectin-like protein in attachment of *Agrobacterium tumefaciens* to carrot suspension culture cells, J. Bacteriol., 174:5999–6003.

Walker-Simmons, M., Hadwiger, L., and Ryan, C. A. (1983). Chitosans and pectic polysaccharides both induce accumulation of the antifungal phytoalexin pisatin in pea pods and antinutrient proteinase inhibitors in tomato leaves, Biochem. Biophys. Res. Commun., 110:194–199.

Walker-Simmons, M., Jin, D., West, C. A., Hadwiger, L., and Ryan, C. A. (1984). Comparison of proteinase inhibitor-inducing activities and phytoalexin elicitor activities of a pure fungal endopolygalacturonase, pectic fragments, and chitosans, Plant Physiol., 76:833–836.

Walker-Simmons, M., and Ryan, C. A. (1984). Proteinase inhibitor synthesis in tomato leaves. Induction by chitosan oligomers and chemically modified chitosan and chitin, Plant Physiol., 76:787–799.

Walker-Simmons, M., and Ryan, C. A. (1985). Proteinase inhibitor I accumulation in tomatovsuspension cultures; induction by plant and fungal cell wall fragments and an extra-cellular polysaccharide secreted into the medium, Plant Physiol., 80:68–75.

Ward, E. R., Payne, G. P., Moyer, M. B., Williams, S. C., Dincher, S. S., Sharkey, K. C., Beck, J. J., Taylor, H. T., Ahl-Goy, P., Meins, F. Jr., and Ryals, J. A. (1991). Differential regulation of β-1,3-glucanase messenger RNAs in response to pathogen infection, Plant Physiol., 96:390–397.

Wessels, J. G. H. (1993). Wall growth, protein excretion and morphogenesis in fungi, Tansley review No. 45, New Phytol., 123:397–413.

Wessels, J. G. H., de Vries, O. M. H. S. A., Asgeirsdottin, S. A., and Schuren, F. H. J. (1991). Hydrophobin genes involved in formation of aerial hyphae and fruit bodies in *Schizophyllum*, Plant Cell, 3:793–799.

West, C. A. (1981). Fungal elicitors of the phytoalexin response in higher plants, Naturwissenschaften, 68:447–467.

White, R. F. (1979). Acetylsalicylic acid (aspirin) induces resistance to tobacco mosaic virus in tobacco, Virology, 99:410–412.

Whitehead, T. M., Dey, P. M., and Dixon, R. A. (1982). Differential patterns of phytoalexin accumulation and enzyme induction in wounded and elicitor treated tissues of *Phaseolus vulgaris*, Planta, 154:156–164.

Wiedhase, R. A., Kramel, H. M., Lehman, J., Leibisch, H. W., Lerbs, W., and Parthier, B. (1987). Methyl jasmonate-induced changes in the polypeptide pattern of senescing barley leaf segments, Plant Sci., 51:177–186.

Wingate, V. P. M., Lawton, M. A., and Lamb, C. J. (1988). Glutathione causes a massive and selective induction of plant defense genes, Plant Physiol., 87:206–210.

Woloshuk, C. P., and Kolatukudy, P. E. (1986). Mechanism by which contact with plant cuticle triggers cutinase gene expression in the spores of *Fusarium solani* f.sp. *pisi*, Proc. Natl. Acad. Sci. USA, 83:1704–1708.

Wynn, W. K. (1976). Appressorium formation over stomates by the bean rust fungus: response to a surface contact stimulus, Phytopathology, 66:136–146.

Wynn, W. K. (1981). Tropic and taxic response of pathogens to plants, Annu. Rev. Phytopathol., 19:237–255.

Xiao, J. Z., Oshima, A., Kamakura, T., Ishiyama, T., and Yamaguchi, I. (1994a). Extracellular glycoprotein(s) associated with cellular differentiation in *Magnaporthe grisea*, Mol. Plant–Microbe Interact., 7:639–644.

Xiao, J. Z., Watanabe, T., Kamakura, T., Oshima, A., and Yamaguchi, I. (1994b). Studies on cellular differentiation of *Magnaporthe grisea*, physiochemical aspects of substratum surfaces in relation to appressorium formation, Physiol. Mol. Plant Pathol., 44:227–236.

Yalpani, N., and Raskin, I. (1993). Salicylic acid: a systemic signal in plant disease resistance, Trends Microbiol., 1:88–92.

Yalpani, N., Silverman, P., Wilson, T. M. A., Kleier, D. A., and Raskin, I. (1991). Salicylic acid is a systemic signal and an inducer of pathogenesis-related proteins in virus-infected tobacco, Plant Cell, 3:809–818.

Yamada, T., Hashimoto, H., Shiraishi, T., and Oku, H. (1989). Suppression of pisatin, phenylalanine ammonia-lyase mRNA, and chalcone synthase mRNA accumulation by a putative pathogenicity factor from the fungus *Mycosphaerella pinodes*, Mol. Plant–Microbe Interact., 2:256–261.

Yamaoka, N., Lyons, P. C., Hipskind, J., and Nicholson, R. L. (1990). Elicitor of sorghum phytoalexin synthesis from *Colletotrichum graminicola*, Physiol. Mol. Plant Pathol., 37:255–270.

Yang, Z., Park, H., Lacy, G. H., and Cramer, C. L. (1991). Differential activation of potato 3-hydroxy-3-methylglutaryl Coenzyme A reductase gene by wounding and pathogen challenge, Plant Cell, 3:397–405.

Yoshikawa, M., Keen, N. T., and Wang, M. C. (1983). A receptor on soybean membranes for a fungal elicitor of phytoalexin accumulation, Plant Physiol., 73:497–506.

Yoshikawa, M., Matama, M., and Masago, H. (1987). Release of a soluble phytoalexin elicitor from mycelial walls of *Phytophthora megasperma* var. *sojae* by soybean tissues, Plant Physiol., 67:1032–1035.

Yoshikawa, M., Takeuchi, Y., and Horino, O. (1990). A mechanism for ethylene-induced disease resistance in soybean: enhanced synthesis of an elicitor-releasing factor, β-1,3-endoglucanase, Physiol. Mol. Plant Pathol., 37:367–376.

Yoshioka, H., Shiraishi, T., Kawamata, S., Nasu, K., Yamada, T., Ichinose, Y., and Oku, H. (1992). Orthovanadate suppresses accumulation of phenylalanine ammonia-lyase mRNA and chalcone synthase mRNA in pea epicotyls induced by elicitor from *Mycosphaerella pinodes*, Plant Cell Physiol., 33:201–204.

Yoshioka, H., Shiraishi, T., Yamada, T., Ichinose, Y., and Oku, H. (1990). Suppression of pisatin production and ATPase activity in pea plasma membranes by orthovanadate, verapamil and a suppressor from *Mycosphaerella pinodes*, Plant Cell Physiol., 31:1139–1146.

Young, D. H., and Kauss, H. (1984). Adhesion of *Colletotrichum lindemuthianum* spores to *Phaseolus vulgaris* hypocotyls and to polystyrene, Appl. Environ. Microbiol., 47:616–619.

Yu, Y. B., and Yang, S. F. (1980). Biosynthesis of wound ethylene, Plant Physiol., 66:281–285.

Zanetti, A., Beauvais, F., Huet, J. C., and Pernollet, J. C. (1992). Movement of elicitins, necrosis-inducing proteins secreted by *Phytophthora* sp., in tobacco, Planta, 187:163–170.

Zhu, J. K., Damsz, B., Kononowioz, R. A. K., Bressan, R. A., and Hasegawa, P. M. (1994). A higher plant extracellular vitronectin-like adhesion protein is related to the translational elongation factor-1α, Plant Cell, 6:393–404.

Zhu, J. K., Shi, J., Singh, U., Wyatt, S. E., Bressan, B. A., Hasegawa, P. M., and Carpita, N. C. (1993). Enrichment of vitronectin- and fibronectin-like proteins in NaCl-adapted plant cells and evidence for their involvement in plasma membrane–cell wall adhesion, Plant J., 3:637–646.

Zocchi, G. (1988). Separation of membrane vesicles from maize roots having different calcium transport activities, Plant Sci., 54:103–107.

Zocchi, G., and Hansen, J. B. (1983). Calcium transport and ATPase activity in a microsomal vesicle fraction from corn roots, Plant Cell Environ., 6:203–209.

Zocchi, G., and Rabotti, G. (1993). Calcium transport in membrane vesicles isolated from maize coleoptiles. Effect of indoleacetic acid and Fusicoccin, Plant Physiol., 101:135–139.

Zook, M. N., and Kuc, J. A. (1991a). Induction of sesquiterpene cyclase and suppression of squalene synthetase activity in elicitor treated or fungal infected potato tuber tissue, Physiol. Mol. Plant Pathol., 39:377–390.
Zook, M. N., and Kuc, J. A. (1991b). The use of a sterol inhibitor to investigate changes in the rate of synthesis of 2,3-oxidosqualene in elicitor-treated potato tuber tissue, Physiol. Mol. Plant Pathol., 39:391–401.
Zuroske, G., Makus, D. J., and Ryan, C. A. (1980). The direction and rate of transport of the proteinase inhibitor inducing factor out of wounded tomato leaves. Plant Physiol., 65:Suppl. p10.

Part II

Molecular Events During Fungal Evasion of Host's Defense Mechanism

2

Cell Wall

I. STRUCTURE OF PLANT CELL WALL

A. Nature of Plant Cell Wall

Plant cell wall is the first barrier and penetration of the cell wall appears to be the first requirement for pathogenesis of fungal pathogens (Pearce and Rutherford, 1981; O'Brien and Leach, 1983; Morris et al., 1989). The cell walls are complex amalgams of carbohydrates (cellulose, hemicellulose, and pectic polysaccharides), proteins, lignin, and incrusting substances such as cutin, suberin, and certain inorganic compounds (Muhlethaler, 1967; Albersheim et al., 1969; Spiro, 1970; Cleland, 1971; Kolattukudy et al., 1981; Northcote, 1985; Showalter and Varner, 1989; Showalter, 1993).

The cell wall is of two types, a thin primary wall and a thicker secondary wall (Albersheim, 1965). The primary wall is found in young, undifferentiated cells that are still growing. It is transformed into a secondary wall after the cell has stopped growing. Primary cell walls of a variety of higher plants appear to have similar structures, but the composition and ultrastructure of secondary walls vary considerably from one cell type to another (Talmadge et al., 1973).

B. Cuticle

Cuticle is the incrusting substance and forms the first layer covering the outer walls of epidermal cells (Martin, 1964; Van den Ende and Linskens, 1974). The cuticle consists of a structural polymer, cutin, impregnated with wax. Cutin is the insoluble polymeric structural component of the cuticle of all aerial parts of plants except periderms (Kolattukudy, 1977; 1980; Espelie et al., 1980). Cutin is composed of mainly two families of monomers: a C_{16} family and a C_{18} family. The predominant component of the C_{16} family is 10,16-dihydroxypalmitic acid and/or its positional isomers in which the midchain hydroxyl group is at C-9, C-8, or C-7; in most cases one isomer predominates. Smaller quantities of 16-hydoxypalmitic acid and palmitic acid are also found in most cases. Major components of the C_{18} family of monomers and 18-hydroxyoleic acid, 18-hydroxy-9,10-epoxystearic acid, and threo-9,10-18-trihydroxystearic acid together with their analogs, contain an additional double bond at C-12 (Walton and Kolattukudy, 1972; Soliday and Kolattukudy, 1977; 1978; Espelie et al., 1979; Kolattukudy, 1970; 1981). In most cases cutin-containing layers have an amorphous appearance, but in some cases they have lamellar appearances (Thomson et al., 1966; Olesen, 1979).

C. Epicuticular Wax

Waxes are embedded in, and protrude from, polymeric cutin (Kolattukudy, 1985; Walton, 1990; Koller, 1991). Cuticular surfaces show characteristic wax crystals. Waxes contain hydrocarbons, long chain esters, free acids, free alcohols, and minor amounts of diol diesters, glycerides, and aldehydes. The major hydrocarbons in nonacosane and major esters are octacosyl esters of C_{14}–C_{32} acids. C_{20} and C_{22} alcohol esters of *trans*-2-docosenoic and tetracosenoic acids are also present. Free acids are C_{14}–C_{32} acids and *trans*-2-docosenoic and tetracosenoic acids. Octacosanol is the principal free alcohol. Hentriacontane-14,16-dione is the β-diketone (Tulloch and Hoffman, 1973). Phenolic esters, monoesters, aldehydes, alkanoic acids, and γ-lactones are also present in waxes (Kolattukudy et al., 1981).

The composition of cuticullar wax varies at different stages of growth (Tulloch, 1973). Wax from different parts of the plants also varies considerably. Alcohols are major components of wax from leaf blades and β-diketones are major components of wax from leaf sheaths of wheat varieties (Tulloch, 1973).

D. Primary Cell Wall

1. Pectic Polysaccharides

The bulk of the primary cell wall consists of pectic polysaccharides (Aspinall, 1970). The polysaccharides are a complex mixture of acidic and neutral polymers characterized by chains of α-1,4-linked galaturonosyl residues in which 2-linked rhamnosyl residues are interspersed. Rhamnogalacturonan is a common feature of all pectic polysaccharides. Pectic galactans and arabinans may vary from plant to plant (Talmadge et al., 1973). Pectic polysaccharides are present in cell walls of all plants (Worth, 1967; Jarvis, 1984).

2. Cellulose

Crystalline cellulose fibers make up an important part of the framework of all walls of plants (Wilson, 1964; Roelofson, 1965; Frey-Wyssling, 1969). In the crystalline cellulose fibers, glucan molecules are bound together by hydrogen bonds. The bonding between approximately 40 glucan chain results in 35 Å diameter threadlike bodies termed elementary fibrils (Frey-Wyssling, 1969).

These fibrils are aggregated into 150–250 Å diameter ropelike structures called microfibrils (Wilson, 1964).

The cellulosic microfibrils are embedded in a matrix of interwoven noncellulosic polysaccharides and proteins (Talbot and Ray, 1992; Carpita and Gibeaut, 1993). The microfibril surfaces are coated with noncellulosic polysaccharides such as xyloglucans, arabinoxylans, and glucomannans that are cellulose-like in their conformations (Iiyama et al., 1994). A proportion of these polysaccharides are hydrogen bonded to the surfaces of cellulose microfibrils and, by virtue of their length, are able to interact with surfaces of more than one microfibril and so act as an adhesive between them (Iiyama et al., 1994).

3. Hemicelluloses

Xyloglucan is the hemicellulose covalently linked to pectic polysaccharides and noncovalently bound to cellulose fibrils of the cell walls. The structure of the polymer is based on a repeating heptasaccharide unit that consists of 4 residues of β-1,4-linked glucose and 3 residues of terminal xylose. A single xylose residue is glycosidically linked to carbon 6 of 3 of the glucosyl residues. This structure consists basically of cellulose-like, β-(1-4)-linked glucan backbone with frequent xylosyl side chains attached to carbon 6 of the glucosyl residues in the backbone. Cellulose fibrils may be held together by hydrogen bonds between oxygens of alternating glycosidic bonds in the one glucan chain and the primary hydroxyl groups at position 6 of glucosyl residues in another chain (Bauer et al., 1973). Besides xyloglucan, other hemicelluloses such as xylans, arabinoxylans, mannans, glucomannans, and galactoglucomannans, also have structures well suited for hydrogen bonding to cellulose chains. (Whistler and Richards, 1970; Blake and Richards, 1971). The structural function of the hemicellulosic polysaccharides is to interconnect the cellulose fibrils and pectic polysaccharides of the wall. This function is based on the ability of the hemicellulosic polysaccharides to bind noncovalently to cellulose and to bind covalently to the pectic polysaccharides (Bauer et al., 1973).

4. Structural Proteins

Plant cell walls contain three major classes of structural proteins: hydroxyproline-rich glycoproteins (HRGP), proline-rich proteins (PRP), and glycine-rich proteins (GRP). These structural proteins contain highly repetitive peptide sequences with characteristic motifs: Ser-Hyp4 in HRGP, Gly-x in GRPs, and Val-Tyr-Lys-Pro-Pro or related proline rich pentamers in PRPs (Roberts and Northcote, 1972; Van Holst and Varner, 1984; Mazau et al., 1988; Keller and Lamb, 1989; Varner and Lin, 1989; Kleis–San Francisco and Tierney, 1990; Sheng et al., 1991). All of the PRPs are characterized by the repeating occurrence of PRO-PRO repeats contained within a variety of other larger repeat units. The PRPs contain approximately equimolar quantities of proline and hydroxyproline (Datta et al., 1989).

HRGPs are a diverse family of cell wall proteins including extensins, hydroxyproline-rich lectins of the solanaceae, and arabinogalactans (Tierney and Varner, 1987; Varner and Lin, 1989). Extensins are rich in hydroxyproline and serine. Other commonly found amino acids are valine, tyrosine, lysine, and histidine. They usually contain the repeating pentapeptide motif Ser-$(Hyp)_4$. Most of the hydroxyproline residues are glycosylated with one to four arabinosyl residues, and some of the serine residues are glycosylated with a single galactose unit. They are basic proteins with isoelectric points of about 10 due to their high lysine content (Biggs and Fry, 1990; Evans et al., 1990; Kawasaki, 1991; Adams et al., 1992; Zhou et al., 1992; Showalter, 1993). Characteristic HRGPs are found only in dicots (Showalter and Rumeau, 1990; Showalter, 1993). In monocots, HRGPs contain threonine (threonine–hydoxyproline-rich glycoprotein [HHRGP]) or histidine (histidine–hydroxypoline-rich glycoprotein [HHRGP]). They are also rich in alanine (Kieliszewski and Lamport, 1987; Kieliszewski et al., 1990). Serine residues may be absent (Caelles et al., 1992).

The carbohydrate component of the dicotyledonous extensins consists of arabinose and galactose. The arabinose is covalently attached to most of the hydroxyproline residues in the form of short arabinoside chains of one to four units in length, while single galactose units are covalently attached to some of the serine residues (Heckman et al., 1988).

The HRGP molecule has a helical conformation stabilized by the carbohydrate moiety. The linear, rod-like molecules may strengthen the cell wall by forming, via isodityrosine residues, an extensively cross-linked insoluble matrix interwoven with the mesh of cellulose micofibrils (Fry, 1982; 1986; Cooper et al., 1984; Stafstrom and Staehelin, 1986a,b; O'Connell et al., 1990).

Repeated peptides with the unique sequence -Ser-Hyp4-, typical of extensin, occur in HRGPs from various plants including melon (Esquerre-Tugaye and Lamport, 1979), tomato (Smith et al., 1984, 1986), bean (Corbin et al., 1987), carrot (Stuart and Varner, 1980; Chen and Varner, 1985a,b; Stafstrom and Staehelin, 1986a,b), and soybean (O'Neill and Selvendran, Averyhart-Fullard et al., 1988). All HRGPs isolated have the hydroxyproline glycosylated by one to four arabinosides (Mazau and Esquerre-Tugaye, 1986) in a way similar to extensin (Lamport and Milles, 1971). HRGPs from different plants are serologically related (Heckman et al., 1988). Antibodies against melon HRGPs specifically recognize HRGPs from tobacco cells (Benhamou et al., 1990).

5. Polymer Composition of Primary Cell Walls of Monocots and Dicots

Seven polymeric components have been detected in sycamore cell walls (Table 1; Talmadge et al., 1973). Other dicotyledonous plants may also have similar composition (Albersheim et al., 1973; Wilder and Albersheim, 1973). However, the cell walls of moncots are different from those of dicots. Like dicots, monocot

Table 1 Polymer composition of sycamore cell walls

Wall component	% Composition
Arabinan	10
3,6-Linked arabinogalactan	2
4-Linked galactan	8
Cellulose	23
Protein	10
Rhamnogalacturonan	16
Tetra-arabinosides (attached to hydroxyproline)	9
Xyloglucan	21

Source: Talmadge et al., 1973.

primary cell walls contain cellulose microfibrils in an amorphous matrix, but the polymers that form this matrix are fundamentally different. The monocot primary wall matrix contains an arabinoxylan as the principal component. No such polymer is present in the primary cell walls of dicots (Whistler and Richards, 1970; Albersheim et al., 1973). The major polymer of dicot cell walls is a xyloglucan but in the monocots it may be absent, or even if present it may not be the major structural component (Burke et al., 1974). However, the pectic polymers may be an integral part of monocot primary cell walls, as they are in dicot cell walls (Clowes and Junifer, 1968; Talmadge et al., 1973).

The primary cell walls of monocots contain from 9 to 14% cellulose, 7 to 18% uronic acids, and 7 to 17% protein. Hydroxyproline accounts for less than 0.2% of the cell walls of monocots (Burke et al., 1974). The major polysaccharide component of all these primary cell walls is an arabinoxylan. The cellulose microfibrils are embedded in heteropolysaccharide matrix. The cell wall arabinoxylan may be linked to cellulose by a paracrystalline interaction similar to that of xyloglucan of dicots (Bauer et al., 1973; Keegstra et al., 1973). The protein present in the monocot cell walls is deficient in hydroxyproline. Arabinogalactans, xyloglucans, and arabinans may also be present in small amounts. Pectic polysaccharides are commonly found (Burke et al., 1974).

E. Secondary Cell Wall

1. Lignin

Lignin is a complex polymer deposited in secondary cell walls (Sarkanen and Ludwig, 1971; Talmadge et al., 1973). HRGP is known to be a matrix for deposition of lignin (Whitmore et al., 1978b; Mazau and Esquerre-Tugaye, 1986). Lignin is a three-dimensional polymer built from three monomers called monolignols. The three building units are p-coumaryl, coniferyl, and sinapyl alcohols, which are polymerized in lignin via free radical production (Gross, 1978). Lignin content and the monomeric composition of lignins may vary with the plant family, the tissue, the developmental stage, and the subcellular localization (Lewis and Yamamoto, 1990).

2. Suberin

Suberin is an insoluble polymeric material attached to the cell walls of periderms, including wound periderms formed on aerial parts of plants (Dean and Kolattukudy, 1977) and endodermis (Espelie and Kolattukudy, 1979a,b). Periderms are formed naturally or due to wounds. The natural and wound periderms consist basically of three tissues: the phellogen, a zone of meristematic cells; the phelloderm, parenchymal cells formed inwardly from the phellogen; and the phellem, the layer of suberized tissue. As the phellem cells mature, they become heavily suberized and are frequently infused with tannins. In woody plants, the epidermis is replaced by the first periderm and subsequently by layers of consecutive periderms (Esau, 1965; Bostock and Stermer, 1989).

Suberin consists of a phenolic matrix attached to the cell wall and aliphatic components attached to the phenolic matrix. The major components of the polymeric aliphatic components are w-hydroxy fatty acids and the corresponding dicarboxylicacids. The aliphatic domains may be embedded in a layer of soluble waxes (Kolattukudy, 1981; 1985). Although the exact composition of phenolic components of suberin is not known, vanillin, p-hydroxybenzaldehyde, syringaldehyde, and cinnamate fractions have been detected (Kolattukudy, 1981). Cross-linking of phenolics forms a polymeric matrix and this matrix is made hydrophobic by attachment of aliphatic domains and by deposition of highly nonpolar waxes into this layer (Kolattukudy et al., 1987).

3. Inorganic Compounds

Silicon accumulates in plant cell wall (Carver et al., 1987; Heath, 1987; Blaich and Grundhofer, 1990). Calcium, magnesium, phosphorus, sulfur, iron, and chlorine have also been detected in cell walls (Bonello et al., 1991). The insoluble calcium may be a component of the pectic materials and may increase the strength of the bonds between adjacent polymer chains by cross-linking them through their acidic groups. The hardening of plant primary walls by cross-linking of pectic polymers has been reported (Akai and Fukutomi, 1980).

II. PENETRATION OF EPICUTICULAR WAXY LAYER BY PATHOGENS

Epicuticular wax is first barrier in the cell wall. Duvatriene diols associated with the cuticular wax of tobacco leaves have been shown to be toxic to fungi (Cruickshank et al., 1977). Virulent pathogens may be able to penetrate the waxy layer. However, the mechanism of degradation of the waxes by the pathogens has not been studied in detail (Kolattukudy, 1985). Utilization of these waxes by the fungal pathogens for their growth and sporulation has, however, been reported. The major components of the lipid extract from uredospores of wheat stripe rust, *Puccinia striiformis*, are β-diketones, n-alcohols, and hydrocarbons (Jackson et al., 1973). The surface lipid extract of the host wheat had a composition qualitatively similar in the major components. Comparison of the β-diketone and free alcohol composition indicated that they were produced by the host transferred to uredospore (Jackson et al., 1973). The presence of the same wax components in the host and pathogen indicates that the pathogen would have utilized its host's wax for its growth.

III. PATHOGENS PRODUCE CUTINASES TO BREACH CUTICLE BARRIER

A. Structure of Cutinases

The structural polymer cutin has been shown to be utilized by many plant pathogens as their sole carbon source. They produce extracellular cutin hydrolyzing enzymes (Shishiyama et al., 1970; Purdy and Kolattukudy, 1973; Baker and Bateman, 1978; Kolattukudy, 1984b,c). *Helminthosporium maydis* race T, *H. carbonum*, *Rhizoctonia solani*, *Pythium aphanidermatum*, *P. arrhenomanes*, *P. ultimum*, *Stemphylium loti*, *Cladosporium cucumerinum*, *Botrytis squamosa*, *Sclerotium rolfsii*, and *Gloeocercospora sorghi* have been reported to produce cutinases (Baker and Baterman, 1978). Cutinase has been isolated and purified from *Helminthosporium sativum*, *Fusarium roseum* Sambucinum, *Ulocladium consortialle* and *Streptomyces scabies* (Lin and Kolattukudy, 1980a), *Fusarium solani* f.sp. *pisi* (Purdy and Kolattukudy 1973; 1975a,b; Shaykh et al., 1977), *F. roseum* Culmorum (Soliday and Kolattukudy, 1976), *Colletotrichum gloeosporioides* (Dickman et al., 1982). *Colletotrichum graminicola*, (Pascholati et al., 1982), *Phytophthora cactorum*, *Colletotrichum capsici*, and *Botrytis cinerea* (Kolattukudy, 1985), *Venturia inaequalis* (Koller and Parker, 1989), *Ascochyta pinodes* and *Ascochyta pisi* (Nasraoui et al., 1990), *Alternaria alternata* (Tanabe et al., 1988), and *Cyphonectria parasitica* (Varley et al., 1992). Two isozymes of cutinase have been purified from *Fusarium oxysporum* f.sp. *pisi* and they were similar in size (25 kDa), amino acid composition, and immunological properties, but cutinase II contained a proteolytic nick near the middle of the polypeptide (Purdy and Kolattukudy, 1975a,b; Soliday and Kolattukudy, 1976). These cutinases are glycoproteins containing 3.5–6.0% carbohydrate (Lin and Kolattukudy, 1976; 1980a,b). The monosaccharides attached to the protein are mannose, arabinose, glucosamine, and glucuronic acid (Soliday and Kolattukudy, 1979; Lin and Kolattukudy, 1980a). The important amino acid residues are serine, threonine, β-hydroxyphenylalanine, and β-hydroxytyrosine (Lin and Kolattukudy, 1976, 1978, 1979).

Venturia inaequalis, the apple (*Malus pumila*) scab pathogen, produces cutinase in culture medium containing cutin. The enzyme is composed of one polypeptide with a molecular mass of approximately 22,000. The enzyme is a glycoprotein with a carbohydrate content of 5.4%. The amino acid composition shows a high content of glycine and nonpolar amino acids (Koller and Parker, 1989). Cutinase enzyme produced by *Colletotrichum gloeosporioides* contains 14–16% carbohydrate and an active serine (Dickman et al., 1982). The presence of a single histidine, a single tryptophan, 1–2 methionine, and a pair of half cysteines is a common feature of cutinases produced by *C. gloeosporioides* and *F. solani* f.sp. *pisi* (Lin and Kolattukudy, 1980a; Koller et al., 1982a; Dickman et al., 1982).

The amino acid composition of fungal cutinases varied (Table 2, Kolattukudy, 1985; Koller and Parker, 1989). In the cutinase from *V. inaequalis* glycine content is higher than that of other cutinases. The number of polar amino acids

Table 2 Amino acid composition of fungal cutinases

Residue	*Venturia inaequalis*	*Colletotrichum gloeosporioides*	*Helminthosporium sativum*	*Fusarium solani*	*Fusarium culmorum*
Aspartic acid	18	18	26	24	21
Glutamic acid	12	17	14	15	16
Serine	16	17	17	20	21
Glycine	42	23	24	27	26
Histidine	2	1	2	1	2
Arginine	6	10	8	14	11
Threonine	12	15	20	10	13
Alanine	20	31	27	23	34
Proline	8	13	12	11	10
Tyrosine	4	11	7	6	9
Valine	14	11	7	6	9
Methionine	2	2	1	1	1
Cysteine	4	2	2	2	4
Isoleucine	10	14	14	9	17
Leucine	10	23	20	18	23
Phenylalanine	8	11	13	6	6
Lysine	8	9	9	5	6
Tryptophan	ND	1	ND	1	1

N.D., Not detected.
Source: Kolattukudy, 1985; Koller and Parker, 1989.

(arginine, proline and tyrosine) is decreased. Cutinases of *C. gloeosporioides* and *H. sativum* have more phenylalanine. Cutinase of *F. solani* contains more arginine but less threonine while that of *F. culmorun* has more alanine. However, the structural elements such as two histidine, two methionine, and four cysteine residues in disulfide linkage are common to all the cutinases (Soliday and Kolattukudy, 1983; Kolattukudy and Koller, 1983; Kolattukudy, 1985; Kolattukudy et al., 1987; Koller and Parker, 1989).

Selective chemical modification of cutinase from *F. solani* f.sp. *pisi* showed that a catalytic triad consisting of one "active serine," one histidine residue, and one carboxyl group is involved in the enzymatic mechanism (Koller and Kolattukudy, 1982). Cutinase is considered a serine esterase (Purdy and Kolattukudy, 1975a,b). Diisopropyl fluorophosphate, an inhibitor of serine hydrolases, strongly inhibited the enzyme and phosphorylation of one active serine residue per cutinase molecule led to full inhibition of the cutinase activity (Koller and Parker, 1989). The enzyme contains one disulfide bridge, which is essential for the activity of the enzyme. The enzyme produced by *C. gloeosporioides* catalyzed hydrolysis of p-nitrophenyl esters of C_4, C_6, C_8, C_{10}, C_{12}, C_{14}, and C_{16} fatty acids at comparable rates (Dickman et al., 1982), but cutinase from *F. solani* f.sp. *pisi* could not hydrolyze the longer chain esters (Lin and Kolattukudy, 1980a).

B. Cutinase Gene

Cutinase cDNA has been cloned from *F. solani* f.sp. *pisi* and the complete nucleotide sequence of the coding region has been done. The cutinase mRNA was 1050 nucleotides in length. An initiation codon and a termination codon were found 702 nucleotides apart in the sequence. This reading frame translated into a protein with a molecular weight of 23,951 (Soliday et al., 1984). Cutinase genes from *C. capsici* and *C. gloeosporioides* have also been cloned and sequenced. Sequence analysis of 5′-flanking region showed that all these three cutinase genes lack consensus TATAA box, while they have a TAAATA box. *C. capsici* and *C. gloeosporioides* had a CAAT box in the 5′-flanking region, but *F. solani* f.sp. *pisi* had no CAAT box. In *F. solani* f.sp. *pisi* there was a CAAG sequence. All the three genes also share consensus core sequence CAGAC associated with transcriptional start site (Ettinger et al., 1987). The promoter for cutinase gene from *F. solani* f.sp. *pisi* is present within a 360 bp 5′-flanking region (Soliday et al., 1989). An inducible promoter has been identified for the cutinase gene of *F. solani* f.sp. *pisi* (Bajar et al., 1991).

A cutinase gene (CUT 1) from *Magnaporthe grisea* has also been cloned (Sweigard et al., 1992b). This gene showed strong sequence similarity to other cutinase genes (Sweigard et al., 1992b). Cutinases from the above four pathogens contain a conserved region (-leu-glu-ala/thr-arg-gln-leu/ser-) in close vicinity to the putative signal cleavage site (Ettinger et al., 1987; Sweigard et al., 1992b).

A cutinase gene from *Alternaria brassicicola* has been cloned and characterized (Yao and Koller, 1994). The cutinase gene was designated CUTAB1. The predicted size of the *Alternaria* cutinase was 22.4 kDa of the full length protein and 20.7 kDa without the putative signal peptide. The total nucleotide sequence was 984 bp long with an open reading frame encoding a protein of 209 amino acids (Yao and Koller, 1994). The conserved sequence (-leu-glu-ala/thr-arg-gln-leu/ser) found in *F. solani* f.sp. *pisi*, *C. gloeosporioides*, *C. capsici*, and *M. grisea* was lacking in *A. brassicicola* cutinase. The N-terminus of *A. brassicicola* cutinase contained a short hydrophobic core, followed by more polar residues and several short and neutral amino acids (-ala-ser-thr-thr-) as putative signal cleavage sites (Yao and Koller, 1994).

C. Variation in Cutinases Produced by Different Pathogens

Variation in cutinases produced by different pathogens has been observed. A cDNA clone of the cutinase gene from a strain of *Colletotrichum capsici* hybridized with genomic DNA from representative strains of *C. graminicola*, *C. gloeosporioides*, and *C. lindemuthianum* but not with DNA from strains of *C. lagenarium* or *C. coccodes* (Ettinger et al., 1987). Comparison of the DNA sequences for cutinase gene of *C. gloeosporioides* and a gene isolated from a strain of *Nectria haematococca* (*Fusarium solani* f.sp. *pisi*) revealed considerable dissimilarity. Only 43% of the amino acids of these two cutinase genes were directly conserved (Ettinger et al., 1987).

There was variation in cutinase produced by a single fungal species also. Two RFLP-defined distinct populations of *Colletotrichum gloeosporioides* isolated from *Citrus* have been reported. Both populations of the strains produced esterases with cutinolytic activity. One population of strains secreted 26 and 19 kDa esterases, whereas, the second population of strains, secreted 24 and 21 kDa esterases. A DNA probe containing the cutinase gene from an isolate of *C. gloeosporioides* hybridized strongly to DNA from the second population of strains and not at all to the first population. The results suggest that distinct cutinase genes may be present in the two types of *C. gloeosporioides* populations (Liyanage et al., 1993).

Cutinases produced by various fungal pathogens vary remarkably in their immunological properties and primary sequences also (Koller, 1990). Optimal pH for hydrolysis of the polymer cutin varied remarkably between cutinases isolated from the fungal pathogens infecting leaf or stem or both. Alkaline pH optima have been reported for the cutinases of pathogens such as *F. solani* f.sp. *pisi* (Purdy and Kolattukudy, 1975b), *F. roseum* f.sp. *culmorum* (Soliday and Kolattukudy, 1976), *F. solani* f.sp. *phaseoli* (Baker and Bateman, 1978), and *Pythium ultimum* (Baker and Bateman, 1978) that infect stem and roots only. *Rhizoctonia solani*, which infects the stem base of bean, has been shown to produce a cutinase with a pH optimum of 8.5 (Trail and Koller, 1990).

Acidic pH optima have been reported for pathogens infecting leaves. Cutinase from *Botrytis cinerea* had a pH optimum of 6.5 (Salinas et al., 1986). Cutin hydrolysis by cutinase of the corn leaf pathogen, *Cochliobolus heterostrophus*, was optimal at pH 6.5 (Trail and Koller, 1990).

The pathogens infecting both stem and leaf (besides other organs) produce cutinases with both acidic and basic pH optima. *Cladosporium cucumerinum*, a stem, leaf, and fruit pathogen of cucumber, has significant activity of both pH 5 and 8.5 (Baker and Bateman, 1978). *Alternaria brassicicola*, the pathogen of *Brassica olearaceae*, infects both leaves and stems and its cutinase shows two distinct pH optima: 7.0 and 9.0 (Trail and Koller, 1990). In fact two distinct forms of cutinase have been detected in culture filtrates of *A. brassicicola* (Trail and Koller, 1990).

The two cutinases A_c and B_a were purified to apparent homogeneity. Cutinase A_c had a cutinolytic pH optimum of 6.5 and cutinase B_a had a cutinolytic pH optimum 8.5. Their molecular weights were 23.0 kDa and 21.0 kDa, respectively. Amino acid compositions were similar but not identical (Trail and Koller, 1993). *Magnaporthe grisea*, the rice blast pathogen produces two different cutinases with pH optima of 6.0 and 9.0, and this pathogen is known to cause leaf blast and neck rot (Sweigard et al., 1992a).

Inoculum from the stem-specific isolate of *R. solani* nonpathogenic on leaves was amended with cutinase purified either from a leaf pathogen *Venturia inaequalis* or from a stem pathogen *F. solani* f.sp. *pisi* (Trail and Koller, 1990). Inoculum amended with cutinase from the leaf-specific pathogen penetrated the bean cuticle and produced disease symptoms whereas inoculum amended with cutinase from the stem-base pathogen remained nonpathogenic on bean leaves (Trail and Koller, 1990). The results suggest a role for cutinase in the expression of tissue specificity.

D. Host Cutin Monomer Triggers Fungal Cutinase

Fungal cutinase is not a constitutive enzyme, but it is inducible. Spores of *F. solani* f.sp. *pisi* generated cutinase only when cutin was present in the medium (Woloshuk and Kolattukudy, 1986). Cutinase could not be detected in the fungal cultures grown in cutin monomer-free medium (Flurkey and Kolattukudy, 1981). Host cutin is insoluble polymer (Lin and Kolattukudy, 1978). If cutin is to induce the fungal cutinase, the insoluble polymer should penetrate the fungal cell. Unless the insoluble polymer is converted into soluble monomer, it is not possible. Hence the fungal pathogen should secrete some basal level of cutinase even in the absence of cutin monomer. Koller et al. (1982a) have demonstrated that the germinating spores of *F. solani* f.sp. *pisi* released a small quantity of cutinase.

This small amount of cutinase may generate soluble cutin monomers that may signal the bulk production of cutinase (Kolattukudy et al., 1989).

The most common host cutin monomers that may act as signals are 10,16-dihydroxyhexadecanoic acid and 9,10,18-trihydroxyoctadecanoic acid (Woloshuk and Kolattukudy, 1986). Production of mRNA for cutinase became detectable within 15 min after spores of *F. solani* f.sp. *pisi* were exposed to cutin (Woloshuk and Kolattukudy, 1986). When nuceli isolated from *F. solani* f.sp. *pisi* culture after different periods of induction with 10,16-dihydroxy-hexadecanoic acid were incubated with [^{32}P] UTP, the amount of label incorporated into cutinase transcripts increased with increasing periods of induction. When nuclei isolated from uninduced (glucose-grown) culture of *F. solani* f.sp. *pisi* were incubated with [^{32}P] UTP, little label was incorporated into cutinase transcripts recovered (Kolattukudy et al., 1989). A protein factor also could be isolated from *F. solani* f.sp. *pisi* mycelium that triggered events required for the selective activation of transcription of cutinase gene (Podilla and Kolattukudy, 1988). When both the host cutin monomer and the fungal protein factor were added together with the nuclei, a major increase in the incorporation of label into cutinase transcript was observed. Neither the protein extract nor the cutin monomer alone had any effect on transcription of cutinase gene (Kolattukudy et al., 1989). These results suggest that the pathogen would have evolved an effective mechanism to sense its contact with the plant surface via the cutin monomers and consequently turn on the genes necessary to penetrate the defensive barrier of the plant.

E. Cutinase is Essential for Cuticle Penetration by Pathogens

1. Ultrastructural Evidence

Fungal pathogens have been observed to penetrate the cuticle. The cuticle at the point of penetration is not depressed inward, suggesting an enzymatic penetration rather than penetration by physical force (Martin, 1964; Van den Ende and Linskens, 1974; Mckeen, 1974; Hashioka, 1975; Murray and Maxwell, 1975; Aist, 1976; Rijkenberg et al., 1980; Staples and Macko, 1980; Stossel et al., 1980; Kritzman et al., 1981; Baker et al., 1982; Kunoh, 1984; Valsangiacomo and Gessler, 1988). Cutinase is released immediately upon contact of *Erysiphe graminis* conidia with a host substratum (Nicholson, 1984; Nicholson et al., 1988; Trail and Koller, 1990; Kunoh et al., 1988; 1990; Nicholson and Epstein, 1991). The release occurs much before appressorium germ tube formation, an event that does not happen earlier than 2 h after contact of the conidium with the host surface (Nicholson et al., 1988).

Ultrastructural evidence for the involvement in cuticular penetration has been presented in many host–pathogen interactions. *F. solani* f.sp. *pisi* spores were placed on pea stem surface and scanned with the electron microscope to determine the progress of penetration of the germinating spores. Treatment of the penetration area with ferritin-conjugated rabbit antibodies prepared against cutinase showed that the penetrating fungus secreted cutinase (Shaykh et al., 1977). Day and Scott (1973) showed the presence of erosion tracks in the barley cuticle underlying hyphae of *E. graminis* f.sp. *hordei*. A liquid was found to be released on the barley leaf surface immediately on contact by conidia of *E. graminis* f.sp. *hordei* (Pascholati et al., 1992). The liquid contained cutinase activity and the presence of cutinase in the conidial exudate was confirmed by enzyme assays using [^{3}H]-cutin as the substrate (Pascholati et al., 1992). Erosion of the host cuticle has been reported in cucumber (Staub et al., 1974) and wheat (Kunoh et al., 1977) also when infected by powdery mildew fungus. Ultrastructural evidence for cuticular penetration by *Alternaria brassicae* in rapeseed (*Brassica napus*) has also been presented (Tewari, 1986).

2. Evidence Using Cutinase Inhibitors

Koller et al., (1991) used an cutinase inhibitor, O-methyl-O-butyl-O-(3,5,6-trichloro-2-pyridyl)phosphate to assess the importance of cutinase in penetration. When the infection droplets containing spores of *Venturia inaequalis* and the inhibitor were incubated on apple leaves, formation of stromata beneath the cuticle was completely prevented. In samples without the inhibitor, subcuticular stromata with a radius of approximately 10 μm were clearly visible after 24 h of germination, and after 96 h subcuticular hyphae had elongated to approximately 100 μm. At this stage, no signs of subcuticular structures beneath appressoria were observed for germ tubes developing in the presence of the inhibitor (Table 3; Koller et al., 1991). The complete lack of subcuticular development indicated a crucial role of cutinase in cuticle penetration.

Several pesticides were found to be potent inhibitors, in vitro, of purified cutinase of *Colletotrichum gloeosporioides*, the causal agent of papaya (*Carica papaya* L.) anthracnose. All of them inhibited cutinase at 10^{-3}–10^{-9} M. At micromolar or lower concentrations they effectively prevented infection of papaya in a laboratory bioassay. None of the compounds was fungitoxic at these concentrations (Dickman et al., 1983). All the inhibitors tested reduced infection of papaya fruit with intact cuticles. To prove further that prevention of infection was due to inhibition of fungal penetration, wound inoculated controls were employed. The results showed that when cuticular barriers were breached by wounding prior to spore inoculation, there was no protection (Dickman et al., 1983).

Diisopropylfluorophosphate (DFP), a serine hydrolase inhibitor, completely inhibited *C. gloeosporioides* cutinase. Rabbit antibodies prepared against *C. gloeosporioides* cutinase (anticutinase) inhibited the cutinase enzyme activity. Both the anticutinase and DFP showed marked suppression of lesion formation in *C. gloeosporioides*-inoculated papaya fruits (Table 4; Dickman et al., 1982).

Table 3 Effect of O-methyl-O-butyl (3,5,6-trichloro-2-pyridyl) phosphate on the infection of apple by *Venturia inaequalis*

Time (h)	Subcuticular stromata (% of germinated conidia)
Control	
24	3.8
48	47.2
72	43.7
96	58.3
Inhibitor	
24	0.0
48	0.0
72	0.0
96	0.0

Source: Koller et al., 1991.

Neither DFP nor anticutinase had any significant effect on either spore germination or mycelial growth. When inoculated fruit surfaces were observed microscopically, where cutinase inhibitors were used, both spore germination and mycelial growth were evident. Hence the lack of lesion formation when cutinase inhibitors were used was not due to the lack of germination of spores or mycelial growth. In fruits whose cuticular layer was breached with mechanical wounding, full-fledged lesions developed in spite of the presence of cutinase inhibitors in the inoculum (Dickman et al., 1982). These observations strongly suggest that the cutinase inhibitors prevent the enzymatic penetration of intact cuticle by the infection hyphae. It can be concluded that *C. gloeosporioides* gains entry into papaya fruit by enzymatic disruption of their cuticular barrier (Dickman et al., 1982).

Table 4 Effect of antibodies and an inhibitor of cutinase on infection of papaya fruit by *Colletotrichum gloeosporioides*

Treatment	Percentage of infection
Control	88
Rabbit preimmune serum	93
Rabbit 1gG from antiserum	0
Diisopropylfluorophosphate	0

Source: Dickman et al., 1982.

DFP and a variety of organophosphates inhibit cutinase produced by *F. solani* f.sp. *pisi*. These organophosphates prevented pathogenic ingress in pea tissues without affecting fungal development (Koller et al., 1982b). Alkylboronic acids are also inhibitors of cutinase and they also prevented penetration of *F. solani* f.sp. *pisi* into pea plants (Carvalho et al., 1984). Incorporation of cutinase inhibitors into the spore suspensions of *V. inaequalis* and *Colletotrichum graminicola* prevented the development of lesions in apple and corn seedlings, respectively (Kolattukudy, 1984a,b,c).

Antibodies were prepared against cutinase purified from *F. solani* f.sp. *pisi* and incorporated into the spore suspension medium. The antibodies prevented cutinase activity and also prevented infection of the pathogen in pea (Maiti and Kolattukudy, 1979).

3. Evidence Using Cutinase-Deficient Fungal Isolates

Virulence of the pathogens has been found to be related to their ability to produce cutinases. Virulence of the isolates of *F. solani* f.sp. *pisi* the pea pathogen was related to their ability to produce cutinase (Koller et al., 1982a). Isolate T-8 was highly virulent whether or not the stem was wounded. On the other hand, isolate T-30 expressed high virulence only when the stem surface was wounded, but much less infection occurred when the stem surface was intact. The isolate T-30 was as virulent as T-8 when the cuticle–cell wall barrier was absent, but it was almost avirulent when the intact barrier layer was present. T-30 produced less cutinase than T-8 during germination (Table 5; Koller et al., 1982a), suggesting a role for cutinase in virulence of the isolates of *F. solani* f.sp. *pisi*.

Restriction fragments of the genomic DNA from isolates T-8 and T-30 were hybridized with the cutinase cDNA probe. T-8 showed a unique restriction fragment that hybridized with the probe and this fragment was not obtained from T-30. The results suggest that only in the isolate T-8, was a copy of the cutinase

Table 5 Cutinase activity released into the extracellular fluid during *Fusarium solani* f.sp. *pisi* spore germination

Isolate	Cutinase activity (^{3}H at min^{-1} released hr^{-1} per 10^{10} spores)	Degrees of virulence: Intact stem surface	Degrees of virulence: Wounded surface
T-8	2,610,000	+++	+++
T-30	3,100	+	+++

+ to +++ indicate increased degrees of virulence.
Source: Koller et al., 1982a.

gene present in a unique location and this gene was highly expressed (Kolattukudy and Soliday, 1985).

Cutinase-deficient mutants of *F. solani* f.sp. *pisi* (the pea pathogen) (Koller et al., 1982c; Dantzig et al., 1986), *C. gloeosporioides* (the papaya pathogen) (Dickman and Patil, 1986) and *Alternaria alternata* (the pear) (*Pyrus communis*) pathogen (Tanabe et al., 1988a,b) showed reduced pathogenicity. Introduction of the cutinase gene into a cutinase-less mutant of *F. solani* restored pathogenicity (Kolattukudy et al., 1989). *Mycosphaerella* sp. can infect papaya fruits only when the cuticular barrier is mechanically breached. Introduction of the cutinase gene from *F. solani* into the wound pathogen *Mycosphaerella* sp. rendered transformants pathogenic on intact surfaces (Dickman et al., 1989).

4. Evidence by Providing Cutinase in the Infection Drops

When the conidia of avirulent isolate T-30 of *F. solani* f.sp. *pisi* were supplemented with cutinase isolated from *F. solani* prior to inoculation of intact pea-stem surfaces, the conidia caused severe infection (Koller et al., 1982a). *Mycosphaerella* sp. is a fungal pathogen of papaya that can enter the fruit only through wounds. However, a very high frequency was observed when the surface of fruit was pretreated for 6 h with purified *C. gloeosporioides* cutinase prior to inoculation. The control, which consisted of cutinase that was inactivated with anticutinase prior to treatment of the fruit, showed an incidence of infection similar to that of the spore suspension alone (water control) (Table 6; Dickman et al., 1983).

5. Evidence Using Systemic Induced Resistance

Systemic resistance can be induced by several biotic and abiotic factors. Systemic resistance is induced against *Colletotrichum lagenarium* on cucumber plants by infiltrating lower leaves of the first true leaf (leaf 1) with a spore suspension of *C. lagenarium* (Xuei et al., 1988). Penetration of *C. lagenarium* through the epidermis was markedly reduced in leaves of immunized (leaves 2 and 3)

Table 6 Effect of cutinase and wounding on the infection of papaya by *Mycosphaerella* sp.

Treatment	Percentage of infection
Control (water)	5
Control (cutinase plus anticutinase)	6
Cutinase	95
Wound	95

Source: Dickman et al., 1983.

compared to control plants (Xuei et al., 1988). The removal of leaf surface from immunized leaves before challenge inoculation with *C. lagenarium* resulted in the decrease of resistance as observed from the increased number of necrotic lesions (Richmond et al., 1979). The observations suggest that leaf cuticle may play an important role in "induced resistance."

To assess the importance of cutin layer in the induced resistance, Huang and Kuc (1995) isolated leaf cutin from leaves 2 and 3 of unimmunized (control) cucumber plants and plants systemically immunized by infiltrating leaf 1 with *C. lagenarium*. The fungus grew better on a medium containing cutin from cucumber (control) leaves as sole carbon source than on a medium containing cutin from leaves of immunized plants (Huang and Kuc, 1995). Esterase activity increased sooner in water droplets containing conidia of *C. lagenarium* retrieved from leaves of control plants as compared with those from leaves of immunized plants (Huang and Kuc, 1995). Hence it appears that the leaves have become resistant as the cutin layer has become difficult to be penetrated by the pathogen. It is possible that in immunized cucumber leaves the degree of cross-links or branching in the cutin would have been increased, rendering the cutin more difficult to be utilized by the fungus. It is also possible that native cutin in immunized leaves is not a good substrate for the esterase enzyme and/or not a good inducer of enzyme production. These results suggest a role for cutinolytic enzymes in penetration of the host tissues.

6. Evidence Using Transformation-Mediated Gene Disruption

Several cutinase-deficient mutants of fungal pathogens have been developed to assess the importance of cutinase in fungal pathogenesis. Stahl and Schaffer (1992) constructed four cutinase-deficient mutants of a highly virulent strain of *Nectria haematococca*, the pea pathogen by transformation-mediated gene disruption. Bioassays of the null mutants and the wild-type showed no difference in the pathogenicity of the fungal strains. Hence it was concluded that the cutinase was not essential for pathogenicity of *N. haematococca*.

The wild-type *N. haematococca* strain (77-2-3) as well as the cutinase deficient mutants 77-75 and 77-102 invaded the pea foot by direct penetration of the outer cell layers of epicotyl, hypocotyl and tap root (Stahl et al., 1994). The infection progressed vertically into the stem by fungal growth into the xylem. There was no difference in intensity of symptoms caused by the wild-type and the mutants. In fact, one of the mutants (77–75) was more virulent than the wild-type. Hence Stahl et al. (1994) concluded that cutinase may not be a virulence factor in *N. haematococca*.

However, this conclusion was challenged by Rogers et al. (1994). They tested the same wild strain 77-2-3 and the same mutant 77-102 for their virulence. They found that the cutinase gene-disrupted mutant consistently showed less virulence than did the wild type. The virulence of the wild type and mutant on pea stems with mechanically breached cuticles was also studied. On the host with mechanically breached cuticle the cutinase gene-disrupted mutant showed the same high degree of virulence as did the wild type. The authors argued that multiple spore levels used for inoculum can detect differences in virulence of the fungus and Stahl and Schafer (1992) have not used multiple spore levels. They (Rogers et al., 1994) believed that cutinase is important for virulence of *N. haematococca*. A critical perusal of the papers published by these two groups of scientists reveals many differences in methodology followed by these groups. Probably cutinase may be one of the several factors responsible for virulence of this pathogen. This conclusion derives support from the work of Oeser and Yoder (1994).

Cochliobolus heterostrophus is a natural pathogen of corn. Its host range could be extended to pea if it was made to overexpress the *N. haematocca* PDA-T9 gene, which encodes pisatin demethylase (PDA) and thus allows the fungus to detoxify the pea phytoalexin pisatin (Schafer et al., 1989). Oeser and Yoder (1994) transformed *C. heterostrophus* with the cutinase encoding gene CUT-T8 from *N. haematococca* with or without PDA-T9 gene. *N. haematococca* cutinase alone had no detectable virulence of *C. heterostrophus*, but in combination with pisatin demethylase cutinase increased virulence (Oeser and Yoder, 1994). Thus cutinase may be one of the several virulence factors. Since a pathogen produces more than one cutinase, gene-disruption transformation studies may not be reliable in some cases.

One of the cutinase genes, CUT 1 of *Magnaporthe grisea*, has been cloned (Sweigard et al., 1992b). Using a one-step strategy to disrupt CUT 1 gene, transformants of *M. grisea* were produced in which CUT 1^+ gene was replaced by a nonfunctional copy constructed in vitro. The Cut 1^- mutants were generated in two strains of *M. grisea*: a strain pathogenic on weeping lovegrass (*Eragrostis curvula*) and barley, and another strain pathogenic on rice. CUT 1 mRNA was not detectable in transformants that contained a disrupted gene. Transformants with a disrupted CUT 1 gene did not produce a cutin-inducible esterase typical of wild-type strains. A single major cutin-inducible esterase was visualized by staining for appropriate enzyme activity on polyacrylamide gels and this band stained much more intensely in gels loaded with proteins from CUT 1 multicopy transformants. This band was absent from proteins of the *cut* 1^- transformant (Sweigard et al., 1992a).

The cut 1^- strains produced reduced levels of cutinase and p-nitrophenyl butyrate (PNB) esterase. Both the disruption transformants showed identical pathogenicity to that of strains with a functional CUT 1 gene in causing rice leaf blast. Wild-type strain produced 14–25 lesions per leaf while the transformant also produced 14–23 lesions per leaf (Sweigard et al., 1992a). The results suggest that the cutinase encoded by CUT 1 is not essential for pathogenesis in leaf blast caused by *M. grisea* (Sweigard et al., 1992a).

However, the above experiments have not provided a conclusive evidence that cutinases are not essential for pathogenesis of *M. grisea*, since transformants in which CUT 1 gene function had been abolished still contained significant activity in standard cutinase assays. It indicates that at least one other gene may encode an enzyme degrading cutin. In polyacrylamide gels, besides a major band representing CUT 1 product, some minor bands with the esterase activity appeared and they were similar in both wild-type strain and the transformant. The extracellular culture filtrates of the wild strain showed four- to fivefold greater activity at pH 9.0 than at pH 6.0. The cut 1^- transformant retained significant amounts of cutinase activity, although this activity was reduced relative to that of the wild-type strain (Table 7, Sweigard et al., 1992a). These data suggested that in addition to the CUT 1 gene product, the culture filtrates contained other enzymes with cutinase activity. These enzymes would have helped the transformant to penetrate the cuticle (Sweigard et al., 1992a).

Further studies may be needed in which transformants, in which all genes encoding all the cutinase enzymes of *M. grisea* have been deleted, are used to assess the importance of cutinase in pathogenesis.

F. Cutinase May Not Be Essential for Pathogenesis

Although cutinase has been shown to be essential for penetration of cuticle in many fungal pathogens, in some pathogens cutinase does not seem to have any role in penetration (Sweigard et al., 1992a). The studies on the process of penetration of the cuticle of apple leaves by *Venturia inaequalis* showed that penetration of the cuticle occurred in both susceptible and resistant varieties (Valsangiacomo and Gessler, 1992). *Magnaporthe grisea* penetrates into host cell wall even in the presence of diisopropylfluorophosphate, an inhibitor of cutinase (Woloshuk et al., 1983). Furthermore tricyclazole, which inhibits penetration process of *M. grisea* into host cell, inhibits penetration of Formvar plastic

Table 7 Cutinase activity in *Magnaporthe grisea* strains

Strain	Cutinase (dpm/h/10 μg protein) pH 6.0	Cutinase (dpm/h/10 μg protein) pH 9.0
Wild-type strain	6743	29047
Cut 1^- strain	3809	21559

Source: Sweigard et al., 1992a.

membranes also by an action that certainly cannot involve cutinase (Woloshuk et al., 1983).

Some pathogens do not produce cutinase but still they penetrate the cuticle efficiently. Other mechanisms such as mechanical pressure may also be involved in cuticle penetration.

IV. MELANIN DEPOSITION IN APPRESSORIA IS REQUIRED FOR PENETRATION OF HOST PLANT CELL WALLS BY FUNGAL PATHOGENS

A. Biosynthesis of Melanin

Some pathogens penetrate cuticle through mechanical force generated by the appressorium and melanin provides the necessary strength and rigidity in the appressorial wall to confine and focus this force (Durell, 1964; Kuo and Alexander, 1967; Bloomfield and Alexander, 1967; Wolkow et al., 1983; Woloshuk et al., 1983; Howard et al., 1991). Several reports have indicated that melanin biosynthesis is required for the penetration of host plant cell walls by appressoria (Suzuki et al., 1982a,b; Kubo et al., 1982a,b, 1983, 1984, 1985, 1986, 1987; Kubo and Furusawa, 1986). 1,8-Dihydroxynaphthalene (DHN) is a fungal melanin precursor (Kubo et al., 1982; Woloshuk and Sisler, 1982; Woloshuk et al., 1983; Wheeler and Bell, 1987; Wheeler and Greenblatt, 1988). DHN melanin is produced at detectable levels in a variety of fungal cell types but the pigment is produced most abundantly just before penetration when a thick melanin layer is deposited in the inner appressorial cell wall (Woloshuk et al., 1983; Wolkow et al., 1983; Sisler, 1986; Kumura et al., 1990). DHN mediates the build up of hydrostatic pressure in the appressorium and this high pressure provides the essential driving force for a mechanical penetration component (Howard and Ferrari, 1989).

The melanins contain protein, carbohydrate, and lipid moieties as well as polymeric nucleus containing quinone, hydroquinone, and semiquinone moieties. The latter polymer is the chromophore (Bell and Wheeler, 1986).

Melanins in cell walls of Basidiomycotina are derived from Y-glutaminyl-3-4-dihydroxy-benzene (GDHB) or catechol, as immediate phenolic precursor of the melanin polymer (Bell and Wheeler, 1986).

In the Ascomycotina and Deuteromycotina, the melanins in cell walls are synthesized from the pentaketide pathway. (+) Scytalone, 1,3,8-trihydroxynaphthalene (1,3,8-THN),-vermelone, and DHN are main intermediates in the melanin pathway. 1,3,6,8-Tetrahydroxynaphthalene (1,3,6,8-THN) has also been suggested as an intermediate (Tokousbalides and Sisler, 1979; Wheeler and Stipanovic, 1979; Ten et al., 1980).

Flaviolin and 2-hydroxyjuglone are major compounds of the pathway, formed by autooxidation of 1,3,6,8-THN and 1,3,8-THN, respectively (Bell et al., 1976b; Stipanovic and Bell, 1977).

Cell-free homogenates of *Verticillium dahliae* metabolized 1,3,6,8-THN, scytalone, 1,3,8-THN, and vermelone as detailed in Table 8 (Wheeler, 1982).

The anaerobic conditions prevented enzymatic-type oxidative reactions and protected the naphtholic compounds from autooxidative reactions. Thus DHN did not form melanin, and 1,3,6,8-THN and 1,3,8-THN were not autooxidized to flaviolin and 2-hydroxyjuglone. Instead, 1,3,6,8-THN and 1,3,8-THN were enzymatically reduced to scytalone and vermelone, respectively, which then were dehydrated enzymatically to 1,3,8-THN and DHN (Wheeler, 1982).

Microsclerotia of three melanin-deficient mutants of *V. dahliae* formed melanin from (+) scytalone, 1,8-dihydroxynaphthalene, catechol and L-3,4-dihydroxyphenylalanine. The melanins formed from (+) scytalone or 1,8-dihydroxynaphthalene resembled wild-type melanin chemically and ultrastructurally. However, the melanins formed from catechol and L-3,4-dihydroxyphenylalanine were different. This suggests that only scytalone and 1,8-dihydroxynaphthalene are natural intermediates of melanin synthesis in *V. dahliae* (Wheeler et al., 1978).

The importance of two reductases and two hydratases in the melanin biosynthetic pathway has been shown by using homogenates of *Pyricularia oryzae* (Wheeler and Greenblatt, 1988). Tricyclazole, chlobenthiazone, fthalide, pyroquilon, and PCBA are the known melanin biosynthesis inhibitors. Reductase activity was measured in the presence and absence of melanin inhibitors with scytalone and 1,3,8-THN. Both substrates were metabolized to DHN when

Table 8 Products formed from melanin substrates by homogenates of *Verticillium dahliae*

Substrate	Product
1,3,6,8-THN	Scytalone
	1,3,8-THN
	Vermelone
	DHN
Scytalone	1,3,8-THN
	Vermelone
	DHN
1,3,8-THN	Vermelone
	DHN
Vermelone	DHN
DHN	None

Source: Wheeler, 1982.

incubated with the fungal extract. All the inhibitors blocked the conversion of 1,3,8-THN to DHN. The fungal extracts also metabolized 1,3,6,8-THN to scytalone and then to DHN. The metabolism of 1,3,6,8-THN to DHN required two reductase reactions and two hydratase reactions. The inhibitors were more effective in preventing the reduction of 1,3,6,8-THN than 1,3,8-THN (Wheeler and Greenblatt, 1988). 1,3,8-THN was not produced from 1,3,6,8-THN in the absence of fungal extract, indicating that it was synthesized enzymatically during the incubation period. Exogenous NADPH increased the levels of DHN production from 1,3,8-THN and 1,3,6,8-THN. Exogenous NADP, NAD, and NADH did not increase DHN production by fungal extracts, suggesting that NADPH alone is required for 1,3,6,8-THN and 1,3,8-THN reductase reactions (Wheeler and Greenblatt, 1988). Dehydratase activity was measured with scytalone as substrate and scytalone was converted mainly to 1,3,8-THN when reaction mixtures were incubated for 20 min. DHN did not accumulate, because the reaction that reduces 1,3,8-THN to vermelone occurred slowly and limited the conversion of 1,3,8-THN to DHN (Wheeler and Greenblatt, 1988).

The reductase and dehydratase enzymes involved in DHN melanin synthesis are unique to this pathway (Wheeler and Stipanovic, 1979; Woloshuk et al., 1980a,b; Wheeler, 1981, 1982, 1983; Yamaguchi et al., 1982, 1983a,b; Buchenauer et al., 1985). The oxidase involved in oxidative polymerization of DHN may be a wall-bound laccase (Bell et al., 1976a; Bell and Wheeler, 1986). In *Colletotrichum lagenarium* reductase and dehydratase involved in melanin biosynthesis are preexisting enzymes in spores or synthesized during 1 h of incubation, and the enzymes are activated after 6 h of incubation, which is the start of appressorium swelling (Kubo et al., 1984).

Conventional genetic analysis of melanin biosynthesis has been carried out in *Cochliobolus miyabeanus* and *Cochliobolus heterostrophus* (Tanaka et al., 1990), and *Pyricularia oryzae* (Chumley and Valent, 1990). The linkage relationship and arrangement of genes involved in melanin biosynthesis were quite different between *Cochliobolus* spp. and *Pyricularia*. In *Cochliobolus* spp., albino gene and reductase genes were closely linked and dehydratase gene was independent of the two genes. In *Pyricularia*, no evident linkage was observed among the three genes. In *Alternaria alternata*, the melanin biosynthetic genes were clustered (Kimura et al., 1990). Many plant pathogens, including *Alternaria brassicicola*, *A. alternata*, *Cochliobolus carbonum*, *C. miyabeanus*, *Sclerotinia sclerotiorum*, *Botrytis cinerea*, *Colletotrichum lindemuthianum*, *Colletotrichum lagenarium*, *Helminthosporium sorokinianum*, *Macrophomina phaseolina*, *Pyricularia oryzae*, *Thielaviopsis basicola*, *Verticillium albo-atrum*, *V. dahliae*, *V. nigrescens*, and *Diplodia gossypina* have been reported to synthesize melanins via DHN pathway (Puhalla, 1975; Wheeler et al., 1976, 1978; Bell et al., 1976a,b; Stipanovic and Bell, 1976; Wheeler and Stipanovic, 1979; Tokousbalides and Sisler, 1978, 1979; Stipanovic and Wheeler, 1980; Wheeler, 1982, 1983; Ten et al., 1980, 1983; Woloshuk et al., 1981, 1983; Wolkow et al., 1983; Buchenauer et al., 1985; Zeun and Buchenauer, 1985; Kubo et al., 1985).

B. Evidence for the Role of Melanin in Host Cell Wall Penetration

1. Melanin-Deficient Mutants

Several melanin-deficient mutants of *Colletotrichum* and *Pyricularia* species have been isolated (Woloshuk et al., 1980; Sisler et al., 1983; Kubo et al., 1983; Bustamam and Sisler, 1987; Kubo and Furusawa, 1991). These melanin-deficient mutants were nonpathogenic, suggesting that melanin biosynthesis is a metabolism essential for pathogenicity in *Colletotrichum* and *Pyricularia* species (Kubo and Furusawa, 1991).

Albino mutants of *Colletotrichum lagenarium* form colorless appressoria and have little penetrating ability by germinating laterally from appressoria (Kubo et al., 1982b, 1983). A cosmid vector pkVB was constructed for isolating genes by complementation of mutations in *C. lagenarium* (Kubo et al., 1991). A genomic DNA library of wild-type *C. lagenarium* was constructed in pKVB. An albino mutant strain was transformed with DNA from this library. Seven melanin-restored transformants were obtained. Albino mutants of *C. lagenarium* formed nonmelanized appressoria and possessed little penetrating ability. However, the transformants formed melanized appressoria with the ability to penetrate as efficiently as in the wild strain (Table 9; Kubo et al., 1991). These transformants also had pathogenicity to the host cucumber plants. The extent of the pathogenicity was the same as in the wild-type strain (Kubo et al., 1991).

Albino mutants of *C. lagenarium* that produced appressoria without melanin pigmentation were obtained by ultraviolet irradiation or N-methyl-N″-nitro-N-nitrosoguanidine treatment (Kubo et al., 1982a). About 70% of appressoria of the parent strain penetrated nitrocellulose membranes after incubation for 72 h. Penetration hyphae growing in the membranes were readily observed. In albino mutants, most appressoria germinated laterally on the surface of membranes and the percentage of penetration into membranes from appressoria was below 10%. However, when spores of mutants were incubated in the presence of 3,4-dihydroxyphenylalanine (DOPA) for 72 h, appressoria with pigment were formed and about 60% of these pigmented appressoria, penetrated the membranes (Kubo et al., 1982a). The results suggest that the decrease in penetration into nitrocellulose membranes in albino mutants is due to the lack of pigmentation of appressoria.

When cucumber leaves were inoculated with a spore suspension of the parent strain of *C. lagenarium*, lesions were observed within 5 days. Pigmented appressoria and elongation of penetration hyphae were also observed. In contrast, when the leaves were inoculated with spores of the albino mutant, lesions were

Table 9 Relationship between appressorial melanin pigmentation and formation of penetration hyphae in various strains of *Colletotrichum lagenarium*

Strain	Appressorial pigmentation	Percentage appressoria with	
		Penetration hyphae	Lateral germination
79215	–	3.0	97.0
104-T	+	89.3	2.1
AC-1	+	83.1	2.7
AC-2	+	70.0	1.1
AC-3	+	70.7	1.2
AC-4	+	81.0	2.5
AC-5	+	71.7	2.8
AC-7	+	70.4	3.0

+, presence, –, absence of pigmentation.
Source: Kubo et al., 1991.

rarely observed. The appressoria were nonpigmented and germinated laterally on the surface of the leaves; and elongation of penetration hyphae was rarely observed (Kubo et al., 1982b). The results suggest that pigmentation of appressoria is essential for host penetration by *C. lagenarium.*

Magnaporthe grisea (anamorph, *Pyricularia oryzae*) variants with buff pigmentation lack pathogenicity. The variants with buff pigmentation which lacked pathogenicity are deficient in the melanin biosynthetic activity (Woloshuk et al., 1980b). Three distinctive phenotypic classes of pigment mutants viz. albino (Alb^-), rosy (Rsy^-), and buff (Buf^-) have been identified by Chumley and Valent (1990). Some of these mutants appeared spontaneously while others were obtained by ultraviolet (UV) mutagenesis. All of them were melanin-deficient. To determine the genetic basis of the pigmentation defects, Alb^-, Rsy^-, and Buf^- mutants were crossed to strains with wild-type pigmentation, and segregation of the melanin-deficient phenotypes was scored among the progeny. In all cases, mutant and wild type phenotypes segregated 1:1 among the progeny, indicating the presence of single gene mutations in the pigment mutants. Pigment mutants caused no visible symptoms when inoculated on rice. Progeny from several crosses between pigment mutants and wild-type strains were tested for pathogenicity. All the pigment-deficient progeny were nonpathogenic while the progeny, which were gray, were pathogenic. Thus, nonpathogenicity and lack of melanin production were due to the same single gene defect (Chumley and Valent, 1990).

The pigment mutants caused no symptoms on intact rice seedlings. However, they caused extensive blast symptoms on wounded seedlings. The lesions produced by pigment mutants on wounded tissue enlarged to the normal size similar to wild-type parent and sporulated normally. These observations led to the conclusion that melanin biosynthesis is required only for initial penetration into the host plant (Chumley and Valent, 1990).

The melanin biosynthetic intermediate scytalone was incorporated into spore suspensions of the different pigment mutants (alb$^-$, rsy$^-$, and buf$^-$) and the mutants were inoculated on rice seedlings. In the presence of scytalone the alb$^-$ strain induced severe blast symptoms; but buf$^-$ and rsy$^-$ strains caused no symptoms even in the presence of scytalone. It shows that only alb$^-$ strain was able to convert scytalone to melanin required for successful penetration of the host's outer barriers (Chumley and Valent, 1990). The buf$^-$ strain may lack 1,3,8-THN [3HN]reductase while rsy$^-$ strain may lack scytalone dehydratase. Alb$^-$ strain may lack an enzyme that converts acetate into 1,3,6,8-THN (Chumley and Valent, 1990).

Wheeler and Greenblatt (1988) obtained buff-colored mutants of *P. oryzae*. Extracts from melanin-deficient buff-colored mutants were tested for both reductase activity and dehydratase activity. They dehydrated scytalone to quantities of 1,3,8-THN similar to those produced by a wild-type control, but the extracts did not have much reductase activity and conversion of 1,3,8-THN to DHN was very much less (Table 10). Scytalone and vermelone did not remain in the mixture containing extracts of the melanin-deficient mutants, since they were completely dehydrated to 1,3,8-THN and DHN, respectively (Wheeler and Greenblatt, 1988). The results suggest that the melanin-deficient buff mutants have a genetic defect that prevents reductase activity and blocks melanin biosynthesis in their hyphae and appressoria. The lack of melanin in appressorial cell walls in the buff strains prevents penetration of the host cell wall (Wheeler and Greenblatt, 1988).

Chida and Sisler (1987) showed that exogenously supplied DHN restored the ability of appressoria of buff mutants of *P. oryzae* with blocked [3HN]reductase activity to penetrate a membrane in vitro. DOPA restored melanization of

Table 10 Conversion of 1,3,8-THN to DHN by extracts of the melanin-deficient strains and a wild-type (P-2) of *Pyricularia oryzae*

Strain	DHN accumulated from 1,3,8-THN (μg/ml)
P-2 (wild)	113
P-2M-1(buff-mutant)	16
P-2M-2(buff-mutant)	16
P-2M-3(buff-mutant)	17

Source: Wheeler and Greenblatt, 1988.

melanin-deficient mutants of *P. oryzae* and led to penetration of host epidermal cell walls (Woloshuk et al., 1980a; Okuno et al., 1983b). DOPA and scytalone restored melanization and penetration of cucumber cell walls and cellulose membranes by the melanin-deficient mutant of *C. lagenarium* (Kubo et al., 1983). These results suggest that melanin is essential for penetration. Wounding allows infection of rice by melanin-deficient mutants, suggesting that penetration is the important factor for pathogenesis of *P. oryzae* (Woloshuk et al., 1983).

2. Melanin Inhibitors

Several "nonfungitoxic" compounds known to control rice blast caused by *Pyricularia oryzae* are inhibitors of melanin biosynthesis in the pathogen (Tokousbalides and Sisler, 1978, 1979; Woloshuk et al., 1980a). They are tricyclazole, pyroquilon, phthalide, coumarin, 2,3,4,5,6-pentachlorobenzyl alcohol; 4,5-dihydro-4-methyltetrazolo (1,5-a) quinazolin-5-one; N-methyl-2-quinolone and s-trizolo (4,3-a) quinolone. Specific epidermal antipenetrant action has been demonstrated for all these inhibitors of melanin biosynthesis (Ishida et al., 1969; Tokousbalides and Sisler, 1978, 1979; Woloshuk et al., 1980a, 1981, 1983; Woloshuk and Sisler, 1982; Chida et al., 1982; Chida and Sisler, 1987; Wheeler and Greenblatt, 1988).

Tricyclazole, at concentrations that inhibit melanization but not mycelial growth, conidial germination, and appressorial formation, prevents penetration of rice epidermis by *P. oryzae* (Woloshuk et al., 1980a,b, 1983; Yamaguchi et al., 1982; Okuno et al., 1983a; Matusra, 1983; Sisler et al., 1983, 1984; Inoue et al., 1984a,b). Tricyclazole inhibits the enzymatic reduction of 1,3,6,8-THN to scytalone and 1,3,8-THN to vermelone. The inhibitor does not prevent the accumulation of 1,3,8-THN or DHN from scytalone or vermelone, respectively. It suggests that it has no apparent inhibitory effect on the enzymatic dehydration of scytalone or vermelone (Wheeler, 1982).

Intact rice plants treated with tricyclazole by soil application were almost completely protected from infection by *P. oryzae* (Woloshuk et al., 1983). The nontreated plants were all severely diseased with numerous, overlapping lesions. With the melanin-deficient mutants P-2m-1 and P-2m-2, there were negligible infections on intact, nontreated rice plants. When the leaf puncture wounds were made, disease occurrence increased on the tricyclazole-treated plants inoculated with the wild-type fungus as well as on the nontreated plants inoculated with the mutants (Table 11; Woloshuk et al., 1983). Twenty-four hours after inoculation of rice stamen filaments, walls of untreated wild-type appressoria were melanized while those of tricyclazole treated wild-type strain and of the melanin-deficient strain (P-2m-2) were nearly colorless and fragile. Penetration was observed at 36 h in cells of filaments inoculated with the untreated wild strain. No penetration was observed by appressoria from the tricyclazole-treated wild strain or from the untreated mutant strain at 36 or 72 h. However, the tricyclazole-treated wild-type

Table 11 Effect of wounding on tricyclazole treated rice plants by wild type and melanin deficient mutants of *Pyricularia oryzae*

P. oryzae strain	Treatment	% infected plants	
		Not wounded	Wounded
Wild	None	100	Not tested
Wild	Tricyclazole	12	100
Mutant (P-2m-1)	None	8	98
Mutant (P-2m-2)	None	12	77

Source: Woloshuk et al., 1983.

and the mutant freely infected damaged host tissue, without appressorial formation (Woloshuk et al., 1983). The metabolite DHN, the fungal melanin precursor, occurs subsequent to the tricyclazole block in the melanin biosynthetic pathway. Supplying 1 μg/ml of DHN to tricyclazole-treated appressoria resulted in partial melanization and epidermal penetrating capacity (Woloshuk et al., 1983). Scytalone, which occurs before tricyclazole block in melanin synthesis, does not restore melanization of the appressoria of *P. oryzae* (Woloshuk and Sisler, 1982) and this treatment did not restore the penetration capability of the fungal pathogens (Kubo et al., 1982b; Woloshuk et al., 1983). DOPA forms a melanin in tricyclazole inhibited appressoria of *P. oryzae* and restores penetration of the host plant (Okuno et al., 1983a).

Melanin appears to provide rigidity to the appressorial wall of *P. oryzae*. The detached appressorial shells maintained their spherical shape, indicating a rigid wall, whereas similar unmelanized appressorial shells of the tricyclazole treated fungus became deformed, indicating a flexible wall. Another indication of greater rigidity of melanized appressorial walls in comparison with nonmelanized walls is the angular impression made in the cuticle by the former compared to the rounded impression made by the latter. Ultrastructural studies of the penetration process suggest that the critical need for wall rigidity is at the basal edge of the thick wall that borders the thin underwall through which turgor forces are exerted on the cuticle. The cuticular penetration by *P. oryzae* was observed to occur near the outer edge of the thin underwall. Failure of tricyclazole-treated appressoria to accomplish cuticular penetration may be due to lack of rigidity at the basal edge of the unmelanized bordering wall (Woloshuk et al., 1983). The portions of the appressorial cell wall of *P. oryzae* juxtaposed to the host plant do not contain melanin, whereas in the rest of the cell wall melanin occurs as dispersed granules (in the inner outer wall) and as dense sheet (in the wall adjacent to the plasmalemma). Appressoria of the fungus exposed to tricyclazole lack the inner electron-dense layer (Woloshuk et al., 1983; Wolkow et al., 1983; Kubo and Furusawa, 1986).

Tricyclazole-treated appressoria of *Colletotrichum lindemuthianum* could not penetrate plant epidermal walls and it has been shown to be due to lack of wall ridigity in the unmelanized appressoria of the fungus. When spores of *C. lagenarium* were incubated at 24°C, dark brown appressoria were formed within 12 h. In the presence of 100 μm tricyclazole, pigmentation of appressoria was completely inhibited and colorless appressoria resembling those of an albino mutant were formed. More than 80% of the appressoria in the parent strain penetrated nitrocellulose membranes during incubation for 72 h (Kubo et al., 1982a). In the presence of 100 μm tricyclazole more than 90% of appressoria germinated laterally on the surface of the membranes, and the percentage of formation of penetration hyphae was less than 5%. When 100 μm tricyclazole was applied after appressorial pigmentation, neither the elongation of penetration hyphae nor halo formation was affected (Kubo et al., 1982a).

Thus several studies have clearly shown the importance of melanin biosynthesis in appressoria of pathogens in penetration of host cell wall.

C. Some Pathogens May Not Require Melanin for Penetration

Some of the pathogens do not require melanin for penetration. Melanin-deficient mutants of *Helminthosporium oryzae* were found to be fully pathogenic (Kubo et al., 1989). *Verticillium dahliae* has no DHN melanin requirement for pathogenicity (Bell et al., 1976a; Bell et al., 1976b; Bell and Wheeler, 1986). It is possible that fungi such as *Fusarium solani* f.sp. *pisi* and *Colletotrichum gloeosporioides*, which penetrate the cuticle via hyphae, may rely primarily on cutinase, whereas fungi such as *P. oryzae* and *Colletotrichum lindemuthianum*, which penetrate the cuticle via an appressorium, may rely primarily on mechanical force provided by melanin (Sweigard et al., 1992a).

V. PATHOGENS PRODUCE PECTIC ENZYMES TO BREACH PECTINACEOUS BARRIER

A. Multiple Pectic Enzymes

Upon penetration through the cuticle, the invading fungus would encounter the pectinaceous barrier (Kolattukudy, 1985). Even when fungi gain entry into plants through stomata and wounds, they would have to degrade the intercellular pectinaceous polymers to initiate infection in the plant (Bateman, 1976; Cooper, 1984; Keon et al., 1987; Crawford and Kolattukudy, 1987). Plant pathogens produce multiple forms of different types of pectic enzymes (Collmer and Keen, 1986). The pectic substances are of two types: pectin (pectinic acid) and pectic acid (polygalacturonic acid) (Vidhyasekaran, 1993). The activity of the pectic enzymes varies in their specificity to the substrate (pectin or pectic acid), in their mechanism to split α-1,4-glycosidic bond (hydrolytic or lyitic), and in the type of cleavage (random or terminal). The pectic enzymes are classified as pectin methylesterase (PME), which converts pectin into polygalacturonic acid; exopolygalacturonase (exo PG) cleaves polygalacturonic acids in a terminal manner, releasing monomeric products (glacturonic acids); and endopolygalacturonase (endo PG), which cleaves polygalacturonic acids in a random manner releasing oligogalacutronic acid. Exo- and endo-pectin methyl galacturonases (PMG) act similarly to polygalacturonases but the preferred substrate is pectin instead of polygalacturonic acid. Both polygalacturonases and polymethylgalaturonases act by hydrolytic action. In contrast, polygalacturonate *trans*-eliminase (PGTE; pectate lyase [PL]) and pectin *trans*-eliminase (PTE; pectin lyase) induce lyitic degradation of the glycosidic linkage, resulting in an unsaturated bond between carbons 4 and 5 of a uronide moiety in the reaction product. The *trans*-eliminases break the glycoside linkage at carbon no. 4 and simultaneously eliminate the H from carbon no. 4, resulting in oligouronides that contain an unsaturated galacturonyl unit. Both exo (terminal) and endo (random) type of reactions are seen in the *trans*-eliminases and they are called exo-PGTE/exo PL, endo-PGTE/endo PL, exo-PTE/exo pectinlyase, and endo-PTE/endo pectinlyase (Vidhyasekaran, 1993). These enzymes not only show preference to substrates but also confer an added advantage to pathogens, because they might be more adaptable to changing environmental conditions and/or because some isoenzymes might be more active or stable within the host plant than others (Hancock, 1976; Byrde and Archer, 1977; Wijesundra et al., 1984; Forster and Rasched, 1985; Peres Artes and Tena, 1990).

The race 0 and race 5 of *Fusarium oxysporum* f.sp. *ciceri*, the chickpea wilt pathogen, produce extracellular polygalacturonase (PG) and pectate lyase (PL) in vitro (Peres-Artes and Tena, 1989). Both the enzymes were purified. The race 0 produces three PG forms (endo-PGIo, endo-PGPIIo, and exo-PGIIIo) and only one PL form (endo-PLo), while the race 5 produces one PG form (exo-PG_5) and two PL forms (endo-PLI_5 and endo-$PLII_5$). The characteristics of these enzymes are given in Table 12 (Peres-Artes and Tena, 1990).

The exo PGs from both races (PGIIIo and PG_5) were apparently the same enzyme. All other enzymes were different from each other (Peres-Artes and Tena, 1990).

Colletotrichum lindemuthianum, the bean pathogen, produces an endo-PG and the enzyme was purified from beta and gamma races of the pathogen (Keon et al., 1990). The molecular weights of the PGs produced by both the races were indistinguishable with similar values (39,500). The isoelectric point of both the enzymes was 9.4. The pH optimum of both enzymes was 5.4. Amino acid analysis of both proteins yielded identical results. No differences were observed in the 36

Table 12 Comparative properties of polygalacturonases (PG) and pectate-lyases (PL) from race 0 and race 5 of *Fusarium oxysporum* f.sp. *ciceri*

	Enzyme						
Property	PGIo	PGIIo	PGIIIo	PG_5	PLo	PLI_5	$PLII_5$
Molecular weight	44,000	44,000	76,000	76,000	25,000	37,000	37,000
Isoelectric point	8.0	6.4	4.2	4.2	> 9.0	> 9.0	—
pH optimum	4.5	4.5	4.5	4.5	9.5	10.5	10.5
Optimum temperature °C	45	45	60	60	45	45	45
Mode of action	endo	endo	exo	exo	endo	endo	endo

Source: Peres-Artes and Tena, 1990.

N-terminal amino acids. Analysis of total carbohydrate content indicated the presence of 7 moles sugar/mole of enzyme for both proteins (Keon et al., 1990). Lorenzo et al (1990) purified endo PGs from α, β, and γ races of *C. lindemuthianum* and all of them exhibited an identical molecular weight of 40,000, each reacted equally with antisera raised against endopolygalacturonases of *Aspergillus niger* and *Fusarium moniliforme.*

Bugbee (1990) purified the enzymes from *Rhizoctonia solani* strain AG 2-2, causing crown and root rot of sugarbeet (*Beta vulgaris*). The fungus produced exo-PG and pectin lyase and the latter was much more active. The pectin lyase showed an endo-mode of cleavage of pectin. No enzyme activity was observed when sodium polypectate was used as the substrate, indicating the absence of pectate lyase. The optimum pH for the pectin lyase was 8.0 (Bugbee, 1990). The molecular weight of the purified pectin lyase was 35 kDa and the isoelectric point was 10.1 (Bugbee, 1990).

Fusarium oxysporum f.sp. *cepae*, the onion (*Allium cepa*) bulb rot pathogen, produced only endo PL and not endo-PG (Holz and Knox-Davies, 1985). Presence of much higher levels of PL than PG activity in infected plant tissues has been reported in *Verticillium* wilt of tomato (Heale and Gupta, 1972; Cooper and Wood, 1980) and *Colletotrichum lindemuthianum*-infected bean (Wijesundara et al., 1984). *Botrytis cinerea* produced both PG and pectinlyase, but PG production was much more than pectin lyase production (Kaile et al., 1991). *B. cinerea* produces multiple forms of polygalacturonoase (Johnston and Williamson, 1992; Johnston et al., 1993; Sharrock and Labavitch, 1994). Endo-pectate lyase has also been reported to be produced by *B. cinerea* (Dilenna et al., 1981; Martinez et al., 1982; Hagerman et al., 1985; Movahedi and Heale, 1990b). Dilenna et al. (1981) reported the secretion of multiple forms of PL by three isolates of *B. cinerea*, with the most virulent producing the greatest number of PL isoenzymes. *Phoma medicaginis* and *Monilinia fructigena* produce PL (Keon, 1985; Pitt, 1988). *Magnaporthe grisea*, the rice blast pathogen, produced pectin methylesterase (PME), pectinlyase, and PG (Bucheli et al., 1990). PME and pectin lyase were further purified. The molecular weights of PME and pectin lyase were 34 kDa and 27 kDa, respectively (Bucheli et al., 1990). *Cladosporium cucumerinum* produced PG in culture (Raa et al., 1977). *Fusarium solani* f.sp. *pisi*, the pea pathogen produces three pectin-degrading enzymes: an acidic hydrolyase, a basic hydrolase, and a pectate lyase. These enzymes have been purified (Crawford and Kolattukudy, 1987). The gene for the lyase has been cloned from the fungus. Northern blot analyses showed the appearance of pectate lyase transcript on the growth of the fungus in the presence of pectin and not in the absence of pectin (Kolattukudy et al., 1989). The molecular mass of the enzyme was 26 kDa. The pH optimum for the lyase was approximately 9.4 and the enzyme was two to three times more active with polygalacturonic acid as substrate than with pectin. The isoelectric point of the pectate lyase was approximately 8.3. The enzyme was an endo-acting enzyme. It exhibited an absolute requirement for calcium. Total poly (A) RNA was isolated from pectin-induced mycelium and translated in vitro. A 29 kDa precursor polypeptide was identified (Crawford and Kolattukudy, 1987).

Two polygalacturonases were detected in the culture filtrate of *Ascochyta pisi*, the leaf and pod spot pathogen of pea. Both the polygalacturonases were partially purified. The first enzyme, PG1, hydrolyzed the polymer in a nonrandom fashion and hence it may be an exopolygalacturonase, whereas PG2 hydrolyzed the polymer randomly and hence it was identified as endopolygalacturonase. The molecular weight of PG1 was 83,000 while that of PG2 was 48,000 (Hoffman and Turner, 1982).

Three polygalacturonases (PG-I, PG-II and PG-III) were detected in the culture filtrate of *Penicillium italicum*, causal organism of blue mold of citrus fruits. PG-I was an exo-enzyme while the other two were endo-enzymes. Exo-PG-I showed higher molecular weight than PG-II and PG-III. Their characters are given in Table 13 (Hershenhorn et al., 1990). All three PGs were detected in infected tissues also (Hershenhorn et al., 1990).

Venturia inaequalis produced exo-PG in culture (Valsangiacomo and Gessler, 1992). The molecular mass of the enzyme was 90,500. Presence of several isoenzymes of the enzyme was detected. The optimum pH of the enzyme was between 5.5 and 6.0. This enzyme did not react with an antiserum produced against a polygalacturonase of *Fusarium moniliforme* (Valsangiacomo and Gessler, 1992). The antiserum produced against a PG of *F. moniliforme* cross-reacts with PGs of a number of pathogens (De Lorenzo et al., 1988). Since this antiserum does not react with the PG of *V. inaequalis*, the PG may not be common to the PGs of other pathogens. This exo-PG caused only limited cell wall degradation (Valsangiacomo and Gessler, 1992).

Table 13 Comparative properties of pectic enzymes produced by *Penicillium italicum*

Property	Exo-PG-I	Endo-PG-II	Endo-PG-III
Percentage of hydrolysis of sodium polypectate at 50% reduction in viscosity	44	3.6	2.4
Molecular weight	60,000	36,000	38,000
Isoelectric point	3.8	7.5	8.0
pH optimum	4.0	4.5	4.5

Source: Hershenhorn et al., 1990.

Gaeumannomyces graminis var. *tritici*, the pathogen causing take-all disease of wheat, produces an endo-PG both in vitro and in vivo (Dori et al., 1990; 1992). A pectate lyase has been purified from *Colletotrichum gloeosporioides*, the causal organism of anthracnose of avocado (*Persea americana*) fruits (Wattad et al., 1994). Its molecular weight was 40 kDa and the optimum pH for activity was 8.9 (Wattad et al., 1994). An endopolygalacturonase produced by *Cryphonectria parasitica*, the chestnut (*Castanea dentata*) blight fungus has been purified and characterized (Gao and Shain, 1994). Its molecular weight was 42 kDa and its isoelectric point was about 8.0. The pH optimum of the enzyme was 4.5–5.0 (Gao and Shain, 1994). *Mycocentrospora acerina*, the fungus responsible for liquorice rot on carrot produced PME, PG, and PL (Lecam et al., 1994b).

Helminthosporium nodulosum, the fingermillet pathogen, and *H. oryzae*, the rice pathogen, produce PME, endo-PG, exo-PG, PGTE, and pectin lyase (Vidhyasekaran, 1972, 1974a,b,c, 1975, 1977, 1978; Sathianathan and Vidhyasekaran, 1978). Multiple pectic enzymes production by several other pathogens have been reported (Vidhyasekaran and Parambaramani, 1973; Muthusamy et al., 1973; Kannaiyan et al., 1975; Sherwood, 1975; Arjunan et al., 1975; Barmore and Brown, 1979; Ramaraj and Vidhyasekaran, 1982, 1986; Dahm and Strzelozyk, 1987, Valsangiacomo et al., 1992).

B. Evidence to Show that Pectic Enzymes Help Pathogens to Penetrate Cell Wall

1. Histologic Evidence

Histological evidence has been provided to show the importance of pectic enzymes in cell wall penetration by fungal pathogens. The gold-complexed lectin from *Aplysia depilans*, an agglutinin with polygalaturonic acid-binding specificity, was used to localize pectin in bean leaf tissues infected with *Colletotrichum lindemuthianum* (Benhamou et al., 1991). As soon as 72 h after inoculation, the host cell walls showed early signs of shredding. At 96 h after inoculation, the fungus penetrated host cell wall. Fungal hyphae were frequently found to be capable of growing within host cell walls and to split them apart. Incubation with the gold-complexed *Aplysia* gonad lectin (AGL) resulted in an irregular deposition of gold particles adjacent to invading hyphae. At 120 h after inoculation, an intense degradation of cell wall components was observed. The intense host wall disintegration in infected area was always accompanied by the release of tiny fragments specifically labeled by gold particles (Benhamou et al., 1991). The results suggested degradation of pectin during cell wall penetration by the pathogen.

The localization of the fungal endopolygalacturonases in *C. lindemuthianum*-infected bean cell walls was studied using polygalacturonase-inhibiting protein (PGIP)–gold complex (Benhamou et al., 1991). Large amounts of gold particles were found over walls of invading hyphae as well as over host cell walls closely neighboring fungal cells. The qualitative evaluation of the label patterns obtained over cells of *C. lindemuthianum* following the PGIP–gold complex revealed that the PG synthesis increased severalfold when the fungus developed in bean leaf tissues. The results demonstrate that *C. lindemuthianum* is capable of producing large amounts of endo PG, causing extensive pectin breakdown (Benhamou et al., 1991).

When tobacco roots were inoculated with zoospores of *Phytophthora parasitica* var. *nicotianae*, penetration of the root epidermis was seen by 24 h after inoculation. By 72 h after inoculation, the fungus had developed through much of the cortex and it was always associated with considerable host wall alterations, ranging from swelling and shredding to complete disruption of the middle lamellar matrices (Benhamou and Cote, 1992). AGL was applied to infected root tissues for studying the pattern of pectin distribution in host cell walls. Labeling with the AGL–gold complex was associated with primary cell walls and middle lamella matrices in noninoculated tobacco root cells. Examination of infected tissues from 48 to 120 h after inoculation showed that labeling with the gold-complexed lectin was markedly reduced in wall areas exhibiting signs of obvious damage (Table 14; Benhamou and Cote, 1992). It indicates that pectin would have been degraded in the infected tissues. The AGL failed to label pectic molecules in primary wall and middle lamella areas near invading hyphae, indicating the release of pectic enzymes by *P. parasitica* var. *nicotianae* (Benhamou and Cote, 1992). Alteration of pectic molecules occurred in wall areas closely adjacent to fungal cells and at a distance from the point of pathogen penetration (Benhamou and Cote, 1992). It suggests that pectic enzymes may have freely diffused extracellularly to facilitate pathogen ingress through loosened middle lamella matrices and host cell walls.

The early demethylation of pectin and degradation of pectin was considered to be the first signs of colonization by *Fusarium oxysporum* f.sp. *dianthi* in carnation (*Dianthus caryophyllus*) cultivars (Niemann et al., 1991). Penetration

Table 14 Density of labeling obtained with the gold-complexed *Aplysia* lectin over primary walls of tobacco root cells infected with *Phytophthora parasitica* var. *nicotianae*

	Density of labeling (number of gold particles per μm^2) over primary wall	
Time after inoculation	Outermost wall layers	Innermost wall layers
Noninoculated tissues	108	100
Inoculated tissues		
72	86	51
96	85	45
120	62	32

Source: Benhamou and Cote, 1992.

of epidermal cells was observed 2–7 h after inoculation with *Pythium ultimum* in cucumber (Cherif et al., 1991). From 24 to 48 h after inoculation, the fungal growth was associated with some changes in the structure of host cell walls such as swelling and decrease in electron density. By 48 h after inoculation marked alterations of the host cell wall involving swelling, disintegration, and loss of the fibrillar wall and middle lamella matrices were observed. AGL was applied to infected cucumber tissues for studying the pattern of pectin distribution during the early stages of colonization. The observations (Table 15) revealed that alteration of pectic macromolecules was an early event and the transmission electron micrographs revealed that the alteration of pectic materials was mainly associated with partial to complete dissolution of middle lamella matrices. The pectin breakdown was not restricted to areas neighboring invading

Table 15 Density (number of gold particles/μm^2) labeling over cell walls of cucumber root and hypocotyl tissues inoculated with *Pythium ultimum*

	Pectic residues	
Time after inoculation (h)	Primary wall	Middle lamella
Control plants	227	327
24	118	307
48	83	144
72–96	41	56

Source: Cherif et al., 1991.

hyphae but also occurred at some distance from the point of fungal attack, suggesting that pectin-degradation enzymes produced by *P. ultimum* freely diffuse extracellularly, thus facilitating pathogen ingress through loosened host walls (Cherif et al., 1991).

2. Evidence Using Antibodies

Fusarium solani f.sp. *pisi*, the pea pathogen, produces pectate lyase. Antibodies against the enzyme were developed. When suspension of the conidia of the fungus was prepared in these antibodies, and inoculated on pea stem, no infection of the pathogen was observed. Penetration of the fungus into the host was very much suppressed by the antibodies (Table 16; Crawford and Kolattukudy, 1987). The results suggest the importance of the PL in pathogenesis of *F. solani* f.sp. *pisi*.

The pectate lyase produced by *Colletotrichum gloeosporioides* (the causal organism of anthracnose of avocado fruits) macerates avocado fruit tissues (Wattad et al., 1994). Antibodies against the enzyme were developed and these antibodies inhibited the enzymatic activity. These antibodies suppressed maceration of the fruit tissue by the enzyme (Wattad et al., 1994). It suggests the importance of the enzyme in disease development in avocado fruit tissue.

3. Evidence Using Pectic Enzyme-Deficient Fungal Isolates

The fungal isolates deficient in pectic enzymes have been shown to be less virulent. The highly virulent isolate of *Fusarium oxysporum* f.sp. *ciceri* produced two forms of endo-PL while the low virulent isolate produced only one form of endo-PL (Peres-Artes and Tena, 1990). Virulent isolates of *Rhizoctonia solani* produced endo-PL while hypovirulent isolates did not produce the enzyme (Marcus et al., 1986). Hypoaggressive isolate of *Mycocentrospora acerina* produces less of PME, PG, and PL in vitro (Lecam et al., 1994b). A hypovirulent strain of *Cyphonectria parasitica* produced less PG compared to the virulent strain (Gao and Shain, 1995). This evidence indicate that pectic enzymes may contribute for virulence of pathogens.

Table 16 Effect of pectate lyase antibodies on the penetration of *Fusarium solani* f.sp. *pisi* into pea stems

Spore suspension in	Frequency of penetration (%)
Water	75
Preimmune IgG	92
Antipectate lyase IgG	10

Source: Crawford and Kolattukudy, 1987.

C. Host Cell Wall Differs in Its Susceptibility to Pectic Enzymes

Efficacy of pectic enzymes to degrade host cell walls may also depend upon the structure of host cell wall (Cooper et al., 1981). The pectic enzymes produced by *Mycocentrospora acerina* hydrolyzed and solubilized the pectins from carrot cultivars (Lecam et al., 1994a). Pectin fractions from the cultivars most (Touchon) and least (Major) susceptible to the fungal infection were incubated with pure endo-PG. After 45 min of reaction, Touchon pectin was hydrolyzed whereas Major pectin was not (Lecam et al., 1994a). Amounts of cell wall material and composition of the pectic fraction were not correlated with cultivar resistance, but a different distribution of methyl groups along the rhamnogalacturonan chains in the different cultivars would have contributed to the difference in behavior of endo-PG on pectin from the Touchon and Major cultivars.

Insoluble forms of pectin (protopectin) are resistant to hydrolysis by pectic enzymes. Susceptibility of strawberries (*Fragaria grandiflora*) to *Botrytis cinerea* was highly correlated with the soluble pectin content of the fruit (Hondelmann and Richter, 1973). The carrot cultivars resistant to *M. acerina* contained more insoluble form of pectin (Lecam et al., 1994a).

D. Cell Wall Proteins Modulate Pectic Enzyme Activity

Some proteins present in the host cell wall modulate pectic enzyme activity in cell walls. Degradability of chickpea cell wall preparations without ionically bound proteins was not related to host resistance or susceptibility to the wilt pathogen, *Fusarium oxysporum* f.sp. *ciceri*. However, walls containing ionically bound proteins from the chickpea susceptible cv.PV-24 were more extensively degraded by the exo-PG and PL forms (PL I and PL II) produced by the pathogen than were similar preparations from the resistant cv.WR 315 (Peres-Artes and Tena, 1990). The results suggest that pectic enzyme activities on plant cell walls can be modulated by certain proteins.

Cell wall proteins are known to inhibit selectively some pectic enzymes (Albersheim and Anderson, 1971; Fisher et al., 1973; Fielding, 1981; Brown and Adikaram, 1982, 1983; Hoffman and Turner, 1982, 1984; Brown, 1984; Lafitte et al., 1984; Barmore and Nguyen, 1985; Cervone et al., 1981, 1986, 1987). The cell wall protein from pea selectively inhibited PG 2 (endopolygalaturonase) and not PG 1 (exopolygalacturonase) produced by *Ascochyta pisi*, the pathogen of pea (Hoffman and Turner, 1982). *A. pisi* produced polymethylgalacturonase and it was also inhibited by the proteinaceous inhibitor that inhibited endopolygalacturonase (Hoffman and Turner, 1982). The inhibitor was purified and it had a molecular weight of 42,000. More than 70% of the inhibitor in pea leaflets was soluble and not bound to the cell walls. The soluble and cell wall bound forms of the inhibitor were distinguishable (Hoffman and Turner, 1982). The proteinaceous inhibitor isolated from bean inhibited only endo-PG produced by *Colletotrichum lindemuthianum*. The exo-PG was not inhibited (Barthe et al., 1981).

A cell-wall-bound inhibitor of PG has been detected in pear fruit. The inhibitor content decreased in the fruit during the process of fruit maturation and ripening. The ripened fruit was highly susceptible to fungal pathogens (Abu-Goukh et al., 1983a,b). On purification two PG inhibitors were identified. Both the inhibitors were identified as proteins and the molecular weight of both was 91,000. Both the inhibitors were effective against PGs produced by *Aspergillus niger* and *Botrytis cinerea* (Abu-Goukh et al., 1983a,b). *Penicillim expansum* could infect immature pear fruits also and the PG produced by the fungus was less inhibited by the proteinaceous inhibitors in the immature fruits (Abu-Goukh and Labavitch, 1983; Abu-Goukh et al., 1983a,b).

Gao and Shain (1995) reported that high amount of a PG inhibitor protein was detected in Chinese chestnut (*Castanea mollissima*) bark, which is resistant to *Cyphonectria parasitica*. The bark of American chestnut (*Castanea dentata*), which is susceptible to the pathogen, contained less inhibitor protein. Lafitte et al. (1984) observed that the isogenic lines of bean that are resistant to *Colletotrichum lindemuthianum* contained higher levels of the polygalacturonase inhibitors than the susceptible ones. The inhibitor was ionically bound to the cell walls of bean hypocotyls and effectively protected cell walls against degradation by PG in vitro. The inhibitor proteins from four bean cultivars have been purified (Lorenzo et al., 1990). The molecular size of the proteins from the four cultivars were indistinguishable by electrophoretic mobility on sodium dodecyl sulfate–polyacrylamide gel electrophoresis (SDS-PAGE). They were of the same molecular size. The abilities of the inhibitors purified from great northern, pinto, red kidney, and small red beans to inhibit the endopolygalacturonases purified from α, β, and γ races of *C. lindemuthianum* was assessed. The three endopolygalacturonases were inhibited to the same extent by the same amount of the inhibitor, regardless of the bean from which the inhibitor was isolated. No significant difference in the rate and pattern of degradation of polygalacturonic acid was observed (Lorenzo et al., 1990). The results suggest that the inhibitor may not be involved in race-cultivar specificity. Anderson et al. (1972) have also reported that the inhibitors from different bean cultivars are equally able to inhibit endopolygalacturonases from different races of *C. lindemuthianum*.

Polygalacturonase inhibitors have been isolated from cucumber (Skare et al., 1975; Bock et al., 1975), avocado (Reymond and Phaff, 1965), onion, sweet paprika (*Capsicum annuum* var. *longum*) and cabbage (*Brassica oleraceae*) (Bock et al., 1975), beans (Albersheim and Anderson, 1971; Anderson and Albersheim, 1972; Fisher et al., 1973; Gobel and Bock, 1978) and several fruit species (Fielding, 1981). *C. lindemuthianum* producd PG in culture but the enzyme could not be detected in infected bean hypocotyls in which a proteinaceous inhibitor of PG could be isolated (Wijesundera et al., 1989). *Cladosporium cucumerinum* also secreted PG in culture, but PG activity was not detected in infected cucumber tissue in which a proteinaceous inhibitor was present (Raa et al., 1977).

Although several lines of evidence have been provided to show that the proteinaceous inhibitors may modulate the action of the pectic enzymes in the host tissues, evidence has also been presented to show that only little of the inhibitor protein is available in the cell walls of host tissues (Hoffman and Turner, 1982, 1984; Turner and Hoffman, 1985). The proteinaceous inhibitor detected in pea leaves effectively inhibited the degradation of polygalacturonate by *Ascochyta pisi*, but it did not affect the degradation of insoluble cell-wall fragments in vitro (Turner and Hoffman, 1985). Oranges (*Citrus sinensis*) (Barmore and Nguyer, 1985), tomato (Brown and Adikaram, 1982), peach (*Prunus persica*), and plum (*Prunus domestica*) fruits (Fielding, 1981) contained proteinaceous inhibitors that effectively inhibited PGs produced by their pathogens such as *Diplodia natalensis*, *Glomerella cingulata*, and *Monilinia* spp. in vitro. However, these pathogens penetrated and macerated all these fruit tissues, suggesting that these inhibitors may not be present in their cell walls in enough quantities to prevent the infection.

When a pathogen produces multiple polygalacturonases, polygalacturonase inhibitor protein may not be sufficient to prevent penetration of fungus into cell wall. One of the polygalacturonases that is not inhibited by the protein may facilitate fungal penetration (Sharrock and Labavitch, 1994).

E. Individual Pectic Enzymes May Not be Sufficient for Cell Wall Penetration and Pathogenicity

Pathogens produce multiple pectic enzymes and single enzyme may not be sufficient to cause disease. *Verticillium dahliae*, the cotton wilt pathogen, produced endo-PG, but endo-PG production by different isolates of *V. dahliae* did not have any relationship to virulence of the isolates (Keen and Erwin, 1971). Mutant strains lacking endo-PG were developed and these also induced severe disease incidence (Puhalla and Howell, 1975; Durrands and Cooper, 1988). PME is produced by *V. dahliae* in culture, but mutant without PME also caused typical symptoms of the disease (Howell, 1976). The pathogen produces PL and pectin lyase also in vitro (Howell, 1976). The mutants deficient for these enzymes also induce all symptoms of the disease (Howell, 1976).

A gene (PGN1) encoding endopolygalacturonase was isolated from *Cochliobolus carbonum* race 1. Genomic and cDNA copies of the gene were isolated and sequenced. The DNA sequence of PGN 1 predicted a polypeptide of 33,992 Da. The PGN 1 gene was required for endo PG production, however, it was not necessary for growth on pectin as sole carbon source. A precise gene disruption mutant in which only the endo PG is affected was developed. When both mutant and wild-type strains were spray-inoculated onto maize leaves, there were no differences in lesion size, lesion number, or rate of disease development. It suggests that endo-PG production is not necessary for pathogenesis of *C. carbonum* (Scott-Craig et al., 1990).

Although the *C. carbonum* mutant does not produce endo-PG, it produces exo-PG (Scott-Craig et al., 1990). *V. dahliae* mutants likewise produce other pectic enzymes (Keen and Erwin, 1971). An avirulent isolate of *F. solani* f.sp. *pisi* became highly virulent when the spore suspension was supplemented with pectinase, PME, cellulase, and cutinase (Koller et al., 1982a). Supplementation by the individual enzymes was not effective, suggesting that all enzymes are essential for penetration and virulence (Koller et al., 1982a). Thus pectic enzymes may be among the several factors that facilitate penetration of the host cell wall and it is quite expected, if one considers the complex nature of plant cell wall.

VI. PATHOGENS PRODUCE CELLULOLYTIC ENZYMES TO BREACH CELL WALL BARRIER

A. Multiple Cellulolytic Enzymes

Cellulose is the major wall polysaccharide and is composed of glucose units in the chain configuration, connected by β-1,4-glycosidic bonds. The enzymatic hydrolysis of cellulose requires the action of endoglucanase (endo β-1,4-glucanase) and exoglucanases (β-1,4-cellobiohydrolase and β-glucosidase) (Yazdi et al., 1990). Several cellulolytic enzymes are known to be produced by pathogens. *Venturia inaequalis*, the apple scab pathogen, produces in culture 12 cellulase isozymes with isoelectric points in the range of 3.7–5.6 (Kollar, 1994). The molecular weights were about 60 kDa for at least five enzymes and about 25 kDa for the five more prominent isozymes. These enzymes could be isolated from apple leaves infected with the pathogen also (Kollar, 1994).

Gaeumannomyces graminis var. *tritici*, the causal agent of take-all disease of wheat, secretes two groups of enzymes that degrade cellulosic polymers (Dori et al., 1995). The first group contains an endoglucanase and β-glucosidase with acidic pIs of 4.0 and 5.6, respectively. The second group contains an endoglucanase and β-glucosidase with basic pIs of 9.3 and > 10, respectively. Acidic and basic groups of endoglucanase and β-glucosidase were also obtained from inoculated wheat roots and they appeared to be similar to those isolated from the culture fluid (Dori et al., 1995). Several other pathogens are known to produce different cellulases in vitro and they have been detected in infected tissues also (Vidhyasekaran, 1974a; Pearson, 1974; Spare et al., 1975; Suzuki et al., 1983; Wagner et al., 1988).

B. Cellulases Are Required for Breaching Cell Wall Barrier

Several studies have provided evidence to show the involvement of cellulolytic enzymes in fungal pathogenesis. Histological studies using tobacco roots infected by *Phytophthora parasitica* var. *nicotianae* demonstrated the involvement of cellulases in fungal pathogenesis (Benhamou and Cote, 1992). The exoglucanase, a β-1,4-D-glucan cellobiohydrolase purified from a cellulase produced by *Trichoderma harzianum*, was used for localizing cellulosic β-1,4-glucans. An intense labeling over the fungus cell walls was observed when incubated with the gold-complexed exoglucanase. Incubation of sections from noninoculated tobacco root tissue resulted in a heavy labeling of primary and secondary cell walls. Between 48 and 72 h after inoculation, the fungus had colonized the epidermis and much of the cortex, causing extensive cell damage and wall alteration. Incubation of these infected tissues with the gold-complexed exoglucanase resulted in the deposition of gold particles over both fungal and plant cell walls. Labeling of the fungal cell walls was similar to that noted over the fungus grown in culture. Great variations in labeling intensity were observed over host cell walls adjacent to invading hyphae (Table 17; Benhamou and Cote, 1992). Labeling decreased in the inoculated tissues, indicating degradation of cellulose (Benhamou and Cote, 1992).

A 95 kDa polypeptide (P-95) was found to be associated with the ability of appressoria of *Colletotrichum lagenarium* to penetrate artificial nitrocellulose membrane (Suzuki et al., 1981). The polypeptide has been suggested to be a cellulase (Suzuki et al., 1982a). Cycloheximide inhibited the synthesis of this polypeptide and in its presence appressoria matured in structure, but lacked the

Table 17 Density of labeling obtained with the gold-complexed *Aplysia* exoglucanase over primary walls of infected tobacco cells inoculated with *Phytophthora parasitica* var. *nicotinae*

	Density of labeling over primary walls (no. of gold particles per μm²)	
Time after inoculation (h)	Outermost wall layers	Innermost wall layers
Noninoculated tissues	182	159
Inoculated tissues		
72	170	87
96	92	39
120	54	12

Source: Benhamou and Cote, 1992.

ability to penetrate cellulose membrane (Suzuki et al., 1981). Appressoria formed in the presence of cycloheximide penetrated into cellulose membranes when the appressoria were pretreated with cellulase (Suzuki et al., 1982a). The results suggest the importance of cellulase in penetration of cellulose membrane.

Three mutants of *C. lagenarium* that lacked their ability to penetrate nitrocellulose membrane were obtained (Katoh et al., 1988). All three mutants synthesized the 95 kDa protein, but they lacked ability to secrete cellulase (Katoh et al., 1988). The studies indicate the importance of secretion of cellulases in host cell wall penetration.

VII. PATHOGENS PRODUCE HEMICELLULASES TO BREACH CELL WALL BARRIER

The important plant cell wall hemicelluloses are araban, xylan, and galactan. Many pathogens, including *Sclerotium rolfsii*, *Verticillium albo-atrum*, *Fusarium oxysporum* f.sp. *lycopersici*, and *Sclerotinia sclerotiorum*, produce high levels of arabanases, xylanases, and galactanases in vitro (Hancock, 1967; Van Etten and Bateman, 1969; Howell, 1975; Baker et al., 1977, 1989; Bauer et al., 1977; Cooper et al., 1978). The cell wall polymers are rapidly degraded in infected tissue that may contain the appropriate enzyme activity (Hancock, 1967; English and Albersheim, 1969; Bateman and Basham, 1976). The matrix of cell wall of potato tubers contains exceptionally high levels (approximately 45%) of β-1,4-galactan. *Phytophthora infestans* and *Phomopsis exiqua*, the pathogens of potato, produce endo-galactanases in vitro and in infected tissues and deplete the galactan (Javis et al., 1981).

Xylans, β-1,4-linked polymers of xylopyranose, are ubiquitous components of cell walls of plants (Darvill et al., 1980). They are the most abundant hemicellulose, constituting up to 40% of the primary cell walls of graminaceous monocotyledons (Burke et al., 1974; Wada and Ray, 1978; Aspinall, 1980). Arabinoxylan is the major noncellulosic component of maize primary cell walls (Kato and Newins, 1984a).

Many fungal pathogens, including *Rhizoctonia solani* (Bateman et al., 1969), *Helminthosporium maydis* (Bateman et al., 1973), *Macrophomina phaseolina* (Dean and Anderson, 1989; Dean et al., 1989), *Fusarium roseum* (Mullen and Bateman, 1975), and *Bipolaris sorokiniana* (Karjalainen et al., 1992), produce xylanase rapidly in high amounts. *Cochliobolus carbonum* secretes xylanase when grown on maize cell walls. On purification three xylanase isoenzymes and a β-xylosidase were detected (Holden and Walton, 1992). Xylanase had a molecular mass of 24,000 and xylanase II and III had molecular masses of 22,000 (Holden and Walton, 1992). Three xylanase genes have been identified in *C. carbonum* (Apel et al., 1993). Xyl 1, the gene for the major xylan-degrading enzyme in *C. carbonum*, has been cloned and sequenced. Xyl 1 encodes two forms of endoxylanase activity (xylanase I and II), which together are responsible for 85–94% of the extracellular xylan-degrading activity of *C. carbonum* when it is grown on corn cell walls or xylan as the sole carbon source (Apel et al., 1993). A mutant specifically lacking a functional copy of this gene was obtained using transformation-mediated gene disruption (Apel et al., 1993). However, growth of the mutant strain was indistinguishable from the wild type in media containing corn cell walls or xylan as the carbon source. It indicates that the residual xylan-degrading activity in the mutant may be sufficient to support wild-type growth rates on xylan. The mutant strain was also pathogenic on corn (Apel et al., 1993). It indicates that xylanase may not be essential for pathogenesis. However, the small but significant amounts of xylanase remaining in the mutant when Xyl 1 is disrupted may be responsible for pathogenicity and hence it cannot be concluded that xylanase is not essential for pathogenicity of *C. carbonum.*

Two xylanases have been purified from *Gaeumannomyces graminis* var. *avenae*, the causal agent of take-all disease of oats (Southern et al., 1993). Their molecular weights were 26.5 kDa and 28 kDa and they had a pI of 10.5. The N-terminal sequences of the two xylanases were identical. These xylanases could be detected in oat and wheat roots infected with *G. graminis* var. *avenae* (Southerton et al., 1993).

Two xylanases have been purified from *Magnaporthe grisea*, the rice blast pathogen. They are secreted when the fungus is grown on rice cell walls as the only carbon source. Their molecular weights were 22 kDa and 33 kDa, and both of them were basic proteins with calculated isoelectric points of 9.95 and 9.71, respectively. The xylanases have been cloned and sequenced (Wu et al., 1995).

Arabinose is a major component of the arabinans. It is also found as a side chain constituent of the xyloglucan, xylan rhamnogalacturonan I, and rhamnogalacturonan II polymers of plant cell walls (McNeil et al., 1984). The removal of the arabinose-containing side chains from these major plant cell wall polymers by arabinofuranosidase might facilitate the digestion of these polymers. *Monilinia fructigena* produces α-L-arabinofuranosidase in vitro and in infected apple fruits. The breakdown product arabinose was detected in apple fruits infected with *M. fructigena* (Laborda et al., 1974; Howell, 1975). A correlation between levels of arabinofuranosidase activity and virulence has been reported in *Monilinia fructigena* (Howell, 1975).

Sclerotinia trifoliorum, a pathogen of legumes, produces an arabinofuranosidase in culture (Rehnstrom et al., 1994). Three arabinofuranosidase-deficient mutants were obtained (Rehnstrom et al., 1994). The pathogenicity of the arabinofuranosidase-deficient mutants was evaluated by inoculating alfalfa and pea stems. These mutants were significantly virulent on the pea stems but not on alfalfa stems. It suggests that arabinofuranosidase may play a role in pathogenicity only under certain conditions (Rehnstrom et al., 1994).

VIII. PROTEASES MAY BE INVOLVED IN DEGRADATION OF CELL WALL PROTEINS

Structural proteins are also important components of the plant cell wall. Proteases may be involved in degradation of plant cell wall proteins (Vidhyasekaran and Parambaramani, 1971; Hislop et al., 1982; Movahedi and Heale, 1990a,b). *Pyrenopeziza brassicae*, the causal agent of light leaf spot of oilseed rape (*Brassica napus* L. ssp *oleifera*) produces a cysteine protease with a molecular weight of 34 kDa (Ball et al., 1991). An ultraviolet (UV)-induced nonpathogenic mutant of the fungus was also deficient in protease production in vitro. The protease mutant was transformed with clones from a genomic library of *P. brassicae* and a transformant obtained showed concomitant restoration of pathogenicity and proteolytic activity in vitro (Ball et al., 1991). The results suggest a role for protease in disease development, but it is not clear whether the protease has degraded plant cell wall protein. It would have degraded membrane proteins to perturb membrane transport functions or would have degraded pathogenesis-related proteins.

IX. PATHOGENS PRODUCE A VARIETY OF ENZYMES TO DEGRADE THE COMPLEX-NATURED CELL WALL

Since plant cell walls contain different polysaccharides, proteins and lipids that are interwoven, a single enzyme may not be able to degrade the cell wall efficiently. A mixture of different enzymes may be necessary (Karr and Albersheim, 1970). *Venturia inaequalis* invades apple leaves by growing between the upper epidermis and the cuticular membrane after penetrating the cuticular membrane without penetrating cells or invading intercellular spaces (Valsangiacomo et al., 1989). Transmission electron microscopic studies indicate a change in the electron density in epidermal cells at the contact point with stroma cells of the fungus. Translucent zones in the cell wall were associated with local degradation since they were present only at the host–pathogen interface (Valsangiacomo et al., 1989). Both commercial enzyme preparations containing cellulase, pectin lyases, polygalacturonases, and hemicelulases and crude extracts from *V. inaequalis* enzymes degraded ^{14}C-labeled cell walls of apple leaves in vitro (Valsangiacomo et al., 1992a). Up to 40% of cell wall materials was degraded by commercial enzymes while only about 8% of the cell wall materials was degraded by crude enzyme extract of *V. albo-atrum*. The crude enzyme extract exhibited only pectinolytic activity. Commercial enzymes were much more effective in cell degradation probably because of the synergistic effect of cellulases, pectinolytic enzymes, and hemicellulases (Valsangiacomo et al., 1992a).

When tobacco roots were inoculated with *Phytophthora parasitica* var. *nicotianae*, cellulose degradation was detected using gold-complexed *Aplysia* exoglucanase (Benhamou and Cote, 1992). The occurrence of gold particles (indicating the presence of undergraded cellulose) over cell wall microfibrils that were severely altered in terms of loosened structure indicated that cellulolytic enzymes were not the first enzymes to be produced by the pathogen, but were likely preceded by pectinases and hemicellulases (Benhamou and Cote, 1992).

X. PATHOGENS ADAPT TO THE NATURE OF THE CELL WALLS OF HOST AND PRODUCE SUITABLE CELL-WALL-DEGRADING ENZYMES

Composition of cell walls of plants varies from plant to plant, particularly between monocots and dicots (Darvill et al., 1980; Cooper, 1983; Selvendran, 1985). Cell walls of monocots contain less than 10% of the pectic polysaccharide content of dicotyledon primary walls (Labavitch and Ray, 1978; Selvendran, 1985). In contrast, primary walls of monocots contain more arabinoxylan and β-1,4-and β-1,3-linked glucans (Burke et al., 1974; Labavitch and Ray, 1978; Darvill et al., 1980, Kato and Nervins, 1984a,b; McNeil et al., 1984). Pathogens of dicots and monocots differ in their ability to produce different enzymes. Pathogens of dicots probably produce more pectic enzymes, while pathogens of monocots produce fewer pectic enzymes and a high amount of other enzymes.

Rhizoctonia cerealis, *Fusarium culmorum*, and *Pseudocercosporella herpotrichoides* are pathogens attacking monocots. When they were grown on wheat seedling cell walls as sole carbon source, they produced a similar spectrum of enzymes and usually in the same sequence (Cooper et al., 1988). Early and high production of arabanase was followed by xylanase, then laminarinase, while low levels of pectinases and cellulases accumulated later. α-L-Arabinosidase appeared much later than arabanase; β-D-galactosidase followed a similar pattern to arabinosidase. Both PG and PL activites were produced by *R. cerealis* and *F. culmorum*, albeit at low levels, but were almost undetectable in cultures of *P. herpotrichoides* (Cooper et al., 1988).

When wheat seedlings were inoculated with *R. cerealis*, only xylanase increased due to infection. Uninfected tissue contained activities of all the glycosidases as well as laminarase and arabanase. Cellulase, endo-glucanase, endo-polygalacturonase, and pectinlyase were almost absent in both healthy and infected tissues (Table 18; Cooper et al., 1988).

In contrast, when *R. cerealis* was inoculated in a dicot (potato) cell walls, PL activity was enhanced above that on wheat seedling cell walls (Cooper et al., 1988). Xylanase activity remained low and β-galactosidase became the predominant activity (Cooper et al., 1988). Xylanases appear to be key enzymes in degrading monocot cell walls (Baker et al., 1977; Cooper, 1983) and pectic enzymes appear to be not very involved in pathogenesis of monocots (Scott-Craig et al., 1990). Pathogens of dicots such as *Rhizoctonia solani*, *Fusarium oxy-*

Table 18 Activities of wall-degrading enzymes in *Rhizoctonia cerealis*-infected wheat seedlings

	Enzyme activities (nmol min^{-1} g^{-1} fresh weight)	
	Healthy	Infected
Xylanase	0	44
Cellulase	0	0
Endo-polygalacturonase	0	0.5
Pectin lyase	0	0.3
Galactanase	0	0
Endo-glucanase	0	0
Laminarinase	177	181
Arabanase	73	77
β-xylosidase	71	66
α-L-Arabinosidase	33	34
β-D-Mannosidase	70	68
β-D-Glucosidase	129	110
α-D-Galactosidase	120	126
β-D-Galactosidase	127	116

Source: Cooper et al., 1988.

sporum, and *F. roseum* produce endo-pectic enzymes predominantly during infection or when grown on host cell walls (Cooper and Wood, 1975; Mullen and Bateman, 1975; Bateman and Basham, 1976).

The pathogen may adapt itself to the cell wall composition of its host and secrete suitable enzymes to degrade the host cell wall. When the monocot pathogen *Rhizoctonia cerealis* was inoculated on a dicot (potato) cell walls, pectin lyase activity was enhanced above that on wheat seedling cell walls, and xylanase activity remained low (Cooper et al., 1988).

XI. PATHOGENS PRODUCE CELL-WALL-DEGRADING ENZYMES IN A SEQUENCE

Linkage of polysaccharides in host cell walls is complex and the linkage has to be broken for effective degradation of cell wall barrier. *Verticillium albo-atrum* and *Fusarium oxysporum* f.sp. *lycopersici*, the wilt pathogens of tomato, synthesize enzymes that are able to act upon most of the major linkages in tomato cell wall polysaccharides: methyl α-4-D-galactopyranuronosyl, α-1,4-D-galactopyranurosyl and/or α-1,3-L-arabinofuranosyl, β-1,4-D-glucopyranosyl, β-1,4-D-galactopyranosyl and β-1,3-D-galactopyranosyl, and/or β-1,6-D-galactopyranosyl (Cooper and Wood, 1975).

The pathogens produce cell-wall-degrading enzymes in sequence. *Verticillium albo-atrum* grown in a medium containing tomato cell walls secreted a range of polysaccharide-degrading enzymes (Cooper and Wood, 1975). Endo-PG increased rapidly after 2 days, followed later by exo-arabinase and endo-PGTE. Endo-xylanase and cellulase were produced at low basal levels up to 6 days, after which production increased rapidly; endo-xylanase productions preceded that of cellulase by 24 h (Cooper and Wood, 1975).

Fusarium oxysporum f.sp. *lycopersici* produced polysaccharidases sequentially in tomato cell wall cultures (Jones et al., 1972). It produced endo-PG, endo-PGTE, cellulase, arabinase, xylanase, and β-galactosidase in sequence (Cooper and Wood, 1975). Sequential production of polysaccharidases by *Colletotrichum lindemuthianum* when grown on cell walls has been reported (English et al., 1971). The first enzyme produced was polygalacturonase followed by arabinase, xylanase, and cellulase (English et al., 1971). Thus the first enzyme to be produced by the pathogens appears to be pectic enzymes. Arabinase, xylanase, and cellulase appear after the pectic enzymes, presumably because their synthesis is induced sequentially as successive substrates become available during progressive degradation of cell walls. A sequential production of enzymes appears to be needed for effective cell wall degradation. Any change in the sequence will make the cell wall resistant to the pathogen.

Production of these enzymes aids in colonization of host tissues by pathogens. However, it is well known that saprophytes like *Aspergillus* and *Penicillium* also produce these enzymes in plenty. Probably production of these enzymes in appropriate site, in appropriate sequence, and in appropriate amount may be important for pathogenesis.

XII. REINFORCEMENT OF HOST CELL WALL DURING FUNGAL INVASION

A. Nature of Cell Wall Reinforcement

The first plant defense against many potential pathogens is the interlocking network of macromolecules in host cell wall. If this barrier is breached, the cell attempts to limit the spread of the infection in several ways. One common response is formation of cell wall appositions, which have been shown to be induced quickly after the onset of fungal penetration, sometimes in a matter of only a few hours (Aist, 1983). These appositions are called papillae and they typically comprise a callose matrix (Hinch and Clarke, 1982; Aist, 1983; Smith and Peterson, 1985; Smart et al., 1986b; Bonhoff et al., 1987; Brammall and Higgins, 1988) variously incorporating pectic materials (El Ghaouth et al., 1994), cellulose, suberin, gums, proteins (including peroxidase enzymes), calcium, and silicon (Aist, 1976b; 1983). Lignification of papillae occurs and it confers extra resistance to penetration (Ride and Pearce, 1979; Edwards and Ayers, 1981; Ride, 1983; Asiegbu et al., 1993, 1994).

Upon penetration by powdery mildew fungi, within the host cell, another reactive material, the collar, is deposited along the haustorial neck. In some cases, this collar may develop to such an extent that the entire hausotrium is encased, thus reducing the haustorial function by acting as a barrier to nutrient uptake (Heath, 1976; Perera and Gay, 1976; Manners and Gay, 1983; Kenji et al., 1985; Stenzei et al., 1985; Godwin et al., 1987; Cohen et al., 1990). The collar is made of two distinct areas, one being amorphous and the other one fibrillar. Neither the amorphous nor the fibrillar material contained chitin, which is of fungal origin. Cellulosic β-1,4-glucans were found to be restricted to the outermost fibrillar layers. The presence of collars is usually associated with poor development of the haustoria (Hajlaoui et al., 1991). The collar may act as a barrier to apoplastic flow in rust fungi (Bushnell and Gay, 1978; Stump and Gay, 1989). In some cases the collar extends to the haustorial body and encases it completely (Heath, 1976). The collar may develop from papillae that form before or during host cell wall penetration (Manners and Gay, 1983). In hop (*Humulus lupulus*) plants infected with *Sphaerotheca fuliginea*, the collar was found to contain callose-like deposits (Cohen et al., 1990).

In the case of powdery mildew fungi the haustorium is surrounded by an extrahaustorial matrix (Bushnell, 1972; Manners and Gay, 1983), which may be composed of a mixture of plant- and fungus-derived compounds (Bracker and Littlefield, 1973; Chong et al., 1985). When *Sphaerotheca pannosa* var. *rosae* was inoculated on rose (*Rosa hybrida*) leaves, fungal growth in the epidermis was associated with the formation of haustoria that appeared multilobed and delimited by an extrahaustorial membrane probably originating from the host plasmalemma. In the extrahaustorial matrix, cellulose and pectin were absent (Hajlaoui et al., 1991).

The extrahaustorial matrix formed in pea leaf cells infected with *Erysiphe pisi* was a fluid that reacted less intensely than the extrahaustorial membrane with polysaccharide agents and was removed by enzyme degrading the host cell wall (Gil and Gay, 1977). The matrix around haustoria of rust fungi contained a mixture of lipids and large amounts of polysaccharides and proteins (Rohringer et al., 1982; Chong et al., 1984, 1985). In the flax (*Linum usitatissimum* L.) rust infections the extrahaustorial matrix contained bound sugars, probably glycoproteins (Coffey and Allen, 1983).

B. Papillae Suppress Fungal Penetration

Papillae may function as a resistance mechanism (Aist and Israel, 1977, 1986; Koga et al., 1980; Tosa and Shishiyama, 1984; Stolzenburg et al., 1984a,b; Smart et al., 1986a,b; Heitefuss and Ebrahim-Nesbat, 1986; Murray and Ye, 1986; Kunoh et al., 1986; Brammal and Higgins, 1988; Cohen et al., 1989). Use of inhibitors of papilla formation in various host–pathogen systems has provided evidence for the role of the papilla in restricting fungal penetration (Vance and Sherwood, 1976; Bird and Ride, 1981; Sherwood and Vance, 1982; Steward and Mansfield, 1985; Gold et al., 1986). When coleoptiles of a barley line resistant to *Erysiphe graminis* f.sp. *hordei* were inoculated with the pathogen, more papillae were formed than in a susceptible barley line. Penetration efficiency of the conidia was also less in the resistant variety and only a few haustoria were formed (Bayles et al., 1990). When the coleoptiles were treated with 2-deoxy-D-glucose (DDG), papillae formation was inhibited in the resistant variety, resulting in more penetration efficiency of the conidia and more haustoria formation (Bayles et al., 1990). Thus when papilla formation is inhibited, penetration is successful. When papilla formation is induced, the plant becomes resistant. Chitosan, when applied as a stem scar treatment, reduced lesion development in bell pepper (*Capsicum annuum* cv. Bellboy) fruit caused by *Botrytis cinerea* (El Ghaouth et al., 1994). In the untreated tissue, deformation of primary cell wall was observed, while in chitosan-treated fruit tissue no cell damage was observed due to fungal inoculation. In chitosan-treated tissue inoculated with the pathogen, papilla formation in host cell was observed (El Ghaouth et al., 1994).

The time-course studies have also demonstrated the role of papillae in restricting fungal penetration (Hachler and Hohl, 1984; Stumm and Gessler, 1986). Delaying or inhibiting papilla formation induces susceptibility (Gold et al., 1986).

Papillae may act as a physical barrier against fungal penetration. In barley varieties resistant to *Erysiphe graminis*, the papillae were larger than in susceptible varieties at the fungal penetration sites (Yokoyama et al., 1991). Oversize papillae have been implicated in host resistance to pathogens (Aist et al., 1976; Smart et al., 1986b; Kunoh, 1990).

The efficacy of papilla formation as a resistant mechanism depends upon its early initiation. A low penetration efficiency of the fungal pathogens was correlated with papillae that were formed in advance of penetration pegs (Aist and Israel, 1977). When papilla deposition started earlier and increased more rapidly, potato plants became resistant to *Phytophthora infestans* (Stromberg and Brishanmar, 1993). Earlier papilla initiation is an important factor for papilla-mediated resistance (Gold et al., 1986; Bayles et al., 1990). The frequency of papilla formation is also important. When water extract of *Reynoutria schaliensis* was sprayed on cucumber leaves, it induced resistance against the powdery mildew fungus *Sphaerotheca fuliginea* (Schneider and Ullrich, 1994). The treatment induced an increased frequency of papillae (Schneider and Ullrich, 1994).

The factors contributing to papilla formation in infected tissues have been studied (Inoue et al., 1992, 1993, 1994a,b). A partially purified aqueous extract from barley seedlings enhanced oversize papilla formation at fungal penetration sites and reduced penetration efficiency (Yokoyama et al., 1991). The extract was referred to as papilla-regulating extract (PRE). Coleoptiles of susceptible barley variety were floated on PRE solution (Inoue et al., 1994a). Papillae were initiated about 23 min earlier in PRE treatments than in controls. Also, papillae were initiated 20 min before penetration peg initiation in the PRE treatments, whereas they were initiated 8 min after penetration peg initiation in the control. Mean papilla diameter at the time of initiation of penetration pegs was significantly greater in PRE treatments than in controls (Inoue et al., 1994a). In PRE treatments, papillae grew 5.1 μM after the initiation of pretreatment pegs, whereas they grew only 1.7 μM during the corresponding period in controls (Inoue et al., 1994a). The results suggest that PRE may be responsible for earlier papilla formation and development.

The PRE appears to contain potassium phosphate. Potassium phosphate, extracted from uninoculated barley leaves, induced papilla-mediated resistance against the powdery mildew fungus *E. graminis* in barley coleoptiles (Inoue et al., 1994b). Formation of resistant, oversize papillae was observed to be induced by $Ca(H_2PO_4)_2$ solution (Aist et al., 1979, Israel et al., 1980; Smart et al., 1986b). Phosphate salts induce local and systemic resistance against northern leaf blight and common rust in maize (Reuveni et al., 1992a,b).

Both PRE-induced resistance and PRE induction of oversize papillae were Ca^{2+} mediated (Inoue et al., 1994b). Calcium and phosphorus have been reported to be highly concentrated in $Ca(H_2PO_4)_2$-induced oversize papillae (Aist et al., 1979; Kunoh et al., 1986; Kunoh, 1990). Calcium and phosphorus are also found in both cytoplasmic aggregate vesicles and developing papillae in epidermal cells of barley leaves (Akutsu et al., 1980).

C. Callose Deposition in Cell Wall

Papillae contain callose and papillae cannot be formed without callose in many host–pathogen interactions (Bayles et al., 1990). All papillae that were formed in barley contained callose (Bayles et al., 1990). Callose is a polysaccharide containing a high proportion of 1,3-β-linked glucose. Callose is a polymer of β-1,3-glucans (Casio et al., 1990). Callose is a minor component of healthy plant tissue. Plants respond to infection by pathogens by the rapid deposition of callose (Schmele and Kauss, 1990).

The paratracheal parenchyma cells adjacent to infected xylem vessels in egg plant (*Solanum melongena*) stem tissues inoculated with *Verticillium albo-*

atrum responded by rapidly sealing off attempted sites of penetration (Benhamou et al., 1995). Electron-opaque, globular structure accumulated in paramural spaces. Large amounts of callose were detected in the electron-opaque globules (Benhamou et al., 1995). Infected epidermal cells of muskmelon showed deposition of callose-like materials around the fungal *Sphaerotheca fuliginea* penetration pegs (Cohen et al., 1990). Cucumber leaves inoculated with *Colletotrichum lagenarium* showed induction of callose-containing papillae (Binder et al., 1989).

D. β-1,3-Glucan Synthase (Callose Synthase) is Involved in Callose Deposition

The UDP-glucose is converted into a β-1,3-glucan, callose by β-1,3-D-glucan synthase (callose synthase) (Hass et al., 1985; Morrow and Lucas, 1986; Hayashi et al., 1987; Lawson et al., 1989). β-1,3-Glucan synthase has been purified from plasma membranes of several host cells (Wasserman and McCarthy, 1986; Fink et al., 1990; Frost et al., 1990; Dugger et al., 1991; Fredrikson et al., 1991; Dhugga and Ray, 1991). The purified β-1,3-glucan synthase from *Beta vulgaris* contains polypeptides at molecular masses of 92, 83, 70, 57, 43, 35, 31/29, and 27 kDa, suggesting the existence of a multisubunit enzyme complex (Wu et al., 1991). A 31 kDa polypeptide has been reported in purified preparations of soybean β-1,3-glucan synthase (Fink et al., 1990). In cotton, eight polypeptides have been found in the purified preparation of β-1,3-glucan synthase (Delmer et al., 1990, 1991).

The plasma membrane-bound callose synthase complex may be a deregulated form of cellulose synthase (Jacob and Northcote, 1985; Delmer, 1987). The enzyme is inactivated by exogenously added phospholipses A2, C, and D (Ma et al., 1991).

E. Callose Synthesis is Activated by Ca^{2+}

Callose synthesis is activated by Ca^{2+} and it has been demonstrated by using chitosan, the fungal elicitor in soybean cells (Kohle et al., 1985). Chitosan induced callose synthesis and the synthesis is immediately stopped when external Ca^{2+} is bound by ethylene glycol-bis-2-aminoethyl ether-N, N′-tetracetate, or cation exchange beads and partly recovers upon restoration of 15 μmol Ca^{2+}. Callose synthesis is observed only when membrane perturbation causing electrolyte leakage from the cells is induced by chitosan. Chitosan treatment does not induce β-1,3-glucan synthase activity in the soybean cells. In the presence of Ca^{2+}, however, the activity of the enzyme is 15–25-fold stimulated in both elicitor-treated and control cells. The β-1,3-glucan synthase activity without Ca^{2+} is not elevated in chitosan-treated cells (Table 19; Kohle et al., 1985).

The first callose formation was detected about 20 min after addition of chitosan. The speed of this response suggests that β-1,3-glucan synthase may be

Table 19 Activity of β-1,3-glucan synthase in chitosin-treated soybean cells and its stimulation by Ca^{2+}

	Callose formed (μg/10 min/50 μl)	
Source of enzyme	$-Ca^{2+}$	$+Ca^{2+}$
Control cells	0.2	4.6
Cells treated with chitosan	0.2	4.4

Source: Kohle et al., 1985.

at the site before callose formation begins, but in an inactive state since this process may be too rapid to involve transcriptional de novo synthesis of enzyme proteins (Kohle et al., 1985). This observation was strengthened by the finding that no increase in total β-1,3-glucan synthase activity was observable in chitosan-treated soybean cells. The enzyme was activated by Ca^{2+} (Kauss et al., 1983; Kauss, 1987a) and this activation may be due to a direct and reversible interaction of Ca^{2+} with the enzyme (Kohle et al., 1985).

Some polycations, including chitosan, poly-L-ornithine, and poly-L-lysine, induce callose synthesis in cells (Kauss, 1987a). Chitosan and other polycations release calcium from suspension-cultured soybean cells (Young and Kauss, 1983). Activation of callose synthesis is associated with a rise in cytoplasmic Ca^{2+} (Kauss, 1990). The activity of the plasma-membrane-located β-1,3-glucan synthase is strictly dependent on Ca^{2+} (Kauss, 1987a,b; Fink et al., 1987; Hayashi et al., 1987). β-1,3-Glucan synthase is localized on the cytoplasmic side of the plasma membrane (Fredrikson and Larsson, 1989). A perturbation leads to elevated cytoplasmic Ca^{2+} (Ohana et al., 1992). A transient increase in cytoplasmic Ca^{2+} concentration may represent the intracellular signal for activation of β-1,3-glucan synthase (Kauss et al., 1989).

When the fungal pathogen attempts to penetrate the cell wall, the Ca^{2+} permeability of the plasma membrane is perturbed (Schemele and Kauss, 1990). Microsomes and plasma membranes, prepared from first leaves of cucumber, exhibited increased specific activity of the Ca^{2+}-regulated β-1,3-glucan synthase. This enzyme activity was latent in epidermal cells of healthy leaves, but activation by attempted penetration by *Colletotrichum legenarium* rendered the cells capable of a rapid production of callose-containing papillae (Schmele and Kauss, 1990). β-1,3-Glucanase required Ca^{2+} for activity with a half-saturation below 1 μM free Ca^{2+}, even in the presence of 5 mM Mg^{2+}. In the absence of Mg^{2+}, Ca^{2+} activation was synergistically enhanced by 50 μM spermine and additively by poly-L-ornithine. The enzyme activity increased up to 10-fold after addition of digitonin. It suggests that β-1,3-glucan synthase is present in right-side-out plasma membrane vesicles and the digitonin increases the permeability of the vesicles, increasing access for substrate and activators to the cytoplasmic membrane surface (Schemele and Kauss, 1990).

Various studies have shown that callose biosynthesis does not involve transcriptional activation of plant defense genes. It is induced by elicitor signals that cause changes in the plasma membrane that lead to specific metabolic processes (Kauss et al., 1990).

F. β-Furfuryl-β-Glucoside is Involved in Callose Synthesis

Elevation of Ca^{2+} does not always lead to callose synthesis and other unknown effectors may be required in addition to Ca^{2+} for elicitation of callose synthesis (Kauss et al., 1991). UDP-glucose: (1-3)-β-glucan synthesis is synergically activated by both Ca^{2+} and a β-glucoside (Hayashi et al., 1987; Callaghan et al., 1988). β-Furfuryl-β-glucoside (FG) is a specific endogenous activator of plant β-1,3-glucan synthase (Ohana et al., 1992) and is present in a variety of plants (Ohana et al., 1991, 1992). However, glycosides are localized within the vacuole (Wagner, 1982) and hence may be inaccessible to the active site of β-1,3-glucan synthase, which is localized on the cytoplasmic side of the plasma membrane (Fredrikson and Larsson, 1989).

In barley suspension-cultured cells FG was found to be sequestered in the vacuole (Ohana et al., 1993). Lowering of cytoplasmic pH led to induction of callose synthesis. Addition of propionic acid to soybean cells elevated the percentage of FG found in the cytoplasmic compartment of soybean cells. Addition of an oligogalacturonide elicitor of DP 10-13 lowered cytoplasmic pH and caused a similar redistribution of FG. When the medium was supplemented with 100 mM $CaCl_2$, the intracellular Ca^{2+} level was elevated. This treatment led to a relative elevation of FG in the cytoplasm. All treatments that led to a relative increase in cytoplasmic FG also led to stimulation of callose synthesis in vivo (Ohana et al., 1993). Supplementation with β-furfuryl alcohol alone stimulated callose synthesis. This treatment elevated 10-fold the overall level of newly synthesized FG in soybean cells. The combination of low pH and supplementation with high levels of external Ca^{2+} led to a dramatic stimulation of callose deposition (Ohana et al., 1993). The results suggest that a relative redistribution of FG between cytoplasm and vacuole is a possible component of the signal transduction pathway for elicitation of callose synthesis in vivo. FG may be the required second signal, in addition to Ca^{2+}, and it should undergo redistribution to elicit callose synthesis. A perturbation leads to elevated cytoplasmic Ca^{2+} and lowered cytoplasmic pH (Quader and Fast, 1990; Matheiu et al., 1991; Ohana et al., 1992). This perturbation may be caused by the fungal penetration.

G. How Do Pathogens Overcome the Papillae and Callose Barrier?

1. Pathogen Delays Papillae Formation

In susceptible hosts, pathogens appear to delay papillae formation. In barley leaves inoculated with *Erysiphe graminis* f.sp. *hordei*, the processes leading to papilla formation are too slow for papillae to be effective as a resistance mechanism (Bayles et al., 1990). When tobacco plant is infected by *Phytophthora parasitica* var. *nicotianae*, cell appositions formed in response to infection, but were expressed too late to account for an effective resistance to fungal colonization (Benhamou and Cote, 1992). Some of the chemicals are known to delay papilla formation and they are used to demonstrate that a delay in papillae formation may enhance penetration efficiency of pathogens. Chlortetracycline treatment delayed papillae formation in barley, resulting in increased penetration efficiency of *E. graminis* f.sp. *hordei* (Gold et al., 1986). DDG treatment delayed the papillae formation induced by *E. graminis* f.sp. *hordei* in barley and resulted in an acceleration of the initiation of the penetration peg and haustorium formation (Bayles et al., 1990).

Papilla formation is induced by Ca^{2+} (Waldmann et al., 1988) and Ca^{2+} chelators induce susceptibility (Gold et al., 1986; Bayles and Aist, 1987). Less rapid triggering of increase in Ca^{2+} delays secretion of papilla precursors (Bayles et al., 1990). The fungal elicitor activates Ca^{2+} signal transduction system, and slower release of elicitor in susceptible interactions would have delayed the action of Ca^{2+}-induced papilla formation. The role of elicitor and Ca^{2} ion in inducing disease resistance has been discussed in Chapter 1.

Phosphate has also been found as a papilla-regulating chemical and the role of phosphorus in signal transduction is known (see Chapter 1). Thus the fungal elicitor may play an important role in papilla formation. Elicitors such as chitosan and β-1,3-glucans are known to induce callose deposition (Kohle et al., 1985; Masuta et al., 1991). Existence of signals or elicitors in induction of papillae formation has been indicated in many studies (Kunoh et al., 1978, 1985; Fric, 1984; Kerby and Somerville, 1989; Scott et al., 1990; Inoue et al., 1994a).

2. Pathogens Are Able to Penetrate Papillae

In compatible interactions, pathogens have been shown to penetrate papillae. When *Heterobasidium annosum* penetrated cells in cortical and endodermal regions of Norway spruce (*Picia abies*), papillae were observed. However, not all hyphae in this region were restricted by papillae, and at a later stage of infection fungal hyphae were able to penetrate papillae (Asiegbu et al., 1994). *Sphaerotheca fuliginea*, the powdery mildew fungus, was inoculated on muskmelon (*Cucumis melo*) plants and basic aniline blue calcofluor double-staining technique was used to locate the penetrations induced by the fungus in the surface of the inoculated leaves (Cohen et al., 1990). Fungal structures (conidia, germ tubes, and hyphae), fluoresced blue under UV epifluorescence microscopy, whereas perforations of the host surface resulting from penetration by the fungus fluoresced intensive yellow due to the accumulation of callose material in the collar region surrounding the penetration peg of the fungus. A total of about 14 perforations per colony resulting from a single conidium were observed at 72 h after inoculation (Cohen et al., 1990). Similar results have been obtained in muskmelon infected with *Pseudoperonospora cubensis* (Cohen et al., 1989) and in hop cultivars infected by *Sphaerotheca humuli* (Godwin et al., 1987).

3. Pathogen Suppresses Callose Synthesis

Pathogen appears to suppress callose synthesis during its penetration process. Wheat line CS 2D/2M and cultivar Thew responded differently to *Puccinia recondita* f.sp. *tritici* strains 104-2,3,6,7,8 and 104-1,2,3,6. Callose deposition was suppressed in compatible interactions while it was abundant in incompatible interactions (Table 20; Southerton and Deverall, 1990). When *Plasmopara lactucae-radicis* was inoculated on susceptible and resistant cultivars of lettuce (*Lactuca sativa*), intercellular runner hyphae and intracellular haustoria could be seen within roots of both the cultivars within 96 h after inoculation. In the resistant cultivar, approximately 94% of the haustoria exhibited intense fluorescence when stained with aniline blue, indicating the deposition of callose, and they appeared to be totally encased in callose (Stanghellini et al., 1993). In the susceptible cultivar, however, haustoria exhibited only a faint fluorescence that was localized around haustorial necks, indicating less amount of deposition of callose (Stanghellini et al., 1993). In *Sphaerotheca fuliginea*–muskmelon interactions callose deposition was observed only around the penetration pegs in the

Table 20 Deposition of callose in wheat mesophyll cells infected with *Puccinia recondita* f.sp. *tritici*

Wheat variety	*Puccinia recondita* f.sp. *tritici* strain	Disease reaction	Callose deposition at times (h) after inoculation			
			20	30	36	48
Lime CS 2D/2m	104-2,3,6,7,8	Compatible	–	–	–	–
	104-1,2,3,6	Incompatible	+	+	+	+
Cultivar Thew	104-1,2,3,6	Compatible	–	–	–	–
	104-2,3,6,7,8	Incompatible	–	–	+	+

–, No deposition; +, deposition as indicated by aniline blue staining.
Source: Southerton and Deverall, 1990.

compatible interactions, while heavy callose deposition was observed in epidermal and mesophyll cells in resistant interactions (Cohen et al., 1990). Less callose deposition appears to favor pathogenesis in susceptible interactions (Cohen and Eyal, 1988). Papillae in susceptible interactions are smaller than those formed in resistant interactions (Stolzenburg et al., 1984a; Gold et al., 1986) and slower deposition of callose deposition has been reported in various susceptible reactions (Zeyen and Bushnell, 1979; Skou, 1985; Gold et al., 1986; Stumm and Gesler, 1986; Skou et al., 1987; Binder et al., 1989).

Less callose deposition appears to be due to inhibition of callose synthesis. Treatment of a genetically resistant lettuce cultivar with 2-deoxy-D-glucose (DDG), an inhibitor of callose synthesis by the plant (Jaffe and Leopold, 1984; Jaffe et al., 1985), results in suppression of callose synthesis and susceptibility to *Plasmopara lactucae-radicis* (Stanghellini et al., 1993). DDG treatment inhibited papillae formation in a barley variety resistant to *Erysiphe graminis* f.sp. *hordei*, resulting in more penetration efficiency of the conidia and more haustoria formation (Bayles et al., 1990).

β-1,3-glucan synthase, the enzyme involved in synthesis of callose, is activated only in resistant interactions. *Colletotrichum lagenarium* inoculation systemically induced resistance in cucurbits against *C. lagenarium*. The induced resistance was associated with an increased ability to prevent penetration of the epidermal cell wall (Jenns and Kuc, 1980; Hammerschmidt and Kuc, 1982; Stumm and Gesler, 1986). Increased β-1,3-glucan synthase activity was detected in the induced resistant cucumber plants challenge-inoculated with *C. lagenarium* (Schmele and Kauss, 1990).

Fungal cell wall components (elicitors) induce callose deposition. Chitosan is a component of cell wall of many pathogenic fungi. When soybean cells were treated with chitosan, the cell walls became highly fluorescent with decolorized aniline blue and this staining property disappeared on preincubation with a mixture of endo-and exo-β-1,3-glucanase, indicating that callose is induced due to the chitosan treatment (Kohle et al., 1983). Detectable callose formation in soybean cells started about 20 min after treatment with chitosan and callose content steadily increased from 1 h onwards (Kohle et al., 1985).

Callose content increased in suspension cultured cells of bean, tobacco, and *Petroselium hortense* treated with chitosan (Kohle et al., 1985). Suspension-cultured rice cells were treated with chitosan. The walls of the treated cells were approximately twice as thick as those of the controls and aniline fluorochrome-positive patches (indicating the presence of callose) appeared around the cell periphery (Masuta et al., 1991).

It is possible that in the susceptible interactions these elicitors would not have been released or released slowly. The possible mechanisms of suppression of elicitor-induced defense reactions of the host has been discussed in detail in Chapter 1.

4. Pathogens May Degrade Callose by Producing β-1,3-Glucanase

Pathogens are known to produce β-1,3-glucanase to degrade callose (β-1,3-glucan). *Fusarium culmorum*, *Pseudocercosporella herpotrichoides*, and *Rhizoctonia cerealis* (Cooper et al., 1988), *Cochliobolus heterostrophus* (Anderson, 1978), *Septoria nodorum* (Lehtinen, 1993), *Claviceps purpurea* (Brockman et al., 1992), and *Cochliobolus carbonum* (Van Hoof et al., 1991) have been reported to produce β-1,3-glucanases in vitro. All plant β-1,3-glucanases with a few exceptions are endo-acting. β-1,3-Glucanase produced by the maize pathogen *Cochliobolus carbonum* is exo-acting and the molecular mass of the enzyme is 63,000 (Van Hoof et al., 1991). The enzyme could digest callose almost completely (Table 21; Van Hoof et al., 1991). Maize papillae contain callose (Hinch and Clarke, 1982) and *C. carbonum* efficiently penetrated the host cell wall, probably by producing β-1,3-glucanase (Van Hoof et al., 1991).

Honey dew of *Claviceps purpurea*-infected rye (*Secale cereale*) plants showed high levels of β-1,3-glucanase activity (Dickerson et al., 1978; Dickerson and Pollard, 1982). The pathogen produces an extracellular endo-β-1,3-glucanase. The enzyme was purified and the molecular weight was about 90 kDa. It had an optimum pH of 4.5 and there was a rapid decline of activity at pH values beyond 5.0. The enzyme was identified as a glycoprotein, containing glucose and/or mannose units. The enzyme was strictly specific for β-1,3-linkages characteristic for callose (Brockmann et al., 1992). During the infection process callose was deposited in several areas of the ryeflower: at the top of the style and in the outer cell walls of stigma trichome cells. Secretion of β-1,3-glucanases at the interface between mycelium and plant tissue may result in degradation of callose (Brockman et al., 1992).

Table 21 Digestion of [^{14}C] callose by β-1,3-glucanase from *Cochliobolus carbonum*

	Ethanol-insoluble radioactive products (ct $min^{-1} \times 10^3$) of protoplast treatment
Before enzyme incubation	72.3
After incubation in	
Buffer alone	57.1
Buffer + glucanase	4.0

Source: Van Hoof et al., 1991.

H. Callose Barrier May Not be Involved in Suppression of Fungal Development

In some host–pathogen interactions, callose may not have much role in suppression of fungal development in the host tissue. 2-Deoxy-D-glucose inhibited callose formation, but did not restrict cowpea rust fungal penetration in the nonhost bean plant (Perumalla and Heath, 1989). It suggests that callose deposition plays no part in restricting the cowpea rust fungus in the nonhost bean plant.

Benhamou (1992) provided convincing evidence that callose deposition in host tissues is of pathogen origin and not of host origin. β-1,3-Glucans are present both in fungal cell wall and host cell wall. Benhamou (1992) used tobacco β-1,3-glucanase tagged with colloidal gold particles as a probe to study the site of accumulation of callose in tobacco plants infected with *Phytophthora parasitica* var. *nicotianae*. Both in culture and in infected tissues the labeling was found more in cell wall of the fungus (Table 22). In the infected host tissues the labeling was detected in the host cell wall, as well as in papilla. Large amounts of β-1,3-glucans were found at sites of attempted penetration such as sieve tube membranes, plasmodesmata, and intercellular spaces (Table 23; Benhamou, 1992).

Thus β-1,3-glucan molecules were mainly associated with the outermost fungal cell layers in infected tissues and glucan molecules were detected in host cell walls adjacent to invading hyphe. It suggests that these molecules may be of pathogen origin. It is possible that they are released from the fungal cell wall through the action of host's β-1,3-glucanases. Based on cytochemical observations, Benhamou (1992) showed that amount of β-1,3-glucans in *P. parasitica* var. *nicotianae* grown in culture decreased several-fold when the fungus developed in

Table 22 Density of labeling obtained with the tobacco β-1,3-glucanase–gold complex over cells of *Phytophthora parasitica* var. *nicotianae*

	Gold particles (μm^{-2}) over cells of *Phytophthora parasitica* var. *nicotianae*	
Structure	In culture	In tobacco root tissues
Cell wall	72.3	28.6
Cytoplasm	5.1	6.1
Organelles	3.4	5.9

Source: Benhamou, 1992.

Table 23 Density of labeling obtained with the tobacco β-1,3-glucanase–gold complex over tobacco root tissues infected by *Phytophthora parasitica* var. *nicotianae*

	Gold particles μm^{-2}				
Tobacco plant	Cell wall	Papilla	Sieve pore	Plasmodesmata	Intercellular space
Healthy	3.5	5.1	24.6	2.9	5.1
Infected	31.4	26.8	41.5	39.1	26.8

Source: Benhamou, 1992.

tobacco root tissues, probably due to hydrolysis of fungal β-1,3-glucans during the course of infection by plant β-1,3-glucanases.

It is known that fungal β-1,3-glucans are elicitors of defense mechanisms and the released glucans may act as elicitors rather than as physical barrier. However, further studies are needed using radiolabeled fungal β-1,3-glucans to differentiate them from host glucans. β-1,3-Glucans may act both as elicitors and physical barrier.

XIII. HYDROXYPROLINE-RICH GLYCOPROTEIN

A. Host Cell Wall Responds to Fungal Invasion by Accumulating HRGP

While the fungal pathogen attempts to penetrate, HRGP accumulates in the host cell wall (Mazau and Esquerre-Tugaye, 1976, 1986, 1987). HRGP increased markedly in the cell walls of melon seedlings during infection by *Colletotrichum lagenarium* (Kimmins and Brown, 1973; Esquerre-Tugaye, 1973; Esquerre-Tugaye et al., 1979; Clarke et al., 1981; Toppan et al., 1982; Esquerre-Tugaye and Mazau, 1984).

When *Colletotrichum lindemuthianum* was inoculated, a few intracellular hyphae became encased in a large deposit of host wall-like material (papillae) and did not develop further. Papillae had a heterogeneous, multilayered structure. In mature papillae, the electron-lucent layers nearest the host plasma membrane were more intensely labeled with immunogold labeling of HRGP than the inner electron-opaque regions. The walls of cells containing papillae were labeled more strongly than cell walls in uninoculated hypocotyl tissue. Gold labeling of cell walls more distant from the site of infection was similar to that of cell walls in uninoculated healthy tissue. The results suggest that HRGPs are present in wall-like papillae produced by plant cells in response to infection by fungi (O'Connell et al., 1990).

HRGP accumulated in many other dicotyldonous plants due to infection with various pathogens, but in monocotyledonous plants, similar increase due to infection was not observed (Table 24; Mazau and Esquerre-Tugaye, 1986).

B. Host Cell Wall Responds to Fungal Invasion by Releasing Ethylene Leading to Accumulation of HRGPs

The accumulation of HRGPs in host cell walls due to fungal invasion appears to depend upon the signal generated by the fungal elicitor and ethylene as second messenger (Esquerre-Tugaye et al., 1979; Toppan et al., 1982). Elicitor from *Colletotrichum lindemuthianum* induced accumulation of HRGP in bean (Bolwell et al., 1985; Bradley et al., 1992). HRGP accumulated in soybeans inoculated with an elicitor from *Phytophthora megasperma* f.sp. *glycinea* (Roby et al., 1985). Elicitor treatment stimulates transcription of genes encoding HRGPs (Corbin et al., 1987; Sauer et al., 1990). Ethylene treatment induces HRGP accumulation in cell walls (Esquerre-Tugaye et al., 1979). When the petioles of melon plants were treated with *C. lagenarium* elicitor, the elicitation of ethylene occurred within the first few hours after the addition of elicitors, while elicitation of HRGP was observed only after 18 h (Roby et al., 1985). An inhibitor of ethylene synthesis, aminoethoxy-vinylglycine (AVG), inhibited synthesis of HRGP (Roby et al.,

Table 24 Changes in levels of HRGP in cell walls of dicoyledonous and monocotyledonous plants infected by various pathogens

Host	Pathogen	HRGP expressed as hydroxyproline levels (μg/100 μg cell wall): Healthy	Infected
Dicots			
Cucumber	*Colletotrichum lagenarium*	0.08	0.78
Bean	*Colletotrichum lindemuthianum*	0.17	0.46
Lucerne	*Colletotrichum trifolii*	0.25	0.49
Melon	*Fusarium oxysporum* f.sp. *melonis*	0.20	0.54
Grapevine	*Plasmopara viticola*	0.16	0.25
Tobacco	*Peronospora tabacina*	0.07	0.18
Monocots			
Rice	*Pyricularia oryzae*	0.08	0.075
Barley	*Erysiphe graminis*	0.065	0.065
Triticum aestivum	*Cercosporella herpotrichoides*	0.063	0.057

Source: Mazau and Esquerre-Tugaye, 1986.

1985). Other specific inhibitors of the ethylene pathway such as 1-canalin inhibited both ethylene and hydroxyproline deposition in the cell wall of melon plants infected with *C. lagenarium* (Toppan et al., 1982). 1-Aminocyclopropane-1-carboxylic acid, the precursor of ethylene, triggered the synthesis of HRGP to the same extent as the elicitor of *C. lagenarium* (Roby et al., 1985). These results suggest that ethylene may be involved in the accumulation of HRGP in plant cell walls due to fungal infection.

Wounding also induced HRGP accumulation (Ludevid et al., 1990; Tagu et al., 1992). Ethylene is known to increase due to wounding (Boller and Kende, 1980; Ecker and Davies, 1987). Ethylene may be the key factor in induction of HRGP accumulation.

Ethylene induces HRGP mRNA accumulation. When mesocotyls of young maize plantlets were exposed to ethylene, HRGP mRNA accumulation was observed (Tagu et al., 1992). A 3 h ethylene treatment produced a 17-fold increase in HRGP mRNA (Tagu et al., 1992). The HRGP gene was detectable at 15 min with a maximum expression between 1 and 2 h (Tagu et al., 1992).

Three species of HRGP mRNAs of sizes 1.5, 1.8, and 4.0 kb have been detected in carrot roots (Ecker and Davis, 1987). Exposure to 10 ppm ethylene caused a dramatic 50–100 fold increase in the level of the 4.0 kb HRGP mRNA. In contrast, the 1.5 kb HRGP transcript decreased in abundance after elicitor treatment. There was only a relatively small increase (two- to fourfold) in the total amount of HRGP homologous sequences present in ethylene-treated versus untreated roots. This was due to the relatively high level of 1.5 kb mRNA compared with the 1.8 kb and 4.0 kb HRGP mRNA in untreated roots. Thus there was a dramatic change in the type of mRNA produced during ethylene treatment. There was a shift in accumulation of HRGP mRNAs from the 1.5 kb mRNA to 4.0 kb mRNA. An additional ethylene-induced mRNA (2.0 kb) was also detected (Ecker and Davis, 1987). The results suggest that the type of HRGP accumulation may also determine disease resistance. Four types of hydroxyproline arabinosides containing 4, 3, 2, and 1 arabinose residues, respectively, were detected in many dicots (Mazau and Esquerre-Tugaye, 1986). Part of the cell wall hydroxyproline was not glycosylated. Only glycosylated hydroxyproline accumulated during infection in all the infected hosts that showed increased levels of hydroxyproline (Mazau and Esquerre-Tugaye, 1986).

C. Host Cell Wall Responds to Fungal Invasion by Strengthening its HRGPs by Glycosylation

Most of the hydroxyproline residues of HRGP have O-glycosidically attached oligosaccharides consisting of L-arabinofuranosyl residues (Lamport, 1967; Ariyama and Kato, 1977). The serine residues have a single galactosyl residue O-glycosidically bound to them (Lamport et al., 1973). The glycosylated hydroxylprolyl and seryl residues are often found in the sequence-Ser-Hyp4, which occurs with a high frequency in wall peptides (Lamport, 1973). The extent of glycosylation of hydroxyproline appears to offer resistance to pathogens.

The glycosylated hydroxyprolyl residues of the wall are linked to arabinose chains and the arabinose residues protect the wall HRGP against proteolysis. The proteolytic enzymes are ineffective in extracting HRGPs without prior deglycosylation of the hydroxyproline residues (Esquerre-Tugaye and Lamport, 1979). Pathogens are known to produce proteolytic enzymes (Pladys and Esquerre-Tugaye, 1994). The increased glycosylation may offer resistance to proteolytic enzymes of pathogens and prevent fungal development in host cell wall.

Pathogens induce glycosylation of HRGP. Infection of muskmelon seedlings by *Colletotrichum lagenarium* causes a 10-fold increase in the amount of cell wall HRGP and the extent of glycosylation of hydroxylation was higher in the fungus-infected muskmelon cell walls (Esquerre-Tugaye and Lamport, 1979). The glycosylation is initiated by a protein: arabinosyl-transferase in the Golgi apparatus of the host cell (Bolwell et al., 1985). Elicitor from *Colletotrichum lindemuthianum* induced rapid induction of the protein/arabinosyl-transferase and increased levels of an arabinosylated hydroxyproline-rich protein in bean cells (Bolwell et al., 1985). It suggests that the pathogen induces glycosylation of HRGP by activating protein: arabinosyl-transferase and the fungal elicitor may be involved in this process.

D. Host Cell Wall Responds to Fungal Invasion by Making HRGPs Insolubilized

HRGPs become insolubilized in plant cell walls when attacked by pathogens (Chen and Varner, 1985). The insolubilization may be caused by the induced cross-linking of HRGPs by intermolecular isodityrosine residues and by a diphenyl ether linkage (Fry, 1982; Chen and Varner, 1985). The increased HRGP cross-linking may lead to a more impenetrable cell wall barrier, thus impending pathogen infection. Cell walls that undergo ultrarapid HRGP cross-linking, are tougher than cell walls of untreated cells (Showalter, 1993).

When bean cells were treated with elicitors isolated from mycelial walls of *Colletotrichum lindemuthianum* a cell wall 35 kDa protein, designated p35, disappeared (Bradley et al., 1992). When soybean cells were treated with elicitors isolated from mycelial walls of *Phytophthora megasperma* f.sp. *glycinea*, a 33 kDa protein disappeared (Bradley et al., 1992). Similar disappearance was observed when treatment with any type of elicitor-active molecules such as the fungal elicitors, salicylic acid, and glutathione was initiated in both bean and soybean cells (Bradley et al., 1992). To assess the causes for disappearance of these proteins in elicitor-treated cells, intensive studies were made using soybean cells. A monoclonal antibody that detects glycoprotein was used as a probe. With this antibody, disappearance of another glycoprotein of 90-110 kDa referred to as p100 was also observed in the walls of elicitor-treated cells. Apart from the disappearance of p33 and p100, no other changes were observed in the pattern of sodium dodecyl sulfate (SDS)-extractable proteins from the walls of elicited cells compared with controls (Bradley et al., 1992).

While the fungal elicitor caused a loss of SDS-extractable immunoreactive material, immunoreactivity was still observed in situ in the walls of elicited cells. The results suggest that p33 and p100 remain in the cell wall, but elicitor induces a change in form that makes these species no longer extractable by SDS. The same results were obtained from parallel experiments with p35 in bean cells (Bradley et al., 1992).

The p33 and p100 were characterized. The N-terminal peptide sequence of soybean p33 comprised a highly repetitive proline rich sequence that contained Val-Tyr-Lys-Pro-Pro repeats and identified as a proline-rich protein (PRP) (Bradley et al., 1992). Although peptide sequence data of p100 have not yet been obtained, it was identified as a HRGP. Both the proteins contained tyrosine abundantly (Bradley et al., 1992).

Treatment of soybean cells with H_2O_2 promoted the loss of SDS-extractable p33 PRP and p100 HRGP from the cell wall. Simultaneous addition of catalase or ascorbate inhibited the effects of fungal elicitor or H_2O_2 (Bradley et al., 1992). It suggests that the fungal elicitor may stimulate H_2O_2-dependent cross-linking of preexisting, soluble forms of p33 PRP and p100 HRGP and that this insolubilization is reflected in their disappearance from the SDS-extractable fractions of the cell wall. The cross-linking of p33/p35 PRP and p100 HRGP in response to fungal elicitor was initiated within 5 min and the response was completed within 20–30 min (Bradley et al., 1992). The rapid burst of H_2O_2 production at the surface of soybean cells within 1–2 min of elicitor treatment (Apostol et al., 1989) would have provided substrate for peroxidase-mediated oxidative cross-linking reactions. Cross-linking could be observed in vitro following addition of H_2O_2 to isolated cell walls, but fungal elicitor was ineffective (Bradley et al., 1992). It suggests that cell wall contains components sufficient for cross-linking, provided that there is a supply of H_2O_2, putatively generated at the plasma membrane. Thus the cross-linking reaction may not be a transcription-dependent process.

E. Host Cell Wall Responds to Fungal Invasion by Enriching HRGP by Lignin Deposition

HRGPs may act as matrices for lignification (Whitmore et al., 1978). When fungal pathogen invades host tissues, HRGPs accumulate and provide a template for the subsequent deposition of lignin (Whitmore, 1978; Ride, 1983). Lignified papillae containing HRGP offer resistance to fungal penetration (Aist, 1976b;

Benhamou et al., 1990a). The association between lignin deposition and HRGP accumulation has been reported in several fungus-infected plants (Mazau and Esquerre-Tugaye, 1986; Corbin et al., 1987). Several cultivars of cucumber were inoculated with *Cladosporium cucumerinum* and simultaneous accumulation of HRGP and lignin was observed in all cultivars that showed resistance to *C. cucumerinum* (Hammerschmidt et al., 1984). These observations suggest that HRGP may have major functions in cell wall architecture and it may play an important role in inhibition of penetration of host cell wall by pathogens.

F. HRGP May Immobilize Plant Pathogens

Extensin, a type of HRGP, immobilizes certain plant pathogens. This agglutination response may result from positively charged extensin molecules interacting ionically with negatively charged surfaces of certain plant pathogens (Leach et al., 1982; Mellon and Helgeson, 1982).

G. How Does Pathogen Overcome HRGP Barrier?

1. Less Accumulation of HRGP in Compatible Interactions

Fungal elicitor induces HRGP accumulation. In some of the susceptible interactions, action of the elicitors may be suppressed as discussed in Chapter 1. When cucumber plants were inoculated with *Cladosporium cucumerinum* HRGP accumulated only in resistant varieties. In all susceptible interactions the accumulation of HRGP in the cell wall was negligible (Table 25; Hammerschmidt et al., 1984).

Table 25 Accumulation of hydroxyproline in cucumber cell walls inoculated with *Cladosporium cucumerinum*

	Hydroxyproline (μg mg^{-1} cell wall)			
Cultivar	Control	Inoculated	% increase	Host response
Marketmore 70	1.7	3.0	74	R
Salty	2.2	3.9	79	R
SMR 58	2.0	3.6	83	R
Victory	2.2	3.2	49	R
Marketer	1.8	1.7	–3	S
Gemini	2.0	2.3	12	S
Shamrock	2.9	3.2	9	S
Straight 8	2.6	2.9	13	S

R, Resistant; S, susceptible.
Source: Hammerschmidt et al., 1984.

The amount of HRGP may not be sufficient for prevention of fungal penetration in the susceptible interactions.

2. Pathogen Overcomes HRGP Barrier by Delaying Accumulation of HRGPs in Host Cell Wall

In many host–pathogen interactions HRGP accumulates almost equally in both compatible and incompatible interactions. In bean, cucumber, and lucern (*Medicago sativa*) plants infected by *Colletotrichum lindemuthianum*, *Fusarium oxysporum* f.sp. *melonis*, and *Colletotrichum trifolii*, respectively, HRGP accumulates almost equally in both susceptible and resistant varieties (Mazau and Esquerre-Tugaye, 1986).

Elicitor from *Colletotrichum lindemuthianum* cell wall caused a marked and prolonged accumulation of three HRGP mRNAs from relatively low basal levels in unelicited cells. The increase in the mRNA level was first observed after a lag of about 4 h. There was a rapid increase in mRNAs between 6 and 12 h after elicitor treatment (Showalter et al., 1985). Changes in the level of HRGP mRNA were studied during race: cultivar-specific interactions between hypocotyls of bean and physiological races of *C. lindemuthianum* (Showalter et al., 1985). In the compatible interaction, the infected cells remained alive and the fungus underwent substantial biotrophic growth. Extensive host cell death occurred, subsequently and spreading anthracnose lesions developed. Marked accumulation of HRGP mRNA occurred later, at about 150 h after inoculation in the necrotrophic phase. In the incompatible interaction, however, when the fungus penetrated the cuticle and the developing hyphae came in contact with the underlying host epidermal cells, a hypersensitive response was observed in about 50 h. HRGP mRNA accumulated 10- and 20-fold above that in equivalent uninfected control hypocotyls during that time (Showalter et al., 1985).

When susceptible cucumber plants were inoculated with *Cladosporium cucumerinum*, HRGP accumulated at 48–72 h after inoculation while similar accumulation was observed at 12–18 h after inoculation in resistant cucumber plants (Hammerschmidt et al., 1984). In the resistant variety the HRGP accumulation was observed at the time of initial penetration by the hyphae while in the susceptible variety the HRGP enhancement was observed when water soaking and necrosis symptoms were observed (Hammerschmidt et al., 1984).

When tomato roots were inoculated with *Fusarium oxysporum* f.sp. *radicis-lycopersici*, cell walls of tomato roots were markedly enriched in HRGPs (Benhamou et al., 1991). HRGPs were found to be present in low amounts in healthy plant cell walls and were not detected in cultured *F. oxysporum* f.sp. *radicis-lycopersici*. The subcellular localization of HRGPs was detected using an antiserum by immunogold cytochemistry. The gold particles were more associated with host cell wall areas adjacent to fungal cells. By 120 h after inoculation, the fungus had ramified through much of the endodermis and had colonized abundantly the vascular stele. Walls of infected inner cortical cells were labeled intensely. Gold particles accumulated greatly over the walls of these highly colonized cells. Wall disorganization was more pronounced in these tissues. Split walls were observed frequently. Gold particles were numerous over both portions of these altered walls (Benhamou et al., 1991). These observations suggest that accumulation of HRGPs occur in cell walls invaded by the pathogen and it was seen mostly during its necrotrophic phase. Thus HRGPs accumulated mostly in highly colonized tissues at the advanced stages of infection. The accumulation of HRGPs in susceptible tomato plants may be a late biochemical event (Benhamou et al., 1991).

Templeton et al. (1990) showed that *Colletotrichum lindemuthianum* infection in an incompatible bean cultivar was characterized by an early (within 4 days) and massive accumulation of HRGP transcript in the epidermal, cortical, and perivascular tissues immediately adjacent to the inoculation site. In the compatible interaction, by 7 days colonization had progressed past the epidermis and throughout the cortical cell layers, but at no time was message accumulation seen in epidermal or cortical cells. Only a late accumulation (7 days after inoculation) of transcript was seen in the perivascular tissue.

These studies suggest that the fungal pathogens may be able to penetrate host cell wall, which does not have high amount of HRGP and delayed accumulation of HRGP only during necrotrophic phase will not be able to suppress the fungal development.

XIV. HOST CELL WALL RESPONDS TO PATHOGEN ATTACK BY ACTIVATING PHENYLPROPANOID METABOLISM

A. Phenylpropanoid Metabolism

In the plant cell walls phenylpropanoid metabolism is activated when the pathogen infects host tissues. In the phenylpropanoid pathway, suberin, lignin, and wall bound phenolics are synthesized (Hahlbrock and Scheel, 1989) as depicted below:

Phenylalanine → Suberin
↓ PAL
Cinnamic acid → Lignin
↓ C_4H
4-Coumaric acid
↓ 4 CL → Other wall bound phenolics
4-Coumaroyl-CoA

PAL—Phenylalanine ammonia-lyase
C_4H—Cinnamate-4-hydroxylase (monooxygenase)
4CL—4-Coumarate : CoA ligase

B. Cell-Wall-Bound Phenolics

1. Host Cell Wall Responds to Fungal Invasion by Activating Enzymes Involved in Synthesis of Wall-Bound Phenolics

PAL, the first enzyme in the phenylpropanoid pathway, was induced in tomato cell cultures, both at the enzyme (Bernards and Ellis, 1991) and mRNA level by *Verticillium albo-atrum* inoculation (Bernards and Ellis, 1991). Maximum enzyme activity occurred by 36 to 48 h postinoculation, concomitant with a marked accumulation of PAL mRNA. Only a low level of PAL and its mRNA was detectable from uninoculated cell cultures (Bernard and Ellis, 1991). Enzymes of general phenylpropanoid metabolism were rapidly induced in suspension-cultured parsley cells treated with elicitor from *Phytophthora megasperma* var. *sojae* (Hahlbrock et al., 1981).

The tomato PAL was purified and the molecular mass of the enzyme was 280 kDa. The PAL was not inhibited by cinnamic, caffeic, quinic, or chlorogenic acids. Two known PAL inhibitors, 2-amino-2-indanephosphate (AIP) and α-aminooxy β-phenylpropanoic acid (AOPP) inhibited the tomato PAL. The accumulation of wall-bound phenolics in inoculated tomato cell cultures was inhibited by the PAL inhibitor AIP (Bernards and Ellis, 1991). When potato leaves were inoculated with *Phytophthora infestans*, rapid increase in the rate of PAL transcription was observed within 1–2 h (Cuypers et al., 1988). Increases in activities of PAL and tyrosine carboxylase in the *P. infestans*-infected tissues led to the synthesis of wall-bound phenolics (Hahlbrock and Scheel, 1989). Within 3–4 h after inoculation with *Erysiphe graminis*, there was an increase in synthesis of cinnamic acid and increase in the level of PAL activity occurred by as early as 2 h after inoculation in barley leaves (Shiraishi et al., 1989). These events corresponded with the time of attempted penetration by the *E. graminis* primary germ tube (Kunoh et al., 1979, 1985a,b; Shiraishi et al., 1989).

Shiraishi et al. (1995) observed increases in PAL activity at two different times in barley cultivars inoculated with *E. graminis*. The first increase in enzyme activity began at 3 h after inoculation and this was followed by a second increase in activity between 12 and 15 h after inoculation. The conidum produces a primary germ tube that attempts penetration beginning about 2 h after inoculation and an appressorium that attempts to penetrate beginning 9–10 h after inoculation. Thus, it appears that the elevation of PAL enzyme levels by hosts is a direct response to attempted penetration by the fungus.

The first detectable increase in PAL mRNA occurred at 30 min after inoculation with *E. graminis* in barley leaves (Shiraishi et al., 1995). This increase in the level of mRNA was considerably earlier than the time of attempted penetration by the fungus. The increase in mRNA may be due to release of elicitor from the fungal cell wall (Shiraishi et al., 1995).

Clark et al. (1994) showed that PAL transcript accumulated in barley cultivars inoculated with *E. graminis*. The initial PAL transcript accumulation occurred between 4 and 6 h and declined to near constitutive levels by 8–10 h. The second peak was from 10 h until 12 h and then declined until 15–18 h. The first increase occurred due to response to the fungal germ tube contact, while the second increase was due to appressorial contact (Clark et al., 1994). The increase in PAL transcripts followed increase in PAL activity (Clark et al., 1994).

The monomeric precursors for the deposition of wall-bound phenolics are hydroxylated and methoxylated cinnamyl alcohols synthesized in two steps from the corresponding coenzyme A esters by cinnomyl-CoA reductase and cinnamyl alcohol dehydrogenase (CAD) (Grisebach, 1981). CAD mRNA appeared within 90 min after elicitor treatment in bean cell cultures (Grand et al., 1987). Only one class of CAD gene has been detected in the bean genome (Walter et al., 1988).

Challenge of parsley cell suspension cultures with fungal elicitors induced the formation of ferulic cell wall esters (Hahlbrock and Scheel, 1989; Pakusch et al., 1989). The incorporation of ferulic acid and related acids into cell walls is a widespread phenomenon and invariably requires activation of S-adenosyl-L-methionine/trans-caffeoyl-CoA 3-0-methyltransferase (CCOAMT), which is responsible for the formation of feruloyl-CoA (Pakusch et al., 1989). CCOAMT has been characterized from elicitor-treated parsley cells (Pakusch et al., 1990). This enzyme possesses a narrow specificity for caffeoyl-CoA (Pakusch and Matern, 1991). CCOAMT is known to be induced de novo in cultured parsley cells upon the addition of elicitors, although noninoculated control cultures already contain fairly high background activity (approximately one-fifth of maximum) and maximal activity is reached within 10–15 h (Pakusch et al., 1989).

CCOAMT is involved in cell wall reinforcement. In cultured parsley cells treated with an elicitor from *Phytophthora megasperma* f.sp. *glycinea*, the enzyme activity is rapidly induced by a transient increase in the rate of de novo transcription (Pakusch and Matern, 1991).

The rapid synthesis of phenolics and their polymerization in the cell wall is generally regulated by p-coumaric hydroxylase, which is extremely pH dependent, and not by de novo enzyme synthesis. Membrane damage leads to decrease in cytoplasmic pH, which activates the hydroxylase (Matern and Kneusel, 1988).

2. Phenolic Deposition in Host Cell Wall in Response to Fungal Invasion

Phenolic compounds are fluorescent and appearance of fluorescent materials in diseased plant tissue is considered to be due to the presence of phenolic materials that accumulate in the tissue as the host attempts to limit the development of the pathogen (Cohen et al., 1990). Autofluorescence of host cells is an early response of host to fungal infection (Kidger and Carver, 1981; Kunoh et al., 1982, 1983, 1985). The autofluorescence occurs early in the interaction of the host and pathogen and is observed in association with attempted penetration of the host cell by the primary germ tube and the appressorium. Localized autofluorescence is a common epidermal cell response to *Erysiphe graminis* appressoria on barley (Aist and Israel, 1986; Kunoh et al., 1982, 1985; Koga et al., 1983), and wheat (Tosa et al., 1990; Carver et al., 1991, 1992a,b). This autofluorescence is localized in limited regions of the host cell wall surrounding fungal germ tube contact sites and in papillae deposited by host epidermal cells as a response to primary and appressorial germtube contact (Aist and Israel, 1986; Mayama and Shishiyama, 1978). PAL inhibitors suppressed accumulation of localized autofluorogens in oats (Carver et al., 1992a), suggesting the accumulation of phenolics.

Tomato cell cultures inoculated with *Verticillium albo-atrum* accumulated up to fivefold higher levels of wall-bound phenolics than were found in uninoculated control cultures (Table 26; Bernards and Ellis, 1991). An array of phenolic compounds were detected in wall preparations of inoculated cell cultures, whereas walls from uninoculated cells yielded only traces of esterified phenolics. A major component in this hydrolyzable phenolic population was sinapic acid (Bernards and Ellis, 1991).

When potato leaves were inoculated with *Phytophthora infestans*, two classes of phenolics accumulated in host cell wall. One class consisted of hydroxybenzoic and hydroxycinnamic acids, such as 4-hydroxybenzoic, 4-coumaric, and ferulic acids. The other class consisted of hydroxycinnamic amides, primarily 4-coumaroyltyramine and feruloyltryramine (Friend, 1981; Clarke 1982). When tomato cell cultures were inoculated with *V. albo-atrum* phenolics accumulated in the cell walls. The analysis of this cell wall-bound material revealed that two populations of phenolic material exist. The first comprises esterified compounds and the second comprises nonbase-labile polymeric material (Bernards and Ellis, 1991). Accumulation of cell wall-bound phenolics in *Fusarium oxysporum* f.sp. *dianthi*-infected carnation stems has been reported (Niemann and Bayer, 1988; Niemann et al., 1991a,b).

Table 26 Accumulation of wall-bound phenolics in *Verticillium albo-atrum*-inoculated tomato cell suspension cultures

	Phenolic accumulation (Å 280/g fresh wt.)	
Time after inoculation (h)	Uninoculated	Inoculated
24	0.49	0.50
36	0.30	1.73
48	0.36	2.25

Source: Bernards and Ellis, 1991.

C. How Does the Pathogen Overcome the Cell-Wall-Bound Phenolics?

1. Pathogen Suppresses Accumulation of Phenolics in Host Cell Wall

When *Sphaerotheca fuliginea*, the powdery mildew fungus, was inoculated on resistant muskmelon plants, the infected epidermal cells emitted autofluorescence from their lumen. Autofluorescence indicates the presence of phenolics. The phenolics were present in cell walls of the resistant variety at 24–96 h after inoculation, but penetrated cells of the susceptible variety did not show any autofluorescence (Cohen and Eyal, 1988; Cohen et al., 1990).

Autofluorescence in cell walls has been shown to be associated only with the manifestations of resistance to *Pseudoperonospora cubensis* in muskmelon (Cohen et al., 1989), *Sphaerotheca humuli* in hop cultivars (Godwin et al., 1987), and *Sphaerotheca pannosa* in rose leaves (Conti et al., 1986). In the susceptible varieties all these pathogens suppressed the accumulation of phenolics in epidermal cell walls (Conti et al., 1986; Godwin et al., 1987; Cohen et al., 1989).

Suppression of PAL has been shown to induce susceptibility in resistant varieties. Oat leaves were infused with α-aminooxy acetic acid (AOA) or α-aminooxy-β-phenylpropionic acid (AOPP), both of which are inhibitory to PAL. The PAL inhibitors suppressed localized autofluorescent host cell responses to contact of germ tube of *E. graminis* f.sp. *avenae* and increased susceptibility to penetration from appressoria (Carver et al., 1992b).

AOPP suppressed the phenolic accumulation in epidermal cell walls in barley and wheat also and this treatment made the hosts susceptibile to their pathogens *E. graminis* f.sp. *hordei* and *E. graminis* f.sp. *tritici*, respectively (Carver et al., 1992c). These observations suggest that in susceptible interactions, phenolic accumulation in plant cell wall is suppressed.

The amount of phenolic accumulation in plant cell wall in susceptible interactions is less. All barley cultivars showed autofluorogenic compound accumulation at 24 h after inoculation with *E. graminis*. In the resistant interactions, higher frequency of intense autofluorescence at appressorium contact sites was observed compared to the susceptible interactions (Clark et al., 1994). When barley leaves were inoculated with a nonpathogen *Erysiphe pisi*, increase in the level of mRNA was observed and the magnitude of PAL mRNA was greatest in response to *E. pisi* while in barley leaves inoculated with the pathogen *E. graminis*, less increase in mRNA was observed (Shiraishi et al., 1995).

2. Pathogen Delays Synthesis of Cell-Wall-Bound Phenolics

Delayed synthesis of cell-wall-bound phenolics at the fungal penetration site has been reported. When potato leaves were inoculated with *Phytophthora infestans*, rapid accumulation of PAL mRNA in a sharply confined area around the fungal penetration site was observed in incompatible interaction while in the compatible interaction PAL mRNA accumulated more slowly and in a more diffuse and further spreading area (Cuypers et al., 1988). In the barley line susceptible to powdery mildew fungus, papillae were formed much later than those in the resistant line (Aist et al., 1988). Chlortetracycline strongly inhibited both resistance to penetration and the incorporation of the light-absorbing component into papillae in the resistant isoline. The penetration efficiency was 5% where papillae contained the light-absorbing component, 9.9% where they did not, and 47% where the amount of light absorbance was intermediate. Autofluorescence and lacmoid staining suggested that the light-absorbing component is rich in phenylpropanoids. The light-absorbing component in papillae may be a molecular component of the resistance mechanism and this phenylpropanoid component is suppressed during successful pathogenesis (Aist et al., 1988).

Yellow autofluorescence (indicating phenolic synthesis) after excitation at 365 nm was emitted by stomatal cells and the cell walls around the necrotic stomata as early as 2 days after inoculation with *Plasmopara viticola* in the resistant *Vitis rotundifolia* (Dai et al., 1995). In contrast, a few stomatal cells with yellow autofluorescence were detected only 8 days after inoculation in lesions on the susceptible *V. vinifera* (Dai et al., 1995). The results suggest that delayed accumulation of phenolics in host cell walls favors pathogenesis.

XV. LIGNIN

A. Host Cell Wall Responds to Fungal Invasion by Increasing Lignification Process

Lignification increased in potato inoculated with *Phytophthora infestans* (Henderson and Friend, 1979), and in *Raphanus japonica* inoculated with *Peronospora parasitica* (Asada and Matsumoto, 1967, 1972; Asada and Kugoh, 1971) and *Alternaria japonica* (Asada and Matsumoto, 1967). The increased lignification is reflected in cucumber, castor bean, and *R. japonica* by enhanced peroxidase (Asada and Matsumoto, 1972; Hammerschmidt et al., 1982; Bruce and Galston, 1989), in muskmelon by increased hydroxycinnamate: CoA ligase (Grand and Rossignol, 1982), in bean by cinnamylalcohol dehydrogenase (CAD) (Grand et al., 1986), in melon by increase in p-coumarate CoA ligase (4CL) (Grand and Rossignol, 1982), and in potato by phenylalanine ammonia-lyase (PAL) (Henderson and Friend, 1979). PAL activity increased by more than 10-fold in pine (*Pinus* sp.) cell cultures after treatment with a fungal elicitor at 24 h after treatment, coinciding with the initiation of cell wall lignification (Campbell and Ellis, 1992). To prove that increased PAL activity is involved, it was shown that treatment of pine cultures with the PAL inhibitor 2-aminoindane-2-phosphonate (AIP) suppressed lignification (Campbell and Ellis, 1992).

Treatment of soybean cotyledons with *Phytophthora megasperma* f.sp. *glycinea* wall glucan elicitor resulted in significant phenolic polymer deposition in the uppermost cell layers as early as 4 h. Phenolic polymer deposition continued rapidly and leveled off by 24 h (Graham and Graham, 1991). The phenolic polymers have been identified as lignin- and suberin-like polymers (Graham and Graham, 1991).

B. Pathogen Suppresses Lignin Deposition

Lignification is suppressed in compatible interactions and only in resistant interactions is lignification predominant (Hijwegen, 1963; Speakman and Lewis, 1978; Ride, 1978, 1980; Ride and Pearce, 1979; Vance et al., 1980; Bird and Ride, 1981; Beardmore et al., 1983; Coffey and Cassidis, 1984; Reisener et al., 1986; Dean and Kuc, 1987; Cadena-Gomez and Nicholson, 1987; Moersbacher et al., 1983, 1989).

Wheat line CS 2D/M and cultivar Thew responded differently to *Puccinia recondita* f.sp. *tritici* strains 104-2,3,6,7,8 and 104-1,2,3,6. Lignin deposition was observed only in resistant interactions (Southerton and Deverall, 1990). Lignin content increased in wheat leaves inoculated with avirulent strain of *Puccinia recondita* f.sp. *tritici*, but no such increase was observed in leaves inoculated with virulent strain (Southerton and Deverall, 1990).

In the susceptible interaction between *Sphaerotheca fuliginea* and muskmelon, the fungal spore developed one or two germ tubes that penetrated into one or two epidermal cells. The penetration zones were surrounded in callose-like material but no autofluorescence nor lignin-like materials were observed in the penetrated epidermal cells. In the resistant varieties, however, the fungus developed a single germtube that induced autofluorescence, callose accumulation, and lignification in the penetrated epidermal cells. In susceptible tissues suppression of phenolics and lignification was observed (Cohen et al., 1990).

The metabolism of radiolabeled aromatic compounds in healthy and *Puccinia graminis tritici*-infected wheat plants was studied. More radiolabel was found in the alcohol insoluble nonhydrolyzable material of infected resistant leaves than in infected susceptible leaves, suggesting that lignification is suppressed in the susceptible reaction of wheat to the stem rust fungus (Rohringer et al., 1967; Fuchs et al., et al., 1967). Histochemical tests indicated the presence of lignin or lignin-like material in wheat cells responding hypersensitively to penetration by haustoria of *Puccinia graminis* f.sp. *tritici* (Tiburzy, 1982).

In wheat plants resistant to *Puccinia graminis tritici*, inoculation with the pathogen resulted in accumulation of lignin in epidermal cells. Histochemical tests for lignin using phloroglucinol and chlorine sulphate were positive in the infected wheat cells, indicating the presence of coniferyldehyde group and syringylpropane units (Tiburzy and Reisener, 1990). Observation of autoradio-

grams of rust-infected leaves by transmission light and fluorescence microscopy showed that the radiolabel of the infiltrated [^{14}C]cinnamic acid was incorporated into autofluorescing necrotic epidermal cells. The incorporated radiolabeled material exhibited the solubility characteristics of lignin. No accumulation of label was found in rust-infected cells of the susceptible wheat cultivar (Tiburzy and Reisener, 1990).

Lignification is common in healthy plants also. The increased lignification observed in resistant varieties appears to be a new type of lignin. A number of histochemical tests such as phloroglucinol–HCl test, toluidinme blue-0-test, chlorine–sulfite test, and modified chlorine–sulfite test (Ride, 1975) are being employed to detect the presence of lignin. When these tests were used to detect the deposition of lignin in infected leaves, additional lignin deposition was observed (Southerton and Deverall, 1989, 1990). The additional lignin observed in walls of wheat leaves due to leaf rust incompatible interaction may be different from that in uninfected leaves because of its green rather than blue–green response to toluidine blue. The failure to react to phloroglucinol may indicate an absence of substituted cinnamaldehyde groups. The additional lignin formed in the incompatible interaction may not be rich with syringic groups because of its failure to stain with chlorine-sulfite (Southerton and Deverall, 1989, 1990). This additional lignin was not detected in susceptible interaction.

C. Pathogen Suppresses Lignin Biosynthetic Enzymes

Phenylalanine ammonia-lyase (PAL), 4-coumarate/CoA ligase (4CL), cinnamyl-alcohol dehydrogenasee (CAD), and peroxidase are the important lignin biosynthetic enzymes. All these enzymes exhibited increased activities at very early stages of infection (8–16 h after inoculation) by *Puccinia graminis* f.sp. *tritici* in both susceptible and resistant wheat lines. Subsequently both PAL and 4CL activities decreased to the level of uninoculated controls in the susceptible interaction. In the resistant interaction, however, their activities continued to increase from 32 h onwards until 6–7 days after inoculation (Moerschbacher et al., 1988). During this period lignification of penetrated resistant host cells were observed (Reisener et al., 1986). In contrast, in the compatible interaction PAL and 4CL activities stayed at the level of uninfected controls or frequently declined below this level (Morschbacher et al., 1988). CAD and peroxidase activities followed the changes in activities of PAL and 4CL (Moerschbacher et al., 1988). Thus the pathogen appears to suppress the increases in all lignin biosynthetic enzymes.

Inoculation of primary roots of *Eucalyptus marginata* (susceptible) and *E. calophylla* (resistant) with *Phytophthora cinnamoni* zoospores caused lesion to develop within 12–16 h (Cahill and McComb, 1992). Lesions in roots of *E. caryophylla* were restricted within 3–4 days and ceased extending, whereas those

in roots of *E. marginata* continued to extend. In the resistant variety PAL activity increased due to infection within 24 h of inoculation. The amount of lignin increased by as much as 53% above control levels due to infection. In the susceptible variety there was no increase in PAL activity when inoculated with the pathogen. The suppression of increase in PAL activity was reflected in unchanged concentrations of lignin in the infected susceptible variety (Cahill and McComb, 1992). To prove the importance of suppression of PAL in suppression of lignin synthesis, the roots of the resistant variety was treated with the PAL inhibitor AOA. This treatment suppressed the synthesis of lignin and phenolics and altered the resistance response of the variety. Typical susceptible symptoms were observed in the resistant variety (Cahill and McComb, 1992). Application of AOA likewise efficiently inhibited the accumulation of lignin in epidermal cells of *Puccinia graminis tritici*-infected wheat variety resistant to the pathogen. This treatment changed the resistant reaction to a susceptible one (Moerschbacher et al., 1990; Tiburzy and Reisener, 1990). In the resistant variety, 82.7% of infection sites were associated with an autofluorescing epidermal cell. In AOA-treated infected tissues, however, only 4% of all infection sites were associated with an autofluorescing epidermal cell (Tiburzy and Reisener, 1990). Application of inhibitors of cinnamyl alcohol dehydrogenase also inhibited lignification in wheat leaves and susceptibility to the rust was induced (Tiburzy and Reisener, 1990).

OH-PAS (2-hydroxyphenyl amino sulphinyl acetic acid, 1,1,-dimethyl ester) is an inhibitor of CAD. The OH-PAS treatment led to increased quantitative susceptibility of the barley lines to *Erysiphe graminis* f.sp. *hordei* as the percentage of appressoria penetrating host cells to form haustoria was approximately doubled. This was associated with reduction in both the frequency and intensity of autofluorescent responses associated with fungal germ tubes (Carver et al., 1994a). Since OH-PAS specifically inhibits CAD (Grand et al., 1985), the results suggest that cinnamyl alcohols or later formed products of the lignin pathway contribute to epidermal cell penetration resistance of barley to *E. graminis* f.sp. *hordei* and suppression of these products induces susceptibility.

When oat leaves were treated with the CAD inhibitor OH-PAS, the leaves become more susceptible to *Erysiphe graminis* f.sp. *avenae* and the treatment reduced localized autofluorescent cell responses (Carver and Zyen, 1993; Carver et al., 1994a,b). These results suggest that suppression of CAD may contribute to suppression of lignin synthesis by the pathogen.

Peroxidase is associated with increased lignin synthesis in many host–pathogen interactions (Hammerschmidt et al., 1982; Smith and Hammerschmidt 1988; Moerschbacher et al., 1988; Kuc, 1990). When the pathogens attack host tissues, peroxidase activity increases in both susceptible and resistant interactions. Two peroxidase isoenzymes with pI 5.2 and pI 8.5 increased three- to fourfold in both susceptible and resistant barley varieties inoculated with *Erysiphe graminis*

f.sp. *hordei* (Kerby and Somerville, 1989). Cadena-Gomez and Nicholson (1987) observed two new isoperoxidase isozymes 24 h after inoculation with *Helminthosporium maydis* and *Colletotrichum graminicola* in both resistant and susceptible maize varieties. In stem rust-infected wheat leaves peroxidase activity increased and two new infection-related peroxidase isozymes were detected 4 days after infection (Holden and Rohringer, 1985). Enhanced peroxidase activity was detected in cucumber inoculated with *Colletotrichum lagenarium* (Hammerschmidt et al., 1982). Papilla formation after inoculation of barley lines with *E. graminis* f.sp. *hordei* has been observed even in susceptible interactions (Kita et al., 1981; Thordal-Christensen and Smedegaard-Petersen, 1988a,b). Peroxidase activity increased during papilla formation and lignin was found to accumulate in the papillae (Moersbacher et al., 1988). In wheat a pathogen-induced peroxidase cDNA clone has been obtained (Schweizer et al., 1989; Rebmann et al., 1991).

Although peroxidases increase in both susceptible and resistant interactions, certain isozymes specifically increase in resistant reactions. Such increase was not observed or the increase was much less in susceptible reactions. Stem infection of tobacco cultivar with *Peronospora tabacina* induced systemic resistance to *P. tabacina* (Ye et al., 1990). Twelve peroxidase isozymes were detected in tobacco cell walls, but two anionic isozymes, P37 and P35 alone increased markedly in induced resistant plant (Table 27; Ye et al., 1990). These two anionic peroxidases appear to be involved in disease resistance and their levels are not elevated in susceptible interactions. These two peroxidases appear to be involved in increased lignification (Ye et al., 1990). Lignin precursors such as coniferyl alcohol have been shown to be preferred substrates of highly anionic peroxidase (Mader et al., 1980).

Several anionic and cationic peroxidases were detected in soybean cotyledons due to wounding (Graham and Graham, 1991). Induction of cationic peroxidases was not correlated with phenolic polymer deposition. Three groups of anionic peroxidases appeared. The group I showed four isozymes and the activity of two of them was enhanced by the *Phytophthora megasperma* f.sp. *glycinea* elicitor, but none of this group of isozymes clearly corresponded to phenolic polymer deposition. The second group of anionic peroxidases was strongly induced by the elicitor as early as 4 h. The induction of the group 2 enzymes correlated very well with the accumulation of phenolic polymers such as lignin and suberin in elicitor-treated tissues (Graham and Graham, 1991).

During infection with *Puccinia graminis* f.sp. *tritici*, the wheat pathogen, two anionic new peroxidase isozymes (IN1 and IN2) appeared in both compatible and incompatible interactions. There were four other preformed peroxidases. The activity of two preformed isoperoxidases of low molecular mass decreased due to infection while the other two preformed isoperoxidases of high molecular mass increased due to infection in both compatible and incompatible interactions. Two

Table 27 Peroxidase isozymes in cell walls of leaves of tobacco plants induced by *Peronospora tabacina*

		Intensity of accumulation of isozymes	
Isozymes	pI	Susceptible plant	Induced resistant plant
P 93	9.3	+	+
P 88	8.8	+	++
P 83	8.3	+	+
P 81	8.1	+	+
P 68	6.8	+	++
P 62	6.2	+	+
P 57	5.7	+	+
P 54	5.4	+	+
P 49	4.9	+	+
P 46	4.6	+	+
P 37	3.7	+	+++
P 35	3.5	+	+++

–, Absence of peroxidase; +, presence of peroxidase; ++, one- to twofold increase in peroxidase; +++, greater than twofold increase in peroxidase.
Source: Ye et al., 1990.

or 3 days after inoculation a new isozyme (IN3) appeared, which was not seen in the compatible interaction. The increase in peroxidase activity continued up to 7 days after inoculation in the incompatible interaction. In contrast, these changes proceeded until 3 days after inoculation and a general decrease in all isozymes was observed until 8 days after inoculation (Flott et al., 1989). After infection activity of basic peroxidases also increased in both compatible and incompatible interactions (Flott et al., 1989). Three groups of basic peroxidases were seen. The group I basic peroxidase could be seen clearly 24 h after inoculation. Generally, this isozyme was less pronounced in the compatible interaction. The group II and group III increased due to infection in both compatible and incompatible interactions. A decrease in peroxidase activity during the compatible interaction 3 days after inoculation could not be observed with the basic isoperoxidases (Flott et al., 1989). However, it is known that basic peroxidases are not involved in the biosynthesis of lignin (Castillo and Greppin, 1986; Schloss et al., 1987).

When barley plants were inoculated with an avirulent strain of *Erysiphe graminis* f.sp. *hordei*, activation of several pathogen-induced genes was observed (Thordal Christensen et al., 1992). Several cDNA clones representing pathogen-induced genes were isolated. Two of the cDNAs were found to be representing peroxidases. One of the cDNA clones (pBH6-301) was characterized. The predicted amino acid sequence revealed a 21 aminoacid signal peptide and a 294 amino acid mature protein (31 kDa). It showed 89% amino acid identity to peroxidase from wheat. The transcript was elevated in response to the avirulent barley powdery mildew fungus as early as 4 h after inoculation. Two maxima in transcript amount seem to occur (i.e., at 5 h and 15–24 h after inoculation) that seem to be correlated with penetration attempts by the fungus. Another transcript (pcD 1311), a putative peroxidase, also showed elevated response to avirulent pathogen. Levels of the pcD 1311 and pBH6-301 transcripts are also elevated in response to inoculation with virulent races. However, the amount induced was much lower than in the incompatible interaction (Thordal Christensen et al., 1992). Thus specific peroxidases involved in lignification appear to be suppressed by pathogens in the compatible interactions.

D. How Does Pathogen Suppress Lignification in Host Cell Wall?

Host cell wall contains material that induces lignification in it (Robertsen, 1986). This host cell wall component is released by the incompatible pathogen, but the compatible pathogen prevents release of the component from the host cell wall. Extracts of cell walls from both resistant and susceptible cucumber varieties elicited lignification (Robertsen, 1986). The hypocotyl segments that had been incubated with cucumber cell wall extracts revealed lignin-like compounds in walls of the cells, similar to the response of resistant hypocotyls to infection by *Cladosporium cucumerinum* (Robertsen, 1986). The elicitors from cucumber cell walls were further purified. The most active fraction from cucumber cell walls contained mainly uronic acid. The most active fraction contained oligomers with 11 galacturonosyl units (Robertsen, 1986).

The elicitors have been detected in cell walls of both resistant and susceptible cucumber hypocotyls. The cell wall pectins may have an important alarm function in plants: if damaged, they can induce defense mechanisms in the surrounding cells (Robertsen, 1986). The dodecamer and tridecamer of galacturonic acid are the most potent elicitors and oligomers with fewer than eight galacturonosyl residues do not have elicitor activity (Nothanagel et al., 1983; Jin and West, 1984; Robertsen, 1986). In the susceptible variety quicker degradation of pectic substances by the pectolytic enzymes produced by the pathogen would have resulted in accumulation of pectic polymer of fewer than eight galacturonosyl residues and hence no lignification would have occurred (Robertsen, 1986).

Fungal cell wall also contains a molecule that elicits lignification. This molecule is released by host enzyme and activation of the host enzyme is under the control of fungal cell wall component. The incubation of insoluble mycelial walls of *Chaetomium globosum* with carrot homogenates released heat-stable soluble factors that stimulated lignification in carrot cells (Kurosaki et al., 1986a).

Table 28 Lignin content in peach bark wounds 3 days after treatment with fungal cell wall extracts and disease development in the treated tissues inoculated with *Leucostoma persooni*

Treatment	Thioglycolic acid–lignin (mg/g)	Canker length (mm)
Cell wall extract from *L. persoonii*	75.7	12.9
Cell wall extract from *L. cincta*	72.6	11.2
Control	41.6	23.4

Source: Biggs and Peterson, 1990.

The activity of these substances was reduced after digestion by chitinase, and they were released by mild hydrolysis with the enzyme. The chitinase activity was induced in cultured carrot cells incubated with the fungal walls, and the soluble fragments liberated from the walls stimulated the biosynthesis of phenolic acids, which are precursors of the lignin synthesized in the cells (Kurosaki et al., 1986a).

A marked increase of chitinase activity was observed in cultured carrot cells incubated with insoluble mycelial walls of *Chaetomium globosum* (Kurosaki et al., 1986b). The enzyme activity increased almost immediately after the mycelial wall treatment. After a lag of 2–4 h, PAL activity increased and it was followed by an increase in phenolic acids (p-hydroxybenzoic acid, caffeic acid, and ferulic acid), which are cinnamate derivatives (Kurosaki et al., 1986b). The results showed that in cultured carrot cells incubated with insoluble mycelial walls, chitinase activity increased first, followed by an increase in PAL. Phenolic acids accumulated in parallel with the induction of PAL activity (Kurosaki et al., 1986b).

Peach canker disease is caused by *Leucostoma persooni* and *L. leucostoma*. The cell wall extracts of both the pathogens induced lignification in peach and induced resistance against the pathogen (Table 28; Biggs and Peterson, 1990).

Thus fungal cell wall component can induce lignification, but the release of this fungal molecule would have been suppressed in susceptible interactions. Further work is necessary to explain this phenomenon.

XVI. SUBERIN

A. Host Cell Wall Responds to Fungal Invasion by Suberization

Suberin is deposited on cell walls of various plant tissues when the pathogen invades host tissues (Kolattukudy, 1981). Suberin is highly resistant to enzymatic degradation by pathogens and hence it is an effective barrier to penetration by fungal pathogens (Kolattukudy, 1981). When potato tubers were inoculated with *Verticillium dahliae*, a rapid increase in the number of layers of suberized cells over that found in controls was observed (Vaughn and Lulai, 1991). Several workers have shown that suberization is responsible for reinforcement of the cell walls, limiting ingress of pathogens into host (Mullick, 1977, 1978; Kolattukudy, 1980; Pearce and Rutherford, 1981; Nnodu et al., 1982; Tippett and Hill, 1984; DeLeeuw, 1985; Espelie and Kolattukudy, 1985; Biggs, 1985a,b; Espelie et al., 1986; Biggs and Miles, 1988; Kolattukudy, et al., 1989).

B. Biosynthesis of Suberin in Pathogen-Inoculated Host Cell Wall

Synthesis of very long chain acids and alcohols, w-hydroxylation, and conversion of the w-hydroxyacids to the corresponding dicarboxylic acids are the major steps involved in the formation of the aliphatic components of suberin (Bolton and Harwood, 1976). w-Hydroxyacid dehydrogenase and an another dehydrogenase are involved in the suberization (Agrawal and Kolattukudy, 1977, 1978a,b). Oxidation of the w-hydroxy acid to dicarboxylic acid, which is unique to suberin, is catalyzed by a specific dehydrogenase induced for suberization (Kolattukudy, 1985). Anionic peroxidase present in cell walls catalyzes polymerization of phenolic monomers and covalent bond formation between cell walls and the phenolic polymer (Whitmore, 1978a; Borchert, 1978; Cottle and Kolattukudy, 1982a,b; Kolattukudy and Soliday, 1985). The production of suberin coatings is dependent on PAL activity (Street et al., 1986). Fungal infection activates all these enzymes (Mohan and Kolattukudy, 1990; Robb et al., 1991; Lee et al., 1992).

PAL induced synthesis of many phenolic acids that are required for synthesis of suberin. Infection with *Fusarium oxysporum* f.sp. *dianthi* increased concentration of vanillic and ferulic acids and induced accumulation of a number of anthranilic acid, amides of p-coumaric, benzoic, salicylic, methoxysalicyclic p-anisic, and β-resorcyclic acids in carnation (Niemann et al., 1991). Cross-linking of such phenolics forms a polymeric matrix. Such a matrix is made hydrophobic by attachment of aliphatic polyester domains and by deposition of highly nonpolar waxes into this layer. The formation of this layer is called suberization (Borchert, 1978; Pearce and Rutherford, 1981; Kolattukudy et al., 1987). The formation of the aromatic matrix is the first step in suberization. This matrix is catalyzed by a highly anionic peroxidase in potato (Cottle and Kolattukudy, 1982a,b; Espelie and Kolattukudy, 1985). This peroxidase has been purified (Espelie and Kolattukudy, 1985) and it was found to be localized in the cell walls of the suberizing cells only (Espelie et al., 1986). This peroxidase cDNA has been cloned and sequenced (Roberts et al., 1988). The peroxidase gene was absent in normal tissue (Kolattukudy et al., 1989).

The aromatic domain of suberin is similar to lignin and the polymerization of the aromatic components of suberin involves an isoperoxidase in a manner similar to that involved in lignin biosynthesis. In potato slices both the time course of appearance and spatial distribution of an anodic isoperoxidase were highly correlated with suberization (Borchert, 1978). The anionic peroxidase has been purified from wound healing potato slices. The molecular weight of the enzyme is 47,000. It is a glycoprotein containing about 17% carbohydrate, approximately one-quarter of which are glucosamine residues (Espelie and Kolattukudy, 1985).

The highly anionic peroxidases of tomato appear to be related to pathogenesis. The basal levels of transcripts of these genes are very low in the root, stem, leaf, and fruit tissues of healthy tomato plants, but expression of these genes is induced to high levels in plants inoculated with fungal pathogens or fungal elicitors (Mohan and Kolattukudy, 1990; Robb et al., 1991).

Two genes encoding homologous isoenzymes of tomato anionic peroxidase have been isolated. These genes, designated *tap* 1 and *tap* 2, display a high degree of sequence identity with each other as well as with potato anionic peroxidase (Roberts and Kolattukudy, 1989). The plasmid pBin 19:α (T1/T2) was engineered to produce high-level, constitutive expression of a composite antisense transcript complementary to the 5′ regions of both *tap* 1 and *tap* 2 mRNAs so that synthesis of peroxidases encoded by both genes can be suppressed. Nineteen transgenic tomato plants expressed the desired 1.4 kb anti-(*tap* 1 and *tap* 2) transcript. These transgenic plants that expressed the anti-(*tap* 1 and *tap* 2) transcript failed to accumulate the anionic peroxidase transcript. However, the level of suberin aliphatics deposited in the wound-healing periderm of the transgenic tomato fruits was unaffected by the absence of the inducible highly anionic peroxidase. The wound periderms of fruits from control and transgenic tomato plants were observed to display similar autofluorescence patterns and intensities in the walls of cells at the wound surface, suggesting that total phenolic deposition was not significantly affected by the loss of anionic peroxidase activity in transgenic plants (Sherf et al., 1993). These results suggest that the expression of highly anionic peroxidase gene is not essential for suberization. However, in view of the multiplicity of peroxidases known to exist in plants, it may be possible that in the antisense transgenic plants another peroxidase has substituted for the abolished activity of the highly anionic peroxidase isozyme. One other nonhomologous peroxidase was induced to high levels in the wound periderm of both non-transformed and anti-(*tap* 1 and *tap* 2)-expressing tomato fruits. This peroxidase did not cross-react with antibodies or DNA probes specific for the highly anionic peroxidase (Sherf et al., 1993). The role of this anionic peroxidase in suberization is yet to be demonstrated.

C. Pathogen Delays Suberin Accumulation

During pathogenesis suberization appears to be delayed in compatible interactions. In tomato, one of the earliest defense responses against *Verticillium albo-atrum* is the coating of xylem vessels and pit membranes with suberin, produced by the xylem parenchyma cells (Street et al., 1986; Robb et al., 1987, 1989, 1991). In the resistant variety, suberization was very rapid, beginning at 8–10 h after inoculation with the pathogen. In sharp contrast, almost none were visible in the susceptible interactions at that time. After 24 h, suberization was visible in the susceptible plants; but it was much less than that in resistant plants (Lee et al., 1992).

When *Verticillium dahliae* infects a resistant potato variety, a complete layer of suberized cells was formed within 2 days after inoculation and on fourth day three or more suberized layers were formed. When tubers of susceptible potato variety were inoculated with the pathogen, however, no suberization was detected in the first few days following inoculation treatment (Vaughn and Luali, 1991). This delay would have helped the pathogen to penetrate host tissues.

The ability of the plants to accumulate suberin also appears to determine susceptibility or resistance. Peach cultivars that accumulate suberin faster after wounding, become resistant to *Leucostoma cincta*, the pathogen causing canker (Biggs, 1989).

D. Pathogen Suppresses Suberin Synthesizing Enzymes

The delay in suberin accumulation appears to be due to suppression of suberin-synthesizing enzymes by the pathogen. PAL, the first enzyme involved in suberin synthesis, increased in both susceptible and resistant tomato plants due to infection with *V. albo-atrum* (Lee et al., 1992). There was an initial increase of about 30% in the PAL mRNA, with a subsequent gradual drop to approximately normal levels in the resistant tomato plant. In contrast, the level in the susceptible plant did not increase and proceeded to drop until after 15 h when it was only 30% of normal (Fig. 1; Lee et al., 1992). The results suggest that the fungal component may suppress or depress PAL mRNA levels in susceptible plants. It also indicates that the fungal pathogen may gain access to its host not simply by passively avoiding defense mechanisms but also by actively inhibiting them (Lee et al., 1992).

Anionic peroxidase is involved in suberization and in the susceptible tomato line. Anionic peroxidase was induced 1 day later than in a near isogenic line resistant to *V. albo-atrum* (Street et al., 1986). Cell suspension cultures from the resistant and susceptible tomato lines were incubated with an elicitor preparation from *V. albo-atrum*. Within 15 min the anionic peroxidase mRNA could be detected in the resistant line. The resistant line responded to the elicitor by

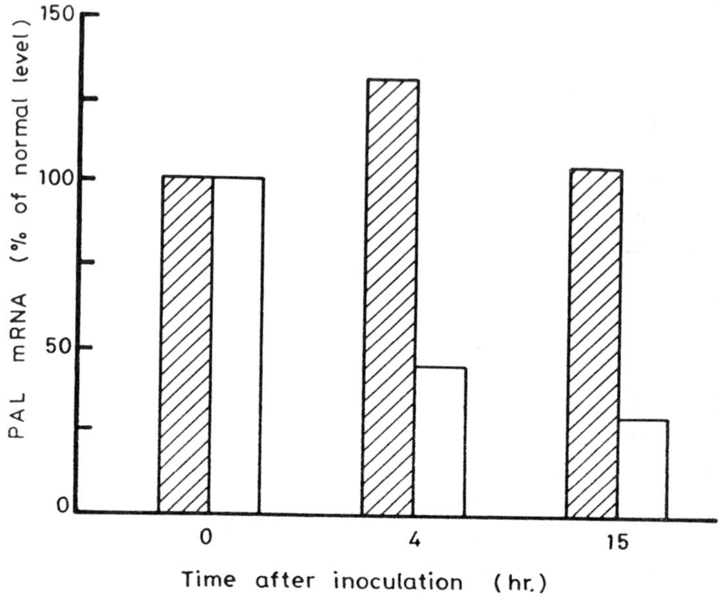

Figure 1 Changes in phenylalanine ammonia-lyase (PAL) mRNA levels in resistant (▨) and susceptible (□) tomato plants infected with *Verticillium albo-atrum*. PAL mRNA levels are expressed relative to the zero time (100%). (From: Lee et al., 1992).

producing high levels of anionic peroxidase mRNA, which reached a maximum by 16 h and then the level decreased. The susceptible line showed only a very little induction (Mohan and Kolattukudy, 1990). The results suggest that the induction of suberization may be due to the fungal elicitor and the elicitor may not be active in the susceptible cultivar or a suppressor may suppress the elicitor activity in the susceptible interaction as discussed in Chapter 1.

E. Pathogens May Penetrate the Suberized Walls of Host Cells

Some fungal pathogens have been reported to penetrate suberized cell walls (Peek et al., 1972; Parameswaran et al., 1976; Peterson et al., 1980). Culture filtrates of *Fusarium solani* f.sp. *pisi* released soluble radioactivity from insoluble-labeled suberin enriched preparations obtained from potato tuber periderm, indicating that the extracellular fluid contained enzymes that degraded the aliphatic and aromatic domains of suberin (Fernando et al., 1984; Kolattukudy, 1985). The extracellular enzyme responsible for releasing the aliphatic components of suberin was isolated in a highly purified state. This esterase, induced as a result of growth of the organism on suberin, was found to be identical to cutinase generated by the same fungus on cutin. They both had identical amino acid composition, molecular weight, kinetic properties, and sensitivity to inhibitors (Fernando et al., 1984).

XVII. DEPOSITION OF MINERAL ELEMENTS IN HOST CELL WALL IN RESPONSE TO FUNGAL INVASION

A. Increased Silicon Deposition May Activate Phenylpropanoid Metabolism and Callose Synthesis in Host Cell Wall

Many mineral elements accumulate in host cell wall when inoculated with pathogens. Silicon deposition at fungal penetration sites has been reported on epidermal cells of barley attacked by *Erysiphe graminis* f.sp. *hordei* (Kunoh et al., 1978; Carver et al., 1987); and on bean mesophyll cells infected by the bean rust fungus (Health, 1979). The infection process of *Sphaerotheca fuliginea* in cucumber was studied using scanning electron microscopy and x-ray micro analysis. The areas of host cell was adjacent to pathogen hyphae had a high concentration of silicon. It appeared that the silicon was being deposited around the points of pathogen penetration of the host tissue (Samuels et al., 1991a,b). Silicon has been reported to accumulate in the papillae of various plants due to infection (Kunoh and Ishizaki, 1975; Sargent and Gay, 1977; Heath, 1981a,b; Koga et al., 1988; Blaich and Grundhofer, 1990; Menzies et al., 1991a).

Silicon has been implicated in the resistance of cucumbers, wheat, barley, beans, and rose to various fungal pathogens (Akai, 1953; Kunoh and Ishizaki, 1975, 1976; Perera and Gay, 1976; Sargent and Gay, 1977; Stumpf and Heath, 1985; Aleshin et al., 1986; Carver et al., 1987; Koga et al., 1988; Leush and Buchenauer, 1989; Jiang et al., 1989). Accumulation of silicon in leaves of cucumber decreases the severity of powdery mildew caused by *Sphaerotheca fuliginea* (Menzies et al., 1991a).

Electron microscopic studies combined with energy-dispersive x-ray microanalysis (EDX) suggested that silicon may act as a mechanical barrier to fungal ingress by accumulation at sites of fungal penetration in *Vitis* and cowpea leaf tissue (Heath, 1981a; Blaich and Grundhofer, 1990). Amendment of nutrient solutions with soluble silicon protected cucumber plants against *Pythium ultimum* (Belanger et al., 1991; Cherif and Belanger, 1992; Cherif et al., 1992a).

Although amendment of nutrient solutions with silicon (Si) protected cucumber plants against *P. ultimum*, the fungal hyphae were able to penetrate silicon-containing cucumber roots and hypocotyls readily (Cherif et al., 1992a). Significant differences in the rate and extent of colonization of +Si and –Si plants by the pathogen were not observed until 48 h after inoculation (Cherif et al.,

1992a). The results suggest that enhanced protection does not result from accumulation of insoluble Si in cell walls.

Cucumber plants grown in hydroponic systems showed increased resistance to *Spherotheca fuliginea* when the nutrient medium was supplemented with soluble silicon in the form of silicate (Menzies et al., 1991a,b; Samuels et al., 1994). The cell walls from –Si plants did not show accumulation of silicon. But cell walls from +Si leaf cells infected with the fungus had silicon-containing sites. Silicon was found in the papillae, in host cell walls, around the haustorial neck, and in deposits between the host cell wall and plasma membrane. However, the haustoria observed in the +Si plants appeared to be healthy despite the presence of the silicon in the adjacent cell wall and papilla (Samuels et al., 1994). This suggests that silicon per se is not directly toxic to the fungus. However, the number of haustoria per colony was significantly reduced due to silicon treatment (Menzies et al., 1991a). At 120 h after inoculation with *S. fuliginea*, colonies had means of 42, 25, and 5 haustoria per colony on low –Si, medium –Si, and high –Si leaf pieces, respectively (Menzies et al., 1991a). The observations suggest that some other defense mechanisms may exist in silicon-treated plants. It is now widely reported that silicon may not prevent pathogen attack directly acting as a mechanical barrier but by indirectly inducing host defense mechanisms (Kunoh and Ishizaki, 1975; Aleshin et al., 1986; Menzies et al., 1991a,b; Samuels et al., 1991b, 1994; Cherif et al., 1992b; Carver et al., 1992a,b).

Silicon may induce phenolic synthesis and induce resistance (Cherif et al., 1992b). Treatment of cucumber plants with silicon markedly stimulated the accumulation of an electron dense, phenolic-like material in host tissues infected with *P. ultimum* (Cherif et al., 1992b). In cucumber leaves autofluorescence (indicating presence of phenolics) was greater in the area around the penetration pegs on high –Si leaf pieces than on low –Si or medium –Si leaf pieces (Menzies et al., 1991a). It suggests greater deposition of phenolics with high –Si treatments of plants. Lignin deposition has also been reported in cucumber plants treated with silicon (Menzies et al., 1991a). Another reaction specific to silicon treatment in the *P. ultimum*-infected cucumber tissues is deposition of thick layers of opaque, osmiophilic material along secondary walls and pit cavities of xylem vessels. This barrier may be due to the suberin nature of the coating material (Cherif et al., 1992b).

Callose deposition was also induced by silicon treatment. The percentage of callose-containing penetration pegs in cucumber tissues inoculated with *S. fuliginea* was 47, 88, and 100% on low, medium, and high –Si leaf pieces (Menzies et al., 1991a). These results suggest that silicon may act as an element capable of inducing the plant to defend itself through the activation of various defense strategies including the impregnation of phenolics, lignin, suberin, and callose in the plant cell walls.

B. Pathogen Suppresses Silicon Deposition

Pathogen may suppress silicon deposition in host cell walls and induce susceptibility. When a nonpathogen of bean (*Uromyces vignae*) was inoculated, penetration of plant cell walls was prevented by the rapid modifications (usually within 12 h of inoculation) of the walls of the bean cells (Heath, 1974). These responses included increased wall refractivity caused by the deposition of silica (Heath, 1979, 1981a,b). When the pathogen (*Uromyces appendiculatus*) was inoculated such response was not seen or occurred too late or at levels too low to stop fungal infection (Fernandez and Heath, 1989).

Elicitors from *U. appendiculatus* induced cell wall autofluorescence (indicating phenolic accumulation) and an increase in refractivity in bean cell walls. In the light microscope, the change in wall refractivity resembled that consistently associated with silica deposition at infection sites of the nonpathogen, *Uromyces vignae*. An electron microscopic study suggested that silica may not be involved in wall refractivity in susceptible host–pathogen interaction (Ryerson and Heath, 1992).

Silicon deposition may not occur in susceptible interaction. *P. ultimum* successfully penetrated cell walls of roots of cucumber plants grown in hydroponic nutrient solutions with or without soluble silicon. Fungal hyphae and penetration sites did not appear to be silicified and were nearly free of silicon accumulation (Cherif et al., 1992b). Similar results were obtained when cucumber leaves were inoculated with *P. ultimum* (Cherif et al., 1992b).

C. Deposition of Calcium in Papillae

Calcium accumulates in papillae during fungal infection. When roots of pine seedlings of a resistant variety were inoculated with *Cylindrocarpon destructans*, papillae were formed. The papillae contained high amount of callose (Table 29; Bonello et al., 1991).

The insoluble calcium observed in the papillae may be a component of pectic materials. Calcium may harden plant primary walls by cross-linking of pectic polymers and confer resistance to pathogen attack (Akai and Fukutomi, 1980). Calcium is a mediator in polysaccharide synthesis (Aist, 1983; Kohle et al., 1985). Calcium is also known to induce callose synthesis in plant cell walls (Kohle et al., 1985; Kauss et al., 1989; Ohana et al., 1992, 1993).

XVIII. CONCLUSION

Plant cell wall is made up of complex defense barriers and molecular events occurring during the fungal attempt to penetrate the barrier are also complex. Cuticle is an insoluble barrier that can be pierced by the fungal hyphae by

Table 29 Concentration of calcium detected in papillae and normal cell walls of pine seedlings inoculated with *Cylindrocarpon destructans*

Root tissue	Calcium content in mM
Inoculated plants	
Papillae	215
Cell wall	41
Uninoculated plants	
Cell wall	31

Source: Bonello et al., 1991.

producing cutinase, which requires a substrate, cutin monomer released from the cuticle by cuticle-sensing cutinase present in fungal cell wall. The released cutin monomer induces bulk cutinases which erode cuticle layer. Melanin accumulates in appressorium, which gives mechanical pressure to penetrate cuticle layer.

Pectin layer is the second layer below cuticle. Pathogens produce pectic enzymes that are inhibited by polygalacturonase inhibitor proteins (PGIP) present in the host cell wall. Partial degradation of pectin results in pectic oligomers that induce synthesis of lignin in cell wall and lignification increases resistance to pathogen penetration. However, the amount of PGIP appears to be less in the cell walls to prevent action of polygalacturonases by the virulent pathogens. Degradation of pectin exposes hemicelluloses that are degraded by the hemicellulases produced by pathogens. The next barrier appears to be cellulose, which is degraded by cellulases produced by pathogens.

When these cell wall barriers are breached by the pathogens, host reinforces its cell wall by forming cell wall appositions and depositing callose, HRGP, and lignin. Chitinase and β-1,3-glucanase activities also increase in the intercellular spaces of the host cell. These host enzymes release elicitors (chitin, β-1,3-glucan) from the fungal cell wall. These elicitors alter cell permeability, resulting in release of Ca^{2+} and β-furfuryl-β-glucoside, which activate 1,3β-glucan synthase resulting in callose accumulation. However, successful pathogens may activate less host chitinase and β-1,3-glucanase, resulting in reduced level of elicitor release. It may lead to delayed deposition or suppression of callose synthesis. Pathogens produce β-1,3-glucanase, which may degrade callose.

Fungal penetration attempts result in release of ethylene, which activates HRGP mRNAs. The fungal elicitor induces H_2O_2 synthesis, which in turn activates peroxidase resulting in cross-linking of HRGPs leading to insolubilization. Some HRGPs immobilize plant pathogens. However, successful pathogens may induce catalase or ascorbic acid synthesis, which may inactivate H_2O_2, which in turn may suppress insolubilization of HRGP. Lignin deposition in HRGP matrix may also be delayed.

During fungal penetration several lignin- and phenol-synthesizing enzymes such as PAL, CAD, CCOAMT, and peroxidase are activated. However, the elicitor from virulent pathogens delayed action of PAL and suppressed the action of other enzymes, also resulting in less lignification and wall bound phenolics. The pathogen delays suberization and produces an enzyme that degrades suberin. Silicon may accumulate in host cell wall and silicon induces synthesis of phenolics and lignin. Successful pathogens suppress silicon accumulation. Calcium also accumulates particularly in the papillae. Calcium induces insolubilization of pectin and callose deposition. However, virulent pathogens suppress calcium accumulation.

The molecular events described are only general events but several exceptions also occur. The virulent pathogens overcome all these cell wall barriers and penetrate the cell successfully. Although these are the conclusions from the work done so far, our knowledge of the mechanism of cell wall penetration by the fungal pathogens still seems to be rudimentary. The complex nature of the cell wall requires a complex number of enzymes for degradation. The important lacuna in our knowledge is the amount of enzyme produced at the site of penetration. If small amount of enzymes alone are produced, the degraded polysaccharides may act as elicitors and induce resistance. If high amounts of these enzymes are produced, tissue disintegration may take place. Cross-linking of several polysaccharides gives strength to the plant cell wall. The enzymes breaking these crosslinks may be vital for degrading cell wall. The mechanism of breaking the cross-links is not yet known.

Most of the hydrolytic enzymes are produced by both pathogen and host. Pectic enzymes and β-1,3-glucanase are commonly produced by host and pathogens. The interaction between the enzymes produced by the host and pathogen is not known. β-1,3-Glucanase of the pathogen may degrade callose and induce susceptibility while β-1,3-glucanase of the host is considered a pathogenesis-related protein (PR-2 protein) and is inhibitory to pathogens. Furthermore, β-1,3-glucanase of the host is known to release elicitors from the fungal cell wall and induce resistance.

Host actively strengthens its cell wall by producing different kinds of defense barriers. The mechanism of induction of these barriers is not yet well understood. Is it because of the elicitors released from fungal cell wall? Elicitors are known to have different actions, eliciting different host defenses. Elicitors to induce cell wall-bound phenolics, callose deposition, lignin deposition, suberin accumulation, silicon, or calcium accumulation have not been identified in many pathogens. In fact, only a few studies have been conducted in this line. Induced lignin seems to be new ones that are not preexisting. Can they be called phytoalexins? New anionic peroxidase isozymes have been detected in induction of suberin. Can they be called as pathogenesis-related (PR-) proteins? Silicon accumulates in host cell walls, but it does not seem to serve as a physical barrier. How does it induce phenolics, lignin, suberin, and callose? Does it serve as an abiotic elicitor? Papilla formation is induced by Ca^{2+}. What, then is the role of Ca^{2+}? Is it an abiotic elicitor or a second messenger?

Pathogen actively suppresses phenolic accumulation, by suppressing PAL mRNA levels. Suppression of CAD leads to susceptibility. It is not known whether a suppressor is produced by the pathogen to suppress this mechanism. Is it a toxin or an inhibitor of an enzyme? How does the suppressor act in suppressing transcription?

It may be useful if we can unravel the phenomenon of active cell wall deposition. By this process we may be able to develop potential tools to manage fungal diseases.

REFERENCES

Abeles, F. B., Hershberger, W. L., and Dunn, L. J. (1988). Hormonal regulation and intracellular localization of a 33-kD cationic peroxidase in excised cucumber cotyledons, Plant Physiol., 89:664–668.

Abu-Goukh, A. A., Greve, L. C., and Labavitch, J. M. (1983a). Purification and partial characterization of "Bartlett" pear fruit polygalacturonase inhibitors, Physiol. Plant Pathol., 23:111–122.

Abu-Goukh, A. A., and Labavitch, J. M. (1983). The in vivo role of "Bartlett" pear fruit polygalacturonase inhibitors, Physiol. Plant Pathol., 23:123–135.

Abu-Goukh, A. A., Strand L. L., and Labavitch, J. M. (1983b). Development-related changes in decay susceptibility and polygalacturonase inhibitor content of "Bartlett" pear fruit, Physiol. Plant Pathol., 23:101–109.

Adams, C. A., Nelson, W. S., Nunberg, A. N., and Thomas, T. L. (1992). A wound-inducible member of the hydroxyproline-rich glycoprotein gene family in sunflower, Plant Physiol., 99:775–776.

Adatia, M. H., and Besford, R. T. (1986). The effect of silicon on cucumber plants grown in recirculating nutrient solution, Ann. Bott., 58:343–351.

Agrawal, V. P., and Kolattukudy, P. E. (1977). Biochemistry of suberization: w-hydroxyacid oxidation in enzyme preparations from suberizing potato tuber disks, Plant Physiol., 59:667–672.

Agrawal, V. P., and Kolattukudy, P. E. (1978a). Mechanism of action of a wound-induced w-hydroxy fatty acid: NADP oxidoreductase isolated from potato tubers (*Solanum tuberosum* L.), Arch. Biochem. Biophys., 191:466–478.

Agrawal, V. P., and Kolattukudy, P. E. (1978b). Purification and characterization of a wound-induced w-hydroxy fatty acid: NADP oxidoreductase from potato tuber disks, Arch. Biochem. Biophys., 191:452–465.

Aist, J. R. (1976a). Cytology of penetration and infection. Encyclopedia of Plant Physiology: Physiological Plant Pathology, Vol. 4 (R. Heitefuss and P. H. Williams, eds.), Springer Verlag, New York, pp. 197–221.

Aist, J. R. (1976b). Papillae and related wound plugs of plant cells, Annu. Rev. Phytopathol., 14:145–163.
Aist, J. R. (1983). Structural responses as resistance mechanisms, Dynamics of the Host Defence (J. A. Bailey and B. J. Deverall, eds.), Academic Press, New York, pp. 33–70.
Aist, J. R., Gold, R. E., Bayles, C. J., Morrison, G. H., Chandra, S., and Israel, H. W. (1988). Evidence that molecular components of papillae may be involved in ml-O-resistance to powdery mildew, Physiol. Mol. Plant Pathol., 33:17–32.
Aist, J. R., and Israel, H. W. (1977). Papilla formation: timing and significance during penetration of barley coleoptiles by *Erysiphe graminis* f.sp. *hordei*, Phytopathology, 67:455–461.
Aist, J. R., and Israel, H. W. (1986). Autofluorescent and ultra-violet absorbing components in cell walls and papillae of barley coleoptiles and their relationship to disease resistance, Can. J. Bot., 64:266–272.
Aist, J. R., Kunoh, H., and Israel, H. W. (1979). Challenge appressoria of *Erysiphe graminis* fail to breach preformed papillae of a compatible barley cultivar, Phytopathology, 69:1245–1250.
Akai, S. (1953). Studies on helminthosporium blight on rice plants. VII. On the relation of silicic acid supply to the outbreak of *Helminthosporium* blight or blast disease of rice plants, Ann. Phytopath. Soc. Japan, 17:109–112.
Akai, S., and Fukutomi, M. (1980). Preformed internal physical barriers, Plant Disease—An advanced Treatise, Vol. V (J. G. Horsfall and E. B. Cowlings, eds.), Academic Press, New York, pp. 135–159.
Akutsu, S., Doi, Y, and Yora, K. (1980). Elementary analysis of papillae and cytoplasmic vesicles formed at the penetration site by *Erysiphe graminis* f.sp. *hordei* in epidermal cells of barley leaves, Ann. Phytopathol. Soc. Japan, 46:667–671.
Albersheim, P. (1965). The structure and function of the cell wall, Plant Biochemistry (J. Bonner and J. Varner, eds.), Academic Press, New York, pp. 151–186.
Albersheim, P., and Anderson, A. J. (1971). Proteins from plant cell walls inhibit polygalacturonases secreted by plant pathogens, Proc. Natl. Acad. Sci. USA, 68:1815–1819.
Albersheim, P., Jones, T. M., and English P. D. (1969). Biochemistry of the cell wall in relation to infective processes, Annu. Rev. Phytopathol., 7:171–194.
Aleshin, N. E., Avakyan, E. R., Dyakun Chak, S. A., Aleshin, D. P., Baryshok, V. P., and Vorokov, M. G. (1986). Role of silicon in resistance of rice to blast, Doklady Akademii Nauk SSSR, 291:499–502.
Anderson, A. J. (1978). Extracellular enzymes produced by *Colletotrichum lindemuthianum* and *Helminthosporium maydis* during growth on isolated bean and corn cell walls, Phytopathology, 68:1585–1589.
Anderson, A. J., and Albersheim, P. (1972). Host pathogen interactions. V. Comparison of the abilities of proteins isolated from three varieties of *Phaseolus vulgaris* to inhibit the endo-polygalacturonase secreted by three races of *Colletotrichum lindemuthianum*, Physiol. Plant Pathol., 2:339–346.
Apel, P. C., Panaccione, D. G., Holden, F. R., and Walton, J. D. (1993). Cloning and targeted gene disruption of XYL1, a β-1,4-xylanase gene from the maize pathogen, *Cochliobolus carbonum*, Mol. Plant-Microbe Interact., 6:467–473.

Apostol, I., Heinstein, P. F., and Low, P. S. (1989). Rapid stimulation of an oxidative burst during elicitation of cultured plant cells, Plant Physiol., 90:109–116.

Araki, F., and Miyagi, Y. (1977). Effects of fungicides on penetration by *Pyricularia oryzae* as evaluated by an improved cellophane method, J. Pestic. Sci., 2:457–461.

Ariyama, Y. K., and Kato, K. (1977). Structure of hydroxyproline arabinoside from tobacco cells, Agric. Biol. Chem., 41:79–81.

Arjunan, G., Vidhyasekaran, P., and Kandaswamy, T. K. (1975). Production of pectolytic enzymes by *Helminthosporium turcicum* in vitro and in vivo, Indian Phytopathol., 28:569–570.

Asada, Y., and Kugoh, T. (1971). Incorporation of tritiated phenylalanine into lignified cell walls of Japanese radish root infected by *Peronospora parasitica*, Ann. Phytopathol. Soc. Japan, 37:311–313.

Asada, Y., and Matsumoto, I. (1967). Formation of lignin in the root tissues of Japanese radish affected by *Alternaria japonica*, Phytopathology, 57:1339–1343.

Asada, Y., and Matsumoto, I. (1972). The nature of lignin obtained from downy mildew-infected Japanese radish root, Phytopathol. Z., 73:208–214.

Asiegbu, F. O., Daniel, G., and Johansson, M. (1993). Studies on the infection of Norway spruce roots by *Heterobasidium annosum*, Can. J. Bot., 71:1552–1561.

Asiegbu, F. O., Daniel, G., and Johansson, M. (1994). Defence related actions of seedling roots of Norway spruce to infection by *Heterobasidium annosum* (Fr.) Bref., Physiol. Mol. Plant Pathol., 45:1–19.

Aspinall, G. O. (1970). Pectins, Plant gums and other plant polysaccharides, The Carbohydrates, Vol. IIB, (W. Pigman and D. Horton, eds.), Academic Press, New York, pp. 515–536.

Aspinall, G. O. (1980). Chemistry of cell wall polysaccharides, The Biochemistry of Plants, Vol. 3 (J. Preiss, ed.), Academic Press, New York, pp. 473–500.

Bajar, A., Podila, G. K., and Kolattukudy, P. E. (1991). Identification of a fungal cutinase promoter that is inducible by a plant signal via a phosphorylated trans-acting factor, Proc. Natl. Acad. Sci. USA, 88:8208–8212.

Baker, C. J., and Bateman, D. F. (1978). Cutin degradation by plant pathogenic fungi, Phytopathology, 68:1577–1584.

Baker, C. J., McCormick, S. L., and Bateman, D. F. (1982). Effects of purified cutin esterase on the permeability and mechanical strength of cutin membranes, Phytopathology, 72:420–423.

Baker, C. J., Whalen, C. H., and Bateman, D. F. (1977). Xylanase from *Trichoderma pseudokoningii*, purification, characterization and effects on isolated plant cell walls, Phytopathology, 67:1250–1258.

Baker, C. J., Whalen, C. H., Korman, R. Z., and Bateman, D. F. (1979). α-L-arabinofuranosidase from *Sclerotinia sclerotiorum*: purification, characterization and effects on plant cell walls and tissues, Phytopathology, 69:789–793.

Ball, A. M., Ashby, A. M., Daniels, M. J., Ingram, D. S., and Johnstone, K. (1991). Evidence for the requirement of extracellular protease in the pathogenic interaction of *Pyrenopeziza brassicae* with oilseed rape, Physiol. Mol. Plant Pathol., 38:147–161.

Barmore, C. R., and Brown, C. E. (1979). Role of pectolytic enzymes and galacturonic acid in citrus fruit decay caused by *Penicillium digitatum*, Phytopathology, 69:675–678.

Barmore, C. R., and Nguyen, T. K. (1985). Polygalacturonase inhibition in rind of Valencia orange infected with *Diplodia natalensis*, Phytopathology, 75:446–449.

Barthe, J. P., Cantenys, D., and Touze A. (1981). Purification and characterization of two polygalacturonases secreted by *Colletotrichum lindemuthianum*, Phytopathol. Z., 100:162–171.

Bateman, D. F. (1976). Plant cell wall hydrolysis by pathogens, Biochemical Aspects of Plant Parasite Relationships (J. Friend and D. R. Threlfall, eds.), Academic Press, London, pp. 79–103.

Bateman, D. F., and Basham, H. G. (1976). Degradation of plant cell walls and membranes by microbial enzymes, Encyclopedia of Plant Physiology New Series, Physiological Plant Pathology, Vol. 4 (R. Heitefuss and P. H. Williams, eds.), Springer-Verlag, New York, pp. 316–355.

Bateman, D. F., Jones, T. M., and Yoder, O. C. (1973). Degradation of corn cell walls by extracellular enzymes produced by *Helminthosporium maydis* race T., Phytopathology, 63:1523–1529.

Bateman, D. F., Van Etten, H. D., English, P. D., Nevins, D. L., and Albersheim, P. (1969). Susceptibility to enzymatic degradation of cell wall from bean plant resistant and susceptible to *Rhizoctonia solani* Kuhn., Plant Physiol., 44:641–648.

Bauer, W. D., Bateman, D. F., and Whalen, C. H. (1977). Purification of an endo-1,4-galactanase produced by *Sclerotinia sclerotiorum*: effects on plant cell walls and tissues, Phytopathology, 67:862–868.

Bauer, W. D., Talmadge, K. W., Keegstra, K., and Albersheim, P. (1973). The structure of plant cell walls. II. The hemicellulose of the walls of suspension-cultured sycamore cells, Plant Physiol., 51:174–187.

Bayles, C. J., and Aist, J. R. (1987). Apparent calcium mediation of resistance of an ml-O barley mutant to powdery mildew, Physiol. Mol. Plant Pathol., 30:337–345.

Bayles, C. J., Ghemawat, M. S., and Aist, J. R. (1990). Inhibition by 2-deoxy-D-glucose of callose formation, papilla deposition, and resistance to powdery mildew in an ml-O barley mutant, Physiol. Mol. Plant Pathol., 36:63–72.

Beardmore, J., Ride J. P., and Granger, J. W. (1983). Cellular lignification as a factor in the hypersensitive resistance of wheat to stem rust, Physiol. Plant Pathol., 22:209–230.

Belanger, R. R., Cherif, M., and Tremblay, J. (1991). Control of *Pythium* root rot in greenhouse grown cucumbers with soluble silicon amendments, Phytopathology, 81:120.

Bell, A. A., Puhalla, J. E., Tolmsoff, W. J., and Stipanovic, R. D. (1976a). Use of mutants to establish (+)-scytalone as an intermediate in melanin biosynthesis by *Verticillium dahliae*, Can. J. Microbiol., 22:787–799.

Bell, A. A., Stipanovic, R. D., and Puhalla, J. E. (1976b). Pentaketide metabolites of *Verticillium dahliae*. Identification of (+)-scytalone as a natural precursor to melanin, Tetrahedron Lett., 32:1353–1356

Bell, A. A., and Wheeler, M. H. (1986). Biosynthesis and functions of fungal melanins, Annul. Rev. Phytopathol., 24:411–451.

Benhamou, N. (1992). Ultrastructural detection of β-1,3-glucans in tobacco root tissues infected by *Phytophthora parasitica* var. *nicotianae* using a gold-complexed tobacco β-1,3-glucanase, Physiol. Mol. Plant Pathol., 41:351–370.

Benhamou N. (1995). Ultrastructural and cytochemical aspects of the response of eggplant parenchyma cells in direct contact with *Verticillium*-infected xylem vessels, Physiol. Mol. Plant Pathol., 46:321–338.

Benhamou, N., and Cote, F. (1992). Ultrastructure and cytochemistry of pectin and cellulose degradation in tobacco roots infected by *Phytophthora parasitica* var. *nicotianae*, Phytopathology, 82:468–478.

Benhamou, N., Chamberland, H., and Pauze, F. J. (1990a). Implications of pectic components in cell surface interactions between tomato root cells and *Fusarium oxysporum* f.sp. *radicis-lycopersici*: a cytochemical study by means of a lectin with polygalacturonic acid-binding specificity, Plant Physiol., 92:995–1003.

Benhamou, N., Lafitte, C., Barthe, J. P., and Esquerre-Tugaye, M. T. (1991). Cell surface interactions between bean leaf cells and *Colletotrichum lindemuthianum*, Plant Physiol., 97:234–244.

Benhamou, N., Mazau, D., and Esquerre-Tugaye, M. T. (1990a). Immunocytochemical localization of hydroxyproline-rich glycoproteins in tomato root cells and *Fusarium oxysporum* f.sp. *radicis-lycopersici*: study of a compatible interaction, Phytopathology, 80:163–173.

Benhamou, N., Mazau, D., Esquerre-Tugaye, M. T., and Asselin, A. (1990b). Immunogold localization of hydroxyproline-rich glycoproteins in necrotic tissue of *Nicotiana tabacum* L. cv. *Xanthi-nc* infected by tobacco mosaic virus, Physiol. Mol. Plant Pathol., 36:129–145.

Bernards, M. A., and Ellis, B. E. (1991). Phenylalanine ammonia-lyase from tomato cell cultures inoculated with *Verticillium albo-atrum*, Plant Physiol., 97:1494–1500.

Biggs, A. R. (1985a). Detection of impervious tissue in tree bark with selective histochemistry and fluorescence microscopy, Stain Technol., 60:299–304.

Biggs, A. R. (1985b). Suberized boundary zones and the chronology of wound response in tree bark, Phytopathology, 75:1191–1195.

Biggs, A. R. (1989). Temporal changes in the infection court after wounding of peach bark and their association with cultivar variation in infection by *Leucostoma persooni*, Phytopathology, 79:627–630.

Biggs, K. J., and Fry, S. C. (1990). Solubilization of covalently bound extensin from *Capsicum* cell walls, Plant Physiol., 92:197–204.

Biggs, A. R., and Miles, N. W. (1988). Association of suberin formation in uninoculated wounds with susceptibility to *Leucostoma cincta* and *L. persoonii* in various peach cultivars, Phytopathology, 78:1070–1074.

Biggs, A. R., and Peterson, C. A. (1990). Effect of chemical applications of peach bark wounds on accumulation of lignin and suberin and susceptibility to *Leucostoma persooni*, Phytopathology, 80:861–865.

Binder, A., Baer, G., Hofmann, C., and Kovats, K. (1989). Mechanisms in systemic induced disease resistance, Signal Molecules in Plants and Plant-Microbe Interactions (B. J. J. Lugtenberg, ed.), Springer-Verlang, Berlin, pp. 267–272.

Bird, P. M., and Ride, J. P. (1981). The resistance of wheat to *Septoria nodorum*: fungal development in relation to host lignification, Physiol. Plant Pathol. 19:289–299.

Blaich, R., and Grundhofer, H. (1990). Mildew attack induces silica incrusts in epidermal cell walls of *Vitis* leaves. Abstracts of the International Congress on Plant-Microbe Interactions, Switzerland, p. 339.

Blake, J. D., and Richards, G. N. (1971). An examination of some methods for fractionation of plant hemicelluloses, Carbohyd. Res., 17:253–268.

Bloomfield, B. J., and Alexander, M. (1967). Melanins and resistance of fungi to lysis, J. Bacteriol., 93:1276–1280.

Bock, W., Dongowski, G., Goebel, H., and Krause, M. (1975). Nachweis der Hemmung mikrobieller Pektin- and Pektatlyase durch pflanzeneigene inhibitoren, Nahrung, 19:411–416.

Boller, T., and Kende, H. (1980). Regulation of wound ethylene synthesis in plants, Nature, 286:259–260.

Bolton, P., and Harwood, J. L. (1976). Fatty acid synthesis in aged potato slices, Phytochemistry, 18:1501–1506.

Bolwell, G. P., Bell, J. N., Cramer, C. L., Schuch, W., Lamb, C. J., and Dixon, R. A. (1985). L-Phenylalanine ammonia-lyase from *Phaseolus vulgaris*: characterization and differential induction of multiple forms from elicitor-treated cell suspension cultures, Eur. J. Biochem., 149:411–419.

Bolwell, G. P., Robbins, M. P., and Dixon, R. A. (1985). Elicitor-induced prolyl hydroxylase from French bean (*Phaseolus vulgaris*), Biochem. J., 229:693–699.

Bonello, P., Pearce, R. B., Watt, F., and Grime, G. W. (1991). An induced papilla response in primary roots on Scots pine challenged in vitro with *Cylindrocarpon destructans*, Physiol. Mol. Plant Pathol., 39:213–228.

Bonhoff, A., Rieth, B., Golecki, J., and Grisenbach, H. (1987). Race cultivar-specific differences in callose deposition in soybean roots following infection with *Phytophthora megasperma* f.sp. *glycinea*, Planta, 1972:101–105.

Bonnen, A. M., and Hammerschmidt, R. (1989a). Cutinolytic enzymes from *Colletotrichum lagenarium*, Physiol. Mol. Plant Pathol., 35:463–474.

Bonnen, A. M., and Hammerschmidt, R. (1989b). Role of cutinolytic enzymes in infection of cucumber by *Colletotrichum lagenarium*, Physiol. Mol. Plant Pathol., 35:475–481.

Borchert, R. (1978). Time course and spatial distribution of phenylalanine ammonia-lyase and peroxidase activity in wounded potato tuber tissue, Plant Physiol., 62:789–793.

Bostock, R. M., and Stermer, B. A. (1989). Perspectives on wound healing in resistance to pathogens, Annu. Rev. Phytopathol., 27:343–371.

Bracker, C. E., and Littlefield, L. J. (1973). Structural concepts of host-pathogen interfaces. Fungal Pathogenicity and the Plant's Response (R. J. W. Byrde and C. V. Cutting, eds.), Academic Press, London, pp. 159–318.

Bradley, D. J., Kjellbom, P., and Lamb, C. J. (1992). Elicitor- and wound-induced oxidative cross-linking of a proline-rich plant cell wall protein: a rapid defense response, Cell, 70:21–30.

Brammall, R. A., and Higgins, V. J. (1988). A histological comparison of fungal colonization in tomato seedlings susceptible or resistant to *Fusarium* crown and root rot disease, Can. J. Bot., 66:915–925.

Brockmann, B., and Smit, R., and Tudzynski, P. (1992). Characterization of an extracellular β-1,3-glucanase of *Claviceps purpurea*, Physiol. Mol. Plant Pathol., 40:191–201.

Brown, A. E. (1984). Relationship of polygalacturonase inhibitor activity to the rate of fungal rot development in apple fruits, Phytopathol. Z., 111:122–132.

Brown, A. E., and Adikaram, N. K. B. (1982). The differential inhibition of pectic enzymes from *Glomerella cingulata* and *Botrytis cinerea* by a cell wall protein from *Capsicum annuum* fruit, Phytopathol. Z., 105:27–38.

Brown, A. E., and Adikaram, N. K. B. (1983). A role for pectinase and protease inhibitors in fungal rot development in tomato fruits, Phytopathol. Z., 106:239–251.

Bruce, R. J., and Galston, A. W. (1989). Elicitation of lignin biosynthesis and isoperoxidase activity by pectic fragments in suspension cultures of castor bean, Plant Physiol., 91:889–897.

Bucheli, P., Doares S. H., Albersheim, P., and Darvill, A. (1990). Host-pathogen interactions. XXXVI. Partial purification and characterization of heat-labile molecules secreted by the rice blast pathogen that solubilize plant cell wall fragments that kill plant cells, Physiol. Mol. Plant Pathol., 36:159–173.

Buchenauer, H., Zeun, R., and Schinzer, U. (1985). Effect of tricyclazole on mycelium growth as well as on development and pigmentation of sclerotia of *Sclerotinia sclerotiorum*, Z. Pflanzenkr. Pflanzenschutz, 92:17–26.

Bugbee, W. M. (1990). Purification and characteristics of pectin lyase from *Rhizoctonia solani*, Physiol. Mol. Plant Pathol., 36:15–25.

Bushnell, W. R. (1972). Physiology of fungal haustoria, Annu. Rev. Phytopathol., 10:151–176.

Bushnell, W. R., and Gay, J. L. (1978). Accumulation of solutes in relation to the structure and function of haustoria of powdery mildews, The Powdery Mildews (D. Spencer, ed.), Academic Press, London, pp. 183–235.

Bustamam, M, and Sister, H. D. (1987). Effect of pentachloronitrobenzene, pentachloroaniline and albinism on epidermal penetration by appressoria of *Pyricularia*, Pestic, Biochem. Physiol., 28:29–37.

Byrde, R. J. W., and Archer, S. A. (1977). Host inhibition or modification of extracellular enzymes of pathogens, Cell Wall Biochemistry Related to Specificity in Host–Plant Interactions, (B. Solheim and J. Raa, eds.), Universitetsforlaget, Tromso, pp. 213–245.

Cadena-Gomez, G., and Nicholson, R. L. (1987). Papilla formation and associated peroxidase activity: A non-specific response to attempted fungal penetration of maize, Physiol. Mol. Plant Pathol., 31:51–67.

Cahill, D. M., and McComb, J. A. (1992). A comparison of changes in phenylalanine ammonia-lyase activity, lignin and phenolic synthesis in the roots of *Eucalyptus calophylla* (field resistant) and *E. marginata* (susceptible) when infected with *Phytophthora cinnamomi*, Physiol. Mol. Plant Pathol., 40:315–332.

Callaghan, T., Ross, P., Weinberger-Ohana, P., and Benziman, M. (1988). β-Glucoside activators of mungbean UDP-glucose: β-glucan synthase. II. comparison of effects of an endogenous β-linked glucolipid and synthetic n-alkyl-D-monoglu-copyranosides, Plant Physiol., 86:1104–1107.

Campbell, M. M., and Ellis, B. E. (1992). Fungal elicitor-mediated responses in pine cell cultures. III. Purification and characterization of phenylalanine ammonia-lyase, Plant Physiol., 98:62–70.

Carvalho, S. C., Allan, C. R., and Kolattukudy, P. E. (1984). Inhibition of cutinase and prevention of fungal penetration into plants by alkyl boronic acids, Am. Chem. Soc. 39th Northwest Region Meet. Program. June 13–15, 1984, Moscow, Idaho, Abstr. 108.

Carver, T. L. W., Robbins, M. P., and Zeyen, R. (1991). Effects of two PAL inhibitors on the susceptibility and localized autofluorescent host cell responses of oat leaves attacked by *Erysiphe graminis* DC, Physiol. Mol. Plant Pathol., 39:269–287.

Carver, T. L. W., Robbins, M. P., Zeyen, R. J., and Dearne, G. A. (1992a). Effects of PAL-specific inhibition on suppression of activated defence and quantitative susceptibility of oats to *Erysiphe graminis*, Physiol. Mol. Plant Pathol., 41:149–163.

Carver, T. L. W., and Zyen, R. J. (1993). Effects of PAL and CAD inhibition on powdery mildew resistance phenomena in cereals, Mechanism of Plant Defense Responses (B. Fritig and M. Legrand, eds.), Kluwer Academic Publishers, Dordrecht, pp. 324–327.

Carver, T. L., Zyen, R. J., and Ahlstrand, G. G. (1987). The relationship between insoluble silicon and success or failure of attempted primary penetration by powdery mildew, *Erysiphe graminis* germlings on barley, Physiol. Mol. Plant Pathol., 31:133–148.

Carver, T. L. W., Zyen, R. J., Bushnell, W. R., and Robbins, M. P. (1994a). Inhibition of phenylalanine ammonia lyase and cinnamyl alcohol dehydrogenase increases quantitative susceptibility of barley to powdery mildew (*Erysiphe graminis* D.C.), Physiol. Mol. Plant Pathol., 44:261–272.

Carver, T. L. W, Zeyen, R. J., Robbins, M. P., and Dearne, G. A. (1992b). Effects of the PAL inhibitor, AOPP, on oat, barley and wheat cell responses to appropriate and inappropriate formae speciales of *Erysiphe graminis* DC., Physiol. Mol. Plant Pathol., 41:397–409.

Carver, T. L. W., Zyen, R. J., Robbins, M. P., Vance, C. P., and Boyles, D. A. (1994b). Suppression of host cinnamyl alcohol dehydrogenase (CAD) and phenylalanine ammonia lyase (PAL) increases oat epidermal cell susceptibility to powdery mildew penetration, Physiol. Mol. Plant Pathol., 44:243–260.

Castillo, F. J., and Greppin, H. (1986). Balance between anionic and cationic extracellular peroxidase activities in *Sedum album* leaves after ozone exposure. Analysis by high-performance liquid chromatography, Physiol. Plant., 68:201–208.

Cervone, F., Anderbahn, T., Coutts, R. H. A., and Wood, R. K. S. (1981). Effects of French bean tissue and leaf protoplasts on *Colletotrichum lindemuthianum* polygalacturonase, Phytopathol. Z., 102:238–246.

Cervone, F., De Lorenzo, G., Degra, L., and Salvi, G. (1986). Interaction of fungal polygalacturonase with plant proteins in relation to specificity and regulation of plant defense response, Recognition in Microbe–Plant Symbiotic and Pathogenic Interactions (B. Lugtenberg, ed.), Springer-Verlag, Berlin, pp. 253–258.

Cervone, F., De Lorenzo, G., Degra, L., Salvi, G., and Bergami, M. (1987). Purification and characterization of a polygalacturonase-inhibiting protein from *Phaseolus vulgaris L.*, Plant Physiol., 85:631–637.

Cervone, F., De Lorenzo, G., Salvi, G, and Camardella, L. (1986). Molecular evolution of fungal polygalacturonase, Biology and Molecular Biology of Plant-Pathogen Interactions (J. Bailey, ed.), Plenum Press, New York, pp. 385–392.

Cervone, F., Scala, A., and Scala, F. (1978). Polygalacturonase from *Rhizoctonia fragariae*:

further characterization of two isoenzymes and their actions towards strawberry tissue, Physiol. Plant Pathol., 12:19–26.

Chen, J., and Varner, J. E. (1985a). Biology and molecular biology of hydroxyproline-rich glycoproteins, Biochemistry of Plants (P. K. Stumpf and E. E. Conn, eds.), Alan L. Liss, New York, pp. 485–520.

Chen, J., and Varner, J. E. (1985b). Isolation and characterization of cDNA clones for carrot extensin and a proline-rich 33-kDa protein, Proc. Natl. Acad. Sci. USA, 82:4399–4403.

Cherif, M., and Belanger, R. R. (1992). Use of potassium silicate amendments in recirculating nutrient solutions to suppress *Pythium ultimum* on long English cucumber, Plant Dis., 76:1008–1011.

Cherif, M., Benhamou, N., and Belanger, R. R. (1991). Ultrastructural and cytochemical studies of fungal development and host reactions in cucumber plants infected by *Pythium ultimum*, Physiol. Mol. Plant Pathol., 39:353–375.

Cherif, M., Benhamou, N., Menzies, J. G., and Belanger, R. R. (1992b). Silicon induced resistance in cucumber plants against *Pythium ultimum*, Physiol. Mol. Plant Pathol., 41:411–425.

Cherif, M., Menzies, J. G., Benhamou, N., and Belanger, R. R. (1992a). Studies on silicon distribution in wounded and *Pythium ultimum* infected cucumber plants, Physiol. Mol. Plant Pathol., 41:371–385.

Chida T., and Sisler, H. D. (1987). Restoration of appressorial penetration ability by melanin precursors in *Pyricularia oryzae* treated with antipenetrants and in melanin-deficient mutants, J. Pestic. Sci., 12:49–55.

Chida, T., Uekita T., Satake, K., Hirano K., Aoki K., and Noguchi, T. (1982). Effect of fthalide on infection process of *Pyricularia oryzae*: observation of the penetration site of appressoria. Ann. Phytopathol. Soc. Japan, 48:58–63.

Chong, J., Harder, D. E., and Rohringer, R. (1984). Cytochemical studies on *Puccinia graminis* f.sp. *tritici* in a compatible wheat host. I. Walls of intercellular hyphal cells and haustorium mother cells, Can. J. Bot., 63:1713–1724.

Chong, J., Harder, D. E., and Rohringer, R. (1985). Cytochemical studies on *Puccinia graminis* f.sp. *tritici* in a compatible wheat host. II. Haustorium mother cell walls at the host cell penetration site, haustorial walls and the extrahaustorial matrix, Can. J. Bot., 64:2561–2575.

Chumley, F. G., and Valent, B. (1990). Genetic analysis of melanin-deficient, nonpathogenic mutants of *Magnaporthe grisea*, Mol. Plant–Microbe Interact., 3:135–143.

Clark, T. A., Zyen, R. J., Smith, A. G., Carver, T. L. W., and Vance, C. P. (1994). Phenylalanine ammonia lyase mRNA accumulation, enzyme activity and cytoplasmic responses in barley isolines differing at M1-a and M1-O, attacked by *Erysiphe graminis* f.sp. *hordei*, Physiol. Mol. Plant Pathol., 44:171–185.

Clarke, D. D. (1982). The accumulation of cinnamic acid amides in the cell walls of potato tissue as an early response to fungal attack, Active Defense Mechanisms in Plants (R. K. S. Wood, ed.), Plenum Press, New York, pp. 321–322.

Clarke, J. A., Lisker, N., Lamport, D. T. A., and Ellingboe, A. E. (1981). Hydroxyproline enhancement as a primary event in the successful development of *Erysiphe graminis* in wheat, Plant Physiol., 67:188–189.

Cleland, R. (1971). Cell wall extension, Annu. Rev. Plant Physiol., 22:197–222.

Coffey, M. D., and Allen, F. H. E. (1983). A quantitative histological and ultrastructural analysis of interactions between the flax rust and near-isogenic host lines varying in their degree on incompatibility, Can. J. Bot., 61:1831–1850.

Coffey, M. D., and Cassidy, D. S. M. (1984). Peroxidase activity and induced lignification in rusted flax interactions varying in their degree of incompatibility, Can. J. Bot., 62:134–141.

Cohen, Y., and Eyal, H. (1988). Epifluorescence microscopy of *Sphaerotheca fuliginea* race 2 on susceptible and resistant genotypes of *Cucumis melo*, Phytopathology, 78:144–148.

Cohen, Y., Eyal, H., and Hanania, J. (1990). Ultrastructure, autofluorescence, callose deposition and lignification in susceptible and resistant muskmelon leaves infected with the powdery mildew fungus *Sphaerotheca fuliginea*, Physiol. Mol. Plant Pathol., 36:191–204.

Cohen, Y., Eyal, H., Hanania, J., and Malik, Z. (1989). Ultrastructure of *Pseudoperonospora cubensis* in muskmelon genotypes susceptible and resistant to downy mildew, Physiol. Mol. Plant Pathol., 34:27–40.

Collendaveloo, J., Legrand, M., and Fritig, B. (1983). Plant disease and the regulation of enzymes involved in lignification-Increased rate of de novo synthesis of the three tobacco-O-methyltransferases during the hypersensitive response to infection by tobacco mosaic virus, Plant Physiol., 73:550–554.

Collmer, A., and Keen, N. T. (1986). The role of pectic enzymes in plant pathogenesis, Annu. Rev. Phytopathol., 24:383–409.

Conti, G. G., Bassi, M., Maffi, D., and Bocci, A. M. (1986). Host-parasite relationship in a susceptible and a resistant rose cultivar inoculated with *Sphaerotheca pannosa*. II. Deposition rates of callose, lignin and phenolics in infected or wounded cells and their possible role in resistance, J. Phytopathology, 117:312–320.

Cooper, R. M. (1983). The mechanisms and significance of enzymic degradation of host cell walls by parasites, Biochemical Plant Pathology (J. A. Callow, ed.), John Wiley, Chichester, pp. 101–135.

Cooper, R. M. (1984). The role of cell wall-degrading enzymes in infection and damage, Plant diseases Infection, Damage and Loss (R. K. S. Wood and G. J. Jellis, ed.), Blackwell Press, Oxford, pp. 13–27.

Cooper, J. B., Chen, J. A., and Varner, J. E. (1984). The glyco-protein component of plant cell walls, Structure, Function, and Biosynthesis of Plant Cell Walls (W. M. Dugger and S. Bartnicki-Garcia, eds.), American Soc. Plant Physiologists, Rockville, MD, pp. 75–78.

Cooper, R. M., Longman, D., Campbell, A., Henry, M., and Lees, P. E. (1988). Enzymic adaptation of cereal pathogens to the monocotyledonous primary wall, Physiol. Mol. Plant Pathol., 32:33–47.

Cooper, R. M., Rankin, B., and Wood, R. K. S. (1978). Cell wall degrading enzymes of vascular wilt fungi. II. Properties and modes of action of polysaccharidases of *Verticillium albo-atrum* and *Fusarium oxysporum* f.sp. *lycopersici*, Physiol. Plant Pathol., 13:101–134.

Cooper, R. M., Wardman, P. A., and Skelton, J. E. M. (1981). The influence of cell walls from host and non-host plants on the production and activity of polygalacturonide-degrading enzymes from fungal pathogens, Physiol. Plant Pathol. 18:239–255.

Cooper, R. M., and Wood, R. K. S. (1975). Regulation of synthesis of cell wall degrading enzymes by *Verticillium albo-atrum* and *Fusarium oxysporum* f.sp. *lycopersici*, Physiol. Plant Pathol., 5:135–156.

Cooper, R. M., and Wood, R. K. S. (1980). Cell wall degrading enzymes of vascular wilt fungi. III. Possible involvement of endo-pectin lyase in *Verticillium* wilt of tomato, Physiol. Plant Pathol., 16:285–300.

Corbin, D. R., Sauer, N., and Lamb, C. J. (1987). Differential regulation of hydroxyproline-rich glycoprotein gene family in wounded and infected plants, Mol. Cell. Biol., 7:4337–4344.

Cosio, E. G., Frey, T., Verduyn, R., Van Boom, J., and Ebel, J. (1990). High affinity binding of a synthetic heptaglucoside and fungal glucan phytoalexin elicitors to soybean membranes, FEBS Lett., 271:223–226.

Cottle, W., and Kolattukudy, P. E., (1982a). Abscisic acid stimulation of suberization: induction of enzymes and deposition of polymeric components and associated waxes in tissue cultures of potato tuber, Plant Physiol., 70:775–780.

Cottle, W., and Kolattukudy, P. E. (1982b). Biosynthesis, deposition and partial characterization of potato suberin phenolics, Plant Physiol., 69:393–399.

Crawford, M. S., and Kolattukudy, P. E. (1987). Pectate lyase from *Fusarium solani* f.sp.pisi: purification, characterization, in vitro translocation of the mRNA, and involvement in pathogenicity, Arch. Biochem. Biophys., 258:196–205.

Cruickshank, I. A. M., Perrin, D. R., and Mandryk, M. (1977). Fungitoxicity of duvatrienediols associated with the cuticular wax of tobacco leaves, Phytopathol. Z., 90:133–146.

Cuypers, B., Schmelzer, E., and Hahlbrock, K. (1988). In situ localization of rapidly accumulated phenylalanine ammonia-lyase mRNA around penetration sites of *Phytophthora infestans* in potato leaves, Mol. Plant–Microbe Interact., 1:157–160.

Dahm, H., and Strezelczyk, E. (1987). Cellulolytic and pectolytic activity of *Cylindrocarpon destructans* (Zin.) Scholt. isolates pathogenic and non-pathogenic to fir (*Abies alba* Mill.) and pine (*Pinus sylvestris* L.), J. Phytopathol., 118:76–83.

Dai, G. H., Andary, C., Mondolot-Cosson, L., and Boubals, D. (1995). Histochemical studies on the interaction between three species of grapevine, *Vitis vinifera*, *V. rupestris* and *V. rotundifolia* and the downy mildew fungus, *Plasmopara viticola*, Physiol. Mol. Plant Pathol., 46:177–188.

Dantzig, A. H., Zuckerman, S. H., and Andonov-Roland, M. M. (1986). Isolation of a *Fusarium solani* mutant reduced in cutinase activity and virulence, J. Bacteriol., 168:911–916.

Darvill, A., McNeil, M., Albersheim, P., and Delmer, D. P. (1980). The primary cell wall of flowering plants, The Biochemistry of Plants, Vol. I, (N. E. Tolbert, ed.), Academic Press, New York, pp. 92–157.

Datta, K., Schmidt, A., and Marcus, A. (1989). Characterization of two soybean repetitive proline-rich proteins and a cognate cDNA from germinated axes, Plant Cell, 1:945–952.

Day, P. R., and Scott, K. J. (1973). Scanning electronmicroscopy of fresh material of *Erysiphe graminis* f.sp. *hordei*, Physiol. Plant Pathol., 3:433–435.

Dean, B. B., and Kolattukudy, P. E. (1977). Biochemistry of suberization: incorporation of

(1-14 C) oleic acid and (1-14 C)-acetate into the aliphatic components in potato tuber disks (*Solanum tuberosum*), Plant Physiol., 59:48–54.

Dean, J. F. D., Gamble, H. R., and Anderson, J. D. (1989). The ethylene biosynthesis-inducing xylanase: its induction in *Trichoderma viride* and certain plant pathogens, Phytopathology, 79:1071–1078.

Dean, R. A., and Kuc, J. (1987). Rapid lignification in response to wounding and infection as a mechanism for induced systemic protection in cucumber, Physiol. Mol. Plant Pathol., 31:69–81.

DeLeeuw, G. T. N. (1985). Deposition of lignin, suberin and callose in relation to the restriction of infection by *Botrytis cinerea* in ghost spots of tomato fruits, Phytopathol. Z., 112:143–152.

Delmer, D. P. (1987). Cellulose biosynthesis, Annu. Rev. Plant Physiol., 38:259–290.

Delmer, D. P., Solomon, M., and Read, S. M. (1991). Direct photolabeling with (32p) UDP-glucose for identification of a subunit of cotton fiber callose synthase, Plant Physiol., 95:556–563.

Delmer, D. P., Solomon, M., Andrawis, A., and Amor, Y. (1990). Cation-induced formation of a macro-glucan synthase complex, Plant Physiol., 93:S-57 (Abstr.).

De Lorenzo, G., Scala, F., Salvi, G., and Cervone, F. (1988). Comparison of immunological reactivity of polygalacturonases from different fungi, G. Botan. Ital. 122:1–6.

Demarty, Y. M., Morvan, D., and Thellier, M. (1986). Calcium and the cell wall, Plant Cell Environ., 7:441–448.

Dhugga, K. S., and Ray, P. M. (1991). A 55-kilodalton plasma membrane-associated polypeptide is involved in β-1,3-glucan activity in pea tissue, FEBS Lett., 278:283–286.

Dickerson, A. G., Mantle, P. G., Nisbet, L. J., and Shaw, B. I. (1978). A role for β-glucanases in the parasitism of cereals by *Claviceps purpurea*, Physiol. Plant Pathol., 12:55–62.

Dickerson, A. G., and Pollard, C. M. D. (1982). Observation on the location of a β-glucanase and an associated β-glucosidase in *Claviceps purpurea* during its development on rye, Physiol. Plant Pathol., 21:179–191.

Dickman, M. B., and Patil, S. S. (1986). Cutinase deficient mutants of *Colletotrichum gloeosporioides* are nonpathogenic to papaya fruit, Physiol. Mol. Plant Pathol., 28:235–245.

Dickman, M. B., Patil, S. S., and Kolattukudy, P. E. (1982). Purification, characterization and role in infection of an extracellular cutinolytic enzyme from *Colletotrichum gloeosporioides* Penz. on *Carica papaya,* Physiol. Plant Pathol., 20:333–347.

Dickman, M. B., Patil, S. S., and Kolattukudy, P. E. (1983). Effects of organophosphorus pesticides on cutinase activity and infection of papayas by *Colletotrichum gloeosporioides*, Phytopathology, 73:1209–1214.

Dickman, M. B., Podila, G. K., and Kolattukudy, P. E. (1989). Insertion of cutinase gene into a wound pathogen enables it to infect intact host, Nature, 342:446–448.

Dilenna, P., Margianno, P., and Magro, P. (1981). Comparative investigation on morphological and physiological features on three isolates of *Botrytis cinerea*, Phytopathol. Z., 100:203–211.

Dori, S., Hershenhorn, J., Solel, Z., and Barash, I. (1992). Characterization of an endopolygalacturonase associated with take-all disease of wheat, Physiol. Mol. Plant Pathol., 40:203–210.

Dori, S., Solel, Z., and Barash, I. (1990). Characterization of extracellular polygalacturonase and cellulases produced by *Gaeumannomyces graminis* var. *tritici* in vitro and in vivo, Fourth International Mycological Congress, p. 227 (Abstr.).

Dori, S., Solel, Z., and Barash, I. (1995). Cell wall-degrading enzymes produced by *Gaeumannomyces graminis* var. *tritici* in vitro and in vivo, Physiol. Mol. Plant Pathol., 46:189–198.

Dugger, W. M., Palmer, R. L., and Black, C. C. (1991). Changes in β-1,3-glucan synthase activity in developing lima bean plants, Plant Physiol., 97:569–573.

Durrands, P. K., and Cooper, R. M. (1988). The role of pectinases in vascular wilt diseases as determined by defined mutants of *Verticillium albo-atrum*, Physiol. Plant Pathol., 32:363–371.

Durrell, L. W. (1964). The composition and structure of walls of dark fungus spores, Mycopathol. Mycol. Appl., 23:339–345.

Edwards, M. C., and Ayers, P. G. (1981). Cell death and cell wall papillae in the resistance of oak species to powdery mildew disease, New Phytol., 89:411–418.

El Ghaouth, A., Arul, J., Wilson, C., and Benhamou, N. (1994). Ultrastructural and cytochemical aspects of the effect of chitosan on decay of bell pepper fruit, Physiol. Mol. Plant Pathol., 44:417–432.

English, P. D., and Albersheim, P. (1969). Host–pathogen interactions. I. A correlation between α-galactosidase production and virulence, Plant Physiol., 44:217–224.

English, P. D., Jurale, J. B., and Albersheim, P. (1971). Host–pathogen interactions. II. Parameters affecting polysaccharide degrading enzyme secretion by *Colletotrichum lindemuthianum* grown in culture, Plant Physiol., 47:1–6.

Esau, K. (1965). Plant Anatomy, Wiley, New York, pp. 767.

Espelie, K. E., Davis, R. W., and Kolattukudy, P. E. (1980). Composition, ultrastructure and function of the cutin- and suberin-containing layers in the leaf, fruit peel, juice-sac and inner seed coat of grape fruit (*Citrus paradisi* Macfed), Planta, 149:498–511.

Espelie, K. E., Dean, B. B., and Kolattukudy, P. E. (1979). Composition of lipid-derived polymers from different anatomical regions of several plant species, Plant Physiol., 64:1089–1093.

Espelie, K. E., Franceshi, V. R., and Kolattukudy, P. E. (1986). Immunocytochemical localization and time course of appearance of an anionic peroxidase associated with suberization in wound healing potato tuber tissue, Plant Physiol., 81:487–492.

Espelie, K. E., and Kolattukudy, P. E. (1979a). Composition of the aliphatic components of 'suberin' from the bundle sheaths of *Zea mays* leaves, Plant Sci. Lett., 15:225–230.

Espelie, K. E., and Kolattukudy, P. E. (1979b). Composition of the aliphatic components of suberin of the endodermal fraction from the first internode of etiolated *Sorghum* seedlings, Plant Physiol., 63:433–435.

Espelie, K. E., and Kolattukudy, P. E. (1985). Purification and characterization of an abscisic acid-inducible anionic peroxidase associated with suberization in potato (*Solanum tuberosum*), Arch. Biochem. Biophys., 240:539–545.

Esquerre-Tugaye, M. T. (1973). Influence d'une maladie parasitaire sur la teneur en hydroxyproline des parois cellulaires d'épicotyles et pétioles de plantes de melon, CR Acad. Sci. Paris, 276:525–528.

Esquerre-Tugaye, M. T., and Lamport, D. T. A. (1979). Cell surfaces in plant–microorganism interactions. I. A structural investigation of cell wall hydroxyproline-rich glycoproteins which accumulate in fungus infected plants, Plant Physiol., 64:314–319.

Esquerre-Tugaye, M. T., Lafitte, C., Mazau, D., Toppan, A., and Touze, A. (1979). Cell surfaces in plant-microorganism interactions. II. Evidence for the accumulation of hydroxyproline-rich glycoproteins in the cell wall of diseased plants as a defense mechanism, Plant Physiol., 64:320–326.

Esquerre-Tugaye, M. T., and Mazau, D. (1984). Effect of a fungal disease on extensin, the plant cell wall glycoprotein, J. Exp. Bot., 25:509–513.

Ettinger, W. F., Thukral, S. K., and Kolattukudy, P. E. (1987). Structure of cutinase gene, cDNA and the derived amino acid sequence from phytopathogenic fungi, Biochemistry, 26:7883–7892.

Evans, I. M., Gatehouse, L. N., Gatehouse, J. A., Yarwood, J. N., Boulter, D., and Croy, R. R. D. (1990). The extensin gene family in oilseed rape (*Brassica napus* L.): characterization of sequences of representative members of the family, Mol. Gen. Genet., 223:273–287.

Fernandez, M., and Heath, M. C. (1991). Interactions of the nonhost Frenchbean plant (*Phaseolus vulgaris*) with parasitic and saprophytic fungi. IV. Effect of preinoculation with the bean rust fungus on growth of parasitic fungi nonpathogenic on bean, Can. J. Bot., 69:1642–1646.

Fernando, G., Zimmerman, W., and Kolattukudy, P. E. (1984). Suberin-grown *Fusarium solani* f.sp. *pisi* generates a cutinase-like esterase which depolymerizes the aliphatic components of suberin, Physiol. Plant Pathol., 24:143–155.

Fielding, A. H. (1981). Natural inhibitors of fungal polygalacturonases in infected tissues, J. Gen. Microbiol., 123:377–381.

Fink, J., Jeblick, W., and Kauss, H. (1990). Partial purification and immunological characterization of 1,3-β-glucan synthase from suspension cells of *Glycine max*, Planta, 181:343–348.

Fink, J., Jeblick, W., Blaschek, W., and Kauss, H. (1987). Calcium ions and polyamines activate the plasma membrane-located 1,3-β-glucan synthase, Planta, 171:130–135.

Fisher, M. L., Anderson, A. J., and Albersheim, P. (1973). Host-pathogen interactions. VI. A single plant protein efficiently inhibits endopolygalacturonases secreted by *Colletotrichum lindemuthianum* and *Aspergillus niger*, Plant Physiol., 51:489–491.

Flott, B. E., Moerschbacher, B. M., and Reisener, H. J. (1989). Peroxidase isoenzyme patterns of resistant and susceptible wheat leaves following stem rust infection, New Phytol., 111:413–421.

Flurkey, W. H., and Kolattukudy, P. E. (1981). In vitro translation of cutinase mRNA: Evidence for a precursor form of an extracellular fungal enzyme, Arch. Biochem. Biophys., 212:154–161.

Forster, H., and Rasched, I. (1985). Purification and characterization of extracellular pectinesterases from *Phytophthora infestans*, Plant Physiol., 77:109–112.

Fredrikson, K., and Larsson, C. (1989). Activation of 1,3-β-glucan synthase by Ca^{2+}, spermine and cellobiose. Localization of activator sites using inside-oat plasma membrane vesicles, Physiol. Plant., 77:196–201.

Fredrikson, T., Kjellbom, P., and Larsson, P. (1991). Isolation and polypeptide composition of 1,3-β-glucan synthase from plasma membranes of *Brassica oleracea*, Physiol. Plant., 81:289–294.

Frey-Wyssling, A. (1969). The structure and biogenesis of native cellulose, Fortschr. Chem. Organ. Naturst., 27:1–30.

Fric, F. (1984). Biochemical changes in barley plants in the preparasitic stage of powdery mildew-barley interactions, Acta Phytopathol. Acad. Sci. Hung., 19:183–191.

Friend, J. (1981). Plant phenolics, lignification and plant disease, Progress in Phytochemistry, Vol. 7 (L. Rheinhold, J. B. Harborne and T. Swain, eds.), Pergamom Press, New York and London, pp. 223–263.

Frost, D. J., Read, S. M., Drake, R. R., Haley, B. E., and Wasserman, B. P. (1990). Identification of the UDP-Glc-binding polypeptide of callose synthase from *Beta vulgaris* L. by photoaffinity labeling with 5-azido-UDP-glucose, J. Biol. Chem., 265:2162–2167.

Fry, S. C. (1982). Isodityrosine, a new cross-linking amino acid from plant cell wall glycoprotein, Biochem. J., 204:449–455.

Fuchs, A., Rohringer, R., and Samborski, D. J. (1967). Metabolism of aromatic compounds in healthy and rust-infected primary leaves of wheat. II. Studies with L-phenylalanine-U-^{14}C-, L-tyrosine-U-^{14}C, and ferulate-U-^{14}C, Can. J. Bot., 45:2137–2153.

Gao, S., and Shain, L. (1994). Characterization of an endopolygalacturonase produced by the chestnut blight fungus, Physiol. Mol. Plant Pathol., 45:169–179.

Gao, S., and Shain, L. (1995). Activity of polygalacturonase produced by *Cryphonectria parasitica* in chestnut bark and its inhibition by extracts from American and Chinese chestnut, Physiol. Mol. Plant Pathol., 46:199–213.

Gil, F., and Gay, J. L. (1977). Ultrastructural and physiological properties of the host interfacial components of haustoria of *Erysiphe pisi* in vivo and in vitro, Physiol. Mol. Plant Pathol., 10:1–12.

Gobel, H., and Bock, W. (1978). Isolierung und charakterisierung eines Pektinase-Inhibitors aus grunen Bohnen (*Phaseolus vulgaris*), Die Nahrung, 22:809–818.

Godwin, J. R., Mansfield, J. W., and Darby, P. (1987). Microscopical studies of resistance to powdery mildew disease in the hop cultivar wye target, Plant Pathol., 36:21–32.

Gold, R. E., Aist, J. R., Hazen, B. E., Stolzeneurg, M. C., Marshall, M. R., and Israel, H. W. (1986). Effects of calcium nitrate and chlortetracycline on papilla formation, ml-O resistance, and susceptibility of barley to powdery mildew, Physiol. Mol. Plant Pathol., 29:115–129.

Graham, M. Y., and Graham, T. L. (1991). Rapid accumulation of anionic peroxidases and phenolic polymers in soybean cotyledon tissues following treatment with *Phytophthora megasperma* f.sp. *glycinea* wall glucan, Plant Physiol., 97:1445–1455.

Grand, C., and Rossignol, M. (1982). Changes in the lignification process induced by localized infection of muskmelons with *Colletotrichum lagenarium*, Plant Sci. Lett., 28:103–110.

Grand, C., Sarni, F., and Boudet, A. M. (1985). Inhibition of cinnamyl-alcohol-dehydrogenase activity and lignin biosynthesis in poplar (*Populus* x *euramericana Dode*) tissues by two organic compounds, Planta, 163:232–237.

Grand, C., Sarni, F., and Lamb, C. J. (1987). Rapid induction by fungal elicitor of the synthesis of cinnamyl-alcohol dehydrogenase, a specific enzyme of lignin synthesis, Eur. J. Biochem., 169:73–77.

Grand, C., Walter, M., Sarni, F., and Lamb, C. J. (1986). Elicitation at the mRNA level of cinnamyl alcohol dehydrogenase, a specific enzyme of lignification in bean cell suspension cultures, International Symposium on Plant Molecular Biology, Strasbourg, July, 1986.

Gregersen, P. L., Collinge, D. B., and Smedegaard-Petersen, V. (1990). Early induction of new mRNAs accompanies the resistance reaction of barley to the wheat pathogen, *Erysiphe graminis* f.sp. *tritici*, Physiol. Mol. Plant Pathol., 36:471–481.

Grisebach, H. (1981). Lignins, The Biochemistry of Plants, Vol. 7 (P. Stump and E. E. Conn, eds.), Academic Press, New York, pp. 457–478.

Hachler, H., and Hohl, H. R. (1984). Temporal and spatial distribution patterns of collar and papillae wall appositions in resistant and susceptible tuber tissue of *Solanum tuberosum* infected by *Phytophthora infestans*, Physiol. Plant Pathol., 24:107–118.

Hagerman, A. E., Blan, D. M., and McClure, A. L. (1985). Plate assay for determining the time of production of protease, cellulase and pectinases by germinating fungal spores, Physiol. Plant Pathol., 17:247–289.

Hahlbrock, K., and Scheel, D. (1989). Physiology and molecular biology of phenylpropanoid metabolism, Annu. Rev. Plant Physiol. Plant Mol. Biol., 40:347–369.

Hahlbrock, K., Lamb, C. J., Purwin, C., Ebel, J., Fautz, E., and Schafer, E. (1981). Rapid response of suspension-cultured parsley cells to the elicitor from *Phytophthora megasperma* var. *sojae*. Induction of the enzymes of general phenylpropanoid metabolism, Plant Physiol., 67:768–773.

Hajlaoui, M. R., Benhamou, N., and Belanger, R. R. (1991). Cytochemical aspects of fungal penetration, haustorium formation and interfacial material in rose leaves infected by *Sphaerotheca pannosa* var. *rosae*, Physiol. Mol. Plant Pathol., 39:341–355.

Hammerschmidt, R., and Kuc, J. (1982). Lignification as a mechanism for induced systemic resistance in cucumber, Physiol. Plant Pathol., 20:61–71.

Hammerschmidt, R., Lamport, D. T. A., and Muldoon, E. P. (1984). Cell wall hydroxyproline enhancement and lignin deposition as an early event in the resistance of cucumber to *Cladosporium cucumerinum*, Physiol. Plant Pathol., 24:43–47.

Hammerschmidt, R., Nuckles, E. M., and Kuc, J. (1982). Association of enhanced peroxidase activity with induced systemic resistance of cucumber to *Colletotrichum lagenarium*, Physiol. Plant Pathol., 20:73–82.

Hancock, J. G. (1967). Hemicellulose degradation in sunflower hypocotyls infected with *Sclerotinia sclerotiorum*, Phytopathology, 57:203–206.

Hancock, J. G. (1976). Multiple forms of endo-pectate lyase formed in culture and in infected squash hypocotyls by *Hypomyces solani* f.sp. *cucurbitae*, Phytopathology, 66:40–45.

Hartley, R. D., and Harris, P. J. (1981). Phenolic constituents of the cell walls of dicotyledons, Biochem. System. Ecol., 9:189–203.

Hartley, R. D., and Jones, E. C. (1977). Lignin–carbohydrate linkages in plant cell walls, Part 5. Phenolic components and degradability of cell walls of grass and legum species, Phytochemistry, 16:1531–1535.

Hashioka, Y. (1975). Ultrastructure of host and nonhost reaction to rice blast fungus, Proc. Jpn. Acad., 51:756–759.

Hass, D., Hackspacher, G., and Franz, G. (1985). Orientation of cell wall β-glucan synthase in plasma membrane vesicles, Plant Sci. 41:1–9.

Heale, J. B., and Gupta, D. P. (1972). Mechanism of vascular wilting induced by *Verticillium albo-atrum*, Trans. Br. Mycol. Soc., 58:19–28.

Heath, M. C. (1984). Light and electron microscope studies of the interactions of host and nonhost plants with the cowpea rust-*Uromyces phaseoli* var. *vignae*, Physiol. Plant Pathol., 4:403–414.

Heath, M. C. (1976). Ultrastructural and functional similarity of the haustorial neckband of rust fungi and the casparian strip of vascular plants, Can. J. Bot., 54:2484–2489.

Heath, M. C. (1979). Partial characterization of the electron-opaque deposits formed in the non-host plant; French bean, after cowpea rust infection. Physiol. Plant Pathol., 15:141–148.

Heath, M. C. (1981a). Insoluble silicon in necrotic cowpea cells following infection with an incompatible isolate of the cowpea rust fungus, Physiol. Plant Pathol., 19:273–276.

Heath, M. C. (1981b). The suppression of the development of silicon-containing deposits in French bean leaves by exudates of the bean rust fungus and extracts from bean tissues, Physiol. Plant Pathol., 18:149–155.

Heckman, J. W., Terhune, B. I., and Lamport, D. T. A. (1988). Characterization of native and modified extensin monomer and oligomer by electron microscopy and gel filtration, Plant Physiol., 86:848–856.

Heitefuss, R., and Ebrahim-Nesbat, F. (1986). Ultrastructural and histochemical studies on mildew of barley (*Erysiphe graminis* DC. f.sp. *hordei* Marchal). III. Ultrastructure of different types of papillae in susceptible and adult plant resistant leaves, J. Phytopathol., 116:358–373.

Henderson, S. J., and Friend, J. (1979). Increase in PAL and lignin-like compounds as race-specific resistance responses of potato tubers to *Phytophthora infestans*, Phytopathol. Z., 94:323–334.

Hershenhorn, J., Manulis, S., and Barash, I. (1990). Polygalacturonases associated with infection of Valencia orange by *Penicillium italicum*, Phytopathology, 80:1374–1376.

Hinch, J. M., and Clarke, A. E. (1982). Callose formation in *Zea mays* as a response to infection with *Phytophthora cinnamomi*, Physiol. Plant Pathol., 21:113–124.

Hijwegen, T. (1963). Lignification, as a mechanism of induced systemic resistance in cucumber, Physiol. Plant Pathol., 20:61–71.

Hislop, E. C., Paver, J. L., and Keon, J. P. R. (1982). An acid protease produced by *Monilinia fructigena* in vitro and in infected apple fruits and its possible role in pathogenesis, J. Gen. Microbiol., 128:799–807.

Hoffman, R. M., and Turner, J. G. (1982). Partial purification of proteins from pea leaflets that inhibit *Ascochyta pisi* endopolygalacturonase, Physiol. Plant Pathol., 20:173–187.

Hoffman, R. M., and Turner, J. G. (1984). Occurrence and specificity of an endopolygalacturonase inhibitor in *Pisum sativum*, Physiol. Plant Pathol., 24:49–59.

Holden, D. W., and Rohringer, R. (1985). Peroxidases and glycosidases in intercellular fluids from noninoculated and rust-affected wheat leaves, Plant Physiol., 79:820–824.

Holden, F. R., and Walton, J. D. (1992). Xylanases from the fungal maize pathogen *Cochliobolus carbonum*, Physiol. Mol. Plant Pathol., 40:39–47.

Holz, G., and Knox-Davies, P. S. (1985). Production of pectic enzymes by *Fusarium oxysporum* f.sp. *cepae* and its involvement in onion bulb rot, Phytopathol. Z., 112:69–80.

Hondelmann, W., and Richter, E. (1973). On the susceptibility of strawberry clones to *Botrytis cinerea* Pers. in relation to pectin quantity and quality in the fruit. Gartenbauwissenschaft, 38:311–314.

Howard, R. J., and Ferrari, M. A. (1989). Role of melanin in appressorium function, Exp. Mycol., 13:403–418.

Howard, R. J., Ferrari, M. A., Roach, D. H., and Money, N. P. (1991). Penetration of hard substrates by a fungus employing enormous turgor pressures, Proc. Natl. Acad. Sci. USA, 88:11281–11284.

Howell, C. R. (1976). Use of enzyme-deficient mutants of *Verticillium dahliae* to assess the importance of pectolytic enzymes in symptom expression of *Verticillium* wilt of cotton, Physiol. Plant Pathol., 9:279–283.

Howell, H. E. (1975). Correlation of virulence with secretion in vitro of three wall-degrading enzymes in isolates of *Sclerotinia fructigena* obtained after mutagen treatment, J. Gen. Microbiol., 90:32–40.

Huang, Q., and Kuc, J. (1995). Cutin, cutinase and esterase as related to the induced systemic resistance of cucumber against *Colletotrichum lagenarium*, Physiol. Mol. Plant Pathol., 46:215–226.

Iiyama, K., Lam, T. B. T., and Stone, B. A. (1994). Covalent cross-links in the cell wall, Plant Physiol., 104:315–320.

Inoue, S., Aist, J. R., and Macko, V. (1992). Induction of both earlier papilla formation and resistance to powdery mildew by a papilla-regulating extract from barley, Phytopathology, 82:1127.

Inoue, S., Macko, V., and Aist, J. R. (1993). Purification and identification of the active component of a papilla-regulating extract (PRE) from barley, Phytopathology, 83:1380.

Inoue, S., Aist, J. R., and Macko, V. (1994a). Earlier papilla formation and resistance to barley powdery mildew induced by a papilla-regulating extract, Physiol. Mol. Plant Pathol., 44:433–440.

Inoue, S., Macko, V., and Aist, J. R. (1994b). Identification of the active component in the papilla-regulating extract, Physiol. Mol. Plant Pathol., 44:441–453.

Inoue, S., Maeda, K., Uematsu, T., and Kato, T. (1984a). Comparison of tetrachlorophthalide and pentachlorobenzyl alcohol with chlobenhiazone and other melanin inhibitors in the mechanism of rice blast control, J. Pestic. Sci., 9:731–736.

Inoue, S., Uematsu, T., and Kato, T. (1984b). Effects of chlobenthiazone on the infection process by *Pyricularia oryzae*, J. Pestic. Sci., 9:689–695.

Ishida, M., Surmi, H., and Oku, H. (1969). Pentachlorobenzyl alcohol, a rice blast control agent, Residue Rev., 25:139–148.

Israel, H. W., Wilson, R. G., Aist, J. R., and Kunoh, H. (1980). Cell wall apposition and plant disease resistance: Acoustic microscopy of papillae that block fungal ingress, Proc. Natl. Acad. Sci. USA, 77:2046–2049.

Jackson, L. I., Dobbs, L., Hildebrand, A., and Yokiel, R. A. (1973). Surface lipids of wheat stripe rust uredospores, *Puccinia striiformis*, compared to those of the host, Phytochemistry, 12:2233–2237.

Jacob, S. R., and Northcote, D. H. (1985). In vitro glucan synthesis by membranes of celery petioles: the role of the membrane in determining the type of linkage formed, J. Cell Sci. Suppl. 2:1–11.

Jaeck, E., Dumas, B., Geoffroy, P., Faver, N., Inze, D., Van Montagu, M., Fritig, B., and Legrand, M. (1992). Regulation of enzymes involved in lignin biosynthesis: induction of O-methyltransferase mRNAs during the hypersensitive reaction of tobacco to tobacco mosaic virus, Mol. Plant–Microbe Interact., 5:294–300.

Jaffe, M. J., Huberman, M., Johnson, J., and Telewski, F. W. (1985). Thigmomorphogenesis: The induction of callose formation and ethylene evolution by mechanical perturbation in bean *Phaseolus vulgaris* cultivar cherokee wax stems, Physiol. Plant., 64:271–279.

Jaffe, M. J., and Leopold, A. C. (1984). Callose deposition during gravitropism of *Zea mays* and *Pisum sativum* and its inhibition by 2-deoxy-D-glucose, Planta, 161:20–26.

Jarvis, M. C. (1984). Structure and properties of pectin gels in plant cell walls, Plant, Cell Environ., 7:153–164.

Jarvis, M. C., Threlfall, D. R., and Friend, J. (1981). Potato cell wall polysaccharides: degradation with enzymes from *Phytophthora infestans*, J. Exp. Bot., 32:1309–1319.

Jenns, A. E., and Kuc, J. (1980). Characteristics of anthracnose resistance induced by localized infection of cucumber with tobacco necrosis virus, Physiol. Plant Pathol., 17:81–91.

Jiang, D., Zeyen, R. J., and Russo, V. (1989). Silicon enhances resistance of barley to powdery mildew, Phytopathology, 79:1198.

Jin, D. F., and West, C. A. (1984). Characteristics of galacturonic acid oligomers as elicitors of casbene synthetase activity in castor bean seedlings, Plant Physiol., 74:989–992.

Johnston, D. J., Ramanathan, V., and Williamson, B. (1993). A protein from immature raspberry fruits which inhibits endopolygalacturonases from *Botrytis cinerea* and other microorganisms, J. Expt. Bolt., 44:971–976.

Johnston, D. J., and Williamson, B. (1992). Purification and characterization of four polygalacturonases from *Botrytis cinerea*, Mycol. Res., 96:343–349.

Jones, T. M., Anderson, A. J., and Albersheim, P. (1972). Host-pathogen interactions. IV. Studies on the polysaccharide-degrading enzymes secreted by *Fusarium oxysporum* f.sp. *lycopersici*, Physiol. Plant Pathol., 2:153–166.

Jorrin, J., and Dixon, R. A. (1990). Stress responses in alfalfa (*Medicago sativa* L.). II. Purification, characterization, and induction of phenylalanine ammonia-lyase isoforms and elicitor-treated cell suspension cultures, Plant Physiol., 92:447–455.

Kaile, A., Pitt, D., and Kuhn, P. J. (1991). Release of calcium and other ions from various plant host tissues infected by different necrotrophic pathogens with special reference to *Botrytis cinerea* Pers., Physiol. Mol. Plant Pathol., 38:275–291.

Kannaiyan, J., Vidhyasekaran, P., and Kandaswamy, T. K. (1975). Production of pectolytic and cellulolytic enzymes by bajra ergot fungus, Indian Phytopath., 28:123–125.

Karjalainen, R., Peltonen, S., Kajander, O., and NikuPaavola, M. L. (1992). Production and characterization of a xylanase from a phytopathogenic fungus *Bipolaris sorokiniana*, Xylans and Xylanases, Progress in Biotechnology, Vol. 7, (J. Visser, G. Beldman, M. A. Kusters van Someren, and A. G. J. Voragen, eds.), Elsevier, Amsterdam, pp. 97–110.

Karr, A. L., and Albersheim, P. (1970). Polysaccharide-degrading enzymes are unable to attack plant cell wall without prior action by a "wall modifying enzyme," Plant Physiol., 46:60–80.

Kato, Y., and Nevins, D. J. (1984a). Enzymic dissociation of Zea shoot cell wall polysaccharides, Plant Physiol., 75:740–765.

Kato, Y., and Nevins, D. J. (1984b). Enzymic dissociation of Zea shoot cell-wall polysaccharides. II. Dissociation of (1-3), (1-4)-β-D-glucan by purified (1-3), (1-4)-β-D-glucan 4 glucanohydrolase from *Bacillus subtilis*, Plant Physiol., 75:745–752.

Katoh, M., Hirose, I., Kubo, Y., Hikichi, Y., Kunoh, H., Furusawa, I., and Shishiyama, J. (1988). Use of mutants to indicate factors prerequisite for penetration of *Colletotrichum lagenarium* by appressoria, Physiol. Mol. Plant Pathol., 32:177–184.

Kauss, H. (1987a). Some aspects of calcium-dependent regulation in plant metabolism, Annu. Rev. Plant Physiol., 38:47–72.

Kauss, H. (1987b). Callose synthase, Regulation durch induzierten Ca^{2+}-Einstrom in Pflanzenzellen, Naturwissenchaften, 74:275–281.

Kauss, H. (1990). Role of the plasma membrane in host/pathogen interaction, The Plant Plasma Membrane—Structure, Function and Molecular Biology (C. Larsson and I. M. Moller, eds.), Springer-Verlag, Berlin, pp. 320–350.

Kauss, H., Jeblick, W., and Domard, A. (1989). The degrees of polymerization and N-acetylation of chitosan determine its ability to elicit callose formation in suspension cells and protoplasts of *Catharanthus roseus*, Planta, 178:385–392.

Kauss, H., Kohle, H., and Jeblick, W. (1983). Proteolytic activation and stimulation by Ca^{2+} of glucan synthase from soybean cells, FEBS Lett., 158:84–88.

Kauss, H., Waldmann, T., Jeblick, N., Euler, G., Ranjeva, D., and Domard, A. (1990). Ca^{2+} is an important but not the only signal in callose synthesis induced by chitosan, saponins, and polyene antibiotics, Signal Molecules in Plant–Microbe Interactions (B. J. J. Lugtenberg, ed.), Springer-Verlag, Berlin, pp. 107–116.

Kauss, H., Waldmann, T., Jeblick W., and Takemoto, J. Y. (1991). The phytotoxin syringomycin elicits Ca^{2+}-dependent callose synthesis in suspension-cultured cells of *Catharanthus roseus*, Physiol. Plant., 81:134–138.

Kawasaki, S. (1991). Extensin secreted into the culture medium of tobacco cells. II. Extensin subfamilies of similar composition, Plant Cell Physiol., 32:721–728.

Keen, N. T., and Erwin, D. C. (1971). Endopolygalacturonase: evidence against involvement in *Verticillium* wilt of cotton, Phytopathology, 61:198–203.

Keller, B., and Lamb, C. J. (1989). Specific expression of a novel cell wall hydroxyproline-rich glycoprotein gene in lateral root initiation, Genes Dev., 3:1639–1646.

Kenji, T., Aist, J. R., and Israel, H. W. (1985). Distribution of hydrolytic enzymes at barley powdery mildew encounter sites: implications for resistance associated with papilla formation in a compatible system, Physiol. Plant Pathol., 27:167–184.

Keon, J. P. R. (1985). Cytological damage and cell-wall modification in cultured apple cells following exposure to pectin lyase from *Monilinia fructigena*, Physiol. Plant Pathol., 26:11–29.

Keon, J. P. R., Byrde, R. J. W., and Cooper, R. M. (1987). Some aspects of fungal enzymes that degrade plant cell walls, Fungal Infection of Plants (G. F. Pegg and P. G. Ayers, eds.), Cambridge University Press, Cambridge, pp. 133–157.

Keon, J. P. R., Waksman, G. W., and Bailey, J. A. (1990). A comparison of the biochemical and physiological properties of a polygalacturonase from two races of *Colletotrichum lindemuthianum*, Physiol. Mol. Plant Pathol., 37:193–206.

Kerby, K., and Somerville, S. (1989). Enhancement of specific intercellular peroxidases following inoculation of barley with *Erysiphe graminis* f.sp. *hordei*, Physiol. Mol. Plant Pathol., 35:323–337.

Kidger, A. L., and Carver, T. L. W. (1981). Autofluorescence in oats infected by powdery mildew, Trans. Br. Mycol. Soc., 76:405–409.

Kieliszewski, M., and Lamport, D. T. A. (1987). Purification and partial characterization of a hydroxyproline-rich glycoprotein in a graminaceous monocot, *Zea mays*, Plant Physiol., 85:823–827.

Kieliszewski, M. J., Leykam, J. F., and Lamport, D. T. A. (1990). Structure of the threonine-rich extensin from *Zea mays*, Plant Physiol., 92:316–326.

Kimmins, W. C., and Brown, R. G. (1973). Hypersensitive resistance: The role of cell wall glycoproteins in vitro localization, Can. J. Bot., 51:1923–1936.

Kimura N., Tsuge, T., Doke, N., and Nishimura, S. (1990). Isolation of genes involved in melanin biosynthesis of *Alternaria alternata*, Ann. Phytopath. Soc. Japan, 56:371.

Kita, N., Toyoda, H., and Shishiyama, J. (1981). Chronological analysis of cytological responses in powdery-mildewed barley leaves, Can. J. Bot., 59:1761–1768.

Kleis-san Francisco, S. M., and Tierney, M. L. (1990). Isolation, and characterization of a proline-rich cell wall protein from soybean seedling, Plant Physiol., 94:1897–1902.

Koga, H., Toyoda, H., Mayama, S., and Shishiyama, J. (1980). Correlation between the deposition of fluorescent compounds in papillae and resistance in barley against *Erysiphe graminis hordei*, Can. J. Bot., 58:536–541.

Koda, H., Toyoda, H., Mayama, S., Shishiyama, J., and Hiura, U. (1983). Characteristic cellular responses as expression of genes for resistance to *Erysiphe graminis* f.sp. *hordei* in barley, Phytopathology, 73:907–910.

Koga, H., Zeyen, R. J., Bushnell, W. R., and Ahlstrand, G. G. (1988). Hypersensitive cell death, autofluorescence, and insoluble silicon accumulation in barley leaf epidermis cells under attack by *Erysiphe graminis* f.sp. *hordei*, Physiol. Mol. Plant Pathol., 32:395–409.

Kohle, H., Jeblick, W., Poten, F., Blaschek, W., and Kauss, H. (1985). Chitosan-elicited callose synthesis in soybean cells as a Ca^{2+}-dependent process, Plant Physiol., 77:544–551.

Kohle, H., Young, D. H., and Kauss., A. (1983). Physiological changes in suspension cultured soybean cells elicited by treatment with chitosan, Plant Sci. Lett., 33:221–230.

Kolattukudy, P. E. (1970). Biosynthesis of cuticular lipids, Annu. Rev. Plant Physiol., 21:163–192.

Kolattukudy, P. E. (1977). Lipid polymers and associated phenolics, their chemistry,

biosynthesis, and role in pathogenesis, Recent Advances in Phytochemistry, Vol. 2 (F. A. Loewus and V. C. Runeckles, eds.), Plenum Press, New York, pp. 185–246.
Kolattukudy, P. E. (1980). Biopolyester membranes of plants. Cutin and suberin, Science, 208:990.
Kolattukudy, P. E. (1981). Structure, biosynthesis and biodegradation of cutin and suberin, Annu. Rev. Plant Physiol., 32:539–567.
Kolattukudy, P. E. (1984a). Biochemistry and function of cutin and suberin, Can. J. Bot., 62:2918–2933.
Kolattukudy, P. E. (1984b). Cutinases from fungi and pollen, Lipases (B. Borgstrom and H. Brockman, eds.), Elsevier, North Holland, Amsterdam, pp. 471–504.
Kolattukudy, P. E. (1984c). Fungal penetration of defensive barriers of plants, Structure, Function and Biosynthesis of Plant Cell Walls (W. M. Dugger and S. Bartnicki-Garcia, eds.), American Society of Plant Physiologists, Rockville, MD, pp. 302–343.
Kolattukudy, P. E. (1985). Enzymatic penetration of the plant cuticle by fungal pathogens. Annu. Rev. Phytopathol., 23:223–250.
Kolattukudy, P. E., Crawford, M. S., Woloshuk, C. P., Ettinger, W. F., and Soliday, C. I. (1987). The role of cutin, the plant cuticular hydroxy fatty acid polymer, in the fungal interaction with plants, Ecology and Metabolism of Plant Lipids (G. Fuller and W. D. Nes, eds.), American Chemical Society, Washington, DC. pp. 152–175.
Kolattukudy, P. E., Espelie, K. E., and Soliday, C. J. (1981). Hydrophobic layers attached to cell walls. Cutin, suberin and associated waxes, Plant Carbohydrates II. Encyclopedia of Plant Physiology New Series, Vol. 13B (W. Tanner and F. A. Loewus, eds.), Springer Verlag, Berlin, pp. 225–254.
Kolattukudy, P. E., Ettinger, W. F., and Sebastian, J. (1987). Cuticular lipids in plant-microbe interactions, The Metabolism, Structure, and Function of Plant Lipids (P. K. Stumpf, J. B. Mudd, and W. D. Nes, eds.), Plenum Press, New York, pp. 473–480.
Kolattukudy, P. E., and Koller, W. (1983). Fungal penetration of the first line defensive barriers of plants, Biochemical Plant Pathology (J. A. Callow, ed.), John Wiley & Sons, Chichester, pp. 79–100.
Kolattukudy, P. E., Podilla, G. K., and Mohan, R. (1989). Molecular basis of the early events in plant-fungus interaction, Genome, 31:342–349.
Kolattukudy, P. E., Podilla, G. K., Roberts, E., and Dickman, M. D. (1989). Gene expression resulting from the early signals in plant-fungus interaction, Molecular Biology of Plant–Pathogen Interactions (B. J. Staskawicz, P. Ahlquist, and O. Yoder, eds.), Alan R. Liss, New York, pp. 87–102.
Kolattukudy, P. E., Purdy, R. E., and Maiti, I. B. (1981). Cutinases from fungi and pollen, Methods Enzymol., 71:652–664.
Kolattukudy, P. E., and Soliday, C. L. (1985). Effects of stress on the defensive barriers of plants, Cellular and Molecular Biology of Plant Stress, Vol. 22 (J. L. Key and T. Kosuge, eds.), University of California at Los Angeles, Symp. Mol. Cell Biology (NS), pp. 381–400.
Kolattukudy, P. E., and Walton, T. J. (1973). The biochemistry of plant cuticular lipids, Progress in the Chemistry of Fats and Lipids, Vol. 13 (R. T. Holman, ed.), Pergamon Press, Oxford, pp. 13:121–175.
Kollar, A. (1994). Characterization of specific induction, activity, and isozyme polymor-

phism of extracellular cellulases from *Venturia inaequalis* detected in vitro and on the host plant, Mol. Plant–Microbe Interact., 7:603–611.

Koller, W. (1991). The plant cuticle: a barrier to be overcome by plant pathogens, The Fungal Spore and Disease Initiation in Plants and Animals, (G. T. Cole and H. C. Hoch, eds.), Plenum Press, New York, pp. 219–246.

Koller, W., Allan, C. R., and Kolattukudy, P. E. (1982a). Role of cutinase and cell wall degrading enzymes in infection of *Pisum sativum* by *Fusarium solani* f.sp. *pisi*, Physiol. Plant Pathol., 20:47–60.

Koller, W., Allan, C. R., and Kolattukudy, P. E. (1982b). Inhibition of cutinase and prevention of fungal penetration into plants by Benomyl—a possible protective mode of action, Pest. Biochem. Physiol., 18:15–25.

Koller, W., Allan, C. R., and Kolattukudy, P. E. (1982c). Protection of *Pisum sativum* from *Fusarium solani* f.sp. *pisi* by inhibition of cutinase with organophosphorous pesticides, Phytopathology, 72:1425–1430.

Koller, W., and Kolattukudy, P. E. (1982). Mechanism of action of cutinase: chemical modification of the catalytic triad characteristic for serine hydrolases, Biochemistry, 21:3083–3090.

Koller, W., and Parker, D. M. (1989). Purification and characterization of cutinase from *Venturia inaequalis*, Phytopathology, 79:278–283.

Koller, W., Parker, D. M., and Becker, C. M. (1991). Role of cutinase in the penetration of apple leaves by *Venturia inaequalis*, Phytopathology, 81:1375–1379.

Kritzman, G., Chet, I., and Gilan, D. (1981). Spore germination and penetration of *Botrytis allii* into *Allium cepa* host, Bot. Gaz., 142:151–155.

Kubo, Y., and Furusawa, I. (1986). Localization of melanin in appressoria of *Colletotrichum lagenarium*, Can. J. Microbiol., 32:280–282.

Kubo, Y., and Furusawa, I. (1991). Melanin biosynthesis: pre-requisite for successful invasion of the plant host by appressoria of *Colletotrichum* and *Pyricularia*, The Fungal Spore and Disease Initiation in Plants and Animals (G. T. Cole and H. C. Hoch, eds.), Plenum Publishing, New York, pp. 205–218.

Kubo, Y., Furusawa, I., and Shishiyama, J. (1987). Relationship between pigment intensity and penetrating ability in appressoria of *Colletotrichum lagenarium*, Can. J. Microbiol., 33:871–873.

Kubo, Y., Furusawa, I., and Yamamoto, M. (1984). Regulation of melanin biosynthesis during appressorium formation in *Colletotrichum lagenarium*, Exp. Mycol., 8:364–369.

Kubo, Y., Nakamura H., Kobayashi, K., Okuno, T., and Furusawa, I. (1991). Cloning of a melanin biosynthetic gene essential for appressorial penetration of *Colletotrichum lagenarium*, Mol. Plant–Microbe Interact., 4:440–445.

Kubo, Y., Suzuki, K., Furusawa, I., Ishida, N., and Yamamoto, M. (1982a). Relation of appressorium pigmentation and penetration of nitrocellulose membranes by *Colletotrichum lagenarium*, Phytopathology, 72:498–501.

Kubo, Y., Suzuki, K., Furusawa, I., and Yamamoto, M. (1982b). Effect of tricyclazole on appressorial pigmentation and penetration from appressoria of *Colletotrichum lagenarium*, Phytopathology, 72:1198–1200.

Kubo, Y., Suzuki, K., Furusawa, I., and Yamamoto, M. (1983). Scytalone as a natural

intermediate of melanin biosynthesis in appressoria of *Colletotrichum lagenarium*, Exp. Mycol., 7:208–215.
Kubo, Y, Suzuki, K., Furusawa, I., and Yamamoto, M. (1985). Melanin biosynthesis as a prerequisite for penetration by appressoria of *Colletotrichum lagenarium*: Site of inhibition by melanin-inhibiting fungicides and their action on appressoria, Pestic. Biochem. Physiol., 23:47–55.
Kubo, Y., Tsuda, M., Furusawa, I., and Shishiyama, J. (1989). Genetic analysis of genes involved in melanin biosynthesis of *Cochliobolus miyabeanus*, Exp. Mycol., 13:77–84.
Kuc, J. (1990). Immunization for the control of plant disease, Biological Control of Soil-Borne Plant Pathogens (D. Hornby, ed.), CAB International, Wallingford, pp. 355–373.
Kumura, N., Tsuge, T., Doke, N., and Nishimura, S. (1990). Isolation of genes involved in melanin biosynthesis of *Alternaria alternata*, Ann. Phytopathol. Soc. Japan, 56:371.
Kunoh, H. (1984). Cytological aspects of penetration of plant epidermis by fungi, Infection Process of Fungi (D. W. Roberts and J. R. Aist, eds.), Conference Report, Rockefeller Foundation, pp. 137–146.
Kunoh, H. (1990). Ultrastructure and mobilization of ions near infection sites, Annu. Rev. Phytopathol., 28:199–207.
Kunoh, H., Aist, J. R., and Hayashimoto, A. (1985). The occurrence of cytoplasmic aggregates induced by *Erysiphe pisi* in barley coleoptile cells before the host cell walls are penetrated, Physiol. Mol. Plant Pathol., 26:199–207.
Kunoh, H., Aist, J. R., and Israel, H. W. (1979). Primary germtubes and host cell penetrations from appressoria of *Erysiphe graminis hordei*, Ann. Phytopath. Soc. Japan, 45:326–332.
Kunoh, H., Aist, J. R., and Israel, H. W. (1986). Elemental composition of barley coleoptile papillae in relation to their ability to prevent penetration by *Erysiphe graminis*, Physiol. Mol. Plant Pathol., 29:69–78.
Kunoh, H., and Ishizaki, H. (1975). Silicon levels near penetration sites of fungi on wheat, barley, cucumber and morning glory leaves, Physiol. Plant Pathol., 5:283–287.
Kunoh, H., and Ishizaki, H. (1976). Accumulation of chemical elements around the penetration sites of *Erysiphe graminis hordei* on barley leaf epidermis. II. Level of silicon in papilla around the haustorial neck, Ann. Phytopathol. Soc. Japan, 42:30–34.
Kunoh, H., Ishizaki, H., and Nakaya, K. (1977). Cytological studies of early stages of powdery mildew in barley and wheat leaves. II. Significance of the primary germtube of *Erysiphe graminis* on barley leaves, Physiol. Plant Pathol., 10:191–199.
Kunoh, H., Kuno, K., and Ishizaki, H. (1985a). Cytological studies of the early stages of powdery mildew in barley and wheat. XI. Autofluorescence and haloes at penetration sites of appressoria of *Erysiphe graminis hordei* and *Erysiphe pisi* on barley coleoptiles, Can. J. Bot., 63:1535–1539.
Kunoh, H., Kuno, K., and Ishizaki, H. (1985b). Cytological studies of the early stages of powdery mildew in barley and wheat. XI. Autofluorescence and haloes at penetration sites of *Erysiphe graminis hordei* on living barley coleoptiles, Physiol. Plant Pathol., 21:373–379.
Kunoh, H., Nicholson, R. L., Yoshioka, H., Yamaoka, N., and Kobayashi, I. (1990).

Preparation of the infection court by *Erysiphe graminis*: degradation of the host cuticle, Physiol. Mol. Plant Pathol., 36:397–407.

Kunoh, H., Takamatsu, S., and Ishizaki, H. (1978). Cytological studies of early stages of powdery mildew in barley and wheat. III. Distribution of residual calcium and silicon in germinated conidia of *Erysiphe graminis* f.sp. *hordei*, Physiol. Plant Pathol., 13:319–325.

Kunoh, H., Tsuuki, T., and Ishizaki, H. (1978). Cytological studies of early stages of powdery mildew in barley and wheat. IV. Direct ingress from superficial primary germ tubes and appressoria of *Erysiphe graminis hordei* on barley leaves, Physiol. Plant Pathol., 13:327–333.

Kunoh, H., Yamamori, K., and Ishizaki, H. (1982). Cytological studies of early stages of powdery mildew in barley and wheat. VIII. Autofluorescence at penetration sites of *Erysiphe graminis hordei* on living barley coleoptiles, Physiol. Plant Pathol., 21:373–379.

Kunoh, H., Yamamori, K., and Ishizaki, H. (1983). Cytological studies of early stages of powdery mildew in barley and wheat. IX. Effect of various inorganic salts on the occurrence of autofluorescence at penetration sites of appressoria of *Erysiphe graminis hordei*, Can. J. Bot., 61:2181–2185.

Kunoh, H., Yamaoka, N., Yoshioka, H., and Nicholson, R. L. (1988). Preparation of the infection court by *Erysiphe graminis*. I. Contact-mediated changes in morphology of the conidium surface, Exp. Mycol., 12:325–335.

Kuo, M. J., and Alexander, M. (1967). Inhibition of the lysis of fungi by melanins, J. Bacteriol., 94:624–629.

Kurosaki, F., Amin, M., and Nishi, A. (1986a). Induction of phytoalexin production and accumulation of phenolic compounds in cultured carrot cells, Physiol. Mol. Plant Pathol., 28:359–370.

Kurosaki, F., Tashiro, N., and Nishi, A. (1986b). Induction of chitinase and phenylalanine ammonia-lyase in cultured carrot cells treated with fungal mycelial walls, Plant Cell Physiol., 27:1587–1591.

Labavitch, J. M., and Ray, P. M., (1978). Structure of hemicellulosic polysaccharides of Avena sativa coleoptile cell walls, Phytochemistry, 17:933–937.

Laborda, F., Archer, S. A., Fielding, A. H., and Byrde, R. J. W. (1974). Studies on the α-L-arabinofuranosidase complex from *Sclerotinia fructigena* in relation of brown rot of apple, J. Gen. Microbiol., 81:151–163.

Lafitte, C., Barthe, J. P., Montillet, J. L., and Touze, A. (1984). Glycoprotein inhibitors of *Colletotrichum lindemuthianum* endopolygalacturonase in near isogenic lines of *Phaseolus vulgaris* resistant and susceptible to anthracnose, Physiol. Plant Pathol., 25:39–53.

Lamport, D. T. A. (1967). Hydroxyproline-O-glycosidic linkage of the plant cell wall glycoprotein extensin, Nature, 216:1322–1324.

Lamport, D. T. A. (1973). The glycopeptide linkage of extensin: 0-D-galactosyl serine and 0-L-arabinosyl hydroxyproline, Biogenesis of the Plant Cell Wall Polysaccharides (F. Loewus, ed.), Academic Press, New York, pp. 149–163.

Lamport, D. T. A., Katona, L., and Roerig, S. (1973). Galactosylserine in extensin, Biochem. J., 133:125–131.

Lawson, S. G., Mason, T. L., Sabin, R. D., Sloan, M. E., Drake, R. R., Haley, D. E., and

Wasserman, B. P. (1989). UDP-Glucose: (1-3)-β-glucan synthase from *Daucus carota* L., Plant Physiol., 90:101–109.

Leach, J. E., Centrell, M. A., and Sequeira, L. (1982). A hydroxyproline-rich bacterial agglutinin from potato: extraction, purification, and characterization, Plant Physiol., 70:1353–1358.

LeCam, B., Massiot, P., Campion, C., and Rouxel, F. (1994a). Susceptibility of carrot cultivars to *Mycocentrospora acerina* and the structure of cell wall polysaccharides, Physiol. Mol. Plant Pathol., 45:139–151.

LeCam, B., Massiot, P., and Rouxel, F. (1994b). Cell wall polysaccharide-degrading enzymes produced by isolates of *Mycocentrospora acerina* differing in aggressiveness on carrot, Physiol. Mol. Plant Pathol., 44:187–198.

Lee, S. W., Nazar, R. N., Powell D. A., and Robb, J. (1992). Reduced PAL gene suppression in *Verticillium*-infected resistant tomatoes, Plant Mol. Biol., 18:345–352.

Lehtinen, U. (1993). Plant cell wall degrading enzymes of *Septoria nodorum*, Physiol. Mol. Plant Pathol., 43:121–134.

Leush, H. J., and Buchenauer, H. (1989). Effects of soil treatments with silica-rich lime fertilizers and sodium trisilicate on the infection of wheat by *Erysiphe graminis* and *Septoria nodorum* depending on the form of N fertilizer, Z. Pflanzenkrankh. Pflanzenschutz., 96:154–172.

Liang, X., Dron, M., Cramer, C. L., Dixon, R. A., and Lamb, C. J. (1989). Differential regulation of phenylalanine ammonia-lyase genes during plant development and by environmental cues, J. Biol. Chem., 264:14486–14492.

Lin, T. S., and Kolattukudy, P. E. (1976). Evidence for novel linkages in a glycoprotein involving β-hydroxyphenylalanine and β-hydroxytyrosine, Biochem. Biophys. Res. Commun., 72:243–250.

Lin, T. S., and Kolattukudy, P. E. (1978). Induction of a polyester hydrolase (cutinase) by low levels of cutin monomers in *Fusarium solani* f.sp. *pisi*, J. Bacteriol., 133:942–951.

Lin, T. S., and Kolattukudy, P. E. (1979). Direct evidence for the presence of β-hydroxyphenylalanine and β-hydroxytyrosine in cutinase from *fusarium solani pisi*, Arch. Biochem. Biophys., 196:225–264.

Lin, T. S., and Kolattukudy, P. E. (1980a). Structural studies on cutinase, a glycoprotein containing novel amino acids and glucuronic acid amide at the N-termimus, Eur. J. Biochem., 106:341–351.

Lin, T. S., and Kolattukudy, P. E. (1980b). Isolation and characterization of a cuticular polyester (cutin) hydrolyzing enzyme from phytopathogenic fungi, Physiol. Plant Pathol., 17:1–15.

Liyanage, H. D., Koller, W., McMillan, R. T., Jr., and Kistler, H. C. (1993). Variation in cutinase from two populations of *Colletotrichum gloeosporioides* from citrus, Phytopathology, 83:113–116.

Lois, R., Dietrich, A., Hahlbrock, K., and Schulz, W. (1989). A phenylalanine ammonia-lyase gene from parsley: structure, regulation and identification of elicitor and light responsive cis-acting elements, EMBO J., 8:1641–1648.

Lorenzo, G. D., Ito, Y., D'Ovidio, R., Cervone, F., Albersheim, P., and Darvill, A. G. (1990). Host-pathogen interactions. XXXVII. Abilities of the polygalacturonase-inhibiting proteins from four cultivars of *Phaseolus vulgaris* to inhibit the endopolygalacturonases from three races of *Colletotrichum lindemuthianum*, Physiol. Mol. Plant Pathol., 36:421–435.

Ludevid, M. D., Ruiz-Avila, L., Vales, M. P., Stiefel, V., Torrent, M., Torne, J. M., and Puigdomenech, P. (1990). Expression of genes for cell-wall proteins in dividing and wounded tissues of *Zea mays* L., Planta, 180:524–529.

Ma, S., Gross, K. D., and Wasserman, B. P. (1991). Developmental regulation of the (1,3)-β-glucan (callose) synthase from tomato, Plant Physiol., 96:664–667.

Maiti, I. B., and Kolattukudy, P. E. (1979). Prevention of fungal infection of plants by specific inhibition of cutinase, Science, 205:507–508.

Manners, J. M., and Gay, J. L. (1983). The host parasite interface and nutrient transfer in biotrophic parasitism, Biochemical Plant Pathology (J. A. Callow, ed.), John Wiley, New York, pp. 165–195.

Marcus, L., Barash, I., Sneh, B., Koltin, Y., and Finkler, A. (1986). Purification and characterization of pectolytic enzymes produced by virulent and hypovirulent isolates of *Rhizoctonia solani* Kuhn, Physiol. Mol. Plant Pathol., 29:325–336.

Martin, J. T. (1964). Role of cuticle in the defense against plant diseases, Annu. Rev. Phytopathol., 2:81–100.

Martinez, M. J., Reyes, F., and Lahoz, R. (1982). Cell wall plant polysaccharide-degrading enzymes in autolysis of *Botrytis cinerea*, Trans. Br. Mycol. Soc., 78:395–403.

Masuta, C., Van den Bulcke, M., Bauw, G., Van Montagu, M., and Caplan, A. B. (1991). Differential effects of elicitors on the viability of rice suspension cells, Plant Physiol., 97:619–629.

Matern, U., and Kneusel, R. E. (1988). Phenolic compounds in plant disease resistance, Phytoparasitica, 16:158–170.

Mathieu, Y., Kurkdjian, A., Xia, H., Guern, J., Koller, A., Spiro, M. D., O'Neill, M., Albersheim, P., and Darvill, A. (1991). Membrane responses induced by oligogalacturonides in suspension-cultured tobacco cells, Plant J., 1:333–343.

Matsuura, K. (1983). Effect of melanin synthesis inhibitors on appressorial function in plant pathogenic fungi, J. Pestic. Sci., 8:379–383.

Mayama S., and Shishiyama, J. (1978). Localized accumulation of fluorescent and U.V.-absorbing compounds at penetration sites in barley leaves infected with *Erysiphe graminis hordei*, Physiol. Plant Pathol., 13:347–354.

Mazau, D., and Esquerre-Tugaye, M. T. (1976). Hydroxyproline et peroxydases dans les parois cellulaires et le cytoplasmic de tiges de melons atteints d'anthracnose, Plant Sci. Lett., 7:119–125.

Mazau, D., and Esquerre-Tugaye, M. T. (1986). Hydroxyproline-rich glycoprotein accumulation in the cell walls of plants infected by various pathogens, Physiol. Mol. Plant Pathol., 29:147–157.

Mazau, D., Rumeau, D., and Esquerre-Tugaye, M. T. (1987). Molecular approaches to understanding cell surface interactions between plants and fungal pathogens, Plant Physiol. Biochem., 25:337–343.

Mazau, D., Rumeau, D., and Esquerre-Tugaye, M. T. (1988). Two different families of hydroxyproline-rich glycoproteins in melon callus, Plant Physiol., 86:540–546.

McKeen, W. E. (1974). Mode of penetration of epidermal cell walls of *Vicia faba* by *Botrytis cinerea*, Phytopathology, 64:461–467.

McNeil, M., Darvill, A. G., Fry, S. C., and Albersheim, P. (1984). Structure and function of the primary cell walls of plants, Annu. Rev. Biochem., 53:625–663.

Mellon, J. E. (1991). Purification and characterization of iso-peroxidases elicited by *Aspergillus flavus* in cotton ovule cultures, Plant Physiol., 95:14–20.

Mellon, J. E., and Helgeson, J. P. (1982). Interaction of hydroxyproline-rich glycoprotein from tobacco callus with potential pathogens, Plant Physiol., 70:401–405.

Menzies, J. G., Ehret, D. L., Glass, A. D. M. and Samuels, A. L. (1991a). The influence of silicon on cytological interactions between *Sphaerotheca fuliginea* and *Cucumis sativus*, Physiol. Mol. Plant Pathol., 39:403–414.

Menzies, J. G., Ehret, D. L., Glass, A. D. M., Helmer, T., Koch, C., and Seywerd, F. (1991b). Effects of soluble silicon on the parasitic fitness of *Sphaerotheca fuliginea* on *Cucumis sativus*, Phytopathology, 81:84–88.

Moerschbacher, B. M., Noll, U., Gorrichon, L., and Reisener, H. J. (1990). Specific inhibition of lignification breaks hypersensitive resistance of wheat to stem rust, Plant Physiol., 93:465–470.

Moerschbacher, B. M., Noll, U. M., Flott, B. E., and Reisener, H. J. (1988). Lignin biosynthetic enzymes in stem rust infected, resistant and susceptible near isogenic wheat lines, Physiol. Mol. Plant Pathol., 33:33–46.

Moerschbacher, B. M., Witte, U., Konigs, D., and Reisener, H. J. (1989). Changes in the level of enzyme activities involved in lignin biosynthesis during the temperature sensitive resistant response of wheat (Sr-6) to stem rust (P6), Plant Sci., 65:183–190.

Mohan, R., and Kolattukudy, P. E. (1990). Differential activation of expression of a suberization-associated anionic peroxidase gene in near-isogenic resistant and susceptible tomato lines by elicitors of *Verticillium albo-atrum*, Plant Physiol., 92:276–280.

Morris, S. C., Forbes-Smith, M. R., and Scriven, F. M. (1989). Determination of optimum conditions for suberization, wound periderm formation, cellular desiccation and pathogen resistance in wounded *Solanum tuberosum* tubers, Physiol. Mol. Plant Pathol., 35:177–190.

Morrow, D. L., and Lucas, W. J. (1986). (1-3)-β-D-glucan synthase from sugar beet. II. Isolation and solubilization, Plant Physiol., 81:171–176.

Movahedi, S., and Heale, J. B. (1990a). Purification and characterization of an aspartic proteinase secreted by *Botrytis cinerea Pers.* ex. *Pers.* in culture and in infected carrots, Physiol. Mol. Plant Pathol., 36:289–302.

Movahedi, S., and Heale, J. B. (1990b). The roles of aspartic proteinase and endo-pectin lyase enzymes in the primary stages of infection and pathogenesis of various host tissues by different isolates of *Botrytis cinerea Pers.* ex. *Pers.*, Physiol. Mol. Plant Pathol., 36:303–324.

Muhlethaler, K. (1967). Ultrastructure and function of plant cell walls, Annu. Rev. Plant Physiol., 18:1–24.

Mullen, J. M., and Bateman, D. F. (1975). Polysaccharide degrading enzymes produced by *Fusarium roseum* "Avenaceum" in culture and during pathogenesis, Physiol. Plant Pathol., 6:233–246.

Mullick, D. B. (1975). A new tissue essential to necrophylactic periderm formation in the bark of four conifers, Can. J. Bot., 53:2443–2457.

Mullick, D. B. (1977). The non-specific nature of defense in bark and wood during wounding, insect, and pathogen attack, Recent Adv. Phytochem., 11:395–441.

Murray, G. M., and Maxwell, D. P. (1975). Penetration of *Zea mays* by *Helminthosporium carbonum*, Can. J. Bot., 53:2872–2883.

Murray, I. D., and Ye, H. (1986). Papilla formation and hypersensitivity at penetration sites and resistance to *Pseudocercosporella herpotrichoides* in winter wheat, Phytopathology, 76:737–744.

Muthusamy, M., Soumini Rajagopalan, C. K., and Vidhyasekaran, P. (1973). Role of pectic and cellulolytic enzymes in pathogenesis of *Pythium aphanidermatum* in tomato seedlings, Madras Agric. J., 60:539–542.

Nasraoui, B., Lepoivre, P., Barthelemy, J. P., and Semal, J. (1990). Evidence of cutinase activity released by *Ascochyta pinodes* and *Ascochyta pisi*, Mededelingen ran de Faculteit Landbouwwetenschappen Rijkssuniversiteit Gent, 55(3a):835–842.

Nicholson, R. L. (1984). Adhesion of fungi to the plant cuticle, Infection Processes in Fungi (D. W. Roberts and J. R. Aist, eds.), Rockefeller Foundation, New York, pp. 74–84.

Nicholson, R. L., and Epstein, L. (1991). Adhesion of fungi to the plant surface, The Fungal Spore and disease Initiation in Plants and Animals (G. T. Cole and H. C. Hoch, eds.), Plenum Press, New York, pp. 3–23.

Nicholson, R. L., Yashioka, H., Yamaoka, N., and Kunoh, H. (1988). Preparation of the infection court by *Erysiphe graminis*. II. Release of esterase enzyme from conidia in response to a contact stimulus, Exp. Mycol., 12:336–349.

Niemann, G. J., and Baayer, R. P. (1988). Involvement of phenol metabolism in resistance of *Dianthus caryophyllus L.* to *Fusarium oxysporum* f.sp. *dianthi*, Neth. J. Plant Pathol., 94:289–301.

Niemann, G. J., Van der Bij, A., Brandt-De-Boer, B., Boon, J. J., and Bayen, R. P. (1991a). Differential response of four carnation cultivars to races 1 and 2 of *Fusarium oxysporum* f.sp. *dianthi* and to *Phialophora cinerescens*, Physiol. Mol. Plant Pathol., 38:117–136.

Niemann, G. J., Van der Kerk, A., Niessen, W. M. A., and Versluis, K. (1991b). Free and cell-wall bound phenolics and other constituents from healthy and fungus-infected carnation (*Dianthus caryophyllus L.*) stems, Physiol. Mol. Plant Pathol., 38:417–432.

Nnodu, E. C., Narrison, M. D., and Parke, R. V. (1982). The effect of temperature and relative humidity on wound healing and infection of potato tubers by *Alternaria solani*, Am. Potato J., 59:297–311.

Nothnagel, E. A., McNeil, M., Albersheim, P., and Dell, A. (1983). Host–pathogen interactions. XXII. A galacturonic acid oligosaccharide from plant cell walls elicits phytoalexins, Plant Physiol., 71:916–926.

O'Brien, V. J., and Leach, S. S. (1983). Investigations into the mode of resistance of potato tubers to *Fusarium roseum* 'Sambucinum,' Am. Potato J., 60:227–233.

O'Connell, R. J., and Bailey, J. A. (1986). Cellular interactions between *Phaseolus vulgaris* and the hemibiotrophic fungus *Colletotrichum lindemuthianum*, Biology and Molecular Biology of Plant–Pathogen Interactions (J. Bailey, ed.), Plenum Press, New York, pp. 39–48.

O'Connell, R. J., Bailey, J. A., and Richmond, D. V. (1985). Cytology and physiology of infection of *Phaseolus vulgaris* by *Colletotrichum lindemuthianum*, Physiol. Plant Pathol., 27:75–98.

O'Connell, R. J., Brown, I. R., Mansfield, J. W., Bailey, J. A., Mazau, D., Rumeau, D., and Esquerre-Tugaye, M. T. (1990). Immunocytochemical localization of hydroxyproline-rich glycoproteins accumulating in melon and bean at sites of resistance to bacteria and fungi, Mol. Plant–Microbe Interact., 3:33–40.

Oeser, B., and Yoder, O. C. (1994). Pathogenesis by *Cochliobolus heterostrophus* transformants expressing a cutinase-encoding gene from Nectria haematococca, Mol. Plant–Microbe Interact., 7:282–288.

Ohana, P., Benziman, M., and Delmer, D. P. (1993). Stimulation of callose synthesis in vivo correlates with changes in intracellular distribution of the callose synthase activator β-furfuryl-β-glucoside, Plant Physiol., 101:187–191.

Ohana, P., Delmer, D. P., Steffens, J. C. Matthews, D. E., Mayer, R., and Benziman, M. (1991). β-Furfuryl-β-glucoside: an endogenous activator of higher plant UDP-Glucose: (1-3)-β-glucan synthase, J. Biol. Chem., 266:13472–13475.

Ohana, P., Delmer, D. P., Volman, G., Steffens, J. C., Matthews, D. E., and Benziman, M. (1992). β-Furfuryl-β-glucoside: an endogenous activator of higher plant UDP-glucose: (1-3)-β-glucan synthase. Biological activity, distribution, and in vitro synthesis, Plant Physiol., 98:708–715.

Okuno, T., Kitamura, Y., and Matsuura, K. (1983a). Mechanism of inhibitory effect of tricyclazole on secondary infection by spores of *Pyricularia oryzae*, J. Pestic. Sci., 8:261–262.

Okuno, T., Matsuura, K., and Furusawa, I. (1983b). Recovery of appressorial penetration by some melanin precursors in *Pyricularia oryzae* treated with tricyclazole and in a melanin deficient mutant, J. Pestic. Sci., 8:357–350.

Olesen, P. (1979). Ultrastructural observations on the cuticular envelope in salt glands of *Frankenia pauciflora*, Protoplasma, 99:1–9.

Pakusch, A. E., Kneusel, R. E., and Matern, U. (1989). S-adenosyl-L-methionine: trans-caffeoyl-CoA 3-0-methyltransferase from elicitor-treated parsley cell suspension cultures, Arch. Biochem. Biophys., 271:488–494.

Pakusch, A. E., Matern, U. (1991). Kinetic characterization of caffeoyl-Coenzyme A-specific 3-0-methyltransferase from elicited parsley suspensions, Plant Physiol., 96:327–330.

Pakusch, A. E., Matern, U., and Schiltz, E. (1990). Elicitor-inducible caffeoyl-CoA 3-0-methyltransferase from *Petroselinum crispum* cell suspensions, purification, partial sequence and antigenicity, Plant Physiol., 95:137–143.

Parameswaran, N., Wilhelm, G. E., and Liese, W. (1976). Ultrastructural aspects of beech bark degradation by fungi, Eur. J. Forens. Pathol., 6:274–286.

Parry, A. L., and Carver, T. L. W. (1986). Relationships between colony development, resistance to penetration and autofluorescence in oats infected with powdery mildew, Trans. Br. Mycol. Soc., 86:355–363.

Pascholati, S. F., Deising, H., Leite, B., Anderson, D., and Nicholson, R.L. (1993). Cutinase and non-specific esterase activities in the conidial mucilage of *Colletotrichum graminicola*, Physiol. Mol. Plant Pathol., 42:37–51.

Pascholati, S. F., Yoshioka, H., Kunoh, H., and Nicholson, R. L. (1992). Preparation of the infection court by *Erysiphe graminis* f.sp. *hordei*: cutinase is a component of the conidial exudate, Physiol. Mol. Plant Pathol., 41:53–59.

Pearce, R. B., and Rutherford, J. (1981). A wound associated suberized barrier to the spread of decay in sapwood of oat (*Quercus robur* L.), Physiol. Plant Pathol., 19:359–369.

Pearson, V. (1974). Virulence and cellulolytic enzyme activity of isolates of *Gaeumannomyces graminis*, Trans. Br. Mycol. Soc., 63:199–202.

Peek, V. R. D., Liese, W., and Parameswaran, N. (1972). Infektion und Abbauder Wulzerinde von Fichte durch *Fomes ammosus*, Eur. J. Forens. Pathol., 2:104–115.

Peng, M., and Kuc, J. (1992). Peroxidase-generated hydrogen peroxide as a source of antifungal activity in vitro and on tobacco leaf disks, Phytopathology, 82:696–699.

Perera, R., and Gay, J. L. (1976). The ultrastructure of haustoria of *Sphaerotheca pannosa* (Wallroth ex Fries) Leveille and changes in infected and associated cells of rose, Physiol. Plant Pathol., 9:57–65.

Peres-Artes, E., and Tena, M. (1989). Pectic enzymes from two races of *Fusarium oxysporum* f.sp. *ciceri*. Enzyme production in culture and enzymatic activity on isolated chickpea cell walls, J. Phytopathology, 124:39–51.

Peres-Artes, E., and Tena, M. (1990). Purification and characterization of pectic enzymes from two races of *Fusarium oxysporum* f.sp. *ciceri* differing in virulence of chickpea (*Cicer arietinum* L.), Physiol. Mol. Plant Pathol., 37:107–124.

Perumalla, C. J., and Heath, M. C. (1989). Effect of callose inhibition of haustorium formation by the cowpea rust fungus in the non-host bean plant, Physiol. Mol. Plant Pathol., 35:375–382.

Peterson, T. A., Mueller, W. C., and Englander, L. (1980). Anatomy and ultrastructure of a rhododendron root-fungus association, Can. J. Bot., 58:2421–2433.

Pitt, D. (1988). Pectin lyase from *Phoma medicaginis* var. *pinodella*, Methods in Enzymology, Vol. 161 (W. A. Wood and S. T. Kellog, eds.), Academic Press, San Diego, pp. 350–354.

Pladys, D., and Esquerre-Tugaye, M. T. (1974). Activité proteo-lytique de *Colletotrichum lagenarium*: comparison de souches de virulence variable et purification, CR Acad. Sci. Paris, 278:743–746.

Podilla, G. K., and Kolattukudy, P. E. (1988). Transcriptional activation of a cutinase gene in isolated fungal nuclei by plant cutin monomers, Science, 242:922–925.

Puhalla, J. E. (1975). Induction, isolation and characterization of mutants of *Verticillium dahliae* with deficiencies in melanin synthesis, Proc. Beltwide Cotton Prod. Res. Conf., p. 26.

Puhalla, J. E., and Howell, C. R. (1975). Significance of endopolygalacturonase activity to symptom expression of *Verticillium wilt* in cotton assessed by the use of mutants of *Verticillium dahliae* Kleb., Physiol. Plant Pathol., 7:147–152.

Purdy, R. E., and Kolattukudy, P. E. (1973). Depolymerization of a hydroxy fatty acid biopolymer, cutin, by an extracellular enzyme from *Fusarium solani* f.sp. *pisi*: isolation and some properties of the enzyme, Arch. Biochem. Biophys., 159:61–69.

Purdy, R. E., and Kolattukudy, P. E. (1975a). Hydrolysis of plant cuticle by plant pathogens. Purification, amino acid composition, and molecular weight of two isozymes of cutinase and a non-specific esterase from *Fusarium solani* f.sp. *pisi*, Biochemistry, 14:2824–2831.

Purdy, R. E., and Kolattukudy, P. E. (1975b). Hydrolysis of plant cuticle by plant pathogens. Properties of cutinase I, cutinase II, and a non-specific esterase isolated from *Fusarium solani pisi*, Biochemistry, 14:2832–2840.

Quader, H., and Fast, H. (1990). Influence of cytosolic pH changes on the organisation of the endoplasmic reticulum in epidermal cells of onion bulb scales: acidification by loading with weak organic acids, Protoplasma, 157:216–224.

Raa, J., Robertson, B., Solheim, B., and Trownso, A. (1977). Cell surface biochemistry related to specificity of pathogenesis and virulence of microorganisms, Cell Wall Biochemistry Related to Specificity in Host–Pathogen Interactions (B. Solheim and J. Raa, eds.), Universitetsforlaget, Tromso, Norway, pp. 11–30.

Ramaraj, B., and Vidhyasekaran, P. (1982). Possible involvement of pectic enzymes in betelvine wilt development, Indian Phytopath., 35:71–72.

Ramaraj, B., and Vidhyasekaran, P. (1986). Inactivation of pectic enzymes produced by *Phytophthora parasitica* var. *piperina* by fungicides, Indian Phytopath., 39:269–270.

Ray, P. M., and Rottenberg, D. A. (1964). Uronic acid constituents of oat-coleoptile cell wall, Biochem. J., 90:646–655.

Rebmann, G., Hertig, C., Bull, J., Mauch, F., and Dudler, R. (1991). Cloning and sequencing of cDNAs encoding a pathogen-induced putative peroxidase of wheat (*Triticum aestivum* L.), Plant Mol. Biol., 16:329–331.

Rehnstrom, A. L., Free, S. J., and Pratt, R. G. (1994). Isolation, characterization and pathogenicity of *Sclerotinia trifoliorum* arabinofuranosidase-deficient mutants, Physiol. Mol. Plant Pathol., 44:199–206.

Reisener, H. J., Tiburzy, R., Kogel, K. H., Moerschbacher, B., and Heck, B. (1986). Mechanisms of resistance of wheat against stem rust in the Sr5/P5 interaction, Biology and Molecular Biology of Plant–Pathogen Interactions (J. A. Bailey, ed.), Springer-Verlag, Berlin, pp. 141–148.

Reuveni, M., Agapov, V., and Reuveni, R. (1992a). Local systemic protection against powdery mildew and growth increase in cucumber plants induced by phosphate salts, Phytopathology, 82:1179.

Reuveni, R., Reuveni, M., and Agapov, V. (1992b). Systemic resistance against northern leaf blight and common rust in maize induced by foliar spray of phosphates, Phytopathology, 82:1179.

Reymond, D., and Phaff, H. J. (1965). Purification and certain properties of avocado polygalacturonase. J. Food Sci., 30:266–273.

Richmond, S., Kuc, J., and Elliston, J. E. (1979). Penetration of cucumber leaves by *Colletotrichum lagenarium* is reduced in plants systemically-protected by previous infection with the pathogen, Physiol. Plant Pathol., 14:329–338.

Ride, J. P. (1975). Lignification in wounded wheat leaves in response to fungi and its possible role in resistance, Physiol. Plant Pathol., 5:125–134.

Ride, J. P. (1978). The role of cell wall alterations in resistance to fungi. Ann. Appl. Biol., 89:302–306.

Ride, J. P. (1980). The effect of induced lignification on the resistance of wheat cell walls to fungal degradation, Physiol. Plant Pathol., 16:187–196.

Ride, J. P. (1983). Cell wall and other structural barriers in defence, Biochemical Plant Pathology (J. A. Callow, ed.), Wiley and Sons, Chichester, U.K., pp. 215–236.

Ride, J. P., and Pearce, R. B. (1979). Lignification and papilla formation sites of attempted penetration of wheat leaves by non-pathogenic fungi, Physiol. Plant Pathol., 15:79–92.

Rijkenberg, F. H. J., de Leeuw, G. T. N., and Verhoeff, K. (1980). Light and electron microscopy studies on the infection of tomato fruits by *Botrytis cinerea*, Can. J. Bot., 58:1394–1404.

Robb, J., Lee, S. W., Mohan, R., and Kolattukudy, P. E. (1991). Chemical characterization of stress-induced vascular coating, Plant Physiol., 97:528–536.

Robb, J., Powell, D. A., and Street, P. F. S. (1987). Time course of wall coating secretion in *Verticillium*-infected tomatoes, Physiol. Mol. Plant Pathol., 31:217–226.

Robb, J., Powell, D. A., Street, P. F. S. (1989). Vascular coating: a barrier to colonization by the pathogen in *Verticillium* wilt of tomato, Can. J. Bot., 67:600–607.

Roberts, E., and Kolattukudy, P. E. (1989). Molecular cloning, nucleotide sequence, and abscisic acid induction of a suberization-associated highly anionic peroxidase, Mol. Gen. Genet., 217:223–232.

Roberts, E., Kutchan, T., and Kolattukudy, P. E. (1988). Cloning and sequencing of cDNA for a highly anionic peroxidase from potato and the induction of its mRNA in suberizing potato tubers and tomato fruits, Plant Mol. Biol., 11:15–26.

Roberts, K., and Northcote, D. H. (1972). Hydroxyproline: observations on its chemical and autoradiographical localization in plant cell wall proteins, Planta, 107:43–51.

Robertsen, R. (1986). Elicitors of the production of lignin-like compounds in cucumber hypocotyls, Physiol. Mol. Plant Pathol., 28:137–148.

Roby, D., Toppan, A., and Esquerre-Tugaye, M. T. (1985). Cell surfaces in plant-microorganism interactions. V. Elicitors of fungal and of plant origin trigger the synthesis of ethylene and of cell wall hydroxyproline-rich glycoprotein in plants, Plant Physiol., 77:700–704.

Roelofson, P. A. (1965). Ultrastructure of the wall in growing cells and its relation to the direction of the growth, Adv. Bot. Res., 2:69–149.

Rogers, L. M., Flaishman, M. A., and Kolattukudy, P. E. (1994). Cutinase gene disruption in *Fusarium solani* f.sp. *pisi* decreases its virulence on pea, Plant Cell, 6:935–945.

Rohringer, R., Chong, J., Kim, W. K., and Harder, D. E. (1982). Cytochemistry of the host/parasite interface and the use of gold markers in cereal/rust interactions, NATO Advanced Study of the Institute of Serology, Serology, A 37:349–350.

Rohringer, R., Fuchs, A., Lunderstaedt, J., and Samborski, D. J. (1967). Metabolism of aromatic compounds in healthy and rust-infected primary leaves of wheat. I. Studies with ^{14}Co2, quinate-U-^{14}C, and shikimate-U-^{14}C as precursors, Can. J. Bot., 45:863–889.

Ryan, C., and Farmer, E. E. (1991). Oligosaccharide signals in plants: a current assessment, Annu. Rev. Plant Physiol., 42:651–674.

Ryerson, D. E., and Heath, M. C. (1992). Fungal elicitation of wall modifications in leaves of *Phaseolus vulgaris* L.cv. *Pinto*. II. Effects of fungal wall components, Physiol. Mol. Plant Pathol., 40:283–298.

Saeki, K., Ishikawa, O., Fukuoka, T., Nakagawa, H., Kai, Y., Yamashita, J., Kasai, N., and Horio, T. (1986). Barley leaf peroxidase: Purification and characterization, J. Biochem., 99:485–494.

Salinas, J., Warnaar, F., and Verhoeff, K. (1986). Production of cutin hydrolyzing enzymes by *Botrytis cinerea* in vitro, J. Phytopathol., 16:299–307.

Samuels, A. L., Glass, A. D. M., Ehret, D. L., and Menzies, J. G. (1991a). Distribution of silicon in cucumber leaves during infection by powdery mildew fungus (*Sphaerotheca fuliginea*), Can. J. Bot., 69:140–146.

Samuels, A. L., Glass, A. D. M., Ehret, D. L., and Menzies, J. G. (1991b). Mobility and deposition of silicon in cucumber plants, Plant, Cell Environ., 14:485–492.

Samuels, A. L., Glass, A. D. M., Menzies, J. G., and Ehret, D. L. (1994). Silicon in cell walls and papillae of *Cucumis sativus* during infection by *Sphaerotheca fuliginea*, Physiol. Mol. Plant Pathol., 44:237–242.

Sargent, C., and Gay, J. L. (1977). Barley epidermal apoplast structure and modification by powdery mildew contact, Physiol. Plant Pathol., 11:195–205.

Sarkanen, K. V., and Ludwig, C. H. (1971). Definition and nomenclature, Lignins—Occurrence, Formation, Structure and Reactions, (K. V. Sarkanen and C. H. Ludwig, eds.), Wiley & Sons, New York, pp. 1–18.

Sathiayanathan, S., and Vidhyasekaran, P. (1978). Involvement of pectolytic enzymes in brown spot disease development in rice, Indian Phytopath., 33:577–580.

Sauer, N., Corbin, D. R., Keller, B., and Lamb, C. J. (1990). Cloning and characterization of a wound-specific hydroxy-proline-rich glycoprotein in *Phaseolus vulgaris*, Plant Cell Environ., 13:257–266.

Schafer, W., Straney, D., Ciuffetti, L., Van Etten, H. D., and Yoder, O. C. (1989). One enzyme makes a fungal pathogen, but not a saprophyte, virulent on a new host plant, Science, 246:247–249.

Schloss, P., Walter, C., and Mader, M. (1987). Basic isoperoxidases in isolated vacuoles of *Nicotiana tabacum* L., Planta, 170:225–229.

Schmele, I., and Kauss, H. (1990). Enhanced activity of the plasma membrane localized callose synthase in cucumber leaves with induced resistance, Physiol. Mol. Plant Pathol., 37:221–228.

Schneider, S., and Ullrich, W. R. (1994). Differential induction of resistance and enhanced enzyme activities in cucumber and tobacco caused by treatment with various abiotic and biotic inducers, Physiol. Mol. Plant Pathol., 45:291–304.

Schweizer, P., Hunziker, W., and Mosinger, E. (1989). cDNA cloning, in vitro transcription and partial sequence analysis of mRNAs from winter wheat (*Triticum aestivum* L.) with induced resistance to *Erysiphe graminis* f.sp. *tritici*, Plant Mol. Biol., 12:643–654.

Scott, K. J., Davidson, A. D., Judidamrongphan, W., Mackinnon, G., and Manners, J. M. (1990). The activation of genes of wheat and barley by fungal pathogens, Austr. J. Bot., 17:229–238.

Scott-Craig, J. S., Panaccione, D. G., Cervone, F., and Walton, J. D. (1990). Endopolygalacturonase is not required for pathogenicity of *Cochliobolus carbonum* on maize, Plant Cell, 2:1191–1200.

Selvendran, R. R. (1985). Development in the chemistry and biochemistry of pectic and hemicellulosic polymers, J. Cell Sci. Suppl. 2:51–88.

Sharrock, K. R., and Labavitch, J. M. (1994). Polygalacturonase inhibitors of Bartlett pear fruits: differential effects on *Botrytis cinerea* polygalacturonase isozymes, and influence on products of fungal hydrolysis of pear cell walls and on ethylene induction in cell culture, Physiol. Mol. Plant Pathol., 45:305–319.

Shaykh, M., Soliday, C. L., and Kolattukudy, P. E. (1977). Proof for the production of cutinase by *Fusarium solani* f.sp. *pisi* during penetration into its host, *Pisum sativum*, Plant Physiol., 60:170–172.

Sherf, B. A., Bajar, A. M., and Kolattukudy, P. E. (1993). Abolition of an inducible highly anionic peroxidase activity in transgenic tomato, Plant Physiol., 101:201–208.

Sherwood, R. T. (1975). Pectin lyase and polygalacturonase production in *Rhizoctonia solani* and other fungi, Phytopathology, 56:279–286.

Sherwood, R. T., and Vance, C. P. (1982). Initial events in the epidermal layer during penetration, The Physiological and Biochemical Basis (Y. Asada, W. R. Bushnell, S. Ouchi and C. P. Vance, eds.), Japan Scientific Societies Press, Tokyo, pp. 27–44.

Shiraishi, T., Yamada, T., Nicholson, R. L., and Kunoh, H. (1995). Phenylalanine ammonia-lyase in barley: activity enhancement in response to *Erysiphe graminis* f.sp. *hordei* (race 1), a pathogen, and *Erysiphe pisi*, a nonpathogen, Physiol. Mol. Plant Pathol., 46:153–162.

Shiraishi, T., Yamaoka, N., and Kunoh, K. (1989). Association between increased phenylalanine ammonia-lyase activity and cinnamic acid synthesis and the induction of temporary inaccessibility caused by *Erysiphe graminis* primary germtube penetration of the barley leaf, Physiol. Mol. Plant Pathol., 34:75–83.

Shishiyama, J., Araki, F., and Akai, S. (1970). Studies on cutin-esterase II. Characteristics of cutin-esterase from *Botrytis cinerea* and its activity on tomato cutin, Plant Cell Physiol., 11:937–945.

Showalter, A. W. (1993). Structure and function of plant cell wall proteins, Plant Cell, 5:9–23.

Showalter, A. M., Bell, J. N., Cramer, C. L., Bailey, J. A., Varner, J. E., and Lamb, C. J. (1985). Accumulation of hydroxyproline-rich glycoprotein mRNAs in response to fungal elicitor and infection, Proc. Natl. Acad. Sci. USA, 82:6551–6555.

Showalter, A. M., and Rumeau, D. (1990). Molecular biology of the plant cell wall hydroxyproline-rich glycoproteins, Organization and Assembly of Plant and Animal Extracellular Matrix (W. S. Adair and R. P. Mecham, eds.), Academic Press, New York, pp. 247–281.

Showalter, A. M., and Varner, J. E. (1989). Plant hydroxy-proline-rich glycoproteins, The Biochemistry of Plants, Vol. 15 (A. Marcus, ed.), Academic Press, New York, pp. 485–520.

Sisler, H. D. (1986). Control of fungal diseases by compounds acting as antipenetrants, Crop. Prot., 5:306–313.

Sisler, H. D., Woloshuk, C. P., and Wolkow, P. M. (1983). Studies on the mode of action of tricyclazole and related compounds, Systemische Fungizide and Antifungale Verbindungen, (H. Lyr and C. Polter, eds.), Akademie-Verlag, pp. 171–176.

Sister, H. D., Woloshuk, C. P., and Wolkow, P. M. (1984). Specific chemical regulation of appressorial function, Tagungsber. Akad. Landwirtschaftswiss, 222:S17–28.

Skare, N. H., Paus, F., and Raa, J. (1975). Production of pectinase and cellulase by *Cladosporium cucumerinum* with dissolved carbohydrates and isolated cell walls of cucumber as carbon sources, Physiol. Plant., 33:299–283.

Skou, J. P. (1985). On the enhanced callose deposition in barley with ml-O powdery mildew resistance genes, Phytopathol. Z., 112:207–216.

Skou, J. P., Jorgensen, J., and Lilholt, U. (1984). Comparative studies on callose formation in powdery mildew compatible and incompatible barley, Phytopathol. Z., 109:147–168.

Smart, M. G., Aist, J. R., and Israel, H. W. (1986a). Structure and function of wall appositions. 1. General histochemistry of papillae in barley coleoptiles attacked by *Erysiphe graminis* f.sp. *hordei*, Can. J. Bot., 64:793–801.

Smart, M. G., Aist, J. R., and Israel, H. W. (1986b). Structure and function of wall appositions. 2. Callose and the resistance of oversize papillae to penetration by *Erysiphe graminis* f.sp. *hordei*, Can. J. Bot., 64:802–804.

Smith, A. K., and Peterson, R. L. (1985). Histochemical features of wall appositions in *Asparagus officinalis* cultivar Viking root meristems infected by *Fusarium oxysporum* f.sp. *asparagi*, Can. J. Plant Pathol., 7:28–36.

Smith, J. A., and Hammerschmidt, R. (1988). Comparative study of acidic peroxidases associated with induced resistance in cucumber, muskmelon and watermelon, Physiol. Mol. Plant Pathol., 33:255–261.

Soliday, C. L., Dickman, M. B., and Kolattukudy, P. E. (1989). Structure of the cutinase gene and detection of promoter activity in the 5′-flanking region by fungal transformation, J. Bacteriol., 171:1942–1951.

Soliday, C. L., Flurkey, W. H., Okit, T. W., and Kolattukudy, P. E. (1984). Cloning and structure determination of cDNA for cutinase, an enzyme involved in fungal penetration of plants, Proc. Natl. Acad. Sci. USA, 81:3939–3943.

Soliday, C. L., and Kolattukudy, P. E. (1976). Isolation and purification of a cutinase from *Fusarium roseum culmorum* and its immunological comparison with cutinases from *F. solani pisi*, Arch. Biochem. Biophys., 176:334–343.

Soliday, C. L., and Kolattukudy, P. E. (1977). Biosynthesis of cutin. w-hydroxylation of fatty acids by a microsomal preparation from germinating *Vicia faba*, Plant Physiol., 59:1116–1121.

Soliday, C. L., and Kolattukudy, P. E. (1978). Midchain hydroxylation of 16-hydroxypalmitic acid by the endoplasmic reticulum fraction from germinating *Vicia faba*, Arch. Biochem. Biophys., 188:338–347.

Soliday, C. L., and Kolattukudy, P. E. (1979). Introduction of O-glycosidically linked mannose into proteins via mannosyl phosphoryl dolichol by microsomes from *Fusarium solani pisi*, Arch. Biochem. Biophys., 197:367–378.

Soliday, C. L., and Kolattukudy, P. E. (1983). Primary structure of the active site region of fungal cutinase, an enzyme involved in phytopathogenesis, Biochem. Biophys. Res. Commun., 114:1017–1022.

Southerton, S. G., and Deverall, B. J. (1989). Histological studies of the expression of the Lr 9, Lr 20 and Lr 28 alleles for resistance to leaf rust in wheat, Plant Pathology, 38:190–199.

Southerton, S. G., and Deverall, B. J. (1990). Histochemical and chemical evidence for lignin accumulation during the expression of resistance to leaf rust fungi in wheat, Physiol. Plant Pathol., 36:483–494.

Southerton, S. G., Osbourn, A. E., Dow, J. M., and Daniels, M. J. (1993). Two xylanases from *Gaeumannomyces graminis* with identical N-terminal amino acid sequence, Physiol. Mol. Plant Pathol., 42:97–107.

Spare, N. H., Paus, F., and Raa, J. (1975). Production of pectinase and cellulase by

Cladosporium cucumerinum with dissolved carbohydrates and isolated cell walls of cucumber as carbon sources, Physiol. Plant., 33:229–233.

Speakman, J. B., and Lewis, B. G. (1978). Limitation of *Gaeumannomyces graminis* by wheat root responses to *Phialophora radicicola*, New Phytol., 80:373–380.

Spiro, R. G. (1970). Glycoproteins, Annu. Rev. Biochem., 39:599–638.

Stafstorm, J. P., and Staehelin, L. A. (1986a). The role of carbohydrate in maintaining extensin in an extended conformation, Plant Physiol., 81:242–246.

Stahl, D. J., and Schafer, W. (1992). Cutinase is not required for fungal pathogenicity on pea, Plant Cell, 4:621–629.

Stahl, D. J., Theuerkauf, A., Heitefuss, R., and Schafer, W. (1994). Cutinase of *Nectria haematococca* (*Fusarium solani* f.sp. *pisi*) is not required for fungal virulence or organ specificity on pea, Mol. Plant–Microbe Interact., 7:713–725.

Stanghellini, M. E., Rasmussen, S. L., and Vandemark, G. J. (1993). Relationship of callose deposition to resistance of lettuce to *Plasmopara lactucae-radicis*, Phytopathology, 83:1498–1501.

Staples, R. C., and Macko, V. (1980). Formation of infection structures as a recognition response in fungi, Exp. Mycol., 4:2–16.

Staub, T., Dahmen, H., and Schwinn, F. J. (1974). Light and scanning electron microscopy of cucumber and barley powdery mildew on host and nonhost plants, Phytopathology, 64:364–372.

Stenzel, K., Steiner, U., and Schonbeck, F. (1985). Effect of induced resistance on the efficiency of powdery mildew haustoria in wheat and barley, Physiol. Plant Pathol., 27:357–367.

Stewart, A., and Mansfield, J. W. (1985). The composition of wall alterations and appositions (reaction material) and their role in the resistance of onion bulb scale epidermis to colonization by *Botrytis allii*, Plant Pathol., 34:25–37.

Stipanovic, R. D., and Bell, A. A. (1976). Pentaketide metabolites of *Verticillium dahliae*. 3. Identification of (-)-3,4-dihydro-3,8-dihydroxy-1 (2H)-naphthalenone [(-)-vermelone] as a precursor to melanin, J. Org. Chem., 41:2468–2469.

Stipanovic, R. D., and Bell, A. A. (1977). Pentaketide metabolites of *Verticillium dahliae*. II. Accumulation of naphthol derivatives by the aberrant-melanin mutant brm-2, Mycologia, 69:164–172.

Stipanovic, R. D., and Wheeler, M. H. (1980). Accumulation of 3,3′-biflaviolin, a melanin shunt product, by tricyclazole-treated *Thielaviopsis basicola*, Pestic. Biochem. Physiol., 13:198–201.

Stoddart, R. W., Barrett, A. J., and Northcote, D. H. (1967). Pectic polysaccharides of growing plant tissues, Biochem. J., 102:194–204.

Stolzenburg, M. C., Aist, J. R., and Israel, H. W. (1984a). The role of papillae in resistance to powdery mildew conditioned by the ml-O gene in barley. I. Correlative evidence, Physiol. Plant Pathol., 25:337–346.

Stolzenburg, M. C., Aist, J. R., and Israel, H. S. (1984b). The role of papillae in resistance to powdery mildew conditioned by the ml-O gene in barley. II. Experimental evidence, Physiol. Plant Pathol., 25:347–361.

Stossel, P., Lazarovits, G., and Ward, E. W. B. (1980). Penetration and growth of compatible races of *Phytophthora megasperma* var. *sojae* in soybean hypocotyl tissues differing in age, Can. J. Bot., 58:2594–2601.

Street, P. F. S., Robb, J., and Ellis, B. E. (1986). Secretion of vascular coating components of xylem parenchyma cells of tomatoes infected with *Verticillium albo-atrum*, Protoplasma, 132:1–11.

Stromberg, A., and Brishammar, S. (1993). A histological evaluation of induced resistance to *Phytophthora infestans* (Mont.) de Bary in potato leaves, J. Phytopathol., 137:15–25.

Stumm, D., and Gessler, C. (1986). Role of papillae in the induced systemic resistance of cucumbers against *Colletotrichum lagenarium*, Physiol. Mol. Plant Pathol., 29:405–410.

Stumpf, M. A., and Gay, J. L. (1989). The haustorial interface in a resistant interaction of *Erysiphe pisi* with *Pisum sativum*, Physiol. Mol. Plant Pathol., 35:519–533.

Stumpf, M. A., and Heath, M. C. (1985). Cytological studies of the interactions between the cowpea rust fungus and silicon-depleted French bean plants, Physiol. Plant Pathol., 27:369–385.

Suzuki, K., Furusawa, I., Ishida, N., and Yamamoto, M. (1981). Protein synthesis during germination and appressorium formation of *Colletotrichum lagenarium* spores, J. Gen. Microbiol., 124:61–69.

Suzuki, K., Furusawa, I., Ishida, N., and Yamamoto, I. (1982a). Chemical dissolution of cellulose membranes as a prerequisite for penetration from appressoria of *Colletotrichum lagenarium*, J. Gen. Microbiol., 128:1035–1040.

Suzuki, K., Furusawa, I., and Yamamoto, M. (1983). Role of chemical dissolution of cellulose membranes in the appressorial penetration by *Colletotrichum lagenarium*, Ann. Phytopath. Soc. Japan, 49:481–487.

Suzuki, K., Kubo, Y., Furusawa, I., and Yamamoto, M. (1982b). Behavior of colorless appressoria in an albino mutant of *Colletotrichum lagenarium*, Can. J. Microbiol., 28:1210–1213.

Sweigard, J. A., Chumley, F. G., and Valent, B. (1992a). Cloning and analysis of CUT1, a cutinase gene from *Magnaporthe grisea*, Mol. Gen. Genet., 232:174–182.

Sweigard, J. A., Chumley, F. G., and Valent, B. (1992b). Disruption of a *Magnaporthe grisea* cutinase gene, Mol. Gen. Genet., 232:183–190.

Tagu, D., Walker, N., Ruiz-Avila, L., Burgess, S., Martinez-Izquierdo, J. A., Leguay, J. J., Netter, P., and Puizdomenech, P. (1992). Regulation of the maize HRGP gene expression by ethylene and wounding. mRNA accumulation and qualitative expression analysis of the promoter by microprojectile bombardment, Plant Mol. Biol., 20:529–538.

Talbott, L. D., and Ray, P. W. (1992). Molecular size and separability features of pea cell wall polysaccharides. Implication for models of primary wall structure, Plant Physiol., 98:357–368.

Talmadge, K. W., Keegstra, K., Bauer, W. D., and Albersheim, P. (1973). The structure of plant cell walls. I. The macromolecular components of the walls of suspension-cultured sycamore cells with a detailed analysis of the pectic polysaccharides, Plant Physiol., 51:158–173.

Tanabe, K., Nishimura, S., and Kohmoto, K. (1988a). Cutinase production by *Alternaria alternata* Japanase pear pathotype and its role in pathogenicity, Ann. Phytopathol. Soc. Japan, 54:483–491.

Tanabe, K., Nishimura, S., and Kohmoto, K. (1988b). Pathogenicity of cutinase- and pectic-enzymes deficient mutants of *Alternaria alternata* Japanese pear pathotype, Ann. Phytopath. Soc. Japan, 54:552–555.

Tanaka, C., Kubo, Y., and Tsuda, M. (1990). Genetic analysis and characterization of *Cochliobolus heterostrophus* colour mutants, Mycol. Res., 95:49–56.

Templeton, M. D., Dixon, R. A., Lamb, C. J., and Lawton, M. A. (1990). Hydroxyproline-rich glycoprotein transcripts exhibit different spatial patterns of accumulation in compatible and incompatible interaction between *Phaseolus vulgaris* and *Colletotrichum lindemuthianum*, Plant Physiol., 94:1265–1269.

Ten, L. N., Stepanichenko, N. N., Mukhamedzhanov, S. J., and Khotyanovich, A. V. (1983). Action of tricyclazole on the bio-synthesis of melanin in some fungi of the genus *Verticillium*, Chem. Nat. Compd. USSR, 19:384–385.

Ten, L. N., Stepanichenko, N. N., Shevtsova, V. M., Mukhamedzhanov, S. Z., Kasyanenko, A. G., and Otroshchenko, O. S. (1980). Metabolites of the pathogenic fungus *Verticillium dahliae*. IX. Pentaketides of mutants and their role in the biosynthesis of melanin, Chem. Nat. Compd., 16:298–302.

Tewari, J. P. (1986). Subcuticular growth of *Alternaria brassicae* in rapeseed, Can. J. Bot., 64:1227–1231.

Thomson, W. W., Platt-Aloia, K. A., and Endress, A. G. (1976). Ultrastructure of oilgland development in the leaf of *Citrus sinensis* L., Bot. Gaz., 137:330–340.

Thordal-Christensen, H., Brandt, J., Cho, B. H., Rasmussen, S. K., Gregersen, P. L., Smedegaard-Petersen, V., and Collinge, D. B. (1992). cDNA cloning and characterization of two barley peroxidase transcripts induced differentially by the powdery mildew fungus *Erysiphe graminis*, Physiol. Mol. Plant Pathol., 40:394–409.

Thordal-Christensen, H., and Smedegaard-Petersen, V. (1988a). Correlation between induced resistance and host fluorescence in barley inoculated with *Erysiphe graminis*, J. Phytopathol., 123:34–46.

Thordal-Christensen, H., and Smedegaard-Petersen, V. (1988b). Comparison of resistance-inducing abilities of virulent and avirulent races of *Erysiphe graminis* f.sp. *hordei* and a race of *Erysiphe graminis* f.sp. *tritici* in barley, Plant Pathology, 37:20–27.

Tiburzy, R. (1982). Fluorescence-microscopical studies on the host-parasite interactions between wheat stem rust and several varieties of wheat, Fitopatol. Brasil., 7:462–463.

Tiburzy, R., and Reisener, H. J. (1990). Resistance of wheat to *Puccinia graminis* f.sp. *tritici*: Association of the hypersensitive reaction with the cellular accumulation of lignin-like material and callose, Physiol. Mol. Plant Pathol., 36:109–120.

Tierney, M. L., and Varner, J. E. (1987). The extensins, Plant Physiol., 84:1–2.

Tippett, J. T., and Hill, T. C. (1984). Role of periderm in resistance of *Eucalyptus marginata* roots against *Phytophthora cinnamomi*, Eur. J. Forens. Pathol., 14:431–439.

Tokousbalides, M. C., and Sisler, H. D. (1978). Effect of tricyclazole on growth and secondary metabolism in *Pyricularia oryzae*, Pestic. Biochem. Physiol., 8:26–32.

Tokousbalides, M. C., and Sisler, H. D. (1979). Site of inhibition by tricyclazole in the melanin biosynthetic path way of *Verticillium dahliae*, Pestic. Biochem. Physiol., 11:64–73.

Toppan, A., Roby, D., and Esquerre-Tugaye, M. T. (1982). Cell surfaces in plant-microorganism interactions. III. In vivo effect of ethylene on hydroxyproline-rich glycoprotein accumulation in the cell wall of diseased plants, Plant Physiol., 70:82–86.

Tosa, Y., Akiyama, T., and Ogura, H. (1990). Cytological aspects of the interactions between formae speciales of *Erysiphe graminis* and genera of gramineous plants, and their evolutionary implications, Can. J. Bot., 68:1249–1253.

Tosa, Y., and Shishiyama, J. (1984). Defense reactions of barley cultivars to an inappropriate forma specialis of the powdery mildew fungus of graminaceous plants, Can. J. Bot., 62:2114–2117.

Trail, F., and Koller, W. (1990). Diversity of cutinases from plant pathogenic fungi: Evidence for a relationship between enzyme properties and tissue specificity, Physiol. Mol. Plant Pathol., 36:495–508.

Trail, F., and Koller, W. (1993). Diversity of cutinases from plant pathogenic fungi: Purification and characterization of two cutinases from *Alternaria brassicicola*, Physiol. Mol. Plant Pathol., 42:205–220.

Tulloch, A. P. (1973). Composition of leaf surface waxes of *Triticum* species: Variation with age and tissue, Phytochemistry, 12:2225–2232.

Tulloch, A. P., and Hoffman, L. L. (1973). Leaf wax of *Triticum aestivum*, Phytochemistry, 12:2217–2223.

Turner, J. G., and Hoffman, R. M. (1985). Effect of the polygalacturonase inhibitor from pea on the hydrolysis of pea cell walls by the endopolygalacturonase from *Ascochyta pisi*, Plant Pathology, 35:54–60.

Valsangiacomo, C., and Gessler, C. (1988). Role of the cuticular membrane in ontogenic and Vf-resistance of apple leaves against *Venturia inaequalis*, Phytopathology, 78:1066–1069.

Valsangiacomo, C., and Gessler, C. (1992). Purification and characterization of an exopolygalacturonase produced by *Venturia inaequalis*, the causal agent of apple scab, Physiol. Mol. Plant Pathol., 40:63–77.

Valsangiacomo, C., Ruckstuhl, M., and Gessler, C. (1992). In vitro degradation of cell walls of apple leaves by pectinolytic enzymes of the scab fungus, *Venturia inaequalis* and by commercial pectinolytic and cellulolytic enzyme preparations, J. Phytopathology, 135:20–27.

Valsangiacomo, C., Wagner, K., Stadler, B., Ruckstuhl, M., Manini-Gessler, P., Michel, M., and Gessler, C. (1989). Aspects of host resistance and pathogenesis in the interaction between *Venturia inaequalis* and apple leaves, Integrated Control of Pome Fruit Disease, Vol. II (C. Gessler, D. J. Butt, and B. Koller, eds.), IOBC Bulletin, pp. 191–204.

Vance, C. P., Kirk, T. K., and Sherwood, R. T. (1980). Lignification as a mechanism of disease resistance, Annu. Rev. Phytopathol., 18:259–288.

Vance, C. P., and Sherwood, R. T. (1976). Cycloheximide treatments implicate papilla formation in resistance of seed canary grass to fungi, Phytopathology, 66:498–502.

van den Ende, G., and Linskens, H. F. (1974). Cutinolytic enzymes in relation to pathogenesis, Annu. Rev. Phytopathol., 12:247–258.

Van Etten, H. D., and Bateman, D. F. (1969). Enzymatic degradation of galactan, galactomannan and xylan by *Sclerotium rolfsii*, Phytopathology, 59:968–972.

Van Holst, G. J., and Varner, J. E. (1984). Reinforced polyproline II conformation in a hydroxyproline-rich cell wall glycoprotein from carrot root, Plant Physiol., 74:247–251.

Van Hoof, A., Leykam, J., Schaeffer, H. J., and Walton, J D. (1991). A single β-1,3-

glucanase secreted by the maize pathogen *Cochliobolus carbonum* acts by an exolytic mechanism, Physiol. Mol. Plant Pathol., 39:259–267.

Varley, D. A., Podila, G. K., and Hiremath, S. T. (1992). Cutinase in *Cryphonectria parasitica*, the chestnut blight fungus. Suppression of cutinase gene expression in isogenic hypovirulent strains containing double-stranded RNAs. Mol. Cell Biol., 12:4539–4544.

Varner, J. E., and Lin, L. S. (1989). Plant cell wall architecture, Cell, 56:231–239.

Vaughn, S. F., and Lulai, E. C. (1991). The involvement of mechanical barriers in the resistance response of a field-resistant and a field-susceptible potato cultivar to *Verticillium dahliae*, Physiol. Mol. Plant Pathol., 38:455–465.

Vidhyasekaran, P. (1972). Nitrogen metabolism of ragi plants in relation to helminthosporiose disease resistance, Phytopathol. Z., 75:223–235.

Vidhyasekaran, P. (1974a). Fingermillet helminthosporiose—a low sugar disease, Z. Pflankranh. Pflanzenschutz, 81:28–38.

Vidhyasekaran, P. (1974b). Possible role of sugars in restriction of lesion development in fingermillet leaves infected with *Helminthosporium tetramera*, Physiol. Plant Pathol., 4:555–565.

Vidhyasekaran, P. (1974c). Possible mode of action of phenolics in inducing disease resistance in ragi plants against *Helminthosporium tetramera*, Indian Phytopath., 27:587–595.

Vidhyasekaran, P. (1975). Role of auxin-phenol complex in fingermillet helminthosporiose disease resistance, Phytopathol. Z., 82:89–99.

Vidhyasekaran, P. (1977). Role of amino acids and amides in fingermillet helminthosporiose incidence. Indian Phytopath., 30:41–46.

Vidhyasekaran, P. (1978). Production of pectolytic enzymes by *Helminthosporium nodulosum* in vitro and in vivo, Indian Phytopath., 31:275–280.

Vidhyasekaran, P. (1988). Physiology of Disease Resistance in Plants, Vol. I, CRC Press, Florida, p. 149.

Vidhyasekaran, P. (1993). Principles of Plant Pathology, CBS Publishers, Delhi, p. 166.

Vidhyasekaran, P., and Parambaramani, C. (1973). Role of pectolytic enzymes in pathogenesis of obligate and facultative parasites causing sorghum diseases, Indian Phytopath., 26:197–209.

Vidhyasekaran, P., Parambaramani, C., and Govindaswamy, C. V. (1971). Role of cellulolytic and proteolytic enzymes in pathogenesis of obligate and facultative parasites causing sorghum diseases, Indian Phytopath., 24:305–309.

Wada, S., and Ray, P. M. (1978). Matrix polysaccharides of oat coleoptile cell walls, Phytochemistry, 17:923–931.

Wagner, G. J. (1982). Compartmentation in plant cells: the role of the vacuole, Recent Adv. Phytochem., 16:1–45.

Wagner, K. L., Hitz-Germann, J. M. S., and Gessler, C. (1988). Cellulolytic ability of the scab fungus, *Venturia inaequalis*, J. Phytopathology, 123:217–221.

Waldmann, T., Jeblick, W., and Kauss, H. (1988). Induced net Ca^{2+} uptake and callose biosynthesis in suspension-cultured plant cells, Planta, 173:88–95.

Walter, M. H., Grima-Pettenati, J., Grand, C., Boudet, A. M., and Lamb, C. J. (1988). Cinnamyl alcohol dehydrogenase, a molecular marker specific for lignin synthesis:

cDNA cloning and mRNA induction by fungal elicitor, Proc. Natl. Acad. Sci. USA, 85:5546–5550.

Walton, T. J. (1990). Waxes, cutin and suberin, Lipids, Membranes and Aspects of Photobiology, Vol. 4 (J. L. Harwood and J. R. Bowyer, eds.), Methods in Plant Biochemistry, Academic Press, New York, pp. 105–108.

Walton, T. J., and Kolattukudy, P. E. (1972). Determination of the structures of cutin monomers by a novel depolymerization procedure and combined gas chromatography and mass spectrometry, Biochemistry, 11:1885–1897.

Wasserman, B. P., and McCarthy, K. J. (1986). Regulation of plasma membrane β-glucan synthase from red beet root by phospholipids. Reactivation of Triton X-100 extracted glucan synthase by phospholipids, Plant Physiol., 82:396–400.

Wattad, C., Dinoor, A., and Prusky, D. (1994). Purification of pectate lyase produced by *Colletotrichum gloeosporioides* and its inhibition by epicatechin: A possible factor involved in the resistance of unripe avocado fruits to anthracnose, Mol. Plant–Microbe Interact., 7:293–297.

Wheeler, M. H. (1981). Melanin biosynthetic enzymes in cell-free homogenates of *Verticillium dahliae* and *Pyricularia oryzae*, Phytopathology, 71:912.

Wheeler, M. H. (1982). Melanin biosynthesis in *Verticillium dahliae*: dehydration and reduction reactions in cell-free homogenates, Exp. Mycol., 6:171–179.

Wheeler, M. H. (1983). Comparisons of fungal melanin biosynthesis in ascomycetous, imperfect and basidiomycetous fungi, Trans. Br. Mycol. Soc., 81:29–36.

Wheeler, M. H., and Bell, A. A. (1987). Melanins and their importance in pathogenic fungi. Curr. Top. Med. Mycol., 2:338–387.

Wheeler, M. H., and Greenblatt, G. A. (1988). The inhibition of melanin biosynthetic reactions in *Pyricularia oryzae* by compounds that prevent rice blast disease, Exp. Mycol., 12:151–160.

Wheeler, M. H., and Stipanovic, R. D. (1979). Melanin biosynthesis in *Thielaviopsis basicola*, Exp. Mycol., 3:340–350.

Wheeler, M. H., Tolmsoff, W. J., Bell, A. A., and Mollenhauer, H. H. (1978). Ultrastructural and chemical distinction of melanins formed by *Verticillium dahliae* from (+) scytalone, 1,8-dihydroxynaphthalene, catechol and L-3,4-dihydroxyphenylalanine, Can. J. Microbiol., 24:289–297.

Whistler, R. L., and Richards, E. L. (1970). Hemicelluloses: The Carbohydrates, Vol. IIA (W. Pigman and D. Horton, eds.), Academic Press, New York, pp. 447–469.

Whitmore, F. W. (1978a). Lignin-carbohydrate complex formed in isolated cell walls of callus, Phytochemistry, 17:421–425.

Whitmore, F. W. (1978b). Lignin-protein complex catalyzed by peroxidase, Plant Science Lett., 13:241–245.

Wijesundra, R. L. C., Bailey, J. A., and Byrde, R. J. W. (1984). Production of pectin lyase by *Colletotrichum lindemuthianum* in culture and in infected bean (*Phaseolus vulgaris*) tissue, J. Gen. Microbiol., 130:285–290.

Wijesundra, R. L. C., Bailey, J. A., Byrde, R. J. W., and Fielding, A. H. (1989). Cell wall degrading enzymes of *Colletotrichum lindemuthianum*: their role in the development of bean anthracnose, Physiol. Mol. Plant Pathol., 34:403–413.

Wilson, K. (1964). The growth of plant cell walls, Int. Rev. Cytol., 17:1–49.

Wilkow, P. M., Sisler, H. D., and Vigil, E. L. (1983). Effect of inhibitors of melanin

biosynthesis on structure and function of appressoria of *Colletotrichum lindemuthianum*, Physiol. Plant Pathol., 22:55–71.
Woloshuk, C. P., and Kolattukudy, P. E. (1986). Mechanism by which contact with plant cuticle triggers cutinase gene expression in the spores of *Fusarium solani* f.sp. *pisi*, Proc. Natl. Acad. Sci. USA, 83:1704–1708.
Woloshuk, C. P., and Sisler, H. D. (1982). Tricyclazole, pyroquilon, tetrachlorophthalide, PCBA, coumarin and related compounds inhibit melanization and epidermal penetration by *Pyricularia oryzae*, J. Pestic. Sci., 7:161–166.
Woloshuk, C. P., Sisler, H. D., and Vigil, E. L. (1983). Action of the antipenetrant, tricyclazole on appressoria of *Pyricularia oryzae*, Physiol. Plant Pathol., 22:245–259.
Woloshuk, C. P., Sisler, H. D., Tokousbalides, M. C., and Dutky, S. R. (1980a). Melanin biosynthesis in *Pyricularia oryzae*. Site of tricyclazole inhibition and pathogenicity of melanin-deficient mutants, Pestic. Biochem. Physiol., 14:256–264.
Woloshuk, C. P., Wolkow, P. M., and Sisler, H. D. (1980b). The effect of three fungicides, specific for the control of rice blast disease on the growth and melanin biosynthesis by *Pyricularia oryzae* Cav., Pestic. Sci., 12:86–90.
Woloshuk, C. P., Wolkow, P. M., and Sisler, H. D. (1981). The effect of three fungicides, specific for the control of rice blast disease on the growth and melanin biosynthesis by *Pyricularia oryzae* Cav., Pestic. Sci., 12:86–90.
Worth, H. J. G. (1967). The chemistry and biochemistry of olefins, dienes and acetylenes via hydroboration, Org. React., 13:1–54.
Wu, A., Harriman, R. W., Frost, D. J., Read, S. M., and Wasserman, B. P. (1991). Rapid enrichment of CHAPS-solubilized UDP-glucose: (1,3)-β-glucan (callose) synthase from *Beta vulgaris* L. by product entrapment, Plant Physiol., 97:684–692.
Wu, S. C., Kauffmann, S., Darvill, A. G., and Albersheim, P. (1995). Purification, cloning and characterization of two xylanases from *Magnaporthe grisea*, the rice blast fungus, Mol. Plant–Microbe Interact., 8:506–514.
Xuei, N. L., Jarlfors, U., and Kuc, J. (1988). Ultrastructural changes associated with induced systemic resistance of cucumber to disease: host response and development of *Colletotrichum lagenarium* in systemically protected leaves, Can. J. Bot., 66:1028–1038.
Yamaguchi, I., Sekido, S., and Misato, T. (1982). The effect of non-fungicidal antiblast chemicals on the melanin biosynthesis and infection by *Pyricularia oryzae*, J. Pestic. Sci.,, 7:523–529.
Yamaguchi, I., Sekido, S., and Misato, T. (1983a). Inhibition of appressorial melanization in *Pyricularia oryzae* by non-fungicidal anti-blast chemicals, J. Pestic. Sci., 8:229–232.
Yamaguchi, I., Sekido, S., Seto, H., and Misato, T. (1983b). Cytotoxic effect of 2-hydroxyjuglone, a metabolite in the branched pathway of melanin biosynthesis in *Pyricularia oryzae*, J. Pestic. Sci., 8:545–550.
Yao, C., and Koller, W. (1994). Diversity of cutinases from plant pathogenic fungi: cloning and characterization of a cutinase gene from *Alternaria brassicicola*, Physiol. Mol. Plant Pathol., 44:81–92.
Yazdi, M. T., Woodward, J. R., and Radford, A. (1990). The cellulase complex of

Neurospora crassa: activity, stability and release, J. Gen. Microbiol., 136:1313–1319.

Ye, X. S., Pan, S. Q., and Kuc, J. (1990). Activity, isozyme pattern, and cellular localization of peroxidase as related to systemic resistance of tobacco to blue mold (*Peronospora tabacina*) and to tobacco mosaic virus, Phytopathology, 80:1295–1299.

Yoder, O. C. (1980). Toxins in pathogenesis, Annu. Rev. Phytopathol., 18:103–109.

Yokoyama, K., Aist, J. R., and Bayles, C. J. (1991). A papilla-regulating extract that induces resistance to barley powdery mildew, Physiol. Mol. Plant Pathol., 39:71–78.

Young, D. H., and Kauss, H. (1983). Release of calcium from suspension-cultured *Glycine max* cells by chitosan, other polycations, and polyamines in relation to effects on membrane permeability, Plant Physiol., 73:698–702.

Zeun, R., and Bucheanauer, H. (1985). Effect of tricyclazole on production and melanin contents of sclerotia of *Botrytis cinerea*, Phytopathol. Z., 112:259–267.

Zeyen, R. J., and Bushnell, W. R. (1979). Papilla response of barley epidermal cells caused by *Erysiphe graminis*: rate and method of deposition determined by microcinematography and transmission electron microscopy, Can. J. Bot., 57:898–913.

3

Lipid Breakdown Products and Active Oxygen Species

I. LIPID PEROXIDATION PRODUCTS

Plant cell membranes consist of mostly lipid and proteins. The membranes are asymmetrical lipid bilayers in which cytoplasmically synthesized proteins are dissolved (Bretscher, 1973). The polyunsaturated octadecanoid fatty acids linoleic acid and linolenic acid are the common constituents of plant membranes (Epton and Deverall, 1968; Hoppe and Heitefuss, 1974). When the plants are infected by pathogens, various lipid breakdown products, including C_6 volatiles, may be formed from linoleic acid and linolenic acid by sequential enzyme steps involving lipoxygenase, hydroperoxide lyase, and hydroperoxide hydratase in the so-called lipoxygenase pathway (Vick and Zimmerman, 1984, 1987a,b). Both hydroperoxide isomerase and hydoperoxide cyclase are known as hydroperoxide dehydrase (Vick and Zimmerman, 1987b).

Plant lipoxygenases (linoleate: oxygen oxidoreductases) are members of a class of nonheme iron-containing dioxygenases that catalyze the addition of molecular oxygen to fatty acids containing a *cis*, *cis* -1,4-pentadiene system to give an unsaturated fatty acid hydroperoxide (Hildebrand, 1989, Siedow, 1991). They are present widely in plants (Hildebrand et al., 1988).

Lipoxygenase (LOX) has been reported in more than 60 species of higher plants (Eskin et al., 1977). Large numbers of LOX genes and isozymes have been reported in plants (Melan et al., 1993). Three isozymes of LOXs have been reported in soybean (Hildebrand et al., 1988). The product of the reaction of linoleic acid with soybean lipoxygenase-1 is almost exclusively 13-hydroperoxy-linoleic acid. With lipoxygenase-2 and -3, roughly equal amounts of the 9- and 13-hydroxy products are obtained (Axelrod, 1974).

LOXs from soybean have been cloned and characterized (Siedow, 1991). LOX-1 from soybean contains 838 amino acids and gives a deduced molecular weight of 94,038 (Shibata et al., 1987). LOX-2 contains 865 amino acids (Mr =

97,053) and LOX-3 contains 857 amino acids (Mr = 96,541) (Shibata et al., 1988; Yenofsky et al., 1988). The derived amino acid sequence for the pea LOX contains 861 amino acids (Mr = 97,628) (Ealing and Casey, 1988).

In rice three LOXs have been isolated from leaves. They have pH optima around 5 (Ohta et al., 1991). Three types of LOXs have been found in rice seeds (Ida et al., 1983; Ohta et al., 1986a,b). The LOXs from seeds have pH optima between pH 5.5 and 7.0 (Galliard and Chan, 1980). The leaf LOX-3 produced preferentially 13-L-hydroperoxy-9,11 (Z,E)-octadecadienoic acid (13-HPODD) from linoleic acid, in contrast with LOX-3 derived from seed (Ohta et al., 1986a). LOXs from rice leaf (Ohta et al., 1991) differed very much from those in seeds (Ida et al., 1983; Ohta et al., 1986b).

Fatty acids in membrane lipids are not good substrates for lipoxygenase. Lipolytic acyl hydrolase (LAH) activity releases fatty acids that LOX can oxidize (Thompson et al., 1987). LAH and LOX generally work in combination because free fatty acids are better substrates for LOX than lipids in membranes (Dhindra et al., 1981; Leshem, 1987). The basal levels of LAH and free fatty acids are relatively high in bean (Sastry and Kates, 1964; Croft et al., 1990). A steady increases in LAH was observed during the resistant host–pathogen interactions in bean (Croft et al., 1990).

Depending on the source of enzyme, LOX activity on linolenic acid produces either 9- or 13-hydroperoxy-linolenic acid or a mixture of both (Matthew and Galliard, 1978). Hydroperoxide lyase catalyzes the breakdown of 13-hydroperoxylinolenic acid to a C_6 product, *cis*-3-hexenal, and a C_{12} product, 12-oxo- *cis*-9- dodecenoic acid. The *cis*-3-hexenal forms *cis*-3- hexenol, *trans*-2-hexenal. The C_{12} product, 12-oxo-*cis*-9-dodecenoic acid, is a precursor of traumatic acid (Croft et al., 1993). The second pathway for the metabolism of lipoxygenase products is initiated by an enzyme, hydroperoxide cyclase, which reacts with 13- hydroperoxy-linolenic acid to give 8-(2-(*cis*-2′-pentenyl)-3-oxo-*cis*-4-cyclopentenyl) octanoic acid, commonly referred to as 12- oxo-phytodienoic acid, which, in turn serves as a precursor for the formation of jasmonic acid (Fig. 1; Siedow, 1991).

II. LIPID HYDROPEROXIDES DECOMPOSITION PRODUCTS

Enzymatic decomposition of lipid hydroperoxides, products of reactions catalyzed by LOX, has been reported in many plants (Vick and Zimmerman, 1987a). Hydroperoxide isomerase from flax seed metabolizes lipid hydroperoxides (Zimmerman, 1966) producing α- and β-ketols.

Another type of isomerase, which converts lipid hydroperoxides to trihydroxy fatty acids, has been isolated from oat (Heimann and Dresen, 1973) and wheat (Graveland, 1970). A hydroperoxide lyase has been reported in banana

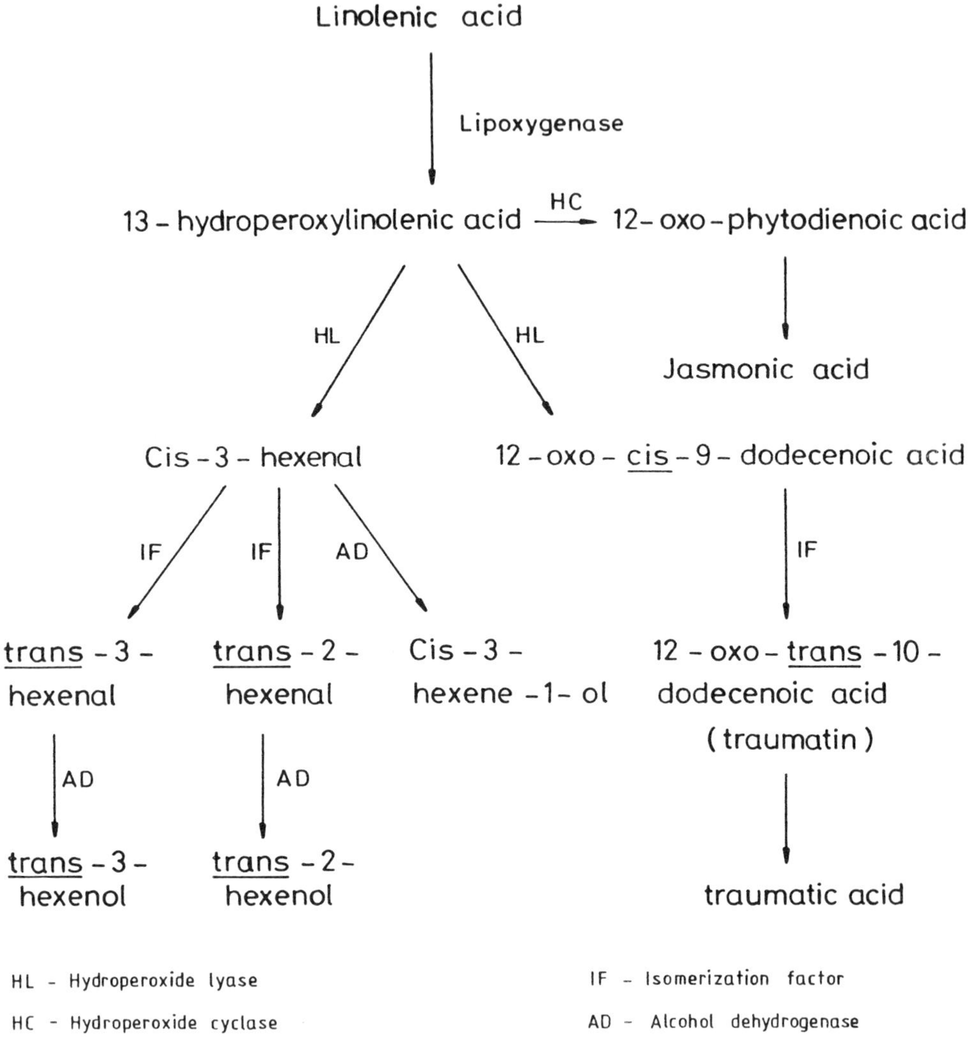

Figure 1 Biochemical pathway of the lipogenase cascade from linolenic acid. (From Siedow, 1991.)

(*Musa* sp.) (Tressl and Drawert, 1973) and watermelon (*Citrullis vulgaris*) (Vick and Zimmerman, 1976). An enzyme, hydroperoxide cyclase, that produces cyclopentenone fatty acids from linolenic acid has been reported in plants (Vick and Zimmerman, 1979). Both hydroperoxide cyclase and hydroperoxide isomerase are considered as a single enzyme (Vick and Zimmerman, 1987a).

An activity that decomposed lipid hydroperoxides was detected in extracts of rice embryos with 9-D-hydroperoxy-10, 12,(E,Z)-octadecadienoic acid (9-LOOH) and 13-L-hydroperoxy-9,11 (Z,E)-octadecadienoic acid (13-LOOH) as substrates. When 9-LOOH was used as the substrate, 9-D-hydroxy-10,12 (E,Z)-octadecadienoic acid (9-LOH) and 9,12,13-trihydroxyoctadec-10-enoic acid (9,12,13-LOH) were formed. When 13-LOOH was the substrate, 13-L-hydroxy-9,11 (Z,E)-octadecadienoic acid (13-LOH) and one more product, which was identified as 9, 10, 13-trihydroxyoctadec-11-enoic acid (9, 10, 13-LOH), were formed. The lipid-decomposing activity was due to two fractions, LHDA-1 and LHDA-2, and these fractions were different from LOX (Ohta et al., 1990).

Soybean LOXs are able to decompose lipid hydroperoxides (Axelrod et al., 1981), but rice LOX-3 has no hydroperoxide-decomposing activity (Ohta et al., 1990).

Enzymatic formation of trihydroxy fatty acids has been reported in wheat (Graveland, 1970) and oat (Heimann and Dresen, 1973). The trihydroxy compound would have been produced from 9-LOOH via 9-hydroxy 12, 13 epoxy octadec-10-enoic acid in rice by an isomerase (Ohta et al., 1990). This isomerase is different from the hydroperoxide isomerases in corn (Gardner, 1970), barley (Yabuuchi, 1976) and flaxseed (Veldink et al., 1970), which produce α-ketol from linoleate hydroperoxide. Oat isomerase generates monohydroxy fatty acid from 9-LOOH or 13-LOOH (Heimann and Dresen, 1973).

III. PRODUCTS OF LIPOXYGENASE ACTIVITY ARE INHIBITORY TO PATHOGENS

13-Hydroxy-*cis*, 9, trans-11-*cis*-15-octadecatrienoic acid, and 9-hydroxy-trans-10, *cis*-12, *cis*-15-octadecatrienoic acid, the LOX products of linolenic acid, inhibit germination of conidia of *Pyricularia oryzae*, the rice blast pathogen (Shimura et al., 1983). The volatile aldehydes hexenal and pentanol are inhibitory to several plant pathogens (Lyr and Basaniak, 1983; Zeringue and McCormick, 1989).

trans-2-Hexenal isolated from leaves of *Ginkgo biloba* inhibits growth of *Monilia fructicola* (Major et al., 1960). *trans*-2-Hexenal produced from wounded cotton leaves inhibited growth of *Aspergillus flavus* (Zeringue and McCormick, 1989). *trans*-2-Hexenal was highly bactericidal at low concentrations, whereas *cis*-3-hexenol was bactericidal only at much higher concentrations (Croft et al., 1993).

Three kinds of hydroperoxides were identified as a result of in vitro lipoxygenase enzyme reaction using the homogenate of blast-infected rice leaves. These three metabolites inhibited germination and germ tube elongation of conidia of *P. oryzae* (Namai et al., 1991). The hydroperoxides correspond to a kind of active oxygen (Namai et al., 1991). Allyl alcohols derived from hydroperoxides are also inhibitory to *P. oryzae* (Kato et al., 1983a,b).

Table 1 Antifungal activities of 9-LOOH and 9-LOH against *Pyricularia oryzae*

	Inhibition (%)			
	9-LOOH (ppm)		9-LOH (ppm)	
	2.5	50	2.5	50
Germination	0	86.0	96.5	100
Germtube growth	38.3	95.2	86.1	100
Appressorium formation	31.9	99.6	—	—

—, No formation of appressoria.
Source: Ohta et al., 1990.

9-Lineolate hydroperoxide (9-LOOH) showed high antifungal activity against *Pyricularia oryzae* at 50 ppm, while the product of 9-LOOH metabolism, 9-LOH, inhibited conidial germination almost completely even at 2.5 ppm (Table 1; Ohta et al., 1990). Thus hydroxy fatty acid is more toxic than hydroperoxide.

Hydroperoxides such as 9-LOOH or 13-LOOH oxidize membrane lipids (Vladimirov et al., 1980) and these hydroperoxides may result in severe damage to the cells of the plant also (Ohta et al., 1990). Antifungal activities of 9-LOH and 9, 12, 13-LOH against *P. oryzae* have been reported (Kato et al., 1983a,b).

13-L-hydroperoxy-9,11 (Z,E)-octadecadienoic acid (13-HPODD) and 13-L-hydroxy-9,11 (Z,E)-octadecadienoic (13-HODD) are the major products of the LOX pathway in *Pyricularia oryzae*-infected rice leaves. Both 13-HPODD and 13-HODD inhibited conidial germination of the fungus at a concentration of 50 μg/ml (Ohta et al., 1991).

LHDA has been shown to convert 13-HPODD (9-HPODD) to 13 HODD (9-HODD) and 9, 10, 13-trihydroxyoctadec-11-enoic acid (9,12,13-trihydroxyoctadec-10-enoic-acid). These decomposed products showed antifungal activity against the rice blast fungus (Kato et al., 1985; Ohta et al., 1990).

IV. LIPOXYGENASE IS INVOLVED IN HOST'S DEFENSE MECHANISMS

The increase in LOX activity has been observed in plants infected with pathogens (Lupu et al., 1980; Ruzicska et al., 1983; Yamamoto and Tani, 1986; Fournier et al., 1986; Ocampo et al., 1986; Keppler and Novacky, 1987; Peever and Higgins, 1989; Croft et al., 1990; Ohta et al., 1990, 1991; Rickauer et al., 1990; Koch et al., 1992). Infection of potato tubers with *Phoma exigua* results in increase in LOX activity (Galliard, 1978). LOX activity increased up to sevenfold in tobacco leaves over a period of 11 days following infection with *Erysiphe cichoracearum* (Lupu et al., 1980). Increase in LOX activity may be due to fatty acids released by free radical attack of membrane lipids during lipid peroxidation (Keppler and Novacky, 1987).

A burst of LOX activity may lead to the production of fatty acid hydroperoxides that are inhibitory to the pathogens (Keppler and Novacky, 1986, Namai et al., 1990; Ohta et al., 1990). LOX and its products can generate active oxygen species (Galliard and Chan, 1980; Veldink and Vliegenthart, 1985). The membrane-bound LOX is capable of generating superoxide anion radicals (Lynch and Thompson, 1984). LOX activity produces singlet oxygen (Anderson, 1989).

LOX activity is involved in induction of defense-related processes. LOX may have a role in signal response coupling of arachidonate elicitation of the defense mechanisms in potato. *Phytophthora* spp. contain fatty acid elicitors such as arachidonic acid and eicosapentaenoic acid (Bostock et al., 1981; Bostock and Stermer, 1989). During the fungal infection, the intrusion of these fatty acids not commonly found in higher plants presents plant LOXs with unusual substrates to produce other highly reactive metabolites that may participate in the induction of defense-related processes (Preisig and Kuc, 1985; Bostock and Stermer, 1989).

The 5-LOX is the principal LOX activity in potato (Galliard and Chan, 1980; Mulliez et al., 1987; Reddanna et al., 1990; Shimizu et al., 1990). LOX inhibitors abolish arachidonic acid (AA) elicitor activity (Preisig and Kuc, 1987). Strong AA elicitor activity was expressed in potato calli that contain normal LOX activity while AA did not elicit responses in LOX-null calli (Vaughn and Lulai, 1992). The structural features in fatty acid elicitors, arachidonic acid, and eicosapentaenoic acid that are required for elicitor activity meet the substrate requirements for the 5-LOX (Bostock et al., 1981; Preisig and Kuc, 1985). These observations suggest that 5-LOX is an intermediate enzyme in the pathway between metabolism and necrosis associated with resistance (Bostock et al., 1981; Preisig and Kuc, 1987).

LOX activity increased above that in untreated potato tuber discs within 30 min after arachidonic acid elicitor treatment, peaked at 1–3 h, and returned to near control levels by 6 h (Bostock et al., 1992). 5-Hydroperoxy-eicosatetraenoic acid (5-HPETE) was the principal product following incubation of these extracts with arachidonic acid. Prior treatment of the discs with abscisic acid, salicylhydroxamic acid, or n-propylgallate suppressed AA induction of the defense response, as well as the arachidonic acid-induced increment in LOX activity. The elicitor-induced increase in 5-LOX activity preceded or temporaly paralelled the arachidonic acid-induced increases in mRNA levels for phenylalanine ammonia-lyase (PAL) and elicitor-responsive isoforms of 3-hydroxy-3-methylglutaryl coenzyme A reductase (HMGR) mRNA were detected within 2 h after treatment, with transcript abundance for both mRNAs attaining a maximum level by 4 h (Bostock et al., 1992). The results suggest that LOX plays a role in the early signaling that leads to expression of defense-related processes (Bostock et al., 1992).

Highly purified preparations of the potato 5-LOX contain additional catalytic activities that convert 5-HPETE to other metabolites, such as leukotriene A4 and lipoxins (Ho and Wong, 1989; Samuelsson et al., 1987). Leukotriene A4 is the first intermediate leading to synthesis of eicosanoids, which are suggested as components of a signaling system (Samuelsson et al., 1987).

LOX-derived fatty acid hydroperoxides are also precursors to molecules such as traumatin and jasmonic acid, which may serve as regulatory molecules (Anderson, 1989; Siedow, 1991). Jasmonates (jasmonic acid and/or its methyl ester) induce specifically the production of a number of proteins in a wide range of plants (Staswick et al., 1991). Jasmonates induce genes encoding phenylalanine ammonia-lyase (Gundlach et al., 1992), chalcone synthase (Creelman et al., 1992b), hydroxyproline-rich cell wall protein (Creelman et al., 1992b), and LOX (Grimes et al., 1992). Jasmonates may also act as signal transduction molecules in the response of plants to pathogen attack (Farmer and Ryan, 1990, 1992; Creelman et al., 1992b; Staswick, 1992). De novo synthesis of jasmonate may be controlled by the availability of free linolenic acid substrate (Farmer and Ryan, 1992). Systemin is a peptide that serves as a systemic signal that releases linolenic acid from membranes after binding to a plasma membrane receptor. Jasmonate synthesized from linolenic acid in turn might activate genes, possibly through another receptor (Farmer and Ryan, 1992). Linolenic acid is the substrate for LOX.

A 2.8 kb, full-length cDNA clone encoding a LOX has been isolated from *Arabidopsis thaliana* (Melan et al., 1993). DNA blot analysis indicated that *Arabidopsis* contains a single LOX 1. LOX 1 mRNA levels increased upon treatment with methyl jasmonate. The induction of the LOX 1 gene may be a regulatory point in the synthesis of jasmonic acid (Melan et al., 1993).

V. HOW DO PATHOGENS OVERCOME LIPOXYGENASE-INDUCED HOST DEFENSE MECHANISMS?

A. Lipoxygenase Activity Markedly Increases Only in Resistant Interaction

LOX activity generally increases in resistant host–pathogen interactions: similar increases are not observed during successful pathogenesis (Croft et al., 1990; 1993). LOX activity has been shown to increase in resistant tissues in oats infected with *Puccinia coronata avenae* (Yamamoto and Tani, 1986), in cucumber infected with *Pseudomonas syringae* pv. *pisi* (Keppler and Novacky, 1986; 1987), and beans infected with *Pseudomonas syringae* pv. *phaseolicola* (Croft et al., 1990). Lipoxygenase activity rapidly increased in rice leaves infected with an incompatible race of *Pyricularia oryzae* (race 131). Although the activity increased in the leaves infected with a compatible race (race 007), the increase in the leaves of the incompatible interaction was much greater than that in the leaves of the compatible interaction (Ohta et al., 1991). Three lipoxygenases (leaf LOX-1,-2,-3) have been isolated from infected and uninoculated leaves. The increase in LOX activity was attributed mainly to the most acidic component, LOX-3. The activity of the leaf LOX-3 was 10 times higher than that in the healthy leaves after inoculation with the incompatible fungus. Activities of the other two LOXs (leaf LOX-1, leaf LOX-2) also increased slightly (Ohta et al., 1991).

When LOX of rice leaves was inhibited by treatment with an inhibitor of the enzyme, the resistant reaction induced by an incompatible race of *P. oryzae* was suppressed, indicating that in susceptible interaction LOX activity is suppressed (Tomita et al., 1984).

Lipid hydroperoxide-decomposing activity (LHDA) also markedly increased in the incompatible *P. oryzae* race-infected rice leaves (Ohta et al., 1991). In contrast, LHDA in leaves inoculated with the compatible race increased only slightly (Ohta et al., 1991). Both LOX and LHDA rapidly increased in rice leaves of a cultivar resistant to *P. oryzae* and similar increase was not observed in susceptible cultivar (Ohta et al., 1990; Namai et al., 1990).

When an elicitor isolated from *Cladosporium fulvum* was injected into tomato leaf tissue, a large increase in levels of lipoxygenase activity was observed at 6 and 12 h after incubation only in resistant cultivar. At 24 h the enzyme activity was not significantly different from the control. No significant differences in peroxidation between control and elicitor-treated tissue were observed 6 h after treatment, but at 12 and 24 h after treatment significantly higher levels of peroxidation were observed in elicitor-treated tissue (Peever and Higgins, 1989). The lipid peroxidation may be due to activated oxygen radicals or to enzymic degradation of membrane lipids by lipoxygenase. However, the first possibility was ruled out since the oxygen radical scavengers α-tocopherol and dabco had no effect on electrolyte leakage induced by the elicitor. The second possibility was strengthened since the lipoxygenase inhibitor piroxicam inhibited the electrolyte leakage induced by the elicitor (Peever and Higgins, 1989). Lipoxygenase-induced electrolyte leakage appears to be found only in resistant cultivar: the host-specific elicitor isolated from *Cladosporium fulvum*-infected tomato leaves induced lipoxygenase and lipid peroxidation only in resistant interaction (Peever and Higgins, 1989).

Inoculation of lower leaf (leaf 1) with *Colletotrichum lagenarium* induced resistance on higher leaves than in cucumber plants. The inducer inoculation induced LOX activity to more than 10 times by day 6 compared to control plants

(Avdiushko et al., 1993). When cucumber leaf 1 was sprayed with dipotassium phosphate or inoculated with tobacco necrosis virus (TNV), resistance against *C. lagenarium* was expressed on leaves above. In all the induced resistant leaves LOX activity increased to a large extent. The amount of LOX protein was significantly increased in leaf 1 of cucumber plants inoculated with *C. lagenarium* or TNV or treated with dipotassium phosphate. The molecular weight of LOX protein was 87 kDa and a degradation product of LOX with molecular weight of 42 kDa was observed in the induced resistant leaves. A LOX isozyme was also detected (Avdiushko et al., 1993). The increase in LOX activity correlated with a decrease in the amount of LOX substrates, linoleic and linolenic acids, and in the total fatty acid fraction in cucumber leaves. The results suggest that lipoxygenase is involved in resistance mechanism and in the susceptible leaves no increase in LOX is observed (Avdiushko et al., 1993).

B. Induction of Lipoxygenases is Delayed in Susceptible Interaction

A delayed induction of LOX activity in the susceptible response compared to the resistant interaction has been reported in several diseases (Lupu et al., 1980; Ruzicska et al., 1983; Yamamoto and Tani, 1986; Ocampo et al., 1986; Keppler and Novacky, 1987; Peever and Higgins, 1989; Rickauer and Higgins, 1989; Ohta et al., 1990; 1991; Croft et al., 1990). LOX mRNA was induced by 3 h and enzyme activity began to increase between 6 and 12 h and had reached maximum levels by 24 to 48 h in bean leaves inoculated with an incompatible bacterial pathogen. In a compatible interaction, LOX mRNA was induced at 12 h and enzyme activity changed only marginally in the first 24 h, then increased steadily up to 72 h, reaching the levels seen in the resistant reaction (Koch et al., 1992). When a cloned avr gene was transferred to a virulent isolate, LOX mRNA rapidly increased similar to that observed for the incompatible strain (Melan et al., 1993). A major difference between susceptibility and resistance to pathogens seems to be in the timing of induction of LOX (Melan et al., 1993). The rapid induction of LOX is under the influence of the avirulence gene of the pathogen. If virulence gene is present, LOX activity induction is suppressed.

C. Pathogen May Induce Antioxidants to Nullify the Action of LOX

When tobacco plants were inoculated with *Erysiphe cichoracearum*, lipoxygenase activity increased and maximum increase was observed on 11th day after inoculation (Lupu et al., 1980). Along with an increase in lipoxygenase activity, there was an increase in the endogenous level of antioxidative activity. Peroxide level was 2.5-fold higher in the infected tissues than in the control leaves (Lupu et al., 1980). It suggested that the pathogen may induce synthesis of antioxidants also along with LOX which may nullify the toxic action of LOX.

D. Lipoxygenase Activity May Not Be Involved in Disease Resistance

Incompatible race of *Phytophthora infestans* strongly and rapidly induces lipid peroxidation in potato tuber discs to an extent that exceeds that in a compatible interaction (Bostock et al., 1986). A corresponding effect on LOX, however, is not observed (Bostock, 1989). The 5-LOX activity in the potato tuber was substantial, even after inoculation with a compatible race (Bostock, 1989) or after abscisic acid treatment, which strongly suppressed the resistant reaction and induced susceptibility to incompatible races (Henfling et al., 1980; Bostock et al., 1992). Hence 5-LOX may not be involved in induction of disease resistance or a discriminating role for the enzyme in disease resistance may reside in its differential recruitment from a sequestered or inactive form (Bostock et al., 1992).

VI. ACTIVE OXYGEN SPECIES

When the pathogens invade host tissues, different oxygen radicals are produced. Superoxide anion (O_2^-) is produced in potato–*Phytophthora infestans* interactions (Jordan and DeVay, 1990), in rice inoculated with *Pyricularia oryzae* (Haga et al., 1986; Averyanov et al., 1987; 1988; Sekizawa et al., 1985; 1987), and in tomato cells treated with the fungal elicitor (Vera-Estrella et al., 1993). The superoxide anion bears an unpaired electron that arises from the one electron reduction of molecular oxygen (Sutherland, 1991).

Extracellular O_2^- in plant cells may be generated in a number of ways. NADH/NADPH is oxidized to NAD/NADP in the presence of peroxidase and Mn^{2+} to reduce molecular oxygen to O_2^- (Gross, 1977, Halliwell, 1978; Stich and Ebermann, 1984). A specific membrane-located NAD(P)H oxidase may catalyze the single electron reduction of molecular oxygen to O_2^- (Doke, 1985; Vianello and Macri, 1989). LOX catalyzes the direct oxygenation of polyunsaturated fatty acids (Saniewski, 1979) and LOX activity produces O_2^- (Lynch and Thompson, 1984; Thompson et al., 1987). The oxidative catabolism of purine leads to synthesis of the ureides, allantoin, and allantoic acid, which constitute major organic nitrogen compounds present in a wide variety of plants (Schubert, 1986). The purine catabolism is catalyzed by xanthine dehydrogenase, leading to uric acid formation. During xanthine oxidation, superoxide radicals are formed (Sandalio et al., 1988).

Singlet oxygen ($^1O_2$) is generated during pathogenesis (Keppler et al., 1989; Salzwedel et al., 1989). Singlet oxygen is an excited state of molecular oxygen that can be generated in a number of ways including the spontaneous dismutation of two O_2^- radicals (Elstner, 1982; Knox and Dodge, 1985).

Hydroxyl radical (OH°) has been detected in *P. infestans*-infected potato leaf tissues (Jordan and DeVay, 1990). This radical has been reported in soybean cotyledons treated with abiotic elicitors (Epperlein et al., 1986). Hydroxyl radicals are derived from O_2^- by the action of superoxide dismutase (SOD) and the iron-catalyzed Haber-Weiss process (Epperlein et al., 1986). H_2O_2 is produced within 5 min of *Verticillium dahliae* elicitor addition in cultured soybean cells (Apostol et al., 1989). The earliest event, which occurred in bean cells treated with *Colletotrichum lindemuthianum* elicitor, was the production of H_2O_2 (Anderson et al., 1991). When suspension cultured white clover (*Trifolium repens*) cells were treated with an abiotic elicitor, H_2O_2 production was observed within 10 min after treatment (Devlin and Gustine, 1992). O_2^- undergoes dismutation to form H_2O_2 (Gross et al., 1977; Anderson, 1989). Uric acid oxidation by uricase also leads to production of H_2O_2 (Tolbert, 1982). Peroxidases may also be possible sources of oxygen radicals in plants (Halliwell, 1978; Albert et al., 1986).

The suggested pathway in the formation of various reactive oxygen species is as follows (Scandalios, 1993):

$$O_2 \xrightarrow{+e} O_2^- \xrightarrow[2H]{+e} H_2O_2 \xrightarrow[2H^+]{+e} OH^\circ$$

Production of active oxygen species has been reported in various plant–pathogen interactions (Doke, 1983a,b; Doke and Chai, 1985; Montalbini and Buonaurio, 1986; Pavlovkin et al., 1986; Keppler and Novacky, 1986, 1987, 1989; Minardi and Mazzucchi, 1988; Chai and Doke, 1987a,b; Moreau and Osman, 1989; Keppler and Baker, 1989).

VII. ACTIVE OXYGEN SPECIES INDUCE RESISTANCE

Oxygen species may be involved in the oxidation of membrane lipids that results in production of several antifungal compounds, initiation of cell wall lignification reaction, direct injury to the pathogens, and signal transduction (Sutherland, 1991).

A. Active Oxygen Species Induce Oxidation of Membrane Lipids

Lipid peroxidation can occur both by enzymic and nonenzymic means. LOX causes lipid peroxidation (Leshem, 1987). Lipid peroxidation can occur by nonenzymic means through the action of various active oxygen species (Svingen et al., 1978; Thompson et al., 1987; Niinomi et al., 1987). Singlet oxygen is highly reactive to oxidize lipids in membranes (Knox and Dodge, 1985). Hydroxyl radicals are derived from O_2^- by the action of superoxide dismutase (Epperlein et al., 1986) and they may induce membrane damage (Adam et al., 1989; Halliwell, 1988; Bielski et al., 1983). OH° attacks membrane lipids in vitro; this in turn initiates a chain reaction of lipid radicals (Halliwell and Gutteridge, 1986). Lipid peroxidation is also known to be due to H_2O_2, perhaps by its effect on peroxidase activation (Keppler and Novacky, 1987; Rogers et al., 1988). Protonation of O_2^- to give its conjugate acid, the perhydroxyl radicals (HO_2^-), can occur in specific cellular microenvironments where there are local concentrations of protons, particularly at the negatively charged surfaces of membranes (Bielski et al., 1983; Fridovich, 1988). The perhydroxyl radical is a stronger oxidizing agent than O_2^- and will attack polyunsaturated fatty acids directly (Bielski et al., 1983; Halliwell, 1988). Thus, LOX activity can lead to the production of various active oxygen species capable of initiating enzyme-independent lipid peroxidation. Once membrane lipid peroxidation has begun lipid peroxy radical (ROO ·) intermediates will propagate the chain reaction among the tightly stacked fatty acids in the lipid bilayer of the membrane (Adam et al., 1989).

B. Active Oxygen Species Are Involved in Induction of Defense Chemicals

Active oxygen species may induce synthesis of various defense chemicals. When suspension cultured cells of tomato were inoculated with a nonspecific elicitor isolated from *Cladosporium fulvum* cell wall, an increase in total phenol content was observed by 1 h of incubation and this increase was relatively constant for 3 h (Vera-Estrella et al., 1993). At least three conspicuous phenolic compounds could be detected in the elicitor-treated cells. One was present at zero time but markedly increased by 3 and 6 h in all treated cells. The other two compounds were not present at zero time and increased between 3 and 6 h in the elicitor-treated cells. (Vera Estrella et al., 1993). To test the possibility that active oxygen species may be involved in the elicitor-induced increase of extracellular phenolic compounds, catalase, SOD, ascorbate, or mannitol was added. SOD is the O_2^- scavenger. Catalase catalyzes the decomposition of H_2O_2. Mannitol is an OH radical scavenger. In the absence of the elicitor, none of these compounds had a marked effect on extracellular phenolic compounds released by the tomato cells. Previous incubation of cells with catalase, SOD, and mannitol prevented the elicitor-induced increase of extracellular phenolic compounds (Vera-Estrella et al., 1993). The results suggest that the active oxygen species may activate the synthesis of defense chemicals.

Digitonin activated the O_2^- generation system, which, in turn activated synthesis of the phytoalexin rishitin in potato tissue. When susceptible potato leaves were treated with digitonin, the leaf tissues became resistant to compatible races of *Phytophthora infestans* (Doke and Chai, 1985). A preinfection infiltration of SOD, which is a scavenger of O_2^-, into potato tissues retarded phytoalexin production (Doke, 1983a). Active oxygen species have been suggested to play an important role in causing phytoalexin accumulation in several host–pathogen interactions (Doke, 1985; Epperlein et al., 1986; Sekizawa et al., 1987; Rogers et al., 1988).

H_2O_2 is reduced by peroxidase to water at the expense of electrons from coniferyl alcohol. The alcohol radical then initiates the lignin chain reaction. Thus H_2O_2 is an important substrate for peroxidases in the oxidation of coniferyl alcohol to initiate the lignification chain reaction (Gross et al., 1977). Lignin synthesis requires H_2O_2 and peroxidases (Van Huytsee, 1987). Indole acetic acid (IAA) inhibits lignin synthesis and H_2O_2 is involved in the metabolism of IAA (Hamilton et al., 1976; Shinshi et al., 1987). H_2O_2-mediated peroxidases catalyze production of hydroxyproline-rich proteins (van Huytsee, 1987). H_2O_2 may also be involved in the synthesis of fungitoxic quinones (van Huytsee, 1987).

C. Active Oxygen Species Cause Direct Injury to Pathogens

All active oxygen species are extremely reactive and cytotoxic. They react with unsaturated fatty acids to cause peroxidation of essential membrane lipids in the plasmalemma or intracellular organelles. Peroxidation damage of the plasmalemma leads to leakage of cellular contents (Scandalios, 1993).

H_2O_2 is a powerful bactericidal agent (Lachmann, 1986) and alone or in combination with cell wall peroxidases it may contribute significantly to repulsion of pathogens (Kim et al., 1988). H_2O_2 is toxic to pathogens (Montalbini, 1991).

Hydroxyl radical is the most potent oxidant known and causes damage to cellular components (Scandalios, 1993). A singlet oxygen quencher, β-carotene, increased bacterial populations in an incompatible host when coinjected into leaves. This suggests the antimicrobial action of singlet oxygen (Roy and Sasser, 1983).

D. Active Oxygen Species are Involved in Signal Transduction

H_2O_2 acts as a second messenger in eliciting the resistance response (Apostol et al., 1989; Heffetz et al., 1990; Anderson et al., 1991). The role of H_2O_2 as a second messenger has been discussed in detail in Chapter 1.

Hydroxyl radical has been suggested to be an important signal in triggering phytoalexin production in soybean cotyledons with abiotic elicitors (Epperlein et al., 1986).

VIII. HOW DOES THE PATHOGEN OVERCOME THE TOXICITY OF ACTIVE OXYGEN SPECIES?

A. Pathogen Suppresses Active Oxygen Species Generation System

Pathogen may suppress the generation of active oxygen species. Superoxide anion was produced much more slowly and in smaller amounts in compatible *Pyricularia oryzae*-rice interaction than in incompatible interaction (Sekizawa et al., 1985; 1987). Production of active oxygen species was initiated within 2 min of treatment when cultivar-specific elicitors isolated from the apoplast of *Cladosporium fulvum*-infected tomato leaves were used on cell suspension cultures derived from a tomato line near isogenic for a gene for resistance to the *C. fulvum* race involved. Cell lines from plants susceptible to this *C. fulvum* race failed to exhibit such oxygen species production (Vera-Estrella et al., 1992).

The incompatible combination of *Cladosporium fulvum* race 4 host-specific elicitor and tomato Cf5 cell lines showed a marked increase in lipid peroxidation within 1 h after elicitor treatment. Similar increase was not observed in compatible combinations. In the incompatible interaction there was an increase in cytochrome c-reducing activity, indicating the production of superoxide (Vera-Extrella et al., 1992). There was no increase in cytochrome c reduction activity in the compatible interaction (Vera-Estrella et al., 1992).

Wounded potato tuber tissues have been found to activate a superoxide anion-generating system almost immediately after penetration of the cells by incompatible but not compatible races of *P. infestans* (Doke, 1983a). The activated system was found to be an O_2^--generating NADPH oxidase (Doke, 1983b).

In potato leaves inoculated with *P. infestans*, activation of O_2^- generation was found to occur before penetration by both compatible and incompatible races, with further O_2^- generation following penetration by incompatible, but not by compatible, races (Chai and Doke, 1983).

In potato tissues inoculated with an incompatible race of *Phytophthora infestans*, NADPH-dependent cytochrome c-reducing activity increased. Similar increase was not observed in compatible interaction. The suppressor isolated from compatible race, but not from incompatible races, of *P. infestans* inhibited the infection-induced NADPH cytochrome c-reducing activity (Doke, 1985). The water-soluble glucans (the suppressor) from compatible races of *P. infestans* specifically inhibited the activation of the NADPH-dependent O_2^--generating reaction in protoplasts (Doke, 1983b). The results suggest that suppression of

O_2^--generating system may be essential for the pathogen to establish the compatible reaction.

B. Pathogen Activates Generation of Scavengers of Oxygen Species

Both enzymic and nonenzymic mechanisms exist in plants to scavenge oxygen species. Ascorbic acid (vitamin C), α-tocopherol (vitamin E), β-carotene, reduced glutathione (GSH), cysteine, hydroquinones, mannitol, flavonoids, dipyridamole, and polyamines are nonenzymic scavengers (Siegel and Halpern, 1965; Tappel, 1965; Morisaki et al., 1982; Law et al., 1983; Drolet et al., 1986; Larson, 1988). Ascorbate peroxidase and GSH reductase are known to scavenge H_2O_2 (Foyer and Halliwell, 1976). Ascorbic acid reacts rapidly with O_2^- and OH^- and helps to remove H_2O_2 by the ascorbate peroxidase reaction (Halliwell, 1982; Abels, 1986). The oxy-radical-bearing fatty acyl chains may partition into the hydrophilic surface of the bilayer so that ascorbic acid interacts directly with them (Lielder et al., 1986). Tocopherol is the major free radical scavenger in thylakoid membranes. It is kept in its reduced form by a nonenzymic reaction with ascorbate. The tocopherol quenches directly lipid peroxy radicals and $^1O_2$ in thylakoid membranes (Krinsky, 1979; Leung et al., 1982). GSH reacts in an enzymatic way through glutathione peroxidase activity (Adam et al., 1989). Catalases and peroxidases remove H_2O_2 very efficiently (Sandalios, 1993). Superoxide dismutases (SODs) scavenge the superoxide anions (Montalbini and Buonaurio, 1986; Moreau and Osman, 1989). Low levels of O_2^- and OH° radicals are normally maintained by the action of SOD and catalase (Galliard and Chen, 1980).

SOD is ubiquitous in plants. SODs are multimeric metalloproteins. There are SODs that contain copper and zinc (Cu/Zn SOD), managenase (Mn SOD), or iron (Fe SOD) (Sandalios, 1993). The plants have multiple forms of SOD (Baum and Scandalios, 1982). Six isozymes of SODs have been reported in maize (Scandalios, 1990). The SOD gene has been cloned from maize (Cannon et al., 1987). Multiple-SOD genes may exist in plants (Scandalios, 1993). The enzymes involved in scavenging oxygen species are mutually protective. O_2^- can inhibit peroxidase and H_2O_2 can inhibit SOD (Fridovich, 1988).

During pathogenesis, pathogens activate synthesis of these scavengers to suppress the defense mechanisms of plants. Ascorbic acid content increased in oat leaves inoculated with highly virulent *Drechslera avenae* (Gonner and Schlosser, 1993). After 110 h the ascorbate content was almost fourfold higher than the content in control leaves and leaves inoculated with the weakly virulent *Drechslera nobleae* (Gonner and Schlosser, 1993). Tocopherol (vitamin E) content also increased more in oat leaves inoculated with *D. avenae* than in those inoculated with *D. nobleae* (Gonner and Schlosser, 1993).

Glutathione content also increased, although only slightly in virulent pathogen-inoculated oat tissue, while there were no marked changes in glutathione content in weakly virulent pathogen-inoculated tissues (Gonner and Schlosser, 1993). Thus all the scavengers of active oxygen species increased in the susceptible interactions in oats (Gonner and Schlosser, 1993).

Catalase activity increases in susceptible interactions to scavenge hydrogen peroxide. Two susceptible and two resistant bean cultivars were inoculated with *Uromyces phaseoli* (Montalbini, 1991). Activity of uricase, the hydrogen peroxide-producing enzyme, increased in both susceptible and resistant cultivars due to infection. An increase in uricase activity was detected 2 days after inoculation in susceptible cultivars, while a considerable increase in uricase was detected 20–22 h after inoculation in the resistant cultivars. Catalase activity was enhanced in the susceptible cultivars at the same extent and simultaneously to uricase (2 days after inoculation). By contrast, the resistant cultivar displayed a catalase increase only 3 days after inoculation with a delay of 2 days compared to uricase activation. The late increase in catalase was also very much less than those observed for uricase. A low catase activation, concomitant with a strong uricase enhancement, may be associated with unscavenged hydrogen peroxide production. During successful pathogenesis, an increase in catalase would have suppressed production of the fungitoxic H_2O_2 (Montalbini, 1991).

Peroxidase in combination with SOD may minimize toxic action of active oxygen species in susceptible interactions. When highly virulent *D. graminea* was inoculated on oat leaves, peroxidase activity increased remarkably and a similar increase was not observed when a weakly virulent *D. nobleae* was inoculated. A large increase in SOD activity was likewise observed 40 h postinoculation with *D. avenae*, while no such increase was observed in oat leaves inoculated with *D. nobleae* (Gonner and Schlosser, 1993).

IX. OXYGEN SPECIES MAY BE INVOLVED IN NECROTIC SYMPTOM DEVELOPMENT RATHER THAN INDUCING RESISTANCE IN SUSCEPTIBLE TISSUES

Bleaching of photosynthetic pigments and cellular injury are characteristic symptoms caused by reactive oxygen species (Rabinowitch and Fridovich, 1983; Kauss, 1990; Rosen and Halpern, 1990) and they are also the characteristic symptoms of several diseases. Hence it is possible that the oxygen species may be involved in symptom development rather than in defense mechanisms.

Drechslera avenae and *Drechslera siccans* cause severe symptoms on oat leaves, while *Drechslera nobleae* and *Drechslera graminea* are only weakly virulent. The technique of spin trapping was applied to study generation of reactive radical intermediates in host–pathogen interactions (Gonner et al., 1993). In this technique, a relatively short-lived radical reacts with a spin trap to yield a more stable radical that can be detected by electron spin resonance (ESR). The spin trapping agent 1,2-dihydroxybenzene-3, 5-disulphonic acid (tiron) forms stable semiquinones or nitroxides by reacting covalently with the unstable oxygen radicals. When the spin-trapping agent tiron was incubated in the intercellular washing fluid of leaflets inoculated with the highly virulent *D. avenae* and *D. siccans*, ESR signal was obtained. Leaves inoculated with the weakly virulent *D. graminea* and *D. nobleae* did not produce a detectable spin adduct after infiltration with tiron. The formation of chlorosis and necrotic lesions coincided with the detection of an ESR signal indicative of free radicals that became apparent 62 h postinoculation with *D. avenae* and 86 h postinoculation with *D. siccans* (Gonner et al., 1993).

When oat leaves were inoculated with virulent pathogens such as *Drechslera avenae* and *D. siccans* and weakly virulent pathogens such as *D. graminea* and *D. nobleae*, decrease in linolenic acid (unsaturated fatty acid) content was observed at 5 days after inoculation only with the virulent pathogens. The amounts of saturated fatty acids, including stearic acid (C18:0) and palmitic acid (C16:0), remained unchanged (Gonner and Schlosser, 1993). Another criterion for detecting lipid peroxidation is the formation of malondialdehyde (MDA). The amount of malondialdehyde increased in leaves of oat inoculated with the highly virulent pathogens *D. avenae* and *D. siccans*. Maximum MDA formation was found to coincide with complete bleaching of the chlorophyll. The inoculation with the weakly virulent *D. nobleae* and *D. graminea* did not cause significant changes. These fungi cause slight symptoms (Gonner and Schlosser, 1993). All these studies suggest that oxygen species may be involved in symptom development rather than inducing resistance.

When the oxygen radicals appear in the intercellular spaces, they may trigger a variety of different forms of cell damage including lipid peroxidation (Kato and Misawa, 1976), inactivation or induction of enzymes (Croft et al., 1990; Hodgson and Raison, 1991), and the loss in structure and function of plasma membranes eventually leading to cell necrosis.

Based on various studies it may be concluded that active oxygen species may occur in both susceptible and resistant interactions. In both cases cell necrosis and death occur. In resistant reactions, cell death may occur much earlier (hypersensitive cell death) while in susceptible interactions cell death may occur at later stages. Early induction of active oxygen species may likewise occur in resistant reaction, while delayed release of active oxygen species may occur in susceptible reaction.

X. ACTIVE OXYGEN SPECIES MAY NOT BE INVOLVED IN HOST DEFENSE MECHANISMS

Although active oxygen species have been suggested to be involved in host defense mechanisms, there are also reports that they may not be involved in the defense mechanism at all. The germination fluid produced by zoospores of *Phytophthora infestans* elicited production of phytoalexins in potato tuber discs (Osman and Moreau, 1985). The elicitor caused an increase in the rate of reduction of cytochrome c in potato tuber discs. However, the superoxide anion was not involved in the reduction of cytochrome c, since very little superoxide was detected in the reaction mixture and addition of superoxide dismutase to the reaction mixtures had very little effect on the rate of reduction of cytochrome c. The results do not support the hypothesis that superoxide is involved in elicitation of phytoalexins in potato tuber discs (Moreau and Osman, 1989).

When suspension-cultured white clover cells were treated with an abiotic elicitor, H_2O_2 production was observed within 10 and 20 min after treatment. Preincubation of plant cells with catalase, which quenches oxidative burst, did not decrease phytoalexin accumulation. The results suggest that the oxidative burst may not be a necessary element of the signaling system for phytoalexin accumulation (Devlin and Gustine, 1992).

Two singlet oxygen quenchers, 1,4-diazabicyclo octane (DABCO) and carotenoid carboxylic acid bixin, did not suppress the hypersensitive reaction induced in suspension cell cultures of tobacco by an incompatible pathogen. The results suggest that singlet oxygen may not be involved in hypersensitive reaction in tobacco cells (Salzwedel et al., 1989). Abiotic elicitors induced phytoalexin production in soybean cotyledons. Both SOD and catalase failed to inhibit this phytoalexin production, implying that O_2^- and H_2O_2 may not be important in the generation of the signal to produce phytoalexin (Epperlein et al., 1986).

XI. CONCLUSION

Plant cell membrane provides many toxic products to offer resistance against fungal pathogens. The fatty acids linoleic and linolenic acids are the common constitutents of plant membranes and various lipid breakdown products are formed through the so-called lipoxygenase pathway. Several enzymes are involved in this pathway and lipoxygenase (LOX) appears to be the key enzyme. Several LOXs have been cloned and characterized. The lipid breakdown products have been shown to be fungitoxic. LOX activity is involved in several defense-related processes. It may have a role in signal response coupling of fatty acid elicitor activity. Its activity leads to synthesis of regulatory molecules such as traumatin and jasmonic acid. Jasmonates are the systemic signal transduction molecules. LOX is involved in phytoalexin, cutin, and suberin synthesis. It is also capable of generating superoxide anion and singlet oxygen.

The active oxygen species (superoxide anion [O_2^-], singlet oxygen [$^1O_2$], hydroxyl radical [OH°]) and H_2O_2 are produced due to fungal infection. Their production is interdependent and also dependent on lipid peroxidation. The earliest event in the process of activation of host defense mechanisms appears to be the generation of active oxygen species, which occurs within few minutes of eicitation. These active oxygen species are cytotoxic and the host cell appears to have a balance between them to escape from the cytotoxicity. Several scavengers of oxygen species also exist in the cells. Catalases and peroxidases remove H_2O_2 efficiently. Superoxide dismutases scavenge the superoxide anion. Nonenzymic scavengers such as ascorbic acid, glutathione, tocopherol, β-carotene, and cysteine may also reduce the toxicity of these active oxygen species.

Active oxygen species induce lipid peroxidation. Singlet oxygen is highly reactive to oxidize lipids. Active oxygen species induce phenolics and phytoalexin synthesis and lignification. Lignification requires H_2O_2 and peroxidase. H_2O_2 is involved in hydroxyproline-rich glycoprotein synthesis. H_2O_2 is a second messenger in induction of phytoalexin synthesis. Hydroxyl radical is also an important signal in triggering phytoalexin production. Superoxide anion activates synthesis of phytoalexins.

Lipid peroxidation and accumulation of active oxygen species increase mostly in the resistant reaction. Successful pathogen appears to suppress the increase in these activities. Scavengers of oxygen species may be activated in the susceptible reaction. It is also possible that both lipid peroxidation and active oxygen species production may not have any role in defense mechanisms and may be involved in cell necrosis only. Their role has been studied more in bacterial diseases and intensive studies are needed in fungal diseases.

REFERENCES

Abeles, F. B. (1986). Plant chemiluminescence, Annu. Rev. Plant Physiol., 37:49–72.

Adam, A., Farkas, T., Somlyai, G., Hevesi, M., and Kiraly, Z. (1989). Consequence of O_2^- generation during a bacterially induced hypersensitive reaction in tobacco: deterioration of membrane lipids, Physiol. Mol. Plant Pathol., 34:13–26.

Albert, F. G., Bennett, L. W., and Anderson, A. J. (1986). Peroxidase associated with the root surface of *Phaseolus vulgaris*, Can. J. Bot., 64:573–578.

Anderson, J. M. (1989). Membrane derived fatty acids as precursors to second messengers, Second Messengers in Plant Growth and Development (W. F. Boss and D. J. Morre, eds.), Alan R. Liss, New York, pp. 182–212.

Anderson, A. J., Rogers, K., Teppers, C., Blee, K., and Cardon, J. (1991). Timing of molecular events following elicitor treatment of plant cells, Physiol. Mol. Plant Pathol., 38:1–13.

Apostol, I., Heinstein, P. F., and Low, P. S. (1989). Rapid stimulation of an oxidative burst during elicitation of cultured plant cells, Plant Physiol., 90:109–116.

Averyanov, A. A., and Lapikova, V. P. (1988). Fungitoxicity determined by active forms of oxygen in excretions of rice leaves, Fiziol. Rastenii, 35:1142–1151.

Averyanov, A. A., Lapikova, B. P., Umnov, A. M., and Dzhavakhiya, V. G. (1987). Generation of superoxide radical by rice leaves in relation to blast resistance, Fiziol. Rastenii, 34:373–379.

Axelrod, B. (1974). Lipoxygenases, ACS Adv. Chem. Ser., 136:324–348.

Axelrod, B., Cheesbrough, T. M., and Laakso, S. (1981). Lipoxygenase from soybeans, Methods Enzymol., 71:441–451.

Baum, J. A., and Scandalios, J. G. (1982). Multiple genes controlling superoxide-dismutase expression in maize, J. Hered., 73:95–100.

Bielski, B. H. J., Arudi, R. L., and Sutherland, M. W. (1983). A study of the reactivity of HO_2/O_2^- with unsaturated fatty acids, J. Biol. Chem., 258:4759–4761.

Bostock, R. M. (1989). Metabolism of lipids containing arachidonic and eicosapentaenoic acids in race-specific interactions between *Phytophthora infestans* and potato, Phytopathology, 79:898–902.

Bostock, R. M., Kuc, J. A., and Laine, R. A. (1981). Eicosapentaenoic and arachidonic acids from *Phytophthora infestans* elicit fungitoxic sesquiterpenes in the potato, Science, 212:67–69.

Bostock, R. M., Schaeffer, D. A., and Hammerschmidt, R. (1986). Comparison of elicitor activities of arachidonic acid, fatty acids and glucans from *Phytophthora infestans* in hypersensitivity expression in potato tuber, Physiol. Mol. Plant Pathol., 29:349–360.

Bostock, R. M., and Stermer, B. A. (1989). Perspectives on wound healing in resistance to pathogens, Annu. Rev. Phytopathol., 27:343–371.

Bostock, R. M., Yamamoto, H., Choi, D., Ricker, K. E., and Ward, B. L. (1992). Rapid stimulation of 5-lipoxygenase activity in potato by the fungal elicitor arachidonic acid, Plant Physiol., 100:1448–1456.

Bretscher, M. S. (1973). Membrane structure: some general principles, Science, 181:622–629.

Cannon, R. E., White, J. A., and Scandalios, J. G. (1987). Cloning of cDNA for maize superoxide dismutase (SOD-2), Proc. Natl. Acad. Sci. USA, 84:179–183.

Chai, H. B., and Doke, N. (1983). Aspects of superoxide anion generation in potato leaf tissues infected by *Phytophthora infestans* and its stimulation by preinfectional treatment with germination fluid, Ann. Phytopath. Soc. Japan, 49:378.

Chai, H. B., and Doke, N. (1987a). Activation of the potential of potato leaf tissue to react hypersensitively to *Phytophthora infestans* by cystospore germination fluid and the enhancement of this potential by calcium ions, Physiol. Mol. Plant Pathol., 30:27–37.

Chai, H. B., and Doke, N. (1987b). Superoxide anion generation: a response of potato leaves to infection with *Phytophthora infestans*, Phytopathology, 77:645–649.

Creelman, R. A., Tierney, M. L., and Mullet, J. E. (1992b). Jasmonic acid/methyl jasmonate accumulate in wounded soybean hypocotyls and modulate wound gene expression, Proc. Natl. Acad. Sci. USA, 89:4938–4941.

Croft, K. P. C., Juttner, F., and Slusarenko, A. J. (1993). Volatile products of the lipoxygenase pathway evolved from *Phaseolus vulgaris* (L.) leaves inoculated with *Pseudomonas syringae* pv. *phaseolicola*, Plant Physiol., 101:13–24.

Croft, K. P. C., Voisey, C. R., and Slusarenko, A. J. (1990). Mechanism of hypersensitive cell collapse: correlation of increased lipoxygenase activity with membrane damage in leaves of *Phaseolus vulgaris* (L.) cv. Red Mexican inoculated with avirulent race 1 cells of *Pseudomonas syringae* pv. *phaseolicola*, Physiol. Mol. Plant Pathol., 36:49–62.

Devlin, W. S., and Gustine, D. L. (1992). Involvement of the oxidative burst in phytoalexin accumulation and the hypersensitive reaction, Plant Physiol., 100:1189–1195.

Dhindra, R. S., Plumb-Dhindra, P., and Thorpe, T. A. (1981). Leaf senescence: correlated with increased levels of membrane permeability and lipid peroxidation, and decreased levels of superoxide dismutase and catalase, J. Exp. Bot., 32:93–101.

Doke, N. (1983a). Involvement of superoxide anion generation in the hypersensitive response of potato tuber tissue to infection with an incompatible race of *Phytophthora infestans* and the hyphal wall components, Physiol. Plant Pathol., 23:345–357.

Doke, N. (1983b). Generation of superoxide anion by potato tuber protoplasts during the hypersensitive response to hyphal wall components of *Phytophthora infestans* and specific inhibition of the reaction by suppressors of hypersensitivity, Physiol. Plant Pathol., 23:359–367.

Doke, N. (1985). NADPH-dependent O_2^- generation in membrane fractions isolated from wounded potato tubers inoculated with *Phytophthora infestans*, Physiol. Plant Pathol., 27:311–322.

Doke, N., and Chai, H. B. (1985). Activation of superoxide generation and enhancement of resistance against compatible races of *Phytophthora infestans* in potato plants treated with digitonin, Physiol. Plant Pathol., 27:323–334.

Drolet, G., Dumbroff, E. B., Legge, R. L., and Thompson, J. E. (1986). Radical scavenging properties of polyamines, Phytochemistry, 25:367–371.

Ealing, P. M., and Casey, R. (1988). Complete amino acid sequence of a pea (*Pisum sativum*) seed lipoxygenase predicted from a near full-length cDNA, Biochem. J., 253:915–918.

Elstner, E. F. (1982). Oxygen activation and oxygen toxicity, Annu. Rev. Plant Physiol., 33:73–96.

Epperlein, M. N., Noronha-Dutra, A. A., and Strange, R. N. (1986). Involvement of the hydroxyl radical in the abiotic elicitation of phytoalexins in legumes, Physiol. Mol. Plant Pathol., 28:67–77.

Epton, H. A. S., and Deverall, B. J. (1968). A biochemical difference between healthy bean leaves resistant and susceptible to the halo blight disease caused by *Pseudomonas phaseolicola*, Ann. Appl. Biol., 61:255–268.

Eskin, N. A. M., Grossman, S., and Pinsky, A. (1977). Biochemistry of lipoxygenase in relation to food quality, Crit. Rev. Food Sci. Nat., 9:1–41.

Farmer, E. E., and Ryan, C. A. (1990). Interplant communication: airborne methyl jasmonate induces synthesis of proteinase inhibitors in plant leaves, Proc. Natl. Acad. Sci. USA, 87:7713–7716.

Farmer, E. E., and Ryan, C. A. (1992). Octadecanoid precursors of jasmonic acid activate the synthesis of wound-inducible proteinase inhibitors, Plant Cell, 4:129–134.

Fournier, J., Pelissier, B., and Esquerre-Tugaye, M. T. (1986). Induction d'une activité lipoxygenase dans les cellules de tabac (*Nicotiana tabacum*) en culture, par des eliciteurs d'ethylene de *Phytophthora parasitica* var. *nicotianae*, C. R. Acad. Sci. Paris, 303:651–654.

Fridovich, I. (1988). The biology of oxygen radicals, Oxygen Radicals and Tissue Injury, (B. Halliwell, ed.), Proc. of a Brook Lodge Symposium, Upjohn, p. 148.

Galliard, T. (1978). Lipolytik and lipoxygenase enzymes in plants and their action in wounded tissues, Biochemistry of Wounded Plant Tissues (G. Kahl, ed.), Walter de Gruyter, Berlin, pp. 155–201.

Galliard, T., and Chan, H. W. S. (1980). Lipoxygenases, The Biochemistry of Plants, Vol. 4 (P. K. Stumpf and E. E. Conn, eds.), Academic Press, New York, pp. 131–161.

Gardner, H. W. (1970). Sequential enzymes of linoleic acid oxidation in corn germ: lipoxygenase and linoleate hydroperoxide isomerase, J. Lipid Res., 11:311–321.

Gonner, V. M., and Schlosser, E. (1993). Oxidative stress in interactions between *Avena sativus* L. and *Drechslera* spp., Physiol. Mol. Plant Pathol., 42:221–234.

Gonner, V. M., Schlosser, E., and Neubacher, H. (1993). Evidence from electron-spin resonance for the formation of free radicals during infection of *Avena sativa* by *Drechslera* spp., Physiol. Mol. Plant Pathol., 42:405–412.

Graveland, A. (1970). Enzymatic oxidation of linoleic acid and glycerol-1-monolinoleate in doughs and flour–water suspensions, J. Am. Oil Chem. Soc., 47:352–361.

Grimes, H. D., Koetje, D. S., and Fransceshi, V. R. (1992). Expression, activity and cellular accumulation of methyl jasmonate-responsive lipoxygenase in soybean seedlings, Plant Physiol., 100:433–443.

Gross, G. G., Janse, C., and Elstner, E. F. (1977). Involvement of malate, monophenols and the superoxide radical in hydrogen peroxide formation by isolated cell walls from horseradish, Planta, 136:271–276.

Gundlach, H., Muller, M. J., Kutchan, T., and Zenk, M. H. (1992). Jasmonic acid is a signal inducer in elicitor-induced plant cell cultures, Proc. Natl. Acad. Sci. USA, 89:2389–2393.

Haga, M., Sekizawa, Y., Ichikawa, M., Hiramatsu, H., Hamamoto, A., Takino, Y., and Ameyama, M. (1986). Nitroblue tetrazolium reduction of rice leaf slices stimulated with blast fungus hyphal component, Agric. Biol. Chem., 50:1427–1436.

Halliwell, B. (1978). Lignin synthesis: the generation of hydrogen peroxide and superoxide by horseradish peroxidase and its stimulation by manganese (II) and phenols, Planta, 140:81–88.

Halliwell, B. (1982). The toxic effects of oxygen on plant tissues, Superoxide Dismutase, Vol. I (L. W. Oberley, ed.), CRC Press, Boca Raton, pp. 89–123.

Halliwell, B. (1988). Oxygen Radicals and Tissue Injury, Proceedings of a Brook Lodge Symposium, Upjohn, p. 748.

Halliwell, B., and Gutteridge, J. M. C. (1986). Iron and free radical reactions: two aspects of antioxidant protection, Trends Biochem. Sci., 11:372–375.

Hamilton, R. H., Meyer, H. E., Burke, R. E., Feung, C. S., and Mumma, R. O. (1976). Metabolism of indole-3-acetic acid, Plant Physiol., 58:77–81.

Heffetz, D., Bushkin, I., Dror, R., and Zick, Y. (1990). The insulinomimetic agents H_2O_2 and vanadate stimulate protein tyrosine phosphorylation in intact cells, J. Biol. Chem., 265:2896–2902.

Heimann, W., and Dresen, P. (1973). About the enzymatic hydroperoxide degradation in cereals, Helv. Chim. Acta, 36:463–469.

Hildebrand, D. F. (1989). Lipoxygenases, Physiol. Plant., 76:249–253.

Hildebrand, D. F., and Grayburn, W. S. (1991). Lipid metabolites—regulators of plant metabolism, Plant Biochemical Regulators (H. Gausman, ed.), Marcel Dekker, New York, pp. 69–95.

Hildebrand, D. F., Hamilton-Kemp, T. R., Leggs, C. S., and Bookjans, G. (1988). Plant lipoxygenases: occurrence, properties and possible functions, Curr. Top. Plant Biochem. Physiol., 7:201–219.

Ho, H. Y., and Wong, P. Y. (1989). The formation of lipoxins and 6,7-dihydrolipoxins by 5-lipoxygenase and reductase partially purified from potato tubers, Adv. Prostaglandin Thromboxane Leukotriene Res., 19:137–140.

Hodgson, R. A. J., and Raison, J. K. (1991). Lipid peroxidation and superoxide dismutase activity in relation to photoinhibition induced by chilling in moderate light, Planta, 185:215–219.

Hoppe, H. H., and Heitefuss, R. (1974). Permeability and membrane lipid metabolism of *Phaseolus vulgaris* infected with *Uromyces phaseoli*. I. Changes in the efflux of cell constituents, Physiol. Plant Pathol., 4:5–9.

Ida, S., Masaki, Y., and Morita, Y. (1983). The isolation of multiple forms and product specificity of rice lipoxygenase, Agric. Biol. Chem., 47:637–641.

Jordan, C. M., and DeVay, J. E. (1990). Lysosome disruption associated with hypersensitive reaction in the potato–*Phytophthora infestans* host–parasite interaction, Physiol. Mol. Plant Pathol., 36:221–236.

Kato, S., and Misawa, T. (1976). Lipid peroxidation during the appearance of hypersensitive reaction in cowpea leaves infected with cucumber mosaic virus, Ann. Phytopathol. Soc. Japan, 42:472–480.

Kato, T., Yamaguchi, T., Abe, N., Uyehara, T., Nakai, T., Kodama, M., and Shiobara, Y. (1985). Structure and synthesis of unsaturated trihydroxy C_{18} fatty acids in rice plant suffering from rice blast disease, Tetrahedron Lett., 26:2357–2360.

Kato, T., Yamaguchi, Y., Uyehara, T., Yokoyama, T., Namai, T., and Yamanaka, S. (1983a). Defense mechanism of the rice plant against rice blast disease, Naturwissenschaften, 70:200–201.

Kato, T., Yamaguchi, Y., Uyehara, T., Yokoyama, T., Namai, T., and Yamanaka, S. (1983b). Self defensive substances in rice plant against rice blast disease, Tetrahedron Lett., 24:4715–4718.

Kauss, H. (1990). Role of the plant plasma membrane in host-pathogen interaction, The Plant Plasma Membrane (C. Larsson and I. M. Moller, eds.), Springer-Verlag, Berlin, pp. 320–351.

Keppler, L. D., and Baker, C. J. (1989). O_2^- initiated lipid peroxidation in a bacteria-induced hypersensitive reaction in tobacco cell suspensions, Phytopathology, 79:555–562.

Keppler, L. D., Baker, C. J., and Atkinson, M. M. (1989). Active oxygen production during a bacteria-induced hypersensitive reaction in tobacco suspension cells, Phytopathology, 79:974–978.

Keppler, L. D., and Novacky, A. (1986). Involvement of membrane lipid peroxidation in the development of a bacterially induced hypersensitive reaction, Phytopathology, 76:104–108.

Keppler, L. D., and Novacky, A. (1987). The initiation of membrane lipid peroxidation during bacteria-induced hypersensitive reaction, Physiol. Mol. Plant Pathol., 30:233–245.

Keppler, L. D., and Novacky, A. (1989). Changes in cucumber cotyledon membrane lipid fatty acids during paraquat treatment and a bacteria-induced hypersensitive reaction, Phytopathology, 79:705–708.

Kim, K. K., Fravel, D. R., and Papavizas, G. C. (1988). Identification of a metabolite produced by *Talaromyces flavus* as glucose oxidase and its role in the biocontrol of *Verticillium dahliae*, Phytopathology, 78:488–492.

Knox, J. P., and Dodge, A. D. (1985). Singlet oxygen and plants, Phytochemistry, 24:889–896.

Koch, E., Meier, B. M., Eiben, H. G., and Slusarenko, A. (1992). A lipoxygenase from leaves of tomato (*Lycopersicon esculentum* Mill.) is induced in response to plant pathogenic pseudomonads, Plant Physiol., 99:571–575.

Krinsky, N. L. (1979). Carotenoid protection against oxidation, Pure Appl. Chemistry, 51:649–660.

Lachmann, P. J. (1986). A common form of killing, Nature, 321:560.

Larson, R. A. (1988). The antioxidants of higher plants, Phytochemistry, 27:969–978.

Law, M. Y., Charles, S. A., and Halliwell, B. (1983). Glutathione and ascorbic acid in spinach (*Spinacia oleracea*) chloroplasts, Biochem. J., 210:899–903.

Leshem, Y. Y. (1987). Membrane phospholipid catabolism and Ca^{2+} activity in control of senescence, Physiol. Plant., 69:551–559.

Leung, H. W., Vang, M. J., and Mavis, R. D. (1982). The cooperation interaction between vitamin E and vitamin C in suppression of peroxidation of membrane phospholipids, Biochem. Biophys. Acta, 641:99–105.

Lielder, C. D., Kling, S. D., and Reed, D. J. (1986). Antioxidant protection of phospholipid bilayers by tocopherol. Control of α-tocopherol status and lipid peroxidation by ascorbic acid and glutathione, J. Biol. Chem., 261:12114–12119.

Lupu, R., Grossman, S., and Cohen, Y. (1980). The involvement of lipoxygenase and antioxidants on pathogenesis of powdery mildew on tobacco plants, Physiol. Plant Pathol., 16:241–248.

Lynch, D. V., and Thompson, J. E. (1984). Lipoxygenase-mediated production of superoxide anion in senescing plant tissue, FEBS Lett., 173:251–254.

Lyr, H., and Basaniak, L. (1983). Alkenals, volatile defense substances in plants, their properties and activities, Acta Phytopathol. Acad. Sci. Hung., 18:3–12.

Major, R. T., Marchini, P., and Sproston, T. (1960). Isolation from *Ginkgo biloba* (L.) of an inhibitor of fungus growth, J. Biol. Chem., 235:3298–3299.

Matthew, J. A., and Galliard, T. (1978). Enzymic formation of carbonyls from linoleic acid in leaves of *Phaseolus vulgaris*, Phytochemistry, 17:1043–1044.

Melan, M. A., Dong, X., Endara, M. E., Davis, K. R., Ausubel, F. M., and Peterman, T. K. (1993). An *Arabidopsis thaliana* lipoxygenase gene can be induced by pathogens, abscisic acid, and methyl jasmonate, Plant Physiol., 101:441–450.

Minardi, P., and Mazzucchi, U. (1988). No evidence of direct superoxide anion effect in hypersensitive death of *Pseudomonas syringae* Van Hall in tobacco leaf tissue, J. Phytopathol., 122:351–358.

Montalbini, P. (1991). Effect of rust infection on levels of uricase, allantoinase, and ureides in susceptible and hypersensitive bean leaves, Physiol. Mol. Plant Pathol., 39:173–188.

Montalbini, P., and Buonaurio, R. (1986). Effect of tobacco mosaic virus infection on levels of soluble superoxide dismutase (SOD) in *Nicotiana tabacum* and *Nicotiana glutinosa* leaves, Plant Sci. Lett., 47:135–143.

Moreau, R. A., and Osman, S. F. (1989). The properties of reducing agents released by treatment of *Solanum tuberosum* with elicitors from *Phytophthora infestans*, Physiol. Plant Pathol., 35:1–10.

Morisaki, N., Stitts, J. M., Bartels-Tomei, L., Milo, G. E., Panganamala, R. V., and Cornwell, D. G. (1982). Dipyridamole: an antioxidant that promotes the proliferation of aorta smooth muscle cells, Artery, 11:88–107.

Mulliez, E., Leblanc, J. P., Girerd, J. J., Rigaud, M., and Chottard, J. C. (1987). 5-Lipoxygenase from potato tubers. Improved purification and physicochemical characteristic, Biochem. Biophys. Acta, 916:13–23.

Namai, T., Kato, T., Yamaguchi, Y., and Togashi, T. (1990). Time-course alteration of lipoxygenase activity in blast-infected rice leaves, Ann. Phytopathol. Soc. Japan, 56:26–32.

Namai, T., Kato, T., Yamaguchi, Y., and Togashi, T. (1991). Cytotoxicity and anti-blast activity of hydroperoxide of linolenic acid produced by in vitro enzyme reaction using blast-infected rice leaves, Ann. Phytopathol. Soc. Japan, 57:339–344.

Niinomi, A., Morimoto, M., and Shimizu, S. (1987). Lipid peroxidation by the (peroxidase/H_2O_2/phenolic) system, Plant Cell Physiol., 28:731–734.

Ocampo, C. A., Moerschbauer, B., and Gramleow, H. J. (1986). Increased lipoxygenase activity is involved in the hypersensitive response of wheat leaf cells infected with avirulent rust fungi or treated with fungal elicitor, Z. Naturforsch., 41C:559–563.

Ohta, H., Ida, S., Mikami, B., and Morita, Y. (1986a). Purification and characterization of rice lipoxygenase component 3 from embryos, Agric. Biol. Chem., 50:3165–3171.

Ohta, H., Ida, S., Mikami, B., and Morita, Y. (1986b). Changes in lipoxygenase components of rice seedlings during germination, Plant Cell Physiol., 27:911–918.

Ohta, H., Shida, K., Peng, Y. L., Furasawa, I., Shishiyama, J., Aibara, S., and Morita, Y. (1990). The occurrence of lipid hydroperoxide decomposing—enzyme activities in rice and the relationship of such activities to the formation of antifungal substances, Plant Cell Physiol., 31:1117–1122.

Ohta, H., Shida, K., Peng, Y. L., Furusawa, I., Shishiyama, J., Aibara, S., and Morita, Y. (1991). A lipoxygenase pathway is activated in rice after infection with the rice blast fungus *Magnaporthe grisea*, Plant Physiol., 97:94–98.

Osman, S. F., and Moreau, R. A. (1985). Potato phytoalexin elicitors in *Phytophthora infestans* spore germination fluids, Plant Science, 41:205–209.

Pavlovkin, J., Novacky, A., and Ullrich-Eberius, C. I. (1986). Membrane potential changes during bacteria-induced hypersensitive reaction, Physiol. Mol. Plant Pathol., 28:125–133.

Peever, T. L., and Higgins, V. J. (1989). Electrolyte leakage, lipoxygenase, and lipid peroxidation induced in tomato leaf tissue by specific and nonspecific elicitors from *Cladosporium fulvum*, Plant Physiol., 90:867–875.

Preisig, C. L., and Kuc, J. A. (1985). Arachidonic acid-related elicitors of the hypersensitive response in potato and enhancement of their activities by glucans from *Phytophthora infestans* (Mont.) de Bary, Arch. Biochem. Biophys., 236:379–389.

Preisig, C. L., and Kuc, J. A. (1987). Inhibition by salicylhydroxamic acid, BW755C, eicosatetraynoic acid, and disulfiram of hypersensitive resistance elicited by arachidonic acid or poly-L-lysine in potato tuber, Plant Physiol., 84:891–894.

Rabinowitch, H. D., and Fridovich, I. (1983). Superoxide radicals, superoxide dismutases and oxygen toxicity in plants, Photochem. Photobiol., 37:679–690.

Rickauer, M., Fournier, J., Pouenat, M. L., Berthalon, E., Bottin, A., and Esquerre-Tugaye, M. T. (1990). Early changes in ethylene synthesis and lipoxygenase activity during defense induction in tobacco cells, Plant Physiol. Biochem., 28:647–653.

Rogers, K. R., Albert, F., and Anderson, A. J. (1988). Lipid peroxidation is a consequence of elicitor activity, Plant Physiol., 86:547–553.

Rosen, G. M., and Halpern, H. J. (1990). Spin trapping biologically generated free radicals: correlating formation with cellular injury, Methods Enzymol., 186:611–621.

Roy, M. A., and Sasser, M. (1983). Evidence for involvement of active oxygen in tobacco resistance to a bacterial pathogen, Phytopathology, 73:968–969.

Ruzicska, P., Gombos, Z., and Farkas, G. (1983). Modification of the fatty acid composition of phospholipids during the hypersensitive reaction in tobacco, Virology, 128:60–64.

Salzwedel, J. L., Daub, M. E., and Huang, J. S. (1989). Effects of singlet oxygen quenchers and pH on the bacterially induced hypersensitive reaction in tobacco suspension cell cultures, Plant Physiol., 90:25–28.

Samuelsson, B., Dahlen, S. E., Lindgren, A., Rouzer, C. A., and Serhan, S. N. (1987). Leukotrienes and lipoxins: structures, biosynthesis, and biological effects, Science, 237:1171–1175.

Sandalio, L. M., Fernandez, V. M., Ruperez, F. L., and Del Rio, L. A. (1988). Superoxide free radicals are produced in glyoxysomes, Plant Physiol., 87:1–4.

Saniewski, M. (1979). Questions about occurrence and possible roles of prostaglandins in the plant kingdom, Acta Horticult., 91:73–81.

Sastry, P. S., and Kates, M. (1964). Lipid components of leaves. V. Galactolipids, cerebrosides, and lecithin of runner-bean leaves, Biochemistry, 3:1271–1280.

Scandalios, J. G. (1990). Response of plant antioxidant defense genes to environmental stress, Adv. Genet., 28:1–41.

Scandalios, J. G. (1993). Oxygen stress and superoxide dismutases, Plant Physiol., 101:7–12.

Schubert, K. R. (1986). Product of biological nitrogen fixation in higher plants: synthesis, transport, and metabolism, Annu. Rev. Plant Physiol., 37:539–574.

Sekizawa, Y., Haga, M., Hirabayashi, E., Takeuchi, N., and Takino, Y. (1987). Dynamic behaviour of superoxide generation in rice leaf tissue infected with blast fungus and its regulation by some substances, Agric. Biol. Chem., 51:763–770.

Sekizawa, Y., Haga, M., Iwata, M., Hamamoto, A., Chihara, C., and Takino, Y. (1985). Probenazole and burst of respiration in rice leaf tissue infected with blast fungus, J. Pesticide Sci., 10:225–231.

Shibata, D., Steczko, J., Dixon, J. E., Andrews, P. C., Hermodson, M., and Axelrod, B. (1988). Primary structure of soybean lipoxygenase-2, J. Biol. Chem., 263:6816–6821.

Shibata, D., Steczko, J., Dixon, J. E., Hermodson, M., Yazdanparast, R., and Axelrod, B. (1987). Primary structure of soybean lipoxygenase-1, J. Biol. Chem., 262:10008–10085.

Shimizu, T., Honda, Z. I., Miki, I., Seyama, Y., Izumi, T., Radmark, O., and Samuelsson, B. (1990). Potato arachidonate 5-lipoxygenase: purification, characterization, and preparation of 5(S)-hydroperoxyeicosatetranoic acid, Methods Enzymol., 187:296–306.

Shimura, M., Mase, S., and Iwata, M. (1983). Anticonidial germination factors induced in the presence of probenazole in infected host leaves. III. Structural elucidation of substances A and C., Agric. Biol. Chem., 47:1983–1989.

Shinshi, H., Mohnen, D., and Meins, F. Jr. (1987). Regulation of a plant pathogenesis-regulated enzyme: inhibition of chitinase and chitinase mRNA accumulation in cultured tobacco tissues by auxin and cytokinin, Proc. Natl. Acad. Sci. USA, 84:89–93.

Siedow, J. N. (1991). Plant lipoxygenase: structure and function, Annu. Rev. Plant Physiol. Plant Mol. Biol., 42:145–188.

Siegel, S. M., and Halpern, L. A. (1965). Effects of peroxide on permeability and their modification by indoles, vitamin E, and other substances, Plant Physiol., 40:792–796.

Staswick, P. E. (1992). Jasmonate, genes and fragrant signals, Plant Physiol., 99:804–807.

Staswick, P. E., Huang, J. F., and Rhee, Y. (1991). Nitrogen and methyl jasmonate induction of soybean vegetative storage protein genes, Plant Physiol., 96:130–136.

Stich, K., and Ebermann, R. (1984). Investigation of hydrogen peroxide formation in plants, Phytochemistry, 23:2719–2722.

Sutherland, M. W. (1991). The generation of oxygen radicals during host plant responses to infection, Physiol. Mol. Plant Pathol., 39:79–93.

Svingen, B. A., O'Neil, F. O., and Aust, S. D. (1978). The role of superoxide and singlet oxygen in lipid peroxidation, Photochem. Photobiol., 28:803–809.

Tappel, A. L. (1965). Free radical peroxidation damage and its inhibition by vitamin E and selenium, Fed. Proc., 24:73–78.

Thompson, J. E., Legge, R. L., and Barber, R. F. (1987). The role of free radicals in senescence and wounding, New Phytol., 105:317–344.

Tolbert, N. E. (1982). Leaf peroxisomes, Ann. New York Acad. Sci., 386:254–268.

Tomita, H., Yamanaka, S., Namai, T., and Kato, T. (1984). Lipid peroxidation in the detached rice sheath tissue infected with rice blast fungus, Ann. Phytopathol. Soc. Japan, 50:107.

Tressl, R., and Drawert, F. (1973). Biosynthesis of banana volatiles, J. Agric. Food Chem., 21:560.

van Huytsee, R. B. (1987). Some molecular aspects of plant peroxidase biosynthetic studies, Annu. Rev. Plant Physiol., 38:205–219.

Vaughn, S. F., and Lulai, E. C. (1992). Further evidence that lipoxygenase activity is required for arachidonic acid-elicited hypersensitivity in potato callus cultures, Plant Sci., 84:91–96.

Veldink, G. A., and Vliegenthart, J. F. G. (1985). Lipoxygenase, nonheme iron-containing enzymes, Adv. Inorg. Biochem., 6:139–161.

Veldink, G. A., Vliegenthart, J. F. G., and Boldingh, J. (1970). The enzymic conversion of linoleic acid hydroperoxide by flax-seed hydroperoxide isomerase, Biochem. J., 120:55–60.

Vera-Estrella, R., Blumwald, E., and Higgins, V. J. (1992). Effect of specific elicitors of *Cladosporium fulvum* on tomato cell suspension cells: evidence for the involvement of active oxygen species, Plant Physiol., 99:1208–1215.

Vera-Estrella, R., Blumwald, E., and Higgins, V. J. (1993). Non-specific glycopeptide elicitors of *Cladosporium fulvum*: evidence for involvement of active oxygen species in elicitor-induced effects on tomato cell suspensions, Physiol. Mol. Plant Pathol., 42:9–22.

Vick, B. A., and Zimmerman, D. C. (1976). Lipoxygenase and hydroperoxide lyase in germinating watermelon seedlings, Plant Physiol., 57:780–788.

Vick, B. A., and Zimmerman, D. C. (1979). Distribution of a fatty acid cyclase system in plants, Plant Physiol., 69:1103–1108.

Vick, B. A., and Zimmerman, D. C. (1984). Biosynthesis of jasmonic acid in plants, Plant Physiol., 75:458–461.

Vick, B. A., and Zimmerman, D. C. (1987a). Oxidative systems for modification of fatty acids: the lipoxygenase pathway, Lipids: Structure and Function, The Biochemistry of Plants, Vol. 9 (P. K. Stumpf, ed.), Academic Press, New York, pp.53–90.

Vick, B. A., and Zimmerman, D. C. (1987b). Pathways of fatty acid hydroperoxide metabolism in spinach leaf chloroplasts, Plant Physiol., 85:1073–1078.

Vladimirov, Y. A., Olenev, V. I., Suslova, T. B., and Cheremisina, Z. P. (1980). Lipid peroxidation in mitochondrial membranes, Adv. Lipid Res., 17:173–249.

Yabuuchi, S., and Amaha, M. (1976). Partial purification and characterization of the linoleate hydroperoxide isomerase from grains of *Hordeum distichum*, Phytochemistry, 15:387–390.

Yamamoto, H., and Tani, T. (1986). Possible involvement of lipoxygenase in the mechanism of resistance of oats to *Puccinia coronata avenae*, J. Phytopathol., 116:329–337.

Yenofsky, R. L., Fine, M., and Liu, C. (1988). Isolation and characterization of a soybean (*Glycine max*) lipoxygenase-3 gene, Mol. Gen. Genet., 211:215–222.

Zeringue, H. J., and McCormick, S. P. (1989). Relationship between cotton leaf-derived volatiles and growth of *Aspergillus flavus*, J. Am. Oil Chem. Soc., 66:581–585.

Zimmerman, D. C. (1966). A new product of linoleic acid oxidation by a flax seed enzyme, Biochem. Biophys. Res. Commun., 23:398–402.

4

Pathogenesis-Related Proteins and Other Antifungal Proteins

I. DEFENSE-RELATED PROTEINS

When the pathogen infects host tissues, a small number of specific genes to produce mRNAs that permit synthesis of a similar number of specific proteins are induced (Vera and Conejero, 1989). Many of these proteins are enzymes such as phenylalanine ammonia-lyase, 4 coumarate-CoA ligase, cinnamic acid-4-hydroxylase, cinnamyl alcohol dehydrogenase, chalcone synthase, and chalcone isomerase (Vidhyasekaran, 1993a) and they are involved in synthesis of low-molecular-weight substances such as phytoalexins, phenols, lignins, tannins, and melanins, which are inhibitory to fungal pathogens (Bell, 1981; Vidhyasekaran, 1988a,b, 1990, 1993a). The other groups of proteins are hydroxyproline-rich glycoproteins and a structurally diverse group of proteins known collectively as pathogenesis-related (PR) proteins (Antoniw and White, 1987; Rigden and Coutts, 1988; Van Loon, 1989). While hydroxyproline-rich glycoproteins may be involved in reinforcement of cell walls to delay fungal penetration, the PR proteins appear to act directly on invading pathogen (Verburg and Huynh, 1991). It is believed that plants are able to defend themselves against fungal pathogens by synthesizing this kind of proteins (Lotan and Fluhr, 1990b; Linthorst, 1991). Several antifungal proteins are also constitutively expressed in plants (Vigers et al., 1991). In this chapter, the role of PR proteins and antifungal proteins in the host defense mechanism and the molecular mechanisms through which the fungal pathogens evade these proteins are discussed.

II. WHAT ARE PR PROTEINS?

PR proteins are soluble and host-coded proteins (Van Loon, 1985; Carr and Klessig, 1989; Vidhyasekaran, 1993b). PR proteins were initially defined as a group of acidic-type proteins found in the extracellular spaces of leaves (Van

Loon and Van Kammen, 1970; Bol et al., 1990). It is now established that most acidic-type PR proteins have basic-type homologues as well (Bol et al., 1990; Eyal and Fluhr, 1991) and they are distributed intracellularly (Van den Bulcke et al., 1989; Keefe et al., 1990). They have unique properties such as high solubility at low pH (Van Loon, 1976) and resistance to many kinds of proteases (Pierpoint, 1983).

PR proteins are induced by a variety of agents (Linthorst, 1991). The infection of plants by several kinds of pathogens such as fungi, bacteria, viruses, and viroids results in the accumulation of PR proteins (Camacho Henriquez and Sanger, 1982b; Van Loon, 1985; Jondle et al., 1989). These proteins have also been shown to accumulate during environmental and chemical stress (Carr and Klessig, 1989; Edreva, 1990). In addition to induction by these stresses, PR proteins are also produced in healthy tissue (Fraser, 1981; Felix and Meins, 1986; Lotan et al., 1989; Neale et al., 1990; Castresana et al., 1990).

III. PR PROTEINS ARE UBIQUITOUS IN PLANTS

Tobacco plants synthesize a large number of PR proteins (Gianinazzi and Kassanis, 1974; Van Loon, 1976, 1977, 1982, 1983; Gianinazzi et al., 1977, 1980; Rohloff and Lerch, 1977; Coutts, 1978; Antoniw and White, 1980, 1983, 1986, 1987; Antoniw et al., 1980, 1985; Pierpoint et al., 1981; Parent and Asselin, 1984; Matsuoka and Ohashi, 1984; Ohashi and Matsuoka, 1985, 1987a; Fortin et al., 1985; Parent et al., 1985, 1988; Oshima et al., 1987; Hogue and Asselin, 1987; Payne et al., 1988, 1989, 1990a; Cote et al., 1989; Van Loon and Gerritsen, 1989; Godiard et al., 1990; Gugler et al., 1990; Kauffmann et al., 1990). These proteins are subdivided into five distinct groups, PR-1, PR-2, PR-3, PR-4, and PR-5 groups, based on enzymatic, structural, and immunological data (Trudel et al., 1989; Bol et al., 1990; Linthorst et al., 1990a; Keefe et al., 1990; Ward et al., 1991; Lawton et al., 1994a; Ohl et al., 1994; Ponstein et al., 1994). A glycine-rich protein (GRP) has also been reported as PR protein in tobacco (Hooft van Huijsduijnen et al., 1986a; Condit and Meagher, 1986; Varner and Cassab, 1986; Van Kan et al., 1988).

In tomato several PR proteins have been identified (Van den Berg-Velthius and De Wit, 1992) and the nomenclature of many of these proteins was based on their molecular mass. The most prominent of the proteins is P14, which has a molecular mass of 14 kDa (Camacho and Sanger, 1982a,b; Nassuth and Sanger, 1986; De Wit et al., 1986; De Wit and Van der Meer, 1986; Fischer et al., 1989). P70 is the second major protein (Christ and Mosinger, 1989). P69, which may be similar to P70 described by Christ and Mosinger (1989), has also been reported (Vera and Conejero, 1988, 1989). Christ and Mosinger (1989) identified 9 more PR proteins, besides P14 and P70 (P69): P80, P36, P31, P30, P26, P21, P18, and P13 in tomato. Granell et al. (1987) detected 10 PR proteins and only one of the

PR proteins was acidic while the other nine were basic. The molecular weight of the major proteins were 14 kDa, 30 kDa, 34 kDa, 36 kDa, and 69 kDa. Two PR proteins, P32 and P34, have been described (Garcia Breijo et al., 1990) and these were identical to PR proteins C7 and C6, which have been described earlier (Granell et al., 1987). PR-P2 (Linthorst et al., 1991) and NP24 are the other PR proteins in tomato (King et al., 1988; Pierpoint et al., 1992). A necrosis-inducing protein has been isolated from the apoplastic fluids of tomato infected with *Cladosporium fulvum*. The protein had a low-molecular mass of 3 kDa and had a high percentage of cysteine residues (Schottens-Toma and De Wit, 1988). Three basic PR proteins (designated P2, P4, and P6) of approximately 15 kDa have been detected in apoplastic fluids obtained from *Cladosporium fulvum*-infected tomato leaves (Joosten et al., 1990). Two other acidic PR proteins with molecular masses of 33 kDa and 35 kDa were detected in the apoplastic fluids of *C. fulvum*-infected leaves (Joosten and De Wit, 1989). A second type of basic 35 kDa protein and an intracellular basic 33 kDa protein have also been purified from *C. fulvum*-infected tomato leaves (Van Kan et al., 1992). Two additional proteins having molecular masses of approximately 63 and 87 kDa have also been detected (Farmer et al., 1992).

In potato eight PR proteins have been detected (Kombrink et al., 1988). The other PR proteins detected in potato are pSTH-2 and pSTH 21 (Matton and Brisson, 1989), *win* 1 and *win* 2 gene-coded proteins (Stanford et al., 1989), *wun* 1 and *wun* 2 gene-coded proteins (Logemann et al., 1988, 1989), prp1 (Sharma et al., 1992), and thaumatin-like (TL) protein (Pierpoint et al., 1990).

In bean three PR proteins (PvPR 1, PvPR 2, and PvPR 3) have been reported (Redolfi et al., 1989; Walter et al., 1990; Sharma et al., 1992). An acidic TL-protein with a molecular weight of 21 kDa has been detected in bean (Sehgal et al., 1991).

Eight pathogenesis-related proteins have been detected in maize leaves. They were designated as PRm proteins and their estimated molecular weights were PRm1 14.2 kDa; PRm2 16.5 kDa; PRm3 and PRm4 25 kDa; PRm5 29 kDa; PRm6a 32 kDa; PRm6b 30.5 kDa; and PRm7 34.5 kDa (Nasser et al., 1988, 1990). Another PR protein has been purified from maize root tissue and named MPR-1. It has a molecular weight of 14,970 (Gillikan et al., 1991). A GRP has also been detected in maize (Gomez et al., 1988; Didierjean et al., 1992). A thaumatin-like (TL) protein, MAI, is reported and its molecular weight is 22.1 kDa (Bryngelsson and Green, 1989). A PR protein resembling PR-1a of tobacco has been detected in maize seeds infected with *Fusarium moniliforme* (Puigdomenech and San Sagundo, 1991). It consists of one acidic and one basic isoform with a molecular mass of 23 and 24 kDa (Garcia-Olmedo et al., 1991; Casacuberta et al., 1991, 1992; Cordero et al., 1992). A 17 kDa PR proteins has also been detected in maize seeds infected with *F. moniliforme* (Roventos et al., 1994).

Pea is reported to accumulate PR proteins coded by the genes such as p149 and p176 (Fristensky et al., 1988) and two cysteine-rich proteins coded by the genes p139 and p1230 with molecular masses of 5 kDa (Chiang and Hadwiger, 1991). Acidic TL-protein with molecular weight of 21 kDa has been detected in leaves of stressed soybean (Graham et al., 1992). In cowpea six PR proteins have been detected (Wagih and Coutts, 1981). Several acid-soluble proteins in powdery mildew pathogen-infected barley leaves have been reported (Bryngelsson et al., 1988). One of them named Hv-1, has been characterized and found to be a thaumatin-like protein (Bryngelsson and Green, 1989). Two PR proteins designated Hv PR-1a and Hv PR-1b have also been detected in barley inoculated with *Erysiphe graminis* f.sp. *hordei* and they showed a high degree of amino acid identity to the PR-1 proteins originally described in tobacco (Bryngelsson et al., 1994). In cucumber leaves infected with *Colletotrichum lagenarium* or *Fusarium oxysporum* f.sp. *cucumerinum*, a new protein with a molecular weight of 16 kDa appeared (Wagih and Coutts, 1982; Gessler and Kuc, 1982).

More than 30 different pathogenesis-related proteins accumulated in spruce roots inoculated with *Pythium* sp. (Sharma et al., 1993). Twelve acidic (B.v. D1 and D12) and eight basic (B.v. R1 to R8) PR proteins have been reported in sugar beet (Fleming et al., 1991; Rousseau-Limouzin and Fritig, 1991). Two of these proteins (D2 and D3) have been characterized and both of them have the same apparent molecular weight of 28 kDa (Fleming et al., 1991). The PR proteins from parsley are acidic and ranged in mass from 17.5 to 19.2 kDa (Somssich et al., 1986) and are named parsley PR 1 proteins (Somssich et al., 1988). PR proteins in rice (Kim and Yoo, 1992), peanut (*Arachis hypogaea*) (Herget et al., 1990), soybean (Roggero et al., 1989), *Solanum dulcamara* (Bronner et al., 1991), melon (Roby et al., 1987), cotton (Singh et al., 1987a), alfalfa (Singh et al., 1987a), rubber (*Hevea brasiliensis*) (Van Parijs et al., 1991), carrot (Singh et al., 1987a), *Citrus medica* (Camacho Henriquez and Sanger, 1982), *Gomphrena globosa* (Redolfi et al., 1982), *Solanum demissum* (Camacho Henriquez and Sanger, 1982), *Allium porum* (Spanu et al., 1989), *Apium graveolens* (Conejero et al., 1979; Camacho Henriquez and Sanger, 1982), *Nicotiana rustica* (Tas and Peters, 1977), *Nicotiana tomentosiformis* (Ahl et al., 1982), *Nicotiana plumbaginifolia* (Gheysen et al., 1990), *Asparagus officinalis* (Warner et al., 1992), *Arabidopsis thaliana* (Samac et al., 1990; Metzler et al., 1991), petunia (*Petunia hybrida*) (Condit and Meagher, 1986), and in several other plants belonging to the families, Aizoaceae, Amaranthaceae, Chenopodiaceae, Cucurbitaceae, Leguminosae, Solanaceae, and Gramineae (Abu Jawdah, 1982; Redolfi, 1983; Redolfi and Cantisani, 1984; Van Loon, 1985; Bol, 1988; Bol et al., 1990a,b; Sehgal et al., 1991) have been reported.

IV. STRUCTURE OF PR PROTEINS

A. PR 1 Proteins

PR proteins show vast diversity in their structure. Attempts have been made to classify them based on their serological relationship and amino acid sequence (Richardson et al., 1987; Pfitzner and Goodman, 1987; Van Loon et al., 1987). The first group of PR proteins is called PR1 proteins. PR1 family in tobacco consists of three acidic polypeptides (PR-1a, PR-1b and PR-1c) that show 90% homology with each other (Pfitzner and Goodman, 1987; Cutt, et al., 1988). P14 is the most predominant PR protein appearing in the intercellular spaces of tomato leaves upon infection by fungi, viruses or viroids (Camacho Henriquez and Sanger, 1982a,b, 1984; De Wit and Van der Meer, 1986; Granell et al., 1987; Fischer et al., 1989b; Vera et al., 1989). P 14 appears to be serologically related to PR-1a, PR-1b, and PR-1c from tobacco (Nassuth and Sanger, 1986).

PR 1a is more acidic than PR 1b and the isoelectric points of PR 1a and PR 1b are 4.3 and 4.8, respectively (Matsuoka and Ohashi, 1984). The N-terminal leader sequences of PR1a and PR1b show typical structure of signal peptides. These are composed of three highly hydrophobic core regions and a hydrophilic β-turn structure near the C terminus. The mature proteins contain six regions (i.e., positions 58-68, 73-78, 93-96, 100-104, 119-129, and 147-153) in which the hydrophobic amino acid residues predominate. Three of the hydrophobic regions (second, fifth, and sixth) have predominant β-sheet structure. The third and fourth hydrophobic regions were calculated to have a helix structure (Matsuoka et al., 1987). The tomato P14 protein has two highly hydrophobic domains predicted as β-sheet in the C terminal region and one highly hydrophobic domain predicted as α-helix structure in the internal region. This homology in the half-side of the C terminus suggests that the half-C terminus side is important for the function of PR-1 proteins (Matsuoka et al., 1987).

P4 and P6 are 15.5 kDa proteins in tomato and they are serologically related to each other and to the PR-1 protein family of tobacco. P4 and P6 represent two differently charged, highly basic isomers of P14 (pI 10.0 and 10.7, respectively) (Joosten et al., 1990). The cDNA sequence data indicate that protein P6 corresponds to the sequence of P14, however, with an extra internal block of five amino acids. An identical block of amino acids occurs in protein P4 and in all three tobacco PR-1 proteins (Van Kan et al., 1992). P4 is a highly related isomer, with only five amino acid substitutes compared to P6 (Van Kan et al., 1992).

The tobacco PR-1 proteins have a basic intracellular counterpart that shows about 65% amino acid similarity to the acidic PR-1 proteins (Cornelissen et al., 1987). Similar intracellular isoform of the basic extracellular proteins P4 and P6 in *Cladosporium fulvum*-infected tomato leaves has not been observed (Joosten et al., 1990).

A PR protein from cowpea appears to be serologically related to P14 of tomato (Nassuth and Sanger, 1986). A polypeptide, MPR-1, serologically related to the tobacco PR-1 family of proteins has been purified from the maize root tissue infected with western corn rootworm and an unidentified fungus. There was 67% amino acid identity between the maize protein MPR-1 and tobacco PR-1a, 66% with PR-1b, and 68% with PR-1c. The tomato PR protein, P14, showed 55% identity with MPR-1 (Gillikin et al., 1991). The maize PR protein, PRm 2, is serologically related to tobacco PR-1b (Nasser et al., 1988). A protein resembling PR-1a of tobacco has been reported in potato (White et al., 1987). Two genomic clones isolated from *Arabidopsis thaliana* hybridized to tobacco PR-1 genes. They did not cross-hybridize with each other, suggesting that there is extensive sequence diversity between the two genes. The sequence of one of these two genomic clones (G19) has been reported. The clone contained a 528-nucleotide open reading frame that begins with a presumptive signal peptide of 43 amino acids. The remaining segment of this open reading frame showed a high level of homology to tobacco PR-1 and to tomato P14 sequence (Metzler et al., 1991).

The biological function of PR 1 group of proteins is not yet known (Lotan and Fluhr, 1990a; Ohshima et al., 1990; Tahiri-Alaoui et al., 1993).

B. β-1,3-Glucanases

β-1,3-glucanase has been detected in many plants (Abeles and Forrence, 1970; Abeles, et al., 1971; Moore and Stone, 1972; Wong and Maclachlan, 1980; Wyatt et al., 1991; Pan et al., 1991b). The group 2 PR proteins show β-1,3-glucanase activity (Felix and Meins, 1985, 1986; Kauffmann et al., 1987; Shinshi et al., 1988; Fritig et al., 1989; Cote et al., 1989; Sock et al., 1990). Several PR proteins with β-1,3-glucanase activity have been detected in tobacco. They are classified into three classes (Payne et al., 1990). The class I enzymes include three related isoforms, characterized by basic pI values, that accumulate predominantly in the vacuole (Van den Bulcke et al., 1989; Keefe et al., 1990). The class II enzymes include the pathogenesis-related proteins PR-2, PR-N, and PR-0, which are acidic proteins found in the extracellular compartment of leaves (Kauffmann et al., 1987; Van den Bulcke et al., 1989). The class III consists of only one enzyme PR-Q′ (Fritig et al., 1990; Payne et al., 1990). It is also an acidic protein that accumulates in the apoplast. It has approximately 55% amino acid identity to both the basic and acidic β-1,3-glucanases of tobacco (Payne et al., 1990).

Ward et al. (1991) proposed three subclasses in class II β-1,3-glucanases. Class IIa comprises PR-2 and PR-N, class IIb comprises PR-0, and class IIc comprises the proteins encoded in cDNA clones GL 153 and G161. Comparisons of cDNA and predicted protein sequences revealed that PR-2 and PR-N are extremely close in sequence, having 99% amino acid identity and greater than 99% DNA sequence similarity. PR-0 is 93% identical with PR-2 and PR-N in amino acid sequence and 94% identical in DNA sequence. GL 153 and GL 161

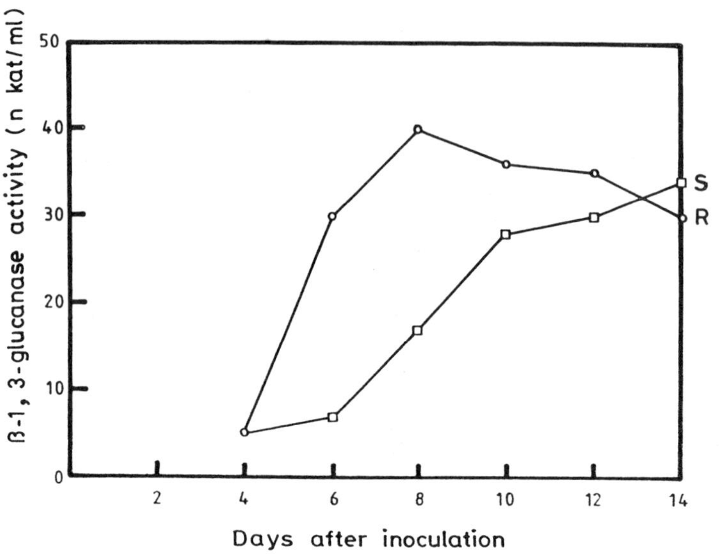

Figure 1 β-1,3-glucanase activity in apoplastic fluids of tomato cultivars of Cf4 (R) and Cf5 (S) after inoculation with race 5 of *Cladosporium fulvum*. R-resistant reaction, S-susceptible reaction. (From Joosten and De Wit, 1989.)

are 89% identical with each other in amino acid sequence and 93% in DNA sequence. These clones are 89–90% identical with PR-2/PR-N at the DNA level and 82–84% identical with PR-2/PR-N at the amino acid level. PR-0 is approximately equally related to PR-2/PR-N, to GL 153 and to GL 161, having an average amino acid identity of 92% and an average DNA sequence identity of 94% to both groups (Ward et al., 1991).

Van den Bulcke et al. (1989) purified three acidic β-1,3-glucanase proteins from tobacco, which they characterized by partial amino acid sequence and designated PR-35, PR-36, and PR-37. Comparison of their amino acid sequence with other PR proteins showed that PR-36 corresponded to PR-2 and PR-37 corresponded to PR-0 (Ward et al., 1991). PR-35 corresponds to the β-1,3-glucanase, PR-Q′ (Payne et al., 1990).

Many isoforms of β-1,3-glucanase have been observed in tobacco (Linthorst et al., 1990a). Several cDNA clones have been isolated from TMV-infected tobacco leaves. Screening of a genomic library of tobacco with labeled acidic β-1,3-glucanase cDNA resulted in six genomic clones with different restriction enzyme maps. Furthermore, the screening of a tobacco genomic library with a cDNA probe corresponding to a basic glucanase resulted in two groups of clones with overlapping inserts, each representing a unique gene. The proteins encoded in clone g19, corresponding to an acidic isozyme and of clone gglb 50, corresponding to a basic enzyme were further studied. The protein encoded in g19 had a molecular weight of 37,763 and was highly acidic (isoelectric point, 4.89). The partial amino acid sequences of the proteins in the cDNA clones of groups c101, c130, c132, and c137 were very similar (> 90% identical) to the one coded in clone g19. It suggests that these clones encode other acidic β-1,3-glucanase isoforms (Linthorst et al., 1990a). The clone c132 was identical to the clone GL 134 described by Ward et al., (1991) and the clone c137 was identical to GL 153 (Ward et al., 1991). The mature tobacco glucanase encoded in clone gglb 50 had a molecular weight of 37,150 and was a neutral, rather than a basic, protein with an isoelectric point of 7.71 (Linthorst et al., 1990a). The tobacco genomic DNA blot analysis showed at least eight distinct bands when a cDNA insert of clone corresponding to acidic glucanase was used as a probe. When a cDNA insert of a clone corresponding to basic glucanase was used as a probe, five or six bands were found to give a hybridization signal. The expression of both types of glucanase genes was highly induced by TMV infection or salicylic acid treatment (Linthorst et al., 1990b).

A β-1,3-glucanase detected in the stigma and style of tobacco flowers was different from other PR-2 β-1,3-glucanases (Cote et al., 1991). Ori et al., (1990) described two cDNAs encoding a class of glucanases specifically in the stylar matrix. One of these, designated Sp41a, is 80% identical to PR-2, 82% identical to PR-0, 88% identical to GL 153, and 85% identical to GL 161 (Ward et al., 1991).

Tomato produces at least three β-1,3-glucanases. One extracellular, slightly acidic 35 kDa β-1,3-glucanase was characterized both at the protein and the cDNA level. A group of cDNA clones encoding a basic 35 kDa protein with high homology to vacuolar tobacco β-1,3-glucanases has been identified. One intracellular, highly basic 33 kDa protein was purified. This protein showed homology to the tobacco PR-Q′ protein (16 identical residues of a total of 24 amino acids sequenced) (Van Kan et al., 1992). However, the tomato 33 kDa β-1,3-glucanase is a basic intracellular protein, whereas the PR-Q′ is extracellular and acidic (Payne et al., 1990; Van Kan et al., 1992). Domingo et al. (1994) have isolated two cDNA clones representing dissimilar β-1,3-glucanase isoforms. Amino acid sequence comparison revealed that they are most similar to β-1,3-glucanases from tobacco, particularly to PR-Q′. They have different isoelectric points: pI 10.5 for the basic isoform and pI 5.2 for the acidic one. They have been designated as Tom PR-Q′a (for the acidic isoform) and Tom PR-Q′b (for the basic isoform).

Two β-1,3-glucanases have been purified from potato leaves infected with *Phytophthora infestans* (Kombrink et al., 1988). In barley two β-1,3-glucanases have been identified (Fincher et al., 1986; Hoj et al., 1989; Jutidamrongpham et al., 1991) and a β-1,3-glucanase has been characterized in soybean (Takeuchi et al., 1990; Ham et al., 1991) and bean (Vogeli et al., 1988).

A β-1,3-glucanase gene has been isolated from apple and characterized (Thimmapuram and Korban, 1994).

Two acidic glucanases and one basic β-1,3-glucanase (36 kDa protein) have been detected in maize (Nasser et al., 1990). All the maize acidic and basic β-1,3-glucanases were serologically closely related. They were serologically related to tobacco acidic β-1,3-glucanase (PR-0) as well as to the bean basic β-1,3-glucanase indicating that the different β-1,3-glucanases from various plant species are serologically closely related (Nasser et al., 1990). Two new PR proteins (28 kDa and 30 kDa) have been detected in the bean variety, 'Saxa.' They possess β-1,3-glucanase activity and they serologically cross-react with tobacco PR-2 and maize PRm-6b proteins (Awade et al., 1989). A single intercellular β-1,3-glucanase has been reported in celery (Krebs and Grumet, 1993). Multiple forms of glucanases have been observed in alfalfa (Maher and Dixon, 1990). The complexity of glucanase gene family has been studied by analysis of genomic DNA isolated from cucumber, tomato, bean, and barley. Multiple bands were observed in all these crops. It suggests that a small gene family encodes the β-1,3-glucanases in a number of plant species (Cote et al., 1991).

The β-1,3-glucanase of pea has 78% identity with a bean basic β-1,3-glucanase (Edington et al., 1991), 62% identity with a *Nicotiana plumbaginifolia* basic β-1,3-glucanase (De Loose et al., 1988), 60% identity with *Nicotiana tabacum* basic β-1,3-glucanase (Shinshi et al., 1988), 57% identity with a soybean β-1,3-glucanase (Takeuchi et al., 1990), 54% identity with a tobacco acidic β-1,3-glucanase (Linthorst et al., 1990a), and 51% identity with barley β-1,3-glucanase (Hoj et al., 1989).

In pea two different β-1,3-glucanases, G1 and G2, have been detected. The molecular weight of G1 is 33,500 while that of G2 is 34,300. The two forms of β-1,3-glucanase differed in their pH optimum, 5.1 and 5.4, respectively. Both were endoglucanases (Mauch et al., 1988a). A β-1,3-glucanase is constitutively expressed in roots of pea seedlings (Chang et al., 1992).

C. Chitinases

The group 3 PR proteins are chitinases (Legrand et al., 1987; Shinshi et al., 1987; Trudel et al., 1989; Flach et al., 1992). Chitinase has been first reported by Grassmann et al. (1934) in extracts of sweet almond (*Prunus amygdalus*). It is reported to be present in seeds of several plants (Powning and Irzykiewicz, 1965). Chitinase has been isolated from bean (Agrawal and Bahl, 1968; Mahe et al., 1992), barley (Kragh et al., 1990), wheat (Ride and Barber, 1990), papaya (Howard and Glazer, 1967), fig (*Ficus benjamina*) (Glazer et al., 1969), pea (Nichols et al., 1980), tomato (Pegg, 1973, Pegg and Young, 1982), and yam (*Dioscorea opposita*) (Tsukamoto et al., 1984; Koga et al., 1992). Chitinases have been characterized in various organs including barley, maize, and wheat grains

(Molano et al., 1979; Roberts and Selitrennikoff, 1988), tomato stem and leaf, bean leaf, pea pod, yam roots and soybean seeds (Pegg and Vessey, 1973; Pegg and Young, 1981, 1982; Pegg, 1988; Boller et al., 1983; Boller, 1985, 1987), barley endosperm (Leah et al., 1987), thorn apple (*Datura stramonium*) leaves (Broeckaert et al., 1988), and oat leaves (Fink et al., 1988).

Six chitinases with molecular weights of 38,000, 38,700, 32,200, 32,600, 34,300, and 33,200 have been identified in potato leaves infected with *Phytophthora infestans*. All isomers are basic proteins with pI values above 7. Most of the isoforms accumulated in intercellular washing fluid after inoculation with *P. infestans* whereas two of them were localized intracellularly. The soluble products formed from [^{3}H]chitin after prolonged incubation with the individual chitinases were oligomers of two to five units (chitobiose, triose, tetraose, and pentose). Free N-acetylglucosamine was detected maximally in 7% of the total products, and acetate was not detectable. Thus, all six enzymes are endochitinases without deacetylase activity. The enzymes showed lysozyme activity (Kombrink et al., 1988). All six chitinases possess similar antigenic sites and they cross-reacted with antisera from bean chitinase (Kombrink et al., 1988). Four isoforms of chitinases that belong to class I chitinases have been identified in potato. The structural features of the potato chitinases included a hydrophobic signal peptide at the N-terminus, a hevein domain which is characteristic of class I chitinases, a proline- and glycine-rich linger region which varies among all potato chitinases, a catalytic domain, and a C-terminal extension (Beerhues and Kombrink, 1994).

Two different molecular forms of chitinase were detected in pea pods inoculated with *Fusarium solani* f.sp. *pisi*. The chitinase, Ch 1, had molecular weight of 33,100 and the other chitinase, Ch 2, had molecular weight of 36,200. Both the chitinases were identified as endochitinases, since they initially catalyzed the release of chitooligosaccharides and with increasing incubation time, both chitinases started to break down the initially released larger oligosaccharides into small units. Hydrolysis by Ch2 ultimately produced a mixture of diacetylchitobiose and triacetylchitotriose. Hydrolysis by Ch1 also produced two saccharides, but it was capable of hydrolyzing triacetylchitotriose into N-acetyl glucosamine and diacetylchitobiose. Ch1 showed approximately 10 times more lysozyme activity than Ch2 (Mauch et al., 1988b).

Four acidic chitinases are found in maize. They can be classified into two groups. The first group (PRm 3 and PRm 4) has the same molecualr weight (25 kDa). Both the chitinases have a similarity in their amino acid composition (Table 1; Nasser et al., 1988) and are serologically related. The second group contains PRm5 and PRm7, which have higher molecular weights (29 and 34.5 kDa, respectively). They are serologically related to each other but have no serological relationships with PRm 3 and PRm 4 (Nasser et al., 1988). One basic chitinase, PRm Ba2 (30 kDa protein) was also detected in maize (Nasser et al., 1990). There was serological relationship between acidic PRm 5 and PRm 7 and

Table 1 Amino acid composition (mol/100 mol) of two chitinases of maize

Amino acid	PRm3	PRm4
Aspartic acid	17.0	16.3
Threonine	5.2	5.5
Serine	7.5	7.1
Glutamic acid	6.4	6.0
Proline	4.8	5.0
Glycine	15.3	16.0
Alanine	9.8	9.0
Cysteine	2.4	2.8
Valine	5.5	5.1
Methionine	0.4	0.4
Isoleucine	5.6	5.5
Leucine	9.0	8.5
Tyrosine	4.4	3.8
Phenylalanine	2.7	3.0
Histidine	0.8	1.0
Lysine	2.0	2.0
Arginine	1.2	3.0

Source: Nasser et al., 1988.

basic PRm Ba 2, but not with the other maize acidic chitinases, PRm 3 and PRm 4 (Nasser et al., 1990).

Two acidic chitinases (PRP and PRQ) and two basic chitinases have been identified in Samsun NN tobacco leaves (Felix and Meins, 1985; Shinshi et al., 1987; Legrand et al., 1987; Kauffmann et al., 1987; Van Loon et al., 1987; Vogeli-Lange et al., 1988; Meins and Ahl, 1989). Basic chitinases were found localized primarily in the central vacuoles of cells (Mauch and Staehlin, 1989). The acidic enzymes were found in the extracellular fluid from leaves, suggesting that they are localized in the apoplastic compartment (Boller, 1987). The partial amino acid sequences of the acidic and basic chitinases show an approximately 70% homology. The molecular weights of the two basic chitinases are 32 kDa and 34 kDa (Hooft van Huijsduijnen et al., 1985, 1987), while those of the two acidic chitinases are in the range of 27–29.5 kDa (Hooft van Huijsduijnen et al., 1985; Jamet and Fritig, 1986). However, both chitinases have homologous C-terminus (Hooft van Huijsduijnen et al., 1987). The acidic and basic chitinases are each encoded by two to four genes in the genome of Samsun NN tobacco (Hooft van Huijsduijnen et al., 1987).

Hooft van Huijsduijnen et al. (1986b) isolated cDNA clones corresponding to several classes of PR proteins from tobacco mosaic virus-infected Samsun NN tobacco leaves. Upon sequencing, two classes of these clones represented by PROB 30 and PROB 3 were found to correspond to acidic and basic chitinases, respectively (Hooft van Huijsduijnen et al., 1987; Shinshi et al., 1987). PROB30 and PROB3 were used to screen a new cDNA library of tobacco (Linthorst et al., 1990a). The tobacco cDNA clones could be classified into two groups. Group 1, represented by clones Tach1 and Tach2, corresponded to PROB 30; the clones in group 2 were represented by clone Tach 3. The sequenced clones contained the complete open reading frames for acidic chitinases. The acidic chitinases encoded by cDNA clones Tach1 and Tach3 corresponded to PR-P and PR-Q of Samsun NN tobacco, respectively, as shown by protein sequence. Both the enzymes were 253 amino acid residues long. Sequence analyses of genomic clone CHA 18 showed that it corresponded to the cDNA sequence encoding PR-P. There was a difference in one nucleotide when this genomic clone was compared to the three PR-P cDNA clones characterized (a G instead of A residue at a position 606). This difference resulted in an amino acid change (cysteine instead of tyrosine at position 199) and suggests the presence of two genes in the tobacco genome encoding PR proteins differing in only one amino acid residue (Linthorst et al., 1990a).

Payne et al. (1990) reported the presence of G residue in several cDNA clones encoding PR-P from Xanthi nc tobacco. The other differences between the Xanthi nc and Samsun NN cDNA clones were a G residue at position 832, where there was a T residue in the Xanthi nc sequence and an extra C residue at position 920 in the Samsun NN cDNA clones. No differences were found between the two PR-Q cDNA sequences from Samsun NN and Xanthi nc (Payne et al., 1990; Linthorst et al., 1990a). Both the acidic chitinases, PR-P and PR-Q from Samsun NN tobacco had a highly hydrophobic NH_2-terminal region, which may probably function as a signal peptide (Linthorst et al., 1990a). This peptide is cleaved between the conserved alanine and glutamine residues at positions 24–25 (Payne et al., 1990). The acidic chitinases lack the cysteine-rich domain (Linthorst et al., 1990a) that is present at the NH_2-terminus of the mature basic enzymes (Lucas et al., 1985).

Trudel et al. (1989) detected four acidic chitinases in Xanthi nc tobacco and three of them corresponded to PR-P, PR-Q and PR-b6c. Chitinase activity of PR-b6c was much higher than the three other activities combined. Four basic chitinases were also identified. One corresponded to PR-b_{13} and other to unidentified proteins. Up to 13 (6 acidic, 7 basic) activities could be detected in Xanthi nc leaf tissue infected with tobacco mosaic virus (Trudel et al., 1989).

The amino-terminal amino acid sequence of bean chitinase has been determined (Lucas et al., 1985). From this sequence a 64-fold degenerate 20-mer oligonucleotide sequence was synthesized. This nucleotide sequence matches, with one substitution, exactly with a cDNA complementary to an ethylene-induced bean chitinase transcript (Broglie et al., 1986; Hedrick et al., 1988). Infection with *Colletotrichum lindemuthianum* resulted in accumulation of chitinase transcripts (Hedrick et al., 1988). Chitinase cDNA sequences hybridized to several genomic fragments, suggesting that there are several chitinase genes in the bean genome (Hedrick et al., 1988). In bean var. 'Saxa,' infection with pathogens or treatment with chemicals such as mercuric chloride induces at least three chitinases (de Tapia et al., 1986; Voisey and Slusarenko, 1989; Awade et al., 1989). They are one basic chitinase, which is located within the vacuole (Boller and Vogeli, 1984), and two acidic isoforms that appear in the extracellular fluid. The two acidic chitinases, P3 and P4, have been characterized. P4-chitinase cDNA clone encodes a 15 residue signal peptide and a mature protein of 255 amino acids (Mr 27300) (Margis-Pinheiro et al., 1991). The complete nucleotide sequence of cDNA encoding P3-chitinase has been determined (Margis-Pinheiro et al., 1993). The 1072 bp P3-chitinase cDNA encodes a mature protein of 268 amino acid residues and the 25 residue NH_2-terminal signal peptide of the precursor (Margis-Pinheiro et al., 1993).

An inducible, acidic chitinase has been purified from cucumber (Metraux et al., 1988; Boller and Metraux, 1988). A cDNA library was constructed from poly $(A)^+$-RNA isolated from cucumber leaves infected with tobacco necrosis virus. The protein encoded by the cDNA sequence contained all the sequenced peptides with agreement at every residue. The deduced protein was 292 amino acids long, which included a mature protein of 267 residues and 25 amino acids at the amino terminus of the protein that are not in the mature protein. No significant homology was found between the chitinase and either the basic chitinase isolated from bean or tobacco (Metraux et al., 1989). A single gene in the cucumber genome encodes this protein (Metraux et al., 1989).

Clones from a cDNA library of *Petunia hybrida* inbred line R27 have been isolated. These petunia cDNA clones encoded an acidic protein similar to the two acidic chitinases of tobacco. The protein had a highly hydrophobic NH_2-terminal region. The petunia chitinase had a molecular mass of 25.1 kDa (Linthorst et al., 1990b). Chitinase-encoding genes have been isolated from rice (Zhu and Lamb, 1991).

Two of the 12 PR proteins (D2 and D3) in sugar beet have been characterized. Both have the same apparent molecular weights (28 kDa) and have a similar, large proportion of acidic amino acids and a smaller proportion of basic amino acids. The first few amino acids from the N-termini of D2 and D3 have been sequenced. Both sequences have high homologies with the deduced amino acid sequences of induced chitinase mRNAs from cucumber and *Arabidopsis thaliana*, suggesting that the two PR proteins are chitinases (Fleming et al., 1991).

Eight different isoforms of chitinases accumulated in spruce roots after infection with *Pythium* sp. (Sharma et al., 1993). *Arabidopsis thaliana* produces a single basic chitinase, encoded by a single-copy gene (Samac et al., 1990). The protein has been purified to homogeneity and its molecular weight is approximately 32 kDa. Its isoelectric point is approximately 8.7. The deduced amino acid sequence of the *Arabidopsis* chitinase showed identity with that of tobacco, potato, and bean (Verburg and Huynh, 1991).

Six isoforms of chitinase have been detected in celery (*Apium graveolens*). Five of them were localized extracellularly and the sixth chitinase was detected in vacuole (Krebs and Grunt, 1993). The chromosomal region encoding the acidic class III chitinase from cucumber has been isolated and characterized. Sequence analysis of a 12,124-bp genomic clone revealed the presence of three complete open reading frames encoding the chitinase gene product. These genes share greater than 90% identity throughout the coding regions, but the homology drops significantly in the 5′ and 3′ flanking regions. The flanking genes CHI1 and CHI 3, although containing features of functional genes, are not expressed. In contrast CHI 2 is induced by pathogens, salicylic acid, and 2′6-dichloroisonicotinic acid (Lawton et al., 1994a). An acidic chitinase has been found to accumulate in sugar beet leaves infected with *Cescospora beticola*. Two isoforms (SE1 and SE2) of 29 kDa were purified to homogenity. One of the isoforms, SE2, is an endochitinase that also exhibits exochitinase activity. The deduced amino acid sequence of SE2 is 58–67% identical to the class III chitinases from cucumber, *Arabidopsis*, and tobacco (Nielsen et al., 1993).

Chitinases have been classified into five structural classes (Shinshi et al., 1990; Collinge et al., 1993; Ohl et al., 1994; Neilsen et al., 1994a,b). Class I chitinases are basic forms with a cysteine-rich N-terminal domain of about 40 amino acid residues, containing 8 conserved cysteines. This region, shown to be a chitin-binding domain, is followed by a variable hinge region and a catalytic domain, the latter comprising about 300 amino acid residues with a highly conserved main structure. Basic forms of chitinases from tobacco (Shinshi et al., 1990), potato (Gaynor, 1988), bean (Broglie et al., 1986), *Arabidopsis* (Samac et al., 1990), and rice (Ilwang et al., 1991; Nishizawa and Hibi, 1991; Zhu and Lamb, 1991) belong to class I and are primarily located in the central vacuole. Class II chitinases are mainly acidic isoforms that resemble class I chitinases, but lack the N-terminal cysteine-rich domain. These isoforms are usually found in the apoplast. This class includes two chitinases from tobacco (Payne et al., 1990), a petunia chitinase (Linthorst et al., 1990b), and a chitinase from barley (Leah et al., 1991). Class III include acidic chitinases from cucumber, tobacco, chickpea, and *Arabidopsis thaliana*, which lack the cysteine-rich domain and have structural homologies to a bifunctional lysozyme/chitinase from *Parthenocissus quinquifolia* (Bernasconi et al., 1987; Metraux et al., 1989; Samac et al., 1990; Vogelsang and Barz, 1993; Lawton et al., 1994a). Class III chitinases are apoplastically located. Basic class III chitinase structurally unrelated to acidic class III chitinase has also been reported in tobacco (Ward et al., 1991).

A 30 kDa basic chitinase has been purified from rapeseed. It is strongly induced in response to pathogen infection (Rasmussen et al., 1992). A chitinase cDNA clone from rapeseed has also been isolated. The cDNA clone, Ch B4, represents the basic chitinase isozyme. This represents a novel group of basic chitinases that shows higher amino acid sequence identity (63%) to the acidic bean chitinase (Margis-Pinheiro et al., 1991) than to the basic class II chitinases (43–48%). The longest open reading frame in Ch B4 encodes a polypeptide of 268 amino acids. This polypeptide consists of a 24 amino acid N-terminal signal peptide, a cysteine-rich domain, and a catalytic domain. Ch B4 has 62% identity to a basic chitinase isolated from sugar beet and 63% identity to an acidic bean chitinase, with a cysteine-rich domain (Margis-Pinheiro et al., 1991). Ch B4 shows low identity to class I and class II chitinases with 43–48% and 35%, respectively. A C-terminal extension is present in all published class I chitinases from dicotyledons (Broglie et al., 1986; Collinge and Slusarenko, 1987; Samac et al., 1990; Shinshi et al., 1990). With a tobacco chitinase gene, it has been found that some C-terminal sequences are necessary for targeting the protein to the vacuole (Neuhaus et al., 1991). No C-terminal extension was found in the rapeseed Ch B4 chitinase, as with the sugar-beet chitinase that has been found to be extracellularly located. Based on the low similarity between the Ch B4 and the other class I and II chitinases, as well as the lack of a C-terminal extension, rapeseed and sugar-beet chitinase may represent a new class of basic chitinases. Collinge et al. (1993) have proposed a fourth class for this type of chitinase. Class IV chitinases resemble the class I chitinases in having a cysteine-rich region with eight cysteines, a hinge region, and a conserved main structure. However, due to four deletions, one in the cysteine-rich domain and three in the catalytic domain, they are about 50 amino acid residues shorter than those of class I (Collinge et al., 1993). The establishment of the class IV chitinases was based on the sequences of three chitinases: the basic sugar beet Ch 4, rapeseed ChB4 chitinases, and the acidic bean PR 4 chitinase (Nielsen et al., 1994b). A maize chitinase has also been identified as class IV chitinase (Hyunh et al., 1992). Two acidic chitinase isoforms SP1 and SP2 have been purified to homogeneity from sugar beet leaves infected with *Cercospora beticola*. SP1 and SP2 were extracellular proteins with an apparent molecular mass of 35 kDa and an approximate pI of 4.2. SP2 chitinase was further characterized by amino acid compositional analysis and partial sequencing. SP2 exhibited significant structural identity with the class IV chitinases from sugar beet, rapeseed, bean, and maize. The major structural difference between SP2 and the other chitinases of the class IV was the longer hinge region in SP2, comprising 22 amino acid residues, with the repeated "TTP" motif (Nielsen et al., 1994b). The acidic SP2 was found extracellularly. Like the other class IV chitinases, SP2 lacked the short C-terminal extension responsible for vacuolar targeting (Nielsen et al., 1994b).

A new class of tobacco chitinases that shows no sequence similarity to the previously identified chitinases (class I–IV) has been recently reported (Ohl et al.,

1994). These chitinases share significant homology with some bacterial exochitinases. They are encoded by a small multigene family and induced by various stresses such as ethylene treatment, wounding, ultraviolet (UV) irradiation and TMV infection. Comparison of a cDNA clone and a genomic clone from two class V chitinases reveals a high degree of sequence identity. Unlike the bacterial chitinases the plant enzymes were endochitinases lacking detectable exochitinase activity (Ohl et al., 1994).

Another type of hydroxyproline-containing chitinases has been reported. The vacuolar class I chitinases A and B in tobacco contain hydroxyprolines (i.e., four to six of the prolyl residues, all localized in the T, P, and G-rich hinge region, and hydroxylated), but they are not glycosylated (Sticher et al., 1992). In contrast, the acidic class IV chitinase of sugarbeet SP2 represents a novel type of hydroxyproline-containing glycoproteins in plants (Nielsen et al., 1994b).

D. PR-4 Group Proteins

Four class II (acidic, extracellular) PR-4 proteins have been reported in tobacco (Pierpoint, 1986; Van Loon et al., 1987; Fritig et al., 1989). A class I PR-4 protein has been recently purified from tobacco (Ponstein et al., 1994). Analysis of the protein and corresponding cDNAs revealed that it contains a C-terminal domain showing homology to class II PR-4 proteins from tobacco. The protein is localized intracellularly and is induced from TMV infection and wounding (Ponstein et al., 1994). A PR protein of tomato, P2, is a basic protein with a molecular weight of 15 kDa and a pI of 10.4. It is serologically related to PR-4 class II protein (Joosten et al., 1990). The PR-4 class I protein also shows homology to the tomato PR protein PR-P2 (Ponstein et al., 1994).

E. Thaumatin-Like Proteins Family

The PR-5 group proteins have close resemblance to a sweet-tasting protein, thaumatin, that occurs in the fruit of the West African shrub *Thaumatococcus danielli* (Cornelissen et al., 1986; Pierpoint et al., 1987). They have a similar sequence of amino acids with 8% cysteine content, a similar molecular weight, a similar circular dichroism spectrum, and an immunological relationship with thaumatin (Cusack and Pierpoint, 1988), but they do not have a marked sweet taste (Pierpoint et al., 1987). The term thaumatin-like (TL) protein is used for this kind of proteins (Pierpoint and Shewry, 1987).

Acidic PR5 protein have been detected in leaf extracts of *Nicotiana tabacum*, *N. glutinosa*, *N. langsdorffii*, and *N. debneyi* plants in stress. *N. glutinosa* produces PR5a while *N. debneyi* expresses PR5b. A hybrid from *N. glutinosa* × *N. debneyi* constitutively expresses both PR5a and PR5b (Pierpoint et al., 1992). An intercellular protein that resembles tobacco PR5 has been purified from potato leaves (Pierpoint et al., 1990). A similar protein named Hv1 has been identified in barley (Bryngelsson and Green, 1989). Acidic TL-proteins with molecular weights near 21,000 have been detected in leaves of stressed soybean (Graham et al., 1992) and Pinto beans (Sehgal et al., 1991). They are reciprocally immunoreactive with PR5 protein from Samsun NN tobacco (Sehgal et al., 1991). In fact, all PR5 proteins are immunologically related (Pierpoint et al., 1992). The PR proteins belonging to PR5 group show a 65% amino acid sequence homology to a maize protein (MAI) that is a bifunctional inhibitor of α-amylase and proteases of insects (Cornelissen et al., 1986; Pierpoint et al., 1987; Singh et al., 1987b; Richardson et al., 1987; Cusack and Pierpoint, 1988).

As well as the above acidic PR5 proteins, a basic intracellular protein in tobacco, homologous to PR5a, has been identified as a TL protein. The protein has been shown to be identical to osmotin, the salt stress protein (Singh et al., 1987b; Grosset et al., 1990; Stintze et al., 1991). Tomato PR protein NP 24 has also been identified as osmotin (King et al., 1988). In tobacco two osmotin-like proteins, osmotin I and II have been described. Both of them have a molecular weight of about 24 kDa (Singh et al., 1987a, 1989). These osmotin-like proteins are synthesized during osmotic adaptation of plant cells in a medium containing osmotic agents (King et al., 1986). When tobacco plants were treated with an incompatible pathogen, the leaf extracts showed strong antifungal activity. The antifungal activity was heat-labile and the active component was identified as a hydrophobic protein with a molecular weight of 24 kDa and a basic isoelectric point. It was called AP 20 (also referred to as AP 24) (Woloshuk et al., 1991b). After purification and determination of the amino acid sequence of the N-terminus of AP 20, this part appeared identical to the corresponding part of osmotin (Woloshuk et al., 1991b). Similar 24 kDa osmotin-like proteins have been isolated from tomato, soybean, millet, carrot, cotton, potato, alfalfa, and bean (Singh et al., 1987a).

The proteins serologically related to osmotin have been reported in the families Fabaceae, Solanaceae, Apiaceae, Malvaceae, and Poaceae (King et al., 1986; Singh et al., 1985, 1987a,b; Pierpoint and Shewry, 1987). Osmotin-like proteins, having amino acid sequence homologous to osmotin, accumulate to substantial levels in cultured tobacco cells (Takeda et al., 1990; Kumar and Spencer, 1992). These osmotin-like proteins are regulated developmentally in intact plants with high levels of expression in roots (King et al., 1986; Singh et al., 1989; Neale et al., 1990; Takeda et al., 1990). They are induced by TMV infection and wounding (Neale et al., 1990; Takeda et al., 1990).

The amino acid sequence indicated that the osmotin-like protein of tobacco is synthesized as a 29 kDa precursor and processed to the 26 kDa mature polypeptide. Sequence analysis also indicated several characteristics of amino acid sequence unique to osmotin-like protein; osmotin-like protein possessed a putative glycosylation site as well as proline–glutamate–serine–theronine rich sequence, neither of which were found in osmotin or tobacco pathogenesis-related protein (PR-S) (Takeda et al., 1991). Sequence comparison revealed that osmotin-like protein was highly homologous to one of the antivirus proteins isolated from the same variety of tobacco (Edelbaum et al., 1990, 1991). High homology suggested that osmotin-like protein might be a precursor of antivirus protein gp22 (Takeda et al., 1991).

When the amino acid sequence of osmotin-like protein of tobacco was compared to sequences of other proteins, the osmotin-like protein was found to be quite similar to osmotin (Singh et al., 1989), tomato protein NP 24 (King et al., 1986), the thaumatin-like PR protein of tobacco, PR-S (Cornelissen et al., 1986), maize α-amylase/trypsin inhibitor (Richardson, et al., 1987), thaumatin (Edens et al., 1982) (76, 77, 60, 54, and 47% identical, respectively) (Takeda et al., 1991).

Yet another new PR protein, TPM 1 has been shown to be induced in tomato leaves after infection with tomato plant macho viroid (TPMV) (Ruiz-Medrano et al., 1992). The gene encoding the TPM1 protein is not expressed in normal healthy conditions, and is strongly induced as a response to viroid infection. The cDNA (TPM1) was 832 bp long and the open reading frame was 714 bp long. This sequence had two polyadenylation signals (Ruiz Medrano et al., 1992). The protein (TPM1) has high homology (about 90%) to tomato osmotin (NP 24) cDNA. The predicted mature N-terminus of TPM 1 is 100% homologous to the N-terminal sequences of the viroid-induced tomato protein P23 and to the fungus-induced tobacco protein with antifungal properties, AP-24, suggesting that all these proteins might be equal (Woloshuk et al., 1991b; Rodrigo et al., 1991; Ruiz-Medrano et al., 1992).

F. PR Proteins Structurally and Serologically Analogous to TL-proteins but Functionally Analogous to β-1,3-Glucanase and Chitinase

An acidic 21-kDa protein, termed PR-4d, has been detected in bean cv Pinto (Mohamed and Sehgal, 1987a,b; Sehgal and Mohamed, 1990). It is secreted into the intercellular fluid (Sehgal and Mohamed, 1990). The bean PR-4d protein shows a strong immunological relationship with tobacco SamsunNN PR-5 protein, which is equivalent to the tobacco Xanthi nc PR-R or tobacco thaumatin-like protein (Sehgal et al., 1991). PR-5 protein of tobacco does not possess any β-1,3-glucanase activity while bean PR-4d protein shows β-1,3-glucanase activity. It suggests that Pinto bean PR-4d and tobacco PR-5 are functionally diverse but structurally analogous (Sehgal et al., 1991).

In another variety of bean, "Saxa," four acidic proteins ranging in molecular weights from 17,000 to 33,500 have been reported (de Tapia et al., 1986). One of these proteins, Saxa PR-4, shows some resemblance to Pinto PR-4d and tobacco PR-5 proteins, but Saxa PR-4 has a larger mass (33.5 kDa) while Pinto PR-4d has a molecular weight of 21 kDa. Saxa PR-4 possesses chitinase activity and accumulates in the intercellular fluid (Awade et al., 1989). The PR-4d tye protein has been detected in *Tetragonia expansa* (Family: Aizoaceae), *Gomphrena globosa* (Amaranthaceae), *Beta vulgaris*, *Chenopodium amaranticolor* and *Chenopodium quinoa* (Chenopodiaceae), *Cucumis sativus* (Cucurbitaceae), *Canavalia ensiformis*, *Phaseolus acutifolius*, *P. angularis*, *P. lunatus*, *P. polyanthus*, *P. trichocarpus*, *P. vulgaris*, *Vigna aconitifolia*, *V. marina*, *V. radiata*, and *V. unguiculata* (cowpea) (Leguminosae), *Datura stramonium*, and *Petunia hybrida* (Solanaceae) (Sehgal et al., 1991).

G. Intracellular Acidic PR Proteins

Another class of PR proteins of unknown function includes parsley PR1-like proteins (Somssich et al., 1988), potato pSTH2 and pSTH21 (Matton and Brisson, 1989; Matton et al., 1990, 1993; Constabel and Brisson, 1992; Constabel et al., 1993), pea p149 (DRRG gene coded protein) (Fristensky et al., 1988; Chiang and Hadwiger, 1990), and bean pvPR 1 and pvPR 2 (Walter et al., 1990; Sharma et al., 1992). Similar proteins have been detected in pea embryos (Barrat and Clark, 1991), soybean cell cultures (Crowell et al., 1992), and birch pollen (Breitender et al., 1989). The acidic PR1 group of PR proteins are all synthesized as precursor proteins with an N-terminal extension of 30 amino acids, which is cleaved during import into the endomembrane system prior to secretion into extracellular spaces. The PR proteins from cultured parsley cells are also acidic and ranged in mass from 17.5 to 19.2 kDa (Somssich et al., 1986). They were named parsley PR1 proteins (Somssich et al., 1988). Sequence analysis of cDNAs coding for these proteins revealed no similarity between tobacco PR1 genes and the three genes coding for parsley PR1-like proteins (Somssich et al., 1988). The parsley PR1-like proteins that are acidic were shown to be intracellular (Somssich et al., 1988). In other cases, only basic PR proteins have been shown to be intracellular. cDNAs homologous to the parsley PR1 cDNA have been cloned from potato (Matton and Brisson, 1989), pea (Fristensky et al., 1988; Iturriaga et al., 1994), bean (Walter et al., 1990), and birch (Betula sp.) pollen (Breitender et al., 1989). In all these cases there was no indication that the predicted protein product of any of these similar genes contained an N-terminal extension characteristic of a signal sequence. Hence it has been suggested that these cDNAs represent a new category of pathogenesis-related genes (Walter et al., 1990). It is suggested that all these proteins may be intracellular and hence designated as intracellular PR proteins (Warner et al., 1992).

A wound-induced cDNA designated AoPR1 (*Asparagus officinalis* pathogenesis-related protein clone1) has been isolated from a suspension of mesophyll cells that had been mechanically isolated from *A. officinalis* seedling by grinding in a mortar and pestle. The cDNA was sequenced and it showed similarities with a pea disease resistance protein, p 149 (27% identity), a potato PR protein, pSTH

2 (34% identity), a parsley PR protein, Pc PR1-1 (32% identity), a birch pollen allergen Betov 1 (36% identity), and the bean elicitor-induced transcripts PvPR1 and PvPR2 (29% identity). Hence AoPR1 of *Asparagus* may also belong to the intracellular PR proteins (Warner et al., 1992).

H. Cysteine-Rich PR Proteins

Two pea genes, p139 and p1230, were induced by infection with *Fusarium solani* f.sp. *pisi*. Nucleotide sequence analysis of p139 and p1230 cDNA revealed open reading frames (ORFs) of 222 and 219 bp, respectively. Both cDNAs contain multiple AATAAA sequences typical of polyadenylation signals. The ORF of p139 encodes 74 amino acids and the ORF of p1230 encodes 73 amino acids. Each of the ORFs contains two domains. The first domain is composed of a long stretch of a hydrophobic residue and the second domain contains all of cysteines. The cysteine-rich domains contain eight cysteine residues, which have the potential to form four disulfide bonds. The molecular masses of the preproteins for p139 and p1230 are 8.2 and 8.0 kDa, respectively. The mature proteins, as indicated by in vitro translation, are 5 kDa ones (Chiang and Hadwiger, 1991). Both the proteins are translocated into the endoplasmic reticulum in vitro (Chiang and Hadwiger, 1991).

A necrosis-inducing peptide has been isolated from the apoplastic fluids of tomato infected with *Cladosporium fulvum*. The secreted peptide had a low molecular mass (3 kDa) and also had a high percentage of cysteine residues; similar to pea p139 and p1230 gene products (Schottens-Toma and De Wit, 1988).

Thionins are a class of basic and cysteine-rich, low-molecular-weight proteins (about 5 kDa) found in a variety of plants. Thionins are predominantly found in seeds but they are also found in leaves and stems. They are toxic to fungal pathogens (Bohlmann and Apel, 1987, 1991; Bohlmann et al., 1988; Garcia-Olmedo et al., 1989; Ebrahim-Nesbat et al., 1989; Apel et al., 1990). They have been shown to permeate fungal cell membranes and inhibit DNA, RNA and protein synthesis (Garcia-Olmedo et al., 1989).

Thionin isolated from barley and other monocots is encoded by a multigenic family. It has eight cysteine residues, which form four disulfide bonds (Ramshaw, 1982; Bohlmann et al., 1988). A novel thionin has been isolated from tobacco flower. The derived amino acid sequence of the central domain of the protein shows homology with γ-thionins (Mendez et al., 1990; Colilla et al., 1990). The thionin protein has an N-terminal domain characteristic of a signal peptide and an acidic C-terminal domain. The spatial pattern of thionin mRNA accumulation suggests that the thionin normally accumulates in highest concentration around the reproductive organs to protect them from potential pathogen attack. The thionin transcripts were inducible in leaves upon fungal pathogen attack (Bohlmann et al., 1988; Ebrahim-Nesbat et al., 1989; Gu et al., 1992).

Similar to thionins, many PR proteins are known to be associated with flower induction (Neale et al., 1990) and some are found in the developing flowers (Lotan et al., 1989). A class of mRNAs encoding thionin accumulated in barley roots 18 h after inoculation with *Drechslera graminea* (Vale et al., 1994).

Radish seeds contain two homologous 5 kDa basic, cysteine-rich proteins, called Rs-AFP1 and Rs-AFP2 which exhibit potent antifungal activity (Terras et al., 1992a). They are related to γ-type thionins (Mendez et al., 1990) and pea pod proteins induced upon fungal attack (Chiang and Hadwiger, 1991). A cDNA encoding the Rs-AFP1 preprotein was isolated form a cDNA library prepared from mRNA of near mature seeds. Transcripts (0.6 kb) hybridizing with a RS-AFP1 cDNA probe were present in near mature and mature seeds, as well as on leaves infected with fungal pathogens, where as such transcripts were barely detectable in healthy uninfected leaves. Hence Rs-AFP-like proteins represent a novel class of pathogenesis-related proteins. Extracts of transgenic tobacco plants expressing Rs-AFP2 showed an up to 16-fold higher antifungal activity against *Fusarium culmorum* relative to extracts from control tobacco plants (Broekaert et al., 1994).

I. Glycine-Rich Proteins

A new protein was found to be induced upon infection with tobacco mosaic virus (Hooft van Huijsduijnen et al., 1986b) or by salicylic acid treatment (Van Kan et al., 1988) in tobacco. Wounding induces glycine-rich protein (GRP) genes in petunia (Condit and Meagher, 1986) and bean (Keller et al., 1989). Abscisic acid and desiccation induce a GRP in maize (Gomez et al., 1988). The expression of some of the GRP genes is organ-specific and developmentally regulated (Oliveira et al., 1990). The sequence of this protein reveals a high percentage of glycine and this GRP may be present in the cell wall (Condit and Meagher, 1986; Varner and Cassab, 1986; Van Kan et al., 1988). GRP gene has been cloned and transgenic tobacco plants expressing GRP constitutively have been obtained (Linthorst et al., 1989). Didierjean et al. (1992) isolated and sequenced two cDNA clones encoding GRPs (CHEM 1 and CHEM 2), in mercuric chloride-treated maize leaves. The gene expression occurred at a basic level in young untreated leaves and the amounts of both mRNAs are significantly increased soon after the mercuric chloride treatment. CHEM 2 GRP is encoded by a small multigene family of approximately five members (Didierjean et al., 1992).

J. *Wun*1, *wun*2, *win*1, and *win*2 Gene-Induced PR Proteins

*Wun*1 isolated from potato is wound- and pathogen-inducible gene (Logemann et al., 1988, 1989). *Wun*1 codes for a small protein of about 18 kDa and its expression is regulated on the transcriptional level (Logemann et al., 1988). To test whether the wound-inducible gene *Wun*1 behaves as a pathogenesis-related gene, the potato leaves were inoculated with *Phytophthora infestans*. *Wun*1 mRNA levels increased severalfold in the infected leaves (Logemann et al., 1989). Construction of *wun*1-chloramphenicol acetyltransferase (CAT) reporter genes was achieved and their expression was analyzed in transgenic tobacco plants. Leaves from greenhouse-grown tobacco plants that were subjected to infection with several pathogens showed a much higher level of CAT activity in nonwounded leaves, indicating that the *wun*1 promoter is activated by pathogen stress in tobacco leaves (Logemann et al., 1989).

In order to obtain information about the protein encoded by *wun*1, the cDNA clone *wun*1-25A2 was sequenced. The *wun*-25A2 consisted of 711 bp with one large open reading frame of 105 amino acids. The large open reading frame coded for the *wun*1 protein. Among the 105 amino acids of the *wun*1 protein, 8 acidic and 11 basic amino acids were present, leading to a slightly basic protein. The size of *wun*1 protein was 12 kDa (Logemann et al., 1989).

The identity of *wun*1 is not yet known. The inducibility of *wun*1 by wounding and by pathogen attack as well as its low molecular weight fulfill some of the criteria for a pathogenesis-related gene (Logemann and Schell, 1989). A comparison of the *wun*1 sequence with all known genes involved in wound or defense reactions showed relatively a high homology (35% homology interrupted by two gaps) to the amino acid sequence of an anionic peroxidase from potato (Logemann and Schell, 1989). The anionic peroxidase is known to be involved in suberization processes in the periderm as a result of wounding (Espelie et al., 1986; Roberts et al., 1988). The activity of *wun*1 was highest in epidermal cells of leaves and stems. Cells containing *wun*1-GUS activity showed more callose synthesis (Sierbertz et al., 1989). A chimeric gene consisting of the *wun*1 promoter, the *wun*1 5′ untranslated region, the reporter gene, GUS and a nopaline synthase 3′ end was used to obtain transgenic tobacco plants. There were similarities in cell-specific expression of the *wun*1 promoter and the formation of callose. The localization of callose in the epidermis and the phloem of wounded tobacco leaves fits the regions where *wun*1 promoter derived GUS activity appears in *wun*1-GUS transgenic tobacco plants (Logeman and Schell, 1989). Callose is involved in plant defense reactions (Kauss, 1989). Enzymes involved in callose formation as well as suberization and lignification are localized in the epidermis and are involved in pathogen defense mechanisms by producing physical barriers (Hahlbrock and Scheel, 1987).

Another wound-induced gene in potato is *wun*2. Its expression is controlled at the level of transcription and the translation of *wun*2 mRNA resulted in the synthesis of 20-kDa protein (Logeman et al., 1988). Wound-induced *win*1 and *win*2 have also been reported (Stanford et al., 1989). The sequence of *win*1 gene was compared with that of the *wun*1 gene. A comparison between the 5′-flanking region of both genes showed two boxes of 85% and 71% homology, respectively, at position –470 to –423 of the *wun* 1 promoter (Siebertz et al., 1989).

K. Proteinase Inhibitors

Although not classically considered as PR proteins (Vigers et al., 1991), proteinase (protease) inhibitors are also induced by pathogens and by application of fungal elicitors (Walker-Simmons and Ryan, 1986; Rickauer et al., 1989) and shown to be inhibitory to fungal pathogens (Roby et al., 1987). When tomato plants are infected by *Phytophthora infestans* proteinase inhibitor increased and the increase was more in resistant varieties than in the susceptible ones (Peng and Black, 1976). Roby et al. (1987b) showed increases in proteinase inhibitor activity in melon plants infected with *Colletotrichum lindemuthianum*. The activity of proteinase inhibitor was more in kernels of smut-resistant wheat cultivars (Yamaleev et al., 1980). Proteinase inhibitors from tomato plants inhibited the activities of proteinases from *Fusarium solani* (Mosolov et al., 1976). The proteinase activities of *C. lindemuthianum* were inhibited by the proteinase inhibitors of bean (Mosolov et al., 1979).. Proteinase inhibitors from potato, mungbean (*Vigna radiata*), beans, chickpea, jack beans (*Canavalia ensiformis*), and adzuki beans (*Vigna angularis*) inhibit many microorganisms (Ryan, 1966; Yoshikawa et al., 1976; Richardson, 1977; Vartak et al., 1980; Svendsen et al., 1984; Seidl et al., 1982; Mosolov and Shulgin, 1986).

Proteinase inhibitor proteins commonly occur in plants (Laskowski and Koto, 1980; Ryan, 1966, 1973, 1981; Cleveland and Black, 1983; Graham et al., 1985; Garcia-Olmedo et al., 1987; Ryan and An, 1988) and they are induced by wounds, insect injury or pathogens (Ryan, 1990). The potato proteinase inhibitor II genes (*pin*2) introduced into rice expressed systemically by wounding or by methyl jasmonate or by abscisic acid (Xu et al., 1993). The proteinase inhibitors strongly inhibit animal and microorganism proteinases, but do not inhibit any known plant proteinases (Walker-Simmons and Ryan, 1977b).

Several nonhomologous families of proteinase inhibitors have been recognized in plants (Garcia-Olmedo et al., 1987; Laskowski and Kato, 1980). Inhibitor families specific for each of the four mechanistic classes of proteolytic enzymes (i.e., serine, cysteine, aspartic and metalloproteases) have been reported. They are soybean trypsin inhibitor (Kuniz) family, Bowman-Birk inhibitor family, barley trypsin inhibitor family, potato inhibitor I family, potato inhibitor II family, squash inhibitor family, ragi 1-2/maize bifunctional inhibitor family, carboxypeptidase A,B inhibitor family, cysteine proteinase inhibitor family (cystatins), and aspartyl proteinase inhibitor family. The majority of known proteinase inhibitors are specific for serine proteinases (Ryan, 1990).

The two kinds of potato proteinase inhibitors (serine proteinase inhibitors) have been detected in many plants. Inhibitor1 was first crystalized from potato tubers by Ryan and Balls (1962), while Bryant et al. (1976) isolated and purified Inhibitor II from potato tubers. Inhibitor I is a potent inhibitor of chymotrypsin while Inhibitor II possesses two reactive sites that are specific for trypsin and chymotrypsin, respectively (Ryan, 1992). These inhibitors have been detected in several plants belonging to Solanaceae (Ryan and Huisman, 1967; Ryan, 1968; Green and Ryan, 1972; Walker-Simmons and Ryan, 1977b; Plunkett et al., 1982; Kuo et al., 1984; An et al., 1989; Pautot et al., 1991), Fabaceae (Brown et al., 1985), Cucurbitaceae (Roby et al., 1987b), and Populus (Bradshaw et al., 1989). Both proteinase inhibitors I and II are located almost entirely in vacuoles of the tomato leaf cells (Shumway et al., 1970, 1972, 1976; Walker-Simmons and Ryan, 1977a).

Proteinase inhibitors, which are known to occur in other organs upon wounding, are found abundantly in potato tubers (Ryan et al., 1976; Sanchez-Serrano et al., 1986, 1987; Cleveland et al., 1987; Pena-Cortes et al., 1988; Stiekema et al., 1988; Keil et al., 1989; Johnson and Ryan, 1990; Palm et al., 1990; Hendricks et al., 1991). Fruit-specific (Margossian et al., 1988; Wingate and Ryan, 1991a) and leaf specific (Lee et al., 1986) tomato inhibitor I genes have been isolated. In *Lycopersicon peruvianum* the inhibitor I gene is both wound-inducibly expressed in leaves and developmentally expressed in unripe fruits (Wingate and Ryan, 1991b). Proteinase inhibitor I protein containing trypsin inhibitor specificity (Wingate et al., 1989) and inhibitor II proteins having both trypsin and chymotrypsin inhibitor specificities (Pearce et al., 1988) accumulate to high levels in unripe fruit of *L. peruvianum*. The levels of both the inhibitor I and inhibitor II proteins diminish as the fruit ripens (Pearce et al., 1988). Both of them are located in the vacuole of the fruit parenchyma cells of *L. peruvianum*.

A genomic clone encoding chymotrypsin inhibitor-2 has been cloned from a barley library and the nucleotide sequence has been obtained (Williamson et al., 1987; Peterson et al., 1991). In tobacco a proteinase inhibitor was strongly induced when infected with tobacco mosaic virus. From its amino acid composition and NH_2-terminal amino acid sequence analysis, it was shown that the inhibitor belongs to the potato inhibitor I family (Geoffroy et al., 1990). The activity of the tobacco inhibitor was found towards serine proteinases of different origins: two animal enzymes (trypsin and chymotrypsin), two bacterial enzymes (the subtilisin from *Bacillus subtilis* and a proteinase from *Streptomyces griseus*) and two fungal proteinases (proteinase K from *Tritirachium album* and an alkaline proteinase from *Aspergillus niger*). The tobacco inhibitor was found to be a potent and a specific inhibitor of the four proteinases of microbial origin. About 0.25 μg inhibitor was sufficient to completely inhibit the activity of 1 μg subtilisin or proteinase K (Geoffroy et al., 1990). The specificity of tobacco inhibitor I for the microbial serine proteinase is strikingly different from inhibitors I from tomato and potato, which are referred to as chymotrypsin inhibitor I due to their high inhibitory activity toward this animal enzyme (Gurusiddaiah et al., 1972; Melville and Ryan, 1972). The accumulation of inhibitor I, which is a specific and a potent inhibitor of microbial proteinases, may represent a defense reaction in the plant (Geoffroy et al., 1990).

Two cysteine inhibitors, designated oryzacystatins I and II, have been purified from rice seeds (Abe et al., 1987, 1991; Chen et al., 1992; Masoud et al., 1993; Benchekroan et al., 1995). cDNA of oryzacystatin I under the control of the cauliflower mosaic virus 35 S promoter was transferred to tobacco and the gene directed the synthesis of the cysteine proteinase inhibitor (Masoud et al., 1993). Requirement of an extracellular cysteine proteinase in the pathogenic interaction of the fungus *Pyrenopeziza brassicae* with oilseed rape has been reported (Ball et al., 1991). Hence cysteine proteinase inhibitor may inhibit the fungal infection. However, the role of cysteine proteinase inhibitor in fungal pathogenesis is yet to be studied.

Two proteinase inhibitors have been isolated from tubers of potato. Based on N-terminal amino acid sequence homologies, they are members of the Kunitz family of proteinase inhibitors. Potato Kunitz inhibitor 2 is an inhibitor of the microbial proteinase subtilisin, while potato Kunitz inhibitor I is a potent inhibitor of the animal pancreatic proteinase trypsin (Walsh and Twitchell, 1991). Cathepsin D Kunitz inhibitor is also commonly detected in potato tubers (Mares et al., 1989). Sporamin is abundantly present in storage roots of sweet potato (Hattori and Nakamura, 1988; Hattori et al., 1989). It is not detectable in organs other than the tuberous roots. Amino acid sequence of sporamin is homologous to Kunitz-type inhibitors of trypsin, chymotrypsin, subtilisin, or α-amylase of legume and cereal seeds (Ohto et al., 1992). It is also homologous to the wound-inducible *win*3 gene product of poplar (Bradshaw et al., 1989) and cathepsin D Kunitz inhibitor of potato tuber, which is also wound-inducible (Suh et al., 1991). Polygalacturonic acid or chitosan induced sporamin in leaf (Ohto et al., 1992). Oligogalacturonide fragments and chitosan are known to induce various defense-related proteins (Walker-Simmons et al., 1983; Bowles, 1990; Ryan and Farmer, 1991). Hence sporamin may also play some defense-related function in sweet potato (Ohto et al., 1992). However, a role for sporamin in fungal pathogenesis is yet to be established.

L. Proteinases

One of the tomato PR proteins, PR 69, has been identified as an alkaline proteinase with a molecular weight of 69,000–70,000 (Vera and Conejero, 1988). It was identified as P70 protein by Christ and Mosinger (1989). It accumulates in tomato leaves after inoculation with races of *Cladosporium fulvum* (De Wit et al., 1986). This protein was found in two compartments (vacuoles and intercellular spaces in a leaf tissue) and it was found to be a potential weapon against attacking pathogens (Vera and Conejero, 1988, 1989). Another proteinase protein has been reported in bean (Van der Wilden et al., 1983).

M. Chitosanases

Grenier and Asselin (1990) have identified chitosanases as antifungal pathogenesis-related proteins in barley, cucumber, and tomato leaves. Chitosanases (EC 3. 2.1.99) (acting on chitosan without activity on chitin) are differentiated from chitinases (acting on chitin without activity on chitosan) by their molecular weight and substrate specificity. Chitosan (poly [1-4]-β-D-glucosamine), the deacetylated chitin, is the substrate for the enzyme. Chitosan is found in various fungal cell walls. Abiotic elicitors like silver nitrate and serine and tobacco mosaic virus induced several PR proteins in barley, cucumber, and tomato. In barley, six chitosanases were detected; four of them were acidic proteins (three at 22 kDa, one at 19 kDa) while the other two were basic ones (one at 22 kDa, one at 19 kDa). Four acidic chitosanases (one at 10 kDa, one at 12 kDa, and two at 14 kDa) were observed in cucumber while one basic chitosanase (24 kDa) alone was detected in tomato. All these chitosanases were detected in intercellular fluids. Three chitosonase activities have been detected in spruce roots infected with *Pythium* sp. (Sharma et al., 1993). Lysis of *Fusarium oxysporum* f.sp. *radicis-lycopersici* spores was observed in stressed barley and tomato intercellular fluid extracts. In tomato the activity lysing spores corresponded to the major chitosanase activity at 24 kDa, which is a basic protein. In barley, the lysis of *F.o.* f.sp. *radicis-lycopersici* spores was detected only with the basic chitosanase. Thus lysis of the *Fusarium* spores has only been observed with basic chitosanases stimulated in stressed tissue. Besides *F.o.* f.sp. *radicis-lycopersici* spores, *Verticillium albo-atrum* and *Ophiostoma ulmi* spores were also lysed by those chitosanases (Grenier and Asselin, 1990).

N. α-Amylases

α-Amylase activity was greatly increased in leaves of tobacco infected with tobacco mosaic virus (Heitz et al., 1991). The basal level measured in healthy leaves was very low. Two α-amylases were purified from TMV-infected leaves and shown to have features in common with well-characterized pathogenesis-related proteins. They were acidic monomers that could be separated upon electrophoresis on basic native gels, and they were found in the apoplastic compartment of the cell. They showed an acidic isoelectric point, a high level of resistance to protease action, a low level of expression in healthy leaves, and similar kinetics of induction during the hypersensitive reaction to TMV (Heitz et al., 1991). Like pathogenesis-related proteins, α-amylase can also be induced by water stress (Jacobsen et al., 1986), by the application of plant hormones (Jacobsen and Higgins, 1982; Hirasawa, 1989), and by senescence (Saeed and Duke, 1990). The strong induction of α-amylases by virus attack and the accumulation of the α-amylases in the apoplast in parallel with other hydrolases, like chitinases and β-1,3-glucanases, suggest a role in the defense reaction.

α-Amylases might be involved in the degradation of substrates released by necrotizing cells, the hydrolysis products being then available to the neighboring cells, which develop the defense reactions (Heitz et al., 1991).

O. Peroxidases

In azuki bean (*Vigna angularis*) seven different proteins (each designated AZ xx according to its molecular mass, xx in kDa) have been shown to accumulate in response to ethylene. Among them an intracellular basic glycoprotein (AZ42) has been detected (Ishige et al., 1991). A cDNA encoding the glycoprotein has been cloned and its complete nucleotide sequence has been determined. An open reading frame (1071 base pairs) in the cDNA encoded a protein of 357 amino acids with a molecular mass of 39.3 kDa. The purified AZ42 had specific peroxidase activity (Ishige et al., 1993a). The amino acid sequence contained three regions that are highly conserved among peroxidases from cucumber, tomato, potato, horseradish (*Armoracia lapathifolia*), turnip (*Brassica campestris*), tobacco, and peanut (Ishige et al., 1993b). The AZ42 is an intracellular glycoprotein (Ishige et al., 1991) and probably localized in the vacuoles (Ishige et al., 1993a).

The mRNA for the basic peroxidase was not present at a detectable level in leaves that had not been treated with ethylene. Hence it appears that ethylene specifically regulates transcription of the gene for the basic isozyme of peroxidase. There were at least two isozymes of acidic peroxidase in both control and ethylene-treated leaves, but the level of the basic isozyme increased only in the treated leaves. The azuki gene of the basic peroxidase was also expressed in response to wounding of tissue and treatment with salicylate (Ishige et al., 1993a). These characteristics of the cationic peroxidase suggest that it may be a PR protein.

Acidic peroxidases have also been considered as PR proteins (Schweizer et al., 1989; Mohan and Kolattukudy, 1990; Ye et al., 1990). Parent and Asselin (1987) showed that the PR proteins of potato include peroxidases. Two tomato anionic peroxidase genes, *tap*1 and *tap*2, are expressed in response to wounding (Roberts et al., 1988), elicitor treatment (Mohan and Kolattukudy, 1990), and fungal attack (Robb et al., 1991; Mohan et al., 1993a,b). They are not constitutively expressed. The two TAP genes are located in tandom on the chromosome. They are highly homologous (about 96%) to each other (Roberts and Kolattukudy, 1989).

Within 6 h of contact, *Drechslera graminea* induced a class of transcripts homologous to a clone coding for peroxidase in wheat rootlets (Vale et al., 1994). Appearance of PR proteins (particularly peroxidase) in roots appears to be reported for the first time by Vale et al. (1994).

P. 1,3;1-4-β-Glucanases

Among plant β-glucanases, endo-1,3-β-glucanases (EC 3.2.1.39) have been described in both monocots and dicots, while endo-1,3;1-4-β-glucanases (EC 3.2.1.73) have been found only in monocots. It is now well known that 1,3-β-glucanases are expressed at low levels in healthy plants and their expression is induced by pathogen attack, salicylic acid, and cytokinin (Mauch and Staehelin, 1989; Memelink et al., 1990; Van den Bulcke et al., 1989). 1,3;1,4-β-Glucanase activity increases at the time of seed germination (Litts et al., 1990; Stuart et al., 1986, 1987). Barley germination 1,3;1,4-β-glucanase genes have been sequenced (Litts et al., 1990; Slakeski et al., 1990). They have extremely G + C-rich coding regions, and a single large 2.5 kb intron in the same position in the signal peptide coding region (Slakeski et al., 1990).

High levels of 1,3;1,4-β-glucanase activity in the germinating caryopses of rice have been reported (Stuart et al., 1987). A rice 1,3;1,4-β-glucanase has been sequenced. The glucanase shows 82% amino acid similarity to the barley 1,3;1,4-β-glucanases. The mature peptide has a calculated molecular mass of 32 kDa. The gene has a large 3145 bp intron in the codon for the 25th amino acid of the signal peptide. This gene has been named Gns1. It is expressed constitutively at relatively low levels in germinating seeds, shoots, leaves, panicles, and callus, but it is expressed at higher levels in roots. Shoots express Gns1 at much higher levels when treated with fungal elicitors derived from the pathogen *Sclerotium oryzae* or from the nonpathogen *Saccharomyces cerevisiae*. Ethylene, cytokinin, and salicylic acid treatment also induce Gns1 (Table 2; Simmons et al., 1992).

Although the physiological role of Gns1 is not yet known, this pattern of expression is consistent with the expression of the PR protein, 1,3-β-glucanase, suggesting that 1,3;1,4-β-glucanase may be a PR protein (Simmons et al., 1992).

Table 2 Relative expression of β-1,3; 1,4-glucanase gene (Gns 1) in rice shoots due to biotic and abiotic treatments

Shoot Treatment	Relative Expression of Gns 1
Control	1.0[a]
Sclerotium oryzae elicitor	23.0
Saccharomyces cerevisiae elicitor	10.0
Ethylene	18.0
Salicylic acid	6.7
Kinetin	14.0

[a]The level of expression in untreated shoots was taken as reference.
Source: Simmons et al., 1992.

V. ANTIFUNGAL PROTEINS

PR proteins accumulate due to fungal infection, some antifungal proteins are constitutively present at high concentrations in seeds and tubers without any fungal infection (Roberts et al., 1988). The antifungal proteins present in seeds include glycosidases (Manners and Marshall, 1973; Roberts and Selitrennikoff, 1986), thionins (Lopez-Fando et al., 1989; Bohlmann and Apel, 1991), permatins (Vigers et al., 1991), and ribosome-inactivating proteins (RIP) (Bowles, 1990; Sinclair et al., 1994). RIP inhibits protein synthesis in target cells by specific RNA, N-glycosidase modification of 28S rRNA (Endo et al., 1988). RIPs do not inactivate "self" ribosomes, but show activity toward ribosomes of distantly related species, including fungal ribosomes (Roberts and Selitrennikoff, 1986b; Stirpe and Hughes, 1989). Barley RIP inhibits growth of fungi in vitro (Leah et al., 1991). Similar to RIP from barley, "tritin" has been isolated from wheat (Coleman and Roberts, 1982).

Logemann et al. (1992) has obtained transgenic tobacco plants that express a barley RIP under the control of the wound-inducible promoter of the potato *wun* gene. The transgenic plants showed increased protection against *Rhizoctonia solani* and the plants expressed the barley RIP upon wounding. Transgenic plants with *wun*1-GUS alone showed no protection against the pathogen (Logemann et al., 1992). In vitro studies with *Trichoderma reesei* and *Fusarium sporotrichoides* showed that the combination of barley RIP and chitinase inhibits fungal growth more efficiently than do either enzyme alone (Leah et al., 1991).

A highly basic protein (isoelectric point higher than 10.5) has been isolated from radish seeds (Terras et al., 1992b). In its unreduced form, it exists as a dimer of a 9 kDa promoter. These characteristics, together with protein sequence data, suggest that the isolated protein is a nonspecific lipid transfer protein (ns-LTP). This protein inhibits fungal growth (Terras et al., 1992b). The involvement of the ns-LTP in the formation of the cutin layer has been suggested (Sterk et al., 1991). This type of protein could confer defense in two ways: indirectly by its involvement in the formation of a mechanical cutin barrier and directly by its intrinsic antifungal activity by following deposition of the transported cutin monomers. ns-LTP is synthesized de novo in the aleurone cells at the onset of water uptake by the seeds and is secreted from the aleurone layer into the surrounding medium (Mundy and Rogers, 1986). Hence ns-LTPs may protect seedlings by secretion from the germinating seeds (Terras et al., 1992b).

Permatins have been detected in seeds (Vigers et al., 1991). Zeamatin, a protein that has inhibitory action against different fungi, has been isolated from corn seeds (Roberts et al., 1988; Roberts and Selitrennikoff, 1990). It exerts antifungal effect by a membrane permeablization mechanism. It induces release of ultraviolet-absorbing material from *Candida albicans*, releases pre-loaded [1-^{14}C]isobutric acid from *Neurospora crassa*, and induces rupture of hyphae of

N. crassa and *C. albicans*. Zeamatin permeablizes fungal membranes causing fungal death. Zeamatin has been purified to homogenity and its molecular weight is 22 kDa (Roberts and Selitrennikoff, 1990).

Zeamatin like proteins have been purified from corn, oats, sorghum, and wheat and the N-terminal sequences for zeamatin and the proteins from all the four crops show remarkable homology. The protein from sorghum is named sormatin, that from oats avenamatin, and that from wheat trimatin (Vigers et al., 1991). All these proteins have the same molecular mass of 22 kDa. Antifungal proteins detected from barley and flax are also 22 kDa proteins. All these proteins also induce hyphal rupture of *N. crassa*, suggesting a common membrane-permeablizing mechanism of action. The N-terminal amino acid sequences of zeamatin, avenamatin, sormatin, and trimatin have homology with the sweet protein thaumatin, PR-5 protein, osmotin, NP 24, and a trypsin/α-amylase inhibitor (Vigers et al., 1991).

Thionins are mostly found as seed proteins. Hordothionins have been detected in the endosperm of barley and purothionins have been isolated from wheat endosperm. Leaf and stem thionins have also been reported. The thionin proteins are subdivided into the α, β, and γ types based on differences in their primary structure (Collila et al., 1990). Thionins are toxic to fungal phytopathogens (Garcia-Olmedo et al., 1989; Apel et al., 1990).

VI. BIOSYNTHESIS OF PR PROTEINS

A. PR Proteins are Expressed at the Level of Transcription and Not at Translation Level

Van Loon (1985) provided evidence to show that PR proteins are expressed not at the level of transcription but at the level of translation. Induction of PR proteins was inhibited by cycloheximide, the inhibitor of protein synthesis on cytoplasmic ribosomes, but not by chloramphenicol, which inhibits protein synthesis on chloroplasts and mitochondria. It suggests that cytoplasmic protein synthesis is necessary for the appearance of PR proteins in tobacco (Van Loon, 1985). Inhibitors of DNA-dependent RNA-synthesis, such as actinomycin D or cordycepin, did not inhibit the induction of PR proteins. It suggests that no transcription is required for the expression of PR proteins. When polyA-mRNA from healthy leaves were translated in vitro, as much PR proteins were synthesized as from mRNA from infected leaves (Carr et al., 1982). These studies suggest that PR-mRNA is constitutively present, but not translated in nonstimulated leaves. Thus, the expression of PR proteins may be controlled at the level of translation (Van Loon, 1985). Matsuoka and Ohashi (1986) demonstrated the putative regulation of PR proteins at the translational level with in vivo pulse-labeling expriments using antibiotics. The synthesis of PR proteins in tobacco leaf discs treated with potassium salicylate for 1 day was inhibited completely by cycloheximide, an inhibitor of protein synthesis in cytoplasmic ribosomes. It suggests that the PR proteins are synthesized de novo on cytoplasmic ribosomes. Actinomycin and α-amanitin are the well known inhibitor of transcription and both of them did not inhibit synthesis of PR proteins. It suggests that transcription is not required for induction of PR proteins. It is possible that untranslatable mRNAs of PR proteins are already present in healthy leaves and are changed to translatable forms by certain stresses like pathogens or salicylic acid treatment (Matsuoka and Ohashi, 1986).

In contrast, others have found that translatable mRNAs for PR-1 proteins are present only in infected or chemically treated leaves but not in healthy leaves (Carr et al., 1985; Van Huijsduijnen et al., 1985; Matsuoka et al., 1985). Northern blot analysis using the cDNAs for PR-1 proteins as probes have shown that the mRNAs for PR 1 proteins are present at a low level in healthy leaves (Van Huijsduijnen et al., 1985, 1986; Pfitzner and Goodman, 1987).

The induction kinetics of the PR1 protein synthesis in tobacco leaves treated with salicylate treatment was studied. There were no detectable mRNAs for PR1 proteins in healthy leaves. There was a lag phase of about 10 h, followed by an almost linear increase in PR1 proteins synthesis until 48 h. The induction kinetics of mRNAs was similar to PR 1 protein synthesis (Matsuoka et al., 1988). The experiments suggest that the synthesis of PR 1 proteins is not regulated at a translational step but at a transcriptional step (Matsuoka et al., 1988).

Acetyl salicylic acid treatment of *Nicotiana tabacum* Xanthi-nc plants induced synthesis of PR1 (16 kDa) proteins. These proteins were not detected in untreated plants. RNA was prepared from acetyl salicylic acid-treated and untreated plants. Rabbit reticulocyte lysates programmed with poly$(A)^+$ RNA from neither treated nor untreated leaves synthesized the 16-kDa protein, but when programmed with poly$(A)^+$ RNA from treated (but not untreated) plants, the wheat germ system synthesized a polypeptide similar in size to PR1 made in vivo. The synthesized product was serologically related to PR1a. PR1a antiserum reacted with PR1b and PR1c also and hence the synthesized product may consist of all three members of the PR1 family. Since functional mRNAs encoding the 16 kDa PR1 are present in acetyl salicylic acid-treated but not in control plants, it has been concluded that induction of PR1 protein synthesis is regulated in part at the level of mRNA accumulation (Carr et al., 1985).

The translation of PR mRNAs in the rabbit reticulocyte lysate system is blocked, but not in the wheat germ extracts. The rabbit reticulocyte lysates, but not wheat germ extracts, contain signal recognition particles (SRPs) that block elongation of signal sequence necessary for contranslational translocation into the endoplasmic reticulum. If the PR1 proteins are initially synthesized as precursors with signal consequences, arrest of polypeptide elongation by signal recognition particles may explain the lack of PR1 synthesis in reticulocyte lysates. The membrane vesicles of dog pancreas microsomes contain docking proteins that release the SRP-mediated elongation arrest and allow translation to continue. In the presence of dog pancreas microsomes, reticulocyte lysates programmed with RNA from acetylsalicylic acid-treated Xanthi-nc leaves synthesized 16 kDa PR1 protein. This in vitro-synthesized PR1 was not produced by reticulocyte lysates programmed with RNA from untreated Xanthi-nc leaves even in the presence of dog pancreas microsomes, again suggesting that PR protein induction is regulated at the level of mRNA accumulation (Carr et al., 1985).

Salicylic acid induced PR proteins in tobacco. Cycloheximide at 0.1 and 1 mg/ml blocked 95 and 99% of protein synthesis, respectively. PR-1a mRNA did not accumulate following treatment with both salicylic acid and cycloheximide. It suggests that continued protein synthesis is required for induction by salicylic acid (Uknes et al., 1993). A single mRNA species of approximately 900 bp was hybridized by the bean PR protein, PvPR3 probe. The PvPR3 transcript was present within uninduced bean cells, and the level subsequently increased slightly within 1 h after elicitor treatment. The mRNA continued to accumulate gradually over a period of 50 h after elicitor treatment (Sharma et al., 1992). Regulation of β-1,3-glucanase has been shown to occur at the mRNA level in barley (Jutidamrongphan et al., 1991), bean (Vogeli et al., 1988), and tobacco (Mohnen et al., 1985). Ethylene, an abiotic elicitor, induced 50–100-fold increase in the level of β-1,3-glucanase mRNA, suggesting that the enzyme is regulated at the level of gene transcription (Takeuchi et al., 1990). A basic β-1,3-glucanase is a major component of the soluble protein of cultured tobacco cells (Felix and Meins, 1986) and has been shown to be downregulated at the mRNA level in the cultured tobacco cells by combinations of the plant hormones auxin and cytokinin (Mohnen et al., 1985; Shinshi et al., 1987; Felix and Meins, 1987). Thiamine treatment induced PR2, N, and O proteins and these β-1,3-glucanases accumulated to high levels in a manner consistent with the observed rise in steady-state levels of glucanase mRNA. The parallel accumulation of PR2, N, and O proteins and mRNA suggests that the β-1,3-glucanases gene expression is regulated at the level of mRNA accumulation (Cote et al., 1991).

The induction of chitinase gene has been found to occur at the level of mRNA accumulation in cucumber (Metraux et al., 1989). In bean plants chitinase is encoded by a multigene family consisting of at least three members (Broglie et al., 1989). One of the genes has been shown to encode an abundant mRNA induced by ethylene (Broglie et al., 1989). Chitinase mRNA synthesis is stimulated in bean cell suspensions by treatment with elicitors from cell wall of *Colletotrichum lagenarium* (Hedrick et al., 1988). It has been demonstrated that modulation of chitinase levels in plants involves the activation of gene transcription (Hedrick et al., 1988).

When bean leaves were treated with ethylene, chitinase and β-1,3-glucanase activities doubled within 2 h and continued to increase over a period of 48 h.

The mRNA activity for both enzymes increased rapidly for the first 2 h after ethylene treatment, and thereafter remained at a high level (Vogeli et al., 1988). The results suggest that chitinase and β-1,3-glucanase are regulated coordinately at the level of mRNA.

Expression of both *wun*1 and *wun*2 genes in wounded potato tubers is controlled at the level of transcription. The translation of *wun*1 and *wun*2 results in the synthesis of 18 and 20 kDa proteins (Logemann et al., 1988).

De novo synthesis of mRNA of PR proteins has been demonstrated in two cultivars of tobacco (Cornelissen et al., 1986; Hooft van Huijsduijnen et al., 1987) and parsley (Somssich et al., 1986). Accumulation of proteinase inhibitors I and II is accompanied by the appearance of translatable mRNAs that code for each (Nelson and Ryan, 1982). The level of mRNAs coding for inhibitors I and II increase rapidly about 4–8 h following wounding and their levels are also transcriptionally regulated (Graham et al., 1986). All the above studies suggest that PR proteins are transcriptionally regulated.

B. Transcriptional Regulation of PR Protein Genes

The controlled expression of many genes depends on the presence of *cis*-acting elements located in the flanking regions of the protein coding genes. Regulatory elements in the immediate vicinity of the transcription sites are called as promoters. They usually comprise conserved transcription signals, such as TATA and CCAAT boxes. The elements controlling gene expression from a greater distance from the RNA start site (up to 30 kb) are referred to as enhancers. Enhancer elements are found upstream as well as downstream from the transcription initiation site of the protein coding regions and are able to exert control in an orientation-independent fashion on the promoter (Beilmann et al., 1992).

To examine the regulation of PR-1 genes, the 5′ flanking regions of the PR-1a gene were joined by a transcriptional fusion to the *Escherichia coli* β-glucuronidase (GUS) gene. Tobacco transgenic plants exhibited considerable GUS activities without any treatment like pathogen attack or chemical stimuli, but these plants did not show any expression of endogenous PR-1 genes. To test whether any heterologous enhancer elements dispersed in the tobacco genome are needed to express PR1 proteins constitutively, domain B of the enhancer of the 35S RNA promoter from cauliflower mosaic virus (CaMV) was fused to PR1/GUS hybrid genes upstream as well as downstream from the RNA start site. These constructs were stably introduced into the tobacco genome. In these transgenic plants, PR-1 synthesis activity was observed. High levels of gene expression from PR-1 promoter were readily obtained when the PR-1 minimal promoter region became associated with an enhancer element. The results suggest that the enhancer element apparently dictates the pattern of gene transcription from the PR-1 promoter (Beilmann et al., 1992).

In contrast, Ohshima et al. (1990) have reported that constructs containing 2.4 kb, 0.9 kb, or 0.3 kb of the PR-1a 5′ flanking region fused to GUS consistently yielded high levels of constitutive expression in transgenic tobacco plants, even exceeding the levels obtained with CaMV 35S RNA promoter. Ohshima et al. (1990) have inserted the PR-1a/GUS constructions in an opposite orientation immediately adjacent to the CaMV 35S enhancer, which is a part of the expression cassette of the selectable marker gene in the binary vector system used for the plant transformation. Hence it is possible that the results obtained by Ohshima et al. (1990) may be due to *cis*-activation of the PR-1 promoter region by the 35S enhancer (Beilmann et al., 1992). The *cis*-acting regulatory elements in the flanking regions of the protein coding genes have been shown to bind *trans*-acting factors that may influence transcription either positively or negatively (Mitchell and Tjian, 1989).

The tissue- and cell type-specific expression of the PR1-GUS gene was expressed in the transgenic tobacco plants. Strong GUS activities were observed in close association with the vascular bundles, namely, the phloem in one transgenic plant. In another transgenic plant high levels of GUS activity were found in the epidermal cell layer of the petiole and the leaf. The data indicate that the expression of the reporter (GUS) gene is not solely directed by the PR-1a promoter region in these plants. It is likely that *cis*-acting regulatory elements located in the chromosomal vicinity of the PR-1/GUS hybrid genes might influence the transcriptional activity of the PR-1 promoter region (Beilmann et al., 1992). A positive *cis*-acting regulatory elements has been defined between –643 and –689 in the PR-1a upstream sequence (Van de Rhee et al., 1990).

Both TMV and salicylic acid induced PR1 proteins in tobacco. To determine the amount of 5′-flanking sequence from the PR-1a sufficient to confer TMV and salicylic acid inducibility, a set of deletions was constructed and fused to the GUS reporter gene. The constructs were transformed into tobacco with *Agrobacterium*. Salicylic acid treatment produced an average induction of greater than fivefold for deletions from –903 to –606. The –600 bp deletion line showed an average induction of 2.3-fold whereas deletion lines –318 to –73 induced 1.3-fold or less. Transgenic tobacco treated with TMV showed strong induction of GUS activity (13-, 10-, 28-, and 12-fold) compared to control treatment for deletion lines –903, –825, –700, and –661, respectively. The deletion lines –600, –318, –222, –150, and –73 had an induction of three-, one-, 1.5-, two- and threefold, respectively. These results indicate that 661 bases of the PR-1a promoter is sufficient for high level induction by TMV or salicylic acid treatment (Uknes et al., 1993).

In experiments designed to identify the *cis*-acting sequences of the PR-1a promoter responsible for induction in tobacco, Ohshima et al. (1990) concluded that as little as 300 bp of 5′ flanking sequence was sufficient to express appropriately a GUS reporter gene. Van de Rhee et al. (1990) using the same PR-1a gene from tobacco and similar reporter gene constructs, however, concluded that induction required at least 685 bp. In the study by Ohshima et al. (1990), an increase of 50% in the level of GUS activity was scored as induced, whereas in the Van de Rhee et al. (1990) study the induction had to be at least fivefold to be scored as induced.

The class I isoforms of tobacco β-1,3-glucanase are encoded by a small gene family. Several genomic clones have been characterized (Linthorst et al., 1990a; Ohme-Takagi and Shinshi, 1990; Sperisen et al., 1991). These isoforms show pathogenesis-related regulation at the mRNA level. GUS reporter gene was used to investigate expression of β-1,3-glucanase B in tobacco leaves due to infection. Promoter deletion analysis was also done to investigate regulation of β-1,3-glucanase B (Vogeli-Lange et al., 1994). The results showed that 1.6 kb of upstream sequence of the β-1,3-glucanase B gene contained elements needed for expression in response to TMV infection. Deletion analyses indicated that β-1,3-glucanase B promoter consists of distinct regions important for high-level gene expression and for regulation. The region from –1452 to –1193 containing two copies of the heptanucleotide AGCCGCC, which is highly conserved in defense-related genes, was necessary for high level expression in leaves (Vogeli-Lange et al., 1994).

By fusing upstream sequences of the PR-5 gene to the GUS reporter gene and analyzing transgenic tobacco plants containing these fusions for systemic induction of GUS activity by TMV, it was found that sequences between –1364 and –718 are involved in TMV induction of PR 5 gene expression (Albrecht et al., 1992).

Chitinase mRNA synthesis is stimulated in bean cell suspensions by treatment with elicitors from cell wall of *Colletotrichum lagenarium* (Hedrick et al., 1988). A chimeric gene was constructed and introduced into bean protoplasts by electroporation. The chimeric gene was composed of a translational fusion between 1.7 kb of 5′ flanking sequences from the chitinase 5B gene and the coding region of the reporter gene GUS, linked to the 3′ untranslated region of nopaline synthase. The chitinase/GUS fusion gene was incorporated into the binary vector pBI101.2 and the resultant binary plasmid was introduced into tobacco plants by *Agrobacterium tumefaciens*-mediated transformation. An analysis of transgenic plants harboring this construct revealed that the GUS is induced when plants are treated with ethylene. Thus, the 5′ flanking region of the chitinase gene alone seems to be sufficient to confer ethylene-regulated expression of the reported gene (Broglie et al., 1989). Analyses of 5′ deletion mutants in the bean protoplast system electroporated with the chitinase gene indicated that DNA sequences located between positions –305 and –236 are important for the elicitor induction of the reporter GUS gene (Roby et al., 1991).

A gene, designated RCH 10, encoding a basic endochitinase, has been isolated from rice. The RCH 10 transcript rapidly accumulates in rice cell suspensions in response to fungal elicitor (Zhu and Lamb, 1991). Treatment of rice cell suspension cultures with a glycan elicitor fraction heat-released from mycelial cell walls of *Phytophthora megasperma* f.sp. *glycinea* causes a marked accumulation of RCH 10 transcripts. The RCH 10 transcripts also accumulate from low basal levels in rice leaf tissue following excision wounding, and that treatment of the excised leaf tissues with fungal elicitor causes a marked induction 24 h after excision (Zhu et al., 1993).

A 1.5 kb promoter fragment from the rice RCH 10 gene that encodes a basic endochitinase inducible by wounding and fungal elicitor was translationally fused to the β-glucuronidase (GUS) reporter gene and transferred to tobacco by *A. tumefaciens*-mediated leaf disc transformation. Wounding of leaves induced GUS activity from low basal levels and addition of fungal elicitor to the wounded tissue caused a further marked activation of the gene fusion (Zhu et al., 1993).

To localize *cis*-elements that specify the stress induction of the RCH 10 promoter, the expression of gene fusions with the promoter 5′ deleted to –326 (BZ 14), –160 (BZ 10), –74 (BZ 74), and –30 (BZ 30) was analyzed. The BZ 14 (–326) and BZ 10 (–160) transformants exhibited wounding and elicitor induction of GUS activity from low basal levels in leaf tissue, with similarfold inductions over basal levels and similar absolute levels of GUS activity in induced tissue as observed in BZ 4 plants containing the full promoter to –1512. Likewise, the kinetics for wounding and elicitor induction of the –326 and –160 promoter deletions were very similar to those observed with the full promoter. BZ 74 (–74) transformants exhibited wounding and elicitor induction of GUS activity from low basal levels in leaf tissue. The results indicate that the presence of a *cis*-element 3′ of –74 is sufficient for stress induction.

To determine whether the proximal promoter region 3′ of –74 involved in stress induction in transgenic tobacco plants also contributed to RCH 10 expression in rice, the expression of RCH 10-GUS gene fusions in transfected rice protoplasts was examined. 5′ deletion to –74 enhanced the level of GUS activity twofold compared with the full length promoter (Zhu et al., 1993). Thus sequences 3′ of –74 were sufficient for strong expression in the protoplasts. It was concluded that sequences of 3′ of –74 are involved in the expression of RCH 10 gene under stress conditions in rice as well as tobacco.

In suspension-cultured tobacco cells, the class I and class II chitinase genes were highly induced by fungal elicitor. To identify the *cis*-regulatory elements involved in the expression of the class I chitinase genes in response to fungal elicitor, the functional analysis of transgenic tobacco calli was performed. The region between –788 and –345 in the 5′-upstream region of the class I chitinase gene was sufficient for elicitor responsive expression. Gel mobility shift assays revealed that tobacco nuclear protein (s) bound specifically to the region between –539 to –518. In addition, the binding activity was present only in the nuclear extracts prepared from elicitor treated tobacco cells, suggesting nuclear factor(s) that interact with region between –539 to –518 are synthesised de novo or are converted to an active binding form during the process of the elicitor signal transduction (Fukuda and Shinshi, 1994). The region between –539 and –518 contains a direct repeat, –533 *GTCAG*AAA*GTCAG* –521. Methylation interference experiments revealed that the repeated sequences are core sequences for the binding of nuclear proteins (s), suggesting that this motif is a candidate for an elicitor-responsive element (Fukuda and Shinshi, 1994).

The 5′ region of the wound-inducible gene *wun* 1 full-length promoter comprised 1201 bp. Several 5′ deletion mutants were fused to GUS and tested in transgenic tobacco plants. Sequences up to –111, including TATA and CAAT box, showed onlly background expression level in transgenic leaves that could not be further stimulated by wounding. The addition of upstream sequences (–571) demonstrated a fourfold higher but still very low GUS expression in nonwounded and in wounded leaves. The addition of further 451 bp (–1022) enhanced the overall expression of 13–370-fold (Siebertz et al., 1989).

Functional analysis of 5′ deletion of potato PR-related gene STH-2 construct revealed an upstream regulatory sequence between –135 and –52 that contains a TGAC motif, and a possible negative regulatory region between –52 and –28. A factor present in nuclear extracts of wounded potato tubers was found to bind specifically to nucleotides located between –135 to –105, suggesting that this region contains important *cis*-regulatory elements (Matton et al., 1993).

Asparagus officinalis intracellular PR1 (AoPR1) gene is expressed in response to wounding and pathogen attack. The AoPR1 putative promoter fragment included DNA sequences coding for the first 31 amino acids of the predicted AoPR1 protein. A probable transcriptional start site 59 bp 5′ to the predicted translational start site has been shown. A putative TATA box with the sequence TATAAAA was present 29 bp upstream of the transcriptional start site. The AoPR1 promoter contains a putative TATA box and 8 bp boxes. A nucleotide motif is also present in the AoPR1 regulatory sequences. These sequences and distances suggest that the longest contiguous stretch of upstream DNA (1132 bp) fragment may contain an active promoter region (Warner et al., 1993).

The ability of the 1132 bp AoPR1 DNA sequence to initiate transcription was evaluated by creating a fusion with a GUS reporter gene using a vector. Transgenic tobacco plants harboring this construct were generated. Two independently transformed plants with a constitutively expressed GUS gene under the control of CaMV 35S promoter were also obtained. Unwounded leaf tissue from AoPR1-GUS transformants showed very little reporter gene expression in contrast to CaMV 35S-GUS plants. In all the AoPR1-GUS plants, GUS activity was induced rapidly and persistently following wounding of leaf tissue demonstrating that the AoPR1 promoter was sufficient to drive wound-inducible expression in transgenic plants. The AoPR1 promoter was responsive to local wound signals and AoPR1 transcript induction was localized to the wound site in asparagus seedlings. The AoPR1 promoter was also responsive to fungal pathogen attack. Infection of 4-week-old AoPR1-GUS tobacco plants with *Botrytis cinerea* resulted in the formation of necrotic lesions. Following histochemical localization of GUS activity, the majority of stained tissue was at the very edges of the lesions and in the vascular tissue near the lesions (Warner et al., 1993). In parsley, following attempted infection by *Phytophthora megasperma*, the parsley PR1 (a homologue of AoPR1) transcript was likewise detectable at high levels at sites of attempted invasion (Schmelzer et al., 1989). In the parsley PR1 promoter, 11 bp motifs have been identified (Meier et al., 1991).

C. PR Proteins are Not Synthesized by Cleavage of Constitutive Proteins

It is now well established that PR proteins are synthesized to a large extent due to stress in plants. Two possibilities may exist. They may be produced by proteolytic cleavage of constitutive proteins that are normally present in healthy leaves or they may be synthesized as large precursors. The PR proteins from tobacco are resistant to trypsin and chymotrypsin as well as to endogenous tobacco proteases. Protease activity strongly increases in hypersensitively reacting tobacco and hence it is possible that the PR proteins may be the stable end-products of proteolysis (Jamet et al., 1985; Van Huijsduijnen et al., 1985). The possibility that PR proteins are produced by cleavage of constitutive proteins was investigated by Jamet et al., (1985). Three days after inoculation with tobacco mosaic virus, tobacco leaves carrying local lesions were each injected with a solution containing 40 μci[^{14}C]leucine and 20 μci[^{14}C]aspartic acid. After 3 h of incubation the leaves were harvested and the proteins were analyzed (experiment A) or reinjected with a solution containing unlabeled leucine and aspartic acid. After a further 16 h (experiment B) or 40 h (experiment C), the leaves were harvested and analyzed. The data (Table 3; Jamet et al., 1985) revealed that the specific radioactivities were lower in experiment B than in A and lower in C than B. This progressive decrease in specific radioactivity of the PR proteins indicates that they were produced by de novo synthesis from a pool of radioactive amino acids, which is later progressively diluted by unlabeled amino acids and not by proteolytic cleavage of constitutive proteins. If it is due to cleavage of constitutive proteins, the 16 or 40 h incubation with unlabeled amino acids should not have diluted the radioactivity (Jamet et al., 1985).

This finding is in contrast to the results obtained by Pierpoint (1983), who observed that PR proteins do not incorporate label when radioactive amino acids are fed to the leaves of hypersensitively reacting tobacco plants. However, Pierpoint (1983) used detached leaves while Jamet et al. (1985) used intact leaves attached to plants. The tissues near the cut end of petiole may act as a sink for

Table 3 Radioactivity of PR proteins labeled by injecting mixtures of [^{14}C]amino acids into infected tobacco leaves

	Experiment		
Experimental Conditions	A	B	C
Period with label (h)	3.0	3.5	3.5
Incubation with nonlabeled amino acids	0.0	16.0	40.0
Results: PR proteins			
1a	1.50	1.10	0.84
1b	0.67	0.62	0.54
1c	1.00	0.55	0.45
2	3.10	2.10	1.30

Source: Jamet et al., 1985.

labeled metabolites (Pratt and Matthews, 1971), so that only a small proportion of that which is fed actually reaches the mesophyll tissues (Jamet et al., 1985).

D. PR Proteins are Synthesized as Larger Precursors

PR proteins may be synthesized as larger precursors. When polyA$^+$ RNA from tobacco leaves treated with potassium salicylate were translated in the rabbit reticulocyte lysate, two translation products were observed and they were immunologically identical to PR1a. However, in the sodium dodecyl sulfate (SDS)–polyacrylamide gel electrophoresis, these two products showed slightly lower mobilities than the radioactive PR1 group produced in tobacco leaf discs. The two translation products were further subjected to two-dimensional gel electrophoresis. These spots separated into at least four bands in the gel electrophoresis, all showing higher isoelectric points. These results suggest that PR1 proteins are synthesized as larger precursors and their shift to higher isoelectric points reflects the involvement of a signal peptide that generally has a basic charge. The in vitro products were translated in the presence of dog pancreatic microsome membrane. The in vitro products were cleaved by the membrane to the same size as mature PR1 proteins. These results clearly show that PR1 proteins are synthesized as larger precursors containing signal peptides (Matsuoka et al., 1988).

Carr et al. (1985) also provided evidences that PR1 proteins are synthesized as large precursors. The PR1 synthesized in vivo in Xanthi-nc tobacco leaves migrated faster in the gel electrophoresis than the in vitro translation product made in the wheat germ cell-free system, suggesting a difference in size of approximately 1 kDa. Van Huijsduijnen et al. (1985) showed that PR proteins are synthesized as large precursors containing signal peptides composed of 30 amino acid residues. The molecular weight of precursors of PR1a and PR1b were calculated to be 15,300 and 15,100, respectively.

Both tobacco and bean β-1,3-glucanases are postulated to be synthesized as larger precursors with an N-terminal signal peptide (Shinshi et al., 1988; Vogeli et al., 1988). Pea β-1,3-glucanase is also likely to be synthesized as a larger precursor with an N-terminal signal peptide (Chang et al., 1982). The complete nucleotide sequence of a bean chitinase gene has been determined. The structural gene contained a single open reading frame comprised of 981 bp that encodes the 301 amino acids of the mature protein and a 26 amino acid signal peptide (Broglie et al., 1989).

Tomato proteinase inhibitor I cDNA encodes a precursor of 111 amino acids that contains an NH_2-terminal propeptide that is removed before maturation of the protein. The propeptide may be 19 amino acids in length (Graham et al., 1985).

E. Processing of Mature Protein

The processing of PR protein in tobacco was studied in detail in respect of β-1,3-glucanase. β-1,3-Glucanase is produced as a 359-residue preproenzyme with an N-terminal hydrophobic signal peptide of 21 residues and a C-terminal extension of 22 residues containing a putative N-glycosylation site. The prepro-β-1,3-glucanase is processed sequentially. First, the N-terminal signal peptide is removed and an oligosaccharide side chain is added. Later, the oligosaccharide and C-terminal extension are removed to give the mature enzyme. The distinctive feature of tobacco β-1,3-glucanase processing is that the proenzyme undergoes a loss of an oligosaccharide side chain and the C-terminal extension. Mature β-1,3-glucanase does not contain N-acetyl glucosamine, which is a constituent of plant N-glycans. This observation and the fact that the only putative N-glycosylation site in the molecules is in the C-terminal extension suggest that processing to the mature form results from the loss of an N-glycopeptide. Processing is important in the intracellular transport and secretion of plant proteins (Shinshi et al., 1988).

Pea β-1,3-glucanase is processed from a larger precursor with an N-terminal signal peptide that is cleaved between Ala-1 and Glu-2. There is a potential intron 3′-splice acceptor site located at just 5′ of Ala-1 (Chang et al., 1992). Pea β-1,3-glucanase is further processed by cleavage of a glycosylated 21 amino acid peptide from its C-terminus (Chang et al., 1992). The enzyme has a single putative glycosylation site, Asn-Xaa-Ser-Thr, starting at Asn-330 in the C-terminal extension, suggesting that processing of the pea β-1,3-glucanase may involve glycosylation. There is a glycine adjacent to a phenylalanine at the C-terminus and processing may involve cleaving its glycosylated C-terminus (Chang et al., 1992). Processing of pea β-1,3-glucanase is similar to that of tobacco basic β-1,3-glucanase. The location of the 3′-splice acceptor site at 5′ of Ala-1 is similar in both the β-1,3,glucanases (Castresana et al., 1990; Ohme-Takagi and Shinshi, 1990). Pea β-1,3-glucanase has significant amino acid homology with the C-terminal regions of tobacco β-1,3-glucanase (Chang et al., 1992). The presence of a glycine adjacent to a phenylalanine at the C-terminus is also common to both the glucanases (Chang et al., 1992).

Tomato proteinase inhibitor I processing shows three domains. These domains are the signal sequences, which spans amino acids 1 through 23, the propeptide spans residues 24 through 42, and the mature protein that begins at residue 43 (Osteryoung et al., 1992). Translation of the inhibitor I precursor results in cleavage of the signal sequence. The signal sequence is cleaved between Ala 23 and Arg 24 of the precursor protein (Osteryoung et al., 1992). The propeptide is retained following cotranslational removal of the signal sequence. Proteolytic process of the inhibitor I propeptide occurs resulting in mature protein (Osteryoung et al., 1992).

VII. SECRETION OF PR PROTEINS

A. Secretory Pathways

PR proteins accumulate in the apoplast (Bol et al., 1990) and vacuole (Keefe et al., 1990). They are synthesized principally in membrane-bound polysomes and mRNAs for PR1 proteins have been found to be associated with membrane-bound polysomes (Carr et al., 1985). It suggests that the PR proteins are secreted from the membranes. The secretory system of plant cells delivers proteins to vacuole, tonoplast, plasma membrane, and the cell wall/extracellular space. In addition, proteins that enter the secretory system may be retained in the endoplastic reticulum (ER) or in various compartments of the Golgi complex. The first step common to the transport of all these proteins is translocation across the ER membrane (Walter and Lingappa, 1986). Once the protein is inside the lumen of the ER, transport depends on two types of information: informational domains that contain specific targeting or retention information and transport competence. Proteins that lack transport competence may be broken down in the secretory system (Pelham, 1989; Klausner and Sitia, 1990). Proteins that have entered the ER and that lack targeting or retention information are secreted via the bulk-flow or default pathway. When chimeric constructs of genes encoding various foreign proteins with the nucleotide sequence for a signal peptide are expressed in plant cells, the resulting proteins are secreted (Lund et al., 1989).

The nature of the bulk flow or default pathway followed by proteins that enter the secretory system by virtue of the presence of a signal peptide suggests that signal peptide of the protein is both necessary and sufficient for efficient secretion (Hunt and Chrispeels, 1991). Entry into the secretory pathway is accompanied by glycosylation in Golgi apparatus (Hunt and Chrispeels, 1991). Glycans help to stabilize protein conformation and protect proteins against breakdown (Faye and Chrispeels, 1989). The glycosylated protein may have a somewhat greater transport competence and stability (Hunt and Chrispeels, 1991). However, glycosylation of chitinase is not required for its efficient secretion from plant cells (Lund and Dunsmuir, 1992). Nonglycosylated forms of the protein are secreted as efficiently as are the glycosylated forms (Sidman, 1981; Kukuruzinska et al., 1987).

B. Targeting of PR Proteins Toward Vacuoles

Vacuoles are part of the endomembrane system that includes the ER, the Golgi, the trans-Golgi network and transitional vesicles mediating transport between the multiple compartments of the endomembrane system (Schroeder et al., 1993). The plant vacuole is a multifunctional organelle essential for numerous functions vital to cellular homeostasis and important for regulatory functions in plant cell growth and development (Boller and Wiemken, 1986). Most of the soluble and membrane-bound secretory proteins enter the secretory pathway by cotranslational insertion into the ER compartment. From there, secretory proteins traverse the Golgi and trans-Golgi network and are transported to the cell surface via the constitutive pathway (Chrispeels, 1991). The proteins targeted towards vacuoles follow the same secretory pathway until they are separated from secretory proteins, probably in the trans-Golgi network, for specific transport to the vacuole. Vacuolar proteins should be sorted from secretory proteins at the branch point of the transport pathway by a sorting machinery that recognizes a sorting signal in the vacuolar protein (Nakamura and Matsuoka, 1993). Some plant proteins contain the vacuolar targeting information either in a C-terminal propeptide, as the basic chitinases from tobacco (Neuhaus et al., 1991b), or in an N-terminal propeptide, as sporamin (Matsuoka and Nakamura, 1991). Many other vacuolar proteins are synthesized without a cleavage propeptide, indicating that the vacuolar targeting information is contained within the mature protein (Chrispeels and Raikhel, 1992). No amino acid sequence homology is found between C-terminal (Bednarek et al., 1990; Neuhaus et al., 1991b) and N-terminal vacuolar sorting signals (Matsuoka and Nakamura, 1991; Holwerda et al., 1992).

A precursor to the basic chitinase of tobacco that is deposited in the vacuole contains a C-terminal extension of 7 amino acids (Nakamura and Matsuoka, 1993). The C-terminal extension is absent in mature proteins as well as in the homologous precursor to the acidic chitinase deposited in the cell wall.

Deletion of this C-terminal extension missorts chitinase, causing it to be secreted into the intercellular space, and fusion of the C-terminal extension to the C terminus of the cucumber secretory form of chitinase through a 3-amino acid linker redirects the fusion protein to the vacuole (Neuhaus et al., 1991b). It suggests that C-terminal extension of tobacco chitinase is necessary and sufficient for vacuolar sorting (Neuhaus et al., 1991b).

Barley lectin is a defense-related protein (Murdock et al., 1990). Barley lectin is synthesized as a preprotein with a N-terminal signal sequence that is cleaved cotranslationally upon translocation in the ER. The preprotein is further modified by glycosylation in the C-terminal propeptide, and during transport or after deposition into the vacuole the glycosylated C-terminal propeptide is removed from the protein. Deletion of C-terminal propeptide leads to secretion of barley lectin (Bednarek and Raikhel, 1991). When the 15 amino acid C-terminal propeptide is fused to the extracellular protein, cucumber chitinase redirects this fusion protein to the vacuole of transgenic tobacco plants (Bednarek and Raikhel, 1991). Hence it was concluded that C-terminal propeptide is necessary to direct barley lectin to the plant cell vacuole (Bednarek and Raikhel, 1991).

Sporamin, a PR protein of sweet potato (*Ipomoea batatas*), is synthesized as a preprotein (Hattori et al., 1985; Nakamura, 1992). The N-terminal region of the sporamin precursor contains a signal sequence for translocation across the ER membrane, followed by an N-terminal propeptide 16 amino acids long. A deletion of N-terminal propeptide from the precursor leads to secretion but the transport to the vacuole is absent. Hence it is proved that N-terminal propeptide is essential for the transport of sporamin to the vacuole (Matsuoka and Nakamura, 1991). When sporamin is expressed in transgenic tobacco plants, the precursor of sporamin is sequentially processed both cotranslationally and posttranslationally and targeted to the vacuole (Matsuoka et al., 1990).

Both barley lectin and sweet potato sporamin were expressed simultaneously in transgenic tobacco plants. Both proteins were targeted quantitatively to the vacuole, indicating that carboxy-terminal and amino-terminal properties are equally recognized by the vacuolar protein-sorting machinery. Both the proteins accumulate in the same vacuole of transgenic tobacco leaf (Schroeder et al., 1993).

The role of C-terminal extension in vacuolar targeting of β-1,3-glucanase has also been reported. After removal of the N-terminal signal peptide, the C-terminal sequence of 22 residues of the basic tobacco β-1,3-glucanase becomes glycosylated and is subsequently cleaved off during vacuolar targeting of the protein (Shinshi et al., 1988). An extracellular acidic β-1,3-glucanase of tobacco lacks this C-terminal extension (Linthorst et al., 1990). A terminal extension with a potential glycosylation site, which is involved in vacuolar targeting, is also present in the C-terminal part of the precursor protein of the basic 35 kDa β-1,3-glucanase of tomato (Van Kan et al., 1992).

VIII. SIGNALS FOR TRANSCRIPTIONAL INDUCTION OF PR PROTEINS

A. Salicylic Acid

Microbial attack or elicitor treatment induces transcription of a battery of genes encoding PR proteins (Lamb et al., 1989; Dixon and Harrison, 1990; Warner et al., 1992) as part of a massive switch in the overall pattern of gene expression (Cramer et al., 1985; Dixon and Harrison, 1990). Wounds or any chemical or environmental stress induces PR proteins (Espelie et al., 1986; Logemann et al., 1988; Logemann and Schell, 1989; Carr and Klessig, 1989). PR proteins are also expressed in plants during normal development in the absence of an applied stress (Zhu et al., 1993).

Several endogenous signals have been identified in plants that are responsible for the induction of PR proteins (Enyedi et al., 1992). Salicylic acid has been identified as a possible endogenous signal molecule (Yalpani et al., 1991). Salicylic acid may be bound by a receptor and the binding may trigger a signal transduction cascade that has an ultimate effect on transcription factors that regulate PR protein expression. A salicylic acid-binding activity has been found in tobacco (Chen and Klessig, 1991).

Salicylic acid induced mRNAs that encode nine different PR proteins in tobacco. They are acidic form of PR1, PR2, PR3, PR4, PR5, basic form of PR1, basic class III chitinase structurally unrelated to PR3, acidic class III chitinase, and PR-Q′. Salicylic acid treatment induces expression of the same set of genes as TMV infection (Ward et al., 1991). When Xanthi-nc tobacco plants were incubated at 32°C instead of 24°C, salicylic acid accumulation was inhibited and it was accompanied by the disappearance of PR proteins (Yalpani et al., 1991).

Malamy et al. (1990) monitored endogenous levels of salicylic acid and PR-1 mRNA in TMV-inoculated resistant Xanthi-nc (NN) and susceptible Xanthi (nn) varieties of tobacco. Salicylic acid levels in resistant, but not in susceptible, tobacco increased almost 50-fold in the TMV-inoculated leaves. In contrast to TMV, salicylic acid was effective in inducing PR proteins in the susceptible Xanthi (nn) tobacco. The results suggest that Xanthi (nn) plants carry the PR genes, and these genes can be activated by treatment with salicylic acid but not with TMV. It is likely that infection of susceptible plants fails to trigger the signal transduction pathway that leads to salicylic acid production, and PR gene expression. Salicylic acid induction of PR-1a depends on 661 bp of the PR-1a promoter (Uknes et al., 1993a). Salicylic acid induced β-1,3-glucanases in tobacco (Mauch and Staehelin, 1989; Memelink et al., 1990; Ohme-Takagi and Shinshi, 1990) and glycine-rich proteins in tobacco (Kan et al., 1988).

Pseudomonas putida produces an enzyme, salicylate hyphoxylase, that degrades salicylic acid to catechol (Lawton et al., 1994b). Transgenic tobacco plants, engineered to express salicylate hydroxylase (NahG) gene from *P. putida*, could not accumulate salicylic acid following pathogen infection (Gaffney et al., 1993). Transgenic *Arabidopsis* plants expressing NahG did not accumulate mRNAs encoding PR proteins, suggesting the importance of salicylic acid in induction of PR proteins (Lawton et al., 1994b).

When lower leaves of tobacco plants are inoculated with tobacco mosaic virus (TMV), systemic resistance against fungal and viral pathogens is induced. Grafting experiments were conducted to assess the role of salicylic acid in systemic signal transduction (Vernooij et al., 1994). Nontransgenic Xanthi-nc tobacco plants and plants from transgenic lines (NahG) that express salicylate hydroxylase at high levels were used for grafting. Four- to 6-week-old scions were grafted onto 6-week-old root stocks. The root stock leaves were inoculated with TMV. The scion leaves were challenge-inoculated with TMV or *Cercospora nicotianae*. The transgenic tobacco rootstocks, although unable to accumulate salicylic acid, were fully capable of delivering a signal that renders nontransgenic scions resistant to further pathogen inoculation. This result indicates that the translocating systemic resistance-inducing signal is not salicylic acid, but resistance could not be induced in scions of NahG/NahG grafted plants by TMV pretreatment of root stocks. Thus NahG rootstocks were capable of producing a systemic signal that induces resistance to a range of pathogens, whereas NahG scion tissues are unable to respond (Vernooij et al., 1994). PR-1, PR-2, and PR-3 genes were strongly expressed in TMV-infected leaves of Xanthi and NahG rootstock. In scion tissue of TMV-pretreated, grafted plants PR-1, PR-2, and PR-3 genes were strongly expressed in Xanthi tissue, irrespective of root stock genotype (Xanthi or NahG). Thus, these PR proteins were only induced in scion tissue showing resistance (Vernooij et al., 1994). The results suggest that the signal requires the presence of salicylic acid in tissues distant from the infection site to induce systemic resistance, but the translocating systemic resistance-inducing signal is not salicylic acid.

B. Ethylene

Ethylene is known as a plant stress hormone. Increased biosynthesis of ethylene has been shown to accompany pathogen attack. Silver nitrate-induced ethylene production leading to accumulation of PR proteins in *Gynura* plants (Belles et al., 1991). Treatment of plants with ethylene results in an increased synthesis of β-1,3-glucanases and chitinases (Roby et al., 1986, 1991; Felix and Meins, 1987; Ecker and Davis, 1987; Boller, 1988; Keefe et al., 1990; Linthorst et al., 1990a; Ishige et al., 1991). Treatment with 1 mM $CoCl_2$, the inhibitor of ethylene production, or with 1.6 mmol/ml norbornadiene in air, the inhibitor of ethylene action, inhibited β-1,3-glucanase accumulation by tobacco callus tissues. The inhibitory effcts were prevented by exogenous ethylene. These observations indicate that ethylene induces β-1,3-glucanase accumulation (Felix and Meins, 1987).

Ozone treatment induces β-1,3-glucanase and chitinase in tobacco (Ernst et al., 1992). Large amounts of ethylene are produced by the ozone-sensitive tobacco cultivar BelW3 after a short ozone exposure, whereas the tolerant cultivar BelB shows no significant induction (Langebartels et al., 1991). Treatment of the ozone-sensitive cultivar BelW3 with ozone markedly increased the mRNA level of basic β-1,3-glucanase, which occurred within 1 h with a transient maximum. Seventeen hours after ozone treatment, the β-1,3-glucanase mRNA level decreased to lower values. The increase of basic chitinase mRNA level was delayed and was less pronounced than that of β-1,3-glucanase. Cultivar BelB, which is tolerant to ozone, showed only a small increase of β-1,3-glucanase mRNA level after the ozone treatment, but its basic chitinase mRNA was strongly induced. The results suggest that the two varieties show a differential accumulation of PR mRNAs upon single ozone pulse (Ernst et al., 1992). Prolonged ozone treatment for 2 days of tobacco BelW3 led to a persistent level of basic β-1,3-glucanase and basic chitinase as well as to an increase of acidic chitinase and PR-1b mRNA levels. The results suggest the importance of ethylene in transcriptional induction of chitinase and β-1,3-glucanase (Ernst et al., 1992). However, Mauch et al. (1984) showed that ethylene may not be a signal for induction of chitinase and β-1,3-glucanase in pea since aminoethoxyvinylglycine, a specific inhibitor of ethylene biosynthesis, did not suppress the synthesis of both chitinase and β-1,3-glucanase.

The induction of acidic type PR proteins may occur through at least two separate signal transduction pathways in tobacco. Induction of PR proteins in tobacco has been shown to be mediated by ethylene (Van Loon, 1985). One induction pathway may be ethylene-dependent in its mode of action as the accumulation of acidic type PR-1 protein is inhibited in the presence of inhibitors of ethylene biosynthesis or action (Felix and Meins, 1987; Vogeli et al., 1988; Lotan and Fluhr, 1990a). The other induction pathway does not require ethylene and is activated by the elicitor xylanase (Lotan and Fluhr, 1990a; Dean and Anderson, 1991). The ethylene-dependent induction was further characterized as requiring light, while ethylene-independent induction did not (Eyal et al., 1992).

The basic PR-1 transcript (prb-1) is readily induced by ethylene in tobacco. In contrast, the acidic-type PR-1 transcript level did not respond directly to ethylene (Eyal et al., 1992) although it was induced with α-aminobutyric acid that depends on ethylene evolution for elicitation (Lotan and Fluhr, 1990b). It suggests that ethylene is necessary but not sufficient for induction of acidic type PR-1 expression. There may be a dual signal requirement for induction of acidic PR-1 via ethylene requirement. One component required is ethylene, as inhibitors of ethylene biosynthesis and action obstruct this induction pathway. The nature of the second component is unknown (Eyal et al., 1992).

Differential effect of light on induction of acidic type and basic-type PR proteins via the ethylene dependent signal transduction pathway in tobacco was studied (Eyal et al., 1992). The elicitor, α-aminobutyric acid, via ethylene did not induce the basic-type prb-1 transcript in the dark. The application of α-aminobutyric acid likewise did not induce accumulation of acidic-type PR-1 transcript in the dark, although it was an efficient inducer in the light. The ethylene-independent elicitor xylanase enhanced the level of basic-type prb-1 transcript in the dark. Xylanase, in contrast to α-aminobutyric acid, was capable of inducing transcript accumulation of acidic-type PR-1 in the dark. To show unequivocally that ethylene-mediated induction requires light, plants were treated with ethylene in the dark and in the light. Accumulation of prb-1 transcript after 6 h in the light was fivefold more than that obtained in the dark. The application of ethylene to leaves in the dark did not significantly enhance dark-induced accumulation of prb-1 transcript (Eyal et al., 1992). The light-dependent responsiveness of prb-1 to ethylene leads us to conclude that the light-requiring step occurs somewhere in the ethylene component of the signal transduction pathway (Eyal et al., 1992).

Ethylene induces PR-1, PR-2, and PR-5 proteins in *Arabidopsis* (Lawton et al., 1994b). Ten PR proteins have been detected in tomato. Nine of them were induced by ethylene while the PR protein C_{10} was not induced by ethylene. A different pathway may exist for C_{10} (Granell et al., 1987). Ethylene decreases ribulose-1,5-bis-phosphate carboxylase/oxygenase (Rubisco) and other soluble proteins while it increases synthesis of PR proteins in tomato. The increased rate of degradation of Rubisco, as well as many other proteins, caused by ethylene may be due to the increased activity of PR-protein P69 (Vera and Conejero, 1990).

There are reports that ethylene may not be responsible for PR proteins in tomato. Silver ions elicit ethylene biosynthesis in tomato plants and induce PR proteins (Conejero et al., 1990). The induction of PR proteins is prevented by specific inhibitors of ethylene biosynthesis such as aminooxyacetic acid and inhibitors of ethylene action like norbornadiene. *Epinastic* is a single-gene tomato mutant derived from a normal tomato variety. The mutant is characterized by constitutive overproduction of ethylene (Fujino et al., 1988). However, no PR proteins are synthesized in the mutant in response to the constitutively elevated ethylene synthesis (Belles et al., 1992). Hence ethylene may not be involved in PR protein synthesis in some cases (Boller, 1990).

C. Systemin

Systemin, a polypeptide detected in tomato leaves, induced synthesis of proteinase inhibitor I and II (Pearce et al., 1991). Cloning and characterization of the cDNA and gene encoding systemin have also been carried out (McGurl et al., 1992). The systemin cDNA was isolated by screening a primary cDNA library synthesized from tomato leaf mRNA with an oligonucleotide corresponding to amino acids 12–18 of systemin, and a second oligonucleotide corresponding to amino acids 1–6 of systemin. One cDNA clone encoding the systemin polypeptide within a larger protein, prosystemin, was obtained. The complete prosystemin mRNA sequence was determined by sequencing the prosystemin gene. The open reading frame was 600 bp encoding a 200-amino acid prosystemin protein. Of the 200-amino acid prosystemin, amino acids 179–196 encode systemin. Prosystemin is encoded by a single gene that consists of 11 exons and 10 introns (McGurl and Ryan, 1992; McGurl et al., 1992).

Prosystemin gene homologs were found in potato also, but not in tobacco and alfalfa. An mRNA species that hybridized to the prosystemin mRNA was also identified in potato. Prosystemin gene itself was wound-inducible. Prosystemin mRNA accumulated in the upper, unwounded leaves of young tomato plants that had been wounded on the lower leaves, demonstrating that prosystemin mRNA, like Inhibitor I mRNA, is systemically wound-inducible. Although Inhibitor I mRNA was absent from the leaves of unwounded tomato plants, a small amount of prosystemin mRNA was detected there, which may provide a continuous supply of systemin and allow the plant to respond immediately to wounding (McGurl et al., 1992).

To determine if the prosystemin gene product is important in the signal transduction pathway leading to the expression of proteinase inhibitor genes in tomato leaves, tomato plants were transformed with a prosystemin antisense gene. In transformed plants with prosystemin antisense gene a complete suppression of the systemic wound induction of proteinase inhibitors was observed. It suggests that systemin is an integral component of the signal transduction system that regulates the synthesis of proteinase inhibitor proteins (McGurl et al., 1992).

D. Jasmonic Acid and Methyl Jasmonate

Jasmonic acid and its methyl ester have been detected in a large number of plant species (Yamane et al., 1991; Meyer et al., 1984; Anderson, 1985; Siedow, 1991; Staswick, 1992) and are involved in the plants' stress metabolism (Parthier, 1991). Jasmonic acid in plants has been shown to originate from linolenic acid (Vick and Zimmerman, 1984). Lipases may be synthesized or activated in response to wounding and linoleic acid released from membranes may initiate the intracellular transduction pathway leading to proteinase inhibitor synthesis (Farmer and Ryan, 1990; Ryan and Farmer, 1991). Free linolenic acid may be converted to jasmonic acid, which may be a signaling component that may be very close to the *trans*-elements that regulate proteinase gene transcription (Farmer et al., 1992). Methyl jasmonate induced accumulation of proteinase inhibitors in tomato (Farmer and Ryan, 1990) and potato (Kim et al., 1992). Jasmonic acid induced accumulation of several polypeptides (Herman et al., 1989; Becker and Apel, 1992) and a prepropolypeptide of thionins, a group of PR proteins (Bohlmann et al., 1988; Apel et al., 1990; Bohlmann and Apel, 1991; Andresen et al., 1992). Jasmonate-induced proteins (JIP) include ribosome-inactivating proteins (RIP) also (Reinbothe et al., 1994a,b; Chaudhry et al., 1994).

Tomato plants exposed to methyl jasmonate vapors contained high levels of mRNAs coding for both inhibitor I and II. These mRNAs were not present in leaves of untreated plants. Methyl jasmonate induced the synthesis of inhibitor I and II proteins and mRNA in a manner similar to wounding, which is known to regulate synthesis of the inhibitor mRNAs transcriptionally (Farmer et al., 1992). Further evidence to prove the transcriptional regulation of proteinase inhibitor synthesis by methyl jasmonate was obtained using transgenic tobacco plants. Transgenic tobacco containing a chimeric gene composed of a wound-inducible inhibitor II gene fused to the chloramphenicol acetyl transferase (CAT) structural gene and terminated with the potato inhibitor II gene 3′ region exhibits wound-inducible CAT activity (Thornburg et al., 1987). Exposure of young transgenic tobacco to methyl jasmonate vapor results in the induction of CAT gene expression in the leaves (Farmer et al., 1992).

Methyl jasmonate induced the accumulation of a proteinase inhibitor in alfalfa leaves that is a wound-inducible member of the Bowman-Birk inhibitor family (Farmer and Ryan, 1990). The induction of proteinase inhibitor by both wounding and methyl jasmonate results in increased alfalfa proteinase inhibitor mRNA levels (Farmer et al., 1992).

E. Peptide

Gianinazzi and Ahl (1983) developed an amphidiploid interspecific hybrid of *Nicotiana glutinosa* × *Nicotiana debneyi* that constitutively produces high levels of PR proteins. They showed that grafting a scion of uninfected *Nicotiana tabacum* cv. Xanthi nc onto a hybrid rootstock leads to the accumulation of PR proteins in the scion. The PR proteins of the hybrid were different from those of Xanthi nc, suggesting that no movement of PR proteins from the root stock occurred. The proteins may be synthesized in the scion leaf tissue following the translocation of an inducing signal from the root stock (Gianinanazzi and Ahl, 1983). Enzymically derived carbohydrate fragments of purified cell wall preparations from hypersensitively responding *Nicotiana* leaves induced resistance when injected into uninfected leaves (Modderman et al., 1985). The interstitial fluid extracts from hypersensitively responding plants that may contain cell wall degradation products induced resistance (Wieringa-Brants and Dekker, 1987).

Further studies were carried out by Gordon-Weeks et al. (1991) to isolate and purify endogenous inducers of PR protein accumulation. The concentrated intercellular fluid from the hybrid *N. glutinosa* × *N. debneyi* or from hypersensitively reacting tobacco necrosis virus-infected *N. debneyi* induced the synthesis of PR 1a and PR 1b in Xanthi nc leaves, whereas intercellular fluid from uninfected Xanthi nc did not. The endogenous inducers of PR proteins could be resolved into components by molecular sieve chromatography. The Peak I material was found to contain PR-1 inducing activity that bound to ConA affinity columns. The presence of ConA-binding activity suggests that some inducing substances are present that contain either glucose or mannose. The activities contained in Peak I were destroyed by sodium metaperiodate, suggesting that carbohydrate moieties are essential for the activity of the larger fraction. The major activity contained in Peak II material, on the other hand, was not absorbed onto ConA affinity columns. It was also not affected by sodium metaperiodate treatment. The compound was a low-molecular-weight substance and could not be inactivated by protease or by boiling. It was positively charged. The active substance may be a small peptide or an amino acid (Gordon-Weeks et al., 1991).

F. Synergistic Effect of Different Signals

Synergistic effect of ethylene and methyl jasmonate in induction of PR proteins has been reported. Ethylene caused osmotin (PR5 protein) mRNA accumulation in tobacco leaf tissues. After 24 h of ethylene treatment, 5-day-old seedlings containing an osmotin promoter-β-glucuronidase (GUS) fusion gene exhibited a fivefold increase in GUS activity. The osmotin promoter was not responsive to treatment with methyl jasmonate alone, but when methyl jasmonate was applied in combination with ethylene, the activity of the osmotin promoter was induced dramatically beyond that seen with either ethylene or methyl jasmonate in the seedlings (Xu et al., 1994). Salicylic acid substantially induced the accumulation of PR-1b protein in Wisconsin 38 tobacco. The combination of salicylic acid and methyl jasmonate induced the accumulation of PR-1b protein to accumulate severalfold more (Xu et al., 1994).

G. Induction of PR Proteins may Require Different Signal Transduction Systems

It is known that different signal transduction systems are involved in plant–microbe interactions. PR proteins may be induced by different signal transduction system rather than by a single transduction system. Ethylene induces PR-1 and PR-5 transcripts in tobacco (Raz and Fluhr, 1993; Xu et al., 1994). The protein kinase C inhibitor 1-(5-isoquinolinylsulfonyl)-2-methylpiperazine (H7) could inhibit the induction of PR-1 transcript by ethylene (Raz and Fluhr, 1993), but H7 was unable to inhibit the induction of PR-5 transcript accumulation by ethylene (Xu et al., 1994). Salicylic acid induced accumulation of PR-1b protein in tobacco, but it could not induce accumulation of PR-5 protein (Xu et al., 1994). These results suggest that two of the PR gene family subgroups, PR-1 and PR-5, are induced through separate signal transduction pathways (Xu et al., 1994).

In maize embryos infected with *Fusarium moniliforme*, a proteinase inhibitor protein (MPI) and another PR-protein (PRms) mRNAs accumulate (Cordero et al., 1994), but treatment with fungal elicitor resulted in an increase in the PRms mRNA level while accumulation of the MPI mRNA level was not observed. Treatment with abscisic acid or methyl jasmonate did not induce PRm mRNA synthesis, while these compounds were effective in induction of the accumulation of the MPI mRNA (Cordero et al., 1994). The results suggest that two distinct signal transduction pathways may exist in triggering induction of PR proteins in maize (Cordero et al., 1994).

Different signal system may affect spatial distribution of PR gene products. When tobacco plants were inoculated with tobacco mosaic virus, acidic PR-5 proteins were not induced outside the stimulated ring around the lesion, whereas others were clearly induced at larger distance. Acidic glucanases (PR-2) and basic chitinases (PR-3) accumulated at a higher level in uninoculated leaves of infected plants than in cells situated outside of the lesions on the inoculated leaf (Heitz et al., 1994). The results suggest that different signals may be involved in induction of different PR proteins.

IX. PR PROTEINS ARE EXPRESSED IN PLANTS DURING NORMAL DEVELOPMENT ALSO

PR proteins are induced in leaf tissues due to fungal, bacterial, viral, and viroid infection. They are also induced in leaves due to various other biotic and abiotic stresses. Under normal growth conditions, the genes coding for PR proteins are quiescent or are expressed at low levels in the leaf tissues (Sehgal et al., 1991). In contrast, PR proteins have been shown to express constitutively in other parts of plants such as roots, stem, latex, flowers, fruits, and seeds. The transcript of the basic-type PR-1 cDNA accumulated in virus-infected or ethephon treated leaves, but it was constituvely present in normal roots of tobacco (Memelink et al., 1987, 1990). PR proteins are known to be associated with flower induction (Neale et al., 1990; Richard et al., 1992) and some are found in the developing flowers (Lotan et al., 1989). In tobacco during normal floral development, a number of β-1,3-glucanase activities were observed in all flower tissues. However, PR2 polypeptides were observed only in sepal tissue. In contrast, a mRNA that hybridized to PR2 cDNA was present in stigma/style tissue and the sepals. Primer extension analysis indicated that the β-1,3-glucanase mRNA and enzyme activities observed in the stigma/style of flowers are attributable to the expression of another class of glucanase genes (Cote et al., 1991). The stylar protein in tobacco flowers has been identified as a glycosylated form of β-glucanase distinct from the forms that accumulated in leaves in response to stress (Ori et al., 1990). In pistil of mature flowers of *Nicotiana alata*, a proteinase inhibitor with partial identity to proteinase inhibitor II of potato and tomato has been isolated (Atkinson et al., 1993). Transcripts encoding a number of PR proteins including chitinase and β-1,3-glucanase have been shown to accumulate in floral organs during normal development (Lotan et al., 1989; Ori et al., 1990). In petunia, an active chitinase is localized in the stigmas of healthy, nonsenescent plants (Leung, 1992). A gene encoding a basic endochitinase has been isolated from rice. A 1.5 kb promoter fragment from the rice chitinase inducible by fungal elicitor was translationally fused to the GUS reporter gene and transferred to tobacco. The chitinase–GUS gene fusion exhibited a characteristic pattern of expression in floral organs. Only low levels of GUS activity were observed in sepals and petals, comparable with the levels in leaves from the same plants; relatively higher levels were found in anther, stigma and ovary extracts. Within anthers there was strong expression of the gene fusion in pollen (Zhu et al., 1993).

Fruit-specific (Margossian et al., 1988; Wingate and Ryan, 1991b) and leaf-specific (Lee et al., 1986) tomato inhibitor I genes have been isolated, but in *Lycopersicon peruvianum* the inhibitor I gene is both wound-inducibly expressed in leaves and developmentally expressed in unripe fruit (Wingate and Ryan, 1991a). Proteinase inhibitor I and inhibitor II proteins accumulate to high levels in unripe fruit of *L. peruvianum*. The levels of both inhibitor I and inhibitor II proteins diminish as the fruit ripens (Pearce et al., 1988). Seeds express PR proteins constitutively. Chitinases have been detected in barley, maize, wheat, rice, and soybean seeds (Roberts and Selitrennikoff, 1988; Leah et al., 1987; Zhu and Lamb, 1991). Inhibitors of the potato proteinase inhibitor I family have been found in barley grains (Svendsen et al., 1982) and in seeds of broadbean (*Vicia faba*) (Svendsen et al., 1984). Zeamatin, sormatin, avenatin, and trimatin have been detected in corn, sorghum, oats, and wheat seeds (Vigers et al., 1991). High levels of 1,3;1,4-β-glucanase activity have been reported in barley and rice seeds (Stuart et al., 1987; Litts et al., 1990; Slakeski et al., 1990). Potato proteinase inhibitor I is expressed in normal potato tubers (Melville and Ryan, 1972).

Constitutive expression of PR proteins in root tissues has been reported widely. β-1,3-Glucanase is constitutively expressed in roots of pea seedlings (Chang et al., 1992). Basic β-1,3-glucanases of tobacco are constitutively expressed in healthy plants in a gradient that increases from the top to bottom of the plant, with the highest expression occurring in the roots (Felix and Meins, 1986; Memelink et al., 1990; Castresana et al., 1990).

Rice chitinase-GUS gene fusion transferred to tobacco expresses during normal development in the absence of an applied stress. High levels of GUS were observed in roots and moderate levels in stems compared with the relatively weak expression in leaves (Zhu et al., 1993). Histochemical analysis showed strong expression of chitinase-GUS in juvenile tissue of apical root tips. In stem, GUS staining was localized to the epidermis and vascular system (Zhu et al., 1993).

The rice chitinase-GUS gene transferred to tobacco-induced GUS activity from low basal levels in tobacco leaves only by the addition of fungal elicitor (Zhu et al., 1993). To localize *cis*-elements that specify the complex developmental regulation in stems and roots and stress induction in tobacco leaves, the expression of the gene fusions with the promoter 5′ deleted to –326, –160, –74, and –30 was analyzed. Functional analysis of the 5′ deletions resulted in different effects on various facets of the overall program of the chitinase promoter activity. Deletion to –160 substantially reduced developmental expression in vegetative organs, but had little effect on the pattern or level of expression in floral organs or in wounded and elicitor-treated leaves. The region from –326 to –160 contains a positive element for development expression in vegetative organs that is not required for other models of expression. Likewise, floral expression is abolished by deletion to –74, indicating the presence of a positive *cis*-element for this mode of expression in the region of –160 to –74. The sequences of 3′ of –74 are sufficient for stress induction. The results suggest that presence of different *cis*-elements is necessary for floral, root, and leaf expression of the chitinase gene in tobacco (Zhu et al., 1993). Hence it is possible that flower and root PR proteins might be diverged subclasses of their leaf homologs (Zhu et al., 1993).

Constitutive expression of PR proteins in floral parts, roots, and epidermis, and vascular system in stems may contribute a permanent form of defense against possible threats from microbes (Memelink et al., 1990). Root tips and epidermal tissues of aerial organs are continually exposed to microbial attack. In addition, many pathogens can rapidly spread throughout the plant if they penetrate into the vascular system. The expression of the PR proteins in these vegetative tissues may represent an anticipatory rather than responsive defense mechanism, to enhance the protection afforded to such vulnerable tissues. Likewise, expression of the PR proteins in stigmas, ovaries, and pollen-containing anthers may anticipate microbial attack of highly exposed and vulnerable tissues or cells critical for the completion of the plant's life cycle (Zhu et al., 1993). It is known that infection of pistils is rare, although the stigmas provide a favorable environment for the growth of pests and pathogens (Kandasamy and Kristen, 1987).

Some PR proteins have been found to be present constitutively in latex of plants. Latex is the cytoplasm of highly specialized cells known as laticifers. The PR proteins are found in the vacuoles (lutoids) of the laticifers (Martin, 1991). Transcripts of PR proteins are expressed at 10–50-fold higher levels in laticifers than in leaves (Kush et al., 1990). Chitinases have been identified in the laticifers of numerous latex-containing species including monocotyledons, dicotyledons, and a gymnosperm (Howard and Glazer, 1967; Glazer et al., 1969; Martin and Gaynor, 1988; Lynn, 1989; Martin, 1991). In many of these species chitinases represent 2–30% of the total soluble protein in the latex (Martin and Gaynor, 1988; Martin, 1991). High levels of chitinases are found in the laticifers of *Hevea brasiliensis*, the rubber tree (Martin, 1991). One acidic and six basic chitinases have been purified from latex of the rubber tree (Martin, 1991). A basic chitinase has been purified from the latex of *H. brasiliensis* (Tata et al., 1983). The N-terminal sequence of that protein has been shown to have homology with an acidic chitinase from *Cucumis sativus* (Joosten and De Wit, 1989) and a chitinase from *Parthenocissus quinquifolia* (Bernasconi et al., 1987). It did not show homology with the basic chitinase from *Phaseolus vulgaris* (Broglie et al., 1986).

A 5 kDa basic protein, hevein, has been isolated from the latex of *H. brasiliensis* (Walujono et al., 1975). It showed a high degree of homology with the N-terminus of wheat germ agglutinin (Raikhel and Wilkins, 1987) and the N-terminus of a basic chitinase from *Phaseolus vulgaris* (Broglie et al., 1986). Hevein is suggested to be a proteolytic fragment of a chitinase (Martin, 1991).

X. ANTIFUNGAL ACTION OF PR PROTEINS

The PR proteins show antifungal activity that has been demonstrated by both in vitro (direct evidence) and in vivo studies (indirect evidence). The PR-protein fractions extracted from infected tomato leaves inhibited both discharge of zoospores from sporangia and the germination of encysted zoospores of *Phytophthora infestans* (Enkerli et al., 1993). The purified PR-1 protein did not show any effect on axenically growing *Chalara elegans* when tested at concentrations ranging from 10 to 300 μg (Tahiri-Alaoui et al., 1993). However, indirect evidence has shown the antifungal nature of PR-1 proteins. Transgenic plants expressing high levels of PR-1a show resistance against *Peronospora tabacina* (Alexander et al., 1992, 1993). PR-1 proteins were found to be associated with the host cell wall outgrowths and papillae produced by tobacco roots in response to *C. elegans* infection. These proteins may increase mechanical strength of these defense-related structures. The strengthening of hypodermal cell walls may be an early host defense strategy (Tahiri-Alaoui et al., 1993). Benhamou et al. (1991a) found that PR-P14 accumulated over papillae in tomato roots infected by *Fusarium oxysporum* f.sp. *radicis-lycopersici* and suggested that PR-P14 may restrict the development of the pathogen by host cell wall modifications.

When PR-P14 and PR-P70 were induced by indole acetic acid treatment, the tomato plants became resistant to *Phytophthora infestans* (Christ and Mosinger, 1989). UV-irradiation induced synthesis of PR proteins in tomato and it resulted in induction of resistance against *P. infestans* (Table 4; Christ and Mosinger, 1989).

Acetyl salicylic acid, kinetin, and abscissic acid treatments did not induce PR protein accumulation and also did not induce resistance in tomato (Christ and Mosinger, 1989).

The PR-2 protein, β-1,3-glucanase has also shown fungitoxicity in vitro. β-1,3-Glucanases (G1 and G2) isolated from pea pods inhibited *Fusarium solani* f.sp. *pisi* in vitro when they were tested at the concentrations of 250 μg/ml. In eggplant infected with *Verticillium albo-atrum*, a purified β-1,3-glucanase was used to study the localization of the enzyme in the infected plant tissues by means of postembedding immunogold labeling (Benhamou et al., 1989). In the infected

Table 4 Relationship between PR proteins and disease resistance in tomato

Treatment	Induction of PR Proteins	Area of necrotic spots (mm^2)
Control	—	46 ± 1
UV-irradiated	P14, P70, P31, P26, P21, P18, P13	6 ± 0

Source: Christ and Mosinger, 1989.

tissues, intense gold deposition occurred on the fungal cell surface. The diffusion of gold particles toward the inside of fungal cells was observed, indicating that the structural integrity of fungal cell walls was generally altered. It suggests that β-1,3-glucanase lyses the fungal cell wall (Benhamou et al., 1989).

Higher constitutive chitinase and β-1,3-glucanase activities were detected in pepper (*Capsicum annuum*) leaves resistant to *Phytophthora capsici* (Mozzetti et al., 1995). *Nicotiana glutinosa* is susceptible to *Cercospora nicotianae*, *Peronospora tabacina*, *Chalara elegans*, and *Phytophthora parasitica* var. *nicotianae*, while *N. debneyi* and the hybrid *N. glutinosa* × *N. debneyi* are resistant to all these diseases (Table 5; Ahl Goy et al., 1992).

Chitinase activity was constitutively expressed to an appreciable extent only in the hybrid, while β-1,3-glucanase was constitutively expressed in both *N. debneyi* and the hybrid, the plants that showed resistance to all the fungal pathogens (Ahl Goy et al., 1992), indicating a greater role for β-1,3-glucanase in the antifungal action.

The purified *Arabidopsis* chitinase effectively inhibited the growth of *Trichoderma reesei* even at as low as 0.5 μg/disc. Growth of several fungi, including *Alternaria solani*, *Fusarium oxysporum*, *Sclerotinia sclerotiorum*, *Gaeumannomyces graminis*, and *Phytophthora megasperma* was not inhibited, even

Table 5 Reaction of *Nicotiana glutinosa*, *N. debneyi*, and their hybrid to fungal pathogens

Plant	*Cercospora nicotianae* (% of infected leaf area)	*Peronospora tabacina* (% of infected leaf area)	*Chalara elegans* (% of infected leaf area)	*Phytophthora parasitica* var. *nicotianae* (% wilted plants)
Nicotiana glutinosa	20	53	92	100
Nicotiana debneyi	0	3	0	50
N. glutinosa × *N. debneyi*	0	0	0	50

Source: Ahl Goy et al., 1992.

when as much as 8 μg purified protein was applied per discs (Verburg and Huynh, 1991). Chitinases purified from thorn-apple, tobacco, and wheat were effective inhibitors of spore germination and hyphal growth of *Trichoderma hamatum* and *Phycomyces blakesleeanus*, but both spore germination and hyphal growth of *Botrytis cinerea* were not affected by these enzymes (Broekhaert et al., 1988). It is suggested that chitin content may vary from fungus to fungus. The ratio of N-acetyl glucosamine to glucose ranges from 2.1:1 for the cell walls of *Morchella crassiper* to a ratio of 1:5.5 for the cell walls of *Candida utilis* (Rokem et al., 1986).

Chitinase, purified to homogeneity from ethylene-treated bean leaves, showed antifungal activity against *Rhizoctonia solani*. The fungal hyphae were markedly reduced in size, and lysis of apical zone was seen within 3 h after treatment with the chitinase. Swelling of the hyphal tips and hyphal distortions were also observed. Disruption of chitin macromolecules preceded cell wall breakdown and protoplasm alteration. Chitin was the main structural component of *R. solani* cell walls (Benhamou et al., 1993).

Chitinase purified from bean inhibited growth of *Trichoderma viride*, the test fungus, even at 2 μg ml. When antibodies were raised against bean chitinase in a rabbit, the antiserum was found to block chitinase activity. It also blocked the inhibitory effect of chitinase on fungal growth (Schlumbaum et al., 1986). Bean seedlings were inoculated with *Uromyces phaseoli*, the rust pathogen. Chitinase was found to be induced about 10-fold in response to infection. Crude extracts from infected leaves produced larger inhibition zones than did extracts from uninfected control leaves, when *T. viride* was used as the test fungus. Adding the antiserum against bean chitinase prevented the formation of an inhibition zone (Table 6; Schlumbaum et al., 1986).

Toyodo et al. (1991) reported that haustoria of *Erysiphe graminis* were

Table 6 Inhibition of *Trichoderma viride* growth by extracts of bean leaves infected with *Uromyces phaseoli*

Treatment (mg/ml growth)	Protein (μg/ml)	Chitinase	Inhibition
Extract from uninfected leaves	7.0	3.0	+
	3.5	1.6	—
Extract from infected leaves	6.6	30	+++
	3.3	15	+++
	1.7	7.5	++
Extract from infected leaves and antiserum	6.6	< 0.05	—

Degrees of inhibition range from +++, strong inhibition to, ++, less inhibition to, —no inhibition.
Source: Schlumbaum et al., 1986.

effectively lysed in the barley epidermal cells by chitinase. Broglie et al. (1991) constructed both transgenic tobacco and *Brassica napa* plants that constitutively express a bean vacuolar chitinase gene under the control of the strong 35S promoter of the cauliflower mosaic virus. The transgenic tobacco plants showed resistance to *Rhizoctonia solani*, the soil-borne pathogen infecting numerous plant species. There was decreased symptom formation. The fungal growth was inhibited by isolated bean chitinase in vitro. In the case of *B. napas* also, significant reduction in fungal growth and delay in disease development were observed. Transgenic tobacco plants expressing chitinase genes have been reported by other workers also to afford protection against *R. solani* (Dunsmuir et al., 1991; Jach et al., 1992; Oppenheim and Chet, 1992; Vierheilig et al., 1993). Transgenic rice plants expressing a class I 35 kDa rice chitinase gene constitutively showed resistance to rice sheath blight pathogen *R. solani* (Lin et al., 1995).

Various biotic agents when inoculated on first leaf of cucumber, increased chitinase activity in first and second leaves. When the second leaf was challenge inoculated with *Colletotrichum lagenarium*, the disease intensity was less. The reduction in disease intensity was very much related to the relative increase in chitinase activity (Table 7; Metraux and Boller, 1986).

Ethylene treatment increased chitinase activity and reduced the disease intensity caused by *C. lagenarium* (Metraux and Boller, 1986).

Table 7 Local and systemic induction of chitinase by pathogens and its effect on disease intensity caused by *Colletotrichum lagenarium*

Treatment	Chitinase activity (μg/g fresh weight)		Successful challenge infections (% of total)
	First leaf	Second leaf	
Experiment I			
Colletotrichum lagenarium	8.4	0.62	41.0
Tobacco necrosis virus	134.0	0.61	19.2
Pseudomonas lachrymans	15.1	0.54	34.0
Pseudoperonospora cubensis	24.6	1.53	12.0
None control	0.15	0.10	72.0
Experiment II			
Tobacco necrosis virus	27.3	1.73	17.0
Nickel chloride	2.5	0.29	72.0
Barium chloride	1.3	0.12	68.0
None control	0.07	0.04	95.0

Source: Metraux and Boller, 1986.

β-1,3-Glucanase and chitinase are simultaneously induced and conferred resistance. Tobacco cv Tn86 was systemically protected against blue mold 21 days after stem inoculation with *Peronospora tabacina*. Both β-1,3-glucanase and chitinase activities were greatly increased in the induced compared to the control Tn86 tobacco plants 21 days after stem inoculation with *P. tabacina* (Table 8; Ye et al., 1990). Chitosan treatment induced chitinase, β-1,3-glucanase and chitosanase in cucumber roots and induced resistance against *Pythium aphanidermatum* (El Ghaouth et al., 1994).

β-1,3-glucanases and chitinases show high antifungal action synergistically. β-1,3-Glucanase 1 (G1), β-1,3-glucanase 2 (G2), chitinase 1 (Ch1), and chitinase 2 (Ch2) were purified from pea pods. Both the β-1,3-glucanases (G1 and G2) inhibited growth of *Fusarium solani* f.sp. *pisi* at 250 μg/ml but did not inhibit *F. solani* f.sp. *phaseoli*, while chitinases Ch 1 and Ch 2 did not inhibit both the fungi. However, a combination of Ch1 and G2 effectively inhibited growth of both the fungi even at the concentrations of 10–30 μg/ml. The results suggest an absolute requirement for both enzyme activities to cause inhibition of the fungi at the concentrations, which may occur in infected pea pod tissues. Treatment of the fungi with purified enzymes caused swelling and lysis of the hyphal tips. In some instances, the fungal tips started to burst within a few minutes after the direct application of the test solutions. Subsequently, the cell contents started to flow, leading to disruption of the cytoplasm remaining in the hyphae. The requirement of β-1,3-glucanase in addition to chitinase to cause lysis indicates that most hyphal tips contain β-1,3-glucan in addition to chitin (Mauch et al., 1988b).

Several classes of chitinases and β-1,3-glucanases have been detected in plants and their fungitoxicity differs. In vitro studies demonstrated that specifically the class I (vacuolar) chitinases and class I β-1,3-glucanases of samsun NN tobacco are potent inhibitors of various fungal species including *Fusarium* and that they act synergistically. The class I chitinases from tobacco inhibits the growth of *Trichoderma viride* and *Alternaria radicina* (Ohl et al., 1994). In contrast, the tobacco class II chitinases and β-1,3-glucanases, which are localized

Table 8 Effect of stem injection with *Peronospora tabacina* on resistance to blue mold and activities of β-1,3-glucanase and chitinase

Treatment	Disease intensity (lesion diameter in mm)	Enzyme activity in units	
		β-1,3-glucanase	Chitinase
Water control	17.2	0.01	7.0
P. tabacina	2.3	1.9	27.0

Source: Ye et al., 1990.

extracellularly, display no antifungal activity in vitro, either alone or in combination with other proteins. The simultaneous expression of class I chitinase, and class I β-1,3-glucanase in transgenic tomato plant resulted in a substantially increased resistance against *Fusarium oxysporum*. However, tomato plants that expressed either of these genes alone were not protected against *Fusarium oxysporum*. These observations suggest that the combined expression of specific chitinases and β-1,3-glucanases enhances fungal resistance in transgenic plants (Logemann et al., 1994).

Transgenic tobacco plants expressing gene encoding rice basic chitinase and tobacco plants expressing alfalfa acidic glucanase were developed. Hybrid plants were generated by crossing transgenic parental lines. Evaluation of disease development in these hybrids, heterozygous for each transgene, and in homozygous selfed progeny, showed that combination of the two transgenes gave substantially greater protection against *Cercospora nicotianae* than either transgene alone. Both rice basic chitinase and alfalfa acidic glucanase transgenes were found to be strongly expressed in the heterozygotes. The results suggest that combinatorial expression of PR 2 and PR 3 gives effective protection against fungal diseases (Zhu et al., 1994).

Tobacco class I chitinase acts synegistically with tobacco class I PR-4 protein inhibiting *Fusarium solani* germlings (Ponstein et al., 1994). Tobacco class I PR-protein likewise acts synergistically with a tobacco class I β-1,3-glucanase against *F. solani* germlings (Ponstein et al., 1994). Tobacco class V chitinases act synergistically with class I β-1,3-glucanase, inhibiting growth of *T. viride* and *Alternaria radicina* (Ohl et al., 1994). A combination of chitinase and ribosome-inactivating protein from barley inhibits fungal growth more efficiently than does either enzyme alone (Leah et al., 1991). Chitinase probably shows more antifungal action when it is combined with other PR proteins.

PR-4 class I protein from tobacco exhibits antifungal activity toward *Trichoderma viride* and *Fusarium solani* by causing lysis of the germ tubes and/or growth inhibition (Ponstein et al., 1994). PR-5 group of proteins contain many antifungal proteins. The tobacco PR protein AP 20 (or AP 24) shows high antifungal activity (Woloshuk et al., 1991b). NP 24, an osmotin-like protein from tomato, inhibits hyphal growth of *Phytophthora infestans* at concentrations greater than 400-nmol (Woloshuk et al., 1991b). Proteinase inhibitors I and II from tomato show antifungal action (Walker-Simmons and Ryan, 1977).

Thionin, the cysteine-rich PR protein from barley and other monocots, is toxic to fungal pathogens (Apel et al., 1990). Chitosanases from cucumber, barley, and tomato lysed *Fusarium oxysporum* f.sp. *radicis-lycopersici*, *Verticillium albo-atrum*, and *Ophiostoma ulmi* spores (Grenier and Asselin, 1990).

It is interesting to note that all the PR proteins, for which antifungal activity in vitro has been shown, are basic proteins (Granier and Asselin, 1990; Woloshuk et al., 1991b; Broglie et al., 1991). Basic isoforms of chitinases and β-1,3-

glucanases purified from *Cladosporium fulvum*-infected tomato plants showed antifungal action against *Trichoderma viride*; but the acidic extracellular enzymes did not show any antifungal activity (Joosten et al., 1995). Acidic isoforms (class II) of the tobacco chitinases (PR-3a and PR-3b) and β-1,3-glucanases (PR-2a, PR-2b and PR-2c) did not show antifungal activity (Sela-Buurlage et al., 1993). No in vitro antifungal activity of acidic class IV chitinase of sugar beet was observed when applied to spore cultures of *Cercospora beticola* grown in wells of microtiter plates (Nielsen et al., 1994b). Transgenic tobacco plants accumulating acidic class III chitinases from sugar beet did not show an increase in resistance against *Cercospora nicotianae* (Nielsen et al., 1993).

XI. HOW DO PATHOGENS OVERCOME PR PROTEINS OF THE HOST?

A. Slower Accumulation of PR Proteins Enables Pathogens to Escape the Antifungal Action of PR Proteins

When the pathogen invades the host tissues, PR proteins accumulate almost immediately (Roby and Esquerre-Tugaye, 1987; Mauch et al., 1988a; Kombrink et al., 1988; Hedrick et al., 1988). However, pathogens try to avoid the antifungal PR proteins by several methods. One such method is that the pathogen delays the accumulation of the PR proteins in the host.

Tomato PR protein P14 belongs to family I of PR protein (PR1 protein). When tomato roots of a susceptible variety are inoculated with *Fusarium oxysporum* f.sp. *radicis-lycopersici*, root colonization starts to appear in the epidermis within 48 h after inoculation and progresses rapidly to reach the vascular stele by 120 h after inoculation (Brammal and Higgins, 1988; Benhamou et al., 1989, 1990, 1991a). The plant reacts to such infection by formation of wall thickenings, plugging of intercellular spaces, and deposition of newly formed structures in xylem vessels (Benhamou et al., 1990). Treatment of sections with PR P14 protein antiserum and gold antibodies resulted in the deposition of gold particles over host cell walls and intercellular spaces. By 48 h after inoculation, the labeling pattern observed over cell walls of the colonized epidermis was mostly similar to the one detected in healthy tissue. A considerable increase in wall labeling was noticeable by 72 h after inoculation. In contrast, in the resistant variety, at 48 h after inoculation treatment of sections with the PR P14 protein antiserum and gold antibodies resulted in the accumulation of gold particles over host cell walls of either colonized or uninvaded cells (Table 9; Benhamou et al., 1991). In this type of interaction also, labeling was preferentially associated with primary walls and secondary thickenings of vessels whereas middle lamellae were nearly free of gold particles. Labeling was also found to occur over intercellular spaces. PR P14 protein does not display a close association with

Table 9 Accumulation of PR P14 protein in tomato cultivars inoculated with *Fusarium oxysporum* f.sp. *radicis-lycopersici*

Time after inoculation (h)	Density of labeling with anti-PR P14 antiserum in tomato root tissue (number of gold particles μm^{-2})			
	Epidermis	Cortex	Endodermis	Vascular elements
48 S	35	32	27	11
R	95	58	47	40
72 S	80	64	36	19
R	177	115	84	69
96 S	99	88	93	65
R	182	111	80	80
120S	140	90	91	89
R	178	112	74	100

S, susceptible cultivar; R, resistant cultivar.
Source: Benhamou et al., 1991.

fungal cell walls but shows predominant accumulation in host cell walls (Table 10; Benhamou et al., 1991).

These histologic observations suggest that the accumulation of PR P14 protein in the resistant variety was detected at 48 h itself while similar accumulation was detected only at 72 h after inoculation in the susceptible variety. The PR P14 protein was present not only in walls of invaded outer cells but also in those of uninvaded inner ones early after inoculation in the resistant variety, indicating that PR P14 accumulation is faster in the resistant variety and slower in the susceptible tomato variety. The slower accumulation of PR P14 protein would have favored the pathogenesis (Benhamou et al., 1991).

Cladosporium fulvum induced synthesis of PR proteins, P14 and P70 in

Table 10 Localized accumulation of PR P14 protein in tomato roots inoculated with *Fusarium oxysporum* f.sp. *radicis-lycopersici*

Structures or Compartments	Density of labeling with anti-PR P14 protein antiserum (particles per μm^2)
Fungal cells	7
Intercellular spaces	51
Papillae	47

Source: Benhamou et al., 1991.

tomato. When compatible race was inoculated, P70 was detectable 8 days after and P14 only 10 days after inoculation. Both the PR proteins, however, appeared even 6 days after inoculation when an incompatible race was inoculated. Seventeen days after inoculation the PR-protein pattern was more strongly evident in extracts from the compatible host (Christ and Mosinger, 1989). De Wit et al. (1986) have also shown delayed accumulation of P14 as an indication of susceptibility in tomato against *Cladosporium fulvum*.

In the compatible *Cladosporium fulvum*–tomato interaction (the tomato genotype Cf5 and fungal race 5), mRNAs encoding the PR proteins PR4 and 6 could be detected only at 6 days after inoculation and they slowly increased to reach a maximum at 12 days after inoculation. In the incompatible interaction, the mRNA levels increased rapidly between 4 and 6 days after inoculation, reaching a maximum 8 days after inoculation. On the 4th day inhibition of the fungal growth was observed in the incompatible interaction while the fungus continued to grow well in the compatible interaction (Van Kan et al., 1992).

Taylor et al. (1990) identified a pathogenesis-related protein, PR 1, in potato. The PR1 mRNA started accumulating about 5 h after inoculation with an incompatible race of *Phytophthora infestans*. A similar increase was observed in the compatible interaction only after about 12 h of inoculation (Taylor et al., 1990). The mRNA accumulated more in an compatible interaction than in an incompatible one at later time points. The delayed accumulation of PR 1 mRNA in a compatible interaction might have favored pathogenesis (Taylor et al., 1990).

When the compatible pathogen *Fusarium solani* f.sp. *pisi* was inoculated, the accumulation of p1230 mRNA that encodes a cysteine-rich PR protein occurred at a slower rate while it accumulated faster in pea tissues inoculated with an incompatible pathogen *F. solani* f.sp *phaseoli* (Chiang and Hadwiger, 1991).

In tobacco β-1,3-glucanase has been shown to be involved in the resistance mechanism to *Peronospora tabacina*. Tobacco plants stem-injected with *P. tabacina* were protected against the disease after challenge with the fungus. Twenty-one days after stem injection, protection was greater than 95% in immunized plants, compared with controls, as determined by leaf area with lesions and by sporulation. β-1,3-Glucanase activity increased in immunized plants prior to challenge, 15 and 21 days after stem injection with *P. tabacina*. These immunized plants were challenged by leaf inoculation 21 days following stem injection. Enzyme activity continued to increase in immunized tobacco 2 and 6 days after challenge. In contrast, β-1,3-glucanase activity was not detected in extracts of control plants prior to challenge. Very low activity was observed in control plants 2 days following challenge. The activity increased by 6 days after inoculation with *P. tabacina* and by that time disease symptoms were already apparent (Tuzun et al., 1988, 1989). The results suggest that although β-1,3-glucanase may suppress the disease development, in successful pathogenesis the

PR protein is induced slowly and high enzyme activity is seen only at the time enough damage has been done by the pathogen (Tuzun et al., 1989).

Changes in isoforms of β-1,3-glucanase in the immunized tobacco plants were also studied (Pan et al., 1991a). Four major isoforms of β-1,3-glucanase (G1, G2, G3, and G4) were detected in the challenged tobacco leaves. The activities of G1 and G2 were higher in leaves of systemically protected plants than in those of the controls during early stages after challenge. G3 was not associated with protection while G4 was of pathogen (*P. tabacina*) origin. In the susceptible control plants symptoms began to appear at 4 days after inoculation with the pathogen and G2 began to accumulate in the control, but its activity was still lower compared to the protected (resistant) plants. Six days after inoculation, the control plants were severely diseased. G1 was detected and the activity of G2 increased markedly in the susceptible control plants. The higher activities of β-1,3-glucanase in the susceptible controls at 6 days after inoculation may be too late to prevent development of *P. tabacina*, formation of lesions, and sporulation of the fungus (Pan et al., 1991a).

The β-1,3-glucanase activity in susceptible leaves, when inoculated with *Erysiphe graminis* f.sp. *hordei*, increased slowly and by the third day after inoculation it increased to 2.8 times the activity of the corresponding control leaves. The enzyme activity increased rapidly to 12.9 times the activity in the corresponding control leaves at 3 days after inoculation (Jutidamrongphan et al., 1991).

Five distinct β-1,3-glucanase isoenzymes in the leaf intercellular fluid were detected in barley. They were named gl_1, (28 kDa), gl_2 (32.5 kDa), gl_3 (33 kDa), gl_4 (34 kDa), and gl_5 (36 kDa). The isoenzymes gl_3 and gl_5 were observed in the uninoculated susceptible variety. The isozymes gl_2, gl_3, and gl_4 were induced after infection of the susceptible variety. By contrast, all the five isoenzymes were much more strongly induced after inoculation in resistant barley lines. The isozyme gl_1 was seen only in the resistant isolines. The enzyme gl_3 was stimulated as early as 1 day after inoculation, and gl_1, gl_2, gl_4, and gl_5 appeared 4 days after inoculation. The induction of gl_2 and gl_4 was much slower in the infected susceptible variety than in the near isogenic resistant lines (Jutidamrongphan, et al., 1991).

Expression of extracellular acidic PR2 protein (β-1,3-glucanase) mRNAs in two tomato genotypes, Cf_4 (incompatible) and cf5 (compatible), infected with *Cladosporium fulvum* race 5 was studied at different days after inoculation (Van Kan et al., 1992). The mRNAs accumulated much faster in incompatible than in compatible interactions. In the incompatible interaction, the mRNA levels increased rapidly between 4 and 6 days after inoculation, reaching a maximum 8 days after inoculation. In the compatible interaction, the mRNAs could be detected 6 days after inoculation and slowly increased to reach a maximum 12 days after inoculation (Van Kan et al., 1992). In the compatible tomato–

Cladosporium fulvum interaction (cf5/race 5), β-1,3-glucanase activity increased slowly compared to the incompatible interaction (Cf5/race 5) (see Fig. 1, p. 270; Joosten and De Wit, 1989).

An antiserum raised against a purified tobacco β-1,3-glucanase was used to study the subcellular localization of β-1,3-glucanase in *Fusarium oxysporum* f.sp. *radicis-lycopersici*-infected tomato plant tissue by means of postembedding immunogold labeling (Benhamou et al., 1989). In the susceptible reaction gold labeling accumulated over host wall areas in the immediate vicinity of fungal cells accompanied by a heavy deposition of gold particles along the fungus pathway, as well as over host wall areas neighboring the channel of penetration. In the resistant variety, although uninvaded by the pathogen, cells of the endodermis, pericycle, and paratracheal parenchyma displayed an intense labeling of their walls. Secondary thickenings of uncolonized xylem vessels were likewise also heavily labeled. This contrasted sharply with susceptible root tissues, in which labeling occurred only over walls of invaded cells. The slow accumulation of β-1,3-glucanase that is also only in the fungus-invaded cells in the susceptible variety is in sharp contrast to that observed in the resistant variety, in which a rapid induction of β-1,3-glucanase even in the uninvaded inner cells far ahead of fungal penetration was observed (Benhamou et al., 1989).

When the underside of leaf 1 of cucumber plants was sprayed with 50 mM K_2HPO_4, chitinase activity increased approximately 3.5-fold within 24 h and eightfold by 40 h, and the leaf became resistant to *Colletotrichum lagenarium*. This treatment induced systemic resistance also and leaf 2 showed about 61% reduced disease intensity when challenge inoculated with the pathogen. The leaf 2 from an induced resistant plant showed higher chitinase activity even at the time of inoculation with the pathogen. The increased activity compared to control susceptible leaf was seen up to 2 days after inoculation, and on the fourth day the chitinase activity increased even in the susceptible control leaf to levels as high as in the induced resistant leaf (Irving and Kuc, 1990). The results suggest that although chitinase activity increased to a great extent in the susceptible leaf, the increase was observed only after 2 days of inoculation. Delayed increase in the chitinase activity would have made the plant susceptible (Irving and Kuc, 1990).

When *Colletotrichum lindemuthianum* spores were inoculated, appressoria were observed after about 30 h. Intracellular primary hyphae emerging from infection vesicles were detected about 60 h after inoculation. Dark brown limited lesion appeared mainly on the veins 92 h after inoculation. Extensive development of primary mycelium was established within 100 h after inoculation. Simultaneously, narrow secondary hyphae branching out of the primary hyphae were visible in brown-colored cells. By 116 h after inoculation, secondary hyphae were well developed in the host cells (Mahe et al., 1992). The chitinase mRNA induction started at 84 h after inoculation and rapidly increased to reach a maximum level at 116 h. The results suggest that the chitinase mRNA is more in the necrotrophic phase and early induction is not observed in the compatible susceptible interaction (Mahe et al., 1992). In the incompatible interaction (host resistant), there was an early increase in chitinase mRNA activity in tissue immediately underlying the site of the spore inoculation and it was correlated with the onset of flecking and expression of hypersensitive resistance. *Cercospora beticola* induced class III acidic, class IV acidic and class IV basic chitinases, and a basic β-1,3-glucanase in sugar beet leaves. The major difference between susceptible and partially resistant plants was a stronger early transient expression in the latter (Nielsen, et al., 1994).

Phoma lingam infection results in a strong induction of chitinase mRNA in rapeseed. In the resistant cultivar, the level of chitinase mRNA was three-fold higher than in the susceptible cultivar 1 day after inoculation. Eight days after inoculation, the chitinase transcript increased approximately 50- and 38-fold in resistant and susceptible cultivars (Rasmussen et al., 1992). When barley cultivars were inoculated with *Erysiphe graminis* f.sp. *hordei*, both chitinase and PR R genes were expressed much earlier in incompatible interactions (Boyd et al., 1994).

Stem injection with *Peronospora tabacina* induced resistance against challenge inoculation with the same pathogen in tobacco. Increased chitinase activity was evident in both immunized and control plants 2 and 6 days after challenge, but lesser enzyme activity was detected in susceptible plants than in immunized plants 2 days after challenge (Fig. 2, Tuzun et al., 1989). The results suggest that delayed increase in chitinase activity allows the pathogen to overcome the defense mechanism (Tuzun et al., 1989).

When *Cladosporium fulvum* was inoculated on a susceptible tomato variety, chitinase activity increased slowly while in the resistant variety the fungus induced rapid induction of chitinase activity (Joosten and De Wit, 1989). β-1,3-Glucanase gene was likewise induced 2–4 days earlier in incompatible interactions than in compatible interactions (Ashfield et al., 1994). The slower accumulation of chitinase would have resulted in susceptibility in cf5/race 5 interaction. The slower rate of accumulation may be due to the presence of race-specific elicitors in the apoplastic fluids, which would not have activated the defense mechanisms in the susceptible interaction (De Wit and Spikman, 1982; De Wit et al., 1986). The presence of a compatible interaction-specific protein has also been reported (Joosten and De Wit, 1989).

Brassica napus is susceptible to *Phoma lingam* while *B. nigra* is resistant. *B. nigra* showed a very rapid production of PR-2, PR-Q, and PR-S due to infection with *P. lingam*. In contrast, in *B. napus* stressed by the fungus, the production of the three PR proteins was delayed or absent (Dixelius, 1994).

Several PR proteins other than chitinases and β-1,3-glucanase also increased in tobacco plants upon infection with *Peronospora tabacina*. In immunized (due to prior inoculation with *P. tabacina*) plants these proteins were

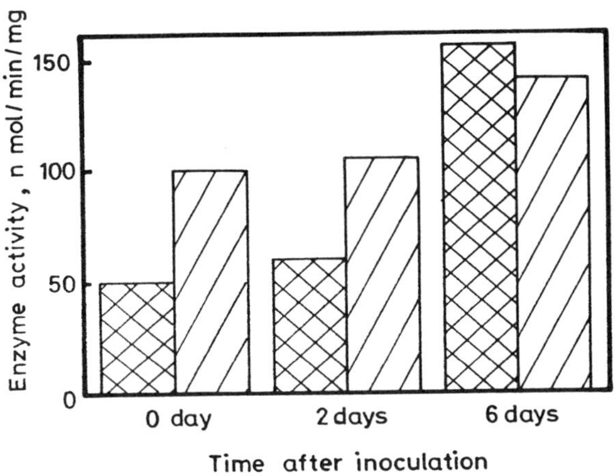

Figure 2 Chitinase activity in susceptible and resistant (immunized) tobacco plants after inoculation with *Peronospora tabacina*. ⊠ -susceptible, ▨ -resistant. (From Tuzun et al., 1989.)

detected at the time of inoculation itself and they continued to increase in immunized plants 2 and 6 days after inoculation, whereas increased amounts were detected in susceptible plants only 6 days after challenge (Tuzun et al., 1989).

All these observations suggest that slower accumulation of PR proteins in the plants may allow the pathogens to escape the antifungal proteins at the critical stage of penetration and establishment in the host. The PR proteins are induced by elicitors present in the fungal cell wall. The major causes for slower accumulation of PR proteins in the susceptible hosts may be delayed release of elicitors from the cell wall of fungal pathogens into host tissues.

When melon plants were inoculated with *Colletotrichum lagenarium*, there was slow increase in chitinase activity. However, when the plants were treated with a fungal elicitor and inoculated with *C. lagenarium* 4 days later, chitinase activity increased at a faster rate than in the plants inoculated with the fungus alone (Roby et al., 1988). The elicitor-treated plants contained up to 42% less chitin than the untreated plants. Chitin content indicates the mycelial content in the infected leaves and hence the results suggest that the early increase in chitinase activity inhibits the fungal development in host tissues (Roby et al., 1988). The results also suggest that there may be slow release of elicitor from the cell wall of the pathogen. By providing elicitor early, the chitinase activity would have increased more quickly.

The other mechanisms involved may be inaction of elicitors in the susceptible hosts or presence of host-specific elicitors that do not act on susceptible hosts or pathogens may have suppressors to suppress action of elicitors, or the susceptible host may have suppressor to suppress action of fungal elicitors or degradation of elicitors by host enzymes as described in detail in Chapter 1.

B. Pathogens May Shed Away from Their Cell Wall the Substrate for the PR Proteins of Enzymatic Nature and Avoid Their Lytic Enzyme Action

Chitin is an important structural component of the cell walls of many plant pathogenic fungi (Wessels and Sietsma, 1981). Chitin is the substrate for PR-3 proteins with chitinase enzyme activities that cause lysis of hyphal tips (Schlumbaum et al., 1986; Broekaert et al., 1988; Mauch et al., 1988). Chitin and chitin oligomers are active elicitors of chitinase activity (Roby et al., 1987). The precise location of the chitin in the infection structures of *Colletotrichum lindemuthianum* during penetration and colonization of bean tissues has been studied (O'Connell and Ride, 1990). Wheat germ agglutinin (WGA) binds strongly to chitin and chitin oligomers and hence WGA, conjugated with collidal gold, was used as a cytochemical probe to detect and localize chitin within the walls of *C. lindemuthianum*. Chitin was detected in *C. lindemuthianum* cell walls and it was absent in the fungal cytoplasm. The lectins, wheat germ agglutinins, and pokeweed mitogen are known to bind with chitin and they bound to conidia, conidial germ tubes, hyaline appressoria, and appressorial germ tubes (O'Connell, 1991). When the fungus was inoculated on the susceptible bean cultivar, WGA-gold bound to the cell walls of germinated conidia indicating the presence of chitin. The cell walls of mature, melanized appressoria also showed the presence of chitin. An electronlucent ingrowth from the appressorial wall, termed the "appressorial cone," surrounded the ventral germ-pore of the appressoria. Both the appressorial cone and the infection peg were not labeled by WGA-gold, and they remained unlabeled at all stages of infection. The cell walls of young, small-diameter intracellular hyphae were also unlabeled, but those of mature, fully expanded vesicles were labeled (O'Connell and Ride, 1990).

The development of *C. lindemuthianum* was studied on the surface of Formvar membranes instead of host surface. The walls of conidial germ-tubes were strongly labeled by WGA-gold. The appressorial germ tubes penetrated Formvar membranes and developed in the water beneath as narrow filamentous hyphae. WGA was found to bind to the walls of young appressorial germ-tubes, unlike the infection pegs and young intracellular hyphae produced during infection. Exclusion of chitin from the hyphal walls of infection peg and young intracellular hyphae appear to be important in pathogenesis: appressorial germ-tubes at an equivalent early stage of development in vitro were labeled by WGA-gold, indicating presence of chitin. The absence of chitin from the fungal wall at the site of initial contact with the host plasma membrane may be important in the establishment of infection by intracellular biotrophs. By excluding chitin from its wall, the fungus may not only resist lysis by host chitinases but also avoid triggering other host defense mechanisms since chitin acts as an elicitor of plant defense reaction also (O'Connell and Ride, 1990).

Similar observations were made in various host–rust interactions. The walls of uredospore-derived substomatal vesicles and basidiospore-derived intraepidermal vesicles contain less chitin than those of appressoria and germ tubes (Holden and Strange, 1984; Kapooria and Mendgen, 1985; Mendgen et al., 1985). The walls of uredospore derived penetration pegs, haustorial necks, and young haustorial bodies appear to lack chitin, while the walls of haustorial mother cells and older haustorial bodies contain chitin (Chong et al., 1981, 1986). The observations suggest that the first fungal structures that contact host cells following intercellular or intracellular penetration lack chitin, and this exclusion is important in pathogenesis (O'Connell and Ride, 1990).

Other supporting evidence for this theory has come from *Rigidoporus lignosus*–rubber tree interaction. *R. lignosus* causes white rot of rubber tree. It attacks roots by means of fast-growing rhizomorphs that contain both thick- and thin-walled mycelium. *R. lignosus* rhizomorphs have been demonstrated to produce specialized infection hyphae (Nicole et al., 1987). Only thin-walled hyphae penetrate the outer root tissue. WGA in conjunction with colloidal gold was used for labeling the chitin oligomers. WGA labeling was substantially reduced over the walls of root-penetrating hyphae compared to that of rhizomorphs. Reduced labeling was likewise found over walls of branching hyphae responsible for host cell wall penetration. Modification or release of chitin from fungal cell walls may thus be related to the establishment of root infection, and to colonization (Nicole et al., 1991; Nicole and Benhamou, 1991).

C. Pathogens May Produce Enzymes That Protect Them from Fungitoxic Action of the PR Protein Chitinase

Chitosan is present along with chitin in the cell wall of fungal pathogens. Chitosan arises mainly by deacetylation of nascent chitin, which is formed by chitin synthase, before the polymer chains aggregate to form fibrils. This mode of synthesis suggests that chitin and chitosan in hyphal walls are not a mixture of homopolymers but represent chitin chains deacetylated to varying degrees (Davis and Bartnicki-Garcia, 1984).

Chitin deacetylase is produced in vitro by *C. lindemuthianum* (Kauss et al., 1983) and *C. lagenarium* (Kauss and Bauch, 1988). When cucumber plants were inoculated with *C. lagenarium*, both chitinase and chitin deacetylase activity increased (Siegrist and Kauss, 1990). The striking increase in chitin deacetylase occurred during lesion development (Table 11; Siegrist and Kauss, 1990), a period in which secondary hyphae presumably spread from cell to cell and through the

Table 11 Chitin deacetylase production after different stages of infection of cucumber leaves with *Colletotrichum lagenarium*

	Chitin deacetylase activity at stages of the infection process		
Experiment	Initial (24–30 h)	Middle (48–72 h)	Late (96–144 h)
I	0.9	0.5	21.4
II	0.4	0.2	7.4
III	1.9	0.3	25.0

Source: Siegrist and Kauss, 1990.

tissue (Xuei et al., 1988). The chitin deacetylase activity in the infected plant tissue was correlated with hyphal growth. This observation was strengthened by the findings that when the cucumber plants were inoculated with tobacco necrosis virus on the first leaf, chitinase activity increased systemically in the second leaves and this increase was associated with reduced lesion formation and chitin deacetylase production by *C. lagenarium*. In other words, chitin deacetylase activity was less in leaf tissues containing less mycelia of the fungus. Chitin deacetylase is known to be secreted in the fungal culture medium. Hence the enzyme may accumulate on the leaf surface and after penetration it may accumulate in the apoplast within the leaf. The chitinase of cucumber is known to be localized in the apoplast (Boller and Metraux, 1988) and so both the enzymes are likely to be present together in the intercellular spaces where the infecting hyphae of *Colletotrichum* are located and first come into contact with the host plasma membrane. The tips of the penetration hypha may synthesize chitin on their outer surfaces. The nascent chitin is a preferred substrate for both chitinase and chitin deacetylase. If deacetylation occurs rapidly, the polymer chains would become less accessible to chitinase and remain intact in the form of partially N-acetylated chitosan polymers. Such a chitosan is likely to remain bound to the hypha and thus sustain the rigidity of the fungal wall. N-acetylation of chitin might, therefore, be a way by which the fungal wall could be partially protected against the fungitoxic action of plant chitinase (Siegrist and Kauss, 1990).

D. Less Elicitor is Released from Pathogen's Cell Wall to Activate Synthesis of PR Proteins

Chitosan treatment induced chitinase activity in celery roots. This chitinase liberated chitin oligosaccharide, N-acetylglucosamine (GlcNAc), from isolated cell walls of compatible race 2 and incompatible race 1 of *Fusarium oxysporum* f.sp. *apii*. Under identical conditions, more GlcNAc was released from race 1 cell walls than from race 2 (Krebs and Grumet, 1993). The results suggest that the ability of the virulent race to invade and infect the host may reside in the cell wall structure that is less accessible to chitinase, the mechanism through which elicitor is released.

E. PR Proteins are Degraded Quickly in Susceptible Host Tissues

When the pathogen establishes infection the PR proteins may be degraded. Pea pods were inoculated with a compatible pathogen, *Fusarium solani* f.sp. *pisi* or with incompatible pathogen, *F. solani* f.sp. *phaseoli*. There was initially little difference in mRNA accumulation between the compatible and incompatible treatments. However, a significant decrease in β-1,3-glucanase mRNA expression was observed in *F. solani* f.sp. *pisi*-inoculated pods between 24 and 48 h after inoculation. In the incompatible reaction, mRNA accumulation remained high for 48 h (Chang et al., 1992). The decrease in mRNA in compatible pathogen inoculated tissues may be due to severe deterioration of the tissue and the breakdown of plant defense system (Chang et al., 1992).

In the susceptible-infected tobacco plants, the isoform of β-1,3-glucanase, G3, decreased when symptoms appeared and subsequently disappeared on 6 days after inoculation with *Peronospora tabacina*. The loss of G3 may be significant in pathogenesis (Pan et al., 1991b).

Tomato PR proteins are degraded by an aspartyl proteinase, a host enzyme constitutively present in healthy and infected leaves at similar levels. However, an acidic pH is required for the activity of this proteinase. Fungal infection leads to acidification of the apoplast and activation of the enzyme activity (Rodrigo et al., 1989). Tobacco leaves were also found to contain an extracellular proteinase that endoproteolytically cleaves tobacco PR proteins. This proteinase was partially purified from tobacco leaves and characterized as an aspartyl proteinase with a pH optimum around pH 3 and a molecular mass of 36–40 kDa. Purified tobacco PR proteins PR-1a, PR-1b, and PR-1c were degraded upon incubation with tobacco intercellular fluids at acidic pH values, remaining practically unattacked at pH values above 3.5. The purified tobacco aspartyl proteinase was able to cleave endoproteolytically tobacco PR-1a, PR-1b, and PR-1c and tomato P1 (p14) at 37°C in citrate–phosphate buffer (pH 2.8). Progressively smaller proteolytic fragments appeared upon prolonged incubation. The degradation products differed in size between the four PR proteins tested. The tomato and tobacco aspartyl proteinase had similar substrate specificities as the pattern of degradation of tomato P1 (p14) by tobacco aspartyl proteinase closely resembled that obtained upon incubation with the tomato 37 kDa aspartyl proteinase (Rodrigo et al., 1991a). The results indicate that these two proteinases are homologous and probably conserved among solanaceae (Rodrigo et al., 1991a). Acidification of the apoplast appears to be important in degradation of PR proteins. It is known that *Sclerotinia sclerotiorum* infection leads to production of oxalic acid in white beans (Tu, 1985).

PR proteins are degraded in elicitor-treated tissues also even in the absence of fungal pathogens. PR proteins in cell-free extracts are resistant to digestion with proteolytic enzymes (Pierpoint, 1983), but in leaf cells they are degraded quickly suggesting the presence or accumulation of an proteolytic enzyme after elicitor treatment. The turnover of PR proteins was measured in tobacco leaf discs by pulse-chase experiments. PR-1a was stable for 72 h and then rapidly degraded, and PR-1b was stable for 40 h and degraded at 40–72 h. PR-1c was gradually degraded until 170 h (Matsuoka and Ohashi, 1986). When tobacco leaves were treated with salicylic acid, synthesis of PR proteins increased after a lag phase of about 8 h. PR 1a synthesis increased almost linearly until 30 h, reached a plateau, and then declined at 50–70 h. Synthesis of PR 1b and PR 1c increased between 8 and 18 h and then immediately declined (Matsuoka and Ohashi, 1986).

F. Fungitoxic PR Proteins are Found Only in Vacuoles and May not Have Antifungal Role in Intercellular Spaces

The central vaucole is considered as a "defense arsenal" (Boller, 1982). Many defense-related proteins like chitinases (Boller and Vogeli, 1984), glucanases (Legrand et al., 1987), osmotin-like proteins (Singh et al., 1987a, Woloshuk et al., 1991b), proteinase inhibitors (Walker-Simmons and Ryan, 1977), proteinases (Nishimura and Beevers, 1979; Wittenbach et al., 1982; Thayer and Huffaker, 1984; Vera and Conejero, 1988), and thionins (Bohlmann et al., 1988) are compartmentalized within the vacuole and all of them are basic in nature.

It has been well established that only intracellular (vacuolar) PR proteins show antifungal activity. Basic chitinases and glucanases that occur intracellularly show antifungal effect while those occurring extracellularly (acidic forms) do not have any antifungal activity (Woloshuk et al., 1991b). Specific enzymatic activity of both basic chitinases and basic β-1,3-glucanases is much higher than that of the acidic isoforms (Kauffmann et al., 1987; Legrand et al., 1987). The osmotin-like protein AP20 that occurs intracellularly in tobacco shows antifungal activity while PR-S, which is very much homologous to osmotin-like proteins, is an extracellular protein and does not have any antifungal activity. The tomato PR protein NP24, a basic osmotin-like protein shows antifungal activity (Woloshuk et al., 1991b). PR-1 proteins also fit into this picture (Woloshuk et al., 1991b). The fungal pathogens, particularly biotrophs, initially develop in the intercellular space and subsequently develop intracellularly in necrotrophic phase. Since only intracellular proteins are fungitoxic, the PR proteins may not have any important role in the defense mechanism of the plants.

Messenger RNA encoding the basic 35 kDa β-1,3-glucanase accumulated very rapidly in both the compatible and incompatible *Cladosporium fulvum*–

tomato interactions. The expression of intracellular β-1,3-glucanase in the compatible interaction almost similar to incompatible interaction suggests that it is not harmful to the fungus (Van Kan et al., 1992). The sites where intracellular chitinases and β-1,3-glucanases accumulate in mesophyll cells of tomato leaves infected by *Cladosporium fulvum* were not significantly different between compatible and incompatible interactions. Chitinase and β-1,3-glucanase accumulated in the cytoplasm of mesophyll cells adjacent to fungal hyphae in both compatible and incompatible interactions (Wubben, 1992). *C. fulvum* grows exclusively in the intercellular space without penetrating tomato cells, and hence intracellular β-1,3-glucanase or chitinase may not interact with fungal hyphae (Van Kan et al., 1992; Wubben, 1992).

The fact that basic vacuolar glucanases may not help plants to defend against fungal pathogens has been demonstrated using antisense transformants (Neuhaus et al., 1992). The chimeric antisense β-1,3-glucanase gene was introduced into *Nicotiana sylvestris* with a Ti-plasmid vector. This vector contained 580 bp of tobacco β-1,3-glucanase gene GLA from position 27 to 608 in the coding sequence inserted in reverse orientation into the polylinker between the CaMV 35S RNA promoter and terminator of pGY 1. The antisense transformants exhibited a 12-fold reduction in β-1,3-glucanase content compared to control plants. Chitinase content was not affected due to antisense transformation (Table 12; Neuhaus et al., 1992).

Three groups of β-1,3-glucanases have been reported in tobacco (Ward et al., 1991a). Extracts of untransformed leaves showed two kinds of proteins with M_r values of about 36 kDa and 33 kDa, corresponding in size to the acidic and basic tobacco isoforms, respectively. The 33 kDa basic β-1,3-glucanase was not found in extracts prepared from leaves of antisense plants, but all the four acidic β-1,3-glucanases were detected in both untransformed and antisense leaves. These results indicate that antisense transformation primarily blocks expression of a class 1 33 kDa β-1,3-glucanase corresponding to the polypeptide encoded by the tobacco gene used to make the antisense construction (Neuhaus et al., 1992).

The antisense transformed plants and untransformed plants were inoculated

Table 12 β-1,3-glucanase and chitinase content of antisense transformant

	Antigen (μg/g fresh weight)	
Plants	β-1,3-glucanase	Chitinase
Empty vector	14.1 ± 0.7	25.9 ± 3.9
Antisense vector	1.0 ± 0.4	40.5 ± 12.2

Source: Neuhaus et al., 1992.

with *Cercospora nicotianae*, the frog eye spot pathogen. The untransformed plants showed a marked increase in both 1,3-glucanase and chitinase relative to mock-inoculated controls. A similar level of induction of chitinase was observed in antisense transformants, but no β-1,3-glucanase antigen was detected in either the mock-infected or infected antisense plants (Table 13; Neuhaus et al., 1992). It shows that antisense transformation effectively blocked induction of basic vacuolar β-1,3-glucanase in response to infection with *C. nicotianae*.

The time course of disease symptoms in the test plants was assessed. There was no consistent difference in the time course of infection between untransformed and antisense transformed plants (Neuhaus et al., 1992). The results suggest that the vacuolar basic β-1,3-glucanases inhibited in the antisense transformant are not important in the successful defense of *N. sylvestris* plants against *C. nicotianae* (Neuhaus et al., 1992).

If the vacuolar basic proteins are targeted into the apoplast, they may induce resistance. Woloshuk et al. (1991a) developed transgenic tobacco plants with the aid of *Agrobacterium tumefaciens* LBA 4404 (pMOG 405). In tobacco plants the osmotin-like protein, AP20 is expressed only in the vacuole. The transgenic plants expressed the altered osmotin gene, eventually resulting in apoplast targeting of the produced osmotin, AP20. In transgenic plants developed with pMOG 404 the osmotin was detected only in the vacuole (Woloshuk et al., 1991a).

The transgenic plants with AP20 targeted to the extracellular space showed more resistance than those with AP20 in vacuole (Table 14; Woloshuk et al., 1991a).

Similar results were obtained with transgenic potato plants. The transgenic plants expressing AP20 with targeting of apoplast (with pMOG 405) showed consistent resistance even 14 days after inoculation with *P. infestans*. The transgenic plants with intracellular target showed only delayed symptom development (Woloshuk et al., 1991a).

Table 13 Effect of *Cercospora nicotianae* infection of β-1,3-glucanase and chitinase antigen content of leaves of antisense transformed and untransformed tobacco plants

		Antigen (μg/g fresh weight)	
Plant	Treatment	β-1,3-glucanase	Chitinase
Transformed	Control	Not detected	11.9
	Infected	Not detected	49.3
Untransformed	Control	1.2	6.8
	Infected	5.6	45.0

Source: Neuhaus et al., 1992.

Table 14 Effect of targeting of AP20 in tobacco lines on their susceptibility to *Phytophthora nicotianae*

Plant line	Leaf area infected by *P. nicotiana*
Non-transgenic	8 discs 100%, 2 discs 60%
pMOG 404 (Vacuole)	7 discs 100%, 3 discs < 25%
pMOG 405 (Apoplast)	1 disc 100%, 2 discs 60%, 7 discs < 25%

Source: Woloshuk et al., 1991a.

PR proteins found intracellularly may be secreted into the intercellular space and by that process they may inhibit fungal growth. Treatment of Samsun NN tobacco leaves with salicylic acid caused an increase of PR 1 proteins. The increase started 1 or 2 days after treatment and continued linearly. PR proteins in the free spaces increased on the 9th day after salicylic acid treatment, while that in the remaining tissue ceased increasing from the sixth day. PR proteins detected in free spaces at the 9th day were about three times the levels of those in the remaining leaf discs, suggesting secretion into the free spaces (Ohashi and Matsuoka, 1987). Although secretion of intracellular PR proteins is possible, the secretion occurs slowly and at 9 days after elicitor treatment. The delayed accumulation would favor the pathogen to escape the defense arsenal.

Although antifungal basic PR proteins occur in vacuoles, decompartmentalization of plant cells has been shown to be a response to pathogen attack (Wilson, 1973; Boller, 1987). Disruption of vacuole will lead to liberation of its hydrolytic constituents to the intracellular spaces, but disruption of vacuole occurs only in late necrotrophic phase. By that time, enough damage would have been caused.

G. Adaptation of Pathogens to PR Proteins

Either slower accumulation or accumulation in lesser amounts may be the major reasons for the pathogens to overcome the antifungal PR proteins in the host. However, it has been shown that even overexpression of chitinases constitutively did not lead to increased resistance of transgenic plants (Neuhaus et al., 1991a). A tobacco basic vacuolar chitinase gene under the control of the CaMV 35S promoter was transferred to the closely related species *Nicotiana sylvestris*. The transgenic plants were shown to accumulate high levels of chitinase in the leaves, but they were found to show only slight increase in resistance to *Cercospora nicotianae*, the frog-eyespot pathogen (Neuhaus et al., 1991a). It is possible that expression of chitinases does not lead to fungal resistance because the fungus has adapted to the defense mechanisms of its host. Basic chitinases and β-1,3-glucanases from tomato showed high antifungal activity, but *Cladosporium fulvum*, the tomato pathogen, was insensitive to these PR proteins (Joosten et al., 1995).

Chitinases from unrelated species in transgenic plants ("new" chitinases) could not be overcome by the invading fungus (Lamb et al., 1992). Tobacco plants expressing bean vacuolar chitinase gene (Roby et al., 1990) show resistance to *Rhizoctonia solani* (Broglie et al., 1991). Potato plants expressing tobacco osmotin gene show delayed development of disease symptoms caused by *Phytophthora infestans*. In contrast, transgenic tobacco plants expressing constitutively their own osmotin gene do not show any increased resistance against *Phytophthora parasitica* var. *nicotianae* (Liu et al., 1994). The chitinase gene (Chi A) from the bacterium *Serratia marcescens* was transferred to tobacco and the transgenic tobacco plants showed resistance to *R. solani* (Jach et al., 1992).

H. PR Proteins May not be Involved in Disease Resistance

There are many reports to indicate that PR proteins are only stress proteins induced due to infection and they may not be involved in host defense mechanism. Nine isoforms of chitinases have been detected in pepper stem tissues after inoculation with *Phytophthora capsici*. All of them except a basic isoform (which also occurred only in a low level in the incompatible interaction) accumulated both in compatible and incompatible interactions (Kim and Huang, 1994). Several isoforms of β-1,3-glucanases accumulated in both the compatible and incompatible interactions in pepper inoculated with *P. capsici* (Kim and Huang, 1994). Accumulation of PR proteins (PR-1, PR-2, PR-3, PR-4 and PR-5) was not a prerequisite for the expression of *Pseudomonas fluorescens*-induced systemic resistance in radish against *Fusarium oxysporum* f.sp. *raphani* (Hoffland et al., 1995).

Different isolates of pathogen induce different kinds of PR proteins in plants. An avirulent isolate of *Phoma lingam* induced PR-Q in *Brassica napus* (susceptible) while it induced PR-2, PR-Q and PR-S in *B. nigra* (Dixelius, 1994). In contrast, the virulent isolate induced PR-2 and PR-S in *Brassica napus* and PR-2 and PR-Q in *B. nigra* (Dixelius, 1994). Thus there was no relationship between a particular PR protein and disease resistance.

XII. CONCLUSION

When fungal pathogens invade the host tissues a small number of genes to produce mRNAs to permit synthesis of a similar number of specific proteins are induced. These proteins, called PR proteins, have been thought to play an important role in the defense mechanisms of plants, because most of them show antifungal action. Both acidic and basic isoforms of these PR proteins have been reported, in many, if not in all of them. The acidic forms accumulate in apoplast while the basic forms accumulate in vacuoles of the host cells. Exceptions have also been reported; a few acidic PR proteins accumulate in vacuoles. Among the PR proteins, basic forms alone have been shown to have fungitoxic action. Pathogens develop in the apoplast in the initial stages and only in the necrotrophic phase do they disrupt the vacuole, releasing the PR proteins into the extracellular space. It suggests that the PR proteins may act only in the later stages of pathogenesis, resulting in symptom suppression. Virulent pathogens delay the accumulation of PR proteins. The delay may be due to delay in signal induction and transcription. It is also possible that delay in elicitor release from the fungal cell wall, degradation of elicitor, action of host specific elicitor, and presence of suppressor would have contributed to the delay in signal induction.

The pathogens would have decreased the pH of the cell sap and in an acidic pH PR proteins are degraded. Host enzymes would have contributed in the degradation of PR proteins. The increase of PR proteins in the host is mostly transitional. Continuous presence of elicitors has been shown to be essential for the induction of PR proteins, but elicitors are degraded by the host cells.

Many PR proteins have been identified as enzymes. Disappearence of the substrates from the infection court would suppress the action of these enzymes. Early exclusion of chitin from the hyphal tip is a typical example of the method by which pathogens escape the fungitoxic action of host chitinases. Adaptation of the pathogens to the host PR proteins may be another method by which the pathogen escapes those toxic actions.

Although several studies have indicated a role for PR proteins in the host defense mechanisms, several questions are unanswered. Every host synthesizes several PR proteins, sometimes more than 20. Several isoforms of these PR proteins also exist. The role of individual PR proteins is being studied intensively. The combined effect of these PR proteins should be looked into. For this purpose, the function of the individual PR proteins should be known. Unfortunately the function of the major PR protein belonging to PR-1 group is not yet known. With the available modern technology, it may be possible to characterize these PR proteins in the near future.

It has been proved in many host–pathogen interactions that pathogens do not simply avoid the host-defense mechanism, but actually suppress it. Intensive studies are needed to find out the mechanism through which the pathogen suppresses the PR proteins. Otherwise it will be a futile exercise for genetic engineers who are trying to clone the PR proteins and obtain transgenic plants to express the defense genes for the management of fungal diseases.

REFERENCES

Abe, K., Emori, Y., Kondo, H., Suzuki K., and Arai, S. (1987). Molecular cloning of a cysteine inhibitor of rice (oryzacystatin): homology and animal cystatins and transient expression in ripening process of rice seeds, J. Biol. Chem., 262:16793–16797.

Abe, K., Kondo, H., Watanabe, H., Emori, Y., and Arai, S. (1991). Oryzacystatins as the first well-defined cystatins of plant origin and their target proteinases in rice seeds, Biomed. Biochim. Acta, 50:4–6.

Abeles, F. B., Bosshart, R. P., Forrence, L. E., and Habig, W. H. (1971). Preparation and purification of glucanase and chitinase from bean leaves, Plant Physiol., 47:129–134.

Abeles, F. B., and Forrence, L. E. (1970). Temporal and hormonal control of β-1,3-glucanase in *Phaseolus vulgaris* L., Plant Physiol., 45:395–400.

Abu Jawdah, Y. (1982). Changes in the soluble proteins of bean leaves upon fungal or viral infections or after injury, Phytopathol. Z., 103:272–279.

Agrawal, K. M. L., and Bahl, O. P. (1968). Glycosidases in *Phaseolus vulgaris*. II, J. Biol. Chem., 234:103–111.

Ahl, P., Antoniw, J. F., White, R. F., and Gianinazzi, S. (1985). Biochemical and serological characterization of b-proteins from *Nicotiana* species, Plant Mol. Biol., 4:31–37.

Ahl, P., Benjama, A., Sanson, R., Gianinazzi, S. (1981). Induction chez le tabac par *Pseudomonas syringae* de nouvells proteines (proteines b) associées au developement d'une resistance nonspecifique à une deuxième infection, Phytopathol. Z., 102:201–202.

Ahl, P., Cornu, A., and Gianinazzi, S. (1982). Soluble proteins as genetic markers in studies of resistance and phylogeny in *Nicotiana*, Phytopathology 72:80–85.

Ahl, P., and Gianinazzi, S. (1982). b-Proteins as a constitutive component in highly (TMV) resistant interspecific hybrids of *Nicotiana glutinosa* and *Nicotiana debneyi*, Plant Sci. Lett., 26:173–181.

Ahl Goy, P., Felix, G., Metraux, J. P., and Meins, Jr. F. (1992). Resistance to disease in the hybrid *Nicotiana glutinosa* × *N. debneyi* is associated with high constitutive levels of β-1,3-glucanase, chitinase, peroxidase and polyphenoloxidase, Physiol. Mol. Plant Pathol., 41:11–21.

Albersheim, P., and Valent, B. (1974). Host–pathogen interactions. VII. Plant pathogens secrete proteins which inhibit enzymes of the host capable of attacking the pathogens, Plant Physiol., 53:684–687.

Albrecht, H., Van de Rhee, M. D., and Bol, J. F. (1992). Analysis of *cis*-regulatory elements involved in induction of a tobacco PR-5 genes by virus infection, Plant Mol. Biol., 18:155–158.

Alexander, D., Glascock, C., Pear, J., Stinson, J., Ahl-Goy, P., Gut-Rella, M., Ward, E., Goodman, R., and Ryals, J. (1992). Systemic acquired resistance in tobacco: use of transgenic expression to study the functions of pathogenesis-related proteins. Sixth International Symp. on Molecular Plant-Microbe Interactions, Seattle, July 11–16, 1992., Abstr. 70.

Alexander, D., Goodman, R. M., Gut-Rella, M., Glascock, C., Weymann, K., Friedrich, L., Maddox, D., Ahl-Goy, P., Luntz, T., Ward, E., and Ryals, J. (1993). Increased tolerance to two oomycete pathogens in transgenic tobacco expressing pathogenesis-related protein 1a, Proc. Natl. Acad. Sci. USA, 90:7327–7331.

An, G., Mitra, A., Choi, H. K., Costa, M. A., An, K., Thornburg, R. W., and Ryan, C. A.

(1989). Functional analysis of the 3′ control region of the potato wound-inducible proteinase inhibitor II gene, Plant Cell, 1:115–122.
Anderson, A. (1978a). Initiation of resistant reponses in bean by mycelial wall fractions from three races of the bean pathogen *Colletotrichum lindemuthiamum*, Can. J. Bot., 56:2247–2251.
Anderson, J. M., (1985). Evidence for phloem transport of jasmonic acid, Plant Physiol., 77:S-75.
Andresen, I., Becker, W., Schluter K., Burges, J., Parthier, B., and Apel, K. (1992). The identification of leaf thionin as one of the main jasmonate-induced proteins of barley (*Hordeum vulgare*), Plant Mol. Biol., 19:193–204.
Antoniw, J. F., Kueh, J., Walkey, D. G. A., and White, R. F. (1981). The presence of pathogenesis-related proteins in callus of Xanthi-nc tobacco, Phytopathol. Z., 101:179–184.
Antoniw, J. F., and Pierpoint, W. S. (1978). The purification and properties of one of the 'b' proteins from virus-infected tobacco plants, J. Gen. Virol., 39:343–350.
Antoniw, J. F., Ritter, C. E., Pierpoint, W. S., and Van Loon, (1980). Comparison of three pathogenesis-related proteins from plants of two cultivars of tobacco infected with TMV, J. Gen. Virol., 47:79–87.
Antoniw, J. F., and White, R. F. (1980). The effects of aspirin and polyacrylic acid on soluble leaf proteins and resistance to virus infection in five cultivars of tobacco, Phytopathol. Z., 98:331–341.
Antoniw, J. F., and White, R. F. (1983). Biochemical properties of the pathogenesis-related proteins from tobacco, Neth. J. Plant Pathol., 89:255–264.
Antoniw, J. F., and White, R. F. (1986). Changes with time in the distribution of virus and PR protein around single local lesions of TMV infected tobacco. Plant Mol. Biol. 6:145–149.
Antoniw, J. F., and White, R. F. (1987). The role of pathogenesis-related proteins, Plant Resistance to Viruses (D. Evered and D. Harnett, eds.), John Wiley & Sons, Chichester, pp. 57–71.
Antoniw, J. F., White, R. F., Barbara D. J., Junes, P., and Longley, A. (1985). The detection of PR (b) proteins and TMV by ELISA in systemic and localized virus infections of tobacco, Plant Mol. Biol. 4:55–60.
Apel, K., Bohlmann, H., and Reimann-Philipp, U. (1990). Leaf thionins, a novel class of putative defense factors, Physiol. Plant., 80:315–321.
Ashfield, T., Hammond-kosack, K. E., Harrison, K., and Jones, J. D. G. (1994). Cf gene-dependent induction of a β-1,3-glucanase promoter in tomato plants infected with *Cladosporium fulvum*, Mol. Plant–Microbe Interact., 7:645–657.
Atkinson, A. H., Heath, R. L., Simpson, R. J., Clarke, A. E., and Anderson, M. A. (1993). Proteinase inhibitors in *Nicotiana alata* stigmas are derived from a precursor protein which is processed into five homologous inhibitors, Plant Cell, 5:203–213.
Awade, A., de Tapia, M., Didierjean, L., and Burkard, G. (1989). Biological function of bean pathogenesis-related (PR3 and PR4) proteins, Plant Sci., 63:121–130.
Awade, A., Metz-Boutigue, M. H., Le Ret, M., Genot, G., Amiri, I., and Burkard, G. (1991). The complete amino acid sequence of the pathogenesis-related (PR2) protein induced in chemically stressed bean leaves, Biochem. Biophys. Acta, 1077:241–244.

Ball, A. M., Ashby, A. M., Daniels, M. J., Ingram, D. S., and Johnstone, K. (1991). Evidence for the requirement of extracellular protease in the pathogenic interaction of *Pyrenopeziza brassicae* with oilseed rape, Physiol. Mol. Plant Pathol., 38:147–161.

Barber, M. S., Bertram, R. E., and Ride, J. P. (1989). Chitin oligosaccharides elicit lignification in wounded wheat leaves, Physiol. Mol. Plant Pathol., 34:3–12.

Barber, M. S., and Ride, J. P. (1989). Purification and properties of a wheat leaf N-acetyl-β-D-hexosaminidase, Plant Science, 60:163–172.

Barrat, D. H. P., and Clark, J. A. (1991). Proteins arising during the late stages of embryogenesis in *Pisum sativum* L., Planta, 184:14–23.

Becker, W., and Apel, K. (1992). Isolation and characterization of a cDNA clone encoding a novel jasmonate-induced protein of barley (*Hordeum vulgare* L.), Plant Mol. Biol., 19:1065–1067.

Bednarek, S., and Raikhel, N. V. (1991). The barley lectin carboxy-terminal propetide is a vacuolar protein-sorting determinant in plants, Plant Cell, 3:1195–1206.

Bednarek, S., and Raikhel, N. V. (1992). Intracellular trafficking of secretory proteins, Plant Mol. Biol., 20:133–150.

Bednarek, S., Wilkins, T. A., Dombrowski, J. E., and Raikhel, N. V. (1990). A carboxyl-terminal propeptide is necessary for proper sorting of barley lectin to vacuoles of tobacco, Plant Cell, 2:1145–1155.

Beerhues, L., and Kombrink, E. (1994). Primary structure and expression of mRNAs encoding basic chitinase and 1,3-β-glucanase in potato, Plant Mol. Biol., 24:353–367.

Beilmann, A., Alblrecht, K., Schultze, S., Wanner, G., and Pfitzner, U. M. (1992). Activation of a truncated PR-1 promoter by endogenous enhancers in transgenic plants, Plant Mol. Biol., 18:65–78.

Bell, A. A. (1981). Biochemical mechanisms of disease resistance, Annu. Rev. Plant Physiol., 32:21–81.

Belles, J. M., Carbonell, J., and Conejero, V. (1991). Polyamines in plants infected by citrus exocortis viroid or treated with silver ions and ethephon, Plant Physiol., 96:1053–1059.

Belles, J. M., Tomero, O., and Conejero, V. (1992). Pathogenesis-related proteins and polyamines in a developmental mutant of tomato, Epinastic, Plant Physiol., 98:1502–1505.

Benchekrown, A., Michaud, D., Nguyen-Quoc, B., Overney, S., Desjardins, Y., and Yelle, S. (1995). Synthesis of active oryzacystatin I in transgenic potato plants, Plant Cell Rep., 14:585–588.

Benhamou, N., Broglie, K., Broglie, R., and Chet, I. (1993). Antifungal effect of bean endochitinase on *Rhizoctonia solani* ultrastructural changes and cytochemical aspects of chitin breakdown, Can. J. Microbiol., 39:318–328.

Benhamou, N., Chamberland, H., and Pauze, P. J. (1990a). Implication of pectic components in cell surface interactions between tomato root cells and *Fusarium oxysporum* f.sp. *lycopersici*: a cytochemical study by means of a lectin with polygalacturonic acid-binding specificity, Plant Physiol., 92:995–1003.

Benhamou, N., Grenier, J., and Asselin, A. (1991). Immunogold localization of pathogenesis-related protein P14 in tomato root cells infected by *Fusarium oxysporum* f.sp. *radicis-lycopersici*, Physiol. Mol. Plant Pathol., 38:237–253.

Benhamou, N., Grenier, J., Asselin, A., and Legrand, M. (1989). Immunogold localization of β-1,3-glucanases in two plants infected by vascular wilt fungi, Plant Cell, 1:1209–1221.

Benhamou, N., Joosten, M. H. A. J., and De Wit, P. J. G. M. (1990b). Subcellular localization of chitinase and of its potential substrate in tomato root tissues infected by *Fusarium oxysporum* f.sp. *radicis-lycopersici*, Plant Physiol., 92:1108–1120.

Benhamou, N., Mazau, D., and Esquerre-Tugaye, M. T. (1990c). Immunocytochemical localization of hydroxyproline-rich glycoproteins in tomato root cells infected by *Fusarium oxysporum* f.sp. *radicis-lycopersici*: study of a compatible interaction, Phytopathology, 80:163–173.

Bernasconi, P., Locher, R., Pilet, P. E., Jolles, J., and Jolles, P. (1987). Purification and N-terminal amino-acid sequence of a basic lysozyme from *Parthenocissus quinquifolia* cultured in vitro, Biochem. Biophys. Acta, 915:254–260.

Bishop, P. D., Pearce, G., Bryant, J. E., and Ryan, C. A. (1984). Isolation and characterization of the proteinase inhibitor-inducing factor from tomato leaves, J. Biol. Chem., 259:13171–13177.

Bohlmann, H., and Apel, K. (1987). Isolation and characterization of cDNAs coding for leaf-specific thionins closely related to the endosperm-specific hordothionin of barley (*Hordeum vulgare* L.), Mol. Gen. Genet., 207:446–454.

Bohlmann, H., and Apel, K. (1991). Thionins, Annu Rev. Plant Physiol. Plant Mol. Biol., 42:227–240.

Bohlmann, H., Clausen, S., Behnke, S., Giese, H., Hiller, C., Reimann-Philipp, U., Schrader, G., Barkholt, V., and Apel, K. (1988). Leaf-specific thionins of barley-a novel class of cell wall proteins toxic to plant pathogenic fungi and possibly involved in the defense mechanism of plants, EMBO J., 7:1559–1565.

Bol, J. F. (1988). Structure and expression of plant genes encoding pathogenesis-related proteins, Plant Gene Research: Temporal and Spatial Regulation of Plant Genes (D. P. S. Verma and R. B. Goldberg, eds.), Springer-Verlag, New York, Wein, pp. 201–221.

Bol, J. F., Linthorst, H. J. M., and Cornelissen, B. J. C. (1990). Plant pathogenesis related proteins induced by virus infection, Annu. Rev. Phytopathol., 28:113–138.

Bol, J. F., and Van Kan, J. A. L. (1988). The synthesis and possible functions of virus-induced proteins in plants, Microbiol. Sci., 5:47–52.

Bol, J. F., Van Kan, J. A. L., and Cornelissen, B. J. C. (1989). Plant defense genes induced by virus infection, Molecular Biology of Plant Pathogen Interactions (B. Staskawicz, P. Ahlquist and O. Yoder, eds.), Alan R. Liss, Inc., New York, pp. 247–257.

Boller, T. (1982). Enzymatic equipment of plant vacuoles, Physiol. Veget., 20:247–257.

Boller, T. (1985). Induction of hydrolases as defense reaction against pathogens, Cellular and Molecular Biology of Plant Stress (J. L. Key and T. Kosuge, eds.), A. R. Liss, New York, pp. 247–262.

Boller, T. (1987). Hydrolytic enzymes in plant disease resistance. Plant Microbe Interactions, Molecular and Genetic Perspectives, Vol. 2 (T. Kosuge and E. W. Nester, eds.), MacMillan, New York, pp. 385–413.

Boller, T. (1988). Ethylene and the regulation of antifungal hydrolases in plants, Oxford Surv. Plant Mol. Cell Biol., 5:145–174.

Boller, T. (1990). Ethylene and plant pathogen interactions, Polyamines and Ethylene: Biochemistry, Physiology and Interactions (H. E. Flores, R. N. Arteca and J. C. Shannon, eds.), American Society of Plant Physiologists, Rockville, MD, pp. 138–145.

Boller, T., Gehri, A., Mauch, F., and Vogeli, U. (1983). Chitinase in bean leaves: induction by ethylene, purification, properties, and possible function, Planta 157:22–31.

Boller, T., and Metraux, J. P. (1988). Extracellular localization of chitinase in cucumber, Physiol. Mol. Plant Pathol., 33:11–16.

Boller, T., and Vogeli, U. (1984). Vacuolar localization of chitinase in bean leaves, Plant Physiol., 74:442–444.

Boller, T., and Wiemken, A. (1986). Dynamics of vacuolar compartmentation, Annu. Rev. Plant Physiol., 37:137–164.

Boyd, L. A., Smith, P. H., and Brown, J. K. M. (1994). Molecular and cellular expression of quantitative resistance in barley to powdery mildew, Physiol. Mol. Plant Pathol., 45:47–58.

Bowles, D. J. (1990). Defense-related proteins in higher plants, Annu. Rev. Biochem., 59:873–907.

Bradshaw, H. D. F., Hollick, J. B., Parsons, T. J., Clarke, H. R. G., and Gordon, M. P. (1989). Systemically wound-responsive genes in poplar trees encode proteins similar to sweet potato sporamins and legume Kunitz trypsin inhibitors, Plant Mol. Biol., 14:51–59.

Brammall, R. A., and Higgins, V. J. (1988). A histological comparison of colonization in tomato seedlings susceptible and resistant to fusarium crown and root rot disease, Can. J. Bot., 66:915–925.

Brederode, F. T., Linthorst, H. J. M., and Bol, J. F. (1991). Differential induction of acquired resistance and PR gene expression in tobacco by virus infection, ethephon treatment, UV light and wounding, Plant Mol. Biol., 17:1117–1125.

Breitender, H., Pettenburger, K., Bito, A., Valenta, R., Kraft, D., Rumpold, H., Scheiner, O., and Breitenbach, M. (1989). The gene coding for the major birch pollen allergen BetvI is highly homologous to a pea disease resistance response gene, EMBO J., 8:1935–1938.

Broekaert, W., Lee, H. I., Kush, A., Chua, N. H., and Raikhel, N. (1990). Wound-induced accumulation of mRNA containing a hevein sequence in laticifers of rubber tree (*Hevea brasiliensis*), Proc. Natl. Acad. Sci. USA, 87:7633–7637.

Broekaert, W. F., Terras, F. R. G., Eggermont, K., Cammue, B. P. A., Kovaleva, V., Raikhel, N. V., Osborn, R. W., Kester, A., Rees, S. B., and Janssens, F. A. (1994). Small cysteine-rice antifungal proteins from radish: role in host defense, J. Cellular Biochemistry Suppl., 18A:86.

Broekaert, W. F., Van Parijs, J., Allen, A. K., and Peumans, W. J. (1988). Comparison of some molecular, enzymatic and antifungal properties of chitinases from thorn-apple, tobacco and wheat. Physiol. Mol. Plant Pathol., 33:319–331.

Broglie, K., Chet, I., Holliday, M., Cressman, R., Biddle, P., Knowlton, S., Mauvais, C. J., and Broglie, R. (1991). Transgenic plants with enhanced resistance to the fungal pathogen *Rhizoctonia solani*, Science, 254:1194–1197.

Broglie, K. E., Biddle, P., Cressman, R., and Broglie, R. (1989). Functional analysis of DNA sequences responsible for ethylene regulation of a bean chitinase gene in transgenic tobacco, Plant Cell, 1:599–607.

Broglie, K. E., Gaynor, J. J., and Broglie, R. M. (1986). Ethylene-regulated gene expression: molecular cloning of the genes encoding an endochitinase from *Phaseolus vulgaris*, Proc. Natl. Acad. Sci. USA, 83:6820–6824.

Broglie, R., Broglie, K., Chet, I., and Benhamou, R. (1991). Transgenic plants with enhanced resistance to *Rhizoctonia solani*, 3rd Int. Congr. Int. Soc. Plant Mol. Biol., p. 1364.

Bronner, R., Westphal, E., and Dreger, F. (1991). Pathogenesis-related proteins in *Solanum dulcamara* L. resistant to the gall mite *Aceria cladophthirus* (Nalepa) (syn: *Eriophyes cladophthirus* Nal.), Physiol. Mol. Plant Pathol., 38:93–104.

Brown, A. E., and Adikaram, N. K. N. (1985). A role for pectinase and protease inhibitors in fungal rot development in tomato fruits. Phytopathol. Z., 106:239–251.

Brown, W. E., Takio, K., Titani, K., and Ryan, C. A. (1985). Wound-induced trypsin inhibitor in alfalfa leaves: Identity as a member of the Bowman-Birk inhibitor family, Biochemistry, 24:2105–2108.

Bryant, J., Green, T., Gurusaddaiah, T., and Ryan, C. A. (1976). Proteinase inhibitor II from potatoes: Isolation and characterization of the isoinhibitor subunits, Biochemistry, 15:3418–3423.

Bryngelsson, T., and Green, B. (1989). Characterization of a pathogenesis-related, thaumatin-like protein isolated from barley challenged with an incompatible race of mildew, Physiol. Mol. Plant Pathol., 35:45–52.

Bryngelsson, T., Gustafsson, M., Ramos Leal, M., and Bartonek, E. (1988). Induction of pathogenesis-related proteins in barley during the resistance reaction to mildew, J. Phytopathol., 123:193–198.

Bryngelsson, T., Sommer-Knudsen, J., Gregersen, P. L., Collinge, D. B., Ek, B., and Thordel-Christensen, H. (1994). Purification, characterization, and molecular cloning of basic PR-1 type pathogenesis-related proteins from barley. Mol. Plant–Microbe Interact, 7:267–275.

Bull, A. T., and Chesters, C. G. C. (1966). The biochemistry of laminarin and nature of laminarinase, Adv. Enzymol., 28:325–364.

Camacho Henriquez, A., and Sanger, H. L. (1982a). Gel electrophoretic analysis of phenol-extractable leaf proteins from different viroid host combinations, Arch. Virol., 74:167–180.

Camacho Henriquez, A., and Sanger, H. L. (1982b). Analysis of acid extractable tomato leaf proteins after infection with a viroid, two viruses and a fungus and partial purification of the 'pathogenesis-related' protein P14, Arch. Virol. 74:181–195.

Camacho Henriquez, A., and Sanger, H. L. (1984). Purification and partial characterization of the major pathogenesis-related tomato leaf protein P14 from tomato spindle tuber viroid (PSTV)-infected leaves, Arch. Virol. 81:263–284.

Carr, J. P., Antoniw, J. F., White, R. R., and Wilson, T. M. A. (1982). Latent messenger RNA in tobacco (*Nicotiana tabacum*), Biochem. Soc. Trans., 10:353–354.

Carr, J. P., Dixon, D. C., and Klessig, D. F. (1985). Synthesis of pathogenesis-related proteins in tobacco is regulated at the level of mRNA accumulation and occurs on membrane bound polysomes, Proc. Natl. Acad. Sci. USA, 82:7999–8003.

Carr, J. P., and Klessig, D. F. (1989). The pathogenesis-related proteins of plants, Genetic Engineering: Principles and Methods, Vol. II (J. K. Setlow, ed.), Plenum Press, New York, pp. 64–109.

Casacuberta, J. M., Raventos, D., Puigdomenech, P., and San Segundo, B. (1992). Expression of the PR-like protein PRms gene in germinating maize embryos, Mol. Gen. Genet., 234:97–104.

Castresana, C., de Carvalho, F., Gheysen, G., Habets, M., Inze, D., and Van Montague, M. (1990). Tissue-specific and pathogen-induced regulation of a *Nicotiana plumbaginifolia* β-1,3-glucanase gene, Plant cell, 2:1131–1143.

Chang, M-M., Hadwiger, L. A., and Horovitz, D. (1992). Molecular characterization of a pea β-1,3-glucanase induced by *Fusarium solani* and chitosan challenge, Plant Mol. Biol., 20:609–618.

Chaudhry, B., Muller-Uri, F., Cameron-Mills, V., Gough, S., Simpson, D., Skriver, K., and Mundy, J. (1994). The barley 60 kDa jasmonate-induced protein (JIP 60) is a novel ribosome-inactivating protein, Plant J., 6:815–824.

Chen, M. S., Johnson, B., Wen, L., Muthukrishnan, S., Kramer, K. J., Morgan, T. D., and Reeck, G. R. (1992). Rice cystatin: bacterial expression, purification, cysteine proteinase inhibitory activity and insect growth suppressing activity of a truncated form of the protein, Protein Exp. Purif., 3:41–49.

Chen, Z., and Klessig, D. F. (1991). Identification of a soluble salicylic acid-binding protein that may function in signal transduction in the plant disease-resistance response, Proc. Natl. Acad. Sci. USA, 88:8179–8183.

Chiang, C. C., and Hadwiger, L. A. (1990). Cloning and characterization of a disease resistance response gene in pea inducible by *Fusarium solani*, Mol. Plant–Microbe Interact., 3:78–85.

Chiang, C. C., and Hadwiger, L. A. (1991). The *Fusarium solani*-induced expression of a pea gene family encoding high cysteine content proteins, Mol. Plant–Microbe Interact., 4:324–331.

Chong, J., Harder, D. E., and Rohringer, R., (1981). Ontogeny of mono- and dikaryotic rust haustoria: cytochemical and ultrastructural studies, Phytopathology, 71:975–983.

Chong, J., Harder, D. E., and Rohringer, R. (1986). Cytochemical studies on *Puccinia graminis* f.sp. *tritici* in a compatible wheat host. II Haustorium mother cell walls at the host cell penetration site, haustorial walls, and the extrahaustorial matrix, Can. J. Bot., 64:2561–2575.

Chrispeels, M. J. (1991). Sorting of proteins in the secretory system, Annu. Rev. Plant Physiol. Plant Mol. Biol., 42:21–53.

Christpeels, M. J. and Raikhel, N. V. (1992). Short peptide domains target proteins to the plant vacuoles, Cell, 68:613–616.

Christ, U., and Mosinger, E. (1989). Pathogenesis-related proteins of tomato: 1. Induction by *Phytophthora infestans* and abiotic inducers and correlations with resistance, Physiol. Mol. Plant Pathol., 35:53–65.

Cleveland, T. E., and Black, L. L. (1983). Partial purification of proteinase inhibitors from tomato plants infected with *Phytophthora infestans*, Phytopathology, 73:664–670.

Cleveland, T. E., Thornburg, R. W., and Ryan, C. A. (1987). Molecular characterization of

a wound-inducible inhibitor I gene from potato and the processing of its mRNA and protein, Plant Mol. Biol., 8:199–207.
Coleman, W. H., and Roberts, W. K. (1982). Inhibitors of animal cells-free protein synthesis from grains, Biochem. Biophys. Acta, 696:239–244.
Colilla, F. J., Rocher, A., and Mendez, E. (1990). Purothionins: amino acid sequence of two polypeptides of a new family of thionins from wheat endosperm, FEBS Lett., 270:191–194.
Collinge, D. B., Kragh, K. M., Mikkelsen, J. D., Nielsen, K. K., Rasmussen, U., and Vad, K. (1993). Plant chitinases, Plant J., 3:31–40.
Collinge, D. B., and Slusarenko, A. J. (1987). Plant gene expression in response to pathogens, Plant Mol. Biol., 9:389–410.
Condit, C. M., and Meagher, R. B. (1986). A gene encoding a novel glycine-rich structural protein of petunia, Nature, 323:178–181.
Conejero, V., Belles, J. M., Garcia-Breijo, F., Garro, R., Hernandez Yago, J., Rodrigo, I., and Vera, P. (1990). Signaling in viroid pathogenesis (R. S. S. Fraser, ed.), Recognition and Response in Plant–Virus Interactions, Vol. 41, Springer Verlag, Berlin, pp. 233–261.
Conejero, V., Picazo, I., and Segado, P. (1979). Citrus exocortis viroid (CEV): protein alterations in different hosts following viroid infection, Virology, 97:454–456.
Constabel, C. P., Bertrand, D., and Brisson, N. (1993). Transgenic potato plants over-expressing the pathogenesis-related STH2 gene show unaltered susceptibility to *Phytophthora infestans* and potato virus X, Plant Mol. Biol., 22:775–782.
Constabel, C. P., and Brisson, N. (1992). The defense-related STH-2 gene product of potato shows race-specific accumulation after inoculation with low concentrations of *Phytophthora infestans* zoospores, Planta, 988:289–295.
Cordero, M. J., Raventos, D., and Segundo, B. S. (1994). Expression of a maize proteinase inhibitor gene is induced in response to wounding and fungal infection: systemic wound-response to a monocot gene, Plant J., 6:141–150.
Cordero, M. J., Raventos, D., and Segundo, B. S. (1992). Induction of PR proteins in germinating maize seeds infected with the fungus *Fusarium moniliforme*, Physiol. Mol. Plant Pathol., 41:189–200.
Cornelissen, B. J. C., Hooft van Huijsduijnen, R. A. M., and Bol, J. F. (1986). A tobacco mosaic virus-induced tobacco protein is homologous to the sweet-tasting protein thaumatin, Nature, 321:531–532.
Cornelissen, B. J. C., Hoof Van Huijsduijnen, R. A. M., Van Loon, L. C., and Bol, J. F. (1986). Molecular characterization of messenger RNAs for pathogenesis-related proteins 1a, 1b and 1c induced by TMV infection of tobacco, EMBO J., 5:37–40.
Cornelissen, B. J. C., Horowitz, J., Van Kan, J. A., Goldberg, R. B., and Bol, J. F. (1987). Structure of tobacco genes encoding pathogenesis-related proteins from the PR-1 group, Nucl. Acids Res., 15:6799–6811.
Cote, F., Cutt, J. R., Asselin, A., and Klessig, D. F. (1991). Pathogenesis-related acidic β-1,3-glucanase genes of tobacco are regulated by both stress and developmental signs, Mol. Plant–Microbe Interact., 4:173–181.
Cote, F., Letarte, J., Grenier, J., Trudel, J., and Asselin, A. (1989). Detection of β-1,3-glucanase activity by native polyacrylamide gel electrophoresis: Application to tobacco pathogenesis-related proteins, Electrophoresis, 10:527–529.

Coutts, R. H. A. (1978). Alterations in the soluble protein pattern of tobacco and cowpea leaves following inoculation with tobacco necrosis virus, Plant Sci. Lett., 12:189–197.

Cramer, C. L., Ryder, T. B., Bell, J. N., and Lamb, C. J. (1985). Rapid switching of plant gene expression by fungal elicitor, Science, 227:1240–1243.

Crowell, D. N., John, M. J., Russell, D., and Amasino, R. M. (1992). Characterisation of a stress-induced developmentally regulated gene family from soybean, Plant Mol. Biol., 18:459–466.

Cusack, M., and Pierpoint, W. (1988). Similarities between sweet protein thaumatin and a pathogenesis-related protein of tobacco, Phytochemistry, 27:3817–3821.

Cutt, J. R., Dixon, D. C., Carr, J. P., and Klessig, D. F. (1988). Isolation and nucleotide sequence of cDNA clones for the pathogenesis-related proteins PR1a, PR1b and PR1c of *Nicotiana tabacum* cv. *Xanthi* nc induced by TMV infection, Nucl. Acids. Res., 16:9861.

Davis, L. L., and Bartinicki-Garcia, S. (1984). Chitosan synthesis by the tandem action of chitin synthetase and chitin deacetylase from *Mucor rouxii*, Biochemistry, 23: 1065–1073.

Dean, J. F. D., and Anderson, J. D. (1991). Ethylene biosynthesis-inducing xylanase II. Purification and physical characterization of the enzyme produced by *Trichoderma viride*, Plant Physiol., 95:316–323.

De Loose, M., Alliotte, T., Gheysen, G., Genetello, C., Gielen, J., Soetaert, P., Van Montagu, M., and Inze, D. (1988). Primary structure of a hormonally regulated β-glucanase of *Nicotiana plumbaginifolia*, Gene, 70:13–23.

de Tapia, M., Bergman, P., Awade, A., and Burkard, G. (1986). Analysis of acid extractable bean leaf proteins induced by mercuric chloride treatment and alfalfa mosaic virus infection. Partial purification and characterization, Plant Sci., 45:167–177.

De Wit, P. J. G. M., Burlage, M. B., and Hammond, K. E. (1986). The occurrence of host-pathogen interaction-specific proteins in the apoplast of *Cladosporium fulvum* (syn. *Fulvia fulva*)-infected tomato leaves, Physiol. Mol. Plant Pathol., 29:159–172.

De Wit, P. J. G. M., and Spikman, G. (1982). Evidence for the occurrence of race and cultivar-specific elicitors of necrosis in intercellular-fluids of compatible interactions of *Cladosporium fulvum* and tomato, Physiol. Plant Pathol., 21:1–11.

De Wit, P. J. G. M., and Van der Meer, F. E. (1986). Accumulation of the pathogenesis-related tomato leaf protein P14 as an early indicator of incompatibility in the interaction between *Cladosporium fulvum* (syn. *Fulvia fulva*) and tomato, Physiol. Mol. Plant Pathol., 28:203–214.

Didierjean, K., Frendo, P., and Burkard, G. (1992). Stress responses in maize: sequence analysis of cDNAs encoding glycine-rich proteins, Plant Mol. Biol., 18:847–849.

Dixelius, C. (1994). Presence of the pathogenesis-related proteins 2, Q and S in stressed *Brassica napus* and *B. nigra* plantlets, Physiol. Mol. Plant Pathol., 44:1–8.

Dixon, R. A., and Harrison, M. J. (1990). Activation, structure, and organization of genes involved in microbial defense in plants, Adv. Genet., 28:165–234.

Domingo, C., Conejero, V., and Vera, P. (1994). Genes encoding acidic and basic class III β-1,3-glucanase are expressed in tomato plants upon viroid infection, Plant Mol. Biol., 24:725–732.

Dunsmuir, P., Howie, W., Newbigin, E., and Suslow, T. (1991). Resistance to *Rhizoctonia solani* in transgenic tobacco expressing the *Serratia marcescens* chitinase gene, 3rd Inst. Cong. Int. Soc. Plant Mol. Biol., p. 1313.

Ebrahim-Nesbat, F., Behnke, S., Kleinhofs, A., and Apel, K. (1989). Cultivar-related differences in the distribution of cell-wall bound thionins in compatible and incompatible interactions between barley and powdery mildew, Planta, 179:203–210.

Ecker, J. R., and Davis, R. W. (1987). Plant defense genes are regulated by ethylene, Proc. Natl. Acad. Sci. USA, 84:5202–5206.

Edelbaum, O., Ilan, N., Grafi, G., Sher, N., Stram, Y., Novick, D., Tal, N., Sela, I., and Rubinstein, M. (1990). Two antiviral proteins from tobacco: purification and characterization by monoclonal antibodies to human β-interferon, Proc. Natl. Acad. Sci. USA, 87:588–592.

Edelbaum, O., Sher, N., Rubinstein, M., Novick, D., Tal, N., Moyer, M., Ward, E., Ryals, J., and Sela, I. (1991). Two antiviral proteins, gp 35 and gp 22, correspond to β-1,3-glucanase and an isoform of PR-5, Plant Mol. Biol., 17:171–173.

Edens, L., Heslinga, L., Klok, R., Ledeboer, A. M., Maat, J., Toonen, M. Y., Visser, C., and Verrips, C. T. (1982). Cloning of cDNA encoding the sweet-tasting plant protein thaumatin and its expression in *Escherichia coli*, Gene, 18:1–12.

Edington, B. V., Lamb, C. J., and Dixon, R. A. (1991). cDNA cloning and characterization of a putative β-1,3-D-glucanase transcript induced by fungal elicitor in bean cell suspension cultures, Plant Mol. Biol., 16:81–94.

Edreva, A. M. (1990). Induction of pathogenesis-related proteins in tobacco leaves by physiological (non-pathogenic) disorders, J. Exp. Bot., 41:701–703.

El Ghaouth, A., Arul, J., Grenier, J., Benhamou, N., Asselin, A., and Belanger, R. (1994). Effect of chitosan on cucumber plants: suppression of *Pythium aphanidermatum* and induction of defense reactions, Phytopathology, 84:313–320.

Endo, Y., Tsurugi, K., and Ebert, R. F. (1988). The mechanism of action of barley toxin: a type I ribosome inactivating-protein with RNA N-glycosidase activity, Biochem. Biophys. Acta, 954:224–226.

Enkerli, J., Gisi, V., and Mosinger, E. (1993). Systemic acquired resistance to *Phytophthora infestans* in tomato and the role of pathogenesis related proteins, Physiol. Mol. Plant Pathol., 43:161–171.

Enyedi, A. J., Yalpani, N., Paul, S., and Raskin, I. (1992). Signal molecules in systemic plant resistance to pathogens and pests, Cell, 70:879–886.

Ernst, D., Schraudner, M., Langebartels, C., and Sandermann, H. Jr. (1992). Ozone-induced changes of mRNA levels of β-1,3-glucanase, chitinase and 'pathogenesis-related' protein 1b in tobacco plants, Plant Mol. Biol., 20:673–682.

Espelie, K. E., Franceschi, V. R., and Kolattukudy, P. E. (1986). Immunocytochemical localization and time course of appearance of an anionic peroxidase associated with suberization in wound-healing potato tuber tissue, Plant Physiol., 81:487–492.

Eyal, Y., and Fluhr, R. (1991). Cellular and molecular biology of pathogenesis-related proteins, Oxford Surveys of Plant Molecular and Cellular Biology (B. J. Miflin, ed.), Oxford University Press, Oxford, pp. 223–254.

Eyal, Y., Sagee, O., and Fluhr, R. (1992). Dark induced accumulation of a basic pathogenesis-related (PR-1) transcript and a light requirement for its induction by ethylene, Plant Mol. Biol., 19:589–599.

Falkenstein, E., Groth, B., Mithofer, A., and Weiler, E. W. (1991). Methyl jasmonate and linolenic acid are potent inducers of tendril coiling, Planta, 185:316–322.

Farmer, E. E., Johnson, R. R., and Ryan, C. A. (1992). Regulation of expression of proteinase inhibitor genes by methyl jasmonate and jasmonic acid. Plant Physiol., 98:995–1002.

Farmer, E. E., and Ryan, C. A. (1990). Interplant communication: Air-borne methyl jasmonate induces synthesis of proteinase inhibitors in plant leaves, Proc. Natl. Acad. Sci. USA, 87:7713–7716.

Faye, L., and Chrispeels, M. J. (1989). Apparent inhibition of β-fructosidase secretion by tunicamycin may be explained by breakdown of the unglycosylated protein during secretion, Plant Physiol., 89:845–851.

Felix, G., and Meins, F. (1985). Purification, immunoassay and characterization of an abundant, cytokinin-regulated polypeptide in cultured tobacco tissues. Evidence the protein is a β-1,3-glucanase, Planta, 164:424–428.

Felix, G., and Meins, F. (1986). Developmental and hormonal regulation of β-1,3-glucanase in tobacco, Planta, 167:206–211.

Felix, G., and Meins, F. (1987). Ethylene regulation of β-1,3-glucanase in tobacco, Planta, 172:386–392.

Fincher, G. B., Lock, P. A., Morgan, M. M., Lingelbach, K., Wettenhall, R. E. H., Mercer, J. F. B., Brandt, A., and Thomsen, K. K. (1986). Primary structure of the (1-3, 1-4)-β-D-glucan 4-glucanohydrolase from barley aleurone, Proc. Natl. Acad. Sci. USA, 83:2081–2085.

Fink, W., Liefland, M., and Mendgen, K. (1988). Chitinase and β-1,3-glucanases in the apoplastic compartment of oat leaves (*Avena sativa* L.), Plant Physiol., 88:270–275.

Fischer, W., Christ, U., Baumgartner, M., Erismann, K. H., and Mosinger, E. (1989). Pathogenesis-related proteins of tomato. II. Biochemical and immunological characterization, Physiol. Mol. Plant Pathol., 35:67–83.

Flach, J., Pilet, P. E., and Jolles, P. (1992). What's new in chitinase research, Experientia 48:701–716.

Fleming, T. M., McCarthy, D. A., White, R. F., Antoniw, J. F., and Mikkelsen, J. D. (1991). Induction and characterization of some of the pathogenesis-related proteins in sugar beet, Physiol. Mol. Plant Pathol., 39:147–160.

Fortin, M. G., Parent, J. G., and Asselin, A. (1985). A comparative study of two groups of b proteins (pathogenesis-related) from the intercellular fluid of *Nicotiana* leaf tissue infected by tobacco mosaic virus, Can. J. Bot., 63:932–937.

Fraser, R. S. S. (1981). Evidence for the occurrence of the "pathogenesis-related" proteins in leaves of healthy tobacco plants during flowering, Physiol. Plant Pathol., 19:69–76.

Fristensky, B., Horivitz, D., and Hadwiger, L. A. (1988). cDNA sequences for pea disease resistance response genes, Plant Mol. Biol., 11:713–715.

Fritig, B., Kauffmann, S., Rouster, J., Dumas, B., Geoffroy, P., Kopp, M., and Legrand, M. (1990). Defense proteins, glycanhydrolases and oligosaccharide signals in plant virus interactions. Recognition and Response in Plant Virus Interactions (R. S. S. Fraser, ed.), NATO ASI Series, Vol. H.41, Springer-Verlag, Berlin, pp. 375–394.

Fritig, B., Rouster, S., Kauffmann, S., Stinzi, A., Geoffroy, P., Kopp, M., and Legrand, M. (1989). Virus-induced glycanhydrolases and effects of oligosaccharide signals on plant virus interactions, Signal Molecules in Plants and Plant–Microbe Interactions (Lugtenberg, B. ed.), Plenum Press, New York, pp. 161–168.

Fujino, D. W., Burger, D. W., Yang, S. F., and Bradford, K. J. (1988). Characterization of an ethylene overproducing mutant of tomato (*Lycopersicon esculentum* Mill, Cultivar VFN 8), Plant Physiol., 88:774–779.

Fukuda, Y., and Shinshi, H. (1994). Characterization of a novel *cis*-acting element that is responsive to a fungal elicitor in the promoter of a tobacco class I chitinase gene, Plant Mol. Biol., 24:485–493.

Gaffney, T., Friedrich, L., Vernooij, B., Negrotto, D., Nye, G., Uknes, S., Ward, E., Kessmann, H., and Ryals, J. (1993). Requirement of salicylic acid for the induction of systemic acquired resistance, Science, 261:754–756.

Garcia Breijo, F. J., Garro, R., and Conejero, V. (1990). $C_7(P_{32})$ and $C_6(P_{34})$ PR proteins induced in tomato leaves by citrus exocortis viroid infection are chitinases, Physiol. Mol. Plant Pathol., 36:249–260.

Garcia-Olmedo, F., Rodriguez-Pelenzuela, P., Hernandez-Lucas, C., Ponz, F., Marana, C., Carmona, M. J., Casacuberta, J. M., Puigdomenech, P., and San Sengundo, B. (1991). A gene coding for a basic pathogenesis-related (PR-like) protein from *Zea mays*. Molecular cloning and induction by a fungus (*Fusarium moniliforme*) in germinating maize seeds, Plant Mol. Biol., 16:527–536.

Garcia-Olmedo, F., Rodriguez-Pelenzuela, P., Hernandez-Lucas, C., Ponz, F., Marana, C., Carmona, M. J., Lopez-Fando, J., Fernandez, J. A., and Carbonero, P. (1989). The thionins: a protein family that includes purothionins, viscotoxins and crambins, Oxford Surv. Plant Mol. Cell Biol., 6:31–60.

Garcia-Olmedo, F., Salcedo, G., Sanchez-Monge, R., Gomez, L., Royo, J., and Carbonero, P. (1987). Plant proteinaceous inhibitors of proteinases and α-amylases, Oxf. Surv. Plant Mol. Cell Biol., 4:275–334.

Gaynor, J. J. (1988). Primary structure of an endochitinase mRNA from *Solanum tuberosum*, Nucl. Acids Res., 16:5210.

Geoffroy, P., Legrand, M., and Fritig, B. (1990). Isolation and characterization of a proteinaceous inhibitor of microbial proteinases induced during the hypersensitive of tobacco to tobacco mosaic virus, Mol. Plant–Microbe Interact., 3:327–333.

Gessler, C., and Kuc, J. (1982). Appearance of a host protein in cucumber plants infected with viruses, bacteria and fungi, J. Exp. Bot., 33:58–66.

Gheysen, G., Inze, D., Soetaert, P., Van Montagu, M., and Castresana, C. (1990). Sequence of a *Nicotiana plumbaginifolia* β(1,3)-glucanase gene encoding a vacuolar isoform, Nucleic Acids Res., 18:6685.

Gianinazzi, S., and Ahl, P. (1983). The genetic and molecular basis of b-proteins in the genus *Nicotiana*, Neth. J. Plant Pathol., 89:275–281.

Gianinazzi, S., Ahl, P., Cornu, A., Scalla, R., and Cassini, R. (1980). First report of host b-protein appearance in response to a fungal infection in tobacco, Physiol. Plant Pathol., 16:337–342.

Gianinazzi, S., and Kassanis, B. (1974). Virus resistance in plants by polyacrylic acid, J. Gen. Virol., 23:1–9.

Gianinazzi, S., Pratt, A., Shewry, P. R., and Miflin, B. J. (1977). Partial purification and preliminary characterization of soluble leaf proteins specific to virus-infected tobacco plants, J. Gen. Virol., 34:345–351.

Gillikin, J. W., Burkhart, W., and Graham, J. S. (1991). Complete amino acid sequence of a polypeptide from *Zea mays* similar to the pathogenesis-related-1 family, Plant Physiol., 96:1372–1375.

Glazer, A. N., Barel, A. O., Howard, J. B., and Brown, D. M. (1969). Isolation and characterization of fig lysozyme, J. Biol. Chem., 244:3583–3589.

Godiard, L., Ragueh, F., Froissard, D., Leguay, J. J., Grosset, J., Chartier, Y., Meyer, Y., and Marco, Y. (1990). Analysis of the synthesis of several pathogenesis-related proteins in tobacco leaves infiltrated with water and with compatible and incompatible isolates of *Pseudomonas solanacearum*, Mol. Plant–Microbe Interact., 3:207–213.

Gomez, J., Sanchez-Martinez, D., Stiefel, V., Riagu, J., Puigdomenech, P., and Pages, M. (1988). A gene induced by the plant hormone abscisic acid in response to water stress encodes a glycine-rich protein, Nature, 334:262–264.

Gordon-weeks, R., White, R. F., and Pierpoint, W. S. (1991). Evidence for the presence of endogenous inducers of the accumulation of pathogenesis-related (PR-1) proteins in leaves of virus-infected tobacco plants and of an interspecific *Nicotiana* hybrid, Physiol. Mol. Plant Pathol., 38:375–392.

Graham, J. S., Burkhart, W., Xiong, J., and Gillikin, J. (1992). Complete amino acid sequence of soybean leaf. P21, Plant Physiol., 98:163–165.

Graham, J. S., Hall, G., Pearce, G., and Ryan, C. A. (1986). Regulation of synthesis of proteinase inhibitors I and II mRNAs in leaves of wounded tomato plants, Planta, 169:399–405.

Graham, J. S., Pearce, G., Merryweather, J., Titani, K., Ericsson, L. H., and Ryan, C. A. (1985). Wound-induced proteinase inhibitors from tomato leaves. I. The cDNA-deduced primary structure of pre-inhibitor I and its post-translational processing, J. Biol. Chem., 260:6555–6560.

Granell, A., Belles, J. M., and Conejero, V. (1987). Induction of pathogenesis-related proteins in tomato by citrus exocortis viroid, silver ion and ethephon, Physiol. Mol. Plant Pathol., 31:83–90.

Grassmann, W., Zechmeister, L., Bender, R., and Toth, G. (1934). Uber die chitin spaltung durch Emulsin-preparate. III. Mitteil uber enzymatosche spaltung fur polysacchariden, Berichte, 67:1–15.

Green, T. R., and Ryan, C. A. (1972). Wound-induced proteinase inhibitor in plant leaves: a possible defense mechanism against insects, Science, 175:776–777.

Grenier, J., and Asselin, A. (1990). Some pathogenesis-related proteins are chitosanases with lytic activity against fungal spores, Mol. Plant–Microbe Interact., 3:401–407.

Grosset, J., Meyer, Y., Chartier, Y., Kauffmann, S., Legrand, M., and Fritig, B. (1990). Tobacco mesophyll protoplasts synthesize 1,3-β-glucanase, chitinases and "osmotins" during in vitro cultures, Plant Physiol., 92:520–527.

Gu, Q., Kawata, C. E., Morse, M. J., Wu, H. M., and Cheung, A. Y. (1992). A flower-specific cDNA encoding a novel thionin in tobacco, Mol. Gen. Genet., 234:89–96.

Gugler, K., Defago, G., and Brodelius, P. (1990). PR-proteins b in *Nicotiana* plants and cell suspension cultures, Physiol. Plant., 79:572.

Gurusiddaiah, S., Kuo, T., and Ryan, C. A. (1972). Immunological comparisons of chymotrypsin inhibitor I among several genera of the Solanaceae, Plant Physiol., 50:627–631.

Hahlbrock, K., and Scheel, D. (1987). Biochemical responses of plants to pathogens, Innovative Approaches to Plant Disease Control (I. Chet, ed.), John Wiley and Sons, pp. 229–254.

Ham, K., Kauffmann, S., Albersheim, P., and Darvill, A. G. (1991). Host-pathogen interactions. XXXIX. A soybean pathogenesis-related protein with β-1,3-glucanase activity releases phytoalexin elicitor-active heat-stable fragments from fungal walls, Mol. Plant–Microbe Interact., 4:545–552.

Harms, T. (1992). Engineering genetic disease resistance into crops: biotechnological approaches to crop protection, Crop Protection, 11:291–306.

Hattori, T., Nakagawa, T., Maeshima, M., Nakamura, K., Asahi, T. (1985). Molecular cloning and nucleotide sequence of cDNA for sporamin; the major soluble protein of sweet potato tuberous roots, Plant Mol. Biol., 5:313–320.

Hattori, T., and Nakamura, K. (1988). Genes coding for the major tuberous root protein of sweet potato: identification of putative regulatory sequence in the 5′-upstream region, Plant Mol. Biol., 11:417–426.

Hattori, T., Yoshida, N., and Nakamura, K. (1989). Structural relationship among the members of a multigene family coding for the sweet potato tuberous root storage proteins, Plant Mol. Biol., 13:563–572.

Hedrick, S. A., Bell, J. N., Boller, T., and Lamb, C. J. (1988). Chitinase cDNA cloning and mRNA induction by fungal elicitor, wounding and infection. Plant Physiol., 86:182–186.

Heitz, T., Fritig, B., and Legrand, M. (1994). Local and systemic accumulation of pathogenesis-related proteins in tobacco plants infected with tobacco mosaic virus. Mol. Plant–Microbe Interact., 7:776–779.

Heitz, T., Geoffroy, P., Fritig, B., and Legrand, M. (1991). Two apoplastic α-amylases are induced in tobacco by virus infection, Plant Physiol., 97:651–656.

Hendricks, T., Vreugdenhil, D., and Stiekema, W. J. (1991). Patatin and four serine proteinase inhibitor genes are differentially expressed during potato tuber development, Plant Mol. Biol., 17:384–394.

Herget, T., Schell, J., and Schreier, P. H. (1990). Elicitor-specific induction of one member of the chitinase gene family in *Arachis hypogaea*, Mol. Gen. Genet., 224:469–476.

Herman, G., Lehmann, J., Peterson, A., Sembdner, G., Weidhase, R. A., and Pathier, B. (1989). Species and tissue specificity of jasmonate-induced abundant proteins, J. Plant Physiol., 134:703–709.

Hirasawa, E. (1989). Auxins induce α-amylase activity in pea cotyledons, Plant Physiol., 91:484–486.

Hoffland, E., Pieterse, C. M. J., Bik, L., and van Pelt, J. A. (1995). Induced systemic resistance in radish is not associated with accumulation of pathogenesis-related proteins, Physiol. Mol. Plant Pathol., 46:309–320.

Hogue, R., and Asselin, A. (1987). Detection of 10 additional pathogenesis-related (b) proteins in intercellular fluid extracts from stressed 'Xanthi-nc' tobacco leaf tissue, Can. J. Bot., 65:476–481.

Hoj, P. B., Hartman, D. J., Morrice, N. A., Doan, D. N. P., and Fincher, G. B. (1989). Purification of (1-3)-β-glucan endohydrolase isoenzyme II from germinated barley and determination of its primary structure from cDNA clone, Plant Mol. Biol., 13:31–42.

Holden, D. W., and Strange, R. W. (1984). Biochemical characteristics of sporelings of *Puccinia graminis* f.sp. *tritici* and protoplasts from wheat isolines of potential significance in specificity, Plant Pathol., 33:233–243.

Holwerda, B. C., Padgett, H. S., and Rogers, J. C. (1992). Efficient vacuolar targeting of proaleurain is mediated by interaction of short contiguous peptide determinants, Plant Cell, 4:307–318.

Hooft van Huijsduijnen, R. A. M., Alblas, S. W., De Rijk, R. H., and Bol, J. F. (1986a). Induction by salicyclic acid of pathogenesis-related proteins and resistance to alfalfa mosaic virus infection in various plant species, J. Gen. Virol., 67:2135–2143.

Hooft van Huijsduijnen, R. A. M., Cornelissen, B. J. C., Van Loon, L. C., Van Boom, J. H., Tromp, M., and Bol, J. F. (1985). Virus-infected synthesis of pathogenesis-related proteins in tobacco, EMBO J., 4:2167–2171.

Hooft van Huijsduijnen, R. A. M., Kauffmann, S., Brederode, F. T., Cornelissen, B. J. C., Legrand, M., Fritig, B., and Bol, J. F. (1987). Homology between chitinases that are induced by TMV infection of tobacco, Plant Mol. Biol., 9:411–420.

Hooft van Huijsduijnen, R. A. M., Van Loon, L. C., and Bol, J. F. (1986b). cDNA cloning of six mRNAs induced by TMV infection of tobacco and characterization of their translation products, EMBO J., 5:2057–2061.

Howard, J. B., and Glazer, A. N. (1967). Studies of the physiochemical and enzymatic properties of papaya lysozyme, J. Biol. Chem., 242:5715–5723.

Huang, J. K., Wen, L., Swegle, M., Tran, H. C., Thin, T. H., Naylor, H. M., Muthukrishnan, S., and Reeck, G. R. (1991). Nucleotide sequence of a rice genomic clone that encodes a class I endochitinase, Plant Mol. Biol., 16:479–480.

Hunt, D. C., and Chrispeels, M. J. (1991). The signal peptide of a vacuolar protein is necessary and sufficient for the efficient secretion of a cytosolic protein, Plant Physiol., 96:18–25.

Hyunh, Q. K., Hironaka, C. M., Levine, E. B., Smith, C. E., Borgmeyer, J. R., and Shah, D. M. (1992). Antifungal proteins from plants. Purification, molecular cloning and antifungal properties of chitinases from maize seed, J. Biol. Chem., 267:6635–6640.

Irving, H. R., and Kuc, J. A. (1990). Local and systemic induction of peroxidase, chitinase and resistance in cucumber plants by K_2HPO_4, Physiol. Mol. Plant Pathol., 37:355–366.

Ishige, F., Mori, H., Yamazaki, K., and Imaseki, H. (1991). The effect of ethylene on the coordinated synthesis of multiple proteins. Accumulation of an acidic chitinase and a basic glycoprotein induced by ethylene in leaves of azuki beans, *Vigna angulasis*, Plant Cell Physiol., 32:681–690.

Ishige, F., Mori, H., Yamazaki, K., and Imasaki, H. (1993a). Identification of a basic glycoprotein induced by ethylene in primary leaves of azuki bean as a cationic peroxidase, Plant Physiol., 101:193–199.

Ishige, F., Mori, H., Yamazaki, K., and Imaseki, H. (1993b). Cloning of a complementary DNA that encodes an acidic chitinase which is induced by ethylene and expression of the corresponding gene, Plant Cell Physiol., 34:103–111.

Iturriaga, E. A., Leech, M. J., Paul Barratt, D. H., and Wang, T. L. (1994). Two ABA-responsive proteins from pea (*Pisum sativum* L.) are closely related to intracellular pathogenesis-related proteins, Plant Mol. Biol., 24:235–240.

Jach, G., Logemann, S., Wolf, G., Oppenheim, A. B., Chet, I., Schell, J., and Logemann,

J. (1992). Expression of a bacterial chitinase leads to improved resistance of transgenic tobacco plants against fungal infection, Biopractice, 1:33–40.
Jacobsen, J. V., Hansen, A. D., and Chandler, P. C. (1986). Water stress enhances expression of an α-amylase gene in barley leaves, Plant Physiol., 80:350–359.
Jacobsen, J. V., and Higgins, T. J. V. (1982). Characterization of the α-amylases synthesized by aleurone layers of Himalaya barley in response to gibberellic acid, Plant Physiol., 70:1647–1653.
Jamet, E., and Fritig, B. (1986). Purification and characterization of 8 of the pathogenesis-related proteins in tobacco leaves reacting hypersensitively to tobacco mosaic virus, Plant Mol. Biol., 6:69–80.
Jamet, E., Kopp, M., and Fritig, B. (1985). The pathogenesis-related proteins of tobacco: their labelling from ^{14}C-amino acids in leaves reacting hypersensitively to infection by tobacco mosaic virus, Physiol. Plant Pathol., 27:29–41.
Johnson, R., and Ryan, C. A. (1990). Wound-inducible potato inhibitor II genes: enhancement of expression by sucrose, Plant Mol. Biol., 14:527–536.
Jondle, D. J., Coors, J. G., and Duke, S. H. (1989). Maize leaf β-1,3-glucanase activity in relation to resistance to *Exserohilum turcicum*, Can. J. Bot., 67:263–266.
Joosten, M. H. A. J., Bergmans, C. J. B., Meulenhoff, E. J. S., Cornelissen, B. J. C., and de Wit, P. J. G. M. (1990). Purification and serological characterization of three basic 15 kD pathogenesis-related proteins from tomato, Plant Physiol., 94:585–591.
Joosten, M. H. A. J., and De Wit, P. J. G. M. (1989). Identification of several pathogenesis-related proteins in tomato leaves inoculated with *Cladosporium fulvum* (syn. *Fulvia fulva*) as 1,3-β-glucanases and chitinases, Plant Physiol., 89:945–951.
Joosten, M. H. A., Verbakel, H. M., Nettekoven, M. E., Van Leeuwen, Van Der Vossen, R. T. M., and De Wit, P. J. G. M. (1995). The phytopathogenic fungus *Cladosporium fulvum* is not sensitive to the chitinase and β-1,3-glucanase defense proteins of its host, tomato, Physiol. Mol. Plant Pathol., 46:45–59.
Jutidamrongphan, W., Andersen, J. B., Mackinnon, G., Manners, J. M., Simpson, R. S., and Scott, K. J. (1991). Induction of β-1,3-glucanase in barley in response to infection by fungal pathogens, Mol. Plant–Microbe Interact., 4:234–238.
Kan, J. A. L., Cornelissen, B. J. C., and Bol, J. F. (1988). A virus-inducible tobacco gene encoding a glycine-rich protein shares putative elements with the ribulose biphosphate carboxylase small subunit gene, Mol. Plant–Microbe Interact., 1:107–112.
Kandasamy, M. K., and Kristen, U. (1987). Developmental aspects of ultrastructure, histochemistry and receptivity of the stigma of *Nicotiana sylvestris*, Ann. Bot., 60:427–437.
Kapooria, R. G., and Mendgen, K. (1985). Infection structures and their surface changes during differentiation in *Uromyces fabae*, Phytopathol. Z., 113:317–323.
Kauffmann, S., Legrand, M., and Fritig, B. (1990). Isolation and characterization of six pathogenesis-related (PR) proteins of Samsun NN tobacco, Plant Mol. Biol., 14:381–390.
Kauffmann, S., Legrand, M., Geoffroy, P., and Fritig, B. (1987). Biological function of 'pathogenesis-related' proteins: four PR proteins of tobacco have β-1,3-glucanase activity, EMBO J., 6:3209–3212.
Kauss, H. (1989). Fluorometric measurement of callose and other 1,3-beta glucans,

Modern Methods of Plant Analysis. Vol. 10 (A. F. Linskens and J. F. Jaction, eds.), Springer, Berlin, pp. 127–137.

Kauss, H., and Bauch, S. (1988). Chitin deacetylase from *Colletotrichum lindemuthianum*, Methods Enzymol., 161B:518–523.

Kauss, H., Jeblick, W., and Young, D. H. (1983). Chitin deacetylase from the plant pathogen *Colletotrichum lindemuthianum*, Plant Sci. Lett., 28:231–236.

Keefe, D., Hinz, U., and Meins, F. Jr. (1990). The effect of ethylene on the cell-type-specific and intracellular localization of β-1,3-glucanase and chitinase in tobacco leaves, Planta, 182:43–51.

Keil, M., Sanchez-Serrano, J. J., and Willmitzer, L. (1989). Both wound-inducible and tuber-specific expression are mediated by the promoter of a single member of the proteinase inhibitor II gene family, EMBO J., 8:1323–1330.

Keller, B., Schmid, J., and Lamb, C. J. (1989). Vascular expression of a bean cell wall glycine-rich protein-β-glucuronidase fusion in transgenic tobacco, EMBO J., 8:1309–1314.

Kim, S. R., Choi, J. L., Costa, M. A., and An, G. (1992). Identification of G-box sequence as an essential element for methyl jasmonate response of potato proteinase inhibitor II promoter, Plant Physiol., 99:627–631.

Kim, Y. J., and Huang, B. K. (1994). Differential accumulation of β-1,3-glucanase and chitinase isoforms in pepper stems infected by compatible and incompatible isolates of *Phytophthora capsici*, Physiol. Mol. Plant Pathol., 45:195–209.

Kim, S. G., and Yoo, J. Y. (1992). Changes in protein patterns resulting from infection of rice leaves with *Xanthomonas oryzae* pv. *oryzae*, Mol. Plant Microbe Interact., 5:356–360.

King, G. J., Hussey, Jr. C. E., and Turner, V. A. (1986). A protein induced by NaCl in suspension cultures of *Nicotiana tabacum* accumulates in whole plant roots, Plant Mol. Biol., 7:441–449.

King, G. J., Turner, V. A., Hussey, C. E., Wurtele, E. S., and Lee, S. M. (1988). Isolation and characterization of a tomato cDNA clone which codes for a salt induced protein, Plant Mol. Biol., 10:401–412.

Klausner, R. D., and Sitia, R. (1990). Protein degradation in the endoplasmic reticulum, Cell, 62:611–614.

Koga, D., Hirata, T., Sueshige, N., Tanaka, S., and Ide, A. (1992). Induction patterns of chitinases in yam callus by inoculation with autoclaved *Fusarium oxysporum*, ethylene, and chitin and chitosan oligosaccharides, Biosci. Biotech. Biochem., 56:280–285.

Kombrink, E., Schroder, M., and Hahlbrock, K. (1988). Several pathogenesis-related proteins in potato are 1,3-β-glucanases and chitinases, Proc. Natl. Acad. Sci. USA, 85:782–786.

Kragh, K. M., Jacobsen, S., and Mikkelsen, J. D. (1990). Induction, purification and characterization of barley leaf chitinase, Plant Sci., 71:55–68.

Krebs, S. L., and Grumet, R. (1993). Characterization of celery hydrolytic enzymes produced in response to infection by *Fusarium oxysporum*, Physiol. Mol. Plant Pathol., 43:193–208.

Kukuruzinska, M. A., Bergh, M. L. E., and Jackson, B. J. (1987). Protein glycosylation in yeast, Annu. Rev. Biochem., 56:915–944.

Kumar, V., and Spencer, M. E. (1992). Nucleotide sequence of an osmotin cDNA from the *Nicotiana tabacum* cv. White Burley generated by the polymerase chain reaction, Plant Mol. Biol., 18:621–622.

Kuo, T., Pearce, G., and Ryan, C. A. (1984). Isolation and characterization of proteinase inhibitor I from etiolated tobacco leaves, Arch. Biochem. Biophys., 230:504–510.

Kush, A., Goyvaerts, E., Chye, M. L., and Chua, N. H. (1990). Laticifer-specific gene expression in *Hevea brasiliensis* (rubber tree), Proc. Natl. Acad. Sci. USA, 87:1787–1790.

Lamb, C. J., Lawton, M. A., Dron, M., and Dixon, R. A. (1989). Signals and transduction mechanisms for activation of plant defenses against microbial attack, Cell, 56:215–224.

Lamb, C. J., Ryals, J. A., Ward, E. R., and Dixon, R. A. (1992). Emerging strategies for enhancing crop resistance to microbial pathogens, Biotechnology, 10:1436–1445.

Langebartels, C., Kerner, K., Leonardi, S., Schraudner, M., Trost, M., Heller, W., and Sandermann, H. (1991). Biochemical plant responses to ozone. 1. Differential induction of polyamine and ethylene biosynthesis in tobacco, Plant Physiol., 95:882–889.

Laskowski, M., and Kato, I. (1980). Protein inhibitors of proteinases, Annu. Rev. Biochem., 49:593–626.

Lawton, K. A., Beck, J., Potter S., Ward, E., and Ryals, J. (1994a). Regulation of cucumber class III chitinase gene expression, Mol. Plant–Microbe Interact., 7:48–57.

Lawton, K. A., Potter, S. L., Uknes, S., and Ryals, J. (1994b). Acquired resistance: signal transduction in *Arabidopsis* is ethylene independent, Plant Cell, 6:581–588.

Lawton, K., Uknes, S., Friedrich, L., Gaffney, T., Alexander, D., Goodman, R., Metraux, J. P., Kessmann, H., Ahl Goy, P., Out Rella, M., Ward, E., and Ryals, J. (1993). The molecular biology of systemic acquired resistance, Mechanisms of Defense Responses in Plants (B. Fritig and M. Legrand, eds.), Kluwer-Academic, Dordrecht, pp. 410–420.

Leah, R., Mikkelsen, J. D., Mundy, J., and Svendsen, I. (1987). Identification of a 28,000 dalton endochitinase in barley endosperm, Carlsberg Res. Commun., 52:31–37.

Leah, R., Tommerup, H., Svendsen, I., and Mundy, J. (1991). Biochemical and molecular characterization of three antifungal proteins from barley seed, J. Biol. Chem., 266:1564–1573.

Lee, J. L., Brown, W. E., Pearce, G., Dreher, T. W., Graham, J. L., Ahern, K. B., Pearson, G. D., and Ryan, C. A. (1986). Molecular characterization and phylogenetic studies of a wound-inducible proteinase inhibitor I gene in *Lycopersicon* species, Proc. Natl. Acad. Sci. USA, 83:7277–7281.

Legrand, M., Kauffman, S., Geoffroy, P., and Fritig, B. (1987). Biological function of pathogenesis-related proteins: four tobacco pathogenesis-related proteins are chitinases, Proc. Natl. Acad. Sci. USA, 84:6750–6754.

Leung, D. W. M. (1992). Involvement of chitinase in sexual reproduction of higher plants, Phytochemistry, 31:1899–1900.

Lin, W., Anuratha, G. S., Datta, K., Potrykus, I., Muthukrishnan, S., and Datta, S. K. (1995). Genetic engineering of rice for resistance to sheath blight, Biotechnology, 13:686–691.

Linthorst, H. J. M. (1991). Pathogenesis-related proteins of plants, Crit. Rev. Plant Sci., 10:123–150.

Linthorst, H. J. M., Danhash, N., Brederode, F. T., Van Kan, J. A. L., De Wit, P. J. G. M., and Bol, J. F. (1991). Tobacco and tomato PR proteins homologous to *Win* and Pro-Hevein lack the "Hevein" domain, Mol. Plant–Microbe Interact, 4:586–592.

Linthorst, H. J. M., Melchers, L. S., Mayer, A., Van Rockel, J. S. C., Cornelissen, B. J. C., and Bol, J. F. (1990a). Analysis of gene families encoding acidic and basic β-1,3-glucanases of tobacco, Proc. Natl. Acad. Sci. USA, 87:8756–8760.

Linthorst, H. J. M., Meuwissen, R. L. J., Kauffman, S., and Bol, J. F. (1989). Constitutive expression of pathogenesis-related proteins PR-1, GPR and PR-S in tobacco has no effect on virus infection, Plant Cell, 1:285–291.

Linthorst, H. J. M., Van Loon, L. C., Van Rossum, C. M. A., Mayer, A., Bol, J. F., Van Roekel, C., Meulenhoff, E. J. S., and Cornelissen, B. J. C. (1990b). Analysis of acidic and basic chitinases from tobacco and *Petunia* and their constitutive expression in transgenic tobacco, Mol. Plant–Microbe Interact., 3:252–258.

Litts, J. C., Simmons, C. R., Karrer, E. E., Huang, N., and Rodriguez, R. L. (1990). The isolation and characterization of a barley 1,3-1,4-β-glucanase gene, Eur. J. Biochem., 194:831–838.

Liu, D., Raghothama, K. G., Hasegawa, P. M., and Bressan, R. A. (1994). Osmotin overexpression in potato delays development of disease symptoms, Proc. Natl. Acad. Sci. USA, 91:1888–1892.

Logemann, J., Jach, G., Tommerup, H., Mundy, J., and Schell, J. (1992). Expression of a barley ribosome-inactivating protein leads to increased fungal protection in transgenic tobacco plants, Biotechnology, 10:305–308.

Logemann, J., Lipphardt, S., Lorz, H., Hauser, I., Willmitzer, L., and Schell, J. (1989). 5′-upstream sequences from the Wun 1 gene are responsible for gene activation by wounding in transgenic plants, Plant Cell, 1:151–158.

Logemann, J., and Schell, J. (1989). Nucleotide sequence and regulated expression of a wound-inducible potato gene (*Wun* 1), Mol. Gen. Genet., 219:81–88.

Logemann, J., Mayer, J. E., Schell, J., and Willmitzer, L. (1988). Differential expression of genes in potato tubers after wounding, Proc. Natl. Acad. Sci. USA, 85:1136–1140.

Logemann, J., Melchers, L. S., Tigelaar, H., Sela-Buurlage, M. B., Ponstein, A. S., van Roekel, J. S. C., Bres-Vloermans, S. A., Dekker, I., Cornelissen, B. J. C., van den Elzen, P. J. M., and Jongedik, E. (1994). Synergistic activity of chitinases and β-1,3-glucanases enhances *Fusarium* resistance in transgenic tomato plants, J. Cell. Biochem. Suppl., 18A:88.

Lopez-Fando, J., Fernandez, J. A., and Carbonero, P. (1989). The thionins: A protein family that includes purothionins, viscotoxins and crambins, Oxford Surv. Plant Mol. Cell Biol., 6:31–60.

Lotan, T., and Fluhr, R. (1990a). Xylanase, a novel elicitor of pathogenesis-related proteins in tobacco uses a non-ethylene pathway for induction, Plant Physiol., 93:811–817.

Lotan, T., and Fluhr, R. (1990b). Function and regulated accumulation of plant pathogenesis-related proteins, Symbiosis, 8:33–46.

Lotan, T., Ori, N., and Fluhr, R. (1989). Pathogenesis-related proteins are developmentally regulated in tobacco flowers, Plant Cell, 1:881–887.

Lucas, J., Henriquez, A. C., Lottspeich, A., Henchen, A., and Sanger, H. L. (1985). Amino

acid sequence of the pathogenesis-related leaf protein p14 in viroid-infected tomato reveals a new type of structurally unfamiliar protein, EMBO J., 4:2745–2749.
Lund, P., and Dunsmuir, P. (1992). A plant signal sequence enhances the secretion of Chi A in transgenic tobacco, Plant Mol. Biol., 18:47–53.
Lynn, K. R. (1989). Four lysozymes from the latex of *Asclepias syriaca*, Phytochemistry, 28:1345–1348.
Mahe, A., Grisvard, J., and Dron, M. (1992). Fungal and plant-specific gene markers to follow the bean anthracnose infection process and normalize a bean chitinase mRNA induction, Mol. Plant–Microbe Interact., 5:242–248.
Maher, E. A., and Dixon, R. A. (1990). Chitinase and glucanase from fungal infected alfalfa, J. Cell Biochem. Suppl., 14E:322.
Malamy, J., Carr, J. P., Klessig, D. F., and Raskin, I. (1990). Salicylic acid: a likely endogenous signal in the resistance response of tobacco to viral infection, Science, 250:1002–1004.
Manners, D. J., and Marshall, J. J. (1973). Some properties of a β-1,3-glucanase from rye, Phytochemistry, 12:547–553.
Mares, M., Meloun, B., Pavlik, M., Kostka, V., and Baudys, M. (1989). Primary structure of cathepsin D inhibitor from potatoes and its structural relationship to soybean trypsin inhibitor family, FEBS Lett., 251:94–98.
Margis-Pinheiro, M., Martin, M., Didierjean, L., and Burkard, G. (1993). Differential expression of bean chitinase genes by virus infection, chemical treatment and UV irradiation, Plant Mol. Biol., 22:659–668.
Margis-Pinheiro, M., Metz-Boutique, M. H., Awade, A., de Tapia, M., le Ret, M., and Burkard, G. (1991). Isolation of a complementary DNA encoding the bean PR4 chitinase: an acidic enzyme with an amino-terminus cysteine-rich domain, Plant Mol. Biol., 17:243–253.
Margossian, L. J., Federman, A. D., Giovannoni, J. J., and Fischer, R. L. (1988). Ethylene-regulated expression of a tomato fruit ripening gene encoding a proteinase inhibitor I with a glutamic residue at the reactive site, Proc. Natl. Acad. Sci. USA, 85:8012–8016.
Martin, M. N. (1991). The latex of *Hevea brasiliensis* contains high levels of both chitinases and chitinases/lysozymes, Plant Physiol., 95:469–476.
Martin, M. N., and Gaynor, J. J. (1988). Expression of chitinase in latex-containing plants, Plant Physiol., 86:S59.
Masoud, S. A., Johnson, L. B., White, F. F., and Reeck, G. R. (1993). Expression of a cysteine proteinase inhibitor (oryzacystatin-I) in transgenic tobacco plants, Plant Mol. Biol., 21:655–663.
Matsuoka, M., Asou, S., and Ohashi, Y. (1985). Transcriptional step is necessary for induction of pathogenesis-related proteins, Proc. Jpn. Acad., 61:486–489.
Matsuoka, M., Asou, S., and Ohashi, Y. (1988). Regulation mechanisms of the synthesis of pathogenesis-related proteins in tobacco leaves, Plant Cell Physiol., 29:1185–1192.
Matsuoka, M., and Ohashi, Y. (1984). Biochemical and serological studies of pathogenesis-related proteins of *Nicotiana* species, J. Gen. Virol., 65:2209–2215.
Matsuoka, M., and Ohashi Y. (1986). Induction of pathogenesis-related proteins in tobacco leaves, Plant Physiol., 80:S505–510.

Matsuoka, K., Matsumoto, S., Hattori, T., Machida, Y., and Nakamura, K. (1990). Vacuolar targeting and posttranslational processing of the precursor to the sweet potato tuberous root storage protein in heterologous plant cells, J. Biol. Chem., 265:19750–19757.

Matsuoka, K., and Nakamura, K. (1991). Propeptide of a precursor to a plant vacuolar protein required for vacuolar targeting, Proc. Natl. Acad. Sci. USA, 88:834–838.

Matsuoka, M., Yamamoto, N., Kano-Murakami, Y., Tanaka, Y., Ozeki, Y., Hirano, H., Kagawa, H., Oshima, M., and Ohashi, Y. (1987). Classification and structural comparison of full-length cDNAs for pathogenesis-related proteins, Plant Physiol., 85:942–946.

Matton, D. P., Bell, B., and Brisson, N. (1990). Nucleotide sequence of a pathogenesis-related gene of potato, Plant Mol. Biol., 14:863–865.

Matton, D. P., and Brisson, N. (1989). Cloning, expression, and sequence conservation of pathogenesis-related gene transcripts of potato, Mol. Plant–Microbe Interact., 2:325–331.

Matton, D. P., Prescott, G., Bertrand, C., Camirand, A., and Brisson, N. (1993). Identification of *cis*-acting elements involved in the regulation of the pathogenesis-related gene STH-2 in potato, Plant Mol. Biol., 22:279–291.

Mauch, F., Hadwiger, L. A., and Boller, T. (1984). Ethylene: symptom, not signal for the induction of chitinase and β-1,3-glucanase in pea pods by pathogens and elicitors, Plant Physiol., 76:607–611.

Mauch, F., Hadwiger, L. A., and Boller, T. (1988a). Antifungal hydrolases in pea tissue. I. Purification and characterization of two chitinases and two β-1,3-glucanases differentially regulated during development and in response to fungal infection, Plant Physiol., 87:325–333.

Mauch, F., Mauch-Mani, B., and Boller, T. (1988). Antifungal hydrolases in pea tissue. II. Inhibition of fungal growth by combinations of chitinase and β-1,3-glucanase, Plant Physiol., 88:936–942.

Mauch, F., and Staehlin, A. (1989). Functional implications of the subcellular localization of ethylene-induced chitinase and β-1,3-glucanase in bean leaves, Plant Cell, 1:447–457.

McGurl, N., Pearce, G., Orozco-Cardenas, M., and Ryan, C. A. (1992). Structure, expression and antisense inhibition of the systemin precursor gene, Science, 255:1570–1573.

McGurl, B., and Ryan, C. A. (1992). The organization of the prosystemin gene, Plant Mol. Biol., 20:405–409.

Meier, I., Hahlbrock, K., and Somssich, I. E. (1991). Elicitor-inducible and constitutive in vivo DNA footprints indicate novel *cis*-acting elements in the promoter of a parsley gene encoding pathogenesis-related protein 1, Plant Cell, 3:309–315.

Meins, F., and Ahl, P. (1989). Induction of chitinase and β-1,3-glucanase in tobacco plants infected with *Pseudomonas tabaci* and *Phytophthora parasitica* var. *nicotianae*, Plant Sci., 61:155–161.

Melville, J. C., and Ryan, C. A. (1972). Chymotrypsin inhibitor I from potatoes. Large scale preparation and characterization of its subunit components, J. Biol. Chem., 11:3445–3453.

Memelink, J., Hoge, J. H. C., and Schilperoort, R. A. (1987). Cytokinin stress changes the

developmental regulation of several defence-related genes in tobacco, EMBO J., 6:3579–3583.
Memelink, J., Linthorst, H. J. M., Schilperoot, R. A., and Hoge, J. H. C. (1990). Tobacco genes encoding acidic and basic isoforms of pathogenesis-related proteins display different expression patterns, Plant Mol. Biol., 14:119–126.
Mendez, E., Moreno, A., Colilla, F., Pelaez, F., Limas, G. G., Mendez, R., Soriano, F., Salinas, M., and de Haro, C. (1990). Primary structure and inhibition of protein synthesis in eukaryotic cell-free system of a novel thionin, γ-hordothionin, from barley endosperm, Eur. J. Biochem., 194:533–539.
Mendgen, K., Lange, M., and Bretschneider, K. (1985). Quantitative estimation of the surface carbohydrates on the infection structures of rust fungi with enzymes and lectins, Arch. Microbiol., 140:307–311.
Metraux, J. P., Ahl Goy, P., Staub, T., Speich, J., Steinemann, A., Ryals, J., and Ward, E. (1991). Induced resistance in cucumber in response to 2,6-dichloro-isonicotinic acid and pathogens, Advances in Molecular Genetics of Plant–Microbe Interactions, Vol. 1, (H. Hennecke and D. P. S. Verma, eds.), Kluwer, Dordrecht, pp. 432–439.
Metraux, J. P., and Boller, T. (1986). Local and systemic induction of chitinase in cucumber plants in response to viral, bacterial and fungal infections, Physiol., Mol. Plant Pathol., 28:161–169.
Metraux, J. P., Burkhart, W., Moyer, M., Dincher, S., Middlesteadt, W., Williams, S., Payne, G., Carnes, M., and Ryals, J. (1989). Isolation of a complementary DNA encoding a chitinase with structural homology to a bifunctional lysozyme/chitinase, Proc. Natl. Acad. Sci. USA, 86:896–900.
Metraux, J. P., Streit, L., and Staub, Th. (1988). A pathogenesis-related protein in cucumber is a chitinase, Physiol. Mol. Plant Pathol., 33:1–9.
Metzler, M. C., Cutt, J. R., and Klessig, D. F. (1991). Isolation and characterization of gene encoding a PR-1 like protein from *Arabidopsis thaliana*, Plant Physiol., 96:346–348.
Meyer, A., Miersch, O., Buttner, C., Dathe, W., and Sembdner, G. (1984). Occurrence of the plant growth regulator jasmonic acid in plants, J. Plant Growth Regul., 3:1–8.
Mitchell, P. J., and Tjian, R. (1989). Transcriptional regulation in mammalian cells by sequence-specific DNA binding proteins, Science, 245:371–378.
Modderman, P. W., Schot, C. P., Klis, F. M., and Wieringa Brants, D. H. (1985). Acquired resistance in hypersensitive tobacco mosaic virus induced by plant cell wall components, Phytopath. Z., 113:165–170.
Mohammed, F., and Sehgal, O. P. (1987a). Alterations in leaf proteins accompanying virus-induced hypersensitive reaction in *Phaseolus vulgaris* L. cv. Pinto, Phytopathology, 77:1765.
Mohammed, F., and Sehgal, O. P. (1987b). Immunoreactivity of viral induced pathogenesis-related proteins, Current Topics in Plant Biochemistry and Physiology, Vol. 6 (D. D. Randall, R. E. Sharp, A. J. Novacky and D. G. Blevins, eds.), University of Missouri, Columbia, p. 172.
Mohan, R., Bajar, A. M., and Kollattukudy, P. E. (1993b). Induction of a tomato anionic peroxidase gene (tap 1) by wounding in transgenic tobacco and activation of *tap 1* GUS and *tap 2*/GUS chimeric gene fusions in transgenic tobacco by wounding and pathogen attack, Plant Mol. Biol., 21:339–354.

Mohan, R., and Kolattukudy, P. E. (1990). Differential activation of expression of a suberization-associated anionic peroxidase gene in near-isogenic resistant and susceptible tomato lines by elicitors of *Verticillium albo-atrum*, Plant Physiol., 92:276–280.

Mohan, R., Vijayan, P., and Kolattukudy, P. E. (1993a). Developmental and tissue-specific expression of a tomato anionic peroxidase (*tap 1*) gene by a minimal promoter, with wound and pathogen induction by an additional 5′-flanking region, Plant Mol. Biol., 22:475–490.

Mohnen, D., Shinshi, H., Felix, G., and Meins, F., Jr. (1985). Hormonal regulation of β-1,3-glucanase messenger RNA levels in cultured tobacco tissues, EMBO J., 4:1631–1635.

Molano, J., Polchek, I., Duran, A., and Cabib, E. (1979). An endochitinase from wheat germ. Activity on nascent and preformed chitin, J. Biol. Chem., 254:4901–4907.

Moore, A. E. and Stone, B. A. (1972). A β-1,3-glucanase hydrolase from *Nicotiana glutinosa*. I. Extraction, purification and physical properties, Biochem. Biophys. Acta, 258:238–247.

Mosolov, V. V., and Shuglin, M. N. (1986). Protein inhibitors of microbial proteinases from wheat, rye and triticale, Planta, 167:595–600.

Mosolov, V. V., Loginova, M. D., Fedurkina, N. V., and Benken, I. I. (1976). The biological significance of proteinase inhibitors in plants, Plant Sci. Lett., 7:77–80.

Mosolov, V. V., Loginova, M. D., Malova, E. I., and Benken, I. I. (1979). A specific inhibitor of *Colletotrichum lindemuthianum* protease from kidney bean (*Phaseolus vulgaris*) seeds, Planta, 144:265–269.

Mozzetti, C., Ferraris, L., Tamietti, G., and Matta, A. (1995). Variation in enzyme activities in leaves and cell suspensions as markers of incompatibility in different *Phytophthora*–pepper interactions, Physiol. Mol. Plant Pathol., 46:95–107.

Mundy, J., and Rogers, J. C. (1986). Selective expression of a probable amylase/protease inhibitor in barley aleurone cells: comparison to the barley amylase/subtilisin inhibitor, Planta, 169:51–63.

Murdock, L. L., Huesing, J. E., Nielsen, S. S., Pratt, R. C., and Shade, R. E. (1990). Biological effects of plant lectins on the cowpea weevil, Phytochemistry, 29:85–89.

Nakamura, K., and Matsuoka, K. (1993). Protein targeting to the vacuole in plant cells, Plant Physiol., 101:1–5.

Nasser, W., De Tapia, M., and Burkard, G. (1990). Maize pathogenesis-related proteins: characterization and cellular distribution of 1,3-β-glucanases and chitinases induced by brome mosaic virus infection or mercuric chloride treatment, Physiol. Mol. Plant Pathol., 36:1–14.

Nasser, W., De Tapia, M., Kauffmann, S., Montasser-Kouhsari, S., and Burkard, G. (1988). Identification and characterization of maize pathogenesis-related proteins. Four maize PR proteins are chitinases, Plant Mol. Biol., 11:529–538.

Nassuth, A., and Sanger, H. L. (1986). Immunological relationship between 'pathogenesis-related' leaf proteins from tomato, tobacco and cowpea, Virus Res., 4:229–242.

Neale, A. D., Wahleithner, J. A., Lund, B., Bonnett, H. T., Kelly, A., Meeks-Wagner, D. R., Peacock, W. J., and Dennis, E. S. (1990). Chitinase, β-1,3-glucanase, osmotin, and extensin are expressed in tobacco explants during flower formation, Plant Cell, 2:673–684.

Nelson, C. E., and Ryan, C. A. (1980). In-vitro synthesis of pre-proteins of two vacuolar compartmented proteinase inhibitors that accumulate in leaves of wounded tomato plants, Proc. Natl. Acad. Sci. USA, 77:1975–1979.
Neuhaus, J. M., Ahl-Goy, P., Hinz, U., Flores, S., and Meins, F. (1991a). High level expression of a tobacco chitinase gene in *Nicotiana sylvestris*. Susceptibility of transgenic tobacco plants to *Cercospora nicotianae*, Plant Mol. Biol., 16:141–151.
Neuhaus, J. M., Flores, S., Keefe, D., Ahl-Goy, P., and Meins, F. Jr. (1992). The function of vacuolar β-1,3-glucanase investigated by antisense transformation: susceptibility of transgenic *Nicotiana sylvestris* plants to *Cercospora nicotianae* infection, Plant Mol. Biol., 19:803–813.
Neuhaus, J. M., Sticher, L., Meins, F. Jr., and Boller, T. (1991b). A short C-terminal sequence is necessary and sufficient for the targeting of chitinases to the plant vacuole, Proc. Natl. Acad. Sci. USA, 88:10362–10366.
Nichols, E. J., Beckman, J. M., and Hadwiger, L. A. (1980). Glycosidic enzyme activity in pea tissue and pea–*Fusarium solani* interactions, Plant Physiol., 66:199–204.
Nicole, M. R., and Benhamou, N. (1991). Ultrastructural localization of chitin in cell walls of *Rigidoporus lignosus*, the white-rot fungus of rubber tree roots, Physiol. Mol. Plant Pathol., 39:415–431.
Nicole, M., Geiger, J. P., and Nandris, D. (1987). Ultrastructural aspects of rubber tree root rot diseases, Eur. J. Forest Pathol., 17:1–10.
Nicole, M., Toppan, A., Geiger, J. P., Roby, D., Nandris, D., and Rio, B. (1991). Defense responses of *Hevea brasiliensis* to elicitors from root rot fungi, Can. J. Bot., 69:1819–1824.
Nielsen, K. K., Bojsen, K., Collinge, D. B., and Mikkelsen, J. D. (1994a). Induced resistance in sugar beet against *Cercospora beticola*: induction by dichloroisonicotinic acid is independent of chitinase and β-1,3-glucanase transcript accumulation, Physiol. Mol. Plant Pathol., 45:89–99.
Nielsen, K. K., Bojsen, K., Roepstorff, P., and Mikkelsen, J. D. (1994b). A hydroxyproline-containing class IV chitinase of sugar beet is glycosylated with xylose, Plant Mol. Biol., 25:241–257.
Nielsen, K. K., Mikkelsen, J. D., Kragh, K. M., and Bojsen, K. (1993). An acidic class III chitinase in sugar beet induction by transgenic tobacco plants, Mol. Plant Microbe Interact., 6:495–506.
Nishimura, M., and Beevers, H. (1979). Hydrolysis of proteins in vacuoles isolated from higher plant tissue, Nature, 277:412–413.
Nishizawa, Y., and Hibi, T. (1991). Rice chitinase gene: cDNA cloning and stress-induced expression, Plant Sci., 76:211–218.
O'Connell, R. J. (1991). Cytochemical analysis of infection structures of *Colletotrichum lindemuthianum* using fluorochrome-labelled lectins, Physiol. Mol. Plant Pathol., 39:189–200.
O'Connell, R. J., and Ride, J. P. (1990). Chemical detection and ultrastructural localization of chitin in cell walls of *Colletotrichum lindemuthianum*, Physiol. Mol. Plant Pathol., 37:39–53.
Ohashi, Y., and Matsuoka, M. (1985). Synthesis of stress proteins in tobacco leaves, Plant Cell Physiol., 26:473–480.
Ohashi, Y., and Matsuoka, M. (1987a). Induction and secretion of pathogenesis-related

proteins by salicylate or plant hormones in tobacco suspension cultures, Plant Cell Physiol., 28:573–580.

Ohashi, Y., and Matsuoka, M. (1987b). Localization of pathogenesis-related proteins in the epidermis and intercellular spaces of tobacco leaves after their induction by potassium salicylate or tobacco mosaic virus infection, Plant Cell Physiol., 28:1227–1235.

Ohl, S., Apotheker-de Groot, M., van der Knaap, J. A., Ponstein, A. S., Sela-Buurlage, M. B., Bol, J. F., Cornelissen, B. J. C., Linthorst, H. J. M., and Melchers, L. S. (1994). A new class of tobacco chitinases homologous to bacterial exo-chininases is active against fungi in vitro, J. Cell. Biochem. Suppl., 18A:90.

Ohme-Takagi, M., and Shinshi, H. (1990). Structure and expression of a tobacco β-1,3-glucanase gene, Plant Mol. Biol., 15:941–946.

Ohshima, M., Itoh, H., Matsuoka, M., Murakami, T., and Ohashi, Y. (1990). Analysis of stress-induced or salicylic acid-induced expression of the pathogenesis-related 1a protein gene in transgenic tobacco, Plant Cell, 2:95–106.

Ohto, M., Nakamura-Kito, K., and Nakamura, K. (1992). Induction of expression of genes coding for sporamin and β-amylase by polygalacturonic acid in leaf-petiole cuttings of sweet potato, Plant Physiol., 99:422–427.

Oliveira, D. E., Seurinck, J., Inze, D., Montagu, M., and Botterman, J. (1990). Differential expression of five *Arabidopsis* genes encoding glycine-rich proteins, Plant Cell, 2:427–436.

Oppenheim, A. B., and Chet, I. (1992). Cloned chitinases in fungal plant-pathogen control strategies, TIBTECH, 10:392–394.

Ori, N., Sessa, G., Lotan, T., Himmelhoch, S., and Fluhr, R. (1990). A major stylar matrix polypeptide (sp 41) is a member of the pathogenesis-related proteins superclass, EMBO J., 9:3429–3436.

Oshima, M., Itoh, H., Matsuoka, M., Murakami, T., and Ohashi, Y. (1990). Analysis of stress-induced or salicylic-induced expression of the pathogenesis-related 1a protein gene in transgenic tobacco, Plant Cell, 2:95–106.

Oshima, M., Matsuoka, M., Yamamoto, N., Tanaka, Y., Kano Murakami, Y., Ozeki Y., Kato, A., Harad, N., and Ohashi, Y. (1987). Nucleic sequence of the PR-1 gene of *Nicotiana tabacum*, FEBS Lett., 225:243–246.

Osteryoung, K. W., Sticher, L., Jones, R. L., and Bennet, A. B. (1992). In vitro processing of tomato proteinase inhibitor I by barley microsomal membranes. A system for analysis of cotranslational processing of plant endomembrane proteins, Plant Physiol., 99:378–382.

Palm, C. J., Costa, M. A., An, G., and Ryan, C. A. (1990). Wound-inducible nuclear gene binds DNA fragments that regulate a proteinase inhibitor II gene from potato, Proc. Natl. Acad. Sci. USA, 87:603–607.

Pan, S. Q., Ye, X. S., and Kuc, J. (1991a). Induction of chitinases in tobacco plants systemically protected against blue mold by *Peronospora tabacina* or tobacco mosaic virus, Phytopathology, 81:970–974.

Pan, S. Q., Ye, R. S., and Kuc, J. (1991b). Association of β-1,3-glucanase activity and isoform pattern with systemic resistance to blue mold in tobacco induced by stem injection with *Peronospora tabacina* or leaf inoculation with tobacco mosaic virus, Physiol. Mol. Plant Pathol., 39:25–39.

Parent, J. C., and Asselin, A. (1984). Detection of pathogenesis-related proteins (PR or b) and other proteins in the intercellular fluid of hypersensitive plants infected with tobacco mosaic virus, Can. J. Bot., 62:564–569.

Parent, J. G., and Asselin, A. (1987). Acidic and basic extracellular pathogenesis-related leaf proteins from 15 potato cultivars, Can. J. Bot., 77:1122–1125.

Parent, J. G., Hogue, R., and Asselin, A. (1985). Glycoproteins and b proteins in intercellular fluid extracts from hypersensitive *Nicotiana* species infected with tobacco mosaic virus, Can. J. Bot., 63:928–931.

Parent, J. G., Hogue, R., and Asselin, A. (1988). Serological relationships between pathogenesis-related leaf proteins from four *Nicotiana* species, *Solanum tuberosum* and *Chenopodium amaranticolor*, Can. J. Bot., 66:199–202.

Parthier, B. (1991). Jasmonates: hormonal regulators or stress factors in leaf senescence? J. Plant Growth Regul., 9:57–63.

Pautot, V., Holzer, F. M., and Walling, L. L. (1991). Differential expression of tomato proteinase inhibitor I and II genes during bacterial pathogen invasion and wounding, Mol. Plant–Microbe Interact., 4:284–292.

Payne, G., Ahl, P., Moyer, M., Harper, A., Beck, J., Meins, F. Jr., and Ryals, J. (1990a). Isolation of complementary DNA clones encoding pathogenesis-related proteins P and Q, two acidic chitinases from tobacco, Proc. Natl. Acad. Sci. USA, 87:98–102.

Payne, G., Middlesteadt, W., Desai, N., Williams, S., Dincher, S., Carnes, M., and Ryals, J. (1989). Isolation and sequence of a genomic clone encoding the basic form of pathogenesis-related protein 1 from *Nicotiana tabacum*, Plant Mol. Biol., 12:595–596.

Payne, G., Middlesteadt, W., Williams, S., Desai, N., Parks, D., Dincher, S., Carnes, M., and Ryals, J. (1988). Isolation and nucleotide sequence of a novel cDNA clone encoding the major form of pathogenesis-related protein R, Plant Mol. Biol., 11:223–234.

Payne, G., Ward, E., Gaffney, T., Ahl Goy, P., Moyer, M., Harper, A., Meins, F., and Ryals, J. (1990b). Evidence for a third structural class of β-1,3-glucanase in tobacco, Plant Mol. Biol., 15:797–808.

Pearce, G., Ryan, C. A., and Liljegren, D. (1988). Proteinase inhibitors I and II in fruit of wild tomato species: transient components of a mechanism for defense and seed dispersal, Planta, 175:527–531.

Pearce, G., Strydom, D., Johnson, S., and Ryan, C. A. (1991). A polypeptide from tomato leaves induces wound-inducible proteinase inhibitor proteins, Science, 253:895–898.

Pegg, G. F. (1973). Chitinase and glucanase activities in *Verticillium albo-atrum*-infected tomato plants, 2nd Int. Cong. Plant Pathol. Minneapolis, Abstr. 968.

Pegg, G. F. (1977). Glucanohydrolases of higher plants: a possible defence mechanism against parasitic fungi, Cell Wall Chemistry Related to Specificity in Host–Pathogen Relationships (Solheim and J. Rau, eds.), Universitetsforlage, Oslo, pp. 304–345.

Pegg, G. F. (1988). Chitinase from tomato *Lycopersicon esculentum*, Methods Enzymol., 161:484–489.

Pegg, G. F. and Vessey, J. C. (1973). Chitinase activity in *Lycopersicon esculentum* and its relationship to the in vivo lysis of *Verticillium albo-atrum* mycelium, Physiol. Plant Pathol., 3:207–222.

Pegg, G. F. and Young, D. H. (1981). Changes in glycosidase activity and their relationship to fungal colonization during infection of tomato by *Verticillium albo-atrum*, Physiol. Plant Pathol., 19:371–382.

Pegg, G. F. and Young, D. H. (1982). Purification and characterization of chitinase enzymes from healthy and *Verticillium albo-atrum* infected tomato plants, and from *V. albo-atrum*, Physiol. Plant Pathol., 21:389–409.

Pelham, H. R. B. (1989). Control of protein exit from the endoplasmic reticulum, Annu. Rev. Cell Biol., 5:1–23.

Pena-Cortes, H., Sanchez-Serrano, J. J., Rocha-Sosa, M., and Willmitzer, L. (1988). Systemic induction of proteinase-inhibitor II gene expression in potato plants by wounding, Planta, 174:84–89.

Peng, J. H. and Black, L. L. (1976). Increased protein inhibitor activity in response to infection of resistant tomato plants by *Phytophthora infestans*, Phytopathology, 66:958–963.

Peterson, D. M., Forde, J., Williamson, M., Rohde, W., and Kreis, M. (1991). Nucleotide sequence of a chymotrypsin inhibitor –2 gene of barley (*Hordeum vulgare L.*), Plant Physiol., 96:1389–1390.

Pfitzner, U. M., and Goodman, H. M. (1987). Isolation and characterization of cDNA clones encoding pathogenesis-related proteins from tobacco mosaic virus-infected plants, Nucleic Acids Res., 15:4449–4465.

Pierpoint, W. S. (1983). The major proteins in extracts of tobacco leaves that are responding hypersensitively to virus infection, Phytochemistry, 22:2691–2697.

Pierpoint, W. S. (1986). The pathogenesis-related proteins of tobacco leaves, Phytochemistry, 25:1595–1601.

Pierpoint, W. S., Gordon-weeks, R., and Jackson, R. J. (1992). The occurrence of the thaumatin-like, pathogenesis related protein, PR-5, in intercellular fluids from *Nicotiana* species and from an interspecific *Nicotiana* hybrid, Physiol. Mol. Plant Pathol., 41:1–10.

Pierpoint, W. S., Jackson, P. J., and Evans, R. M. (1990). The presence of a thaumatin-like protein, a chitinase and a glucanase among the pathogenesis-related proteins of potato (*Solanum tuberosum*), Physiol. Mol. Plant Pathol., 36:325–338.

Pierpoint, W. S., and Shewry, P. R. (1987). Amino acid homologies suggest functions for pathogenesis-related proteins, Oxford Surv. Plant Mol. Cell Biol., 4:337–342.

Pierpoint, W. S., Robinson, N. P., and Leason, M. B. (1981). The pathogenesis-related proteins of tobacco: their induction by viruses in intact plants and their induction by chemicals in detached leaves, Physiol. Plant Pathol., 19:85–97.

Pierpoint, W. S., Tatham, A. S., and Pappin, D. J. C. (1987). Identification of the virus-induced protein of tobacco leaves that resembles the sweet-protein thaumatin, Physiol. Mol. Plant Pathol., 31:291–298.

Plunkett, G., Senear, D. F., Zuroske, G., and Ryan, C. A. (1982). Proteinase inhibitors I and II from leaves of wounded tomato plants. Purification and properties, Arch. Biochem. Biphys., 213:463–472.

Ponstein, A. S., Bres-Vloemans, A. A., Sella-Buurlage, M. B., Cornelissen, B. J. C., and Melchers, L. S. (1994). The "missing" class I PR-4 protein from tobacco exhibits antifungal activity, J. Cell. Biochem. Suppl., 18A:90.

Powning, R. F., and Irzykiewicz, H. (1965). Studies on the chitinase systems in bean and other seeds, Comp. Biochem. Physiol., 14:127–133.

Pratt, M. J., and Mathews, R. E. F. (1971). Non-uniformities in the metabolism of excised leaves and leaf discs, Planta, 99:21–36.

Puigdomenech, P., and San Sagundo, B. (1991). A gene coding for a basic pathogenesis-related (PR-like) protein from *Zea mays*. Molecular cloning and induction by a fungus (*Fusarium moniliforme*) in germinating maize seeds, Plant Mol. Biol., 16:527–536.

Raikhel, N. V., and Wilkins, T. A. (1987). Isolation and characterization of a cDNA encoding wheat germ agglutinin, Proc. Natl. Acad. Sci. USA, 84:6745–6749.

Ramshaw, J. A. M. (1982). Structures of plant proteins, Nucleic Acids and Proteins in Plants. (I. D. Boulter and B. Parthier, eds.) I. Structure, Biochemistry and Physiology of Proteins. Springer-Verlag, Berlin. pp. 229–290.

Ramshaw, J. A. M. (1982). Structures of plant proteins. Nucleic acids and proteins in plants. I. Structure, Biochemistry and Physiology of Proteins. (I. D. Boulter and B. Parthier, eds.), Springer-Verlag, Berlin, pp. 229–290.

Rasmussen, U., Bojsen, K., and Collinge, D. B. (1992). Cloning and characterization of a pathogen-induced chitinase in *Brassica napus*, Plant Mol. Biol., 20:277–287.

Raz, V. and Fluhr, R. (1993). Ethylene signal is transduced via protein phosphorylation events in plants, Plant Cell, 5:523–530.

Redolfi P. (1983). Occurrence of pathogenesis-related (b) and similar proteins in different plant species, Neth. J. Plant Pathol., 89:245–254.

Redolfi, P. and Cantisani, A. (1984). Preliminary characterisation of new soluble proteins in *Phaseolus vulgaris* cv. Saxa reacting hypersensitively to viral infection, Physiol. Plant Pathol., 25:9–19.

Redolfi, P., Cantisani, A., and Migheli, Q. (1989). The pathogenesis-related proteins of *Phaseolus* bean, J. Phytopathol., 126:193–203.

Redolfi, P., Veechiati, M., and Gianinazzi, S. (1982). Changes in the soluble leaf protein constitution of *Gomphrena globosa* during the hypersensitive reaction to different viruses, Phytopathol. Z., 103:48–54.

Reinbothe, S., Mollenhauer, B., and Reinbothe, C. (1994a). JIPs and RIPs: The regulation of plant gene expression by jasmonates in response to environmental cues and pathogens, Plant Cell, 6:1197–1209.

Reinbothe, S., Reinbothe, C., Lehmann, J., Becker, W., Apel, K., and Parthier, B. (1994b). JIP 60, a methyl jasmonate-induced ribosome-inactivating protein involved in plant stress reactions, Proc. Natl. Acad. Sci. USA, 91:7012–7016.

Richard, K.,, Montserrat, A., Hoebeke, J., Meeks-Wagner, R., and Tran Thanh Van K. (1992). Immunological evidence of thaumatin-like proteins during tobacco floral differentiation, Plant Physiol., 98:337–342.

Richardson, M. (1977). The proteinase inhibitors of plants and microorganisms, Phytochemistry, 16:159–169.

Richardson, M., Valdes-Rodriguez, S., and Blanco-Labra, A. (1987). A possible function of a thaumatin and a TMV-induced protein suggested by homology to a maize inhibitor, Nature, 327:432–434.

Rickauer, M., Fournier, J., and Esquerre-Tugaye, M. T. (1989). Induction of proteinase

inhibitors in tobacco cell suspension culture by elicitors of *Phytophthora parasitica* cv. *nicotianae*, Plant Physiol., 90:1065–1070.
Ride, J. P., and Barber, M. S. (1990). Purification and characterization of multiple forms of endochitinase from wheat leaves, Plant Sci., 71:185–197.
Rigden, J., and Coutts, R. (1988). Pathogenesis-related proteins in plants, Trends Genet., 4:87–89.
Robb, J., Lee, S. W., Mohan, R., and Kolattukudy, P. E. (1991). Chemical characterization of stress-induced vascular coating, Plant Physiol., 97:528–536.
Roberts, E., and Kolattukudy, P. E. (1989). Molecular cloning, nucleotide sequence and abscisic acid induction of a suberization-associated highly anionic peroxidase, Mol. Gen. Genet., 217:223–231.
Roberts, E., Kutchan, T., and Kolattukudy, P. E. (1988). Cloning and sequencing of cDNA for a highly anionic peroxidase from potato and the induction of its mRNA in suberizing potato tubers and tomato fruits, Plant Mol. Biol., 11:15–26.
Roberts, W. K., Laue, B. E., and Selitrennikoff, C. P. (1988). Antifungal proteins from plants, Ann. N.Y. Acad. Sci., 544:141–151.
Roberts, W. K., and Selitrennikoff, C. P. (1986a). Isolation and partial characterization of two antifungal proteins from barley, Biochem. Biophys. Acta, 880:161–170.
Roberts, W. K., and Selitrennikoff, C. P. (1986b). Plant proteins that inactivate foreign ribosomes, Biosci. Rep., 6:19–29.
Roberts, W. K., and Selitrennikoff, C. P. (1988). Plant and bacterial chitinases differ in antifungal activity, J. Gen. Microbiol., 134:169–176.
Roberts, W. K., and Selitrennikoff, C. P. (1990). Zeamatin, an antifungal protein from maize with membrane-permeabilizing activity, J. Gen. Microbiol., 136:1771–1778.
Roby, D., Broglie, K., Cressman, R., Biddle, P., Chet, I., and Broglie, R. (1990). Activation of bean chitinase promoter in transgenic tobacco plants by phytopathogenic fungi, Plant Cell, 2:999–1007.
Roby, D., Broglie, K., Gaynor, J., and Broglie, R. (1991). Regulation of a chitinase gene promoter by ethylene and elicitors in protoplasts, Plant Physiol., 97:433–439.
Roby, D., and Esquerre-Tugaye, M. T. (1987). Purification and some properties of chitinases from melon plants infected by *Colletotrichum lagenarium*, Carbohydr. Res., 165:93–104.
Roby, D., Gadelle, A., and Toppan, A. (1987a). Chitin oligosaccharides as elicitors of chitinase activity in melon plants, Biochem. Biophys. Res. Commun., 143:885–892.
Roby, D., Toppan, A., and Esquerre-Tugaye, M. T. (1986). Cell surfaces in plant–microorganism interactions. VI. Elicitors of ethylene from *Colletotrichum lagenarium* trigger chitinase activity in plants, Plant Physiol., 81:228–233.
Roby, D., Toppan, A., and Esquerre-Tugaye, M. T. (1987b). Cell surfaces in plant microorganism interactions. VIII. Increased proteinase inhibitor activity in melon plants in response to infection by *Colletotrichum lagenarium* or to treatment with an elicitor fraction from this fungus, Physiol. Mol. Plant Physiol., 30:453–460.
Roby, D., Toppan, A., and Esquerre-Tugaye, M. T. (1988). Systemic induction of chitinase activity and resistance in melon plants upon fungal infection or elicitor treatment, Physiol. Mol. Plant Pathol., 33:409–417.
Rodrigo, I., Vera, P., and Conejero, V. (1989). Degradation of tomato pathogenesis-related proteins by an endogenous aspartyl endoproteinase, Eur. J. Biochem., 184:663–669.

Rodrigo, I., Vera, P., Frank, K., and Conejero, V. (1991a). Identification of the viroid-induced tomato pathogenesis-related (PR) protein P23 as the thaumatin-like protein NP24 associated with osmotic stress, Plant Mol. Biol., 16:931–934.

Rodrigo, I., Vera, P., Van Loon, L. C., and Conejero, V. (1991b). Degradation of tobacco pathogenesis-related proteins, Plant Physiol., 95:616–622.

Roggero, P., and Pennazio, S. (1989). The extracellular acidic and basic pathogenesis-related proteins of soybean induced by viral infection, J. Phytopathol., 127:274–280.

Rohloff, H., and Lerch, B. (1977). Soluble leaf proteins in virus infected plants and acquired resistance I. Investigations on *Nicotiana tabacum* var. "Samsun" and "Samsun NN," Phytopath. Z., 89:306–316.

Rokem, J. S., Klein, D., Toder, H., and Zomer, E. (1986). Degradation of fungal cell walls taking into consideration the polysaccharide composition, Enzyme Microbiol. Technol., 8:588–592.

Rousseau-Limouzin, M., and Fritig, B. (1991). Induction of chitinases, 1,3-β-glucanases and other pathogenesis-related proteins in sugar beet leaves upon infection with *Cercospora beticola*, Plant Physiol. Biochem., 29:105–117.

Roventos, D., Cordero, M. J., and Segundo, B. S. (1994). Fungal-induced synthesis of pathogenesis-related proteins in germinating maize embryos, Physiol. Mol. Plant Pathol., 45:349–358.

Ruiz-Medrano, R., Jimenez-Moraila, B., Herrara-Estrella, L., and Rivera-Bustamante, R. F. (1992). Nucleotide sequence of an osmotin-like cDNA induced in tomato during viroid infection, Plant Mol. Biol., 20:1199–1202.

Ryan, C. A. (1966). Chymotrypsin inhibitor I from potatoes; reactivity with mammalian, plant, bacterial and fungal proteinases, Biochemistry, 5:1592–1596.

Ryan, C. A. (1968). An inducible protein in tomato and potato leaflets, Plant Physiol., 43:1880.

Ryan, C. A. (1973). Proteolytic enzymes and their inhibitors in plants, Annu. Rev. Plant Physiol., 24:173–196.

Ryan, C. A. (1981). Proteinase inhibitors, Biochemistry of Plants Vol. 6 (A. Marcas, ed.), Academic Press, New York, pp. 351–370.

Ryan, C. A. (1990). Protease inhibitors in plants: genes for improving defenses against insects and pathogens, Annu. Rev. Phytopathol., 28:425–449.

Ryan, C. A. (1992). The search for the proteinase inhibitor-inducing factor, PIIF, Plant Mol. Biol., 19:123–133.

Ryan, C. A., and An, G. (1988). Molecular biology of wound-inducible proteinase inhibitors in plants, Plant Cell Envir., 11:345–349.

Ryan, C. A., and Balls, A. K. (1962). An inhibitor of chymotrypsin from *Solanum tuberosum* and its behaviour toward trypsin, Proc. Natl. Acad. Sci. USA, 48:1839–1844.

Ryan, C. A., and Farmer, E. E. (1991). Oligosaccharide signals in plants: a current assessment. Annu. Rev. Plant Physiol. Mol. Biol., 42:651–674.

Ryan, C. A., and Huisman, O. C. (1967). Chymotrypsin inhibitor I from potatoes. A transient component in leaves of young potato plants, Nature, 214:1047.

Ryan, C., Pearce, G. K., and Kunkel, R. (1976). Variability in the concentration of three heat stable proteinase inhibitor proteins in potato tubers, Am. Potato J., 53:443–455.

Saeed, M., and Duke, S. H. (1990). Amylases in pea tissues with reduced chloroplast density and/or function, Plant Physiol., 94:1813–1819.

Samac, D. A., Hironaka, C. M., Yallaly, P. E., and Shah, D. M. (1990). Isolation and characterization of the genes encoding basic and acidic endochitinase in *Arabidopsis thaliana*, Plant Physiol., 93:907–914.

Sanchez-Serrano, J. J., Keil, M., O'Connor, A., Schell, J., and Willmitzer, L. (1987). Wound-inducible expression of a potato proteinase inhibitor II gene in transgenic tobacco plants, EMBO J., 6:303–306.

Sanchez-Serrano, J. J., Schmidt, R., Schell, J., and Willmitzer, L. (1986). Nucleotide sequence of proteinase inhibitor II encoding cDNA of potato (*Solanum tuberosum*) and its mode of expression, Mol. Gen. Genet., 203:15–20.

Schlumbaum, A., Mauch, F., Vogeli, U., and Boller, T. (1986). Plant chitinases are potent inhibitors of fungal growth, Nature, 324:365–367.

Schmelzer, E., Kruger-Lebus, S., and Hahlbrock, K. (1989). Temporal and spatial patterns of gene expression around sites of attempted fungal infection in parsley leaves, Plant Cell, 1:993–1001.

Schottens-Toma, I. M. J., and De Wit, J. G. M. (1988). Purification and primary structure of a necrosis-inducing peptide from the apoplastic fluids of tomato infected with *Cladosporium fulvum* (syn: *Fulvia fulva*), Physiol. Mol. Plant Pathol., 33:59–67.

Schroeder, M. R., Borkhsenious, O. N., Matsuoka, K., Nakamura, K., and Raikhel, N. V. (1993). Colocalization of barley lectin and sporamin in vacuoles of transgenic tobacco-plants, Plant Physiol., 101:451–458.

Schweizer, P., Hunziker, W., and Mosinger, E. (1989). cDNA cloning, in vitro transcription and partial sequence analysis of mRNAs from winter wheat (*Triticum aestivum* L.) with induced resistance to *Erysiphe graminis* f.sp. *tritici*, Plant Mol. Biol., 12:643–654.

Sehgal, O. P., Rieger, R., and Mohamed, F. (1991). Induction of bean PR-4d-type protein in divergent plant species after infection with tobacco ringspot virus and its relationship with tobacco PR-5, Phytopathology, 81:215–219.

Sehgal, O. P., and Mohamed, F. (1990). Pathogenesis-related proteins, Plant Viruses—Pathology, Vol. II (C. L. Mandahar, ed.), CRC Press, Boca Raton, FL, pp. 65–83.

Seidl, D. S., Abreu, H., and Jaffe, W. G. (1982). Purification of a subtilisin inhibitor from black bean seeds, FEBS Lett., 92:245–250.

Seidl, D. S., Lorenzo, P., Pineli, B. E., and Tovar, J. (1988). Subtilisin inhibitors in legume seeds, Biol. Chem. Hoppe-Seyler, 369:69–73.

Sela-Buurlage, M. B. Ponstein, A. S., Bres-Vloemans, S. A., Melchers, L. S., Van Den Elzen, P. J. M., and Cornelissen, B. J. C. (1993). Only specific tobacco (*Nicotiana tabacum*) chitinases and β-1,3-glucanases exhibit antifungal activity, Plant Physiol., 101:857–863.

Sharma, P., Borja, D., Stougaard, P., and Lonneborg, A. (1993). PR-proteins accumulating in spruce roots infected with a pathogenic *Pythium* sp. isolate include chitinases, chitosanases and β-1,3-glucanases, Physiol. Mol. Plant Pathol., 43:57–67.

Sharma, Y. K., Hinojos, C. M., and Mehdy, M. C. (1992). cDNA cloning, structure and expression of a novel pathogenesis-related protein in bean, Mol. Plant–Microbe Interact., 5:89–95.

Shinshi, H., Mohnen, D., and Meins, F. (1987). Regulation of a plant-pathogenesis related enzyme: Inhibition of chitinase and chitinase mRNA accumulation in cultured tobacco tissues by auxin and cytokinin, Proc. Natl. Acad. Sci. USA, 84:89–93.

Shinshi, H., Neuhaus, J. M., Ryals, J., and Meins, F. (1990). Structure of a tobacco endochitinase gene: evidence that different chitinase genes can arise by transposition of sequences encoding a cysteine-rich domain, Plant Mol. Biol., 14:357–368.

Shinshi, H., Wenzler, H., Neuhaus, J. M., Felix, G., Hofsteenge, J., and Meins, F. (1988). Evidence for N- and C-terminal processing of a plant defense-related enzyme: Primary structure of tobacco prepro-β-1,3-glucanase, Proc. Natl. Acad. Sci. USA, 85:5541–5545.

Showalter, A. M., Bell, J. N., Cramer, C. L., Bailey, J. A., Varner, J. E., and Lamb, C. L. (1985). Accumulation of hydroxyproline-rich glycoprotein mRNAs in response to fungal elicitor and infection, Proc. Natl. Acad. Sci. USA, 82:6551–6555.

Shumway, L. K., Cheng, V., and Ryan, C. A. (1972). Vacuolar protein in apical and flower-petal cells, Planta, 106:279–290.

Shumway, L. K., Rancour, J. M., and Ryan, C. A. (1970). Vacuolar protein bodies in tomato leaf cells and their relationship to storage of chymotrypsin inhibitor I protein, Planta, 93:1–14.

Shumway, L. K., Yang, V., and Ryan, C. A. (1976). Evidence for the presence of proteinase inhibitor I in vacuolar protein bodies of plant cells, Planta, 129:161–165.

Sidman, C. (1981). Differing requirements for glycosylation in the secretion of related glycoproteins is determined neither by the producing cell nor by the relative number of oligosaccharide units, J. Biol. Chem., 256:9374–9376.

Siebertz, B., Logemann, J., Willmitzer, L., and Schell, J. (1989). *Cis*-Analysis of the wound-inducible promoter *Wun* 1 in transgenic tobacco plants and histochemical localization of its expression, Plant Cell, 1:961–968.

Siedow, J. N. (1991). Plant lipoxygenase: structure and function, Annu. Rev. Plant Physiol., 42:651–674.

Siegrist, J., and Kauss, H. (1990). Chitin deacetylase in cucumber leaves infected by *Colletotrichum lagenarium*, Physiol. Mol. Plant Pathol., 36:267–275.

Simmons, C. R., Litts, J. C., Huang, H., and Rodriguez, R. L. (1992). Structure of a rice β-glucanase gene regulated by ethylene, cytokinin, wounding, salicylic acid and fungal elicitors, Plant Mol. Biol., 18:33–45.

Sinclair, L. J., Carzaniga, R., Harris, N., Fordham-Skelton, A. P., and Croy, R. R. D. (1994). Synthesis and targeting of saponins, type-1 ribosome-inactivating proteins from *S. officinalis* (soapwort), J. Cell. Biochem. Suppl., 18A:111.

Singh, N. K., Bracker, C. A., Hasegawa, P. M., Handa, A. K., Buckel, S., Hermodson, M. A., Pfankoch, E., Regnier, F. E., and Bressan, R. A. (1987a). Characterization of osmotin: a thaumatin-like protein associated with osmotic adaptation in plant cells, Plant Physiol., 85:529–536.

Singh, N. K., Handa, A. K., Hasegawa, P. M., and Bressan, R. A. (1985). Proteins associated with adaptation of cultured tobacco cells to NaCl, Plant Physiol., 79:126–137.

Singh, N. K., Lahosa, P. C., Handa, A. K., Hasegawa, P. M., and Bressan, R. A. (1987b). Hormonal regulation of protein synthesis associated with salt tolerance in cells, Proc. Natl. Acad. Sci. USA, 84:739–743.

Singh, N. K., Nelson, D. E., Kuhn, D., Hasegawa, P. M., and Bressan, R. A. (1989).

Molecular cloning of osmotin and regulation of its expression by ABA and adaptation to low water potential, Plant Physiol., 90:1096–1101.

Slakeski, N., Baulcombe, C. C., Devos, K. M., Ahluwalia, B., Doan, D. N. P., and Fincher, G. B. (1990). Structure and tissue specific regulation of genes encoding barley (1-3,1-4)-β-glucan endohydrolases, Mol. Gen. Genet., 224:437–449.

Sock, J., Rohringer, R., and Kang, Z. (1990). Extracellular β-1,3-glucanases in stem rust-affected and abiotically stressed wheat leaves. Immunocytochemical localization of the enzyme and detection of isozymes in gels by activity staining with dye-labeled laminarin, Plant Physiol., 94:1376–1389.

Somssich, I. E., Schmelzer, E., Bollmann, J., and Hahlbrock, K. (1986). Rapid activation by fungal elicitor of genes encoding "pathogenesis-related" proteins in cultured parsley cells, Proc. Natl. Acad. Sci. USA, 83:2427–2430.

Somssich, I. E., Schmelzer, E., Kawalleck, P., and Hahlbrock, K. (1988). Gene structure and in situ transcript localization of pathogenesis-related protein I in parsley, Mol. Gen. Genet., 213:93–98.

Spanu, P., Boller, T., Ludwig, A., Wiemken, A., Faccio, A., and Bonfante-Fasolo, P. (1989). Chitinase in roots of mycorrhizal *Allium porrum*: regulation and localization, Planta, 177:447–455.

Sperisen, C., Ryals, J., and Meins, F. Jr. (1991). Comparison of cloned genes provides evidence for intergenomic exchange of DNA in the evolution of a tobacco glucan endo-1,3-β-glucosidase gene family, Proc. Natl. Acad. Sci. USA, 88:1820–1824.

Stanford, A., Bevan, M., and Northcote, D. (1989). Differential expression within a family of novel wound-induced genes in potato, Mol. Gen. Genet., 215:200–208.

Staswick, P. E. (1992). Jasmonate, genes, and fragrant signals, Plant Physiol., 99:804–807.

Sterk, P., Booij, H., Schellekens, G. A., and Van Kammen, A., De Vries, S. C. (1991). Cell-specific expression of the carrot EP,2 lipid transfer protein gene, Plant Cell, 3:907–921.

Sticher, L., Hofsteenge, J., Milani, A., Neuhaus, J. M., and Meins, F. Jr. (1992). Vacuolar chitinases of tobacco: a new class of hydroxyproline-containing proteins, Science, 257:655–657.

Stiekema, W. J., Heidekamp, F., Dikse, W. G., Van Beckum, J., De Haan, Ten Bosch, G., and Louwerse, J. D. (1988). Molecular cloning and analysis of four potato tuber mRNAs, Plant Mol. Biol., 11:255–269.

Stintzi, A., Heitz, T., Kauffmann, S., Legrand, M., and Fritig, B. (1991). Identification of a basic pathogenesis-related thaumatin-like protein of virus-infected tobacco as osmotin, Physiol. Mol. Plant Pathol., 38:137–146.

Stripe, F., and Hughes, R. C. (1989). Specificity of Ribosome-inactivating proteins with RNA N-glycosidase activity, Biochem. J., 262:1001–1002.

Stuart, I. M., Loi, L., and Fincher, G. B. (1986). Development of (1-3, 1-4)-β-glucan endohydrolase isoenzymes in isolated scutella and aleurone layers of barley (*Hordeum vulgare*), Plant Physiol., 80:310–314.

Stuart, I. M., Loi, L., and Fincher, G. B. (1987). Immunological comparison of (1-3; 1-4) β-glucan endohydrolases in germinating cereals, J. Cereal Sci., 6:45–52.

Suh, S. G., Stiekema, W. J., and Hannapel, D. J. (1991). Proteinase-inhibitor activity and wound-inducible gene expression of the 22 kDa potato-tuber proteins, Planta, 184:423–430.

Svendsen, I., Biosen, S., and Hejgaard, J. (1982). Amino acid sequence of serine protease inhibitor CI-I from barley. Homology with barley inhibitor CI-2, potato inhibitor I, and leech eglin, Carlsberg Res. Commun., 47:45–53.

Svendsen, I., Hejgaard, J., and Chavan, J. K. (1984). Subtilisin inhibitor from seeds of broadbean (*Vicia faba*): purification, amino acid sequence and specificity of inhibition, Carlsberg Res. Commun., 49:493–502.

Tahiri-Alaoui, A., Dumas-Gaudot, E., and Gianinazzi, S. (1993). Immunocytochemical localization of pathogenesis-related PR-1 proteins in tobacco root tissues infected in vitro by the black root rot fungus *Chalara elegans*, Physiol. Mol. Plant Pathol., 42:69–82.

Takeda, S., Sato, F., Ida, K., and Yamada, Y. (1990). Characterization of polypeptides that accumulate in cultured *Nicotiana tabacum* cells, Plant Cell Physiol., 31:215–221.

Takeda, S., Sato, F., Ida, K., and Yamada, Y. (1991). Nucleotide sequence of a cDNA for osmotin-like protein from cultured tobacco cells, Plant Physiol., 97:844–846.

Takeuchi, Y., Yoshikawa, M., Takeba, K., Tanaka, K., Shibata, D., and Horino, O. (1990). Molecular cloning and ethylene induction of mRNA encoding a phytoalexin elicitor-releasing factor, β-1,3-endoglucanase, in soybean, Plant Physiol., 93:673–682.

Tas, P. W. I., and Peters, D. (1977). The occurrence of a soluble protein (E_1) in cucumber cotyledons infected with plant viruses, Neth. J. Plant Pathol., 83:5–12.

Tata, S. J., Beintema, J. J., and Balabaskaran, S. (1983). The lysozyme of *Hevea brasiliensis* latex: isolation, purification, enzyme kinetics and partial amino acid sequence, J. Rubber Res. Inst. Malays., 31:35–48.

Taylor, J. L., Fritzemeier, K. H., Hauser, I., Kombrink, E., Rohwer, F., Schroder, M., Strittmatter, G., and Hahlbrock, K. (1990). Structural analysis and activation by fungal infection of a gene encoding a pathogenesis-related protein in potato, Mol. Plant–Microbe Interact., 3:72–77.

Terras, F. R. G., Goderis, I. J., Van Leuven, F., Vanderleyden, J., Cammue, B. P. A., and Broekaert, W. F. (1992a). In vitro antifungal activity of a radish (*Raphanus sativus* L.) seed protein homologous to nonspecific lipid transfer protein. Plant Physiol., 100:1055–1058.

Terras, F. R. G., Schoofs, H. M. E., De Bolle, M. F. C., Van Leuven, F., Rees, S. B., Vanderleyden, J., Cammue, B. P. A., and Brockaert, W. F. (1992b). Analysis of two novel classes of plant antifungal proteins from radish (*Raphanus sativus* L.) seeds, J. Biol. Chem., 267:15301–15309.

Thayer, S. S., and Huffaker, R. C. (1984). Vacuolar localization of endoproteinases EP1 and EP2 in barley mesophyll cells, Plant Physiol., 75:70–73.

Thimmapuram, J., and Korban, S. S. (1994). Isolation and characerization of a β-1,3-glucanase gene in apple. J. Cell. Biochem. Suppl., 18A:91.

Thornburg, R. W., An, G., Cleveland, T. E., Johnson, R., and Ryan, C. A. (1987). Wound-inducible expression of a potato inhibitor II-chloramphenicol acetyltransferase gene fusion in transgenic tobacco plants, Proc. Natl. Acad. Sci. USA, 84:744–748.

Toyoda, H., Matsuda, Y., Yamaga, T., Ikeda, S., Morita, M., Tamai, T., and Ouchi, S. (1991). Suppression of powdery mildew pathogen by chitinase microinjected into barley coleoptile epidermal cells, Plant Cell Rep., 10:217–220.

Trudel, J., Audy, P., and Asselin, A. (1989). Electrophoretic forms of chitinase activity in

Xanthi-nc tobacco, healthy and infected with tobacco mosaic virus, Mol. Plant–Microbe Interact., 2:315–324.

Tsukamoto, T., Koga, D. A., Ishibashi, T., Horino-Matsushige, M., Yasishita, K., and Imoto, T. (1984). Purification and some properties of chitinases from Yam, *Dioscorea opposita*, Thumb. Agric. Biol. Chem., 48:931–939.

Tu, J. C. (1985). Tolerance of white bean *Phaseolus vulgaris* to white mold (*Sclerotinia sclerotiorum*) associated with tolerance to oxalic acid, Physiol. Plant Pathol., 26:111–117.

Tuzun, S., Rao, M. N., Vogeli, U., Schardl, C. L., and Kuc, J. (1988). Early accumulation of chitinases, β-1,3-glucanases and other b-proteins in tobacco immunized against blue mold, Phytopathology, 78:1556.

Tuzun, S., Rao, M. N., Vogeli, U., Schardl, C. I., and Kuc, J. (1989). Induced systemic resistance to blue mold: Early induction and accumulation of β-1,3-glucanases, chitinases and other pathogenesis-related proteins (b-proteins) in immunized tobacco, Phytopathology, 79:979–983.

Uknes, S., Dincher, S., Friedrich, L., Negrotto, D., Williams, S., Thompson-Taylor, H., Potter, S., Ward, E., and Ryals, J. (1993a). Regulation of pathogenesis-related protein-1a gene expression in tobacco, Plant Cell, 5:159–169.

Vale, G. P., Torrigiani, E., Gatti, A., Delogu, G., Porta-Puglia, A., Vannacci, G., and Cattivelli, L. (1994). Activation of genes in barley roots in response to infection by two *Drechslera graminea* isolates, Physiol. Mol. Plant Pathol., 44:207–215.

Van de Rhee, M. D., Van Kan, J. A. L., Gonzalez-Jaen, M. T., and Bol, J. F. (1990). Analysis of regulatory elements involved in the induction of two tobacco genes by salicylate treatment and virus infection, Plant Cell, 2:357–366.

Van den Berg-Velthuis, G. C. M., and De Wit, P. J. G. M. (1992). Differential accumulation of mRNAs encoding extracellular and intracellular PR proteins in tomato induced by virulent and avirulent races of *Cladosporium fulvum*, Plant Mol. Biol., 20:513–527.

Van den Bulcke, M., Bauw, G., Castresana, C., Van Montague, M., and Vande Kerckhove, J. (1989). Characterization of vacuolar and extracellular β-1,3-glucanases of tobacco: evidence for a strictly compartmentalized plant defense system, Proc. Natl. Acad. Sci. USA, 86:2673–2677.

Van der Wilden, J. H. L., Seger, M. J., and Chrispeels, M. J. (1983). Cell wall of *Phaseolus vulgaris* leaves contain the azocoll-digesting proteinase, Plant Physiol., 73:576–578.

Van Kan, J. A. L., Cornelissen, B. J. C., and Bol, J. F. (1988). A virus-inducible gene encoding a glycine-rich protein shares putative regulatory elements with the ribulose bisphosphate carboxylase small subunit gene, Mol. Plant–Microbe Interact., 1:107–112.

Van Kan, J. A. L., Joosten, M. H. A. J., Wagemakers, C. A. M., Van den Berg-Velthius, G. C. M., and de Wit, P. J. G. M. (1992). Differential accumulation of mRNAs encoding intercellular and intracellular PR proteins in tomato induced by virulent and avirulent races of *Cladosporium fulvum*, Plant Mol. Biol., 20:513–527.

Van Loon, L. C. (1976). Specific soluble leaf proteins in virus-infected tobacco plants are not normal constituents, J. Gen. Virol., 30:375–379.

Van Loon, L. C. (1977). Induction by 2-chloroethylphosphonic acid of viral like lesions, associated proteins and systemic resistance in tobacco, Virology, 80:417–420.

Van Loon, L. C. (1982). Regulation of changes in proteins and enzymes associated with active defense against virus infection, Active Defense Mechanisms in Plants (R. K. S. Wood, ed.), Plenum Press, New York, pp. 247–273.

Van Loon, L. C. (1983). The induction of pathogenesis-related proteins by pathogens and specific chemicals, Neth. J. Plant Pathol., 89:265–273.

Van Loon, L. C. (1985). Pathogenesis-related proteins, Plant Mol. Biol., 4:111–116.

Van Loon, L. C. (1989). Stress proteins in infected plants, Plant–Microbe Interactions: Molecular and Genetic Perspectives, Vol. 3 (T. Kosuge and F. W. Nester, eds.), McGraw-Hill, New York, pp. 198–237.

Van Loon, L. C., and Gerritsen, Y. A. M. (1989). Localization of pathogenesis-related proteins in infected and non-infected leaves of Samsun NN tobacco during the hypersensitive reaction to tobacco mosaic virus, Plant Sci., 63:131–140.

Van Loon, L. C., Gerritsen, Y. A. M., and Ritter, C. E. (1987). Identification, purification and characterization of pathogenesis-related proteins from virus-infected Samsun NN tobacco leaves, Plant Mol. Biol., 9:593–609.

Van Loon, L. C., and Van Kammen, A. (1970). Polyacrylamide disc electrophoresis of the soluble leaf proteins from *Nicotiana tabacum* var. Samsun NN. Changes in proteins constitution after infection with tobacco, Virology, 40:199–201.

Van Parijs, J., Broekaert, W. F., Goldstein, I. J., and Peumans, W. J. (1991). Hevein—an antifungal protein from rubber-tree (*Hevea brasiliensis*) latex, Planta, 183:258–264.

Varnar, J. E., and Cassab, G. I. (1986). A new protein in petunia, Nature, 323:110–112.

Vartak, H. G., Rele, M. V., and Jagannathan, V. (1980). Proteinase inhibitors from *Vigna unguiculata* subsp. *cylindrica* III. Properties and kinetics of inhibitors of papain, subtilisin, and trypsin, Arch. Biochem. Biphys., 204:134–140.

Vera, P., and Conejero, V. (1988). Pathogenesis-related proteins in tomato. P69 as an alkaline endoproteinase, Plant Physiol., 87:58–63.

Vera, P., and Conejero, V. (1989). The induction and accumulation of the pathogenesis-related P69 proteinase in tomato during citrus exocortis viroid infection and after chemical treatments, Physiol. Mol. Plant Pathol., 34:323–334.

Vera, P., and Conejero, V. (1990). Effect of ethephon on protein degradation and the accumulation of 'pathogenesis-related' (PR) proteins in tomato leaf discs, Plant Physiol., 92:227–233.

Vera, P., Hernandez-Yago, J., and Conejero, V. (1989). Pathogenesis-related P1 (P14) Protein. Vacuolar and apoplastic localization in leaf tissue from tomato plants infected with citrus exocortis viroid: in vitro synthesis and processing, J. Gen. Virol., 70:1933–1942.

Vera, P., Yago, J. H., and Conejero, V. (1989). Immunogold localization of the citrus exocortis viroid-induced pathogenesis-related proteinase P69 in tomato leaves, Plant Physiol., 91:119–123.

Verburg, J. G., and Huynh, Q. K. (1991). Purification and characterization of an antifungal chitinase from *Arabidopsis thaliana*, Plant Physiol., 95:450–455.

Vernooj, B., Friedrich, L., Morse, A., Reist, R., Kolditz-Jawhar, R., Ward, E., Uknes, S., Kessmann, H., and Ryals, J. (1994). Salicyclic acid is not the translocated signal responsible for inducing systemic acquired resistance but is required in signal transduction, Plant Cell, 6:6959–6965.

Vick, B. A., and Zimmerman, D. C. (1984). Biosynthesis of jasmonic acid or by several plant species, Plant Physiol., 75:458–461.

Vidhyasekaran, P. (1988a). Physiology of Disease Resistance in Plants, Vol. I, CRC Press, Boca Raton, FL, pp. 149.

Vidhyasekaran, P. (1988b). Physiology of Disease Resistance in Plants, Vol. II, CRC Press, Boca Raton, FL, pp. 128.

Vidhyasekaran, P. (1990). Physiology of Disease Resistance in Field Crops, Today and Tomorrow's Publishers, New Delhi, pp. 137.

Vidhyasekaran, P. (1993b). Defense genes for crop disease management, Genetic Engineering, Tissue Culture, and Molecular Biology for Crop Pest and Disease Management (P. Vidhyasekaran, ed.), Daya Publishing House, Delhi, pp. 17–30.

Vidhyasekaran, P. (1993b). Chitinase gene and pathogenesis-related proteins for crop disease management, Genetic Engineering, Tissue Culture, and Molecular Biology for Crop Pest and Disease Management (P. Vidhyasekaran, ed.), Daya Publishing House, Delhi, pp. 41–63.

Vierheilig, H., Alt, M., Neuhaus, J. M., Boller, T., and Wiemken, A. (1993). Colonization of transgenic *Nicotiana sylvestris* plants, expressing different forms of *Nicotiana tabacum* chitinase, by the root pathogen *Rhizoctonia solani* and by the mycorrhizal symbiont *Glomus mosseae*, Mol. Plant–Microbe Interact., 6:261–264.

Vigers, A. J., Roberts, W. K., and Selitrennikoff, C. P. (1991). A new family of plant antifungal proteins, Mol. Plant–Microbe Interact., 4:315–323.

Vogeli, U., Meins, F., and Boller, T. (1988). Co-ordinated regulation of chitinase and β-1,3-glucanase in bean leaves, Planta, 174:364–372.

Vogeli-Lange, R., Frundt, C., Hart, C. M., Nagy, F., and Merias, F. Jr. (1994). Developmental, hormonal and pathogenesis-related regulation of the tobacco class I β-1,3-glucanase B promoter, Plant Mol. Biol., 25:299–311.

Vogeli-Lange, R., Hansen-Gehri, A., Boller, T., and Meins, F. Jr. (1988). Induction of the defense-related glucanohydrolases, β-1,3-glucanase and chitinase, by tobacco mosaic virus infection of tobacco leaves, Plant Sci., 54:171–176.

Vogelsang, R., and Barz, W. (1993). Purification, characterization and differential regulation of a β-1,3-glucanase and two chitinases from chickpea (*Cicer arietinum L.*), Planta, 189:60–69.

Voisey, C. R., and Slusarenko, A. J. (1989). Chitinase mRNA and enzyme activity in *Phaseolus vulgaris* (L.) increase more rapidly in response to avirulent than to virulent cells of *Pseudomonas syringae* pv. *phaseolicola*, Physiol. Mol. Plant Pathol., 35:403–412.

Wagih, E. E., and Coutts, R. H. A. (1981). Similarities in the soluble protein profiles of leaf tissue following either a hypersensitive reaction to virus infection or plasmolysis, Plant Sci. Lett., 21:61–69.

Wagih, E. E., and Coutts, R. H. A. (1982). Comparison of virus-elicited and other stresses on the soluble protein fraction of cucumber cotyledons, Phytopathol. Z., 104:364–374.

Walker-Simmons, M., Hadwiger, L., and Ryan, C. A. (1983). Chitosans and pectic polysaccharides both induce the accumulation of the antifungal phytoalexin pisatin in pea pods and antinutrient proteinase inhibitors in tomato leaves, Biochem. Biophys. Res. Commun., 110:194–199.

Walker-Simmons, M., and Ryan, A. (1977a). Immunological identification of proteinase inhibitors I and II in isolated tomato leaf vacuoles, Plant Physiol., 60:61–63.

Walker-Simmons, M., and Ryan, C. A. (1977b). Wound-induced accumulation of trypsin inhibitor activities in plant leaves: survey of several plant genera, Plant Physiol., 59:437–439.

Walker-Simmons, M., and Ryan, C. A. (1986). Proteinase inhibitor I accumulation in tomato suspension cultures: Induction by plant and fungal cell wall fragments and an extracellular polysaccharide secreted into the medium, Plant Physiol., 80:68–71.

Walsh, T. A., and Twitchell, W. P. (1991). Two Kunitz type proteinase inhibitors from potato, Plant Physiol., 97:15–18.

Walter, M. H., Liu, J. W., Grand, C., Lamb, C. J., and Hess, D. (1990). Bean pathogenesis-related (PR) proteins deduced from elicitor-induced transcripts are members of a ubiquitous new class of conserved PR proteins including pollen allergens, Mol. Gen. Genet., 222:353–360.

Walter, P., and Lingappa, R. (1986). Mechanism of protein translocation across the endoplasmic reticulum membrane, Annu. Rev. Cell Biol., 2:499–516.

Walujono, K., Schooma, R. A., and Beintema, J. J. (1975). Amino acid sequence of hevein, Proc. International Rubber Conference, Kuala Lumpur, pp. 518–531.

Ward, E. R., Payne, G. P., Moyer, M. B., Williams, S. C., Dincher, S. S., Sharkey, K. C., Beck, J. J., Taylor, H. T., Ahl-Goy, P., Meins, F. Jr., and Ryals, J. A. (1991). Differential regulation of β-1,3-glucanase messenger RNAs in response to pathogen infection, Plant Physiol., 96:390–397.

Warner, S. A. J., Scott, R., and Draper, J. (1992). Characterization of a wound induced transcript from the monocot asparagus that shares similarity with a class of intracellular pathogenesis-related (PR) proteins, Plant Mol. Biol., 19:555–561.

Warner, S. A. J., Scott, R., and Draper, J. (1993). Isolation of an asparagus intracellular PR gene (AoPR1) wound responsive promoter by the inverse polymerase chain reaction and its characterization in transgenic tobacco, Plant J., 3:191–201.

Wessels, J. G. H., and Sietsma, J. H. (1981). Fungal cell walls: a survey, Encyclopedia of Plant Physiology, New Series Vol. 13 B, (W. Tanner and F. A. Loewus, eds.), Springer, New York, pp. 352–394.

White, R. F., Rybicki, E. P., Van Wechmar, M. B., Dekker, J. L., and Antoniw, J. F. (1987). Detection of PR-1 type proteins in Amaranthaceae, Chenopodiaceae, Gramineae and Solanaceae by immunoelectroblotting, J. Gen. Virol., 68:2043–2048.

Wieringa-Brants, D. H., and Dekker, W. C. (1987). Induced resistance in hypersensitive tobacco against tobacco mosaic virus by infection of intercellular fluid from tobacco plants with systemic acquired resistance, J. Phytopathol., 118:165–170.

Williamson, M. S., Forde, J., Buxton, B., and Kreis, M. (1987). Nucleotide sequence of barley chymotrypsin inhibitor-2 (CI-2) and its expression in normal and high-lysine barley, Eur. J. Biochem., 115:99–106.

Wilson, C. L. (1973). A lysosomal concept for plant pathology, Annu. Rev. Phytopathol., 11:247–272.

Wingate, V. P. M., Broadway, R. M., and Ryan, C. A. (1989). Isolation and characterization of a novel developmentally regulated proteinase inhibitor I protein and cDNA from the fruit of a wild tomato species of tomato, J. Biol. Chem., 264:17734–17738.

Wingate, V. P. M., Franceschi, V. R., and Ryan, C. A. (1991). Tissue and cellular localization of proteinase inhibitors I and II in the fruit of the wild tomato *Lycopersicon peruvianum* (L.) *Mill.*, Plant Physiol., 97:490–495.

Wingate, V. P. M., and Ryan, C. A. (1991a). A novel fruit expressed trypsin inhibitor I gene from a wild species of tomato, J. Biol. Chem., 266:5814–5818.

Wingate, V. P. M., and Ryan, C. A. (1991b). Uniquely regulated proteinase inhibitor I gene in a wild tomato species. Inhibitor I family gene is wound-inducible in leaves and developmentally regulated in fruit, Plant Physiol., 97:496–501.

Wittenbach, V. A., Lin, W., and Hebert, R. (1982). Vacuolar localization of proteases and degradation of chloroplast in mesophyll protoplast from sensing primary wheat leaves, Plant Physiol., 69:82–102.

Woloshuk, C. P., Melchers, L. S., Cornelissen, B. J. C., Meulenhoff, E. J. S., Sela-Buurlage, M. B., and Van Den Elzen, P. J. M. (1991a). New antifungal preparations, process for making such preparations, process for obtaining plants with decreased susceptibility to fungi, International Patent Application, PCT/NL 91/0089, pp. 44.

Woloshuk, C. P., Meulenhoff, J. S., Sela-Buurlage, M., Van Den Elzen, P. J. M., and Cornelissen, B. J. C. (1991b). Pathogen-induced proteins with inhibitory activity towards *Phytophthora infestans*, Plant Cell, 3:619–628.

Wong, Y., and Maclachlan, G. A. (1980). (1-3)-β-D-glucanases from *Pisum sativum* seedlings III. Development and distribution of endogenous substrates, Plant Physiol., 65:222–228.

Wubben, J. P. (1992). Subcellular localization of plant chitinases and 1,3-β-glucanases in *Cladosporium fulvum* (Syn. *Fulvia fulva*)-infected tomato leaves, Physiol. Mol. Plant Pathol., 41:23–32.

Wyatt, S. E., Pan, S. Q., and Kuc, J. (1991). β-1,3-glucanase, chitinase, and peroxidase activities in tobacco tissues resistant and susceptible to blue mould as related to flowering, age and sucker development, Physiol. Mol. Plant Pathol., 39:433–440.

Xu, D., McElroy, D., Thornburg, R. W., and Wu, R. (1993). Systemic induction of a potato *pin* 2 promoter by wounding, methyl jasmonate, and abscisic acid in transgenic rice plants, Plant Mol. Biol., 22:573–588.

Xu, Yi., Chang, Pi-F. L., Liu, D., Narasimhan, M. L., Raghothama, K. G., Hasegawa, P. M., and Bressan, R. A. (1994). Plant defense genes are synergistically induced by ethylene and methyl jasmonate, Plant Cell, 6:1077–1085.

Xuei, X. L., Jarlfors, U., and Kuc, J. (1988). Ultrastructural changes associated with induced systemic resistance of cucumber to disease: host response and development of *Colletotrichum lagenarium* in systemically protected leaves, Can. J. Bot., 66:1028–1038.

Yalpani, N., Silverman, P., Wilson, T. M. A., Kleier, D. A., and Raskin, I. (1991). Salicylic acid is a systemic signal and an inducer of pathogenesis-related proteins in virus-infected tobacco, Plant Cell, 3:809–818.

Yamaleev, A. M., Mikhsionov, U. Kh., Isaev, R. F., Yamaleeva, A. A., and Krivchenko, V. I. (1980). The activity of inhibitors of protease and the resistance of wheat to the causal agent of hard smut, Skkh. Biol., 15:143–144.

Yamane, H., Takagi, H., Abe, H., Yokota, T., and Takahashi, N. (1981). Identification of jasmonic acid in three species of higher plants and its biological activities, Plant Cell Physiol., 22:689–697.

Ye, X. S., Pan, S. Q., and Kuc, J. (1990). Association of pathogenesis-related proteins and activities of peroxidase, β-1,3-glucanase and chitinase with systemic induced resistance to blue mould of tobacco but not to systemic tobacco mosaic virus, Physiol. Mol. Plant Pathol., 36:523–536.

Yoshikawa, M., Iwasake, T., Fujii, M., and Oogaki, M. (1976). Isolation and some properties of a subtilisin inhibitor from barley, J. Biochem., 79:765–773.

Zhu, Q., Doerner, P. W., and Lamb, C. J. (1993). Stress induction and developmental regulation of a rice chitinase promoter in transgenic tobacco, Plant J., 3:203–212.

Zhu, Q., and Lamb, C. J. (1991). Isolation and characterization of a rice gene encoding a basic chitinase, Mol. Gen. Genet., 226:289–296.

Zhu, Q., Maher, E. A., Masoud, S., Dixon, R. A., and Lamb, C. J. (1994). Enhanced protection against fungal attack by constitutive co-expression of chitinase and glucanase genes in transgenic tobacco, Biotechnology, 12:807–812.

5

Phytoalexins

I. WHAT ARE PHYTOALEXINS?

Phytoalexins are low-molecular-weight antimicrobial compounds that are both synthesized by and accumulated in cells after exposure to microorganisms (Paxton, 1981). For a compound to be properly classified as a phytoalexin it should also satisfy the criteria of de novo synthesis in response to infection and accumulation to antimicrobial concentrations in the area of infection (Dixon and Harrison, 1990; VanEtten et al., 1995).

Phytoalexins constitute a chemically heterogeneous group of substances and more than 150 phytoalexins have been detected in plants infected with pathogens. The phytoalexins belong to three major groups (phenylpropanoids, terpenoids, and fatty acid derivatives) and belong to various classes of compounds: isoflavonoids, stilbenes, coumarins, dihydrophenanthrenes, sesquiterpenes, diterpenes, polyacetylenes, and other compounds (Ebel, 1986).

The important isoflavonoid phytoalexins are glyceollins in soybean, maackiiain and pisatin in pea, phaseollin in bean, and medicarpin in alfalfa (Oku et al., 1975a; Pueppke and VanEtten, 1976; Shiraishi et al., 1977; Dixon et al., 1983; Ebel et al., 1984). The stilbenoid phytoalexins are piceatannol in sugarcane (*Saccharum officinarum*) (Brinker and Seigler, 1991, 1993), resveratrol in peanut, and stilbene in *Pinus* and in plants belonging to Vitaceae (Kindl, 1985; Hoos and Blaich, 1988). The coumarin phytoalexins are umbelliferone in plants belonging to Umbelliferae and scopoletin and ayapin in sunflower (*Helianthus annuus*) (Tal and Robeson, 1986). In parsley a number of furanocoumarins and dihydropyranocoumarin have been identified as phytoalexins (Scheel et al., 1986). Hircinol and orchinol in *Orchis* belong to the dihydrophenanthrenes group (Fritzemeier and Kindl, 1983). The terpenoid phytoalexins are ipomeamarone (furanosesquiterpene) in sweet potato (Fujita and Ashai, 1985), lubimin and rishitin in potato (Shih et al., 1973; Brindle et al., 1985), casbene in castor bean (diterpene) (West et al., 1985; Moesta and West, 1985), and lettucenin A in lettuce (Tamara et al., 1988; Bennett et al., 1994). The fatty acid derivatives are falcarindiol (polyacetylene) in tomato and wyerone acid (furanoacetylene) in broadbean, safynol and dehydrosafynol in *Carthamus tinctorius* (Tietjen and Matern, 1984).

In one and the same plant different kinds of phytoalexins have been reported. The polyacetylenes falcarinol and falcarindiol and the sesquiterpene rishitin have been reported in tomato (DeWit and Kodde, 1981). In broad bean the furanoacetylene, wyerone acid, and the isoflavonoid medicarpin have been detected (Hargreaves et al., 1977).

Phytoalexins have been reported mostly in dicotyledons (Bailey, 1982; Bailey and Mansfield, 1982). In a few monocots such as rice (Cartwright et al., 1981) and oat (Mayama et al., 1981), phytoalexins have been reported. Isoflavonoids are common in the Leguminosae (Ingham, 1982). Sesquiterpenoids are common in the Solanaceae. Stilbenes have been detected in Vitaceae and Leguminosae (Bailey and Mansfield, 1982).

II. BIOSYNTHESIS OF PHENYLPROPANOID PHYTOALEXINS

A. Biosynthesis of Phytoalexins in Bean

1. Biosynthetic Pathway

Biosynthetic pathways of phytoalexins in beans have been studied in detail. Phaseollin, phaseollidin, and kievitone are the important phytoalexins in bean. Phaseollinisoflavan, coumestrol, and genistein are the other phytoalexins reported in bean (Rathmell and Bendall, 1971; Bailey and Burden, 1973; Biggs, 1975; Dixon and Bendall, 1978b). Increases in the activity of a number of biosynthetic enzymes and concomitant accumulation of phytoalexins have been reported in bean (Bailey, 1974; Dixon and Bendall, 1978; Robbins and Dixon, 1984; Cramer et al., 1985a). Phenylalanine ammonia-lyase (PAL) is the first enzyme in the phenyl propanoid pathway. Cinnamic acid 4-hydroxylase (Ca4H), 4 coumarate: CoA ligase (4CL), chalcone synthase (CHS), and chalcone isomerase (CHI) are the key enzymes in biosynthesis of isoflavonoids (Figure 1; Dixon, 1986; Collinge and Slusarenko, 1987; Dixon and Harrison, 1990).

2. Phenylalanine Ammonia-Lyase

The first enzyme in phenylpropanoid metabolism, PAL is present constitutively in all stages of plant development (Hahlbrock and Grisebach, 1979; Knogge et al., 1987; Jahnen and Hahlbrock, 1988a,b; Hahlbrock and Scheel, 1989). The enzyme is a tetramer and contains two active sites per molecule (Bolwell et al., 1985). Multiple tetrameric forms differing in Km, size, and pI of the subunits have been obtained from purified PAL preparations (Bolwell et al., 1985). Multiple mRNA species for the enzyme have been reported (Edwards et al., 1985; Dixon,

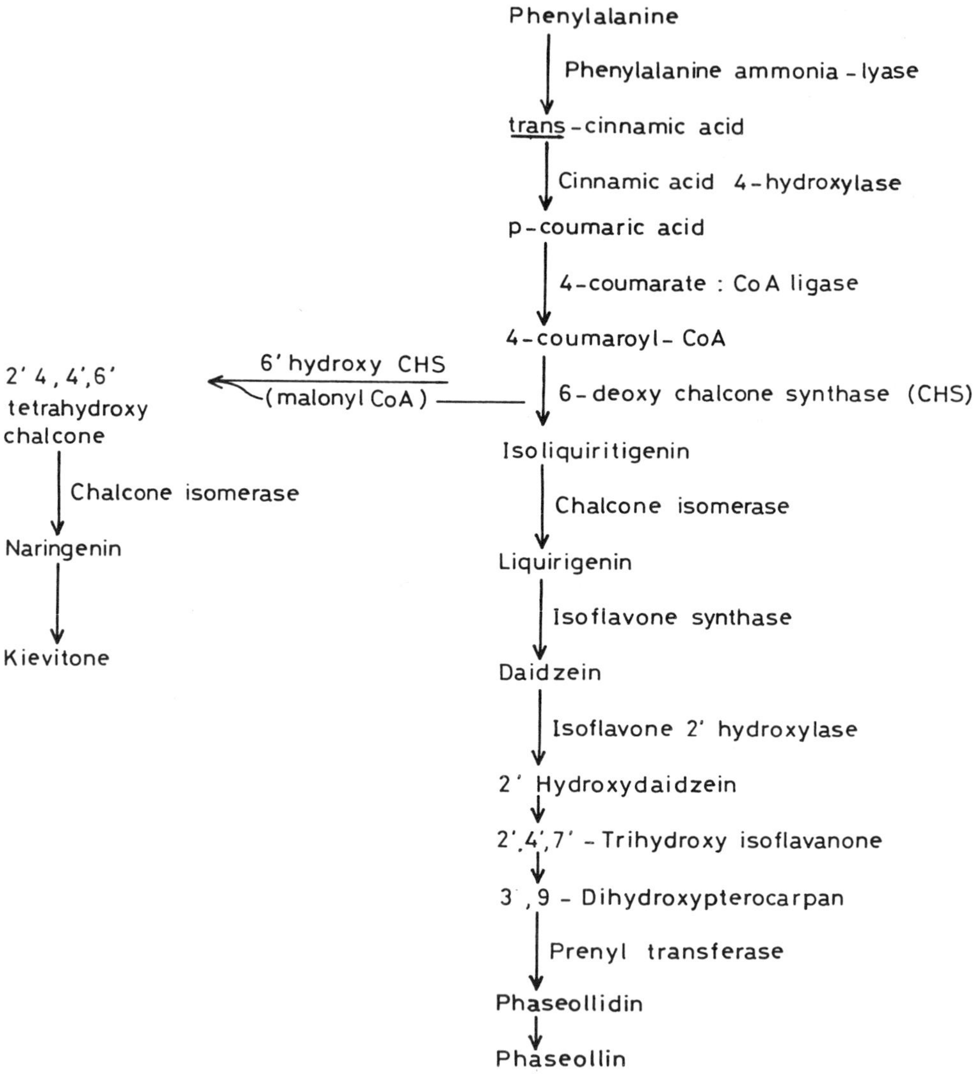

Figure 1 Biosynthesis of phaseollin, phaseollidin, and kievitone in bean. (From Dixon, 1986; Collinge and Slusarenko, 1987; Dixon and Harrison, 1990.)

1986; Cramer et al., 1989). PAL is encoded in bean by a family of divergent classes of genes (Dixon et al., 1986; Hahlbrock and Scheel, 1989; Mavandad et al., 1990). Several PAL poly-peptides have been observed following in vitro translation of mRNA isolated from elicited bean cell cultures (Bolwell et al., 1985). Three divergent classes of PAL gene have been cloned in bean (Cramer et al., 1989). The nucleotide sequences of gPAL2 and gPAL3 have been determined. These genes contain open reading frames encoding polypeptides of 712 and 710 amino acids, respectively. Single introns in the two genes (1720 bp in gPAL2, 447 bp in gPAL3) occur in identical positions. They show 59% nucleotide sequence similarity in exon I, 74% similarity in exon II, and extensive divergence in the intron, 5′ and 3′ flanking sequences (Cramer et al., 1985b; 1989). The full sequence of PAL1 gene has not yet been obtained (Dixon and Harrison, 1990).

Fungal infection results in de novo synthesis of PAL (Lawton et al., 1983a;b). PAL is rapidly synthesized de novo and accumulated in bean cell cultures due to elicitor treatment also (Cramer et al., 1985b). Transcriptional activation of the genes encoding PAL has been observed within 5 min of elicitor treatment (Lamb et al., 1986).

PAL mRNA was almost completely absent from control bean cells but rapidly accumulated following *Colletotrichum lindemuthianum* elicitor treatment. Increases in PAL mRNA could be observed 30 min after elicitation and maximal accumulation occurred 3–4 h after elicitor treatment. Subsequently, the mRNA decayed rapidly to relatively low levels. PAL mRNA did not accumulate in unelicited control cells. Accumulation and decay of PAL mRNA in total and polysomal RNA fractions exhibited very similar kinetics. There was a close correlation between changes in hybridizable mRNA, translatable mRNA activity, and PAL synthesis in vivo. Maximal stimulation coincided with the period of most rapid increase in PAL enzyme activity (Cramer et al., 1985b; Edwards et al., 1985;). Increased mRNA activity encoding PAL has been observed in the total RNA, polysomal RNA, and poly A^+ RNA-fractions (Lawton et al., 1983b). cDNA clones to bean PAL have been obtained (Edwards et al., 1985).

The genes of gPAL1 and gPAL2 are activated by fungal elicitor in cell cultures of bean cultivar Canadian Wonder while gPAL3 is not activated (Cramer et al., 1989; Liang et al., 1989a). In suspension cultures of bean cultivar Immuna, all three PAL genes are elicitor inducible and particularly PAL3 transcripts were strongly induced (Ellis et al., 1989). Transgenic tobacco plants expressing PAL2 gene have been developed (Liang et al., 1989b; Bevan et al., 1989). In these plants, the gene was developmentally expressed in vascular tissue (Liang et al., 1989b). Thus the PAL genes appear to express variably in different bean cultivars or in different plants.

3. Cinnamic acid 4-hydroxylase

Cinnamic acid 4-hydroxylase (CA4H) activity is induced before the accumulation of phytoalexin in bean cell cultures treated with fungal elicitor (Dixon and Bendall, 1978; Bolwell et al., 1985a).

4. 4-Coumarate: CoA ligase

4-Coumarate: CoA ligase (4CL) activity increased in cell suspension cultures of French bean treated with elicitor before the accumulation of phaseollin (Dixon and Bendall, 1978). Messenger RNA of 4CL was induced before the increase in the enzyme activity was observed (Hahlbrock et al., 1981; Ragg et al., 1981; Hahlbrock et al., 1983).

5. Chalcone synthase

The first committed step in the biosynthesis of isoflavonoids is catalyzed by chalcone synthase (CHS) (Hahlbrock and Grisebach, 1979; Heller and Forkmann, 1988). Substrates of the CHS reaction are the major product of general phenylpropanoid metabolism, 4-coumaroyl-CoA, and the product of the acetyl CoA carboxylase reaction, malony,-CoA. Three acetate units from malonyl-CoA are converted into the C_{15} intermediate 2′, 4, 4′, 6′-tetrahydroxy chalcone (Ebel, 1986; Dixon, 1986; Hahlbrock and Scheel, 1989). CHS enzyme is necessary for formation of phytoalexins in French bean tissues (Dixon and Bendall, 1978; Dixon et al., 1981; Dewick et al., 1982).

CHS mRNA was almost completely absent in control bean cells but was rapidly induced following *Colletotrichum lindemuthianum* elicitor treatment. Increases in CHS mRNA could be observed 30 min after elicitation and maximal induction occurred about 3 h after elicitor treatment. Subsequently, the mRNA rapidly decayed to relatively low levels. There was a close correlation between the kinetics of induction of translatable mRNA activity and CHS synthesis in vivo (Ryder et al., 1984).

The increase in CHS mRNA levels was due to de novo synthesis (Cramer et al., 1985b). Increased levels of mRNA are followed by the accumulation of isoflavonoid-derived phytoalexins in elictor-treated bean cells and infected plants (Lawton et al., 1983a, b; Ryder et al., 1984; 1987; Bell et al., 1984; 1986; Ebel, 1986). Induction of CHS transcription in response to fungal elicitor is observed within 5–10 min (Lamb et al., 1986; Lawton and Lamb, 1987). It suggests that an efficient signal pathway may operate elicitor perception by putative receptors at the plasma membrane. Efficient induction of CHS activity was observed only at higher elicitor concentrations (Lawton et al., 1983a), and there may be some form of posttranslational control over the rate of accumulation of CHS activity (Lawton et al., 1980; 1983a).

Differential de novo synthesis of 9 CHS polypeptides has been observed in elicitor-treated bean cell cultures as well as in hypocotyls inoculated with spores of *Colletotrichum lindemuthianum,* prior to accumulation of isoflavonoid phytoalexins in bean (Ryder et al., 1987).

Seven CHS genes have been detected in bean genome (Ryder et al., 1987). Clustering of these genes have been observed and they may be found in the same genomic clones (Dixon, 1986; Ryder et al., 1987). Two of these genes, CHS 14 and CHS 17, are arranged 12 kb apart in opposite transcriptional orientations (Ryder et al., 1987). CHS genes show different patterns of expression in bean plants. CHS 8 transcripts were present in roots, stems, and floral buds whereas the transcript levels in fully expanded leaves and mature flowers were too low to be detected. CHS 4 transcript was present in high levels in floral buds (Schmid et al., 1990). The induction of CHS 8 transcripts was observed in elicitor-treated bean cell suspension cultures and in wounded hypocotyls (Ryder et al., 1987). Transcripts specific for CHS 1 gene increased very slowly in response to *C. lindemuthianum* elicitor, reaching maximum only after 24 h (Ellis et al., 1989), but total CHS activity increased even within 1.5 h after elicitor treatment. This increase was preceded by an increase in total CHS transcripts reaching maximum levels 4 h postelicitation. The elicitor induced accumulation of phaseollin and kievitone in cell cultures of bean cv. Imuna. At 22.5 h postelicitation, kievitone was the major phytoalexin present, whereas phaseollin began to predominate after 50 h. The elevated activities of the CHS 1 from 24 to 36 h preceded the major phase of accumulation of phaseollin. The results suggest that CHS 1 may be specifically involved in phaseollin biosynthesis (Ellis et al., 1989). The bean contains a separate 6′-deoxy-chalcone synthase involved in the synthesis of phaseollin (Dewick et al., 1982). Two different chalcone synthases may be involved in biosynthesis of kievitone and phaseollin (Dixon, 1986).

The CHS 8 promoter containing 1250 bp upstream of the transcription start site and 162 bp of the untranslated leader sequence extending to 1 bp 5′ of the translation site was fused to the β-glucuronidase (GUS) coding sequence in a vector pBI 101.1 (Schmid et al., 1990). In addition, the complete CHS 8 gene was ligated into the polylinker of the binary vector BIN 19. Both constructs were transferred to tobacco. Mature leaf tissue of transgenic plants containing the CHS 8–GUS gene fusion contained very low levels of GUS activity. Localized wounding of the tissue of a detached mature leaf caused induction of GUS activity in the rings of tissue immediately adjacent to the wound sites. Weak staining at some sites was observed 16 h after wounding, while strong GUS staining was observed at all wound sites after 48–72 h. However, immediate application of fungal elicitor to the wound site caused a marked induction of CHS 8 promoter after 16–24 h compared with the response of wounding alone. Strong GUS staining was observed in tissue immediately adjacent to the site of *Phytophthora megasperma* f.sp. *glycinea* elicitor application (within 2–3 mm) and weaker staining in tissue up to 10–15 mm distant. Activation of the CHS 8 promoter at a distance was also observed in transgenic tobacco leaves infiltrated locally with an incompatible isolate of *Pseudomonas syringae* or treated with potassium oxalate. Induction of GUS activity was observed in leaf epidermal, mesophyll, and vascular tissue within this zone (Schmid et al., 1990). The results suggest that bean CHS 8 is activated by the fungal elicitor and this activation is found even in tobacco. It is interesting that tobacco also has CHS gene, but these genes are not induced by any stress (Dixon et al., 1983).

Bean CHS 15 is also elicitor-inducible (Dixon et al., 1988). A number of deletions of the bean CHS 15 promoter with 5′ end points of –326, –173, –130, –72, and –19 were fused to the bacterial chloramphenicol acetyl transferase (CAT) gene and introduced into soybean protoplasts by electroporation. The –326 construct was strongly induced by elicitor from *Colletotrichum lindemuthianum* (Dron et al., 1988). The construct was only weakly expressed in the soybean protoplasts without elicitor treatment (Dron et al., 1988). Deletion to –173 results a 2.5-fold increase in expression in response to elicitor. Deletion of the bean CHS promoter to –72 almost totally abolished CAT expression, whereas the –130 deletion gives expression similar to that of the –326 construct (Dron et al., 1988). The results suggest that the region between –130 and TATA box-containing sequences is essential for transcriptional activation (Dixon and Harrison, 1990).

These experiments were repeated in alfalfa protoplasts and entirely different results were obtained. The deletion to –173 resulted in decreased expression, suggesting that the region may contain both positive and negative elements whose operation depends on the cell-type environment (Dixon and Harrison, 1990).

The 336 bp upstream of the transcription start site in CHS 15 gene has been found to be sufficient to confer tissue-specific and elicitor responsiveness to the GUS reporter gene in transgenic tobacco (Stermer et al., 1990). When it is fused to the bacterial CAT gene, it drives CAT expression in an elicitor-inducible manner on electroporation of the CHS-CAT nopaline synthase (NOS) 3′ construct into protoplasts of soybean, tobacco, or alfalfa (Dron et al., 1988; Choudhary et al., 1990). A longer promoter from the CHS 8 gene is also inducible by bacterial infection of transgenic tobacco plants harboring a CHS 8–GUS chimeric gene (Stermer et al., 1990; Dixon and Harrison, 1990).

Functional analysis of the promoter of CHS 15 gene, one of a family of bean genes encoding CHS, which is inducible by fungal elicitor or reduced glutathione in electroporated soybean and alfalfa protoplasts, has defined the location of both positive and negative *cis*-acting elements, which affect quantitatively the expression of the gene, and sequences essential for elicitor inducibility within –326 bp of the transcription start site (Dron et al., 1988; Stermer et al., 1990; Harrison et al., 1991a; Lawton et al., 1991). A silencer is located between positions –326 and –173 in CHS 15 promoter (Dron et al., 1988; Harrison et al., 1991b).

The functional architecture of the CHS 15 promoter was dissected by a novel homologous plant in vitro transcription initiation system in which whole-cell and nuclear extracts from suspension-cultured soybean cells direct accurate and efficient transcription from an immobilized promoter template (Arias et al., 1993). 5′ deletion from –130 to –72 abolished CHS 15 transcript by the soybean whole-cell extract. Preincubation of the soybean whole-cell extract with CHS 15 sequences from –4 to +105 had little effect on subsequent transcription of the immobilized CHS 15 promoter, indicating that this region of the promoter does not contain *cis*-elements that bind *trans* factors in the soybean extract essential for CHS 15 transcription. In contrast, depletion of the extract by preincubation with CHS 15 sequences from either positions –132 to –80 or from positions –80 to –40 as *trans* competitors essentially abolished activity, indicating that these specific regions contain *cis* elements that bind *trans* factors in the soybean whole-cell extract that are essential for transcription from the CHS 15 promoter in vitro (Arias et al., 1993). Assay of the effects of depletion of the soybean whole-cell extract by preincubation with small regions of the CHS 15 promoter or defined *cis* elements showed that *trans* factors that bind to G-box (CACGTG, –74 to –69) and H-box (CCTACC, –61 to –56 and –121 to –126) *cis* elements, respectively, make major contributions to the transcription of the CHS 15 promoter in vitro. Both *cis* element/*trans* factor interactions in combination are required for maximal activity (Arias et al., 1993).

Combination of H-box (CCTACC (N)7-CT) and G-box (CACGTG) *cis* elements has been shown to be necessary for feed-forward stimulation of a chalcone synthase promoter by the phenylpropanoid-pathway intermediate p-coumaric acid (Loake et al., 1992).

CHS 8 and CHS 15 were each fused upstream of the coding region of the β-glucuronidase (GUS) reporter gene and transformed into tobacco. Short wave ultraviolet (UV) light or mercuric chloride induced the expression of both gene fusions in transgenic tobacco leaves. Application of oxalate induced the expression of the CHS 8-GUS, but not the CHS 15-GUS gene fusion. The data suggest the operation of several distinct mechanisms for stress activation of defense genes (Stermer et al., 1990).

The presence of binding sites for bean and alfalfa nuclear proteins within these regulatory elements has been demonstrated (Harrison et al., 1991a; b; Lawton et al., 1991).

Sequences homologous to the CHS *cis*-acting elements are also found in the promoter of a bean PAL gene, which is coordinately induced with CHS at the onset of phytoalexin synthesis (Cramer et al., 1985a; 1989a).

6. Chalcone Isomerase

Chalcone isomerase (CHI) from bean catalyzes the stereospecific isomerization of both 2′,4,4′-trihydroxy chalcone (isoliquiritigenin) and 2′,4,4′,6′-tetrahydroxy chalcone to yield the corresponding (–) flavanones (Dixon et al., 1982; 1983b).

A single isomerase enzyme may be involved in the biosynthesis of both kievitone and phaseollin in bean, while two chalcone synthases may be required for it (Dixon, 1986). CHI appears to be encoded by a single gene (Dixon et al., 1986; Mehdy and Lamb, 1987). CHI in bean is a monomeric protein of Mr 27,000 (Robbins and Dixon, 1984).

CHI activity is usually high in unelicited bean cell cultures and its activity increased due to elicitor treatment (Dixon and Bendall, 1978; Dixon and Lamb, 1979; Robbins et al., 1985). In the induction of CHI in *Colletotrichum lindemuthianum*-infected or elicitor-treated cells, de novo synthesis of this enzyme has been shown (Dixon et al., 1983c; Robbins and Dixon, 1984; Cramer et al., 1985a; b). A significant proportion of elicitor-induced enzyme activity may be due to the activation of the preexisting enzyme (Dixon et al., 1983c; Robbins and Dixon, 1984). Elicitor may induce the formation of both active and inactive CHI molecules (Dixon et al., 1983c; Robbins and Dixon, 1984). In *Petunia hybrida* CHI is not involved in synthesis of phytoalexins (van Tunen et al., 1988), while CHI in bean is involved in synthesis of phytoalexins. These two enzymes have different substrate specificities. The bean CHI catalyzes the isomerization of chalcones both with and without a 6′ hydroxyl group, whereas the two CHI isozymes of *Petunia* require a 6′ hydroxyl activity (Dixon et al., 1983; Blyden et al., 1989). The isoflavonoid phytoalexins of bean lack a hydroxyl group at the 6′ position, chalcone numbering (Dixon et al., 1983).

In bean hypocotyls infected with *C. lindemuthianum,* the magnitude and timing of CHI induction paralleled those of the preceding biosynthetic enzymes (Dixon, 1986).

7. Other Enzymes

Isoflavone synthetase (= isoflavone synthase), isoflavone 2′-hydroxylase, and prenyl transferase are difficult to purify. However, their increased activities lead to induction of phytoalexin synthesis (Dixon et al., 1983; Dixon and Bolwell, 1986; Biggs et al., 1987; Bless and Barz, 1988; Welle and Grisebach, 1988a; b; Dixon and Harrison, 1990; Tiemann et al., 1991).

8. Coordinated Induction of Enzymes

The induction of phytoalexin synthesis appears to be associated with coordinated induction of relevant biosynthetic enzymes (Barz et al., 1988; Daniel et al., 1988; Dalkin et al., 1990a). Coordinate regulation of PAL, 4-hydroxy cinnamate CoA ligase, and CHS has been reported (Whitehead et al., 1982; Dixon et al., 1983; Lawton et al., 1983a; Robbins et al., 1985). PAL, CHS, and CHI are coordinately regulated in bean cell cultures treated with the elicitor (Cramer et al., 1985a; Dixon, 1986).

CHI mRNA is coordinately induced with PAL and CHS mRNAs in elicitor-treated bean cell cultures (Mehdy and Lamb, 1987). CHS mRNA activity was higher than PAL at later times (Dixon and Lamb, 1979; Robbins and Dixon, 1984; Dixon et al., 1986).

The appearance of increased activities of phenylpropanoid biosynthetic enzymes and levels of their transcripts in plant suspension cultures is both rapid and transient, implying tight control of both initation and cessation of transcription (Dixon and Harrison, 1990). Cinnamic acid, the immediate product of the PAL reaction, may be involved in the downregulation of the phenylpropanoid pathway (Edwards et al., 1990). Exogenous additions of cinnamic acid prevent the induction of PAL enzyme activity in bean cell suspension cultures (Dixon et al., 1980; Bolwell et al., 1986). The rapid degradation of PAL activity in elicitor-stimulated bean cells following exogenous addition of *trans*-cinnamic acid has been reported (Shields et al., 1982; Walter and Hahlbrock, 1984). Exogenously added cinnamic acid affects the rate of PAL synthesis by inhibiting appearance of PAL transcripts (Bolwell et al., 1988). Cinnamic acid may increase the rate of removal of active PAL enzyme by inducing synthesis of a proteinaceous inhibitor of the enzyme (Bolwell et al., 1986).

When bean cells were treated with *C. lindemuthianum* elicitor, all three PAL genes were induced. Concentrations of cinnamic acid above 10^{-4} M inhibited appearance of all three PAL (PAL1, PAL2 and PAL3) transcripts (Mavandad et al., 1990). L-α-Aminooxy-3-phenyl propionic acid (AOPP) is a potent and specific inhibitor of PAL activity in vivo. Blocking cinnamic acid production in elicitor-treated cells with AOPP resulted in increased levels of each of the three individual PAL transcripts. It suggests that the cinnamic acid may act as a component in a regulatory feedback system operating at the level of phenyl propanoid gene transcription (Mavandad et al., 1990).

CHS is also inhibited by application of relatively high concentrations (10^{-4}–10^{-3} M) of *trans*-cinnamic acid at which concentrations PAL is also inhibited (Bolwell et al., 1988; Mavandad et al., 1990). CHS appears to be regulated by cinnamic acid at the transcriptional level and not posttranslationally as in the cause of PAL (Bolwell et al., 1988). The inhibition of PAL activity by the AOPP results in a delayed superinduction of transcripts encoding CHS (Bolwell et al., 1988).

The effect of various cinnamic acid derivatives on bean CHS 15 expression was assessed in electroporated alfalfa protoplasts. *Trans*-p-coumaric acid, the direct product of the CA 4-hydroxylase, stimulated expression of CHS gene up to 4.5-fold at 5×10^{-4} M. Cinnamic acid at high concentrations inhibited CHS gene activity. Caffeic and sinapinic acids did not inhibit CHS gene activity (Table 1; Loake et al., 1991).

Cinnamic acid at low concentrations stimulated the expression of bean CHS 15 promoter in the alfalfa protoplast system. At high concentrations it inhibited the expression of the promoter (Loake et al., 1991). The stimulatory effects of

Table 1 Effects of cinnamic acid derivatives on CHS-chloramphenicol acetyltransferase (CAT) and nopaline synthase (NOS) chimeric gene expression in electroporated alfalfa protoplasts

	Relative CAT activity (%) at concentrations (M) below		
Compound	0	5×10^{-6}	5×10^{-4}
trans-cinnamic acid	100	119	13
trans-p-coumaric acid	100	126	456
trans-caffeic acid	100	105	96
trans-sinapinic acid	100	104	113

Source: Loake et al., 1991.

low cinnamic acid concentrations may result from metabolism to *trans*-p-coumaric acid (Loake et al., 1991).

The stimulation of expression of the CHS 15 promoter by low concentrations of cinnamic acid and *trans*-p-coumaric acid appears to occur through specific *cis*-acting sequences because this stimulation was totally abolished by deletion of the promoter to position –130. The effects of this deletion appear to be due to the removal of a binding site for a *trans*-acting factor because electroporation of the sequence from –183 to –130 in *trans* with the plasmid pCHC1 also results in inhibition of stimulation by cinnamic acid and p-coumaric acid, presumably by competition for this factor (Loake et al., 1991).

Thus cinnamic acid synthesized due to the activation of PAL may induce the activation of the enzyme leading to synthesis of p-coumaric acid which in turn activates CHS. The coordinated induction of PAL and CHS induces CHI mRNA synthesis (Mehdy and Lamb, 1987). This kind of coordinated expression leads to synthesis of the phytoalexins. If there is any difference in timing of this coordination there may be accumulation of cinnamic acid or p-coumaric acid, which in turn may inhibit PAL, CHS and CHI (Hahlbrock and Scheel, 1989).

Several elicitor-active fractions have been obtained from culture filtrate of *C. lindemuthianuum*. All of them induced mRNA activities encoding PAL, CHS, and CHI (Hamdan and Dixon, 1987). The sequence CCTACC (N) CT is found at positions –61 to –47, –154 to –140, and in the opposite orientation on the opposite strand, –122 to –136, in the bean CHS 15 promoter. The same sequence occurs five times in the promoter of the bean gPAL2 gene, which is coinduced with CHS in response to elicitor (Dixon and Harrison, 1990). Sequences between –130 and the TATA box are essential for elicitor regulation of both the enzymes (Dixon and Harrison, 1990).

The different phytoalexins of bean are also induced in sequence. Kievitone accumulates rapidly to a maximum level of approximately 40 nmol/g fresh weight at between 8 and 12 h after treatment with the elicitor isolated from *C. lindemuthianum*. However, phaseollin is present in very low levels up to 24 h postelicitation, but then it accumulates to levels in excess of 500 nmol/g fresh weight at around 48 h postelicitation (Robbins et al., 1985).

B. Biosynthesis of Phytoalexins in Soybean

Four isomeric pterocarpan phytoalexins, glyceollins I–IV, have been characterized in soybean (Partridge and Keen, 1977; Lyne and Mulheiru, 1978). Due to *Phytophthora megasperma* infection or elicitor treatment these phytoalexins accumulate following increases in all the enzymes of general phenylpropanoid metabolism and of isoflavonoid biosynthesis as well as enzymes specifically involved in later steps of glyceollin biosynthesis (Figure 2; Zahringer et al., 1978; 1979; Leube and Grisebach, 1983; Hagmann et al., 1984; Kimpel and Kosuge, 1985; Grab et al., 1985; Bonhoff et al., 1986a,b; Habereder et al., 1989).

Fungal infection results in de novo synthesis of PAL and the increased synthesis results from increased mRNA activity in soybean (Borner and Grisebach, 1982; Lawton et al., 1983b). The importance of PAL activity in glyceollin biosynthesis has been demonstrated. PAL activity is inhibited in soybean roots by treatment with L-amino-oxy-3-phenylpropionic acid (AOPP) or R-(1-amino 2-phenyl-ethyl) phosphonic acid (APEP). This treatment completely abolished glyceollin induction and changed the phenotype of the incompatible to compatible (Moesta and Grisebach, 1982).

The increase in glyceollin biosynthesis is accompanied by increased 4CL in cell suspension cultures and cotyledons of soybean due to fungal elicitor treatment (Ebel, 1979; Hahlbrock et al., 1980; Ebel et al., 1984). Increase in mRNA encoding 4CL was observed in soybean hypocotyls infected with *Phytophthora megasperma* sp. *glycinea* and it was followed by the accumulation of glyceollin (Schmelzer et al., 1984).

Several enzymes of 4CL have been reported in soybean. The activity of isoenzyme 2 alone increased due to elicitor treatment and this isoenzyme has been shown to be involved in flavonoid biosynthesis (Hahlbrock and Grisebach, 1979; Hille et al., 1982).

CHS activity increased in soybean due to fungal infection. CHS has been purified from soybean (Welle and Grisebach, 1987; Welle et al., 1991). CHS genes from soybean have been cloned and characterized (Dangl et al., 1989; Wingender et al., 1989; Akada et al., 1990; 1991; Estabrook and Gopalan, 1991). Among the three CHS genes, only one gene (chs 1) was found to be transcribed after elicitor treatment (Wingender et al., 1989). Two soybean CHS genes (*chs* 1 and *chs* 3) are tightly linked in a 5′ to 5′ orientation (Wingender et al., 1989). The

Phenylalanine

↓ Phenylalanine ammonia-lyase (PAL)

trans-cinnamic acid

↓ Cinnamic acid 4-hydroxylase (CA4H)

p-coumaric acid

↓ 4-coumarate : CoA ligase (4 CL)

4-coumaroyl-CoA

↓ Chalcone synthase (CHS)

Isoliquiritigenin

↓ Chalcone isomerase (CHI)

Liquiritigenin

↓ Isoflavone synthase (IFS)

Daidzein

↓ Isoflavone 2- hydroxylase (IFH)

2' Hydroxydaidzein

↓

2',4',7 -Trihydroxy isoflavanone

↓

3, 9 - Dihydroxypterocarpan

↓ pterocarpan 6a-hydroxylase

3, 6a, 9 - trihydroxypterocarpan

↓ pterocarpan dimethyl allyl transferase

Glyceollin I, II and III

Figure 2 Biosynthesis of glyceollins in soybean. (From Bonhoff et al., 1986a,b.)

5′ region of *chs* 1 of soybean was subjected to deletion analysis with the help of chimeric CHS-neomycin phosphotransferase (NPT II)/GUS gene constructs. This analysis delimited the sequences necessary for elicitor inducibility to –175 and –134 of the *chs* 1 promoter (Wingender et al., 1990).

Parsley cells accumulate furanocoumarins (not flavonoids) upon pathogenic attack and the endogenous CHS is not responsive to elicitor treatment (Schulze-Lefert et al., 1989a; b). The same sequences necessary for elicitor inducibility of *chs* 1 promoter in soybean were able to direct elicitor inducibility in parsley protoplasts (Wingender et al., 1990). It suggests that the *cis*-acting elements involved in plant defense are functionally conserved between species because they seem to be recognized by the appropriate *trans*-acting factors in a heterologous host (Wingender et al., 1990).

Glyceollin synthesis is accompanied by increased transcription and activity of PAL and CHS, followed by CHI (Borner and Grisebach, 1982; Schmelzer et al., 1984; Bonhoff et al., 1986a; Esnault et al., 1987). CHI catalyzes the stereospecific isomerization of a chalcone to give a (2S)-flavanone: the immediate precursor of both flavones and isofavones (Ebel, 1986). The reductase involved in the biosynthesis of 6′-deoxychalcone, the flavanones naringenin, and its 5-deoxy derivative are formed in soybean (Welle and Grisebach, 1989). An isoflavone synthetase has been shown to be involved in the biosynthetic pathway (Hagmann and Grisebach, 1984). The isoflavone genistein is formed from naringenin through a two-step 2,3-aryl migration (Hagmann and Grisebach, 1984; Kochs and Grisebach, 1986). A 2,3-aryl migrates with a flavonoid intermediate (Hahlbrock and Grisebach, 1975; Dewick, 1982). Enzymic arrangement proceeds with a flavonone as substrate. A NADPH- and dioxygen-dependent rearrangement of (2S)-naringenin (5,7,4′-trihydroxy-flavanone) to genistein (5,7,4′-trihydroxyflavonone) has been demonstrated (Hagmann and Grisebach, 1984).

The isoflavones daidzein and genistein are present constitutively in large quantities as conjugates in all seedling organs of soybean cultivars. The conjugates have been identified as the 7-0-glucosyl-isoflavones. The conjugates are hydrolyzed to free daidzein and high levels of glyceollin subsequently accumulate. The glyceollin biosynthesis may not be solely dependent on the induction of enzymes of early phenylpropanoid and flavonoid metabolism, but may depend partly on the rapid release of the isoflavone aglycones from their conjugates and/or in the later steps of glyceollin biosynthesis (Graham et al., 1990).

Addition of daidzein to infection sites on hypocotyls results in a large increase in daidzein accumulation but not in the levels of glyceollins (Table 2; Morris et al., 1991). It indicates that metabolism of daidzein to the glyceollins is a rate-limiting step and that, following infection, the supply of substrates through the phenylpropanoid pathway is more than adequate for glyceollin biosynthesis with accumulation of daidzein and its glucosides after infection occuring as a consequence (Morris et al., 1991).

Table 2 Accumulation of isoflavonoids in soybean hypocotyls inoculated with *Phytophthora megasperma* f.sp. *glycinea* with and without an exogenous supply of daidzein

	Compound (μg/g fresh wt.)		
Compound	Control	Pathogen	Pathogen + daidzein
Daidzein	2.5	109	2730
Glyceollins	0	4668	4285

Source: Morris et al., 1991.

Induction of several of the later enzymes in glyceollin biosynthesis has been reported in infected soybean roots (Benhoff et al., 1986a, b). A pterocarpan 6a-hydroxylase has been shown to convert 3,9-dihydroxypterocarpan sterospecifically to a 3,6a,9-trihydroxypterocarpan (glycinol) in the presence of NADPH and dioxygen in soybean (Hagmann et al., 1984). A dimethyl-allylpyrophosphate, 3,6a,9-trihydroxypterocarpan dimethylallyltransferase, catalyzes the formation of the 4- as well as 2-dimethylallyl-trihydroxypterocarpan (Zahringer et al., 1979; Leube and Grisebach, 1983, Welle and Grisebach, 1988b). The products of the dimethylallyl transfer may be the intermediates in the biosynthesis of the various glyceollin isomers in soybean (Ebel, 1986).

C. Biosynthesis of Phytoalexins in Alfalfa

1. De Novo Synthesis of Enzymes

The pterocarpan medicarpin is the major phytoalexin of alfalfa (Latunde-Dada and Lucas, 1985; Baker et al., 1988; Dalkin et al., 1990a). The biosynthetic pathway of medicarpin is presented in Figure 3 (Dalkin et al., 1990a). When alfalfa-cultured cells are treated with elicitor from *C. lindemuthianum*, the phytoalexin medicarpin and related compounds accumulate. The phytoalexin accumulation results from rapid and extensive increases in the extractable activities of PAL, 4 CL, CHS, CHI, and isoflavone-0-methyl transferase (Dalkin et al., 1990a). PAL activity increased approximately fourfold, reaching maximum values 10–14 h after elicitor treatment (Kessmann et al., 1990a). PAL activity showed rapid increase but the induction was only transient with a quick decline (Dalkin et al., 1990a).

Three isoforms of PAL have been purified from alfalfa cell suspension cultures treated with the fungal elicitor (Jorrin and Dixon, 1990). However, it has been suggested that there may be more than four PAL genes in alfalfa (Gowri et al., 1991). A single clone (APAL 1) that encoded a full-length PAL cDNA frame encoding a 725 amino acid polypeptide, 96 bp 5′-untranslated leader, and 128 bp

Phenylalanine

↓ phenylalanine ammonia-lyase (PAL)

trans-cinnamic acid

↓ cinnamic acid 4-hydroxylase (CA4H)

p-coumaric acid

↓ 4-coumarate CoA ligase (4CL)

4-coumaroyl-CoA

↓ chalcone synthase (CHS)

2′,4,4′-trihydroxychalcone

↓ chalcone isomerase (CHI)

4′,7-dihydroxy flavanone

↓ isoflavone synthase (IFS)

Daidzein

↓ isoflavone 4′-O-methyltransferase (=Daidzein O-methyl transferase (IFMT = DOMT)

Formononetin

↓ isoflavone-2′-hydroxylase (IFOH)

2′-Hydroxyformononetin

↓ isoflavone reductase (IFR)

Vestitone

↓ pterocarpan synthase (PTS)

Medicarpin

Figure 3 Biosynthesis of medicarpin in alfalfa. (From Dalkin et al., 1990a.)

3′-noncoding region has been isolated. The deduced amino acid sequence was 86.5%, similar to that of PAL 2 gene of bean, and encoded a polypeptide of Mr 78865. A second PAL cDNA species has also been isolated whose 3′-untranslated region was 86% identical to that of APAL 1 (Gowri et al., 1991). Of the PAL species for which complete deduced polypeptide sequences are available, alfalfa PAL is the largest (Gowri et al., 1991).

The PAL species from alfalfa, bean, rice, sweet potato, and parsley differ mostly at their amino acid and carboxy termini. Internal to these regions, alfalfa and bean PAL show considerable amino acid sequence similarity. In bean, parsley, and rice PAL genes, the single intron splits an arginine codon at a position equivalent to between nucleotides 524 and 525 in the alfalfa APAL 1 sequence (amino acid 143). This is one of the most highly conserved regions, comprising 32 identical amino acids in the alfalfa, bean, parsley, and sweet potato PAL coding regions (Gowri et al., 1991).

PAL activity was induced 50-fold in alfalfa cell suspension cultures within 12 h of exposure to elicitor. PAL transcripts accumulated rapidly and massively within 2 h after elicitation. By 12 h after elicitation, the transcript levels had started to decline, reaching near basal levels by 30 h after elicitation, and a similar trend was observed for the activity of the enzyme (Gowri et al., 1991). Similar results have been reported by Dalkin et al. (1990a). PAL activity showed rapid increase but the induction was only transient with a quick decline (Dalkin et al., 1990a). CA4H activity also increased fourfold (Kessmann et al., 1990a), but this increase was also only transient (Dalkin et al., 1990a). The presence of at least 6 elicitor-inducible CHS isopolypeptides has been reported in alfalfa cell suspension cultures (Dalkin et al., 1990a). Five distinct CHS cDNAs (CHS 1, 2, 4, 8, and 9) have been isolated from an alfalfa cell suspension culture (Dalkin et al., 1990a) after exposure to an elicitor from *C. lindemuthianum*. At the nucleotide level, CHS 2 was 78% identical to the bean CHS 17. The individual alfalfa CHS clones were more closely related to one another than was any one of them to the bean CHS. All cDNAs have TAG as the stop codon. The lengths of the 3′-untranslated regions were 135, 130, 131, 139, and 124 nucleotides for CHS 1, 2, 4, 8, and 9 respectively. The first 120 nucleotides of the 3′-untranslated regions of CHS 4 was 87% identical to that of CHS 8. Comparison of the deduced amino acid sequences from all five alfalfa CHS clones showed extensive sequence conservation, although many of the amino acid differences between the individual CHS cDNA were not conservative changes. The calculated isoelectric points of the 5 CHS proteins ranged from 5.7 to 6.1 (Junghans et al., 1993).

Exposure of an alfalfa cell suspension culture to a fungal elicitor resulted in a rapid and transient increase in CHS transcripts. Maximum transcript levels were attained around 4 or 5 h after elicitation and a large increase had occurred as early as 1 h after elicitation (Junghans et al., 1993). CHS activity rose approximately 10-fold to a maximum value 19 h postelicitation (Kessmann et al., 1990a).

There are many CHS genes in healthy alfalfa plants and some of the CHS genes may have close physical linkage (Junghans et al., 1993). As a whole the CHS gene family was primarily expressed in the lower parts of the alfalfa plant (roots, nodules, and lower stems). Moderate expression was also seen in upper stems and petioles with very low expression in leaves and mature flowers. Specific CHS 2, 10, and 7 transcripts were expressed exclusively in roots. CHS 1, 2, 5, and 10 transcripts were not present at any stage of floral development (Junghans et al., 1993).

The alfalfa CHS 2 differs from the consensus sequence derived from the sequences of CHS from *Antirrhinum majus,* barley, maize, *Petunia hybrida, Magnolia liliflora,* and *Ranunculus acer* by only 16 out of 389 amino acids (Niesbach-Klosgen et al., 1987; Junghans et al., 1993). Included in a conserved region of 28 amino acids near the C-terminus of the CHS open reading frame from alfalfa and all other species is the sequence MSSACV, which contains cysteine 341, which is an essential feature of the CHS active site (Rohde et al., 1991).

The transcripts corresponding to the five alfalfa clones (CHS 1, 2, 4, 8, and 9) were not expressed in healthy leaves and were induced in leaves in response to *Phoma medicaginis* infection (Junghans et al., 1993). When alfalfa leaves were inoculated with *Phoma medicaginis,* CHS transcripts were induced rapidly within 2 h and the elevated transcript level was maintained up to 72 h after inoculation (Junghans et al., 1993). The alfalfa CHS genes appear to have close physical linkage (Junghans et al., 1993).

CHI catalyzes the isomerization of the chalcone to give a (2S)-flavanone (Hagmann and Grisebach, 1984). CHI shows approximately fourfold increases within 10–14 h in alfalfa suspension cultured cells due to elicitor treatment (Kessmann et al., 1990a). CHI activity remained elevated up to 48 h postelicitation (Dalkin et al., 1990a).

The membrane-associated cytochrome P_{450} isoflavone synthase (IFS) activity was strongly induced, reaching maximum activities at 11 h after elicitor treatment (Kessmann et al., 1990a, Dalkin et al., 1990a). Isoflavone 4′-0-methyl transferase, IFMT (= daidzein 0-methyl transferase, DOMT) increased rapidly and remained elevated up to 48 h postelicitation (Dalkin et al., 1990a).

The alfalfa cells contained both daidzein and genistein-0-methyl transferase, and activity against genistein was induced in a similar manner to the DOMT activity. The alfalfa DOMT and genistein 0-methyl transferase activites may be functions of the same enzyme (Dalkin et al., 1990a). Isoflavone 2′-hydroxylase activity was strongly induced and the maximum activity was observed at 14 h after elicitor treatment (Kessmann et al., 1990a). Isoflavone reductase (IFR) and pterocarpan synthase were also induced due to the fungal elicitor treatment (Dalkin et al., 1990a). IFR has been cloned from alfalfa (Paiva et al., 1991) and chickpea (Tiemann et al., 1991).

Treatment of alfalfa cell suspension cultures with elicitor from *C. lindemu-*

thianum results in the rapid and transient induction of 11 enzyme activities involved in the biosynthesis of the isoflavonid phytoalexin medicarpin (Dalkin et al., 1990a; Kessmann et al., 1990a). PAL and CHS mRNAs were synthesized with a time lag of less than 1 h. 4 CL mRNA appeared at 1 h time lag. CHS was the major newly synthesized enzyme (Dalkin et al., 1990b).

Most of the induced enzymes started declining very soon. By 12 h PAL transcript levels started declining. CHS transcripts declined rapidly after 3 h postelicitation and IFR transcripts declined 8 h after elicitation (Gowri et al., 1991). However caffeic acid 0-methyl transferase (COMT) transcripts remained at elevated levels up to 48 h after elicitation. IFR and COMT enzyme activities remained at high levels up to 48 h after elicitation (Gowri et al., 1991). These observations indicate that transcriptional activation, mRNA turnover, and enzyme stability may all be sites for regulation of phytoalexin biosynthetic enzyme production in response to elicitation (Gowri et al., 1991).

2. Rapid Release of Phytoalexins from Their Conjugates

The isoflavanoid conjugates medicarpin-3-0-glucoside-6″-0-malonate (MGM), afrormosin-7-0-glucoside (AG), and afrormosin-7-0-glucoside-6″-0-malonate (AGM) have been isolated from cell suspension cultures of alfalfa. They were the major constitutive secondary metabolites in the cells. Due to *Colletotrichum lindemuthianum* elicitor treatment, rapid accumulation of medicarpin, with maximum levels at 14 h has been observed. Following the maximum of medicarpin accumulation, peak levels of MGM were observed at 24 h. The MGM accumulated during the period in which free medicarpin was declining (12–48 h), suggesting that the phytoalexin was being actively glucosylated. However, the decline in medicarpin content during this period (800–320 nmol/g fresh weight) was not accompanied by a directly corresponding increase in MGM (40–100 nmol/g), suggesting that the disappearance of medicarpin can only be partially explained by the formation of the malonyl glucoside (Kessmann et al., 1990b). The concentration of AG and AGM were not affected by treatment of cells with fungal elicitor (Kessmann et al., 1990b).

Treatment of alfalfa cells with elicitor together with 100 μm L-α-amino-oxy-β-phenylpropionic acid (AOPP) reduced the incorporation of [^{14}C]phenylalanine into isoflavonoids to control levels for the first 4 h of treatment. To determine whether elicited alfalfa cells could potentially mobilize MGM into free medicarpin, changes in the specific activity of [^{14}C]medicarpin were compared in cells treated with elicitor alone or elicitor plus AOPP. During the first 4 h after addition of AOPP, during which period the inhibitor greatly decreased incorporation from [^{14}C]phenylalanine into medicarpin, similar quantities of medicarpin appeared in cells treated with AOPP and in cells treated with elicitor alone (Table 3; Kessmann et al., 1990b). Since the specific activity of ^{14}C in this medicarpin was much lower than in the medicarpin accumulating at this period in the absence

Table 3 Effect of AOPP on medicarpin synthesis in elicited alfalfa cell cultures

	[^{14}C]medicarpin[a] (per g fresh weight)			
	Elicited		Elicited + AOPP	
Time post-elicitation	dpm	nmol	dpm	nmol
0	0	1	0	3
2	1071	18	137	18
4	2808	20	107	25

[a][^{14}C]Medicarpin is expressed as nmol of phytoalexin and radioactivity (dpm) incorporated from L-(U-^{14}C) phenylalanine, at times shown.
Source: Kessmann et al., 1990b.

of AOPP, it was concluded that the medicarpin can originate from a preformed precursor such as MGM in elicited cells in which the flux into the isoflavonoid pathway is blocked (Kessmann et al., 1990b).

The isoflavonoid conjugates may act as a source of isoflavonoid metabolites that may be mobilized under stress conditions if the availability of carbon sources for phenyl-propanoid biosynthesis becomes rate limiting; for example, in the late stages of an infection that depleted phenylalanine pools (Kessmann et al., 1990b). A pathway for the biosynthesis of isoflavonoid conjugates in alfalfa has been suggested, as shown in Figure 4 (Kessmann et al., 1990b).

In unelicited alfalfa cells, the major products are afrormosin conjugates. This suggests that under such conditions, the first enzyme specific for medicarpin biosynthesis, formononetin 2′-hydroxylase, is rate limiting, and that flow occurs instead through the 6-hydroxylase, the constitutive pathway (Kessmann et al., 1990b). Isoflavone 2′-hydroxylase activity is undetectable in unelicited alfalfa cells, but is induced approximately 5–10-fold on treatment with elicitor (Kessmann et al., 1990a). It will activate the synthesis of medicarpin through inducible pathway (Fig. 4; Kessmann et al., 1990b).

3. Biosynthesis of Phytoalexins in Sorghum

When sorghum tissues were inoculated with *Colletotrichum graminicola,* rapid accumulation of two flavonoid phytoalexins has been observed (Nicholson et al., 1987; 1988; Yamaoka et al., 1990). The phytoalexins have been identified as 2-deoxyanthocyanidins, apigeninidin, and luteolinidin (Nicholson et al., 1987).

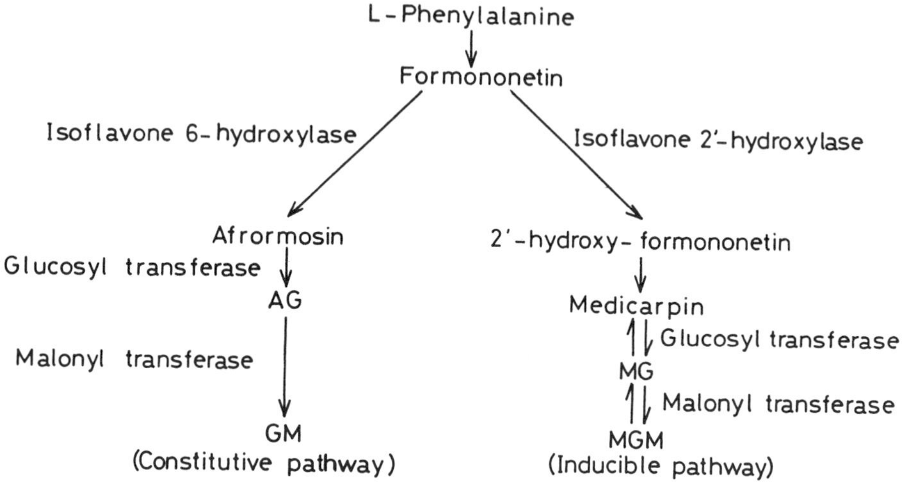

Figure 4 The biosynthetic pathway for isoflavonoid conjugates in alfalfa cell cultures. (From Kessmann et al., 1990b.)

Inoculation with the pathogen does not result in changes in activities of either PAL or 4-hydroxycinnamic acid. CoA ligase, the first and last enzymes of the phenylpropanoid pathway (Nicholson et al., 1987).

When sorghum mesocotyls were inoculated with *C. graminiola,* an increase in chalcone synthase activity was detected as early as 6 h after inoculation. The accumulation of sorghum phytoalexins apigeninidin and luteolinidin occurred as early as 9 h and there was at least a time differential of 3 h between the initial increase in CHS and the appearance of the phytoalexins (Lue et al., 1989). The infection did not cause an increase in the level of PAL or tyrosine ammonia-lyase (TAL) activity (Lue et al., 1989). The results suggest that the coordinate regulation of the three enzymes, PAL, hydroxycinnamate CoA ligase, and CHS, does not occur in response to *C. graminicola* inoculation in sorghum (Lue et al., 1989).

E. Biosynthesis of Phytoalexins in Parsley

The potential phytoalexins in parsley are the linear furanocoumarins, marmesin and psoralen, their coumarin precursor, umbelliferone, and the methoxylated psoralen derivatives, xanthotoxin, bergapten, and isopimpinellin (Tietjen et al., 1983a; Scheel et al., 1986). They accumulate in parsley leaves inoculated with

Phytophthora megasperma f.sp. *glycinea* (Scheel et al., 1986; Jahnen and Hahlbrock, 1988a). But only umbelliferone and marmesin are not detectable in uninfected, healthy parsley leaves.

3-Deoxy-D-arabino-heptulosonate 7-phosphate synthase (DAHP synthase) is the first enzyme of the shikimate pathway that is common to the synthesis of the three aromatic amino acids: phenylalanine, tryptophan, and tyrosine. These amino acids serve dual pruposes as substrates in protein synthesis and in the synthesis of secondary products such as phytoalexins (Henstrand et al., 1992). Parsley cell cultures treated with an elicitor show a twofold increase of DAHP synthase (McCue and Conn, 1989).

Elicitor from *Phytophthora megasperma* f.sp. *glycinea* causes a rapid accumulation of DAHP synthase mRNA in parsley cells grown in suspension cultures (Henstrand et al., 1992). The elicitor-induced accumulation of DAHP synthase mRNA reached maximal levels within 2 h and remained elevated for at least 24 h after elicitor treatment. Actinomycin D, an inhibitor of transcription, causes a significant decrease in DAHP synthase mRNA accumulation (Henstrand et al., 1992). PAL mRNA also accumulated after elicitor treatment. The expression pattern showed a similar increased rate of accumulation between 45 and 60 min after treatment with elicitor (Henstrand et al., 1992). It is possible that DAHP synthase transcription is responding to a depletion of phenylalanine pools caused by increased PAL enzyme activity (Henstrand et al., 1992). PAL, CA4H, and 4 CL appear to be the enzymes involved in biosynthesis of furanocoumarins (Figure 5; Collinge and Slusarenko, 1987).

Increases in mRNA activities for PAL and 4 CL were followed by accumulation of furanocoumarins in parsley cells treated with fungal elicitor (Hahlbrock et al., 1981; Kuhn et al., 1984). 4 CL mRNA synthesis was at its maximum at 1.5 h in parsley cells treated with the elicitor (Chappell and Hahlbrock, 1984). Increases in mRNAs of these enzymes were shown to be due to de novo transcription (Chappell and Hahlbrock, 1984). Different PAL and 4 CL genes have been found to express in parsley due to fungal elicitor treatment (Kuhn et al., 1984; Hahlbrock et al., 1984; 1986; Kombrink et al., 1986; Douglas et al., 1987).

PAL is encoded by a family of four genes, and out of the four genes three appear to be activated by fungal elicitors (Lois et al., 1989a). The parsley PAL 1 has an open reading frame encoding a 716 amino acid polypeptide subunit, with a single intron of approximately 800 bp flanked with AG/GT consensus sequences (Lois et al., 1989a). *Cis*-acting elements have been identified and DNA sequences within these elements are conserved in the promoters of at least one PAL gene (PAL 1) (Lois et al., 1989b).

4 CL is a monomeric enzyme and occurs in two isoforms in parsley, each of which is encoded by a single-copy gene (Douglas et al., 1987; Lozoya et al., 1988). The two genes are similar in exon–intron structure and nucleotide sequence (Douglas et al., 1987). The exceptions are a small (54-bp) and a larger

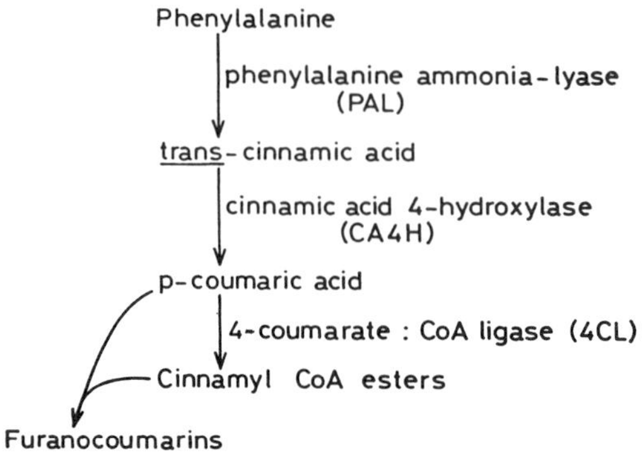

Figure 5 Biosynthesis of furanocoumarins in parsley. (From Collinge and Slusarenko, 1987.)

(600-bp) insertion in the second intron of one 4 CL gene relative to the other. Both the genes were activated in elicitor-treated cells (Douglas et al., 1987). The two genes encode 4 CL isoenzymes that differ in three amino acid residues (Lozoya et al., 1988). 4CL mRNA is rapidly and transiently accumulated in small, confined areas around fungal penetration sites in infected parsley leaves (Hahlbrock and Scheel, 1989).

CHS is the most abundant of all biosynthetically related proteins in parsley (Kruezaler and Hahlbrock, 1973; Kruezaler et al., 1978, Heller and Hahlbrock, 1980; Schroder and Schafer, 1980; Schulze-Lefert et al., 1989a,b). CHS gene occurs in only one copy in parsley with two alleles. The two alleles differ in size by a 927-bp transposon-like insertion in the promoter region of one allele to the other (position –538) (Herrmann et al., 1988). The CHS gene is located on the long arm of one of a small group of submetacentric chromosomes (Hermann et al., 1988).

The biosynthetic steps of furanocoumarin biosynthesis in parsley appear to include the intermediate formation of glucosides or glucose esters. These may be cleaved by cell wall- or plasma membrane-associated glucosidases as part of the excretion mechanism (Hahlbrock and Scheel, 1989). Elicitor treatment rapidly induces UDP-glucose: cinnamate, UDP-glucose: 4-coumarate, and UDP-glucose: ferulate 0-glucosyl transferases (Hahlbrock and Scheel, 1989). Dimethylallyl diphosphate: umbelliferone dimethylallyl transferase catalyzes the prenylation of umbelliferone to give demethylsuberosin (Tietjen and Matern, 1983), which is converted to (+) marmesin by marmesin synthase (Hamerski and Matern, 1988; Hahlbrock and Scheel, 1989). Marmesin is a substrate for psoralen synthase (Wendorff and Matern, 1986). Psoralen is the precursor for all linear furanocoumarins (Brown, 1985; Hauffe et al., 1986). Psoralen synthase catalyzes the oxidative removal of the hydroxypropyl residue from (+) marmesin to yield psoralen (Hauffe et al., 1986). The two methyl transferases S-adenosylmethionine: xanthotoxol 0-methyltransferase (XMT) catalyzes the formation of xanthotoxin and S-adenosylmethionine:bergaptol 0-methyl-transferase (BMT) catalyzes the methylation of bergaptol and 5-hydroxyxanthotoxin to bergapten and isopimpinellin, respectively (Tietjen et al., 1983). The elicitor induces dimethylallyl diphosphate:umbelliferone dimethylallyl transferase, marmesin synthase, psoralen synthase, XMT, and BMT in cultured parsley cells (Hauffe et al., 1986; Hahlbrock and Scheel, 1989). The BMT mRNA activation and enzyme accumulation occurred slowly in infected parsley tissue (Scheel et al., 1986; 1987).

F. Biosynthesis of Phytoalexins in Pea

Pisatin, the phytoalexin of pea, is synthesized almost in the same pathway of phaseollin biosynthesis in bean. Increases in the activity of a number of biosynthetic enzymes and concomitant accumulation of the phytoalexin pisatin have been reported in pea (Loschke et al., 1981; Sweigard et al., 1986).

Fungal infection results in de novo synthesis of PAL and the increased synthesis results from increased mRNA activity in pea (Hahlbrock et al., 1981). PAL from pea has been cloned (Kawamata et al., 1992). CHS is encoded by a multigene family in pea (Harker et al., 1990). Three members of the CHS multigene family, CHS 1, CHS 2, and CHS 3, have been detected in pea. CHS 1 and CHS 3 are expressed in both petal and root tissue, whereas CHS 2 expression was detected only in root tissue. The expression of all three CHS genes is induced by an abiotic elicitor (Harker et al., 1990).

Pea possesses an inducible methyltransferase that catalyzes the terminal step in the biosynthesis of pisatin (Sweigard et al., 1986). When pea seedlings were infected with *Nectria haematococca* or *Aphanomyces euteiches* or treated with the abiotic elicitor, $CuCl_2$, the methyltransferase activity increased severalfold. This increase was followed by accumulation of pisatin (Sweigard et al., 1986).

The methyltransferase activity in pea is dependent on 6a-hydroxymaackiain (Sweigard et al., 1986). 6a-Hydroxymaackiain is incorporated more efficiently than maackiain and much more efficiently than pterocarpin (Banks and Dewick, 1982; Sweigard et al., 1986). The high degree of substrate discrimination suggests that the 6a-hydroxymaackiain-dependent reaction is catalyzed by a specific enzyme for pisatin biosynthesis, rather than by a fortuitous activity of a nonspecific methyltransferase. Furthermore, the enzyme is induced by treatments that elicit pisatin biosynthesis (Sweigard et al., 1986).

G. Biosynthesis of Phytoalexins in Chickpea

In chickpea also medicarpin is a major phytoalexin. The enzymes catalyzing isoflavonoid conjugation have been shown to be constitutively expressed glucosyl and malonyl transferases with high activity that exhibit strict specificity for their respective aglycones/conjugates (Barz et al., 1989). The conjugated isoflavonoids that accumulate are derivatives of precursors of medicarpin rather than the phytoalexin itself (Barz et al., 1989). An isoflavone-0-methyltransferase has been purified from chickpea cell cultures and is active specifically toward daidzein and genistein (Wengenmayer et al., 1974).

III. STILBENES

Stilbenoids are derived from the cinnamate–malonate pathway (Grisebach and Ebel, 1978; Kindl, 1985). The biosynthesis of stilbenes takes place via cinnamoyl-CoA derivatives (Figure 6; Fritzemeir and Kindl, 1981). Stilbene synthase catalyzes the formation of stilbenes and cinnamic acid CoA esters and malonyl-CoA are the substrates for the enzyme (Rolfs et al., 1981; Fritzemeir and Kindl, 1981; Kindl, 1985). PAL, CA4H, 4CL, and stilbene synthase are involved in the formation of the stilbene skeletons in leaves of Vitaceae. Large increases and subsequent decreases in the activities of PAL, CA4H, and stilbene synthase occurred concomitantly. The coordinated changes in enzyme activities were characterized by a maximum after 15 h. The phytoalexin resveratrol accumulated following the increase in the enzyme activities (Fritzemeir and Kindl, 1981). In the case of stilbene synthase, more than a hundredfold increase in activity was found and the enzyme was almost absent in nonelicited leaves (Fritzemeir and Kindl, 1981). To prove the importance of stilbene synthase in the phytoalexin synthesis, stilbene synthase from peanut has been cloned. The transgenic tobacco calli containing the stilbene synthase gene, when treated with the elicitors produced resveratrol (Hain et al., 1990; Kishore and Somerville, 1993). Several stilbene synthase genes may be present in the peanut genome (Schroder et al., 1988).

Synthesis of resveratrol involves two condensation steps of acetate units from malonyl-CoA with 4-coumaroyl CoA catalyzed by resveratrol synthase (Schoppner and Kindl, 1984). Resveratrol synthase is induced by treatment of cultured peanut cells with elicitor (Vornam et al., 1988). Resveratrol synthase genes have been cloned and their molecular structure has been studied (Schroder et al., 1988; Lanz et al., 1990).

Stilbene synthases (STS) often use the same substrates as CHS, but the products are different (Figure 7; Fliegmann et al., 1992).

cDNA clones and genomic clones for STS show 70–75% homology to CHS sequences (Schroder et al., 1988). CHS and STS are closely related both in structure and function (Schroder and Schroder, 1990). They have extensive overall similarities of the protein sequences and a strictly conserved amino acid sequence motif (GVLFGFGPGLT) is present in these proteins (CHS/STS family signature) (Bairoch, 1991). Both CHS and STS catalyze the stepwise condensation of three acetate units (from malonyl Co-A) to a starter molecule (4-coumaroyl Co-A and cinnamoyl-CoA) (Fleigmann et al., 1992). A cysteine at a

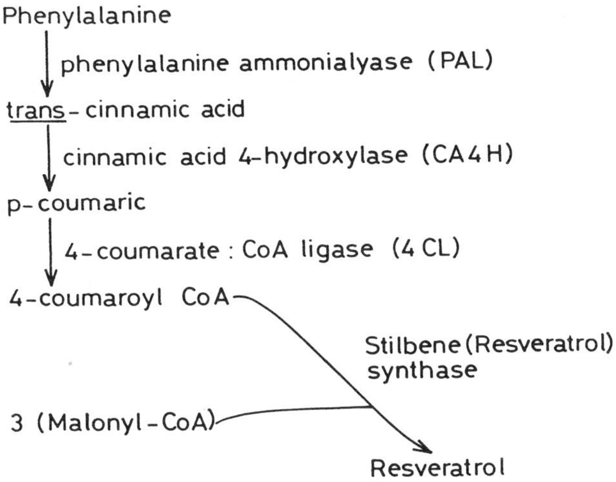

Figure 6 Biosynthesis of resveratrol in peanut. (From Fritzemeir and Kindl, 1981.)

position strictly conserved in both proteins is essential to enzyme activity, and this cysteine most likely represents the active site for the condensing reaction (Lanz et al., 1991). These two enzymes differ in their catalytic functions. STS performs an additional decarboxylation that removes a carbon atom that remains in the chalcone product of CHS (Schroder and Schroder, 1990).

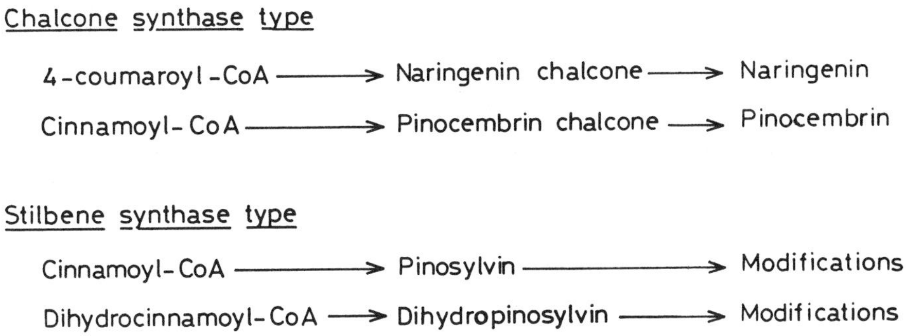

Figure 7 Chalcone synthase and stilbene synthase type enzyme activities in *Pinus sylvestris*. (From Fliegmann et al., 1992.)

Pinus sylvestris possesses CHS as well as STS. It contains the stilbenes pinosylvin, dihydropinosylvin, and their derivatives (Kindl, 1985; Gorham, 1989). Pinosylvin is synthesized with cinnamoyl-CoA as starter by the enzyme pinosylvin synthase and dihydropinosylvin is produced from dihydrocinnamoyl-CoA by the action of dihydropinosylvin synthase (Gorham, 1989). *P. sylvestris* contains pinocembrin, via a CHS-type reaction (Sandermann et al., 1989).

CHS and STS genes were cloned from *P. sylvestris*. CHS was active not only with 4-coumaroyl-CoA (to naringenin chalcone), but also with cinnamoyl-CoA (leading to pinocembrin). The STS preferred dihydrocinnamoyl-CoA to cinnamoyl-CoA as substrate and hence the STS may be dihydropinosylvin synthase. The STS had 73.2% identity with the CHS from *P. sylvestris* and only 65.3% with a STS from peanut (Fliegmann et al., 1992). Both CHS and STS were induced due to stress (Fliegmann et al., 1992). STS was found to be induced in *P. sylvestris* due to fungal infection (Gehlert et al., 1990). Grapevine stilbene synthase cDNA is only slightly different from CHS mRNA (Melchoir and Kindl, 1990).

The heartwood of pine trees contains considerable amounts of the stilbenes pinosylvin and pinosylvin monomethyl ether. In young plants of *Pinus sylvestris*, these stilbenes are almost absent. Due to fungal attack, the stilbene formation is greatly increased and pinosylvin becomes the preferentially formed phytoalexin (Gehlert et al., 1990).

Most species of the genus *Pinus* synthesize pinosylvin and pinosylvin monomethyl ether (Rudloff and Jorgensen, 1963; Schoppner and Kindl, 1979). The biosynthetic pathway of hydroxy stilbenes originates from L-phenylalanine and PAL, an acyl-CoA synthetase, and stilbene synthase are involved (Kindl, 1985). Stilbene synthase activity is very important in the synthesis of pinosylvin and pinosylvin monomethyl ether (Gehlert et al., 1990).

A unique synthase acting on cinnamoyl-CoA is the key enzyme of the stilbenes in *Pinus*. Stilbene synthase (pinosylvin-forming) was purified and this preparation was free of chalcone synthase activity. Its molecular weight is 43,000. Multiple forms of the enzyme differing in isolectric points between pH 5.1 and 4.6 have been identified. Inoculation of pine seedlings with *Botrytis cinerea* leads to a 38-fold increase in stilbene synthase activity within 1 day (Gehlert et al., 1990).

Cultured cells of peanut when treated with elicitor from *Phytophthora cambivora* produced hydroxystilbenes primarily resveratrol and derivatives of 5-[3′,4′-dihydroxyphenylethenyl) benzene-1,3-diol (astringenin or piceatannol) and the maximum induction was observed between 2 and 12 h after the elicitor treatment. Two enzymes on the pathway to stilbenes (i.e., PAL and STS) showed a dramatic increase. The maximum increase in PAL activity was observed at 3 h after elicitation, while the peak in STS activity was observed at 5 h (Steffens et al., 1989). PAL activity peaked 2 h prior to STS activity and this was highly characteristic in peanut cells.

IV. DIHYDROXYPHENANTHRENE PHYTOALEXINS

Dihydroxyphenanthrene phytoalexins are also derived from the cinnamate–malonate pathway (Kindl, 1985). The biosynthetic pathway of 9,10-dihydrophenanthrenes involves dihydro-3-coumaroyl-CoA, dihydrostilbene (bibenzyl) synthase, and oxidative coupling of a dihydrostilbene intermediate to yield the phytoalexins, hircinol, and orchinol in orchids (Fritzmeier and Kindl, 1983) and hircinol in plants belonging to Dioscoreaceae (Fritzmeier et al., 1984).

V. SESQUITERPENOID PHYTOALEXINS

A. Biosynthesis of Phytoalexins in Potato

In potato three sesquiterpenoid phytoalexins (rishitin, lubimin, and solavetivone) are synthesized due to fungal infection (Grisebach and Ebel, 1978; Coolbear and Threlfall, 1985). Their biosynthetic pathway is shown in Figure 8 (Coolbear and Threlfall, 1985).

The enzymes involved may be isopentenyl pyrophosphate isomerase, farnesyl pyrophosphate (FPP) synthase, FPP cyclase, 3- hydroxy-3 methylglutaryl coenzyme A reductase, and various oxidative enzymes (Coolbear and Threlfall, 1985; Oba et al., 1985; Stermer and Bostock, 1987; Zook and Kuc, 1991; Stermer et al., 1991). Monooxygenases are involved in the later stages of the biosynthesis of sesquiterpenoid phytoalexins in potato (Brindle et al., 1985). Radiolabeled isopentenyl pyrophosphate was found to be incorporated into the sesquiterpenoid phytoalexin lubimin in potato tissues treated with elicitor from *Phytophthora infestans* (Coolbear and Threlfall, 1985).

B. Biosynthesis of Phytoalexins in Sweet Potato

When *Ceratocystis fimbriata* infects sweet potato, the activities of enzymes responsible for the conversion of mevalonte to isopentylpyrophosphate increase and this increase is correlated with the production of furanoterpenoid phytoalexin ipomeamarone and structurally related compounds (Suzuki et al., 1975; Oba et al., 1976; Brindle and Threlfall, 1983; Fujita and Ashai, 1985). The fungal infection induces 3-hydroxy-3-methylglutaryl-CoA reductase (HMGR), which catalyzes the NADPH-dependent reduction of 3-hydroxy-3-methylglutaryl-CoA to mevalonate in sweet potato (Oshima and Uritani, 1968; Suzuki et al., 1975; Uritani, 1978). HMGR activity in diseased tissue was largely associated with

Acetate
↓
Mevalonate
↓
Isopentenyl pyrophosphate
↓
Farnesyl pyrophosphate
↓
Eudesmane cation
↓
Vetaspirane cation
↓
Solavetivone
↓
Isolubimin
↓
Dihydrolubimin
↓
Lubimin
↓
Hydroxylubimin
↓
Risitin

Figure 8 Biosynthesis of sesquiterpenoid phytoalexins in potato. (From Coolbear and Threlfall, 1985.)

microsomes. The enzyme activity preceded the terpene formation and reached a maximum in 2 days (Suzuki et al., 1975). The phytoalexin accumulation has also been correlated with activity increases in pyrophosphomevalonate decarboxylase, and dehydroipomeamarone reductase (Suzuki et al., 1975. Oba et al., 1976; Ito et al., 1979; Takeuchi et al., 1980; Inoue et al., 1984). Pyrophosphomevalonate decarboxylase was greatest in the 12–24 h period after inoculation with *C. fimbriata.* The synthesis of the enzyme protein was required for the induction of terpene production in sweet potato root tissue (Oba et al., 1976).

C. Biosynthesis of Phytoalexins in Tobacco

The elicitor-induced synthesis of sesquiterpenoids in tobacco cell suspension cultures (Chappell and Nable, 1987) is correlated with the induction of two enzymes of sesquiterpenoid biosynthetic pathway: HMGR and sesquiterpene cyclase (Vogeli and Chappell, 1988). The induction of sesquiterpenoid accumulation was correlated with a suppression of sterol biosynthesis, an isoprenoid branch pathway that will be competing with sesquiterpene biosynthesis between these two pathways. The suppression of sterol biosynthesis was correlated with a suppression of squalene synthetase, the first committed enzyme for this biosynthetic pathway (Vogeli and Chappell, 1988). There was coordinated inhibition of squalene synthetase and induction of enzymes of sesquiterpenoid phytoalexin biosynthesis in tobacco (Threlfall and Whitehead, 1988; 1990).

The elicitor from *Phytophthora parasitica* cell wall induced both accumulation of capsidiol and browning while pectolyase from *Aspergillus japonicus* induced only browning in tobacco cell suspension cultures. The elicitor from *P. parasitica* induced more HMGR than the level found in control cell cultures. Incubation of the cell cultures with pectolyase did not induce HMGR enzyme activity significantly, but both the fungal wall elicitor and pectolyase induced PAL and sesquiterpene cyclase and suppressed squalene synthetase activities in tobacco cell suspension cultures (Table 4, Chappell et al., 1991). The studies have demonstrated the importance of inducible HMGR activity for elicitor-stimulated sesquiterpenoid biosynthesis. Tobacco cell suspension cultures incubated with pectolyase did not exhibit a transient induction of HMGR activity, nor did they accumulate sesquiterpenoids. This was in spite of the fact that all the necessary biosynthetic machinery beyond HMGR was apparently induced, as well as other responses such as culture browning and induction of PAL (Chappell et al., 1991).

Table 4 Effects of *Phytophthora parasitica* elicitor and pectolyase on various enzyme activities in tobacco cell suspension cultures

	Enzyme activities (nmol/mg protein/h)			
Treatment	Phenylalanine ammonia-lyase	Sesquiterpene cyclase	Squalene synthetase	HMGR
None	75	0.0	6.7	2.7
Phytophthora parasitica elicitor	550	6.8	1.2	46.0
Pectolyase	760	2.8	1.9	3.7

Source: Chappell et al., 1991.

Mevinolin, a competitive inhibitor of HMGR, inhibited capsidiol accumulation in elicitor-treated tobacco cell suspension cultures (Chappell and Nable, 1987).

To prove the importance of sesquiterpene cyclase in synthesis of sesquiterpenoid phytoalexins, a sesquiterpene cyclase, trichodiene synthase, from *Fusarium sporotrichoides* has been introduced into tobacco. The fungal enzyme was active in the leaves of transgenic plants and trichodiene was detected at a level of 5–10 ng/g fresh weight (Hohn and Ohlrogge, 1991).

D. Biosynthesis of Phytoalexin in Cotton

Gossypol is a dimeric sesquiterpenoid in cotton and is synthesized from mevalonic acid. Geranyl pyrophosphate, neryl-pyrophosphate, and the isomers of farnesyl pyrophosphate are the precursors for the phytoalexin (Heinstein et al., 1970).

VI. DITERPENOID PHYTOALEXINS

A. Biosynthesis of Phytoalexins in Castor Bean

Casbene, a macrocyclic diterpene hydrocarbon phytoalexin is produced in castor bean inoculated with *Rhizopus stolonifer* or treated with its enzyme (Sitton and West, 1975; Lee and West, 1981; Bruce and West, 1982). The enzymes catalyzing the last four steps of casbene biosynthesis are isopentenyl pyrophosphate isomerase, geranyl transferase (farnesyl pyrophosphate synthetase), farnesyl transferase (geranyl geranyl pyrophosphate synthetase), and casbene synthetase (Dudley et al., 1985; 1986; Dueber et al., 1978; Lee and West, 1981; Collinge and Slusarenko, 1987; Lois and West, 1990). The activities of farnesyl transferase and casbene synthetase are quite low in extracts of healthy tissues (Dudley et al., 1986). Geranyl transferases have been partially purified and characterized from plant sources (Ogura et al., 1968; Green and West, 1974; Widmaier et al., 1980; Allen and Banthorpe, 1981). Geranyl–geranyl pyrophosphate synthetase, which catalyzes a farnesyl transferase reaction with farnesyl pyrophosphate as the allylic substrate has been studied in carrot (Nandi and Porter, 1964), pumpkin (*Cucurbita pepo*) (Ogura et al., 1972; Shinka et al., 1975) and tomato (Spurgeon et al., 1984).

Farnesyl transferase has been purified from extracts of castor bean seedlings that were elicited by exposure for 10 h to *R. stolonifer* spores. The molecular weight of the enzyme was about 72,000 (Dudley et al., 1986). Farnesyl pyrophosphate in combination with isopentenyl pyrophosphate was the most effective substrate for the production of geranylgeranyl pyrophosphate. Dimethyl-allyl pyrophosphate was not utilized by the enzyme. One peak of farnesyl transferase activity and two peaks of geranyl transferase activity (farnesyl pyrophosphate synthetase) from extracts of elicited castor bean seedlings were resolved by ion exchange chromatography. It suggests that the pathway for geranylgeranyl pyrophosphate synthesis in elicited castor bean seedlings involves the successive actions of two enzymes: a geranyl transferase that utilizes dimethylallyl-pyrophosphate and isopentenyl pyrophosphate as substrates and a farnesyl transferase that utilizes the farnesyl pyrophosphate produced in the first step and isopentenyl pyrophosphate substrates (Dudley et al., 1986).

Casbene synthetase has been purified from castor bean seedlings infected with *Rhizopus stolonifer* (Moesta and West, 1985). The molecular weight of casbene synthetase is 59,000 (Moesta and West, 1985). No casbene synthetase mRNA has been detected in uninfected seedlings. Casbene synthetase activity preceded the formation of cyclic diterpene casbene in castor bean seedlings infected with the pathogen (Sitton and West, 1975; Dueber et al., 1978; Stekoll and West, 1978; Lee and West, 1981; West et al., 1985). Messenger RNA for casbene synthetase increased in castor bean seedlings treated with the elicitor of the pathogen within 6 hr of treatment (Moesta and West, 1985).

B. Biosynthesis of Phytoalexins in Rice

Momilactones A and B, oryzalexin A-D, and several other antifungal agents have been reported to accumulate in rice leaves infected with blast pathogen *Pyricularia oryzae* (Cartwright et al., 1981; Kono et al., 1984; Akatsuka et al., 1985; Matsuyama and Wakimoto, 1985; 1988; Sekido et al., 1986; Kodama et al., 1988). The momilactones are oxygenated derivatives of 9 βH- pimara-7,15-diene-19,6-β-olide and oryzalexins are oxygenated derivatives of a stereorically different pimaradiene, *ent*-sandaracopimara-8,(14),15-diene (Ren and West, 1992).

Momilactones are formed by oxygenation of 9βH-pimara- 7,15-diene, which is generated by a two-step cyclization of geranylgeranyl pyrophosphate via the bicyclic intermediate, 9,10- syn-copalyl pyrophosphate. The oryzalexins are formed by oxygenation of *ent*-sandaracopimaradiene, which originates in a two-step cyclization with *ent*-copalyl pyrophosphate as the bicyclic intermediate (Figure 9; Ren and West, 1992). The precursors of the rice phytoalexins have been identified as *ent*-sandara-copimara-9(14),15-diene and 9βH-pimara-7,15-diene (Wickham and West, 1992).

The diterpene hydrocarbon synthase (cyclase) with geranylgeranyl pyrophosphate as the substrate may induce synthesis of diterpene phytoalexins in rice (Ren and West, 1992). The cell wall preparation from the rice pathogen *Fusarium moniliforme* as well as chitin induced the cyclase enzyme activity. However, the activity was first detectable only after about 24 h, and the maximum activity was seen at about 40 h. The existence of a long period of time lag suggests that the induction of diterpene hydrocarbon synthase activity might be a secondary event dependent on some primary action of the elicitor treatment. A significant increase in chitinase activity was evident 12 h after addition of chitin to the rice culture.

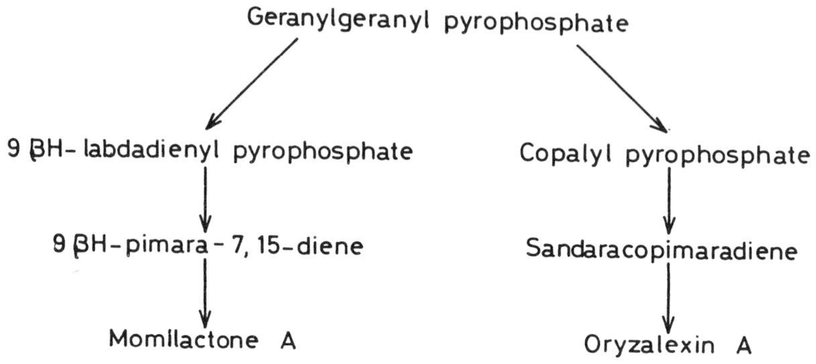

Figure 9 Pathways of diterpene phytoalexins in rice.

The production of chitinase may be an early event and phytoalexin biosynthesis may occur later. The chitinase produced by the plant may release water-soluble chitin fragments from the fungal cell wall and the soluble chitin may elicit phytoalexin synthesis by activating cyclase activity (Ren and West, 1992).

VII. SITE OF SYNTHESIS OF PHYTOALEXINS

Phytoalexin may be synthesized by living cells undergoing attack by the pathogen. In sorghum three phytoalexins of the deoxyanthocyanidin class have been reported to accumulate in response to fungal infection (Nicholson et al., 1987; Hipskind et al., 1990). The important phytoalexins are luteolinidin and apigeninidin (Snyder et al., 1991). These phytoalexins are pigmented and they first accumulate in the cell that is mostly undergoing attack by *Colletotrichum graminicola* (Snyder and Nicholson, 1990). Subcellular vesicle-like inclusions appear in the host cell on which the appressorium had formed, migrate to the site of appressorium attachment, coalesce, and ultimately burst releasing their contents into the host cell itself (Snyder and Nicholson, 1990).

The synthesized phytoalexins may be secreted from the cells. Phytoalexin accumulates in potato-cell suspension cultures inoculated with either an incompatible or compatible race of *Phytophthora infestans* (Table 5; Brindle et al., 1983). The large proportion of each phytoalexin in the culture medium suggests that the phytoalexins diffused into the medium from the cells in which they were synthesized. It is possible that the phytoalexins are synthesized in healthy living cells and then move out and accumulate in the adjoining necrotic tissue (Brindle et al., 1983).

Table 5 Phytoalexin content of potato-cell suspension cultures inoculated with *Phytophthora infestans*

	Phytoalexin (μg)					
Phytophthora infestans race	Lubimin		Rishitin		Solavetivone	
	Cells	Medium	Cells	Medium	Cells	Medium
Incompatible	225	622	19	44	95	174
Compatible	197	685	50	163	73	141

Source: Brindle et al., 1983.

VIII. PHYTOALEXINS ARE FUNGITOXIC

Phytoalexins have been reported to be highly fungitoxic. Casbene is inhibitory toward the growth of *Aspergillus niger* even at 10 μg/ml (Dueber et al., 1978). The phytoalexins are more inhibitory than their precursors. The fungitoxicity of medicarpin, the phytoalexin of alfalfa, and its biosynthetic intermediates daidzein, formononetin, 2′-hydroxyformononetin, and vestitone has been tested against different pathogens of alfalfa: *Phytophthora megasperma* f.sp. *medicaginis, Phoma medicagninis,* and *Nectria haematococca* (Blount et al., 1992). Neither daidzein nor formononetin significantly inhibited mycelial growth of any of these fungi, even at 0.5 mM (Miao and VanEtten, 19912). Medicarpin and vestitone inhibited mycelial growth of *Phoma medicaginis* while 2′-hydroxyformononetin, vestitone, and medicarpin inhibited growth of *Phytophthora megasperma* f.sp. *medicaginis.* Medicarpin alone inhibited growth of *N. haematoccoa* and biosynthetic precursors of medicarpin did not inhibit the growth of the fungus (Table 6; Blount et al., 1992).

In *Cassia obtusifolia* a flavonoid phytoalexin identified as 2-(p-hydroxyphenoxy)-5,7-dihydroxychromone was induced due to infection with *Alternaria cassiae* (Sharon and Gressel, 1991). It inhibited growth of *A. cassiae* at 0.3 millimolar concentration (Sharon et al., 1992b). Phytoalexins inhibit spore germination (Bailey and Deverall, 1971; Khan and Milton, 1975; Skipp and Bailey, 1976; Ward and Stoessl, 1977). Maackiain inhibits fungal germtube elongation (Higgins, 1978; Lucy et al., 1988).

Pathogens produce toxins responsible for disease symptom development. Phytoalexins may suppress toxin production by the pathogens. *Fusarium sporotrichoides* is a pathogen of parsnip (*Pastinaca sativa*). It produces trichothecenes, which are potent phytotoxins. There was a correlation between virulence of the fungal isolates and their ability to produce toxins in vitro (Manka et al., 1985).

Table 6 Relative growth of fungal pathogens in the presence of medicarpin or its biosynthetic precursors

	Relative growth in the presence of				
Fungus	Medicarpin	Vestitone	2′-hydroxy formononetin	Formononetin	Daidzein
Phytophthora megasperma f.sp. *medicaginis*	−44.6	−52.5	−44.0	+12.6	+6.5
Phoma medicaginis	−42.1	−44.2	−7.0	+5.7	+11.3
Nectria haematococca	−62.3	−17.6	−7.3	−1.6	−0.3

Source: Blount et al., 1992.

The toxins could be isolated from diseased plant tissues (Miller et al., 1985). Xanthotoxin and angelicin are the furanocoumarin phytoalexins in parsnip (Johnson et al., 1973; Desjardins et al., 1989b). Trichothecene toxin production by *F. sporotrichoides* is completely inhibited by furanocoumarins at concentrations well below those reported to accumulate in infected parsnips (Desjardins et al., 1988).

In *F. sporotrichoides*-infected parsnip root tissues xanthotoxin and angelicin accumulated to high levels, whereas trichothecene toxin was present only at low levels. Trichodiene, the precursor of the toxin, which was not detectable in liquid culture, accumulated. In contrast, in microwave-cooked parsnip root infected with the fungus very low levels of furanocoumarins and trichodiene were detected, and the level of trichothecene toxin was comparable with that observed in liquid cultures (Desjardins et al., 1989b). The results suggest that the furanocoumarins may inhibit biosynthesis of trichothecene from trichodiene in the parsnip roots infected with *F. sporotrichoides*.

IX. HOW DO PATHOGENS OVERCOME THE ANTIFUNGAL PHYTOALEXINS?

A. Pathogen Detoxifies Phytoalexins

The pathogens have been reported to detoxify phytoalexins of the host (Lappe and Barz, 1978; Yoshikawa et al., 1979). Pea pathogens are known to detoxify pisatin, the pea phytoalexin (Nonaka, 1967; Wit-Elshove and Fuchs, 1971). *Nectria haematococca* is a pathogen of pea that demethylates pisatin. The phenolic product, 3,6a-dihydroxy-8,9-methylenedioxypterocarpan, is less toxic than pisatin (VanEtten et al., 1975, 1982; Fuchs et al., 1980). All virulent isolates of *N. haematococca* are able to demethylate pisatin (VanEtten et al., 1980; Tegtmeier and VanEtten, 1982; Kistler and VanEtten, 1984a,b; VanEtten and Matthews, 1984). These isolates are also highly tolerant to pisatin in vitro. Isolates unable to demethylate pisatin are low in virulence and sensitive to pisatin.

A series of crosses were made between isolates of *Nectria haematococca* mating population VI that differed in sensitivity to the pea phytoalexin pisatin. The progeny from some crosses were of two discrete classes of sensitivity to pisatin. All progeny in the tolerant class were able to demethylate pisatin; none in the more pisatin-sensitive class could demethylate this phytoalexin. Of the 345 analyzed all of the moderately or highly virulent progeny were tolerant of pisatin and able to demethylate it. Virulent isolates of *N. haematococca* are more sensitive to pisatin, have demethylating ability, and induce less pisatin in infected tissue (Table 7; Tegtmeier and VanEtten, 1982).

Ascochyta pisi degrades the pisatin to products that are less toxic to the pisatin-sensitive fungus *Glomerella cingulata* (Uehara, 1964; Sanz Platero and Fuchs, 1978). The highly virulent isolates of pathogens of pea *A. pisi* and *Rhizoctonia solani* demethylate pisatin more rapidly than the less virulent ones (VanEtten et al., 1989). *Phoma pinodella* and *Mycospherella pinodes*, the other pea pathogens, also demethylate pisatin (Delserone and VanEtten, 1987). Pisatin demethylation appears to be necessary for tolerance to pisatin and for virulence on pea for the pathogens (VanEtten et al., 1980; Tegtmeier and VanEtten, 1982).

Pisatin demethylation is catalyzed by a microsomal NADPH-dependent cytochrome P-450 (Matthews and VanEtten, 1983). Cytochrome P-450 monooxygenases are multicomponent enzyme systems. The components are cytochrome P-450 (Matthews and VanEtten, 1983; Shoun et al., 1983) and NADPH-cytochrome P-450 reductase (Coon and Persson, 1980; Desjardins et al., 1984). Cytochrome P-450 and NADPH-cytochrome P-450 reductase are the key components in microsomal monooxygenase systems of animals, plants, and fungi. Cytochrome P-450-dependent monooxygenases detoxify a wide variety of drugs,

Table 7 Differential sensitivity of the pea pathogen *Nectria haematococca* isolates to pisatin

Isolate	Lesion length (cm)	Inhibition by pisatin (%)	Demethylating ability	Pisatin µg/cm^3 lesion tissue
T_{30}	11.5	16	+	96
T_8	9.5	15	+	115
T_{95}	2.4	55	–	505
T_{213}	2.9	47	–	551

Source: Tegtmeier and VanEttern, 1982a.

insecticides, and other xenobiotic compounds (Coon and Persson, 1980; Matthews and VanEtten, 1983; Desjardins et al., 1984).

The cytochrome P-450 and NADPH-cytochrome P-450 reductase of *N. haematococca* have been purified. Upon sodium dodecyl sulfate-polyacrylamide gel electrophoresis, the reductase fraction contained one major band with a molecular weight of 84,000. The partially purified cytochrome P-450 fraction contained a number of minor bands and three major bands of molecular weights 52,000, 56,000, and 58,000. Thus multiple forms of P-450 may exist in *N. haematococca* (Desjardins and VanEtten, 1986). *N. haematococca* may possess a family of P-450 cytochromes with at least two subfamilies (VanEtten et al., 1989).

NADPH-cytochrome P-450 reductase was found in all *N. haematococca* isolates tested, whether or not they are able to demethylate pisatin (Desjardins et al., 1984). Reductase from both demethylating and nondemethylating *N. haematococca* isolates is competent in reconstituting demethylation (Desjardins et al., 1984). The reductase has been purified and appears to be the same in both demethylating and nondemethylating isolates (Scala et al., 1988). It is possible that the amount of reductase present in microsomes of *N. haematococca* may limit the rate of pisatin demethylation (Desjardins and VanEtten, 1986). The inability of some isolates to demethylate pisatin appears to be due to the absence of a suitable cytochrome P-450 rather than to a lack of functional reductase (Desjardins et al., 1984).

Pisatin demethylase activity was lost when the reductase was removed by adsorption on 2′,5′-ADP-agarose. The demethylase activity of reductase-free fraction could be restored by adding a reductase preparation purified approximately 100-fold from microsomes of *N. haematococca* isolate 74-8-1 that does not demethylate pisatin. The results suggest that pisatin demethylation requires NADPH-cytochrome C reductase for activity (Desjardins et al., 1984).

Pisatin demethylation is strongly induced by pisatin. No other compound was effective in inducing pisatin demethylation (VanEtten and Barz, 1981). The demethylase activity in microsomal preparations from pisatin-induced mycelium of *Ascochyta pisi* and *Mycosphaerella pinodes* also showed a substrate preference for pisatin (VanEtten et al., 1989).

Different regulatory phenotypes for pisatin demethylating isolates of *N. haematococca* have been reported (VanEtten and Matthews, 1984). Some isolates demethylate pisatin only at a low rate ("non-inducible" isolates, designated PDA^n), while others show much greater activity when pretreated with substrate ("inducible" isolates, designated PDA^i). The inducible, high-activity PDA^i isolates showed detectable demethylase activity after 6–8 h pretreatment with substrate, while the noninducible, low-activity PDA^n isolates demethylated pisatin appreciably after 48 h only (Kistler and VanEtten, 1984b).

A cross between PDA^i and PDA^n field isolates resulted in some progeny that lacked demethylase activity altogether (Cross 44) (Cowling and VanEtten,

1986). Twenty of 100 ascospores were of the PDA$^-$ (without pisatin demethylase activity) phenotype, indicating that the determinants of the parental demethylating phenotypes are nonallelic. Crosses between PDA$^-$ progeny and the PDA^n parent isolate resulted in 1:1 segregation for the parental demethylating phenotypes, indicating that these strains differed at a single gene. Crosses between a PDA^i parent isolate and PDA$^-$ progeny of cross 44 showed an approximate 2:1:2 ratio for PDA^i:PDA^n:PDA$^-$. This segregation pattern indicated a more complex control of pisatin-demethylating ability in the PDA$^-$ isolate. Further studies with crosses between the two naturally occurring isolates (parents of cross 44) suggest that three loci determined pisatin demethylating ability. A pair of alleles at one locus (designated pda-1) determines the presence or absence of the PDA^i phenotype. Alleles at two other loci (pda-2 and pda-3) each can confer the PDA^n phenotype. Negative alleles at every locus result in PDA$^-$ individuals (Kistler and VanEtten, 1984b). Each locus may represent a distinct structural gene for a pisatin demethylase enzyme (Kistler and VanEtten, 1984b).

PDA$^-$ isolates could be deficient in any one of the biochemical components of pisatin demethylase, because of the absence of a structural gene for such components, or they could have a PDA$^-$ phenotype as a result of other genes affecting the expression of these structural genes. Segregation patterns in the cross between two PDA$^+$ field isolates indicated that the PDA$^-$ phenotype required the negative allele of each of three genes. These genes appeared to be homologous in function, since the positive allele of any one was sufficient for the PDA$^+$ phenotype. This may be structural genes for cytochrome P-450 isozymes capable of accepting pisatin as a substrate (Kistler and VanEtten, 1984b).

Attempts were made to identify functional classes of pda genes by searching for PDA$^+$ recombinants among the *N. haematococca* ascospore progeny of crosses between PDA$^-$ progeny. No PDA$^+$ recombinants were detected from 99 crosses of PDA$^-$ isolates when analyzed by this method, nor were any detected among 181 ascospore progeny assayed individually. Complementary heterokaryons for the PDA$^+$ phenotype were not formed from mixtures of 37 PDA$^-$ isolates when they were grown together on a pisatin-amended medium that was inhibitory to the PDA$^-$ isolates. Thus no evidence was found to suggest that the PDA$^-$ phenotype can be caused by genes with different functions in *N. haematococca* (Cowling and VanEtten, 1986).

N. haematococca may possess multiple Pda genes (Kistler and VanEtten, 1984b). Six Pda genes have been identified in different isolates of *N. haematococca* (VanEtten et al., 1989). These genes behave like independent structural genes in that the positive allele of any one confers the ability to demethylate pisatin (VanEtten et al., 1989). Pda 1 gene has been cloned and the 5′ untranslated region has been sequenced (Straney and VanEtten, 1994).

VanEtten et al. (1989) could identify three kinds of *N. haematococca* isolates based on the time required for a noninduced culture to metabolize pisatin (lag period) and the rate of demethylation induced. These phenotypes were designated Pda^{SH} (short lag, high induced activity), Pda^{SM} (short lag, moderate activity), and Pda^{LL} (long lag, low activity). Only genes for the Pda^{SH} and Pda^{SM} phenotypes were linked to virulence. Pda^{LL} field isolates and progeny were no more virulent than Pda^{-} isolates. In a cross between Pda^{SH} and Pda^{SM} isolates, the relative virulence was strongly correlated with their level of inducible demethylase activity (VanEtten et al., 1989).

Genetic analyses indicated that Pda^{SM} and Pda^{LL} field isolates of *N. haematococca* are encoded by different genes (Mackintosh et al., 1989). The gene responsible for Pda^{SM} phenotype may be Pda 4, whereas the Pda^{LL} gene may represent another allele, Pda 3-2 (Mackintosh et al., 1989). Among hundreds of ascospore progeny representing numerous recombinational events, all Pda^{-} and pda^{LL} ascospore isolates of *N. haematococca* have been essentially nonpathogenic on pea (Mackintosh et al., 1989). No combination of genes had been obtained that could replace the dependence of this fungus on a gene conferring the Pda^{SH} or Pda^{SM} phenotype for pathogenicity on pea (Mackintosh et al., 1989). It suggests that adequate production of this enzyme is required for pathogenicity and the genes conferring this property are therefore pathogenicity genes (Mackintosh et al., 1989).

A fragment of *N. haematococca* DNA was found to confer on *Aspergillus nidulans* the ability to demethylate pisatin. The transformed *A. nidulans* showed increased tolerance to pisatin (Weltring et al., 1988). This fragment encodes a single 1.8 kb transcript. A Pda^{-} isolate of *N. haematococca* was transformed with this DNA fragment. The PDA^{+} transformants were more tolerant of pisatin and they showed significantly more virulence towards pea (Ciuffetti et al., 1988). A nonpathogen of pea, *Cochliobolus heterostrophus* (maize pathogen), was transformed with the Pda^{+} gene from the pea pathogen, *N. haematococca* and the transformed maize pathogen became pathogenic on pea, producing larger lesions (Schafer et al., 1988). Overexpression of Pda gene from *N. haematococca* caused increased virulence to pea of several independently generated *C. heterostrophus* transformants (Oeser and Yoder, 1994). These studies have established that degradation in pea is needed for pathogenesis by fungal pathogens.

Maackiain and medicarpin are the phytoalexins produced in many leguminous plants. *Ascochyta rabiei* cleaves maackiain and medicarpin by three different ways. A reductive cleavage of ring C producing isoflavans, hydroxylation of ring A at position 1a resulting in production of 1a-hydroxydienones and 9-0-methylation are the three different initial reactions in the metabolism of maackiain and medicarpin (Kraft et al., 1987). The enzyme responsible for the first reaction, reductive cleavage of the pterocarpan to isoflavan has been purified. It is a soluble NADPH-dependent reductase with a strong preference for medicarpin and maackiain as substrates (Hohl and Barz, 1987). The enzyme preparations also catalyzed the 1a-hydroxylation and 9-0-demethylation (Hohl and Barz, 1987).

The 1a-hydroxylation by enzyme preparations obtained from *A. rabiei* is O_2^- dependent (Hohl and Barz, 1987). *Sclerotinia trifoliorum* degrades maackiain (MacFoy and Smith, 1982).

N. haematococca degrades maackiain and medicarpin by three alternative reactions. The products correspond to oxygen attack at positions 6a, 1a, and 11a (Denny and VanEtten, 1982) and all these reactions may be oxygenase-catalyzed (Denny and VanEtten, 1982). The degraded products are less toxic than medicarpin and maackiain. The most virulent isolates of *N. haematococca* were relatively tolerant of maackiain and medicarpin and almost all of these isolates could metabolize maackiain and medicarpin by at least one route (Denny and VanEtten, 1981; Lucy et al., 1988). Three genes controlling maackiain metabolism have been reported in *N. haematococca* (Miao and VanEtten, 1987). *Mak* 1 and *Mak* 2 confer 1a-hydroxylation and *Mak* 3 confers 6a-hydroxylation (Miao et al., 1986). The enzymes encoded by *Mak* 2 and *Mak* 3 are specific for maackiain while *Mak* 1 may be identical to Pda 6 (Miao et al., 1986).

Maackiain and pisatin detoxification are required for pathogenicity of *N. haematococca* on chickpea and pea respectively, and each reaction can be controlled independently (VanEtten et al., 1989).

Phaseollin (Cruickshank and Perrin, 1963; Perrin, 1964), phaseollidin (Perrin et al., 1974), phaseollinisoflavan, and kievitone (Burden et al., 1972; Bailey and Burden, 1973) are the important phytoalexins in French bean. *Fusarium solani* f.sp *phaseoli* metabolizes all the four phytoalexins of bean, (Smith et al., 1980; Zhang and Smith, 1983; Kuhn et al., 1977; Kuhn and Smith, 1978, 1979). The resulting metabolites are less toxic than their parent phytoalexins (Kuhn et al., 1977; Kuhn and Smith, 1978; Smith et al., 1980, 1981; VanEtten et al., 1982). Phaseollin is detoxified by 1a-hydroxylation by an intracellular oxygenase (Kistler and VanEtten, 1981). Isotope incorporation studies have provided direct evidence for participation of the monooxygenase in metabolism of phaseollin by *F. solani* f.sp. *phaseoli* (Kistler and VanEtten, 1981). Phaseollidin is also detoxified by hydration of its isopentenyl side chain (Smith et al., 1980, 1981). Phaseollinisoflavan is metabolized to a product that is less fungitoxic (Zhang and Smith, 1983; Wictor-Orlandi and Smith, 1985).

When phaseollinisoflavan was added to the liquid cultures of *F. solani* f.sp. *phaseoli*, its concentration decreased for 21 h and thereafter no phaseollinisoflavan could be recovered. Conversely, a new compound, designated M-1, was detected in the culture medium after 9 h and increased steadily. The metabolite M-1 was less toxic to the fungus compared to the phytoalexin. Both M-1 and phaseollinisoflavan could be detected in bean hypocotyls inoculated with the pathogen indicating that they would have been metabolized in vivo in cells (Wictor-Orlandi and Smith, 1985). Both phaseollin and its metabolite 1a-hydroxyphaseollone were detected in bean hypocotyls infected with *F. solani* f.sp. *phaseoli* (VanEtten et al., 1982, 1989). Kievitone and its metabolite kievitone hydrate have also been detected in bean infected with the pathogen (Kuhn and Smith, 1978).

Kievitone hydratase is produced constitutively by *F. solani* f.sp. *phaseoli*. The enzyme induces hydration of isopentenyl side chain in kievitone resulting in the formation of kievitone hydrate (Kuhn and Smith, 1978, 1979; Smith et al., 1982). Kievitone hydratase is an acidic glycoprotein (Cleveland and Smith, 1983). The enzyme does not show any substrate specificity. Phaseollinisoflavan induced kievitone hydratase levels approximately 10-fold higher than kievitone (VanEtten et al., 1989).

The highly virulent isolates of *F. solani* f.sp. *phaseoli* produced more kievitone hydratase (Smith et al., 1984). Two mutants of the pathogen produced less kievitone hydratase and showed less tolerance to kievitone. They were also less pathogenic on bean (Smith et al., 1984). These mutants were also less tolerant to phaseollin and phaseollinisoflavan (Choi et al., 1987).

The potato pathogen, *Gibberella pulicaris* degrades both lubimin and rishitin, the phytoalexins of potato (Gardner et al., 1988; Desjardins and Gardner, 1989; Desjardins et al., 1989a). Seven metabolites of lubimin have been reported (Gardner et al., 1988; Desjardins et al., 1989a). The lubimin may be metabolized to 1,5-dihydrolubimin or to 2-dehydrolubimin. Both are converted to isolubimin and then to cyclodehydroisolubimin (Desjardins et al., 1989a). Both 1,5-dihydrolubimin and isolubimin are toxic to *G. pulicaris*, while cyclodehydroisolubimin and two later tricyclic metabolites are not toxic to the fungus (Desjardins et al., 1989a).

All isolates of *G. pulicaris* that were highly virulent on potato degraded both lubimin and rishitin relatively rapidly and were relatively tolerant to them. The isolates that were less tolerant, metabolized the phytoalexins more slowly and they were low in virulence (Desjardins et al., 1989a). There may be at least two genes in *G. pulicaris* that confer high tolerance to and rapid metabolism of rishitin (Desjardins and Gardner, 1989).

Botrytis cinerea is known to detoxify phytoalexins from a number of plants (Mansfield, 1980). The pathogenicity of isolates of *B. cinerea* towards French bean was found to be more closely correlated with their ability to metabolize phaseollin than with their ability to tolerate the phytoalexin (Van Den Heuvel, 1976).

Pterostilbene (*trans* 3,5-dimethoxy-4′-hydroxystilbene) and resveratrol (*trans* 3,5,4′-trihydroxystilbene) are constitutive components of the woody parts of many species of the Vitaceae (Hart, 1981; Pool et al., 1981). However, these compounds are only produced in the leaves and fruits after fungal infection and hence they are considered as phytoalexins (Barlass et al., 1987; Dercks and Creasy, 1989). Pterostilbene has been identified as a phytoalexin in grapevine (*Vitis vinifera*) leaves infected with *Plasmopara viticola* (Langcake et al., 1979) although it is found in healthy grape berries of *Vitis vinifera* (Pezet and Pont,

1988). Pterostilbene is highly toxic to *Botrytis cinerea* (Pezet and Pont, 1990; Pont and Pezet, 1990). The protein extract of culture filtrates of *Botrytis cinerea* caused a rapid change in the UV spectra when added to solutions of resveratrol or pterostilbene (Pezet et al., 1991), indicating that the chemical structure of the stilbenes was being modified. Degradation of the stilbenes was confirmed by HPLC analysis. After 10 min enzyme action, 100% of the pterostilbene and 94.3% of the resveratrol had been degraded to yield a number of products. No degradation occurred in the absence of the protein extract of culture filtrate (Pezet et al., 1991).

Pterostilbene solution lost its lethal property completely after the treatment with the protein extract of culture filtrate (Table 8; Pezet et al., 1991).

Inhibitors of laccase, such as sodium diethyldithiocarbamate and sodium azide, inhibited the activity of the protein extract. This inhibition was also observed using thiourea, an OH^- scavenger and inhibitor of hydroxylation, indicating that hydroxylation is an important step in hydroxystilbene oxidation (Pezet et al., 1991). The results suggest that *B. cinerea* produces a hydroxystilbene-degrading enzyme and the enzyme, stilbene oxidase, is a laccase. Stilbene oxidase oxidizes both pterostilbene and resveratrol to produce nontoxic products (Pezet et al., 1991).

Botrytis cinerea requires two signals for maximal laccase production. One is provided by phenolics produced by the plants, and the second by decomposition products of pectin that also signal the extent of breakdown of host tissues (Marbach et al., 1984, 1985). Oxidation of resveratrol facilitates pathogenesis of *Botrytis cinerea* in grapevine (Hoos and Blaich, 1988, 1990). A positive correlation has been reported between stilbene production and resistance of grapevine varieties to *B. cinerea* (Langcake et al., 1979; Langcake and McCarthy, 1979; Stein and Hoos, 1984). *B. cinerea* produces large amount of laccase (polyphenol oxidase) (Dubernet and Ribereau-Gayon, 1973) secreted from the cytoplasm of its mycelial cells through the walls and mucilage sheath (Hodson et al., 1987). *B. cinerea* produces laccase isozymes (Zouari et al., 1987).

The furanocoumarin concentration in parsnip roots changed sharply at the leading edge of infection by *Fusarium sporotrichoides*. The concentration of total

Table 8 Detoxication of pterostilbene by the culture filtrate of *Botrytis cinerea*

Stilbene	Inhibition of germination (%)	Dead conidia (%)
Pterostilbene	100	100
Pterostilbene + protein extract of culture filtrate	0	0

Source: Pezet et al., 1991.

furanocoumarins in advance of infection was from 0.4 to 1.0 μmoles of fresh weight, which was enough to slow but less than that necessary to inhibit fungal growth completely in vitro. Furanocoumarin concentrations were highest in tissue sections collected several millimetres behind the leading edge of infection. Furanocoumarin levels in tissue sections closer to original inoculum were very much less (Desjardins et al., 1989).

Both xanthotoxin and angelicin were fungistatic, but not fungicidal to *F. sporotrichoides*. Cultures of the fungus kept for more than 2 weeks at furanocoumarin concentrations completely inhibitory to growth had the same growth rate as control cultures when transferred to agar media without furanocoumarins. The fungus was able to metabolize xanthotoxin, completely, added to 24 h liquid cultures to a final concentration of up to 5 mM (Desjardins et al., 1989).

B. Pathogen Induces Accumulation of Phytoalexins Slowly in the Susceptible Host

Incompatible pathogens induce rapid accumulation of phytoalexins (Barlass et al., 1987; Bavaresco and Eibach, 1987; Dercks and Creasy, 1989; Elgersma and Liem, 1989). In contrast, pathogens delay the accumulation of phytoalexins during their successful pathogenesis. Glyceollins I, II, and III accumulate slower in susceptible soybean cultivar inoculated with *Phytophthora megasperma* f.sp. *glycinea* compared to that in the resistant variety (Bhattacharyya and Ward, 1986). Glyceollin accumulates in resistant soybean roots at 2 h after inoculation with the pathogen but not until 12 h in susceptible roots (Hahn et al., 1985). In the incompatible interaction glyceollin begins to accumulate at 6–7 h and reaches Ec_{90} (0.6 mM for glyceollin) 8 h after inoculation, and at this time the fungal growth appeared to slow or stop (9–12 h after inoculation). In the compatible interaction, glyceollin levels within newly invaded tissue layers did not exceed the Ec_{90} value at any time after inoculation (Yoshikawa et al., 1978).

The importance of early induction of phytoalexins to suppress the pathogen development has been shown by Albersheim and Valent (1978). The soybean hypocotyls were treated with the fungal elicitor 6 h before inoculation with the pathogen *P. megasperma* f.sp *glycinea* and this treatment induced resistance against the pathogen. The elicitor treatment induced phytoalexin synthesis, and at the time of inoculation with the pathogen the elicitor-induced phytoalexin level was already sufficient for containment of the fungus (Albersheim and Valent, 1978).

The delay in synthesis of phytoalexins in the infected susceptible host may be due to delayed stimulation of activity of phytoalexin biosynthetic enzymes. In the susceptible soybean tissues infected with *P. megasperma* f.sp. *glycinea* the activity of phytoalexin biosynthetic enzymes remains the same as, or was only slightly higher than, in the uninoculated controls during the experimental period of 2–8 h after inoculation. In contrast, in the incompatible interaction there was an early stimulation of the various enzyme activities starting at 2–4 h after inoculation with accumulation of phytoalexins (Hahn et al., 1985).

When soybean hypocotyls were inoculated with *P. megasperma* f.sp. *glycinea* mRNAs for both PAL and CHS were rapidly produced even within 3 h following inoculation in a resistant variety, but in the susceptible variety only slight enhancement of mRNAs was observed at that time. In soybean roots inoculated with *P. megasperma* f.sp. *glycinea*, an increase in the activity of dihydroxypterocarpan 6a-hydroxylase, an enzyme that catalyzes the production of the immediate precursor of glyceollin, is observed 6 h earlier in resistant than in susceptible roots (Bonhoff et al., 1986a,b). The observations indicate that the production of mRNAs for enzymes leading to phenylpropanoid biosynthesis in soybean cotyledons is delayed in response to infection with compatible pathogen, but it is an early response in an incompatible interaction (Esnault et al., 1987). Delayed production of CHS mRNA in a susceptible *Colletotrichum lindemuthianum*–bean interaction has been reported (Bell et al., 1984). CHS mRNA was produced at a slower rate in the susceptible interaction in soybean also (Schmelzer et al., 1984).

The delayed release of precursors of phytoalexin may also delay the synthesis of phytoalexins. The precursors of glyceollin, daidzein conjugates, are present at higher levels in a soybean cultivar susceptible to *P. megasperma* f.sp. *glycinea* compared to that in a resistant variety. In incompatible reactions, there was a marked and immediate reduction in levels of the conjugates. These fell to less than half their original levels by 24 h and to nearly nondetectable levels within just 48 h. Concurrent with this, there was an accumulation of free daidzein and glyceollin. In contrast, the compatible reaction was characterized by a delayed disappearance of the daidzein conjugates, a nearly stoichiometric accumulation of daidzein, and little to no accumulation of glyceollin. In the compatible reactions, only low levels of glyceollin were synthesized despite the accumulation of very large amounts of free daidzein. It suggests that by the time the isoflavones are released, the majority of cells in the susceptible infected tissues are no longer capable of glyceollin biosynthesis. The delayed release of daidzein from its conjugates may be important in pathogenesis of the pathogen (Graham et al., 1990).

C. Pathogen Induces Diffusion of Phytoalexins in Susceptible Host

In resistant interactions defense gene-activated products may accumulate in a narrow zone around the infection site. In susceptible interactions the gene product may be diffused. PAL activation was delayed, more diffuse, and nontransient in the susceptible *Phytophthora infestans*–potato interactions than in the resistant interactions. PAL mRNA accumulated in a narrow band of cells immediately surrounding the hypersensitive lesion. PAL messenger in this sharply defined zone was at high levels by 3 h and declined to near control levels by 6 h. Immunohistochemical localization of the fungus demonstrated that this very high activation of PAL involved the apparent transmission of a signal immediately in advance of the fungal hyphae (Cuypers et al., 1988).

The existence of sharply restricted zone of cells expressing defense genes has been reported in parsley also. Within this zone, rapid induction of PAL, 4 CL, and other defense-related genes was observed in incompatible interactions (Graham and Graham, 1991).

D. Pathogen Suppresses Accumulation of Phytoalexins in Susceptible Host

Several phytoalexins (oryzalexin A, B, C, and D) have been detected in rice leaves infected with *Pyricularia oryzae* (Iwakuma et al., 1990). When rice leaves were inoculated with incompatible race of *P. oryzae*, the phytoalexin production was detected at 36–48 h after inoculation at which time insertion of the infection hyphae had started. At this time, phytoalexin was not detected in rice leaves inoculated with virulent pathogen. In general the appearance of the phytoalexin was 12–24 h faster in incompatible interaction than in compatible interaction (Iwakuma et al., 1990). This race specificity disappeared when heat-killed spores were applied and the phytoalexin appeared at 48 h in both incompatible and compatible interactions. Hyphal wall components (elicitors) from both compatible and incompatible races of *P. oryzae* induced similar amount of phytoalexins. The results suggest that the compatible pathogen may suppress the accumulation of phytoalexins while its cell wall component may be able to induce phytoalexins faster. The race specificity in phytoalexin elicitation is observed when living spores are applied and not with the treatments with heat-killed spores or the hyphal cell wall components. A suppressor or suppressing system in living fungal cells may exist (Iwakuma et al., 1990).

The induction of phytoalexin in barley leaves by an incompatible race of *Erysiphe graminis* f.sp. *hordei* was suppressed by preliminary inoculation with a compatible race (Oku et al., 1975b, 1980). Pycnospore germination fluid of *Mycosphaerella pinodes*, the pathogen of pea, contains both elicitors and suppressors for the accumulation of pisatin in pea (Shiraishi et al., 1978; Oku et al., 1987). The elicitors induce PAL gene in pea while the suppressors suppress it (Hiramatsu et al., 1986). Similar elicitors and suppressors have been isolated from *P. megasperma* f.sp. *glycinea*, the soybean pathogen (Ziegler and Pontzen, 1982).

When pine seedlings were inoculated with *Botrytis cinerea*, stilbene synthase activity was induced about 38-fold in the initial stages of infection. However, after 24 h, the increase in the enzyme activity started declining and by about 72 h its activity reached to its near basal level (Gehlert et al., 1990). The phytoalexin content also started increasing after the fungal infection and the maximum content was observed at about 20 h after fungal inoculation. Subsequently the phytoalexin content declined very rapidly (Gehlert et al., 1990) and the decline may be due to reduced activity of stilbene synthase (Gehlert et al., 1990).

The suppressor isolated from the pea pathogen, *M. pinodes* delayed the accumulation of PAL and CHS mRNAs by 3 h in pea. There was a 6 h delay in the increase of PAL enzyme activity, and a 6–9 h suppression of pisatin accumulation due to suppressor treatment (Yamada et al., 1989).

The suppressor isolated from *M. pinodes* suppressed ATPase activity in pea plasma membrane. The ATPase activity in pea plasma membrane was unaffected by the addition of elicitor. Orthovanadate suppressed ATPase activity similar to the fungal suppressor (Yoshioka et al., 1990). Orthovanadate (Yoshioka et al., 1990) and the fungal suppressor (Yamada et al., 1989) suppressed the accumulation of pisatin. The accumulation of PAL- or CHS-mRNA is also delayed due to addition of orthovanadate (Yoshioka et al., 1992). The inhibition of ATPase activity in pea plasma membrane in vitro seems to be closely correlated with the suppression of production of pisatin in vivo (Yoshioka et al., 1990). The effect of the suppressor on host defense reactions seems to be resulted from its inhibition of the ATPase in the host plasma membrane, because the ATPase in the plasma membranes is thought to be a "master enzyme" responsible for many important and fundamental functions of the cells (Serrano, 1989).

E. Pathogen Induces Only Less Amount of Phytoalexins in Susceptible Host

Plants may respond to pathogens producing phytoalexins. The differences between resistance and susceptibility may be quantitative rather than qualitative. The susceptible plants may possess the machinary necessary for induction of phytoalexins, but it may not be activated in sufficient magnitude to restrict the infection (Kuc, 1983). When a cultivar of tobacco resistant to *Phytophthora nicotianae* var. *nicotianae* was inoculated with the pathogen the sesquiterpenoid phytoalexins (capsidiol, phytuberin, rishitin, and phytuberol) accumulated to higher level when compared to that in a susceptible variety (Nemestothy and Guest, 1990).

When the tobacco variety susceptible to *P. nicotianae* var. *nicotianae* was fed with fosetyl-Al, the plant became resistant. This treatment induced accumulation of all the four phytoalexins including capsidiol in the susceptible variety (Nemestothy and Guest, 1990). Mevinolin completely inhibited phytoalexin accumulation 24 h after inoculation in both susceptible and resistant varieties either with or without fosetyl-Al treatment. The inhibitor treatment increased the susceptibility of both fosetyl-Al-treated and untreated leaves of both cultivars (Table 9; Nemestothy and Guest, 1990). These observations suggest that the accumulation of phytoalexins is important in determining susceptibility.

Lucerne callus lines differed in their reaction to the pathogen *Verticillium albo-atrum*. Callus line 4 showed resistance while callus line 16 was susceptible. Medicarpin accumulated in the callus lines after inoculation with *V. albo-atrum*. The amounts of the phytoalexin extracted from callus line 4 were consistently higher than those found in callus line 16 (Latunde-Dada and Lucas, 1985). When the resistant callus line 4 was treated with 0.1 mM glyphosate, the medicarpin production in response to infection with *V. albo-atrum* was suppressed. This treatment converted the callus line 4 into a susceptible one (Latunde-Dada and Lucas, 1985). The herbicide glyphosate specifically inhibits the formation of chorismate from shikimate in the shikimic acid pathway (Amrhein et al., 1980; Hollander and Amrhein, 1980) and this leads to reduced production of phenylpropanoids including medicarpin.

The susceptible soybean leaves inoculated with *P. megasperma* f.sp. *glycinea* accumulate less glyceollin than the resistant soybean leaves (Bhattacharyya and Ward, 1985; Keen, 1986; Ebel and Grisebach, 1988). The accumulation of the phytoalexin glyceollin in soybean inoculated with *P. megasperma* f.sp. *glycinea* or *Pseudomonas syringae* pv. *glycinea* was suppressed due to the herbicide glyphosate treatment and this resulted in suppression of resistance mechanism in soybean resistant cultivars (Holliday and Keen, 1982; Keen et al., 1982).

A fluorescent phytoalexin, 6,7-dimethoxycoumarin (DMC) accumulates in sour orange (*Citrus aurantium*) inoculated with *Phytophthora citrophthora* (Musumeci and Oliveira, 1975; Afek et al., 1986; Afek and Szteinberg, 1988). The accumulation of DMC in infected stems of different citrus species was positively correlated with their degree of resistance to *P. citrophthora* (Afek et al., 1986; Afek and Szteinberg, 1988; Sulistyowati et al., 1990). Soil salinity

Table 9 Relationship between phytoalexin content and susceptibility to *Phytophthora nicotianae* var. *nicotianae* in tobacco

	Phytoalexin mg/g fresh weight				Lesion diameter (mm)			
	Susceptible variety		Resistant variety		Susceptible variety		Resistant variety	
Treatment	–fosetyl –Al	+fosetyl –Al	–fosetyl –Al	+fosetyl –Al	–fosetyl –Al	+fosetyl –Al	–fosetyl –Al	+fosetyl –Al
Control	50	107	134	114	27	12	16	4
Mevinolin	9	9	6	8	28	25	35	12

Source: Nemestothy and Guest, 1990.

caused reduced accumulation of the phytoalexin, 6,7-dimethoxycoumarin in the troyer citrange (*Poncirus trifoliata* × *Citrus sinensis*) and increased susceptibility of plant tissues to invasion by *P. citrophthora* (Sulistyowati and Keane, 1992). Two phytoalexins (1,3-dion-5-octyl-cyclopentaene[tsibulin 1d] and 1,3-dion-5-hexyl-cyclopen-taene[tsibulin 2d]) when onion bulbs were inoculated with the nonpathogen, accumulated to a very high level. When a pathogen of onion, *Botrytis allii* was inoculated, the accumulation of the phytoalexins was almost negligible (Dmitriev et al., 1990).

When susceptible and resistant chickpea cultivars were inoculated with *Ascochyta rabiei*, increase in medicarpin was negligible in the susceptible cultivar and the second phytoalexin maackiain could not be detected at all. In contrast, the resistant cultivar showed accumulation of large amounts of both the phytoalexins (Weigand et al., 1986).

All these studies indicate that reduced accumulation of phytoalexins may confer susceptibility in plants. It is possible that reduced release of elicitors from the fungal cell wall may be the cause for the reduced accumulation of phytoalexins during successful pathogenesis. When soybean hypocotyls were treated with metalaxyl, defense mechanisms were activated against *Phytophthora megasperma* f.sp. *glycinea* (Lazarovits and Ward, 1982; Ward, 1984). The defense mechanisms included the accumulation of high levels of the phytoalexin glyceollin (Ward et al., 1980; Lazarovits and Ward, 1982). When soybean hypocotyls were inoculated with metalaxyl-tolerant isolates of *P. megasperma* f.sp. *glycinea*; severe disease symptoms appeared with low levels of glyceollin. This contrasted with restricted brown lesions with high glyceollin content that developed following inoculation with the metalaxyl-sensitive isolates. Comparisons were made of the release of elicitors of glyceollin into culture fluids of a metalaxyl sensitive and a tolerant isolate following addition of metalaxyl to the culture medium. The elicitor activity in culture filtrates of the metalaxyl-sensitive isolate increased with increasing metalaxyl concentration (Cahill and Ward, 1989). There were no comparable increases in elicitor activity in culture fluids from the metalaxyl-tolerant isolate, regardless of the metalaxyl concentration in the medium (Table 10; Cahill and Ward, 1989). The results suggest that in virulent pathogen elicitor is not released from its cell wall to induce the defense mechanisms (Cahill and Ward, 1989).

F. Phytoalexins May be Degraded by the Susceptible Host Itself

Phytoalexins synthesized during fungal infection may not persist for a long time in host cells. Phytoalexins are not necessarily end products of plant metabolism and their concentration may decrease due to transformations catalyzed by the tissue itself (Stoessl et al., 1977; Gay, 1985). When carnation cells were treated

Table 10 Effect of metalaxyl on production of elicitors in vitro by a metalaxyl-sensitive and a metalaxyl-tolerant isolate of *Phytophthora megasperma* f.sp. *glycinea*

Concentration of metalaxyl (μg/ml)	Elicitor activity (mean μg glyceollin/μg carbohydrate per hypocotyl) Metalaxyl-sensitive isolate	Metalaxyl-tolerant isolate
0	16.6	22.2
1.0	27.6	24.0
10.0	39.8	24.8

Source: Cahill and Ward, 1989.

with an elicitor preparation from *Phytophthora parasitica*, the phytoalexin dianthalexin synthesis began after a lag phase of 14 h and increased to a maximum after 24 h after treatment. Thereafter, the concentration in the cells decreased steadily and 6 days after treatment only small amounts were detectable (Gay, 1985). Healthy cell suspensions of potato metabolized exogenously supplied rishitin and lubimin to 13-hydroxyrishitin and unindentifiable lubimin metabolites, respectively (Ward et al., 1977).

The elicitor isolated form *Ascochyta rabiei* induced the phytoalexins medicarpin and maackiain in chickpea. After 60–90 h after elicitor treatment the content of the phytoalexins declined rapidly (Kessmann and Barz, 1986). The precursor of these phytoalexins, formononetin, also declined rapidly after 36–40 h (Kessmann and Barz, 1986). Degradation of the phytoalexins may be due to catabolic reactions catalyzed by hydrolases and peroxidases of the host (Kessmann and Barz, 1986).

G. Phytoalexins May Not Have Any Role in Defense Mechanisms of Plants

In some host–pathogen interactions, phytoalexins may not have any role in disease resistance. Accumulation of phytoalexins may be only a metabolic process activated by a stress. Phytoalexins may accumulate in both susceptible and resistant varieties due to infection. In pea the pathogen *Fusarium solani* f.sp. *pisi* induced more pisatin than nonpathogen *F. solani* f.sp. *phaseoli* (Kendra and Hadwiger, 1987). Inoculation with *F. solani* f.sp. *pisi* induced PAL severalfold in pea and it was much higher than that induced by the nonpathogen *F. solani* f.sp. *phaseoli* (Loschke et al., 1981).

Three phytoalexins have been detected in sorghum after inoculation with pathogens or nonpathogens (Nicholson et al., 1987, 1988; Hipskind et al., 1990). They are 3-deoxyanthocyanidins and have been identified as apigeninidin, luteolinidin (Nicholson et al., 1987), and a caffeic acid ester of arabinosyl-5-0-apigeninidin (Hipskind et al., 1990). The deoxyanthocyanidin phytoalexins accumulate rapidly in both sorghum mesocotyls and leaves following attempted fungal infection by pathogens and nonpathogens (Nicholson et al., 1987, 1988).

X. CONCLUSION

When a fungal pathogen invades host tissues, de novo synthesis of phytoalexins occurs in host cells within a few hours of attempted fungal ingress. The synthesized phytoalexins may be secreted from the cells and may be accumulating in the infection court. These phytoalexins are fungitoxic and may eliminate the invading fungus. Different kinds of phytoalexins may be produced in the same host and are synthesized in different pathways. Several enzymes have to be coordinately induced to produce these phytoalexins. Phenylalanine ammonia-lyase (PAL), cinnamic acid 4-hydroxylase(CA4H), 4-coumarate:CoA ligase (4 CL), isoflavone synthase (IFS), isoflavone 2′-hydroxylase (IFH), and prenyl transferase are the key enzymes in the synthesis of isoflavonoid phytoalexins. Methyl transferases catalyze the terminal step in the biosynthesis of certain phytoalexins, such as pisatin and medicarpin. Stilbene synthase is the terminal enzyme in resveratrol biosynthesis. Pyrophosphomevalonate decarboxylase, 3-hydroxy-3-methylglutaryl-CoA reductase (HMGR), and sesquiterpene cyclase are the key enzymes in the synthesis of terpenoids. Farnesyl transferase and casbene synthetase are involved in the synthesis of casbene. Several other enzymes have been shown to be involved in biosynthesis of phytoalexins.

Proper coordination of all these enzymes are necessary in the synthesis of phytoalexins. Activation of a particular enzyme alone is not sufficient for phytoalexin synthesis. Although a significant increase in CHS mRNA synthesis and CHS activity was noticed in soybean infected with *Phytophthora megasperma* f.sp. *glycinea*, concomitant increase in the phytoalexin, glyceollin, had not been observed (Ebel et al., 1984; Bonhoff et al., 1986a,b; Dhawale et al., 1989).

The different phytoalexins of the host may also be produced in sequence. Kievitone accumulates rapidly while phaseollin is synthesized much later in bean. Hence, the role of these phytoalexins during the infection process has to be cautiously interpreted. Successful pathogens may have the ability to overcome the toxic action of phytoalexins. In the first instance the pathogens may suppress the de novo synthesis of these phytoalexins. Reduced release of elicitors from the fungal cell wall may result in reduced induction of phytoalexins. Defense gene activation may be suppressed, probably by the action of suppressor. Accumulation of defense-related enzymes in the infection court may be diffused. Pathogens produce enzymes that detoxify the phytoalexins. Monooxygenases are the common enzymes produced by various pathogens that detoxify phytoalexins including pisatin, phaseollin, phaseollinisoflavan, phaseollidin, kievitone, maackiain, and medicarpin. Laccases produced by pathogens degrade phytoalexins such as pterostilbene and resveratrol.

Phytoalexins have been detected mostly in dicots. The causal role for phytoalexins in disease resistance mechanism could not be established in some host–pathogen interactions. However, it appears that the gene that confers the ability to overcome the toxic effect of phytoalexins may be one of the pathogenicity genes of a number of pathogens.

REFERENCES

Afek, U., and Szteinberg, A. (1988). Accumulation of seoparone, a phytoalexin associated with resistance of citrus to *Phytophthora citrophthora*, Phytopathology, 78:1678–1682.

Afek, U., Szteinberg, A., and Carmely, S. (1986). 6,7-Dimethoxycoumarin, a citrus phytoalexin conferring resistance against *Phytophthora* gummosis, Phytochemistry, 25:1855–1856.

Akada, S., Kung, S. D., and Dube, S. K. (1990). The nucleotide sequence of gene 3 of the soybean chalcone synthase multigene family, Nucleic Acids Res., 18:19.

Akada, S., Kung, S. D., and Dube, S. K. (1991). The nucleotide sequence of gene 1 of the soybean chalcone synthase multigene family, Plant Mol. Biol., 16:751–752.

Akatsuka, T., Kodama, O., Sekido, H., Kono, Y., and Takeuchi, S. (1985). Novel phytoalexins (oryzalexins A, B and C) isolated from rice blast leaves infected with *Pyricularia oryzae*. Part I: Isolation, characterization and biological activities of oryzalexins, Agric. Biol. Chem., 49:1689–1694.

Albersheim, P., and Valent, B. S. (1978). Host–pathogen interactions in plants. Plants, when exposed to oligosaccharides of fungal origin, defend themselves by accumulating antibiotics, J. Cell Biol., 78:627–643.

Allen, B. F., and Banthorpe, D. V. (1981). Partial purification and properties of prenyltransferase from *Pisum sativum*, Phytochemistry, 20:35–40.

Amrhein, N., Deus, B., Gehrke, P., and Steinrucken, H. C. (1980). The site of the inhibition of the shikimate pathway by glyphosate. II. Interference of glyphosate with chorismate formation in vivo and in vitro, Plant Physiol., 66:830–834.

Arias, J. A., Dixon, R. A., and Lamb, C. J. (1993). Dissection of the functional architecture of a plant defense gene promoter using a homologous in vitro transcription initiation system, Plant Cell, 5:485–496.

Bailey, J. A. (1974). The relationship between symptom expression and phytoalexin accumulation in hypocotyls of *Phaseolus vulgaris* infected with *Colletotrichum lindemuthianum*, Physiol. Plant Pathol., 4:477–488.

Bailey, J. A. (1982). Mechanisms of phytoalexin accumulation, Phytoalexins, (J. A. Bailey and J. W. Mansfield, eds.), Blackie Press, Glasgow and London, pp. 289–318.

Bailey, J. A., and Burden, R. S. (1973). Biochemical changes and phytoalexin accumulation in *Phaseolus vulgaris* following cellular browning caused by tobacco necrosis virus, Physiol. Plant Pathol., 3:171–177.

Bailey, J. A., and Deverall, B. J. (1971). Formation and activity of phaseollin in the interaction between bean hypocotyls (*Phaseolus vulgaris*) and physiological races of *Colletotrichum lindemuthianum*, Physiol. Plant Pathol., 1:435–449.

Bailey, J. A., and Mansfield, J. W. (1982). Phytoalexins, Blackie Press, Glasgow and London, pp. 334.

Bairoch, A. (1991). PROSITE: a dictionary of sites and patterns in proteins, Release 7.00, Medical Biochemistry Department, University of Geneva, Switzerland.

Baker, C. J., O'Neill, N. R., and Bauchan, G. R. (1988). Production of phytoalexin in race-clone interactions of alfalfa and *Colletotrichum trifolii*, Phytopathology, 78:1590.

Banks, S. W., and Dewick, P. M. (1982). Biosynthesis of the 6a-hydroxypterocarpan phytoalexin pisatin in *Pisum sativum*, Phytochemistry, 21:2235–2242.

Barlass, M., Miller, R. M., and Douglas, T. J. (1987). Development of methods for screening grapevines for resistance to infection by downy mildew. II. Resveratrol production, Am. J. Enol. Viticult., 38:65–68.

Barz, W., Bless, W., Daniel, S., Gunia, W., Hinderer, W., Jaques, U., Kessmann, H., Meier, D., Tiemann, K., and Wittkampf, U. (1989). Elicitation and suppression of isoflavones and pterocarpan phytoalexin in chickpea (*Cicer arietinum* L.) cell cultures, Secondary Metabolism of Plant Cell Cultures II, (W. G. K. Kurz, ed.), Springer-Verlag, Berlin, pp. 208–218.

Barz, W., Daniel, S., Hinderer, W., Jaques, U., Kessmann, H., Koster, J., Otto, C., and Tiemann, K. (1988). Elicitation and metabolism of phytoalexins in plant cell cultures, Ciba Found. Symp., 137:178–198.

Bavaresco, L., and Eibach, R. (1987). Investigations on the influence of N fertilizer on resistance to powdery mildew (*Oidium tuckeri*) and downy mildew (*Plasmopara viticola*) and on phytoalexin synthesis in different grapevine varieties, Vitis, 26:192–200.

Bell, J. N., Dixon, R. A., Bailey, J. A., Rowell, P. M., and Lamb, C. J. (1984). Differential induction of chalcone synthase mRNA activity at the onset of phytoalexin accumulation in compatible and incompatible plant-pathogen interactions, Proc. Natl. Acad. Sci. USA, 81:3384–3388.

Bell, J. N., Ryder, T. B., Wingate, V. P. M., Bailey, J. A., and Lamb, C. J. (1986). Differential accumulation of plant defense gene transcripts in a compatible and an incompatible plant-pathogen interaction, Mol. Cell Biol., 6:1615–1623.

Bennett, M. H., Gallagher, M. D. S., Bestwick, C. S., Rossiter, J. T., and Mansfield, J. W. (1994). The phytoalexin response to lettuce to challenge by *Botrytis cinerea*, *Bremia lactucae* and *Pseudomonas syringae* pv. *phaseolicola*, Physiol. Mol. Plant Pathol., 44:321–333.

Bevan, M., Shufflebottom, D., Edwards, K., Jefferson, R., and Schuch, W. (1989). Tissue- and cell-specific activity of a phenylalanine ammonia-lyase promoter in transgenic plants, EMBO J., 8:1899–1906.

Bhattacharyya, M. K., and Ward, E. W. B. (1985). Differential sensitivity of *Phytophthora megasperma* f.sp. *glycinea* isolates to glyceollin isomers, Physiol. Plant Pathol., 27:299–310.

Bhattacharyya, M. K., and Ward, E. W. B. (1986). Resistance, susceptibility and accumulation of glyceollins I-III in soybean organs inoculated with *Phytophthora megasperma* f.sp. *glycinea*, Physiol. Mol. Plant Pathol., 29:227–237.

Biggs, R. (1975). Post-infectional compounds from the French bean *Phaseolus vulgaris*: isolation and identification of genistein and 2′4′,5-7-tetrahydroxyisoflavone, Austr. J. Chem., 28:1389–1392.

Biggs, D. R., Welle, R., Visser, F. R., and Grisebach, H. (1987). Dimethylallylpyrophosphate: 3,9-dihydroxypterocarpan 10-dimethylallyl transferase from *Phaseolus vulgaris*, FEBS Lett., 220:223–226.

Bless, W., and Barz, W. (1988). Isolation of pterocarpan synthase, the terminal enzyme of pterocarpan phytoalexin biosynthesis in cell suspension cultures of *Cicer arietinum*, FEBS Lett., 235:47–50.

Blount, J. W., Dixon, R. A., and Paiva, N. L. (1992). Stress responses in alfalfa (*Medicago sativa* L.). XVI. Antifungal activity of medicarpin and its biosynthetic precursors; implications for the genetic manipulation of stress metabolites, Physiol. Mol. Plant Pathol., 41:333–349.

Blyden, E. R., Dixon, R. A., Lamb, C. J., Vriglandt, E., van Tunen, A. J., and Mol, J. N. M. (1989). Comparative molecular studies of chalcone isomerases, J. Cell Biochem. Suppl., 13d:274.

Bolwell, G. P., Bell, J. N., Cramer, C. L., Schuch, W., Lamb, C. J., and Dixon, R. A. (1985). L-phenylalanine ammonia-lyase from *Phaseolus vulgaris*. Characterisation and differential induction of multiple forms from elicitor-treated cell suspension cultures, Eur. J. Biochem., 149:411–419.

Bolwell, G. P., Cramer, C. L., Lamb, C. J., Schuch, W., and Dixon, R. A. (1986). L-phenylalanine ammonia-lyase from *Phaseolus vulgaris*. Modulation of the levels of active enzyme by trans-cinnamic acid, Planta, 169:97–107.

Bolwell, G. P., Mavandad, M., Millar, D. J., Edwards, K. J., Schuch, W., and Dixon, R. A. (1988). Inhibition of mRNA levels and activities by *trans*-cinnamic acid in elicitor-induced bean cells, Phytochemistry, 27:2109–2117.

Bolwell, G. P., Robbins, M. P., and Dixon, R. A. (1985a). Metabolic changes in elicitor-treated bean cells. Enzymic responses associated with rapid changes in cell wall components, Eur. J. Biochem., 148:571–578.

Bonhoff, A., Loyal, R., Ebel, J., and Grisebach, H. (1986a). Race: cultivar specific induction of enzymes related to phytoalexin biosynthesis in soybean roots following infection with *Phytophthora megasperma* f.sp. *glycinea*, Arch. Biochem. Biophys., 246:149–154.

Bonhoff, A., Loyal, R., Feller, K., Ebel, J., and Grisebach, H. (1986b). Further investigations of race: cultivar-specific induction of enzymes related to phytoalexin biosynthesis in soybean roots following infection with *Phytophthora megasperma* f.sp. *glycinea*, Biol. Chem. Hoppe. Seyler, 367:797–802.

Borner, H., and Grisebach, H. (1982). Enzyme induction in soybean infected by *Phytophthora megasperma* f.sp. *glycinea*, Arch. Biochem. Biophys., 217:65–71.

Brindle, P. A., Coolbear, T., Kuhn, P. J., and Threlfall, D. R. (1985). An $^{18}O_2$ study of the biosynthesis and metabolism of rishitin, Phytochemistry, 24:1219–1222.

Brindle, P. A., Kuhn, P. J., and Threlfall, D. R. (1983). Accumulation of phytoalexins in potato-cell suspension cultures, Phytochemistry, 22:2719–2721.

Brindle, P. A., and Threlfall, D. R. (1983). The metabolism of phytoalexins, Biochem. Soc. Trans., 11:516–522.

Brinker, A. M., and Seigler, D. S. (1991). Isolation and identification of piceatannol as a phytoalexin from sugarcane, Phytochemistry, 30:3229–3232.

Brinker, A. M., and Seigler, D. S. (1993). Time course of piceatannol accumulation in resistant and susceptible sugarcane stalks after inoculation with *Colletotrichum falcatum*, Physiol. Mol. Plant Pathol., 42:169–176.

Brown, S. A. (1985). Recent advances in the biosynthesis of coumarins, The Biochemistry of Plant Phenolics (C. F. Van Sumere and P. J. Lea, eds.), Clarendon Press, Oxford, pp. 257–270.

Bruce, R. J., and West, C. A. (1982). Elicitation of casbene synthetase activity in castor bean. The role of pectic fragments of the plant cell wall in elicitation by a fungal endopolygalacturonase, Plant Physiol., 69:181–188.

Burden, R. S., Bailey, J. A., and Dawson, G. W. (1972). Structures of three new isoflavnoids from *Phaseolus vulgaris* infected with tobacco necrosis virus, Tetrahedron Lett., 41:4175–4178.

Cahill, D. M., and Ward, E. W. B. (1989). Effects of metalaxyl on elicitor activity, stimulation of glyceollin production and growth of sensitive and tolerant isolates of *Phytophthora megasperma* f.sp. *glycinea*, Physiol. Mol. Plant Pathol., 35:97–112.

Cartwright, D. W., Langcake, P., Pryce, R. J., Leworthy, D. P., and Ride, J. P. (1981). Isolation and characterization of phytoalexins from rice as momilactones A and B, Phytochemistry, 20:535–537.

Chappell, J., and Hahlbrock, K. (1984). Transcription of plant defense genes in response to UV light or fungal elicitor, Nature, 311:76–78.

Chappell, J., and Nable, R. (1987). Induction of sesquiterpenoid biosynthesis in tobacco cell suspension cultures by fungal elicitor, Plant Physiol., 85:469–473.

Chappell, J., Vonlanken, C., and Vogeli, U. (1991). Elicitor-inducible 3-hydroxy-3-methylglutaryl coenzyme A reductase activity is required for sesquiterpene accumulation in tobacco cell suspension cultures, Plant Physiol., 97:693–698.

Choi, G. H., Richey, M. G., Turbek, C. S. M., and Smith, D. A. (1987). Comparative characteristics of a kievitone hydratase-producing isolate of *Fusarium solani* f.sp. *phaseoli* and of two strains deficient in this enzyme, Physiol. Mol. Plant Pathol., 30:215–224.

Choudhary, A. D., Kessmann, H., Lamb, C. J., and Dixon, R. A. (1990). Stress responses in alfalfa (*Medicago sativa* L.). IV. Expression of defense gene constructs in electroporated suspension cell protoplasts, Plant Cell Rep., 9:42–46.

Ciuffetti, L. M., Weltring, K. M., Turgeon, B. G., Yoder, O. C., and VanEtten, H. D. (1988). Transformation of *Nectria haematococca* with a gene for pisatin demethylating activity, and the role of pisatin detoxification in virulence, J. Cell Biochem., 12C:278.

Cleveland, T. E., and Smith, D. A. (1983). Partial purification, and further characterization, of kievitone hydratase from cell-free culture filtrates of *Fusarium solani* f.sp. *phaseoli*, Physiol. Plant Pathol., 22:129–142.

Collinge, D. B., and Slusarenko, A. J. (1987). Plant gene expression in response to pathogens, Plant Mol. Biol., 9:389–410.

Coolbear, T., and Threlfall, D. R. (1985). The biosynthesis of lubimin from (I-^{14}C) isopentenyl pyrophosphate by cell-free extracts of potato tuber tissue inoculated with an elicitor preparation from *Phytophthora infestans*, Phytochemistry, 24:1963–1971.

Coon, M. J., and Persson, A. V. (1980). Microsomal cytochrome P-450: A central catalyst in detoxication reactions, Enzymatic Basis of Detoxication Vol. I (W. B. Jakoby, ed.), Academic Press, New York, pp. 117–133.

Cowling, W. A., and VanEtten, H. D. (1986). Genetic evidence for a single functional deficiency in isolates of *Nectria haematococca* unable to demethylate pisatin, Can. J. Bot., 64:350–354.

Cramer, C. L., Bell, J. N., Ryder, T. B., Bailey, J. A., Schuch, W., Bolwell, G. P., Robbins, M. P., Dixon, R. A., and Lamb, C. J. (1985a). Co-ordinated synthesis of phytoalexin biosynthetic enzymes in biologically-stressed cells of bean (*Phaseolus vulgaris* L.), EMBO J., 4:285–289.

Cramer, C. L., Edwards, K., Dron, M., Liang, X., Dildine, S. L., Bolwell, G. P., Dixon, R. A., Lamb, C. J., and Schuch, W. (1989). Phenylalanine ammonia-lyase gene organization structure, Plant Mol. Biol., 12:367–383.

Cramer, C. L., Ryder, T. B., Bell, J. N., and Lamb, C. J. (1985b). Rapid switching of plant gene expression induced by fungal elicitor, Science, 227:1240–1243.

Cruickshank, I. A. M., and Perrin, D. R. (1963). Phytoalexins of the Leguminosae. Phaseollin from *Phaseolus vulgaris* L., Life Sciences, 2:680–682.

Cuypers, B., Schmelzer, E., and Hahlbrock, K. (1988). In situ localization of early accumulated phenylalanine ammonia-lyase mRNA around penetration site of *Phytophthora infestans* in potato leaves, Mol. Plant–Microbe Interact., 1:157–160.

Dalkin, K., Edwards, R., Edington, B., and Dixon, R. A. (1990a). Stress responses in alfalfa (*Medicago sativa* L.). I. Induction of phenylpropanoid biosynthesis and hydrolytic enzymes in elicitor-treated cell suspension cultures, Plant Physiol., 92:440–446.

Dalkin, K., Jorrin, J., and Dixon, R. A. (1990b). Stress responses in alfalfa (*Medicago sativa* L.). VII. Induction of defence related mRNAs in elicitor-treated cell suspension cultures, Physiol. Mol. Plant Pathol., 37:293–307.

Dangl, J. L., Hahlbrock, K., and Schell, J. (1989). Regulation of structure of chalcone synthase genes, Cell Culture and Somatic Cell Genetics of Plants, Vol. 6 (I. K. Vasil and J. Schell, eds.), Academic Press, San Diego, California, pp. 155–174.

Daniel, S., and Barz, W. (1990). Elicitor-induced metabolic changes in cell cultures of chickpea (*Cicer arietinum* L.) cultivars resistant and susceptible to *Ascochyta rabiei*. II. Differential induction of chalcone synthase-mRNA activity and analysis of in vitro translated protein patterns, Planta, 182:279–286.

Daniel, S., Hinderer, W., and Barz, W. (1988). Elicitor-induced changes of enzyme activities related to isoflavone and pterocarpan accumulation in chickpea (*Cicer arietinum* L.) cell suspension cultures, Z. Naturforsch. C. Biosci., 43C:536–544.

Delserone, L. M., and VanEtten, H. D. (1987). Demethylation of pisatin by three fungal pathogens of *Pisum sativum*, Phytopathology, 77:116.

Denny, T. P., and VanEtten, H. D. (1981). Tolerance by *Nectria haematococca* MP VI of the chickpea (*Cicer arietinum*) phytoalexins medicarpin and maackiain, Physiol. Plant Pathol., 19:419–437.

Denny, T. P., and VanEtten, H. D. (1982). Metabolisma of the phytoalexins medicarpin and maackiain by *Fusarium solani*, Phytochemistry, 21:1023–1028.

Dercks, W., and Creasy, L. L. (1989). The significance of stilbene phytoalexins in the *Plasmopara viticola*–grapevine interaction, Physiol. Mol. Plant Pathol., 34:189–202.

Desjardins, A. E., and Gardner, H. W. (1989). Genetic analysis in *Gibberella pulicaris*, Mol. Plant–Microbe Interact., 2:26–34.

Desjardins, A. E., Gardner, H. W., and Plattner, R. D. (1989a). Detoxification of the potato phytoalexin lubimin by *Gibberella pulicaris*, Phytochemistry, 28:432–437.

Desjardins, A. E., Matthews, D. E., and VanEtten, H. D. (1984). Solubilization and reconstitution of pisatin demethylase, a cytochrome P-450 from the pathogenic fungus *Nectria haematococca*, Plant Physiol., 75:611–616.

Desjardins, A. E., Plattner, R. D., and Spencer, G. F. (1988). Inhibition of trichothecene toxin biosynthesis by naturally occurring shikimate aromatics, Phytochemistry, 27:767–771.

Desjardins, A. E., Spencer, G. F., Plattner, R. D., and Beremand, M. N. (1989b). Furanocoumarin phytoalexins, trichothecene toxins, and infection of Pastinaca sativa by *Fusarium sporotrichoides*, Phytopathology, 79:170–175.

Desjardins, A. E., and VanEtten, H. D. (1986). Partial purification of pisatin demethylase, a cytochrome P-450 from the pathogenic fungus *Nectria haematococca*, Arch. Microbiol., 144:84–90.

Dewick, P. M. (1982). Isoflavonoids, The Flavonoids: Advances in Research (J. B. Harborne and T. J. Mabry, eds.), Chapman and Hall, London, pp. 535–640.

Dewick, P. M., Steele, M. J., Dixon, R. A., and Whitehead, I. M. (1982). Biosynthesis of isofavonoid phytoalexins: incorporation of sodium (1,2-13$_C$) acetate into phaseollin and kievitone, Z. Naturforschung, 37C:363–368.

DeWit, P. J. G. M., and Kodde, E. (1981). Induction of polyacetylenic phytoalexins in *Lycopersicon esculentum* after inoculation with *Cladosporium fulvum*, Physiol. Plant Pathol., 18:143–148.

Dhawale, S., Souciet, G., and Kuhn, D. N. (1989). Increase of chalcone synthase mRNA in pathogen-inoculated soybeans with race-specific resistance is different in leaves and roots, Plant Physiol., 91:911–916.

Dixon, R. A. (1986). The phytoalexin response: elicitation, signalling and control of host gene expression, Biol. Rev., 61:239–291.

Dixon, R. A., and Bendall, D. S. (1978). Changes in the levels of enzymes of phenylpropanoid and flavonoid synthesis during phaseollin production in cell suspension cultures of *Phaseolus vulgaris*, Physiol. Plant Pathol., 13:295–306.

Dixon, R. A., and Bolwell, G. P. (1986). Modulation of the phenylpropanoid pathway in bean (*Phaseolus vulgaris*) cell suspension cultures, Secondary Metabolism in Plant Cell Cultures (P. Morris, A. H. Scragg, A. Stafford, and M. W. Fowler, eds.), Cambridge University Press, London and New York, pp. 89–102.

Dixon, R. A., Bailey, J. A., Bell, J. N., Bolwell, G. P., Cramer, C. L., Edwards, K., Hamdan, M. A. M. S., Lamb, C. J., Robbins, M. P., Ryder, T. B., and Schuch, W. (1986). Rapid changes in gene expression in response to microbial elicitation, Phil. Trans. R. Soc. Lond. B., 314:411–426.

Dixon, R. A., Blyden, E. R., Robbins, M. P., van Tunen, A. J., and Mol, J. N. M. (1988). Comparative biochemistry of chalcone isomerases, Phytochemistry, 27:2801–2808.

Dixon, R. A., Browne, T., and Ward, M. (1980). Modulation of L-phenylalanine ammonia-lyase by pathway intermediates in cell suspension cultures of dwarf French bean (*Phaseolus vulgaris* L.), Planta, 150:279–285.

Dixon, R. A., Dey, P. M., and Lamb, C. J. (1983a). Phytoalexins: enzymology and molecular biology, Adv. Enzymol. Rel. Areas Mol. Biol., 55:1–136.

Dixon, R. A., Dey, P. M., Lawton, M. A., and Lamb, C. J. (1983b). Phytoalexin production in French bean. Intercellular transmission of elicitation in cell suspension cultures and hypocotyl sections of *Phaseolus vulgaris*, Plant Physiol., 71:251–256.

Dixon, R. A., Dey, P. M., Murphy, D. L., and Whitehead, I. M. (1981). Dose responses for *Colletotrichum lindemuthianum* elicitor-mediated enzyme induction in French bean cell suspension cultures, Planta, 151:272–280.

Dixon, R. A., Dey, P. M., and Whitehead, I. M. (1982). Purification and properties of chalcone isomerase from cell suspension cultures of *Phaseolus vulgaris*, Biochemica et Biophysica Acta, 715:25–33.

Dixon, R. A., Gerrish, C., Lamb, C. J., and Robbins, M. P. (1983c). Elicitor-mediated induction of chalcone isomerase in *Phaseolus vulgaris*, Planta, 159:561–569.

Dixon, R. A., and Harrison, M. J. (1990). Activation, structure and organization of genes involved in microbial defense in plants, Adv. Genet., 28:165–234.

Dixon, R. A., and Lamb, C. J. (1979). Stimulation of de novo synthesis of L-phenylalanine ammonia-lyase in relation to phytoalexin accumulation in *Colletotrichum lindemuthianum* elicitor-treated cell suspension cultures of French bean (*Phaseolus vulgaris*), Biochem. Biophys. Acta, 586:453–464.

Dixon, R. A., and Lamb, C. J. (1990). Molecular communications in interactions between plants and microbial pathogens, Annu. Rev. Plant Physiol. Plant Mol. Biol., 41:339–367.

Dmitriev, A. P., Tverskoy, L. A., Kozlovsky, A. G., and Grodzinsky, D. M. (1990). Phytoalexins from onion and their role in disease resistance, Physiol. Mol. Plant Pathol., 37:235–244.

Douglas, C., Hoffmann, H., Schluz, W., and Hahlbrock, K. (1987). Structure and elicitor or U.V.-light-stimulated expression of two 4-coumarate:CoA ligase genes in parsley, EMBO J., 6:1189–1195.

Dron, M., Clouse, S. D., Lawton, M. A., Dixon, R. A., and Lamb, C. J. (1988). Glutathione and fungal elicitor regulation of a plant defense gene promoter in electroporated protoplasts, Proc. Natl. Acad. Sci. USA, 85:6738–6742.

Dubernet, M., and Ribereau-Gayon, P. (1973). Les 'polyphenoloxydases' du raisin sain et du raisin parasite par *Botrytis cinerea*, CR Acad. Sci., Paris, Serie D, 277:975–978.

Dudley, M. W., Dueber, M. T., and West, C. A. (1986). Biosynthesis of the macrocyclic diterpene casbene in castor bean (*Ricinus communis* L.) seedlings. Changes in enzyme levels induced by fungal infection and intracellular localisation of the pathway, Plant Physiol., 81:335–342.

Dudley, M. W., Green, T. R., and West, C. A. (1985). Biosynthesis of the macrocyclic diterpene casbene in castor bean (*Ricinus communis* L.) seedlings. The purification and properties of farnesyltransferase from elicited seedlings, Plant Physiol., 81:343–348.

Dueber, M. T., Adolf, W., and West, C. A. (1978). Biosynthesis of the diterpene phytoalexin casbene. Partial purification and characterisation of casbene synthetase from *Ricinus communis*, Plant Physiol., 62:598–603.

Ebel, J. (1979). Elicitor-induced phytoalexin synthesis in soybean (*Glycine max*), Regulation of Secondary Product and Plant Hormone Metabolism (M. Luckner and K. Schreiber, eds.), Pergamon Press, Oxford, pp. 155–161.

Ebel, J. (1984). Induction of phytoalexin synthesis in plants following microbial infection or treatment with elicitors. Bioregulators: Chemistry and Uses, American Chemical Society, Washington, DC, USA, pp. 257–271.

Ebel, J. (1986). Phytoalexin synthesis: the biochemical analysis of the induction process, Annu. Rev. Phytopathol., 24:235–264.

Ebel, J., and Grisebach, H. (1988). Defense strategies of soybean against the fungus *Phytophthora megasperma* f.sp. *glycinea*: a molecular analysis, Trends Biol. Sci., 13:23–27.

Ebel, J., Schmidt, W. E., and Loyal, R. (1984). Phytoalexin synthesis in soybean cells: elicitor induction of phenylalanine ammonia-lyase and chalcone synthase mRNAs and correlation with phytoalexin accumulation, Arch. Biochem. Biophys., 232:240–248.

Edwards, K., Cramer, C. L., Bolwell, G. P., Dixon, R. A., Schuch, W., and Lamb, C. J. (1985). Rapid transient induction of phenylalanine ammonia-lyase mRNA in elicitor-treated bean cells, Proc. Natl. Acad. Sci. USA, 82:6731–6735.

Edwards, R., Mavandad, M., and Dixon, R. A. (1990). Metabolic fate of cinnamic acid in elicitor treated cell suspension cultures of *Phaseolus vulgaris*, Phytochemistry, 29:1867–1873.

Elgersma, L. M., and Liem, J. I. (1989). Accumulation of phytoalexins in susceptible and resistant near-isogenic lines of tomato infected with *Verticillium albo-atrum* or *Fusarium oxysporum* f.sp. *lycopersici*, Physiol. Mol. Plant Pathol., 34:545–555.

Ellis, J. S., Jennings, A. C., Edwards, L. A., Lamb, C. J., and Dixon, R. A. (1989). Defense gene expression in elicitor-treated cell suspension cultures of French bean cv. Imuna, Plant Cell Rep., 8:504–507.

Esnault, R., Chibbar, R. N., Lee, D., Van Huystee, R. B., and Ward, E. W. B. (1987). Early differences in production of mRNAs for phenylalanine ammonia-lyase and chalcone synthase in resistant and susceptible cultivars of soybean inoculated with *Phytophthora megasperma* f.sp. *glycinea*, Physiol. Mol. Plant Pathol., 30:293–297.

Estabrook, E. M., and Gopalan, S. (1991). Differential expression of phenylalanine ammonia-lyase and chalcone synthase during soybean nodule development, Plant Cell, 3:299–308.

Fliegmann, J., Schroder, G., Schanz, S., Britsch, L., and Schroder, J. (1992). Molecular analysis of chalcone and dihydropinosylvin synthase from Scots pine (*Pinus sylvestris*), and differential regulation of these and related enzyme activities in stressed plants, Plant Mol. Biol., 18:489–503.

Fritzemeier, K. H., and Kindl, H. (1981). Coordinate induction by UV light of stilbene synthase, phenylalanine ammonia-lyase and cinnamate 4-hydroxylase in leaves of vitaceae, Planta, 151:48–52.

Fritzemeier, K. H., and Kindl, H. (1983). 9,10-Dihydrophenanthrenes as phytoalexins of orchidaceae. Biosynthetic studies in vitro and in vivo providing the route from L-phenylalanine to dihydro-m-coumaric acid, dihydrostilbene and dihydrophenanthrenes, Eur. J. Biochem., 133:545–550.

Fritzemeier, K. H., Kindl, H., and Schlosser, E. (1984). Two different pathways leading to phenanthrenes and 9,10-dihydrophenanthrenes of the genus *Dioscorea*, Z. Naturforsch., 39C:217–221.

Fuchs, A., de Vries, F. W., and Platero Sanz, M. (1980). The mechanism of pisatin degradation by *Fusarium oxysporum* f.sp. *pisi*, Physiol. Plant Pathol., 16:119–133.

Fujita, M., and Asahi, T. (1985). Different intracellular localization of two cytochrome P-450 systems, ipomeamarone-15-hydroxylase and cinnamic acid 4-hydroxylase, in sweet potato root tissue infected with *Ceratocystis fimbriata*, Plant Cell Physiol., 26:389–395.

Gardner, H. W., Desjardins, A. E., Weisleder, D., and Plattner, R. A. (1988). Biotransformation of the potato phytoalexin, lubimin, by *Gibberella pulicaris*. Identification of major products, Biochem. Biophys. Acta, 966:347–356.

Gay, L. (1985). Phytoalexin formation in cell cultures of *Dianthus caryophyllus* treated by an extract from the culture medium of *Phytophthora parasitica*, Physiol. Plant Pathol., 26:143–150.

Gehlert, R., Schoppner, A., and Kindl, H. (1990). Stilbene synthase from seedlings of *Pinus sylvestris*: purification and induction in response to fungal infection, Mol. Plant–Microbe Interact., 3:444–449.

Gorham, J. (1989). Stilbenes and phenanthrenes, Methods in Plant Biochemistry, Vol. 1 (P. M. Dey and J. B. Harbone, eds.), Academic Press, London, pp. 159–196.

Gowri, G., Paiva, N. L., and Dixon, R. A. (1991). Stress responses in alfalfa (*Medicago sativa* L.). 12. Sequence analysis of phenylalanine ammonia-lyase (PAL) cDNA clones and appearance of PAL transcripts in elicitor-treated cell cultures and developing plants, Plant Mol. Biol., 17:415–429.

Grab, D., Loyal, R., and Ebel, J. (1985). Elicitor-induced phytoalexin synthesis in soybean cells. Changes in the activity of chalcone synthase mRNA and the total population of translatable mRNA, Arch. Biochem. Biophys., 243:523–529.

Graham, T. L., and Graham, M. Y. (1991). Cellular coordination of molecular responses in plant defense, Mol. Plant–Microbe Interact., 4:415–422.

Graham, T. L., Kim, J. E., and Graham, M. Y. (1990). Role of constitutive isoflavone conjugates in the accumulation of glyceollin in soybean infected with *Phytophthora megasperma*, Mol. Plant–Microbe Interact., 3:157–166.

Green, T. R., and West, C. A. (1974). Purification and characterization of two forms of geranyl transferase from *Ricinus communis*, Biochemistry, 13:4720–4729.

Grisebach, H., and Ebel, J. (1978). Phytoalexins, chemical defense substances of higher plants? Angew. Chem. Int. Ed. Engl., 17:635–647.

Habereder, H., Schroder, G., and Ebel, J. (1989). Rapid induction of phenylalanine ammonia-lyase and chalcone synthase mRNAs during fungus infection of soybean (*Glycine max*) roots or elicitor treatment of soybean cell cultures at the onset of phytoalexin synthesis, Planta, 177:58–65.

Hagmann, M., and Grisebach, H. (1984). Enzymatic rearrangement of flavonone to isoflavone, FEBS Lett., 175:199–202.

Hagmann, M. L., Heller, W., and Grisebach, H. (1984). Induction of phytoalexin synthesis in soybean. Stereospecific 3,9-dihydroxypterocarpan 6a-hydroxylase from elicitor-induced soybean cell cultures, Eur. J. Biochem., 142:127–131.

Hahlbrock, K., Boudet, A. M., Chappell, J., Kreuzaler, F., Kuhn, D. N., and Ragg, H. (1983). Differential induction of mRNAs by light and elicitor in cultured plant cells,

Structure and Function of Plant Genomes, (O. Cifferi and L. Dure, eds.), NATO ASI Series, Chapman and Hall, London, pp. 15–23.

Hahlbrock, K., Chappell, J., and Kuhn, D. N. (1984). Rapid induction of mRNAs involved in defense reactions in plants, The Genetic Manipulation of Plants and its Application to Agriculture (P. J. Lea and G. R. Stewart, eds.), Clarendon Press, Oxford, pp. 171–182.

Hahlbrock, K., Cuypers, B., Douglas, C., Fritzemeier, K. H., Hoffman, H., Rohwer, F., Scheel, D., and Schluz, W. (1986). Biochemical interactions of plants with potentially pathogenic fungi, Recognition in Microbe–Plant Symbiotic Interactions (B. Lugtenberg, ed.), Springer-Verlag, Berlin, pp. 311–323.

Hahlbrock, K., and Grisebach, H. (1975). Biosynthesis of flavonoids, The Flavonoids (J. B. Harborne, T. J. Mabry and H. Mabry, eds.), Chapman and Hall, London, pp. 866–915.

Hahlbrock, K., and Grisebach, H. (1979). Enzymic controls in the biosynthesis of lignin and flavonoids, Annu. Rev. Plant Physiol., 30:105–130.

Hahlbrock, K., Lamb, C. J., Purwin, C., Ebel, J., Fautz, E., and Schafer, E. (1981). Rapid response of suspension-cultured parsley cells to the elicitor from *Phytophthora megasperma* var. *sojae*. Induction of the enzymes of general phenylpropanoid metabolism, Plant Physiol., 67:768–773.

Hahlbrock, K., and Scheel, D. (1989). Physiology and molecular biology of phenylpropanoid metabolism, Annu. Rev. Plant Physiol. Plant Mol. Biol., 40:347–369.

Hahlbrock, K., Schroder, J., and Vieregge, J. (1980). Enzyme regulation in parsley and soybean cell cultures, Adv. Biochem. Engin., 18:39–60.

Hahn, M. G., Bonhoff, A., and Grisebach, H. (1985). Quantitative localization of the phytoalexin glyceollin I in relation to fungal hyphae in soybean roots infected with *Phytophthora megasperma* f.sp. *glycinea*, Plant Physiol., 77:591–601.

Hain, R., Bieseler, B., Kindl, H., Schroder, G., and Stocker, R. (1990). Expression of a stilbene synthase gene in *Nicotiana tabacum* results in synthesis of the phytoalexin resveratrol, Plant Mol. Biol., 15:325–335.

Hamdan, M. A. M. S., and Dixon, R. A. (1987). Differential patterns of protein synthesis in bean cells exposed to elicitor fractions from *Colletotrichum lindemuthianum*, Physiol. Mol. Plant Pathol., 31:105–121.

Hamerski, D., and Matern, U. (1988). Elicitor-induced biosynthesis of psoralens in *Ammi majus* L. suspension cultures. Microsomal conversion of demethylsuberosin into (+) marmesin and psoralen, Eur. J. Biochem., 171:369–375.

Hargreaves, J. A., Mansfield, J. W., and Rossall, S. (1977). Changes in phytoalexin concentrations in tissues of the broad bean plant (*Vicia faba* L.), following inoculation with species of *Botrytis*, Physiol. Plant Pathol., 11:227–242.

Harker, C. L., Ellis, T. H. N., and Coen, E. S. (1990). Identification and genetic regulation of the chalcone synthase multigene family in pea, Plant Cell, 2:185–194.

Harrison, M. J., Choudhary, A. D., Kooter, J., Lamb, C. J., and Dixon, R. A. (1991a). Stress-responses in alfalfa (*Medicago sativa* L.). 8. *Cis*-elements and *trans*-acting factors for the quantitative expression of a bean chalcone synthase gene promoter in electroporated alfalfa protoplasts, Plant Mol. Biol., 16:877–890.

Harrison, M. J., Lawton, M. W., Lamb, C. J., and Dixon, R. A. (1990). Effects of trans-cinnamic acid on expression of the bean phenylalanine ammonia-lyase gene family, Plant Physiol., 94:671–680.

Harrison, M. J., Lawton, M. A., Lamb, C. J., and Dixon, R. A. (1991b). Characterization of a nuclear protein that binds to three elements within the silencer region of a bean chalcone synthase gene, Proc. Natl. Acad. Sci. USA, 88:2515–2519.

Hart, J. M. (1981). Role of phytostilbenes in decay and disease resistance, Annu. Rev. Phytopathol., 19:437–458.

Hauffe, K. D., Hahlbrock, K., and Scheel, D. (1986). Elicitor-stimulated furanocoumarin biosynthesis in cultured parsley cells: S-adenosyl L-methionine: bergaptol and S-adenosyl-L-methionine: xanthotoxol 0-methyltransferases, Z. Naturforsch., 41C:228–239.

Heinstein, P. F., Herman, D. L., Tove, S. B., and Smith, F. H. (1970). Biosynthesis of gossypol. Incorporation of mevalonate-2-^{14}C and isoprenyl pyrophopshates, J. Biol. Chem., 245:4658–4665.

Heller, W., and Forkmann, G. (1988). Biosynthesis, The Flavonoids (J. B. Harbone, ed.), Chapman and Hall Press, London, pp. 399–425.

Heller, W., and Hahlbrock, K. (1980). Highly purified "flavanone synthase" from parsley catalyzes the formation of naringenin chalcone, Arch. Biochem. Biophys., 200:617–619.

Henstrand, J. M., McCue, K. F., Brink, K., Handa, A. K., Herrmann, K. M., and Conn, E. E. (1992). Light and fungal elicitor induce 3-Deoxy-D-arabino-heptulosonate 7-phosphate synthase mRNA in suspension cultured cells of parsley (*Petroselinum crispum* L.), Plant Physiol., 98:761–763.

Herrmann, A., Schluz, W., and Hahlbrock, K. (1988). Two alleles of the single-copy chalcone synthase gene in parsley differ by a transposon-like element, Mol. Gen. Genet., 212:93–98.

Higgins, V. J. (1978). The effect of some pterocarpanoid phytoalexins on germtube elongation of *Stemphylium botryosum*, Phytopathology, 68:339–345.

Hille, A., Purwin, C., and Ebel, J. (1982). Induction of enzymes of phytoalexin synthesis in cultured soybean cells by an elicitor from *Phytophthora megasperma* f.sp. *glycinea*, Plant Cell Rep., 1:123–127.

Hipskind, J. D., Hanau, R., Leite, B., and Nicholson, R. L. (1990). Phytoalexin synthesis in sorghum identification of an apigeninidin acyl ester, Physiol. Mol. Plant Pathol., 36:381–396.

Hiramatsu, I., Ichinose, Y., Shiraishi, T., Oku, H., and Ouchi, S. (1986). Regulation of pisatin biosynthesis in pea leaves by elicitor and suppressor produced by *Mycosphaerella pinodes*, Ann. Phytopathol. Soc. Japan., 52:53–58.

Hodson, M. J., Bar-Nun, N., and Mayer, A. M. (1987). Cytochemical localization of laccase in *Botrytis cinerea*, Trans. Br. Mycol. Soc., 89:572–575.

Hohl, B., and Barz, W. (1987). Partial characterization of an enzyme from the fungus *Ascochyta rabiei* for the reductive cleavage of pterocarpan phytoalexins to 2′-hydroxy-isoflavans, Z. Naturforsch., 42C:897–901.

Hohn, T. M., and Ohlrogge, J. B. (1991). Expression of a fungal sesquiterpene cyclase gene in transgenic tobacco, Plant Physiol., 97:460–462.

Hollander, H., and Amrhein, N. (1980). The site of inhibition of the shikimate pathway by glyphosate. I. Inhibition by glyphosate of phenylpropanoid synthesis in buckwheat (*Fagopyrum esculentum* Moench), Plant Physiol., 66:823–829.

Holliday, M. J., and Keen, N. T. (1982). The role of phytoalexins in the resistance of soybean leaves to bacteria: effect of glyphosate on glyceollin accumulation, Phytopathology, 72:1470–1474.

Hoos, G., and Blaich, R. (1988). Metabolism of stilbene phytoalexins in grapevines: oxidation of resveratrol in single-cell cultures, Vitis, 27:1–12.

Hoos, G., and Blaich, R. (1990). Influence of resveratrol on germination of conidia and mycelial growth of *Botrytis cinerea* and *Phomopsis viticola*, J. Phytopathol., 129:102–110.

Ingham, J. L. (1982). Phytoalexins from the Leguminosae, Phytoalexins (J. A. Bailey and J. W. Mansfield, eds.), Blackie Press, Glasgow, pp. 21–80.

Inoue, H., Oba, K., Ando, M., and Uritani, I. (1984). Enzymatic reduction of dehydroipomeamarone to ipomeamarone in sweet potato root tissue infected by *Ceratocystis fimbriata*, Physiol. Plant Pathol., 25:1–8.

Ito, R., Oba, K., and Uritani, I. (1979). Mechanism for the induction of 3-hydroxy-3-methylglutaryl coenzyme A reductase in mercury (II) chloride-treated sweet potato root tissue, Plant Cell Physiol., 20:867–874.

Iwakuma, T., Ataka, Y., Matsuyama, N., and Wakimoto, S. (1990). Phytoalexin production elicited by *Pyricularia oryzae* infection and its hyphal wall-component treatment in rice leaves, Ann. Phytopathol. Soc. Japan., 56:665–670.

Jahnen, W., and Hahlbrock, K. (1988a). Cellular localization of nonhost resistance reactions of parsley (*Petroselinum crispum*) to fungal infection, Planta, 173:197–204.

Jahnen, W., and Hahlbrock, K. (1988b). Differential regulation and tissue-specific distribution of enzymes related to phenylpropanoid branch pathways in developing parsley seedlings, Planta, 173:453–458.

Johnson, C., Brannon, D. R., and Kuc, J. (1973). Xanthotoxin: a phytoalexin of *Pastinaca sativa* root, Phytochemistry, 12:2961–2962.

Jorrin, J., and Dixon, R. A. (1990). Stress responses in alfalfa (*Medicago sativa* L.). II. Purification, characterization, and induction of phenylalanine ammonia-lyase isoforms from elicitor-treated cell suspension cultures, Plant Physiol., 92:447–455.

Junghans, H., Dalkin, K., and Dixon, R. A. (1993). Stress responses in alfalfa (*Medicago sativa* L.). 15. Characterization and expression patterns of members of a subset of the chalcone synthase multigene family, Plant Mol. Biol., 22:239–253.

Kawamata, S., Yamada, T., Tanaka, Y., Sriprasertsak, P., Kato, H., Ichinose, Y., Shiraishi, T., and Oku, H. (1992). Molecular cloning of phenylalanine ammonia-lyase cDNA from *Pisum sativum*, Plant Mol. Biol., 20:167–170.

Keen, N. T. (1986). Phytoalexins and their involvement in plant disease resistance, Iowa State J. Res., 60:477–499.

Keen, N. T., Holliday, M. J., and Yoshikawa, M. (1982). Effects of glyphosate on glyceollin production and the expression of resistance to *Phytophthora megasperma* f.sp. *glycinea* in soybean, Phytopathology, 72:1467–1470.

Kendra, D. F., and Hadwiger, L. A. (1987). Calcium and calmodulin may not regulate the disease resistance and pisatin formation responses of *Pisum sativum* to chitosan or *Fusarium solani*, Physiol. Mol. Plant Pathol., 31:337–348.

Kessmann, H., and Barz, W. (1986). Elicitation and suppression of phytoalexin and

isoflavone accumulation in cotyledons of *Cicer arietinum* L. as caused by wounding and by polymeric components from the fungus, *Ascochyta rabiei*, J. Phytopathol., 117:321–335.
Kessmann, H., Choudhary, A. D., and Dixon, R. A. (1990a). Stress responses in alfalfa (*Medicago sativa* L.). III. Induction of medicarpin and cytochrome P 450 enzyme activities in elicitor-treated cell suspension cultures and protoplasts, Plant Cell Rep., 9:38–41.
Kessmann, H., Edwards, R., Geno, P. W., and Dixon, R. A. (1990b). Stress responses in alfalfa (*Medicago sativa* L.). V. Constitutive and elicitor-induced accumulation of isoflavonoid conjugates in cell suspension cultures, Plant Physiol., 94:227–232.
Khan, F. Z., and Milton, J. M. (1975). Phytoalexin production by lucerne (*Medicago sativa* L.) in response to infection by *Verticillium*, Physiol. Plant Pathol., 7:179–187.
Kimpel, J. A., and Kosuge, T. (1985). Metabolic regulation during glyceollin biosynthesis in green soybean hypocotyls, Plant Physiol., 77:1–7.
Kindl, H. (1985). Biosynthesis of stilbenes, Biosynthesis and Biodegradation of Wood Components (T. Higuchi, ed.), Academic Press, New York, pp. 349–377.
Kishore, G. M., and Somerville, C. R. (1993). Genetic engineering of commercially useful biosynthetic pathways in transgenic plants, Curr. Opin. Biotechnol., 4:152–158.
Kistler, H. C., and VanEtten, H. D. (1981). Phaseollin metabolism and tolerance in *Fusarium solani* f.sp. *phaseoli*, Physiol. Plant Pathol., 19:257–271.
Kistler, H. C., and VanEtten, H. D. (1984a). Regulation of pisatin demethylation in *Nectria haematococca* and its influence on pisatin tolerance and virulence, J. Gen. Microbiol., 130:2605–2613.
Kistler, H. C., and VanEtten, H. D. (1984b). Three non-allelic genes for pisatin demethylation in the fungus *Nectria haematococca*, J. Gen. Microbiol., 130:2595–2603.
Knogge, W., Kombrink, E., Schmelzer, E., and Hahlbrock, K. (1987). Occurrence of phytoalexins and other putative defense-related substances in uninfected parsley plants, Planta, 174:279–287.
Kochs, G., and Grisebach, H. (1986). Enzymic synthesis of isoflavones, Eur. J. Biochem., 155:311–318.
Kodama, O., Suzuki, T., Miyakawa, J., and Akatsuka, T. (1988). Ultraviolet-induced accumulation of phytoalexins in rice leaves, Agric. Biol. Chem., 52:2469–2473.
Kombrink, E., Bollmann, J., Hauffe, K. D., Knogge, W., Scheel, D., Schmelzer, E., Somssich, I., and Hahlbrock,, K. (1986). Biochemical responses of non-host plant cells to fungi and fungal elicitors, Biology and Molecular Biology of Plant–Pathogen Interactions (J. A. Bailey, ed.), Springer-Verlag, Berlin, pp. 253–262.
Kono, Y., Takeuchi, S., Kodama, O., and Akatsuka, T. (1984). Absolute configuration of oryzalexin A and structures of its related phytoalexins isolated from rice blast leaves infected with *Pyricularia oryzae*, Agric. Biol. Chem., 48:253–255.
Kraft, B., Schwenen, L., Stockl, D., and Barz, W. (1987). Degradation of the pterocarpan phytoalexin medicarpin by *Ascochyta rabiei*, Arch. Microbiol., 147:201–206.
Kruezaler, F., and Hahlbrock, K. (1973). Flavonoid glycosides from illuminated cell suspension cultures of *Petroselinum hartense*, Phytochemistry, 12:1149–1152.
Kruezaler, F., Light, R. J., and Hahlbrock, K. (1978). Flavonone synthase catalyzes CO_2 exchange and decarboxylation of maonyl-CoA, FEBS Lett., 94:175–178.
Kuhn, D., Chappell, J., Boudet, A., and Hahlbrock, K. (1984). Induction of phenylalanine

ammonia-lyase and 4-coumarate: CoA ligase mRNAs in cultured plant cells by UV light or fungal elicitor, Proc. Natl. Acad. Sci. USA, 81:1102–1106.

Kuhn, P. J., and Smith, D. A. (1978). Detoxification of the phytoalexin, kievitone, by *Fusarium solani* f.sp. *phaseoli*, Ann. Appl. Biol., 89:362–366.

Kuhn, P. J., and Smith, D. A. (1979). Isolation from *Fusarium solani* f.sp. *phaseoli* of an enzymic system responsible for kievitone and phaseollidin detoxification, Physiol. Plant Pathol., 14:179–190.

Kuhn, P. J., Smith, D. A., and Ewing, D. F. (1977). 5,7,2′,4′-Tetrahyroxy-8-(3″-hydroxy-3″-methyl-butyl) isoflavanone, a metabolite of kievitone produced by *Fusarium solani* f.sp. *phaseoli*, Phytochemistry, 16:296–297.

Lamb, C. J., Corbin, D. R., Lawton, M. A., Sauer, N., and Wingate, V. P. M. (1986). Recognition and response in plants: pathogen interactions, Recognition in Microbe–Plant Symbiotic Interactions, (B. Lugtenberg, ed.), Springer-Verlag, Berlin, pp. 333–344.

Langcake, P., Cornford, C. A., and Pryce, R. J. (1979). Identification of pterostilbene as a phytoalexin from *Vitis vinifera* leaves, Phytochemistry, 18:1025–1027.

Langcake, P., McCarthy, W. V. (1979). The relationship of resveratrol production to infection of grapevine leaves by *Botrytis cinerea*, Vitis, 18:244–253.

Lanz, T., Schroder, G., and Schroder, J. (1990). Differential regulation of genes for resveratrol synthase in cell cultures of *Arachis hypogaea*, Planta, 181:169–175.

Lanz, T., Tropf, S., Marner, F. J., Schroder, J., and Schroder, G. (1991). The role of cysteines in polyketide synthases; site-directed mutagenesis of resveratrol and chalcone synthases, two key enzymes in different plant-specific pathways, J. Biol. Chem., 266:9971–9976.

Lappe, U., and Barz, W. (1978). Degradation of pisatin by fungi of the genus, Fusarium, Z. Naturforsch., 33C:301–302.

Latunde-Dada, A. O., and Lucas, J. A. (1985). Involvement of the phytoalexin medicarpin in the differential response of callus lines of lucerne (*Medicago sativa*) to infection by *Verticillium albo-atrum*, Physiol. Plant Pathol., 26:31–42.

Lawton, M. A., Dean, S. M., Dron, M., Kooter, J. M., Kragh, K., Harrison, M. J., Yu, L., Tanguay, L., Dixon, R. A., and Lamb, C. J. (1991). Silencer region of a chalcone synthase promoter contains multiple binding sites for a factor, SBF-1 closely related to GT-1, Plant Mol. Biol., 16:235–240.

Lawton, M. A., Dixon, R. A., Hahlbrock, K., and Lamb, C. J. (1983a). Rapid induction of phenylalanine ammonia-lyase and of chalcone synthase synthesis in elicitor-treated plant cells, Eur. J. Biochem., 129:593–601.

Lawton, M. A., Dixon, R. A., Hahlbrock, K., and Lamb, C. J. (1983b). Elicitor induction of mRNA activity: rapid effects of elicitor on phenylalanine ammonia-lyase and chalcone synthase mRNA activities in bean cells, Eur. J. Biochem., 130:131–139.

Lawton, M. A., Dixon, R. A., and Lamb, C. J. (1980). Elicitor modulation of the turnover of L-phenylalanine ammonia-lyase in French bean cell suspension cultures, Biochem. Biophys. Acta, 633:162–175.

Lawton, M. A., and Lamb, C. J. (1987). Transcriptional activation of plant defense gene by fungal elicitor, wounding and infection, Mol. Cell Biol., 7:335–341.

Lazarovits, G., and Ward, E. W. B. (1982). Relationship between localized glyceollin accumulation and metalaxyl treatment in the control of *Phytophthora* rot in soybean hypocotyls, Phytopathology, 72:1217–1221.

Lee, S. C., and West, C. A. (1981). Polygalacturonase from *Rhizopus stolonifer*, an elicitor of casbene synthetase activity in castor bean (*Ricinus communis* L.) seedlings, Plant Physiol., 67:633–639.

Leube, J., and Grisebach, H. (1983). Further studies on induction of enzymes of phytoalexin synthesis in soybean and cultured soybean cells, Z. Naturforsch. C. Biosci., 38:730–735.

Liang, X., Dron, M., Schmid, J., Dixon, R. A., and Lamb, C. J. (1989b). Developmental and environmental regulation of a phenylalanine ammonia-lyase-β-glucuronidase gene fusion in transgenic tobacco plants, Proc. Natl. Acad. Sci. USA, 86:9284–9288.

Liang, X, Dron, M., Cramer, C. L., Dixon, R. A., and Lamb, C. J. (1989a). Differential regulation of phenylalanine ammonia-lyase genes during plant development and by environmental cues, J. Biol. Chem., 264:14486–14492.

Loake, G. J., Choudhary, A. D., Harrison, M. J., Lamb, C. J., and Dixon, R. A. (1991). Phenylpropanoid pathway intermediates regulate transient expression of a chalcone synthase gene promoter, Plant Cell, 3:829–840.

Loake, G. J., Faktor, O., Lamb, C. J., and Dixon, R. A. (1992). Combination of H-box (CCTACC(N)$_7$CT) and G-box (CACGTG) *cis* elements is necessary for feed-forward stimulation of a chalcone synthase promoter by the phenylpropanoid-pathway intermediate p-coumaric acid, Proc. Natl. Acad. Sci. USA, 89:9230–9234.

Lois, R., Dietrich, A., and Hahlbrock, K. (1989a). A phenylalanine ammonia-lyase gene from parsley: Structure, regulation and identification of elicitor and abiotic phytoalexin inducers, Physiol. Plant Pathol., 23:163–173.

Lois, R., Dietrich, A., Hahlbrock, K., and Schulz, W. (1989b). A phenylalanine ammonia-lyase gene from parsley: structure, regulation, and identification of elicitor and light-responsive *cis*-acting elements, EMBO J., 8:1641–1648.

Lois, A. F., and West, C. A. (1990). Regulation of expression of the casbene synthetase gene during elicitation of castor bean seedlings with pectic fragments, Arch. Biochem. Biophys., 276:270–277.

Loschke, D. C., Hadwiger, L. A., Schroder, J., and Hahlbrock, K. (1981). Effects of light and of *Fusarium solani* on synthesis and activity of phenylalanine ammonia-lyase in peas, Plant Physiol., 68:680–685.

Lozoya, E., Hoffmann, H., Douglas, C., Schulz, W., Scheel, D., and Hahlbrock, K. (1988). Primary structures and catalytic properties of isoenzymes encoded by the two 4-coumarate: CoA ligase genes in parsley, Eur. J. Biochem., 176:661–667.

Lucy, M. C., Matthews, P. S., and VanEtten, H. D. (1988). Metabolic detoxification of the phytoalexins maackiain and medicarpin by *Nectria haematococca* field isolates: relationship to virulence on chickpea, Physiol. Mol. Plant Pathol., 33:187–199.

Lue, W. L., Kuhn, D., and Nicholson, R. L. (1989). Chalcone synthase activity in sorghum mesocotyls inoculated with *Colletotrichum graminicola*, Physiol. Mol. Plant Pathol., 35:413–422.

Lyne, R. L., and Mulheirn, L. J. (1978). Minor pterocarpinoids of soybean, Tetrahedon Lett., 34:3127–3128.

MacFoy, C. A., and Smith, I. M. (1982). Degradation of maackiain by a cell-free extract from *Sclerotinia trifoliorum*, Plant Pathol., 31:195–199.

Mackintosh, S. F., Matthews, D. E., and VanEtten, H. D. (1989). Two additional genes for pisatin demethylation and their relationship to the pathogenicity of *Nectria haematococca* on pea, Mol. Plant–Microbe Interact., 2:354–362.

Manka, M., Visconti, A., Chelkowski, J., and Bottalico, A. (1985). Pathogenicity of *Fusarium* isolates from wheat, rye and triticale towards seedlings and their ability to produce trichothecenes and zearalenone, Phytopathol. Z., 113:24–29.

Mansfield, J. W. (1980). Mechanisms of resistance to *Botrytis*, Biology of *Botrytis* (J. R. Coley-Smith, K. Verhoeff and W. R. Jarvis, eds.), Academic Press, London, New York, pp. 181–218.

Marbach, I., Harel, E., and Mayer, A. M. (1984). Molecular properties of extracellular *Botrytis cinerea* laccase, Phytochemistry, 23:2713–2717.

Marbach, I., Harel, E., and Mayer, A. M (1985). Pectin, a second inducer for laccase production by *Botrytis cinerea*, Phytochemistry, 24:2559–2561.

Matsuyama, N., and Wakimoto, S. (1985). Purification and characterization of anti-blast substance, S-1, formed mainly in blast-resistant lower rice leaves, Ann. Phytopathol. Soc. Jpn., 51:498–500.

Matsuyama, N., and Wakimoto, S. (1988). Isolation and identification of diterpenoid antiblast substance produced in the blast-infected rice leaves, Nippon Shokubutsu Bjori Gakkaiho, 54:183–188.

Matthews, D. E., and VanEtten, H. D. (1983). Detoxification of the phytoalexin pisatin by a fungal cytochrome P-450, Arch. Biochem. Biophys., 224:494–505.

Mavandad, M., Edwards, R., Liang, X., Lamb, C. J., and Dixon, R. A. (1990). Effects of *trans*-cinnamic acid on expression of the bean phenylalanine ammonia-lyase gene family, Plant Physiol., 94:671–680.

Mayama, S., Tani, T., Matsuura, Y., Ueno, T., and Fukami, H. (1981). The production of phytoalexins by oat in response to crown rust, *Puccinia coronata* f.sp. *avenae*, Physiol. Plant Pathol., 19:217–226.

McCue, K., and Conn, E. E. (1989). Induction of 3-deoxy-D-arabino-heptulosonate 7-phosphate synthase activity by fungal elicitor in cultures of *Petroselinum crispum*, Proc. Natl. Acad. Sci. USA, 86:7374–7377.

Mehdy, M. C., and Lamb, C. J. (1987). Chalcone isomerase cDNA cloning and mRNA induction by fungal elicitor, wounding and infection, EMBO J., 6:1527–1533.

Melchior, F., and Kindl, H. (1990). Grapevine stilbene synthase cDNA only slightly differing from chalcone synthase cDNA is expressed in *Escherichia coli* into a catalytically active enzyme, FEBS Lett., 268:17–20.

Miao, V., Lucy, M., and VanEtten, H. D. (1986). Detoxification of the phytoalexin maackiain is important for virulence of *Nectria haematococca* on chickpea, J. Cell Biochem., 10C:25.

Miao, V., and VanEtten, H. D. (1987). Inheritance of maackiain metabolizing ability in *Nectria haematococca*, Can. J. Plant Pathol., 9:283.

Miao, V. P. W., and VanEtten, H. D. (1992). Three genes for metabolism of the phytoalexin maackiain in the plant pathogen *Nectria haematococca*: meiotic instability and relationship to a new gene for pisatin demethylase, Appl. Environ. Microbiol., 58:801–808.

Miller, J. D., Young, C. J., and Sampson, D. R. (1985). Deoxynivalenol and fusarium head blight resistance in spring cereals, Phytopathol. Z., 113:359–367.

Moesta, P., and Grisebach, H. (1982). L-2-Aminooxy-3-phenyl-propionic acid inhibits phytoalexin accumulation in soybean with concomitant loss of resistance against *Phytophthora megasperma* f.sp. *glycinea*, Physiol. Plant Pathol., 21:65–70.

Moesta, P., and West, C. A. (1985). Casbene synthetase: regulation of phytoalexin biosynthesis in *Ricinus communis* L. seedlings. Purification of casbene synthetase and regulation of its biosynthesis during elicitation, Arch. Biochem. Biophys., 238:325–333.

Morris, P. F., Savard, M. E., and Ward, E. W. B. (1991). Identification and accumulation of isoflavonoids and isoflavone glucosides in soybean leaves and hypocotyls in resistance responses to *Phytophthora megasperma* f.sp. *glycinea*, Physiol. Mol. Plant Pathol., 39:229–244.

Musumeci, M. R., and Oliveira, A. R. (1975). Accumulation of phenols and phytoalexins in citrus tissues inoculated with *P. citrophthora* (Sm. & Sm.) Leonian, Summa Phytopathol., 1:275–282.

Nandi, D. L., and Porter, J. W. (1964). The enzymatic synthesis of geranylgeranyl pyrophosphate by enzymes of carrot root and pig liver, Arch. Biochem. Biophys., 105:7–19.

Nemestothy, G. S., and Guest, D. I. (1990). Phytoalexin accumulation, phenylalanine ammonia-lyase activity and ethylene biosynthesis in fosetyl-Al treated resistant and susceptible tobacco cultivars infected with *Phytophthora nicotinae* var. *nicotinae*, Physiol. Mol. Plant Pathol., 37:207–219.

Nicholson, R. L., Jamil, F. F., Snyder, B. A., Lue, W. L., and Hipskind, J. (1988). Phytoalexin synthesis in the juvenile sorghum leaf, Physiol. Mol. Plant Pathol., 33:271–278.

Nicholson, R. L., Kollipara, S. S., Vincent, J. R., Lyon, P. C., and Cadena-Gomez, G. (1987). Phytoalexin synthesis in the sorghum mesocotyl in response to infection by pathogenic and nonpathogenic fungi, Proc. Natl. Acad. Sci. USA, 84:5520–5524.

Niesbach-Klosgen, U., Barzen, E., Bernhardt, J., Rhode, W., Schwarz-Sommer, Z., Reif, H. J., Wienand, U., and Saedler, H. (1987). Chalcone synthase genes in plants: a tool to study evolutionary relationships, J. Mol. Evol., 26:213–225.

Nonaka, F. (1967). Inactivation of pisatin by pathogenic fungi, Agric. Bull. Saga Univ., 24:109–121.

Oba, K., Kondo, K., Doke, N., and Uritani, I. (1985). Induction of 3-hydroxy-3-methyl-glutaryl CoA reductase in potato tuber after slicing, fungal infection or chemical treatment, and some properties of the enzyme, Plant Cell Physiol., 26:873–880.

Oba, K., Tatematsu, H., Yamashita, G., and Uritani, I. (1976). Induction of furano-terpene production and formation of the enzyme system from mevalonate to isopentenyl pyrophosphate in sweet potato root tissue injured by *Ceratocystis fimbriata* and by toxic chemicals, Plant Physiol., 58:51–56.

Oeser, B., and Yoder, O. C. (1994). Pathogenesis by *Cochliobolus heterostrophus* transformants expressing a cutinase-encoding gene from *Nectria haematococca*, Mol. Plant–Microbe Interact., 7:282–288.

Ogura, K., Nishino, T., and Seto, S. (1968). The purification of prenyltransferase and isopentenyl pyrophosphate isomerase of pumpkin fruit and some of their properties, J. Biochem., 64:197–203.

Ogura, K., Shinka, T., and Seto, S. (1972). Purification and properties of geranylgeranyl pyrophosphate synthetase from pumpkin fruit, J. Biochem., 72:1101–1108.

Oku, H., Ouchi, S., Shiraishi, T., and Baba, T. (1975a). Pisatin production in powdery mildewed pea seedlings, Phytopathology, 65:1263–1267.

Oku, H., Ouchi, S., Shiraishi, T., Komoto, M., and Oki, K. (1975b). Phytoalexin activity in barley powdery mildew, Ann. Phytopathol. Soc. Japan., 41:185–191.

Oku, H., Shiraishi, T., and Ouchi, S. (1987). Role of specific suppressors in pathogenesis of *Mycosphaerella* species, Molecular Determinants of Plant Diseases, (S. Nishimura, C. P. Vance, and N. Doke, eds.), Japan Scientific Society Press, Tokyo; Springer-Verlag, Berlin, pp. 145–156.

Oku, H., Shiraishi, T., Ouchi, S., and Ishiura, M. (1980). A new determinant of pathogenicity in plant disease, Naturwissenchaften, 67:310.

Oshima, K., and Uritani, I. (1968). Enzymatic synthesis of a β-hydroxy-3-methylglutaric acid derivative by a cell-free system from sweet potato with black rot, J. Biochem., 63:617–625.

Paiva, N. L., Edwards, R., Sun, Y., Hrazdina, G., and Dixon, R. A. (1991). Stress responses in alfalfa (*Medicago sativa* L.). XI. Molecular cloning and expression of alfalfa isoflavone reductase, a key enzyme of isoflavonoid phytoalexin biosynthesis, Plant Mol. Biol., 17:653–667.

Partridge, J. E., and Keen, N. T. (1977). Soybean phytoalexins: Rates of synthesis are not regulated by activation of initial enzymes in flavonoid biosynthesis, Phytopathology, 67:50–55.

Paxton, J. D. (1981). Phytoalexins—a working redefinition, Phytopathol. Z., 101:106–109.

Perrin, D. R., Biggs, D. R., and Cruickshank, I. A. M. (1974). Phaseollidin, a phytoalexin from *Phaseolus vulgaris*: induction, physicochemical properties and antifungal activity, Aust. J. Chem., 27:1607–1611.

Perrin, D. R. (1964). The structure of phaseolin, Tetrahedron Lett., 33:29–35.

Pezet, R., and Pont, V. (1988). Mise en évidence de pterostilbene dans les grappes de *Vitis vinifera*, Plant Physiol. Biochem., 26:603–607.

Pezet, R., and Pont, V. (1990). Ultrastructural observations of pterostilbene fungitoxicity in dormant conidia of *Botrytis cinerea* Pers., J. Phytopathol., 129:19–30.

Pezet, R., Pont, V., and Hoang-Van, K. (1991). Evidence for oxidative detoxication of pterostilbene and resveratrol by a laccase-like stilbene oxidase produced by *Botrytis cinerea*, Physiol. Mol. Plant Pathol., 39:441–450.

Pont, V., and Pezet, R. (1990). Relation between the chemical structure and the biological activity of hydroxystilbenes against *Botrytis cinerea*, J. Phytopathol., 130:1–8.

Pool, R. M., Creasy, L. L., and Frackelton, A. S. (1981). Resveratrol and the viniferins, their application to screening for disease resistance in grape breeding programs, Vitis, 20:136–145.

Pueppke, S. G., and VanEtten, H. D. (1976). Accumulation of pisatin and three additional antifungal pterocarpans in *Fusarium solani*-infected tissues of *Pisum sativum*, Physiol. Plant Pathol., 8:51–61.

Ragg, H., Kuhn, D. N., and Hahlbrock, K. (1981). Coordinated regulation of 4-coumarate: CoA ligase and phenylalanine ammonia-lyase mRNAs in cultured plant cells, J. Biol. Chem., 256:10061–10065.

Rathmell, W. G., and Bendall, D. S. (1971). Phenolic compounds in relation to phytoalexin biosynthesis in hypocotyls of *Phaseolus vulgaris*, Physiol. Plant Pathol., 1:351–362.

Ren, Y. Y., and West, C. A. (1992). Elicitation of diterpene biosynthesis in rice (*Oryza sativa* L.) by chitin, Plant Physiol., 99:1169–1178.

Robbins, M. P., Bolwell, G. P., and Dixon, R. A. (1985). Metabolic changes in elicitor-treated bean cells. Selectivity of enzyme induction in relation to phytoalexin accumulation, Eur. J. Biochem., 148:563–569.

Robbins, M. P., and Dixon, R. A. (1984). Induction of chalcone isomerase in elicitor-treated bean cells. Comparison of rates of synthesis and appearance of immunodetectable enzyme, Eur. J. Biochem., 145:195–202.

Rohde, W., Dorr, S., Salamini, F., and Becker, D. (1991). Structure of a chalcone synthase gene from *Hordeum vulgare*, Plant Mol. Biol., 16:1103–1106.

Rolfs, C. H., Fritzemeier, K. H., and Kindl, H. (1981). Cultured cells of *Arachis hypogea* susceptible to induction of stilbene synthase (resveratrol-forming), Plant Cell Rep., 1:83–85.

Rudloff, E. von., and Jorgensen, E. (1963). The biosynthesis of pinosylvin in the sapwood of *Pinus resinosa* AIT, Phytochemistry, 2:297–304.

Ryder, T. B., Cramer, C. L., Bell, J. N., Robbins, M. P., Dixon, R. A., and Lamb, C. J. (1984). Elicitor rapidly induces chalcone synthase mRNA in *Phaseolus vulgaris* cells at the onset of the phytoalexin response, Proc. Natl. Acad. Sci. USA, 81:5724–5728.

Ryder, T. B., Hedrick, S. A., Bell, J. N., Liang, X., Clouse, S. D., and Lamb, C. J. (1987). Organization and differential activation of a gene family encoding the plant defense enzyme chalcone synthase in *Phaseolus vulgaris*, Mol. Gen. Genet., 210:219–233.

Sandermann, H. Jr., Schmitt, R., Heller, W., Rosemann, D., and Langebartels, C. (1989). Ozone-induced early biochemical reactions in conifers, Acid Deposition: Sources, Effects, and Controls (J. W. S. Longhurst, ed.), British Library, London, pp. 243–254.

Sanz Platero, de M., and Fuchs, A. (1978). Degradation of pisatin, an antimicrobial compound produced by *Pisum sativum* L., Phytopathol. Medit., 17:14–17.

Scala, F., Matthews, D., Costa, M., and VanEtten, H. D. (1988). Immunochemical relatedness of fungal NADPH-cytochrome P-450 reductases and their ability to reconstitute pisatin demethylase activity, Exp. Mycol., 12:377–385.

Schafer, W., VanEtten, H. D., and Yoder, O. C. (1988). Transformation of the maize pathogen *Cochliobolus heterostrophus* to a pea pathogen by insertion of a PDA gene of *Nectria haematococca*, J. Cell Biol., 12C:291.

Scheel, D., Dangl, J. L., Douglas, C., Hauffe, K. D., Hermann, A., and Hahlbrock, K. (1987). Stimulation of phenylpropanoid pathways by environmental factors, Plant Molecular Biology (D. von Wettstein and N. H. Chua, eds.), Plenum Press, New York, pp. 243–251.

Scheel, D., Hauffe, K. D., Jahnen, W., and Hahlbrock, K. (1986). Stimulation of phytoalexin formation in fungus-infected plants and elicitor-treated cell cultures of parsley, Recognition in Microbe–Plant Symbiotic and Pathogenic Interactions (B. Lugtenberg, ed.), Springer-Verlag, Berlin, pp. 325–331.

Schmelzer, E., Borner, H., Grisebach, H., Ebel, J., and Hahlbrock, K. (1984). Phytoalexin synthesis in soybean (*Glycine max*). Similar time course of mRNA induction in

hypocotyls infected with a fungal pathogen and in cell cultures treated with fungal elicitor, FEBS Lett., 172:59–63.
Schmelzer, E., Kruger-Lebus, S., and Hahlbrock, K. (1989). Temporal and spatial patterns of gene expression around sites of attempted fungal infection in parsley leaves, Plant Cell, 1:993–1001.
Schmid, J., Doerner, P. W., Clouse, S. D., Dixon, R. A., and Lamb, C. J. (1990). Developmental and environmental regulation of a bean chalcone synthase promoter in transgenic tobacco, Plant Cell, 2:619–631.
Schoppner, A., and Kindl, H. (1979). Stilbene synthase (pinosylvin synthase) and its induction by UV-light, FEBS Lett., 108:349–352.
Schroder, G., Brown, J. W. S., and Schroder, J. (1988). Molecular analysis of resveratrol synthase, cDNA, genomic clones and relationship with chalcone synthase, Eur. J. Biochem., 172:161–169.
Schroder, J., and Schafer, E. (1980). Radio-iodinated antibodies, a tool in studies on the presence and role of inactive enzyme forms: regulation of chalcone synthase in parsley cell suspension cultures, Arch. Biochem. Biophys., 203:800–808.
Schroder, J., and Schroder, G. (1990). Stilbene and chalcone synthases: related enzymes with key functions in plant-specific pathways, Z. Naturforsch., 45C:1–8.
Schulze-Lefert, P., Becker-Andre, M., Schulz, W., Hahlbrock, K., and Dangl, J. L. (1989a). Functional architecture of the light-responsive chalcone synthase promoter from parsley, Plant Cell, 1:707–714.
Schulze-Lefert, P., Dangl, J., Becker-Andre, M., Hahlbrock, K., and Schulz, W. (1989b). Inducible in vivo DNA footprints define sequences necessary for UV light activation of the parsley chalcone synthase gene, EMBO J, 8:651–656.
Sekido, H., Endo, T., Suga, R., Odama, O., Akatsuka, T., Kono, Y., and Takeuchi, S. (1986). Oryzalexin D (3,7-dihydroxy-(+)-sandaracopimaradiene), a new phytoalexin isolated from blast infected rice leaves, J. Pestic. Sci., 11:369–372.
Serrano, R. (1989). Structure and function of plasma membrane ATPase, Annu. Rev. Plant Physiol. Plant Mol. Biol., 40:61–94.
Sharon, A., Amsellem, Z., and Gressel, J. (1992a). Glyphosate suppression of an elicited defense response: increased susceptibility of *Cassia obtusifolia* to a mycoherbicide, Plant Physiol., 98:654–659.
Sharon, A., Ghirlando, R., and Gressel, J. (1992b). Isolation, purification, and identification of 2-(p-hydroxyphenoxy)-5,7-dihydroxychromone: a fungal induced phytoalexin from *Cassia obtusifolia*, Plant Physiol., 98:303–308.
Sharon, A., and Gressel, J. (1991). Elicitation of a flavanoid phytoalexin accumulation in *Cassia obtusifolia* by a mycoherbicide: estimation by $AlCl_3$ spectrofluorimetry, Pestic. Biochem. Physiol., 41:142–149.
Shields, S. E., Wingate, V. P. M., and Lamb, C. J. (1982). Dual control of phenylalanine ammonia-lyase production and removal by its product cinnamic acid, Eur. J. Biochem., 123:389–395.
Shih, M., Kuc, J., and Williams, E. B. (1973). Suppression of steroid glycoalkaloid accumulation as related to rishitin accumulation in potato tubers, Phytopathology, 63:821–826.
Shinka, T., Ogura, K., and Seto, S. (1975). Comparative specificity of geranylgeranyl

pyrophosphate synthetase of *Micrococcus lysodeikticus* and pumpkin, J. Biochem., 78:1177–1181.

Shiraishi, T., Oku, H., Ouchi, S., and Tsuji, Y. (1977). Local accumulation of pisatin in tissues of pea seedlings infected by powdery mildew fungi, Phytopathol. Z., 88:131–135.

Shiraishi, T., Oku, H., Yamashita, M., and Ouchi, S. (1978). Elicitor and suppressor of pisatin induction in spore germination fluid of pea pathogen, *Mycosphaerella pinodes*, Ann. Phytopathol. Soc. Japan, 44:659–665.

Shoun, H., Sudo, Y., Seto, Y., and Beppu, T. (1983). Purification and properties of a cytochrome P-450 of a fungus, *Fusarium oxysporum*, J. Biochem., 94:1221–1229.

Sitton, D., and West, C. A. (1975). Casbene: an antifungal diterpene produced in cell-free extracts of *Ricinus communis* seedlings, Phytochemistry, 14:1921–1925.

Skipp, R. A., and Bailey, J. A. (1976). The effect of phaseollin on the growth of *Colletotrichum lindemuthianum* in bioassays designed to measure fungitoxicity, Physiol. Plant Pathol., 9:253–263.

Smith, D. A., Harrer, J. M., and Cleveland, T. E. (1981). Simultaneous detoxification of phytoalexins by *Fusarium solani* f.sp. *phaseoli*, Phytopathology, 71:1212–1215.

Smith, D. A., Harrer, J. M., and Cleveland, T. E. (1982). Relation between production of extracellular kievitone hydratase by isolates of *Fusarium* and their pathogenicity on *Phaseolus vulgaris*, Phytopathology, 72:1319–1323.

Smith, D. A., Kuhn, P. J., Bailey, J. A., and Burden, R. S. (1980). Detoxification of phaseollidin by *Fusarium solani* f.sp. *phaseoli*, Phytochemistry, 19:1673–1675.

Smith, D. A., Wheeler, H. E., Banks, S. W., and Cleveland, T. E. (1984). Association between lowered kievitone hydratase activity and reduced virulence to bean in variants of *Fusarium solani* f.sp. *phaseoli*, Physiol. Plant Pathol., 25:135–147.

Snyder, B. A., Leite, B., Hipskind, J., Butler, L. G., and Nicholson, R. L. (1991). Accumulation of sorghum phytoalexins induced by *Colletotrichum graminicola* at the infection site, Physiol. Mol. Plant Pathol., 39:463–470.

Snyder, B. A., and Nicholson, R. L. (1990). Synthesis of phytoalexins in sorghum as a site specific response to fungal ingress, Science, 248:1637–1639.

Spurgeon, S. L., Sathyamoorthy, N., and Porter, J. W. (1984). Isopentenyl pyrophosphate isomerase and prenyltransferase from tomato fruit plastids, Arch. Biochem. Biophys., 1230:446–454.

Steffens, M., Ettl, F., Kranz, D., Kindl, H. (1989). Vanadate mimics effects of fungal cell wall in eliciting gene activation in plant cell cultures, Planta, 177:160–168.

Stein, U., and Hoos, G. (1984). Induktions- and Nachweis-methoden fur stilbene bei Vitaceen, Vitis, 23:179–194.

Stekoll, M., and West, C. A. (1978). Purification and properties of an elicitor of castor bean phytoalein from culture filtrates of the fungus *Rhizopus stolonifer*, Plant Physiol., 61:38–45.

Stermer, B. A., and Bostock, R. M. (1987). Involvement of 3-hydroxy-3-methylglutaryl coenzyme A reductase in the regulation of sesquiterpenoid phytoalexin synthesis in potato, Plant Physiol., 84:404–408.

Stermer, B. A., Edward, L. A., Edington, B. V., and Dixon, R. A. (1991). Analysis of elicitor-inducible transcripts encoding 3-hydroxy-3-methylglutaryl coenzyme A reductase in potato, Physiol. Mol. Plant Pathol., 39:135–145.

Stermer, B. A., Schmid, J., Lamb, C. J., and Dixon, R. A. (1990). Infection and stress activation of bean chalcone synthase promoters in transgenic tobacco, Mol. Plant–Microbe Interact., 3:381–388.

Stoessl, A., Robinson, J. R., Rock, G. L., and Ward, E. W. B. (1977). Metabolism of capsidiol by sweet pepper tissue: some possible implications for phytoalexin studies, Phytopathology, 67:64–66.

Straney, D. C., and VanEtten, H. D. (1994). Characterization of the PDA 1 promoter of *Nectria haematococca* and identification of a region that binds a pisatin-responsive DNA binding factor, Mol. Plant–Microbe Interact., 7:256–266.

Sulistyowati, L., and Keane, P. J. (1992). Effect of soil salinity and water content on stem rot caused by *Phytophthora citrophthora* and accumulation of phytoalexin in citrus rootstocks, Phytopathology, 82:771–777.

Sulistyowati, L., Keane, P. J., and Anderson, J. W. (1990). Accumulation of the phytoalexin, 6,7-dimethoxycoumarin, in roots and stems of citrus seedlings following inoculation with *Phytophthora citrophthora*, Physiol. Mol. Plant Pathol., 37:451–461.

Suzuki, H., Oba, K., and Uritani, I. (1975). The occurrence and some properties of 3-hydroxy-3-methylglutaryl coenzyme A reductase in sweet potato roots infected by *Ceratocystis fimbriata*, Physiol. Plant Pathol., 7:265–276.

Sweigard, J. A., Matthews, D. E., and VanEtten, H. D. (1986). Synthesis of the phytoalexin pisatin by a methyltransferase from pea, Plant Physiol., 80:277–279.

Takeuchi, A., Scott, K. J., Oba, K., and Uritani, I. (1980). Phytopathological chemistry of sweet potato with black rot and injury. Part 135. Possible role of the cytosol pathway of acetyl CoA supply in terpene biosynthesis in sweet potato infected with *Ceratocystis fimbriata*, Plant Cell Physiol., 21:917–922.

Tal, B., and Robeson, D. J. (1986). The induction by fungal inoculation of ayapin and scopoletin biosynthesis in *Helianthus annuus*, Phytochemistry, 25:77–79.

Tamara, S., Hanawa, F., Harada, Y., and Muzutani, J. (1988). A fungitoxin inducibly produced by dandelion leaves treated with cupric chloride, Agric. Biol. Chem., 52:2947–2948.

Tanaka, Y., Matsuoka, M., Yamamoto, N., Ohashi, Y., Kano-Murakami, Y., and Ozeki, Y. (1989). Structure and characterization of a cDNA clone for phenylalanine ammonia-lyase from cut-injured roots of sweet potato, Plant Physiol., 90:1403–1407.

Tegtmeier, K. J., and VanEtten, H. D. (1982). The role of pisatin tolerance and degradation in the virulence of *Nectria haematococca* on pea. A genetic analysis, Phytopathology, 72:608–612.

Threlfall, D. R., and Whitehead, I. M. (1988). Co-ordinated inhibition of squalene synthetase and induction of enzymes of sesquiterpenoid phytoalexin biosynthesis in cultures of *Nicotiana tabacum*, Phytochemistry, 27:2567–2580.

Threlfall, D. R., and Whitehead, I. M. (1990). Redirection of terpenoid biosynthesis in elicitor-treated plant cell suspension cultures, Plant Lipid Biochemistry Structure and Utilization (P. J. Quinn, and J. L. Harwood, eds.), Portland Press, London, pp. 344–346.

Tiemann, K., Inze, D., Van Montagu, M., and Barz, W. (1991). Pterocarpan phytoalexin biosynthesis in elicitor-challenged chickpea (*Cicer arietinum* L.) cell cultures.

Purification, characterization and cDNA cloning of NADPH: isoflavone oxido-reductase, Eur. J. Biochem., 200:751–757.
Tietjen, K. G., and Hunkler, D., and Matern, U. (1983). Differential response of cultured parsley cells to elicitors form two nonpathogenic strains of fungi. I. Identification of induced products as coumarin derivatives, Eur. J. Biochem., 131:401–407.
Tietjen, K. G., and Matern, U. (1983). Differential response of cultured parsley cells to elicitors from two non-pathogenic strains of fungi. 2. Effects of enzyme activities, Eur. J. Biochem., 131:409–413.
Tietjen, K. G., and Matern, U. (1984). Induction and suppression of phytoalexin synthesis in cultured cells of safflower *Carthamus tinctorius* L. by metabolites of *Alternaria carthami* Chowdhury, Arch. Biochem. Biophys., 229:136–144.
Uehara, K. (1964). Relationship between host specificity of pathogen and phytoalexin, Ann. Phytopathol. Soc. Jpn., 29:103–110.
Uritani, I. (1978). Bichemistry of host response to infection, Progress in Phytochemistry (L. Reinhold, J. B. Harborne and T. Swain, eds.), Pergamon Press, Oxford, pp. 29–63.
Van Den Heuvel, J. (1976). Sensitivity to, and metabolism of, phaseollin in relation to the pathogenicity of different isolates of *Botrytis cinerea* to bean (*Phaseolus vulgaris*), Nether. J. Plant Pathol., 82:153–160.
VanEtten, H. D., and Barz, W. (1981). Expression of pisatin demethylating ability in *Nectria haematococca*, Arch. Microbiol., 129:56–60.
VanEtten, H. D., and Matthews, P. S. (1984). Naturally occurring variation in the inducibility of pisatin demethylating activity in *Nectria haematococca* mating population VI, Physiol. Plant Pathol., 25:149–160.
VanEtten, H. D., Matthews, D. E., and Matthews, P. S. (1989). Phytoalexin detoxification: importance for pathogenicity and practical implications, Annu. Rev. Phytopathol., 27:143–164.
VanEtten, H. D., Matthews, D. E., and Smith, D. A. (1982). Metabolism of phytoalexins, Phytoalexins, (J. A. Bailey and J. W. Mansfield, eds.), Blackie and Son Ltd., Glasgow, pp. 181–217.
VanEtten, H. D., Matthews, P. S., Tegtmeier, K. J., Dietert, M. F., and Stein, J. I. (1980). The association of pisatin tolerance and demethylation with virulence on pea in *Nectria haematococca*, Physiol. Plant Pathol., 16:257–268.
VanEtten, H. D., Pueppke, S. G., and Kelsey, T. C. (1975). 3,5a-Dihydroxy-8,9-methylenedioxypterocarpan as a metabolite of pisatin produced by *Fusarium solani* f.sp. *pisi*, Phytochemistry, 14:1103–1105.
VanEtten, H. D., Sandrock, R. W., Wasmam, C. C., Soby, S. D., Mellusky, K., and Wang, P. (1995). Detoxification of phytoanticipins and phytoalexins by phytopathogenic fungi, Can. J. Bot., 73(Suppl. 1):S518–S525.
van Tunen, A. J., Koes, R. E., Spelt, C. E., van der Krol, A. R., Stuitje, A. R., and Mol, J. N. M. (1988). Cloning of the two chalcone flavanone isomerase genes from *Petunia hybrida*: coordinate, light-regulated and differential expression of flavonoid genes, EMBO J., 7:1257–1263.
Vogeli, U., and Chappell, J. (1988). Induction of sesquiterpene cyclase and suppression of squalene synthase activity in plant cell culture treated with fungal elicitor, Plant Physiol., 88:1291–1296.

Vornam, B., Schon, H., and Kindl, H. (1988). Control of gene expression during induction of cultured peanut cells: mRNA levels, protein synthesis and enzyme activity of stilbene synthase, Plant Mol. Biol., 10:235–243.

Walter, M. H., and Hahlbrock, K. (1984). Cinnamic acid induces PAL mRNA and inhibits mRNA translation in cultured parsley cells, Plant Physiol. (Suppl), 75:155.

Ward, E. W. B. (1984). Suppression of metalaxyl activity by glyphosate: evidence that host defence mechanisms contribute to metalaxyl inhibition of *Phytophthora megasperma* f.sp. *glycinea* in soybeans, Physiol. Plant Pathol., 25:381–386.

Ward, E. W. B., Lazarovits, G., Stossel, P., Barrie, S. D., and Unwin, C. H. (1980). Glyceollin production associated with control of phytophthora rot of soybeans by the systemic fungicide, metalaxyl, Phytopathology, 70:738–740.

Ward, E. W. B., and Stoessl, A. (1977). Phytoalexins from potatoes: evidence for the conversion of lubimin to 1,5-dihydrolubimin by fungi, Phytopathology, 67:468–471.

Ward, E. W. B., Stoessl, A., and Stothers, J. B. (1977). Metabolism of the sesquiterpenoid phytoalexins capsidiol and rishitin to their 13-hydroxy derivatives by plant cells, Phytochemistry, 16:2024–2025.

Weigand, F., Koster, J., Weltzien, H. C., and Barz, W. (1986). Accumulation of phytoalexins and isoflavone glucosides in a resistant and a susceptible cultivar of *Cicer arietinum* during infection with *Ascochyta rabiei*, J. Phytopathol., 115:214–221.

Welle, R., and Grisebach, H. (1987). Purification and properties of chalcone synthase from cell suspension cultures of soybean, Z. Naturforsch., 42C:1200–1206.

Welle, R., and Grisebach, H. (1988a). Induction of phytoalexin synthesis in soybean: Enzymatic cyclization of prenylated pterocarpans to glyceollin isomers, Arch. Biochem. Biophys., 263:191–198.

Welle, R., and Grisebach, H. (1988b). Isolation of a novel NADPH-dependent reductase which coacts with chalcone synthase in the biosynthesis of 6′-deoxychalcone, FEBS Lett., 236:221–225.

Welle, R., and Grisebach, H. (1989). Phytoalexin synthesis in soybean cells: Elicitor induction of reductase involved in biosynthesis of 6′-deoxychalcone, Arch. Biochem. Biophys., 272:97–102.

Welle, R., Schroder, G., Schiltz, E., Grisebach, H., and Schroder, J. (1991). Induced plant responses to pathogen attack: analysis and heterologous expression of the key enzyme in the biosynthesis of phytoalexins in soybean (*Glycine max* L. Merr. cv. Harosoy 63), Eur. J. Biochem., 196:423–430.

Weltring, K. M., Turgeon, B. G., Yoder, O. C., and VanEtten, H. D. (1988). Isolation of a phytoalexin-detoxifying gene from the plant pathogenic fungus *Nectria haematococca* by detecting its expression in *Aspergillus nidulans*, Gene, 68:335–344.

Wendorff, H., and Matern, U. (1986). Differential response of cultured parsley cells to elicitors from two non-pathogenic strains of fungi. Microsomal conversion of (+) marmesin into psoralen, Eur. J. Biochem., 161:391–398.

Wengenmayer, H., Ebel, J., and Grisebach, H. (1974). Purification and properties of a S-adenosylmethionine isoflavone 4′-0-methyltransferase from cell suspension cultures of *Cicer arietinum* L., Eur. J. Biochem., 50:135–143.

West, C. A., Moesta, P., Jin, D. F., Lois, A. F., and Wickham, K. A. (1985). The role of pectic fragments of the plant cell wall in the response to biological stress, Cellular and Molecular Biology of Plant Stress (J. Key and T. Kosuge, eds.), Alan R. Liss Press, New York, pp. 335–349.

Whitehead, I. M., Dey, P. M., and Dixon, R. A. (1982). Differential patterns of phytoalexin accumulation and enzyme induction in wounded and elicitor-treated tissues of *Phaseolus vulgaris*, Planta, 154:156–164.

Wickham, K. A., and West, C. A. (1992). Biosynthesis of rice phytoalexins: identification of putative diterpene hydrocarbon precursors, Arch. Biochem. Biophys., 293:320–332.

Wictor-Orlandi, E. A., and Smith, D. A. (1985). Metabolism of the phytoalexin, phaseollinisoflavan, by *Fusarium solani* f.sp. *phaseoli*, Physiol. Plant Pathol., 27:197–207.

Widmaier, R., Howe, J., and Heinstein, P. (1980). Prenyltransferase from *Gossypium hirsutum*, Arch. Biochem. Biphys., 200:609–616.

Wingender, R., Rohrig, H., Horicke, C., and Schell, J. (1990). *Cis*-regulatory elements involved in ultraviolet light regulation and plant defense, Plant Cell, 2:1019–1026.

Wingender, R., Rohrig, H., Horicke, C., Wing, D., and Schell, J. (1989). Differential regulation of soybean chalcone synthase genes in plant defence, symbiosis and upon environmental stimuli, Mol. Gen. Genet., 218:315–322.

Wit-Elshove, A. de., and Fuchs, A. (1971). The influence of the carbohydrate source on pisatin breakdown by fungi pathogenic to pea (*Pisum sativum*), Physiol. Plant Pathol., 1:17–24.

Yamada, T., Hashimoto, H., Shiraishi, T., and Oku, H. (1989). Suppression of pisatin, phenylalanine ammonia-lyase mRNA, and chalcone synthase mRNA accumulation by a putative pathogenicity factor from the fungus *Mycosphaerella pinodes*, Mol. Plant–Microbe Interact., 2:256–261.

Yamaoka, H., Lyons, P. C., Hipskind, J., and Nicholson, R. L. (1990). Elicitor of sorghum phytoalexin synthesis from *Colletotrichum graminicola*, Physiol. Mol. Plant Pathol., 37:255–270.

Yoshikawa, M., Yamauchi, K., and Masago, H. (1978). Glyceollin: its role in restricting fungal growth in resistant soybean hypocotyls infected with *Phytophthora megasperma* var. *sojae*, Physiol. Plant Pathol., 12:73–82.

Yoshikawa, M., Yamauchi, K., and Masago, H. (1979). Biosynthesis and biodegradation of glyceollin by soybean hypocotyls infected with *Phytophthora megasperma* var. *sojae*, Physiol. Plant Pathol., 14:157–169.

Yoshioka, H, Shiraishi, T., Kawamata, S., Nasu, K., Yamada, T., Ichinose, Y., and Oku, H. (1992). Orthovanadate suppresses accumulation of phenylalanine ammonia-lyase mRNA and chalcone synthase mRNA in pea epicotyls induced by elicitor from *Mycosphaerella pinodes*, Plant Cell Physiol., 33:201–204.

Yoshioka, H., Shiraishi, T., Yamada, T., Ichinose, Y., and Oku, H. (1990). Suppression of pisatin production and ATPase activity in pea plasma membranes by orthovanadate, verapamil and a suppressor from *Mycosphaerella pinodes*, Plant Cell Physiol., 31:1139–1146.

Zahringer, U., Ebel, J., and Grisebach, H. (1978). Induction of phytoalexin synthesis in soybean. Elicitor-induced increase in enzyme activities of flavonoid biosynthesis and incorporation of mevalonate into glyceollin, Arch. Biochem. Biophys., 188:450–455.

Zahringer, U., Ebel, J., Mulheirn, L. J., Lyne, R. L., and Grisebach, H. (1979). Induction of phytoalexin synthesis in soybean. Dimethylallylpyrophosphate: tri-hydroxy-pterocarpan dimethylallyl transferase from elicitor-induced cotyledons, FEBS Lett., 101:90–92.

Zhang, Y., and Smith, D. A. (1983). Concurrent metabolism of the phytoalexins phaseollin, kievitone and phaseollinisoflavan by *Fusarium solani* f.sp. *phaseoli*, Physiol. Plant Pathol., 23:89–100.

Ziegler, E., and Pontzen, R. (1982). Specific inhibition of glucan-elicited glyceollin accumulation in soybeans by an extracellular mannan-glycoprotein of *Phytophthora megasperma* f.sp. *glycinea*, Physiol. Plant Pathol., 20:321–331.

Zook, M. N., and Kuc, J. A. (1991). Induction of sesquiterpene cyclase and suppression of squalene synthetase activity in elicitor treated or fungal infected potato tuber tissue, Physiol. Mol. Plant Pathol., 39:377–390.

Zouari, N., Romette, J. L., and Thomas, D. (1987). Purification and properties of two laccase isoenzymes produced by *Botrytis cinerea*, Appl. Biochem. and Biotechnol., 15:213–225.

6

Phytoanticipins

I. WHAT ARE PHYTOANTICIPINS?

Besides phytoalexins, which are antimicrobial compounds synthesized de novo when microorganisms invade them, most plants synthesize toxic compounds as part of normal development also and such compounds are called phytoanticipins. Phytoanticipins are low-molecular-weight antimicrobial compounds that are present in plants before challenge by microorganisms or are produced after infection solely from preexisting constituents (VanEtten et al., 1995). Phenolics, glucosinolates, cyanogenic glucosides, saponins, alkaloids, dienes, and antimetabolites are the important phytoanticipins found in plants.

II. PHENOLICS

Vidhyasekaran (1988a) described the occurrence of many kinds of phenolics in plants. Monophenols, di-, and tri-hydroxy phenols, phenolic acids, pterocarpans, isoflavans, isoflavones, isoflavanones, glucosides of isoflavonoids, furanocoumarins, and anthocyanidins are the important antifungal phenolics in plants. Catechol, protocatechuic acid, chlorogenic acid, and caffeic acid have been reported in many plants (Johnson and Schaal, 1957). Ferulic, p-coumaric and syringic acids are commonly found in wheat cultivars (Southerton and Deverall, 1990). Chlorogenic acid is the major phenol found in roots of potato plants (Patil et al., 1964). Catechol and protocatechuic acid are found in onion (Walker and Stahmann, 1955). Scopolin, scopoletin, chlorogenic acid, and O-coumaric acid have been reported in tobacco (Reuveni and Cohen, 1978). Salicylic acid is a phenolic widely found in plants (Malamy et al., 1992; Yalpani and Raskin, 1993). Catechin is a predominant phenolic in cotton (Hunter, 1974). The main constitutive phenolics of chickpea are the isoflavones biochanin A and formononetin, which occur as aglycones, the 7-O-glucosides and the isoflavone 7-O-glucoside-6″-O-malonate esters (Koster et al., 1983; Kramer et al., 1984). The isoflavonoids, daidzein and malonyl-daidzein, glucosides of daidzein and genistin, and malonyl-genistin and glucosides of genistein are present in soybean (Graham et al., 1990; Graham and Graham, 1991; Graham 1991). Only low levels of isoflavone glucosides have been detected in soybean leaves (Fett, 1984; Cosio et al., 1985; Morris et al., 1991). However, in some plants such as *Trifolium pratense* (Jones, 1979; McMurray, 1986) and chickpea (Weigand et al., 1986), high levels of isoflavone glucosides have been reported. Phloretin (Noveroske et al., 1964), sieboldin (Hunter, 1971), chlorogenic acid (Oydvin and Richardson, 1987), and flavan-3-ol- (Treutter and Feucht, 1990) are the important phenolics in apple. Several isoflavonoids have been detected in *Phaseolus vulgaris* (Woodward, 1980), *Phaseolus aureus* (O'Neill, 1983; O'Neill et al., 1983), *Phaseolus mungo* (Adesanya et al., 1984), *Phaseolus coccineus* (Adesanya et al., 1985), and *Phaseolus lunatus* (O'Neill et al., 1986). They are isoflavones, isoflavanones, pterocarpans, and isoflavans. Anthocyanins are glycosides of anthocyanidins and the final products of flavonoid metabolism. They are commonly found in corn leaves (Hammerschmidt and Nicholson, 1977). Furanocoumarins occur in many plants, especially in species of the Umbelliferae, Leguminosae, Rutaceae, and Moraceae (Pathak et al., 1962). Furanocoumarins are heterocyclic compounds derived from coumarin by the addition of a furan ring at the 6, 7 positions (linear furanocoumarins) or 7, 8 positions (angular furanocoumarins). The furanocoumarins angelicin, bergapten, psoralen, trimethylpsoralen, xanthotoxin, iso-pimpinellin, oxypeucedanin and sphondin have been identified in celery plants (Scheel et al., 1963; Floss et al., 1969; Ashwood-Smith et al., 1985; Chaudhury et al., 1985; Ataga et al., 1993).

A. Toxicity of Phenolics to Pathogens

Toxicity of different phenolics varies and sensitivity of pathogens to the phenolics also varies. Differential sensitivity to monophenols and di- and tri-hydroxy phenols has been reported among *Phytophthora* spp. (Table 1; Casares et al., 1986). Salicylic acid, p-coumaric acid, vanillic acid, and pyrogallol were highly inhibitory to *P. cambivora* and not to *P. cinnamomi*. Cinnamic acid inhibited both the fungi effectively. In general, *P. cinnamomi* was less sensitive to the phenols than *P. cambivora* (Casares et al., 1986).

Effect of different phenolics in inhibiting growth of *Phytophthora cinnamomi* varied (Table 2; Cahill and McComb, 1992). Cinnamic acid, p-coumaric acid, vanillic acid, ferulic acid, p-hydroxybenzoic acid, caffeic acid, and coumarin completely inhibited growth of the fungus at 10 mM, while several other phenolics did not inhibit the growth appreciably (Cahill and McComb, 1992).

The isoflavones even at low concentrations inhibit the fungi in vitro (Naim et al., 1974; Johnson et al., 1976; Adesanya et al., 1986; Weidenborner et al., 1990). The isoflavone biochanin A and its dehydroderivatives (isoflavanones) show high antifungal activity against *Rhizoctonia solani* and *Sclerotium rolfsii* at a concen-

Table 1 Relative growth of *Phytophthora cinnamomi* and *P. cambivora* under the effect of phenols

	Growth (%)[a]	
Phenols	*Phytophthora cinnamomi*	*Phytophthora cambivora*
Benzoic acid	97	82
Cinnamic acid	0	0
Phenol	98	76
Guaiacol	81	90
p-OH-Benzaldehyde	97	84
p-OH-Benzoic acid	110	78
Salicylic acid	73	33
P-coumaric acid	100	53
Tyrosine	97	90
Vanillin	79	76
Vanillic acid	94	50
Ferulic acid	84	70
Syringic acid	83	69
Pyrocatechol	73	66
Resorcinol	92	90
Hydroquinone	97	78
Protocatechualdehyde	92	62
Protocatechuic acid	95	87
Caffeic acid	97	78
Pyrogallol	82	29
Phloroglucinol	98	100
Gallic acid	105	92
Tannic acid	82	62

[a]Growth in medium without phenols was taken as 100%.
Source: Casares et al., 1986.

tration of about 0.8 mM. Genistein isoflavan and other isoflavans with two hydroxyl groups and one methoxy group are also fungitoxic (Weidenborner et al., 1990).

Genistein and biochanin A are the naturally occurring isoflavones. Their dihydroderivatives (isoflavanones) and per-hydrogenated isoflavones (isoflavans) were tested for their antifungal activity. Unsubstituted isoflavonoids (isoflavone, isoflavanone, isoflavan) possess weak antifungal activity. Only after the introduction of suitable substituents in particular positions did these isoflavanoid structures become fungitoxic. Two hydroxyl groups at C-5 and C-7 and one methoxy group at C-4′ are conducive to a high antifungal activity (Weidenborner et al., 1990).

Table 2 Effect of phenolic compounds on the growth of *P. cinnamomi* in vitro

Phenolic compound	Growth (%) control
Cinnamic acid	0
p-Coumaric acid	0
Vanillic acid	0
Ferulic acid	0
Hydroquinone	18
p-Hydroxy benzoic acid	0
Caffeic acid	0
Vanillin	42
Ellagic acid	38
Gallic acid	77
Gentisic acid	21
Coumarin	0
Protocatechuic acid	80
(+) Catechin	85
(–) Epicatechin	85
Chlorogenic acid	77
Phloroglucinol	71

Source: Cahill and McComb, 1992.

A comparison of the effectiveness of the isoflavones genistein and biochanin reveals that the presence of a methoxy group instead of the hydroxyl group at C-4′ bestows a high activity against *R. solani* and *S. rolfsii* (Weidenborner et al., 1990). Similar results were obtained against *Cercospora beticola* and *Monilinia fructicola* (Johnson et al., 1976).

The fungitoxicity of isoflavones, isoflavanones, pterocarpans, and isoflavans was tested against *Cladosporium cucumerinum* (Table 3; Adesanya et al., 1986). Pterocarpans and isoflavans are more fungitoxic than other classes of isoflavonoids (Adesanya et al., 1986). The presence of dimethylallyl side chain enhances fungitoxic activity. In the isoflavone group, phaseoluteone is more active in inhibiting spore germination than its nonprenylated counterpart 2′-hydroxygenistein. 2,3-Dehydrokievitone is also more active than 2′-hydroxygenistein against *C. cucumerinum* (Adesanya et al., 1986).

Differential sensitivity of *Aphanomyces euteiches*, *Fusarium solani*, and *Rhizoctonia solani* to pterocarpans and related isoflavonoids has been reported (VanEtten, 1967). The presence of a dimethylallyl side chain in the isoflavone, isoflavanone, and pterocarpan series enhances fungitoxicity (Harborne and Ingham, 1978; Adesanya et al., 1986). Dimethylallyl side chain improves lipophili-

Table 3 Inhibition of spore germination in *Cladosporium cucumerinum* by isoflavonoids

Compound	Minimum inhibitory concentration (μg/ml)
Isoflavones	
Daidzein	50–75
Genistein	50–75
Isoprunetin	50–75
2′-Hydroxydaidzein	25–50
2′-Hydroxygenistein	50–75
2′-Methoxygenistein	50–75
2′-Hydroxyisoprunetin	25–50
2,3-Dehydrokievitone	25–50
2,3-Dehydrokievitol	> 100
Phaseoluteone	10–25
Isoflavanones	
Dalbergioidin	75–100
Isoferreirin	10–25
5-Deoxykievitone	25–50
2′,4′,5,7-Tetramethylkievitone	> 100
2′,4′,5-Trimethylkievitone	25–50
Kievitol	> 100
3′(y,y-dimethylallyl)-kievitone	25–50
Cyclokievitone	50–75
2′,4′,5-Trimethylcyclokievitone	25–50
Pterocarpans	
Demethylmedicarpin	25–50
Isomedicarpin	10–25
Isoflavan	
Demethylvestitol	25–50
Coumestan	
Aureol	25–50

Source: Adesanya et al., 1986.

city and hence aids penetration through the fungal cell wall (Harborne and Ingham, 1978). Isopentenyl substituent or methylation improves lipophilicity of the parent isoflavonoid (Adesanya et al., 1986). At least one phenolic group is required for fungitoxicity (Adesanya et al., 1986).

It is likely that the plant uses all the active compounds it produces in a concerted effort to combat fungal invasion. Twenty isoflavonoids have been isolated from *Phaseolus vulgaris*. It is possible that each compound may have a different function. The isoflavonoids are able to attack fungi by affecting membrane porosity, by inhibiting certain fungal enzymes, or by affecting DNA transcription (Adesanya et al., 1986).

B. Phenolics Are Involved in Suppression of Fungal Pathogenesis

The presence of phenolic compounds in plants or their synthesis in response to infection, have been associated with resistance against fungal pathogens (Farkas and Kiraly, 1962). Treatment of susceptible tomato plants with quinic acid or phenylalanine increased their phenolic content and their resistance to *Fusarium oxysporum* f.sp. *lycopersici* (Carrasco et al., 1978).

Twelve-day-old seedlings of cotton were more resistant than 5- or 6-day-old seedlings to infection by *Rhizoctonia solani.* At 24 h after inoculation, the concentrations of phenolic compounds, predominantly catechin, were greater in infected 6-day-old seedlings than in healthy seedlings. The increase was greater in the older plants. Oxidized catechin was highly inhibitory to the polygalacturonase produced by the pathogen (Hunter, 1978). Production and oxidation of catechin, in response to infection by *R. solani* appears to be partially responsible for the resistance of older seedlings (Hunter, 1974). When tomato plants were fed with catechol, a marked accumulation of total phenols was observed and it resulted in suppression of disease symptom expression after infection by *Fusarium oxysporum* f.sp. *lycopersici* (Retig and Chet, 1974). It suggests the importance of phenolics in disease resistance.

Oxidized phenolics are much more toxic. Peroxidase and polyphenoloxidase (PPO) are capable of oxidizing phenolic compounds. In tomato plants resistant to *Fusarium oxysporum* f.sp. *lycopersici*, peroxidase activity increased significantly 24 h after inoculation, whereas in susceptible plants a similar increase was found 24 h later. A rapid initial increase in PPO activity in the resistant plant, followed by a very high activity during a period of 12–24 h after inoculation was observed. No increase in activity after infection was found in the susceptible plants. Susceptible plants treated with Ethephon showed an increase in resistance to *Fusarium*. Peroxidase and PPO activities were higher in Ethephon-treated than in control plants. The results suggest that peroxidase and PPO may oxidize phenolics and such oxidation products may inhibit the pathogen development (Retig, 1974).

C. Some Phenolics May Act as Signal Molecules in Inducing Disease Resistance

Salicylic acid is a widely occurring phenolic in plants (Malamy et al., 1992; Enyedi et al., 1992; Yalpani and Raskin, 1993). Salicylic acid acts as a systemic signal molecule in inducing resistance against diseases (Malamy et al., 1990; Metraux et al., 1990; Rasmussen et al., 1991; Yalpani et al., 1991). The role of salicylic acid in inducing resistance has been described in detail in Chapters 1 and 4.

D. Some Phenolics May Act as Antioxidants

Some of the phenolics are known to act as antioxidants and induce resistance. Phenolic antioxidants are present in plant lipophylic regions (Hudson and Mahgoub, 1980; Takahama, 1985). The soluble phenolic flavan-3-ol epicatechin is an antioxidant and acts as a trap for free radicals (Mehta and Seshadri, 1958).

In the avocado cultivars susceptible to *Colletotrichum gloeosporioides*, symptoms appeared in the harvested fruits on the same or one day after fruit softening while in the resistant cultivars symptoms appeared only after 4–10 days after fruit softening (Prusky et al., 1988). Epicatechin concentration in peel of the fruits of the susceptible variety decreased to 12 μg/g fresh weight concomitantly with fruit softening. In the resistant cultivar, 10 days after harvest, when the fruit was completely soft, the epicatechin concentration was still 456 μg/g fresh weight. Four days later, epicatechin decreased to 15 μg/g fresh weight (Table 4; Prusky et al., 1988). Avocado fruits contain 1-acetoxy-2-hydroxy-4-oxo-heneicosa-12,15-diene (diene) (Prusky et al., 1982, 1983). When the susceptible fruits became soft, concentration of the diene had decreased to 120 μg/g fresh weight and symptoms appeared. In resistant fruits, the diene content was still 238 μg/g fresh weight at fruit softening; and it decreased to 159 μg/g fresh weight when symptoms appeared, 4 days later (Table 4; Prusky et al., 1988). Thus epicatechin and diene appear to be involved in disease resistance.

The diene inhibits mycelial growth and spore germination at 790 μg/ml

Table 4 Relationship between epicatechin and diene concentrations in avocado fruits and susceptibility to *Colletotrichum gloeosporioides*

	Susceptible cultivar				Resistant cultivar			
Days after harvest	Firmness (kg)	Epicatechin (μg/g)	Diene (μg/g)	Symptom appearance	Firmness (kg)	Epicatechin (μg/g)	Diene (μg/g)	Symptom appearance
1	10.7	433	1420	–				
2					12.3	519	620	–
4	5.2	344	540	–				
6	0.5	12	120	+				
7					8.5	591	821	–
14					0.5	15	159	+

Source: Prusky et al., 1988.

(Prusky et al., 1982). The diene is degraded by lipoxygenase extracted from avocado peel. The diene compound functions as a substrate for lipoxygenase. The specific activity of lipoxygenase in peel extracts increases by 80% coincident with a rapid decrease of the diene in fruit peel (Prusky et al., 1983).

Epicatechin inhibits lipoxygenase in vitro (Marcus et al., 1988). It may act in vivo as a regular of membrane-found lipoxygenase (Marcus et al., 1988). The concentration of epicatechin in the avocado peel was inversely correlated with lipoxygenase activity and decreased significantly when lipoxygenase increased. The enzyme protein content ranged from 10 to 30 μg/g fresh weight of peel, and was not directly related to the enzyme activity. It suggests that the changes in lipoxygenase activity did not result from changes in the enzyme amount, but are closely related to changes in the concentration of its inhibitor, epicatechin (Karni et al., 1989). Treatments with epicatechin and other antioxidants delayed the decrease of the antifungal diene (Prusky, 1988), α-Tocopherol acetate, an antioxidant, inhibits oxidation of linoleic acid and the avocado diene by lipoxygenases from avocado. Infiltration of *C. gloeosporioides*-inoculated fruits with α-tocopherol acetate partially eliminates reduction of the diene compound during fruit ripening and fungus development in the peel is delayed (Prusky et al., 1983). These results suggest that epicatechin may induce resistance by inhibiting lipoxygenase, which may degrade the antifungal diene into a nontoxic one. Antifungal diene decrease is regulated by lipoxygenase activity, which, in turn, is regulated by a decrease in the concentration of the endogenous antioxidant, epicatechin (Prusky et al., 1985, 1988, 1991; Karni et al., 1989). When avocado fruits were exposed to CO_2 for 24 h, level of diene as well as epicatechin increased while there was inhibition of lipoxygenase activity (Prusky et al., 1991).

Inoculation of freshly harvested avocado fruits with a mutant strain of *Colletotrichum magna* inhibited subsequent decay development by the pathogen *C. gloeosporioides*. The mutant strain induced higher levels of epicatechin that lasted in ripening fruits. It suggests that enhanced resistance of avocado fruits to *C. gloeosporioides* may be due to the induction of epicatechin that inhibits oxidation of the antifungal diene (Prusky et al., 1994).

E. How Does Pathogen Overcome the Antifungal Phenolics?

1. Pathogen Degrades Phenolics to Nontoxic Products

Successful pathogens have been shown to be potent degraders of plant defense compounds, particularly phenolics. *Fusarium* spp. degrade isoflavones of chickpea (Willeke and Barz, 1982a,b). Biochanin A (5,7-dihydroxy-4′-methoxyisoflavone), its 7-O-glucoside, and the biochanin A-7-O-glucoside-6″-O-malonate are the main phenolic constituents of chickpea (Koster et al., 1983). *Fusarium javanicum* (Willeke and Barz, 1982a) and *Fusarium solani* (Willeke et al., 1983) degrade biochanin A. The catabolic route first involves dihydrobiochanin A (5,7-dihydroxy-4′-methoxyisoflavone) and then leads via 3-(p-methoxyphenyl)-4,6-diketo-5,6-dihydro-4H-pyran to phenylacetic acids derived from the side chain phenyl ring and carbon atoms 2 and 3 of the heterocycle. O-Demethylation of biochanin A has also been reported to occur (Willeke and Barz, 1982a; Willeke et al., 1983).

Ascochyta rabiei rapidly degraded biochanin A within 1–3 h after incubation in the nutrient medium, and transient formation of two isoflavones pratensein and genistein was seen (Kraft and Barz, 1985). 3′-Hydroxylation and O-demethylation reactions led to the isoflavones pratensein, genistein, and orobol, which were rapidly further degraded. Dihydrogenistein and p-hydroxyphenylacetic were also identified as catabolites (Kraft and Barz, 1985). Biochanin A-7-O-glucoside was degraded, leading to aglycone and pratensein. Biochanin A-7-O-glucoside-6″-O-malonate was very slowly degraded without subsequent accumulation of catabolites (Kraft and Barz, 1985). Removal of malonic acid from biochanin A-7-O-glucoside-6″-malonate by action of an esterase has been shown to occur with *Fusarium javanicum* (Schlieper et al., 1984). The secondary spread of fungi that cause foliar disease results when sporulation occurs on the leaf surface and conidia are dispersed to previously uninfected tissue. Some fungi produce conidia in a water-soluble mucilage and, in this case, conidium dispersal depends on both water and wind since liquid water is typically involved in the movement of propagules away from the site of sporulation onto new, healthy tissues. Sporulation occurs within a zone of necrotic tissue that may contain high levels of phenols produced during lesion restriction. Toxic phenols may be exuded from the lesions and may exist as normal components of the phylloplane. Analysis of the material that leached into water droplets on necrotic lesions showed the presence of significant levels of phenols within 1 h, (Nicholson et al., 1989) whereas phenols could not be detected in leachate from green leaf tissue. The leachate from necrotic lesions contained p-coumaric and ferulic acids. These phenolic acids inhibited conidial germination in vitro (Nicholson et al., 1989). The phenols released into the water that carries conidia to new infection sites may inhibit conidium germination and germtube elongation resulting in suppression of disease spread (Nicholson et al., 1989).

Conidia of *Colletotrichum graminicola*, the causal agent of corn anthracnose, are embedded in a mucilaginous matrix (Nicholson and Moraes, 1980; Ramadoss et al., 1985). The mucilage contains a proline-rich protein fraction capable of selectively binding to condensed tannins and other polyphenolic compounds (Nicholson et al., 1986). The phenols could inhibit conidium germination only when the mucilage was removed from conidia but not when mucilage was added back to the washed conidia (Nicholson et al., 1986). The mucilage contained a nonspecific esterase as well as a β-glucosidase (Nicholson and Moraes, 1980; Ramadoss et al., 1985). The enzymes found in the mucilage may cleave the phenolic esters and glycosides, freeing the aglycones and making them more available for binding to the extracellular proline-rich proteins of the mucilage (Nicholson et al., 1986). As a consequence their toxicity would be reduced. Thus the proline-rich proteins of the spore mucilage may protect conidia from toxic phenols that accumulate in the water necessary for conidium dispersal and secondary spread of the fungus (Nicholson et al., 1989).

2. Pathogen Suppresses Synthesis of Phenolics in Plants

The pathogens may suppress increased synthesis of phenolics in susceptible tissues. Uninfected roots of *Eucalyptus calophylla* resistant to *Phytophthora cinnamomi* contained more than three times the amount of phenolic substances as those of *E. marginata* (susceptible to the pathogen). A rapid increase in the concentration of total phenolics in roots of *E. calophylla* followed inoculation with the pathogen. The level increased up to 97% above control levels within 72 h after inoculation. In roots of *E. marginata* there were only small increases in the concentration of total phenolics (Cahill and McComb, 1992).

A correlation between accumulation of phenolic compounds and resistance in alfalfa stems inoculated with *Colletotrichum trifolii* has been reported (Baker et al., 1989). In quantitative high-performance liquid chromatographic (HPLC) studies several distinct peaks were found to correlate with resistant interactions. In general the accumulation of fluorescent and ultraviolet- (UV) absorbing compounds in alfalfa stem tissue depended on both the phenotype of the clone and race of the pathogen. RR clones (resistant to race 1 and 2 of *C. trifolii*) accumulated high levels of these compounds in response to both races. RS clones (resistant to race 1 and susceptible to race 2) accumulated high levels only in response to race 1. SS clones (susceptible to both races) did not accumulate comparable levels of these compounds (Baker et al., 1989). Thus in susceptible interactions phenolic synthesis does not seem to be accelerated.

3. Pathogen Suppresses Phenol Biosynthetic Enzymes

PAL activity increased rapidly to a very high level in roots of *Eucalyptus calophylla*, a species resistant to *Phytophthora cinnamomi* after infection with the pathogen. In a susceptible species, however, *E. marginata*, the activity of PAL increased only slightly after inoculation with the fungus in the first 24 h. Subsequently a rapid decline in the PAL activity was observed and after 96 h after inoculation PAL activity declined even up to 77% of control levels (Cahill and McComb, 1992). The importance of suppression of PAL in reduction in phenolic synthesis was shown by using an inhibitor of PAL: aminooxyacetic acid (AOA). The increase in phenolic synthesis was suppressed when the resistant *E. calophylla* was treated with AOA (Table 5; Cahill and McComb, 1992). AOA treatment made the resistant *E. calophylla* into a susceptible one (Cahill and

Table 5 Effect of aminooxyacetate (AOA) on phenolic concentration after inoculation of roots of *Eucalyptus calophylla* with *Phytophthora cinnamomi*

	Total phenolic concentration (mg p-coumaric acid equivalent/g fresh wt.)	
Treatment	Uninoculated	Inoculated
Control	1.69	2.71
AOA	1.76	1.36

Source: Cahill and McComb, 1992.

McComb, 1992) and it suggests that suppression of phenolic synthesis leads to susceptibility.

PAL may play a key role in the development of resistance to pathogens in plants (Ralton et al., 1989; Ward et al., 1989). Changes in PAL activity and transcription generally do not occur in compatible interactions (Esnault et al., 1987; Bhattacharyya and Ward, 1988). The switching-off of PAL genes appears to be an important early step in the development of pathogens in susceptible hosts (Cahill and McComb, 1992).

4. Pathogen Suppresses Phenolic Metabolism by its Suppressor Molecule

Pathogens are known to produce suppressors. The culture broth of *Ascochyta rabiei*, the chickpea pathogen, showed inhibitory activity on isoflavone accumulation in chickpea cotyledons. The inhibitory activity was found in a glycoprotein fraction that was called "suppressor." The inhibitory activity of the suppressor was observed in the concomitant decrease in the accumulation of all phenolic constituents of chickpea (Table 6; Kessmann and Barz, 1986).

5. Pathogen Suppresses Phenolic Metabolism by Producing Toxins

Toxins produced by pathogens may also suppress phenolic biosynthesis in plants. When rice leaves were treated with toxin produced by *Helminthosporium oryzae* (Vidhyasekaran et al., 1986) phenolic content decreased and PAL activity was suppressed (Table 7; Vidhyasekaran et al., 1992). When *H. oryzae* was inoculated, phenolic content of rice leaves decreased at 24 h after inoculation. A decrease in PAL activity was observed on the third day. When an avirulent isolate that is not

Table 6 Suppression of phenolic accumulation in chickpea cotyledons by the suppressor of *Ascochyta rabiei*

	Isoflavonoids (nmol/g fresh wt.)			
Treatment	Formononetin	Biochanin A	Formonetin 7-O-glucoside-6″-malonate	Biochanin 7-O-glucoside-6″-malonate
Control	37	104	140	57
Suppressor	30	78	45	27

Source: Kessmann and Barz, 1986.

able to produce the toxin was inoculated on the rice leaves, there was no decrease in phenolic content or reduction in PAL activity. The results suggest that the toxin produced by the pathogen may suppress the phenolic metabolism of plants, resulting in severe incidence of the disease (Vidnyasekaran et al., 1992).

6. Pathogen Suppresses Oxidation of Phenolics by Inhibiting Polyphenol Oxidase

Alternaria alternata induces chlorotic symptoms in mungbean. Tentoxin is a cyclic tetrapeptide produced by the fungus and the toxin induces chlorosis (Templeton, 1972; Duke et al., 1980, 1982). Mungbean plants showed high levels of PPO. PPO was absent from plastids of tentoxin-treated mungbeans. Mixed control and tentoxin-treated extracts showed no inhibition of PPO activity nor did direct addition of tentoxin solutions to assay mixtures of control preparations, indicating that neither tentoxin nor a substance induced in tentoxin-treated plants was a PPO inhibitor. Tentoxin reduced PPO activity by 90% in ivy-leaf morning

Table 7 Suppression of phenolic content and phenylalanine ammonia-lyase activity in rice leaves due to treatment with *Helminthosporium oryzae* toxin

Treatment	Phenolics (μg/g fresh tissue)	Phenylalanine ammonia-lyase (nmol cinnamic acid produced/ g fresh tissue/h)
Untreated control	1924	918
Toxin	865	597

Source: Vidhyasekaran et al., 1992.

glory, a sensitive species. Tentoxin had no effect on PPO from soybean, a species insensitive to tentoxin (Table 8; Vaughn and Duke, 1981a).

Tentoxin appears to be a specific inhibitor of chloroplast protein incorporation in mungbean. Control plastids and plastids recovering from a tentoxin treatment always contain the nuclear-coded protein polyphenol oxidase (PPO), yet affected plastids totally lacked this enzyme activity (Vaughn and Duke, 1981a, 1982). It suggests that tentoxin neither interferes with synthesis nor acts as an inhibitor of PPO, but instead affects its activation or incorporation into the plastid.

Electrophoretic separation of thylakoid proteins from control mungbeans followed by incubation in aerated dihydroxyphenylalanine solutions gave a strong band of activity at 36.5 kDa. Electrophoresis of tentoxin-treated material revealed no band at 36.5 kDa (Vaughn and Duke, 1981a). Cytochemical studies revealed that PPO activity was present in the control plastids along the photosynthetic lamellae when no activity was noted in the tenxotin-treated plastids. All these studies indicated that tentoxin-treated material lacked any plastidic PPO. Integration of PPO into the plastid may occur by incorporation of this protein into the small vesicles found along the plastid envelope. Only in apical plastids were envelope-associated areas of PPO-staining noted, suggesting that this may be the vehicle for entry of PPO into the developing plastid. This process of apparent PPO integration did not occur in tentoxin-treated apical plastids (Vaughn and Duke, 1981a).

Immunocytochemical studies with *Vicia faba* indicated that an immuno-

Table 8 Polyphenol oxidase activities of tentoxin-treated plant extracts

Treatment + plant material	Polyphenoloxidase activity as change of absorbance (min/g fresh weight)
Mungbean leaves	
Control	190
Tentoxin	4
Mungbean leaves	
Control plus extract of tentoxin treated	185
Control plus tentoxin in vitro	168
Tentoxin	0
Morningglory cotyledons	
Control	123
Tentoxin	15
Soybean seedlings	
Control	38
Tentoxin	40

Source: Vaughn and Duke, 1981a.

logically recognizable PPO is present at the envelope but not incorporated into the plastids in tentoxin-treated *V. faba* plants (Vaughn and Duke, 1984). Sodium dodeyl sulfate-polyacrylamide gel electrophoresis (SDS-PAGE) of whole leaf extracts of control and tentoxin-treated *V. faba*, with subsequent staining for PPO, revealed one stained band at 40 kDa in the control extracts. No stained band was noted at any position of the gel in the tentoxin-treated extracts (Vaughn and Duke, 1984b). Other parts of the gel not stained for PPO activity were blotted to nitrocellulose and reacted with monospecific antibody to *Vicia* PPO. In both control and tentoxin-treated extracts a single band at 40 kDa was observed. The observations suggest that although the PPO of tentoxin-treated plants is inactive, there is no significant difference between the molecular weight of the inactive and active PPO forms (Vaughn and Duke, 1984). All of these studies indicate that PPO accumulates in the plastid envelope membranes in tentoxin-affected cells and that PPO in these treated plants is not processed to an active protein.

7. Phenolics are Fungitoxic but Do Not Accumulate to Fungitoxic Levels during Pathogenesis

Phenolics may accumulate during infection but even the increased concentrations may not be sufficient to inhibit pathogens. In an incompatible interaction there were large accumulations of the malonyl and neutral glucosides of daidzein and genistein at 24–36 h after inoculation with *Phytophthora megasperma* f.sp. *glycinea* in soybean leaves (Table 9; Morris et al., 1991). No growth inhibition of *P. megasperma* f.sp. *glycinea* could be demonstrated on medium that contained up to 200 μg/ml daidzein, or formononetin, or genistein (Morris et al., 1991). Constitutive levels of isoflavonoids and their glucosides both in leaves and hypocotyls of soybean were found to be too low to contribute to resistance based on their fungitoxicity (Fett, 1984; Fett and Jones, 1984; Cosio et al., 1985; Graham, 1991; Morris et al., 1991).

F. Phenolics May Not Have Any Role in Host's Defense Mechanisms

Several studies indicate that phenolics may not play an important role in host's defense mechanisms. Two types of single-gene-based resistance to the leaf rust fungus (*Puccinia recondita* f.sp. *tritici*) have been reported in wheat. The type, based on the Lr 28 allele, is rapid, being effective at about 24 h after inoculation while the other type, based on the Lr 20 allele, is slower, being effective at about 40 h from inoculation (Southerton and Deverall, 1989). The rust strain 104-2,3,6,7,8 is virulent on Lr 28 gene cultivar while 104-1,2,3,6 is avirulent on it. In contrast, the first strain is avirulent on Lr 20 gene cultivar while the latter strain is highly virulent on it. Three soluble phenolic acids were detected in extracts

Table 9 Accumulation of isoflavonoids and isoflavone glucosides in leaves of a resistant soybean cultivar inoculated with *Phytophthora megasperma* f.sp. *glycinea* (μg/g fresh weight)

	Healthy leaves (control)	Infected leaves
Daidzein	< 5	< 5
Daidzin	20	152
6″-0-malonyl-7-0-glucosyl-daidzein	36	308
Genistein	< 1	4
Genistin	34	102
6″-0-malonyl-7-0-glucosyl-genistein	68	228
Glyceitin-7-0-β-glucoside	6	61
Formononetin	–	11
Biochanin A	< 5	< 5

Source: Morris et al., 1991.

from healthy and virulent and avirulent isolates-inoculated leaves of both cultivars (Table 10; Southerton and Deverall, 1990).

There were only slight differences in ferulic, p-coumaric, and syringic acid levels between control leaves and leaves of both wheat cultivars inoculated with either rust strain (Southern and Deverall, 1990). Attempts to affect expression of

Table 10 Changes in phenolic acid content of wheat leaves inoculated with *Puccinia recondita* f.sp. *tritici*

	Phenolic acid content of leaves (μg/g dry weight)					
	Lr28			Lr20		
Phenolic acid	Control	Race 104-2,3,6,7,8 (virulent)	Race 104-1,2,3,6 (avirulent)	Control	Race 104-1,2,3,6 (virulent)	Race 104-2,3,6,7,8 (avirulent)
Ferulic acid	661	611	656	1250	1297	1321
p-Coumaric acid	24	23	23	69	68	69
Syringic acid	119	141	150	81	78	86
Total soluble	804	775	829	1400	1443	1476

Source: Southerton and Deverall, 1990.

resistance by application of inhibitors of phenylalanine ammonia-lyase (PAL) such as α-aminooxyacetate (AOA) and 2-aminoindan-2-phosphoric acid (AIP) were not successful and both wheat cultivars remained resistant to avirulent rust strains (Southerton and Deverall, 1990). The results suggest that phenolic acids may not be involved as a defense mechanism in wheat plants against the leaf rust fungus.

It appears that, in the case of wheat stem rust also, phenolic is not involved in host's resistance mechanism observed in resistant varieties. In contrast, there were decreases in free phenolic acid content in both susceptible and resistant varieties at 2 and 5 days after inoculation with *Puccinia graminis* f.sp. *tritici* (Chigrin et al., 1973). Both the susceptible and resistant chickpea cultivars accumulate almost identical amounts of the phenolic conjugates when inoculated with *Ascochyta rabiei* (Weigand et al., 1986).

Phloretin (Noveroske et al., 1964) and sieboldin (Hunter, 1971) are the preformed phenolics in apple that have been related to resistance against the scab fungus, *Venturia inaequalis*. Phloridzin was metabolized to phloretin by a crude enzyme extract from apple leaves and further oxidation products inhibited germination of conidia of *V. inaequalis* (Noveroske et al., 1964). However, phloridzin, phloretin and sieboldin have been reported to be noninhibitory to the fungal growth in culture (Noveroske et al., 1964; Hunter, 1971). The presence of phloridzin and sieboldin in progeny from Cox's Orange Pippin and *Malus floribunda* is not linked to their resistance (Noveroske et al., 1964; Hunter, 1971).

Chlorogenic acid was suggested to be involved in resistance against *V. inaequalis* in apple (Oydvin and Richardson, 1987). However, the observations were not consistent (Sierotzki and Gessler, 1993). It has also been reported that the flavan-3-ol in apple leaves of different cultivars was correlated positively with resistance (Treutter and Feucht, 1990). However, the content of flavan-3-ols in the leaves of apple is highly dependent on the age of the leaf. The first three leaves had the highest content, with decreasing amounts in older leaves. A positive correlation between preformed flavan-3-ol and ontogenetic resistance of apple trees against *V. inaequalis* was not evident. The flavan-3-ol contents of different apple cultivars varied from year to year. There was no correlation between flavan-3-ol content and resistance groups found in various crosses (Sierotzki and Gessler, 1993). No association between production of polyphenols in the phloem of stems of *Eucalyptus marginata* and the ability of *Phytophthora cinnamomi* to colonize tissues could be observed (Tippett et al., 1989).

Even the enzymes involved in biosynthesis of phenolics may not have any role in the defense system of the host. PAL and tyrosine ammonia–lyase (TAL) are the key enzymes in the phenylpropanoid pathway and may be involved in phenolic synthesis (Burrell and Rees, 1974). Extracts of rice leaves susceptible to *Pyricularia oryzae* had higher activities of PAL than extracts of resistant leaves and the PAL activity increased in both susceptible and resistant leaves inoculated with the pathogen, but the activity of inoculated resistant leaves was much lower than that from comparable inoculated susceptible leaves. No differences were found in TAL activity between extracts of uninfected and inoculated plants. The results suggest that increases in PAL or TAL may not be part of the mechanisms of resistance but may be secondary effects of infection (Burrell and Rees, 1974).

Phenolics may accumulate more in the susceptible interactions, probably as degradation products. Histological examination of the interaction between *Rhynchosporium secalis* and susceptible and resistant cultivars of barley revealed that cells collapsed and phenolics accumulated in advance of hyphae in susceptible cultivar only. No cell collapse or accumulation of phenolic substances took place in the resistant cultivar after inoculation with the fungus (Jorgensen et al., 1993). The results suggest that phenolics may not be involved in disease resistance, but they may be only as products accumulating due to cell necrosis and cell death.

III. GLUCOSINOLATES

A. Biosynthesis of Glucosinolates

Glucosinolates are sulfur-containing glycosides and are commonly found in the family Cruciferae (Hooker et al., 1945; Heaney and Fenwick, 1989; Porter et al., 1991). They are comprised of a common glycoside moiety and a variable side chain (Mithen, 1992). Three major classes of glucosinolates have been recognized in *Brassica*. The first possesses side chains that may have aliphatic alkenyl or hydroxyalkenyl groups. The second has side chains that contain an indolyl group and the third one contains aralkyl side chains (Mithen, 1992).

The major glucosinolates found in *Brassica* species are sinigrin, gluconapin, glucobrassicanapin, progoitrin, hapoleiferin, glucoiberverin, glucoraphanin, hydroxyglucobrassicin, glucobrassicin, methoxyglucobrassicin, neoglucobrassicin, gluconasturtium, and glucotropaeolum (Mithen et al., 1987b; Mithen, 1992; Mithen and Magrath, 1992). The content of individual glucosinolates varies widely between different *Brassica* plants. *Brassica oleracea* and *B. cretica* subsp. *cretica* contain more gluconapin, while *B. rupestris* does not contain gluconapin and contains only glucoiberin. *B. oleracea* var. *capitate* contains mostly glucobrassicin and sinigrin and methoxyglucobrassicin are also present in appreciable amounts in it. *B. depranensis* contains glucoiberverin and glucoiberin while *B. macrocarpa*, *B. insularis*, and *B. cretica* subsp. *laconica* contain more sinigrin (Mithen et al., 1987b). *B. rapa* contains gluconapin and glucobrassicanapin (Mithen et al., 1987b). Glucobrassicin-1-sulphonate has been isolated from *Isatis tinctoria* (Elliott and Stowe, 1971).

Glucosinolates are derived from amino acids along a biosynthetic pathway that consists of two parts. Initially there is a chain elongation phase, which is followed by the biosynthesis of the specific part of the glusinolate molecule (Underhill et al., 1973). The side chains are derived from amino acids such as methionine, tryptophan, and phenylalanine (Underhill et al., 1973). The indolyl glucosinolates are derived from tryptophan (Mithen and Magrath, 1992). The glycone moiety is developed through a complex series of general nitrogenous and sulfur-containing intermediates (Underhill et al., 1973).

Glucosinolates are hydrolyzed by an endogenous enzyme called myrosinase (thioglucoside glucohydrolase). The hydrolysis products include nitriles, isothiocyanates, oxazolidinethiones, and thiocyanates (Fenwick et al., 1983; Mithen, 1992). The enzyme at low pH (3–4) hydrolyzes glucobrassicin to 3-indoleacetonitrile (IAN), glucose, sulfate, hydrogen sulfide and sulphur hydrogen sulfide and sulphur (Thimann and Mahadevan, 1964). Isothiocyanates are derived from the hydrolysis of alkenyl glucosinolates such as sinigrin and gluconapin. Thiocyanate ion is derived from indole glucosinolate hydrolysis while vinyloxazolidine-2-thione is derived from the hydrolysis of the hydroxyalkenyl glucosinolate progoitrin (Mithen et al., 1987a). Indolyl glucosinolates may produce nitrile derivatives also. Under acidic conditions the predominant hydrolytic products from alkenyl and hydroxyalkenyl glucosinolates are nitriles rather than isothiocyanates (Mithen, 1992). Aldoxime is an intermediate in pathway in the biosynthesis of aliphatic glucosinolates as well as in the biosynthesis of cyanogenic glycosides (Underhill et al., 1973).

Myrosinase, the enzyme that hydrolyzes glucosinolates, occurs in multiple forms. They are all glycoproteins with two to four subunits and a molecular weight between 125 and 153 kDa (Mithen, 1992). Myrosinase may be associated within a group of histochemically distinct cells known as myrosin cells (Thagstad et al., 1991). The myrosinase is associated with the tonoplast-like membrane that surrounds the myrosin grains within the myrosin cells. Iverson et al. (1979) have demonstrated myrosinase activity within tissues that lack myrosin cells. Hence myrosinase and myrosin grains may occur within many cells. Myrosinase may be associated with membrane while glucosinolates may be located within the vacuole (Matile, 1980; Luthy and Matile, 1984). Ascorbic acid promotes the activity of myrosinase (Mithen, 1992).

B. Toxicity of Glucosinolates to Fungal Pathogens

Glucosinolates are not toxic to fungal pathogens, but the hydrolysis products of glucosinolates are fungitoxic (Rawlinson et al., 1985; Mithen et al., 1987a; Doughty et al., 1991). 1-Methoxyglucobrassicin and sinigrin did not inhibit growth of *Leptosphaeria maculans*, the pathogen that causes stem canker disease of oilseed rape in vitro (Mithen et al., 1986). However, all glucosinolates (sinigrin, gluconapin, glucobrassicin, and 1-methoxyglucobrassicin) reduced fungal growth when added to cultures containing myrosinase. Progoitrin was not fungitoxic even in the presence of myrosinase. Sinigrin in the presence of myrosinase had the most pronounced effect. Volatile products from sinigrin and gluconapin were strongly inhibitory to *L. maculans* and no growth occurred on petri plates exposed to these products (Mithen et al., 1986). The results suggest that hydrolysis products of glucosinolates are fungitoxic rather than glucosinolates themselves, and toxicity of these hydrolysis products vary among glucosinolates.

Mithen et al. (1986) showed that hydrolysis products of glucobrassicin are highly fungitoxic. Further antifungal activity of the hydrolysis products was attributable to the products possessing an indole nucleus and not to the thiocyanate ion. Of the indole compounds, indolyl-3-carbinol had the greatest activity (Table 11; Mithen et al., 1986).

The volatile isothiocyanate products such as allyl-(or 2-propenyl-) and 3-butenyl isothiocyanates formed from the glucosinolates, sinigrin, and gluconapin, respectively, have toxic activity against many fungi (Walker et al., 1937; Pryor et al., 1940; Davis, 1964; Drobnica et al., 1967; Hartill, 1978). Isothiocyanates are toxic towards *Peronospora parasitica* (Greenhalgh and Mitchell, 1976), and *Mycosphaerella brassicae* (Hartill and Sutton, 1980). The germination of resting spores of *Plasmodiophora brassicae* was severely inhibited by allyl and β-phenethyl-isothiocyanate concentrations as low as 5 μg/ml. Davis (1964) showed that β-phenethyl isothiocyanate inhibited growth of *Fusarium oxysporum* at a concentration of 30 μg/ml.

Glucosinolates may serve as source for synthesis of phytoalexins in oilseed rape leaves (Conn et al., 1988). The phytoalexin cyclobrassinin accumulates in oilseed rape leaves due to infection with *Alternaria brassicae*. This compound is derived from tryptophan and indolyl glucosinolates are also derived from tryptophan. Hence it has been suggested that the phytoalexin is derived from

Table 11 Inhibition of *Leptosphaeria maculans* by glucobrassicin and its degradation products

Treatment	Growth (% of control)
No glucosinolate (control)	100.0
Glucobrassicin	95.0
Glucobrassicin + myrosinase	14.7
Thiocyanate ion	106.3
Indolyl-3-carbinol	7.9
Indole-3-acetonitrile	71.6
Diindolylmethane	72.1

Source: Mithen et al., 1986.

indolyl glucosinolates (Hanley and Parsley, 1990). The phytoalexin is inhibitory to pathogens.

C. How Does the Pathogen Overcome Toxicity of Glucosinolates?

1. Concentration of Glucosinolates May be Less in Susceptible Tissues

The plants susceptible to pathogens have been shown to contain fewer glucosinolates. Genotypes in both *Brassica oleracea* and *B. rapa* susceptible to *L. maculans* show less glucosinolate content than the resistant genotypes (Table 12; Mithen et al., 1987a).

High levels of resistance to *Peronospora parasitica* have been found in wild members of the *B. oleracea* group (Greenhalgh and Mitchell, 1976). Wild populations contain a higher total glucosinolate content than cultivars (Mithen et al., 1987b). The cabbage variety susceptible to *P. parasitica* contained lower concentrations of allyl isothiocyanate than the susceptible cultivars (Greenhalgh and Mitchell, 1976).

2. Glucosinolates May not be Involved in Disease Resistance Unless the Tissue is Damaged

Glucosinolates, unless hydrolyzed, are not toxic to pathogens. Hydrolysis requires high activation of myrosinases in plants. Similar activation may not occur unless the leaf tissue is highly damaged, as occurs with insect injury (Mithen and Magrath, 1992). Volatile derivatives of the glucosinolate sinigrin are produced on mechanical maceration of *Brassica* leaf tissue. Walker and Stahmann (1955) concluded that glucosinolates may not be important in the resistance of crucifers to *Plasmodiophora brassicae*, since hydrolysis of a glucoside occurs only when host cells are severely damaged, but there was intimate contact of the plasmodium with the cytoplasm of the invaded cell indicating less or no damage of host cells.

Table 12 Glucosinolate content of genotypes of *Brassica oleracea* and *B. rapa*, susceptible or resistant to *Leptosphaeria maculans*

Brassica species	Genotypes	Total glucosinolates (μmol/g fresh wt.)
B. oleracea	Susceptible	1.16
	Resistant	9.42
B. rapa	Susceptible	0.92
	Resistant	7.94

Source: Mithen et al., 1987a.

Although it is known that fungi can also produce myrosinase (Reece et al., 1958; Ohtsuru et al., 1969), it has not yet been established that enough myrosinase is produced in infected leaves to release sufficient fungitoxic hydrolytic products (Mithen, 1992).

In many host–pathogen interactions, no relationship between glucosinolates and disease resistance has been observed. The rape cultivar Bienvenu contains high glucosinolate content while the rape cultivar Cobra contains very low glucosinolate content. The cultivar Cobra had a reduced activity of myrosinase also (Porter et al., 1990), but both the cultivars were equally susceptible to *Alternaria brassicae* (Doughty et al., 1991). Inoculation with *A. brassicae* increased the concentration of glucosinolates in rape leaves by 4–10-fold, with the greatest increases in aromatic and indolyl glucosinolates. Both the cultivars Cobra and Bienvenu responded similarly to infection (Doughty et al., 1991).

Synthetic lines of *B. napus* were derived by combining the genomes of *B. atlantica* and *B. oleracea* var. *alboglabra*, which were, respectively, resistant and susceptible to *Leptosphaeria maculans* with a susceptible line of *B. rapa* (Mithen and Magrath, 1992). Resistance was expressed in the synthetic lines containing the genome of *B. atlantica*. The leaves of the synthetic *Brassica napus* line 235, derived from *B. atlantica* were highly resistant to *L. maculans*. The synthetic *B. napus* line 235 contained predominantly glucobrassicanapin and gluconapin, with smaller amounts of sinigrin and low levels of indolyl glucosinolates. The synthetic line 233, which is susceptible to *L. maculans*, contained smaller amounts of alkenyl glucosinolates than line 235 but higher levels of the indolyl glucosinolate glucobrassicin. The F_1 hybrid between these lines contained intermediate levels of both alkenyl and indolyl glucosinolates (Mithen and Magrath, 1992). Due to infection glucosinolate content was very much reduced in both susceptible and resistant lines. Those from the resistant synthetic *B. napus* line 235 contained predominantly alkenyl glucosinolates while those from the susceptible *B. napus* line 233 had predominantly indolyl glucosinolates with only trace levels of alkenyl glucosinolates. Within the F_2 population of a cross between the synthetic lines, there was a segregation of both disease resistance and glucosinolates, but there was no correlation between the levels of alkenyl or indolyl glucosinolates and disease resistance (Mithen and Magrath, 1992). Thus it appears that resistance genes are not related to the glucosinolate profile.

Mithen (1992) suggested that different types of hydrolysis of glucosinolates may occur in susceptible and resistant interactions. For example, in susceptible interactions, nitriles and indole acetic acid may be formed (Butcher et al., 1974). In contrast, in incompatible interactions, the unstable isothiocyanate may be formed and subsequently antifungal compounds such as indole carbinol and its derivatives and, potentially, the indolyl phytoalexins such as brassilexin may be formed (Rouxel et al., 1989). However, this possibility is yet to be proved with experimental evidence.

IV. CYANOGENIC GLUCOSIDES

Some of the plants contain hydrocyanic acid (Vidhyasekaran, 1988b). Young sorghum plants contain abundant hydrocyanic acid (Vidhyasekaran et al., 1971). Sudan grass (*Sorghum vulgare* var. *sudanensis*), flax, and birdsfoot trefoil are the other important cyanogenic plants. Hydrocyanic acid is liberated from nitrile glucosides and it is released only when the cells are injured (Vidhyasekaran et al., 1971). Cyanide is toxic to most organisms because it inhibits respiration in mitochondria (Fry and Evans, 1977). It is toxic to fungi (Vidhyasekaran, 1988b). The growth of *Fusarium oxysporum* f.sp. *lini*, the flax wilt fungus, was strongly inhibited in vitro at 135 μg/ml hydrocyanic acid (Trione, 1960). Hydrocyanic acid was toxic to germination of uredospores of *Puccinia purpurea*, the sorghum pathogen, even at 15 μg/ml (Vidhyasekaran et al., 1971).

Stemphylium loti is a pathogen of birdsfood trefoil (*Lotus corniculata*) while *S. botryosum*, *S. sarcinaeforme*, and *S. consortiale* are nonpathogens of trefoil. *S. loti* is more tolerant to HCN than the nonpathogens. An enzyme has been suggested to be the cause for tolerance of *S. loti* to hydrocyanic acid (Fry and Millar, 1969).

Fungi capable of causing diseases in cyanogenic plants are known to produce cyanide hydratase (Fry and Evans, 1977; Wang et al., 1992; Cluness et al., 1993). Cyanide hydratase is a substrate-inducible fungal enzyme capable of converting cyanide to formamide (Fry and Evans, 1977). Cyanide hydratase of the sorghum pathogen *Gloeocercospora sorghi* has been characterized. It was cyanide-inducible and functioned as an aggregated protein consisting of 45-kDa polypeptides (Wang et al., 1992). The corresponding gene has been cloned (Wang and VanEtten, 1992).

V. SAPONINS

Saponins have been implicated in disease resistance in some plants. The saponin avenacin occurs in oat roots. The resistance of oats to nonpathogens has been attributed to the presence of avenacin (Maizel et al., 1964). Avenacin is a mixture of four major compounds (avenacins A-1, A-2, B-1, and B-2), which are all glycoconjugates containing β-linked glucosyl residues (Crombie et al., 1984a,b, 1986a). Avenacin A-1 is the most abundant of the four in oat roots (Crombie and Crombie, 1986). *Gaeumannomyces graminis* is a pathogen of oats and wheat while *G. graminis* var. *tritici* can infect only wheat and not oats. Avenacin inhibits the growth of *G. graminis* var. *tritici* in culture while *G. graminis* var. *avenae* isolates are less sensitive to avenacin (Crombie et al., 1986b). These isolates deglucosylate avenacin A-1 to less toxic derivatives (Turner, 1961; Crombie et al., 1986b) by producing avenacinase with β-glucosidase activity (Turner, 1961). Other oat-infecting fungi, *Fusarium avenaceum* and *Phialophora* sp., also degrade avenacin (Crombie et al., 1986b). *G. graminis* var. *tritici* isolates produce less avenacinase (Crombie et al., 1986b).

The effect of avenacin A-1 on the growth of *G. graminis* var. *avenae* and *G. graminis* var. *tritici* isolates was studied (Osbourn et al., 1991). Growth of all the *G. graminis* var. *tritici* was completely inhibited at 5 μg avenacin A-1 per ml of agar, while *G. graminis* var. *avenae* isolates were still able to grow in the presence of higher levels of avenacin A-1 (20–40 μg/ml) (Osbourn et al., 1991). *G. graminis* var. *tritici* isolates did not produce avenacinase, while *G. graminis* var. *avenae* isolates produced significant amount of avenacinase. When culture filtrates of *G. graminis* var. *avenae* were incubated with avenacin A-1, deglucosylation of the substrate to both the mono- and bis-deglucosylated forms occurred. *G. graminis* var. *tritici* had only a very weak ability (less than 1/100 of that present in *G. graminis* var. *avenae* isolates) to generate the monodeglucosylated form. The bisdeglucosylated form was never observed when *G. graminis* culture filtrate preparations were incubated with avenacin A-1. The ability to generate the bisdeglucosylated form may be particularly important in overcoming the toxic effects of avenacin during infection of oat roots, since it has been shown that while *mono*-deglucosylated avenacin A-1 has considerably reduced toxicity, the bisdeglucosylated form is even less toxic (Crombie et al., 1986b).

Although *G. graminis* var. *avenae* grew well in the presence of avenacin A-1, there was no relationship between avenacinase-producing ability of the isolates and their virulence (Table 13; Osbourn et al., 1991). Although the isolate Gg 178 was the most pathogenic isolate on oats, it produced only less avenacinase. The isolate Hornby 61 is the least virulent, but it produced high amount of avenacinase. The results indicate that the ability to produce avenacinase may be an absolute requirement during early infection of oat roots (since var. *tritici* isolates lacking ability to produce avenacinase could not infect oat roots) but once

Table 13 Relationship between virulence and avenacinase production by *Gaeumannomyces graminis* var. *avenae* isolates

G. graminis var. *avenae* isolate	Oat seedlings showing symptoms (%)[a] 0	+	++	+++	Avenacinase activity (μmol/min) mycelium (g × 10^3)
Hornby 61	71	29	0	0	11.9
Hornby 174	47	47	6	0	3.4
Gg 178	0	0	58	42	4.4

[a]Symptoms based on the following scale: 0, no symptoms; +, lesions visible on roots; ++, lesions on roots extensive; +++, lesions on roots as well as browning of leaf sheath.
Source: Osbourn et al., 1991.

within the root cortex the subsequent extent of infection may depend on other factors. Avenacin is restricted to the outer cell layers of oat roots only (Osbourn et al., 1991).

Avenacinase activity was shown to reside in a single polypeptide species of 100 kDa. This protein was able to generate both the mono- and bisdeglucosylated forms of avenacin A-1 (Osbourn et al., 1991).

Specific mutations have been made by transformation-mediated gene disruption procedures that eliminate the ability of *G. graminis* var. *avenae* to detoxify avenacin (VanEtten et al., 1995). The mutants lost their pathogenicity on oats (VanEtten et al., 1995). The importance of avenacin in disease resistance is further demonstrated by the observation that one diploid oat species, *Avena longiglumis*, which lacked detectable levels of avenacin, was susceptible to infection by the nonpathogen of oat *G. graminis* var. *tritici* (Osbourn et al., 1994). Thus avenacin degradation appears to be important in pathogenesis.

VI. STEROID ALKALOID

Alkaloids also contribute to disease resistance in some plants. α-Tomatine is a steroidal glycoalkaloid found in tomato. It is highly toxic to fungi. However, tomato pathogens are less sensitive to α-tomatine than the nonpathogens of tomato (Arneson and Durbin, 1968). The minimum molar concentration of tomatine that completely inhibits mycelial growth of pathogens of tomato such as *Septoria lycopersici*, *Alternaria solani*, *Phytophthora infestans*, *Fusarium oxysporum* f.sp. *lycopersici*, and *Verticillium albo-atrum* ranges from 0.01 to 0.85, while concentration of tomatine required to inhibit mycelial growth of nonpathogens of tomato such as *Cercospora beticola*, *Alternaria tenuis*, *Septoria linicola*, *F. oxysporum* f.sp. *conglutinans*, and *Helminthosporium turcicum* ranged from 0.00013 to 0.0020 M (Arneson and Durbin, 1968). Tomatine is degraded by a glucosidase produced by the tomato pathogen *S. lycopersici* (Arneson and Durbin, 1967). *S. lycopersici* detoxified tomatine both in vitro and in infected tomato leaves (Arneson and Durbin, 1967). Hence it appears that for successful pathogenesis the pathogen should be able to degrade tomatine in tomato.

VII. DIENES

Dienes are antifungal substances found in plants. Avocado fruits contain 1-acetoxy-2-hydroxy-4-oxo-heneicosa-12,15-diene, which has been suggested as the basis of resistance to fruit decay (Prusky et al., 1982, 1983). The concentration of the preformed diene is enhanced when the fruit is inoculated with *Colletotrichum gloeosporioides* (Prusky et al., 1990). When avocado fruits were exposed to 30% carbon dioxide for 24 h, level of the antifungal compound increased to a greater extent. This increase in diene level resulted in a delay in symptom development due to *C. gloeosporioides* infection (Prusky et al., 1991). Four different preformed antifungal compounds have been detected in the peel of avocado fruits (Sivanathan and Adikaram, 1989). The diene has been isolated from avocado leaves also (Chang et al., 1976). It inhibited mycelial growth as well as spore germination at 790 μg/ml (Prusky et al., 1982).

Pathogens could not infect avocado fruits when diene content is more, but diene content decreases to subfungitoxic concentrations during fruit ripening, permitting the development of disease symptoms caused by *C. gloeosporioides* (Prusky et al., 1982). The decrease in concentration of the diene has been reported to depend on the lipoxygenase-catalyzed oxidation of the compound (Prusky et al., 1985, 1988; Karni et al., 1989).

VIII. ANTIMETABOLITES

Antimetabolites may play an important role in inducing resistance against fungal pathogens. Amino acid analogues, particularly canavanine, ethionine, and p-fluorophenylalanine, inhibited rust development in wheat leaves. The inhibitions were reversed by comparable levels of arginine (Samborski and Forsyth, 1960). Increased resistance to *Venturia inaequalis* following injection of the host with analogues of normal metabolites has also been reported (Kirkham and Flood, 1956; Kuc et al., 1969). These studies suggest the importance of antimetabolites inducing resistance. Antimetabolites may act by preventing the metabolite from entering the cell rather than by inhibiting its utilization (Samborski and Forsyth, 1960).

Some nonprotein amino acids act as antimetabolites inhibiting fungal growth (Rosenthal, 1991). Canavanine (2-amino-4-(guanidinooxy)-butyric acid) does not occur naturally in protein (Ho and Shen, 1966). It is a naturally occurring structural analogue of arginine. The compound has been found in the subfamily Papilionidae of the Fabaceae (Rosenthal, 1990). Canavanine competes with arginine for binding onto transfer ribonucucleic acid (tRNA) (Allende and Allende, 1964). Canavanine suppresses histone biosynthesis in fungi (Childs et al., 1971).

L-canalin (L-α-amino-y-(aminooxy)-n-butyric acid) is hydrolytically cleaved by arginase from the nonprotein amino acid L-canavanine. L-canalin inhibits the action of pyridoxal phosphate-dependent enzymes (Rahiala et al., 1971; Rosenthal, 1981). Inhibition of pyridoxal phosphate-dependent enzymes prevents ethylene biosynthesis in higher plants (Murr and Yang, 1975; Hodges and Campbell, 1994). L-canaline decreases biosynthesis of endogenous ethylene and subsequent loss of chlorophyll by *Poa pratensis* leaf blades infected by *Bipolaris sorokiniana* (Hodges and Campbell, 1994). L-canavanine affects the development of some fungi by inhibiting hyphal growth and sporulation (Samborski and Forsyth, 1960; Shepherd and Mandyk, 1964; Toshikazu et al., 1971). Hyphal growth of *B. sorokiniana* is inhibited by L-canaline (Hodges and Campbell, 1994).

BABA (DL-3-amino-n-butanoic acid, β-aminobutyric acid) is a nonprotein amino acid found in some plants (Rosenthal, 1982; Gamliel and Katan, 1992). It may interfere with normal amino acid and/or protein synthesis (Cohen and Gisi, 1994). Application of BABA induces resistance against *Phytophthora infestans* in tomato (Cohen, 1994b; Cohen et al., 1994; Cohen and Gisi, 1994), *Aphanomyces euteiches* in pea (Papavizas, 1964), and *Peronospora tabacina* in tobacco (Cohen, 1994a). These results suggest that nonprotein amino acids may play an important role in disease resistance and fungal pathogenesis. However, the role of the antimetabolites during fungal pathogenesis has not yet been studied in detail.

IX. CONCLUSION

Phenolics are ubiquitous in plants. They are highly fungitoxic. The oxidized phenolics are much more inhibitory. Polyphenol oxidases oxidize phenolics to quinones and hence they may be involved in the defense mechanisms of the plants. Fungitoxicity of the phenolics toward different pathogens varies. The phenolics may affect fungi by altering membrane porosity, or by inhibiting enzymes, or by affecting DNA transcription. Some phenolics serve as signal molecules in inducing resistance. They are also antioxidants and may induce resistance by their action on lipid metabolism.

Successful pathogens have been shown to be able to degrade the phenolics into nontoxic ones. The pathogens produce enzymes that metabolize the phenolics. The pathogens may not induce the synthesis of phenolics in the invaded tissues. Absence of specific elicitors or presence of suppressors in the fungal cell wall may be the important cause for noninduction of defense chemicals. It is also possible that the pathogen may suppress the biosynthesis of phenolics, inhibiting the enzymes involved in the biosynthetic pathway. The pathogen may produce toxins that may suppress the phenolic synthesis. Successful pathogens may produce toxins that may suppress the host's polyphenol oxidases.

Although several kinds of phenolics are found in plants, they may not accumulate to fungitoxic levels during pathogenesis. In many host–pathogen interactions it has been demonstrated that phenolics may not have any role in the host's defense mechanisms. In some cases, phenolics accumulate more during pathogenesis, probably due to large-scale cell necrosis. Thus these phenolics may be only degradation products and may not be involved in the host's defense mechanisms. Several studies indicate that phenolics may not be solely responsible for inducing resistance, although they may also contribute to the host's defense mechanism. Successful pathogens overcome the fungitoxic phenolics by several methods and cause the disease.

Preformed antifungal compounds other than phenolics are also present in certain plants. Among them glucosinolates are common in plants belonging to the family Cruciferae. Only hydrolysis products of glucosinolates are fungitoxic. Glucosinolates may be located within the vacuole while myrosinase, the enzyme that hydrolyzes glucosinolates, may be associated with cell membranes. Hence tissue maceration seems to be essential for the synthesis of hydrolysis products of glucosinolates. Although some fungi are known to produce myrosinase, its role in infected plants has not been studied. Probably some insect pest injury may induce synthesis of hydrolytic products of glucosinolates and induce resistance against pathogens. However, experimental evidence is lacking to prove this possibility.

Cyanogenic glucosides are commonly found in some plants. Hydrocyanic acid is also released only when the cells are injured. Pathogens detoxify hydrocyanic acid by producing cyanide hydratase. Some saponins play an important role in disease resistance. Avenacin seems to be responsible for resistance in oats to nonpathogens. The oat pathogen is capable of producing avenacinase, which detoxifies avenacin. The steroid glycoalkaloid tomatine has been shown to be responsible for resistance of tomato to its nonpathogens. Tomatine is detoxified by a glucosidase produced by pathogens of tomato. Dienes are also fungitoxic substances and they are also degraded by oxidation catalyzed by lipoxygenase. Fungal infection would have accelerated lipoxygenase activity, which in turn detoxifies dienes.

Nonprotein amino acids are also common in plants and they may act as antimetabolites. Some of the antimetabolites have been shown to induce systemic resistance against disease. However, the present knowledge about antimetabolites is only rudimentary. Intensive studies may help us to devise effective control measures against diseases using antimetabolites.

REFERENCES

Adesanya, S. A., O'Neill, M. J., and Roberts, M. F. (1984). Induced and constitutive isoflavonoids in *Phaseolus mungo* Leguminosae, Z. Naturforschung, 39C:881–893.

Adesanya, S. A., O'Neill, M. J., and Roberts, M. F. (1985). Isoflavonoids from *Phaseolus occineus* L., Phytochemistry, 24:2699–2706.

Adesanya, S. A., O'Neill, M. J., and Roberts, M. F. (1986). Structure-related fungitoxicity of isoflavonoids, Physiol. Mol. Plant Pathol., 29:95–103.

Allende, C. C., and Allende, J. E. (1964). Purification and substrate specificity of arginyl-ribonucleic acid synthetase from rat liver, J. Biol. Chem., 239:1102.

Arneson, P. A., and Durbin, R. D. (1967). Hydrolysis of tomatine by *Septoria lycopersici*: a detoxification mechanism, Phytopathology, 57:1358–1360.

Arneson, P. A., and Durbin, R. D. (1968). The sensitivity of fungi to α-tomatine, Phytopathology, 58:536–537.

Ashwood-Smith, M. J., Ceska, O., and Chaudhury, S. K. (1985). Mechanism of photosensitivity reactions of diseased celery, Br. Med. J., 290:1249.

Ataga, A. E., Epton, H. A. S., and Frost, R. R. (1993). Effect of virus infection on the concentration of furanocoumarins in celery (*Apium graveolens* L. var. *dulce* Mill, D. C.), Physiol. Mol. Plant Pathol., 42:161–168.

Baker, C. J., O'Neill, N. R., and Tomerlin, J. R. (1989). Accumulation of phenolic compounds in incompatible clone/race interactions of *Medicago sativa* and *Colletotrichum trifolii*, Physiol. Mol. Plant Pathol., 35:231–241.

Bhattacharyya, M. K., and Ward, E. W. B. (1988). Phenylalanine ammonia-lyase activity in soybean hypocotyls and leaves following infection with *Phytophthora megasperma* f.sp. *glycinea*, Can. J. Bot., 66:18–23.

Burrell, M. M., and Rees, T. Ap. (1974). Metabolism of phenylalanine and tyrosine by rice leaves infected by *Piricularia oryzae*, Physiol. Plant Pathol., 4:497–508.

Butcher, D., El-Tigani, S., and Ingram, D. (1974). The role of indole glucosinolates in the clubroot disease of the Cruciferae, Physiol. Plant Pathol., 4:127–140.

Cahill, D. M., and McComb, J. A. (1992). A comparison of changes in phenylalanine ammonia-lyase activity, lignin and phenolic synthesis in the roots of *Eucalyptus calophylla* (field resistant) and *E. marginata* (susceptible) when infected with *Phytophthora cinnamomi*, Physiol. Mol. Plant Pathol., 40:315–332.

Carrasco, A., Boudet, A. M., and Marigo, G. (1978). Enhanced resistance of tomato plants to *Fusarium* by controlled stimulation of their natural phenolic production, Physiol. Plant Pathol., 12:225–232.

Casares, A., Melo, E. M. P. F., Ferraz, J. F. P., and Ricardo, C. P. P. (1986). Differences in ability to *Phytophthora cambivora* and *P. cinnamomi* to dephenolize lignin, Trans. Brit. Mycol. Soc., 87:229–235.

Chang, C., Isogai, A., Kamikado, S., Murakoshi, A., Sakurai, A., and Tamura, S. (1975). Isolation and structure elucidation of growth inhibitors for silkworm larvae from avocado leaves, Agric. Biochem., 39:1167–1168.

Chaudhury, S. K., Ceska, O., Warrington, P. J., and Ashwood-Smith, M. J. (1985). Increased furanocoumarin content of celery during storage. J. Agric. Food Chem., 33:1152–1157.

Chigrin, V. V., Rozum, L. V., and Zaprometov, M. N. (1973). Phenolcarboxylic acids and lignin in leaves of resistant and sensitive varieties of spring wheat during infection with stem rust, Soviet Plant Physiol., 20:801–806.

Childs, E. A., Ayres, J. C., and Koehler, P. E. (1971). Differentiation in *Aspergillus flavus* as influenced by L-canavanine, Mycologia, 63:181–184.

Cluness, M. J., Turner, P. D., Clements, E., Brown, D. T., and O'Reilly, C. (1993). Purification and properties of cyanide hydratase from *Fusarium lateritium* and analysis of the corresponding *chy* 1 gene, J. Gen. Microbiol., 139:1807–1815.

Cohen, Y. (1994a). 3-Aminobutyric acid induces systemic resistance against *Peronospora tabacina*, Physiol. Mol. Plant Pathol., 44:273–288.

Cohen, Y. (1994b). Local and systemic control of *Phytophthora infestans* in tomato plants by DL-3-amino-n-butanoic acids, Phytopathology, 84:55–59.

Cohen, Y., and Gisi, U. (1994). Systemic translocation of ^{14}C-DL-3-aminobutyric acid in tomato plants in relation to induced resistance against *Phytophthora infestans*, Physiol. Mol. Plant Pathol., 45:441–456.

Cohen, Y., Niderman, T., Mosinger, E., and Fluhr, R. (1994). β-aminobutyric acid induces the accumulation of pathogenesis-related proteins in tomato (*Lycopersicon esculentum* L.) and resistance to late blight infection caused by *Phytophthora infestans*, Plant Physiol., 104:59–66.

Conn, K. I., Tewari, J. P., and Dahiya, J. S. (1988). Resistance to *Alternaria brassicae* and phytoalexin-elicitation in rapeseed and other crucifers, Plant Sci., 56:21–25.

Cosio, E. G., Weissenbock, G., and MeClure, J. W. (1985). Acifluorfen-induced isoflavonoids and enzymes of their biosynthesis in mature soybean leaves, Plant Physiol., 78:14–19.

Crombie, L., Crombie, W. M. L., and Whiting, D. A. (1984a). Isolation of avenacins A-1, A-2, B-1 and B-2 form oat roots: structures of their aglycones, the avenestergenins, J. Chem. Soc. Chem. Commun. 244.

Crombie, L., Crombie, W. M. L., and Whiting, D. A. (1984b). Structures of the four avenacins, oat root resistance factors to take-all disease, J. Chem. Soc. Chem. Commun. 246.

Crombie, L., Crombie, W. M. L., and Whiting, D. A. (1986a). Structures of the oat root resistance factors to take-all disease, avenacins A-1, A-2, B-1 and B-2 and their companion substances, J. Chem. Soc. Perkins Trans., 1:1917–1922.

Crombie, W. M. L., and Crombie, L. (1986). Distribution of avenacins A-1, A-2, B-1 and B-2 in oat roots: their fungicidal activity towards take-all fungus, Phytochemistry, 25:2069–2073.

Crombie, W. M. L., Crombie, L., Green, J. B., and Lucas, J. A. (1986b). Pathogenicity of the take-all fungus to oats: in relationship to the concentration and detoxification of the four avenacins, Phytochemistry, 25:2075–2083.

Davis, D. (1964). Host fungitoxicity in selective pathogenicity of *Fusarium oxysporum*, Phytopathology, 54:290–293.

Doughty, K. J., Porter, A. J. R., Morton, A. M., Kiddle, G., Bock, C. H., and Wallsgrove, R. (1991). Variation in the glucosinolate content of oilseed rape (*Brassica napus* L.) leaves. II. Response to infection by *Alternaria brassicae* (Berk.) Sacc., Ann. Appl. Biol., 118:469–477.

Drobnica, I., Zemanova, M., Nemec, P., Antos, K., Kristian, P., Stullerova, A., Knoppova, V., and Nemec, P. Jr. (1967). Antifungal activity of isothiocyanates and their analogues, Appl. Microbiol., 15:701–717.

Duffy, S. S., and Felton, G. (1991). Enzymatic antinutritive defenses of the tomato plant against insects, Naturally Occurring Pest Bioregulators, (P. Heldin, ed.), American Chemical Society, Washington, DC., pp. 167–197.

Duke, S. O., Paul, R. N., and Wickliff, J. N. (1980). Tentoxin effects on ultrastructure and greening of ivyleaf morning glory (*Ipomoea hederacea* var. *hederacea*) cotyledons, Physiol. Plant., 49:27–36.

Duke, S. O., Wickliff, J. L., Vaughn, K. C., and Paul, R. N. (1982). Tentoxin does not cause chlorosis in greening mungbean by inhibiting photophosphorylation, Physiol. Plant., 56:387–398.

Elliott, M. C., and Stowe, B. B. (1971). Indole compounds related to auxins and goitrogens of woad (*Isatis tinctoria* L.), Plant Physiol., 47:366–372.

Enyedi, A. J., Yalpani, N., Silverman, P., and Raskin, I. (1992). Localization, conjugation and function of salicylic acid in tobacco during the hypersensitive response to tobacco mosaic virus, Proc. Natl. Acad. Sci. USA, 89:2480–2484.

Esnault, R., Chibbar, R. N., Lee, D., Van Huystee, R. B., and Ward, E. W. B. (1987). Early differences in production of mRNAs for phenylalanine ammonia-lyase and chalcone synthase in resistant and susceptible cultivars of soybean inoculated with *Phytophthora megasperma* f.sp. *glycinea*, Physiol. Mol. Plant Pathol., 30:293–297.

Farkas, G. L., and Kiraly, Z. (1962). Role of phenolic compounds in the physiology of plant diseases and disease resistance, Phytopath. Z., 44:105–150.

Fenwick, G. R., Heaney, R. K., and Mullin, W. J. (1983). Glucosinolates and their breakdown products in food and food plants, CRC Crit. Rev. Food Sci. Nutrit., 18:123–201.

Fett, W. F. (1984). Accumulation of isoflavonoids and isoflavone glucosides after inoculation of soybean leaves with *Xanthomonas campestris* pv. *campestris* and a study of their role in resistance, Physiol. Plant Pathol., 24:303–320.

Fett, W. F., and Jones, S. B. (1984). Stress metabolite accumulation, bacterial growth and bacterial immobilization during host and nonhost responses of soybean to bacteria, Physiol. Plant Pathol., 25:277–296.

Floss, H. G., and Guenther, H., and Hadwiger, L. A. (1969). Biosynthesis of furanocoumarin in diseased celery, Phytochemistry, 8:585.

Fry, W. E., and Evans, P. H. (1977). Association of formamide hydro-lyase with fungal pathogenicity to cyanogenic plants, Phytopathology, 67:1001–1006.

Fry, E. W., and Millar, R. L. (1969). Cyanide tolerance in *Stemphylium loti*, Phytopathology, 61:494–500.

Gamliel, A., and Katan, J. (1992). Influence of seed and root exudates on fluorescent pseudomonads and fungi in solarized soil. Phytopathology, 82:320–327.

Golbeck, J. H., and Cammarata, K. V. (1981). Spinach thylakoid polyphenoloxidase, isolation, activation, and properties of the native chloroplast enzyme, Plant Physiol., 67:977–984.

Graham, T. L. (1991). Flavonoid and isoflavonoid distribution in developing soybean seedling tissues and in seed and root exudates, Plant Physiol., 95:594–603.

Graham, T. L., and Graham, M. Y. (1991). Glyceollin elicitors induce major but distinctly different shifts in isoflavonoid metabolism in proximal and distal soybean cell populations, Mol. Plant–Microbe Interact., 4:60–68.

Graham, T. L., Kim, J. E., and Graham, M. Y. (1990). Role of constitutive isoflavone conjugates in the accumulation of glyceollin in soybean infected with *Phytophthora megasperma*, Mol. Plant–Microbe Interact., 3:157–166.

Greenhalgh, J. R., and Mitchell, N. D. (1976). The involvement of flavour volatiles in the resistance to downy mildew of wild and cultivated forms of *Brassica oleracea*, New. Phytol., 77:391–398.

Gregory, P., Tingey, W. M., Ave, D. A., and Bouthyette, P. Y. (1986). Potato glandular trichomes: a physiochemical defense mechanism against insects, Natural Resistance of Plants to Pests: Roles of Allelochemicals (M. B. Green and P. A. Heldin, eds.), American Chemical Society, Washington, DC., pp. 160–167.

Hammerschmidt, R., and Nicholson, R. L. (1977). Resistance of maize to anthracnose: effect of light intensity on lesion development, Phytopathology, 67:247–250.

Hanley, A. H., and Parsley, K. R. (1990). Identification of 1-methoxyindolyl-3-methyl isothiocyanate as an indole glucosinolate breakdown product, Phytochemistry, 29:769–771.

Harborne, J. B., and Ingham, J. (1978). Biochemical aspects of coevolution of higher plants with their fungal parasites, Biochemical Aspects of Coevolution in Higher Plants (J. B. Harborne, ed.), Academic Press, London, pp. 343–405.

Hartill, W. F. T. (1978). The role of allyl isothiocyanate in the development of 'ringspot' epidemics in brassicas, Proceedings, 3rd International Congress of Plant Pathology, Abstr., p. 217.

Hartill, W. F. T., and Sutton, P. G. (1980). Inhibition of germination of *Mycosphaerella brassicicola* ascospores on young cabbage and cauliflower leaves, Ann. Appl. Biol., 96:153–161.

Heaney, R. K., and Fenwick, G. R. (1980). Glucosinolates in *Brassica* vegetables. Analysis of 22 varieties of Brussel sprout (*Brassica oleracea* var. *gemmifera*), J. Sci. Food Agric., 31:785–793.

Ho, H., and Shen, C. (1966). Studies on the distribution of canavanine in Taiwan plants and its metabolic change during germination, J. Agr. Chem. Soc. Cec. Sp. Iss. 23.

Hodges, C. F., and Campbell, D. A. (1994). Inhibition of endogenous ethylene biosynthesis by L-canaline in leaves of *Poa pratensis* infected by *Bipolaris sorokiniana*, Physiol. Mol. Plant Pathol., 44:301–309.

Hooker, W. J., Walker, J. C., and Link, K. P. (1945). Effects of two mustard oils on *Plasmodiophora brassicae* and their relation to clubroot, J. Agr. Res., 70:63–78.

Hudson, B. J. F., and Mahgoub, S. E. O. (1980). Naturally occurring antioxidants in leaf lipids, J. Sci. Food Agric., 31:646–650.

Hunter, L. D. (1971). Is sieboldin a factor in apple scab resistance? Report of East Malling Research Station for 1970, pp. 134.

Hunter, R. E. (1974). Inactivation of pectic enzymes by polyphenols in cotton seedlings of different ages infected with *Rhizoctonia solani*, Physiol. Plant Pathol., 4:151–159.

Hunter, R. E. (1978). Effects of catechin in culture and in cotton seedlings on growth and polygalacturonase activity in *Rhizoctonia solani*, Phytopathology, 68:1032–1036.

Iverson, T. H., Baggerind, C., and Betavag, T. (1979). Myrosin cells in Brassicaceae roots, Z. Pflanzenphysiol., 94:143–154.

Johnson, G., Maag, D. D., Johnson, D. K., and Thomas, R. D. (1976). The possible role of phytoalexins in the resistance of sugarbeet (*Beta vulgaris*) to *Cercospora beticola*, Physiol. Plant Pathol., 8:225–230.

Johnson, G., and Schaal, L. A. (1957). Chlorogenic acid and other orthodihydric phenols in scab-resistant Russet Burbank and scab-susceptible Triumph potato tubers of different maturities, Phytopathology, 47:253–255.

Jolley, R. L., Robb, D. A., and Mason, H. S. (1969). The multiple forms of mushroom tyrosinase, J. Biol. Chem., 244:1593–1599.

Jones, R. (1979). The destruction of β-glucosidase activity by oven-drying and its effect on formononetin estimation in red clover, J. Sci. Food Agric., 30:243–245.

Jorgensen, H. J. L., Neergaard, E. De., and Smedegaard-Petersen, V. (1993). Histological examination of the interaction between *Rhynchosporium secalis* and susceptible and resistant cultivars of barley, Physiol. Mol. Plant Pathol., 42:345–358.

Karni, L., Prusky, D., Kobiler, I., Bar-Shira, E., and Kobiler, D. (1989). Involvement of epicatechin in the regulation of lipoxygenase activity during activation of quiescent *Colletotrichum gloeosporioides* infections of ripening avocado fruits, Physiol. Mol. Plant Pathol., 35:367–374.

Kessmann, H., and Barz, W. (1986). Elicitation and suppression of phytoalexin and isoflavone accumulation of cotyledons of *Cicer arietinum* L. as caused by wounding and by polymeric components from the fungus *Ascochyta rabiei*, J. Phytopathol., 117:321–335.

Kirkham, D. S., and Flood, A. E. (1956). Inhibition of *Venturia* spp. by analogues of host metabolites, Nature, 178:422–423.

Koster, J., Strack, D., and Barz, W. (1983). High performance liquid chromatographic separation of isoflavones and structural elucidation of isoflavone-7-O-glucoside-6″-malonates from *Cicer arietinum*, Planta Med., 48:131–135.

Kowalski, S. P., Bamberg, J., Tingey, W. M., and Steffens, J. C. (1990). Insect resistance in the wild potato *Solanum berthaultii*: inheritance of glandular trichome polyphenol oxidase, J. Hered., 81:475–478.

Kraft, B., and Barz, W. (1985). Degradation of the isoflavone Biochanin A and its glucoside conjugates by *Ascochyta rabiei*, Appl. Environ. Microbiol., 50:45–48.

Kramer, R. P., Hindore, H., Hem, C. J., Kallage, J. J., and Zilliken, F. (1984). Antifungal activity of soybean and chickpea isoflavones and their reduced derivatives, Phytochemistry, 23:2203–2205.

Kuc, J., Barnes, E., Daftsios, A., and Williams, E. B. (1959). The effect of amino acids on susceptibility of apple varieties to scab, Phytopathology, 49:313–315.

Luthy, H., and Matile, P. (1984). The mustard oil bomb rectified analysis of the subcellular organisation of the myrosinase system, Physiol. Pflanzen., 179:5–12.

Maizel, J. V., Burkhardt, H. J., and Mitchell, H. K. (1964). Avenacin, an antimicrobial substance isolated from *Avena sativa*: I. Isolation and antimicrobial activity, Biochemistry, 3:424–426.

Malamy, J., Carr, J. P., Klessig, D. F., and Raskin, I. (1990). Salicylic acid: a likely endogenous signal in the resistance response of tobacco to viral infection, Science, 250:1002–1004.

Malamy, J., Hennig, J., and Klessig, D. E. (1992). Temperature-dependent induction of salicylic acid and its conjugates during the resistance response to tobacco mosaic virus infection, Plant Cell, 4:359–366.

Marcus, L., Prusky, D., and Jacoby, B. (1988). Purification and characterization of avocado lipoxygenase, Phytochemistry, 27:323–327.

Martin, J. H. (1931). Hydrocyanic acid content of different parts of the sorghum plant, J. Am. Soc. Agron., 30:725–734.

Matile, P. (1980). Die Senfolbaombe: zur kompartimentierung des Myrosinasesystems, Biochem. Physiol. Pflanzen., 175:722–731.

McMurray, C. H., Laidlaw, A. S., and McElroy, M. (1986). The effect of plant development and environment on formononetin concentration in red clover (*Trifolium pratense* L.), J. Sci. Food Agric., 37:333–340.

Mehta, A. C., and Seshadri, T. R. (1958). Flavonoids as antioxidants, J. Sci. Indus. Res., 18B:24.

Metraux, J. P., Signer, H., Ryals, J., Ward, E., Wyss-Benz, M., Gaudin, J., Raschdorf, K., Schmid, E., Blum, W., and Inverardi, B. (1990). Increase in salicylic acid at the onset of systemic acquired resistance in cucumber, Science, 250:1004–1006.

Mithen, R. (1992). Leaf glucosinolate profiles and their relationship to pest and disease resistance in oilseed rape, Euphytica, 63:71–83.

Mithen, R. F., Lewis, B. G., and Fenwick, G. R. (1986). In vitro activity of glucosinolates and their products against *Leptosphaeria maculans*, Trans. Br. Mycol. Soc., 87:433–440.

Mithen, R. F., Lewis, B. G., Heaney, R. K., and Fenwick, G. R. (1987a). Resistance of *Brassica* species to *Leptosphaeria maculans*, Trans. Br. Mycol. Soc., 88:525–531.

Mithen, R. F., Lewis, B. G., Heaney, R. K., and Fenwick, G. R. (1987b). Glucosinolates of wild and cultivated *Brassica* species, Phytochemistry, 26:1969–1973.

Mithen, R. F., and Magrath, R. (1992). Glucosinolates and resistance to *Leptosphaeria maculans* in wild and cultivated *Brassica* species, Plant Breeding, 108:60–68.

Morris, P. F., Savard, M. E., and Ward, E. W. B. (1991). Identification and accumulation of isoflavonoids and isoflavone glucosides in soybean leaves and hypocotyls in resistance responses to *Phytophthora megasperma* f.sp. *glycinea*, Physiol. Mol. Plant Pathol., 39:229–244.

Murr, D. P., and Yang, S. F. (1975). Inhibition of in vivo conversion of methionine to ethylene by L-canaline and 2,4-dinitrophenol, Plant Physiol., 55:79–82.

Naim, M., Gestener, B., Zilkah, S., Birk, Y., and Bondi, A. (1974). Soybean isoflavones. Characterization, determination, and antifungal activity, J. Agric. Food Chem., 22:806–810.

Nicholson, R. L., Butler, L. G., and Asquith, T. N. (1986). Glycoproteins from *Colletotrichum graminicola* that bind phenols: implications for survival and virulence of phytopathogenic fungi, Phytopathology, 76:1315–1318.

Nicholson, R. L., Hipskind, J., and Hanau, R. M. (1989). Protection against phenol toxicity by the spore mucilage of *Colletotrichum graminicola*, an aid to secondary spread, Physiol. Mol. Plant Pathol., 35:243–252.

Nicholson, R. L., and Moraes, W. B. C. (1980). Survival of *Colletotrichum graminicola*: importance of the spore matrix, Phytopathology, 70:255–261.

Noveroske, F. L., Kuc, L., and Williams, E. B. (1964). Oxidation of phloridzin and phloretin related to resistance of *Malus* to *Venturia inaequalis*, Phytopathology, 54:92–97.

Ohtsuru, M., Tsuruo, I., and Hata, T. (1969). Studies on fungus myrosinase, Part 1. Production, purification and some characteristics, Agric. Biol. Chem., 33:1309–1314.

O'Neill, M. J. (1983). Aureol and Phaseol, two new coumestans from *Phaseolus aureus* Roxb., Z. Naturforschung., 38C:698–700.

O'Neill, M. J., Adesanya, S. A., Roberts, M. F., and Pantry, I. R. (1986). Novel inducible isoflavonoids from the lima bean *Phaseolus lunatus* L., Phytochemistry, 25:1315–1322.

Osbourn, A. E., Clarke, B. R., Dow, J. M., and Daniels, M. J. (1991). Partial characterization of avenacinase from *Gaeumannomyces graminis* var. *avenae*, Physiol. Mol. Plant Physiol., 38:301–312.

Osbourn, A. E., Clarke, B. R., Lunness, P., Scot., P. R., and Daniels, M. J. (1994). An oat species lacking avenacin is susceptible to infection by *Gaeumannomyces graminis* var. *tritici*, Physiol. Mol. Plant Pathol., 45:457–467.

Oydvin, J., and Richardson, D. G. (1987). A paper chromatography survey of the phenol content of apple peel from scab resistant and susceptible genotype, Nor. J. Agric. Sci., 1:7–13.

Papavizas, F. G. (1964). Greenhouse control of *Aphanomyces* root rot of peas with aminobutyric acid and methylaspartic acid, Plant Dis. Reptr., 48:537–541.

Pathak, M. A., Daniels, F., and Fitzpatrick, T. B. (1962). The presently known distribution of furanocoumarins (psoralens) in plants, J. Invest. Dermatol., 39:225–239.

Patil, S. S., Powelson, R. L., and Young, R. A. (1964). Relation of chlorogenic acid and free phenols in potato roots to infection by *Verticillium albo-atrum*, Phytopathology, 54:531–535.

Porter, A. J. R., Kiddle, G., and Wallsgrove, R. M. (1990). Disease resistance and myrosinase activity in single and double-low varieties of oilseed rape (*Brassica napus*), Plant Physiol., 93S:54.

Porter, A. J. R., Morton, A. M., Kiddle, G., Doughty, A. J., and Wallsgrove, R. M. (1991). Variation in the glucosinolate content of oilseed rape (*Brassica napus*) leaves. 1. Effect of leaf age and position, Ann. Appl. Biol., 118:461–467.

Prusky, D. (1988). The use of antioxidants to delay the onset of anthracnose and stem end decay in avocado fruits after harvest, Plant Dis., 72:381–384.

Prusky, D., Freeman, S., Rodriguez, R. J., and Keen, N. T. (1994). A nonpathogenic mutant strain of *Colletotrichum magna* induces resistance to *C. gloeosporioides* in avocado fruits, Mol. Plant–Microbe Interact., 7:326–333.

Prusky, D., Karni, L., Kobiler, I., and Plumbley, R. A. (1990). Induction of the antifungal diene in unripe avocado fruits: effect of inoculation with *Colletotrichum gloeosporioides*, Physiol. Mol. Plant Pathol., 37:425–435.

Prusky, D., Keen, N. T., and Eaks, I. (1983). Further evidence for the involvement of a preformed antifungal compound in the latency of *Colletotrichum gloeosporioides* in unripe avocado fruits, Physiol. Plant Pathol., 22:189–198.

Prusky, D., Keen, N. T., Sims, J. J., and Midland, S. (1982). Possible involvement of an antifungal compound in latency of *Colletotrichum gloeosporioides* on unripe avocado fruits, Phytopathology, 72:1578–1582.

Prusky, D., Kobiler, I., and Jacoby, B. (1988). Involvement of epicatechin in cultivar susceptibility of avocado fruits to *Colletotrichum gloeosporioides* after harvest, Phytopathol. Z., 123:140–146.

Prusky, D., Kobiler, I., Jacoby, B., Sims, J. J., and Midland, S. L. (1985). Inhibition of avocado lipoxygenase: their possible relationship with the latency of *Colletotrichum gloeosporioides*, Physiol. Plant Pathol., 27:269–279.

Prusky, D., Plumbley, R. A., and Kobiler, I. (1991). Modulation of natural resistance of avocado fruits to *Colletotrichum gloeosporioides* by CO_2 treatment, Physiol. Mol. Plant Pathol., 39:325–334.

Pryor, D. E., Walker, J. C., and Stahmann, M. A. (1940). Toxicity of allyl-isothiocyanate vapor to certain fungi, Am. J. Bot., 28:30–38.

Rahiala, E. L., Kekoma, M., Janne, J., Raina, A., and Raiha, N. C. R. (1971). Inhibition of pyridoxal enzymes by L-canaline, Biochim. Biophys. Acta, 227:337–343.

Ralton, J. E., Howlett, B. J., Clarke, A. E., Irwin, J. A. G., and Imrie, B. (1989). Interaction of cowpea with *Phytophthora vignae*: inheritance of resistance and production of phenylalanine ammonia-lyase as a resistance response, Physiol. Mol. Plant Pathol., 32:89–103.

Ramadoss, C. S., Ulig, J., Carlson, D. M., Butler, L. G., and Nicholson, R. I. (1985). Composition of the mucilaginous spore matrix of *Colletotrichum graminicola*, a pathogen of corn, sorghum and other grasses, J. Agric. Food Chem., 33:728–732.

Rasmussen, J. B., Hammerschmidt, R., and Zook, M. N. (1991). Systemic induction of salicylic acid accumulation in cucumber after inoculation with *Pseudomonas syringae* pv. *syringae*, Plant Physiol., 97:1342–1343.

Rawlinson, C. J., Muthyala, G., Poole, V. A., Cayley, G. R., Hulme, P. J., and Pickett, J. A. (1985). Mustard oils fungicides and disease, Rothamstead J. Exp. Station Report for 1984, pp. 124–125.

Reece, E. T., Clapp, R. C., and Mandels, M. (1958). A thioglucosidase in fungi, Arch. Biochem. Biophys., 75:228–242.

Retig, N. (1974). Changes in peroxidase and polyphenoloxidase associated with natural and induced resistance of tomato to *Fusarium* wilt, Physiol. Plant Pathol., 4:145–150.

Retig, N., and Chet, I. (1974). Catechol-induced resistance of tomato plants to *Fusarium* wilt, Physiol. Plant Pathol., 4:469–475.

Reuveni, M., and Cohen, Y. (1978). Growth retardation and changes in phenolic compounds, with special reference to scopoletin, in mildewed and ethylene-treated tobacco plants, Physiol. Plant Pathol., 12:179–189.

Rosenthal, G. A. (1981). A mechanism of L-canaline toxicity, Eur. J. Biochem., 114:301–304.

Rosenthal, G. A. (1982). Plant Nonprotein Amino and Imino acids: Biological, Biochemical and Toxicological Properties, Academic Press, London, pp. 273.

Rosenthal, G. A. (1990). Metabolism of L-canavanine and L-canaline in leguminous plants, Plant Physiol., 94:1–3.

Rosenthal, G. A. (1991). The biochemical basis for the deleterious effects of L-canavanine, Phytochemistry, 30:1055–1058.

Rouxel, T., Sarnignet, A., Kollman, A., and Bosquet, J. F. (1989). Accumulation of a phytoalexin in *Brassica* spp. in relation to a hypersensitive reaction of *Leptosphaeria maculans*, Physiol. Mol. Plant Pathol., 34:507–517.

Samborski, D. J., and Forsyth, F. R. (1960). Inhibition of rust development on detached wheat leaves by metabolites, antimetabolites, and enzyme poisons, Can. J. Bot., 38:467–476.

Scheel, L. D., Perone, V. B., Larkin, R., and Kupel, R. (1963). The isolation and characterization of two phytotoxic furanocoumarins (psoralens) from diseased celery, Biochemistry, 2:1127–1131.

Schlieper, D., Komossa, D., and Barz, W. (1984). Degradation of the isoflavone biochanin A-7-O-glucoside-6″-malonate and phenylacetic acids by *Fusarium javanicum*, Z. Naturforsch. Sect. C., 39:882–887.

Shepherd, C. J., and Mandryk, M. (1964). Effects of metabolites and antimetabolites on the sporulation of *Peronospora tabacina* Adam. on tobacco leaf disks, Austr. J. Biol. Sci., 17:878–891.

Sierotzki, H., and Gessler, C. (1993). Flavan-3-ol content and the resistance of Malus x domestica to *Venturia inaequalis* (Cke.) Wint, Physiol. Mol. Plant Pathol., 42:291–297.

Sivanathan, S., and Adikaram, N. K. B. (1989). Biological activity of four antifungal compounds in immature avocado, J. Phytopathol., 125:97–109.

Southerton, S. G., and Deverall, B. J. (1989). Histological studies of the expression of the Lr 9, Lr 20 and Lr 28 alleles for resistance to leaf rust in wheat, Plant Pathology, 38:190–199.

Southerton, S. G., and Deverall, B. J. (1990). Changes in phenolic acid levels in wheat leaves expressing resistance to *Puccinia recondita* f.sp. *tritici*, Physiol. Mol. Plant Pathol., 37:437–450.

Steffens, J. C., and Walters, D. S. (1991). Biosynthesis of glucose esters by glandular trichomes of *Lycopersicon pennellii*, Naturally Occurring Pest Bioregulators (P. A. Heldin, ed.), American Chemical Society, Washington, DC., pp. 136–149.

Takahama, U. (1985). Inhibition of lipoxygenase-dependent peroxidation by quercitin: mechanism of antioxidative function, Phytochemistry, 24:1443–1446.

Templeton, G. E. (1972). *Alternaria* toxins related to pathogenesis in plants, Microbial Toxins, Vol. 8 (S. J. Ajl, S. Nadis and A. Ciegler, eds.), Academic Press, New York, pp. 169–192.

Thagstad, O. P., Evjen, K., and Bones, A. (1991). Immunogold-EM localisation of myrosinase in Brassicaceae, Protoplasma, 161:85–93.

Thimann, K. V., and Mahadevan, S. (1964). Nitrilase. I. Occurrence, preparation and general properties of the enzyme, Arch. Biochem. Biophys., 105:133–141.

Tippett, J. T., McGrath, J. F., and Hill, T. C. (1989). Site and seasonal effects on susceptibility of *Eucalyptus marginata* to *Phytophthora cinnamomi*, Austr. J. Bot., 37:481–490.

Toshikazu, T., Irikura, H., and Naito, N. (1971). Inhibition of uredosorus formation of *Puccinia coronata* by plant growth regulators and antimetabolites, Tech. Bull. Fac. Agric. Kagawa Univ., 23:42–50.

Treutter, D., and Feucht, W. (1990). The pattern of flavan-3-ols in relation to scab resistance of apple cultivars, J. Hort. Sci., 65:511–517.

Trione, J. (1960). The HCN content of flax in relation to flax wilt resistance, Phytopathology, 50:482–486.

Turner, E. M. (1961). An enzymic basis for pathogen specificity in *Ophiobolus graminis*, J. Expt. Bot., 12:169–175.

Underhill, E. W., Wetter, L. R., and Chisholm, M. D. (1973). Biosynthesis of glucosinolates, Biochem. Soc. Symp., 38:303–326.

VanEtten, H. D. (1967). Antifungal activity of pterocarpans and other selected isoflavonoids, Phytochemistry, 15:655–659.

VanEtten, H. D., Sandrock, R. W., Wasmann, C. C., Soby, S. D., McCluskey, K., and Wang, P. (1995). Detoxification of phytoanticipins and phytoalexins by phytopathogenic fungi, Can. J. Bot., 73(Suppl I):S518–S525.

Vaughn, K. C., and Duke, S. O. (1981a). Tentoxin-induced loss of plastidic polyphenol oxidase, Physiol. Plant., 53:421–428.

Vaughn, K. C., and Duke, S. O. (1981b). Tissue localization of polyphenol oxidase in *Sorghum*, Protoplasma, 108:319–327.

Vaughn, K. C., and Duke, S. O. (1982). Tentoxin effects on sorghum: the role of polyphenol oxidase, Protoplasma, 110:48–53.

Vaughn, K. C., and Duke, S. O. (1984a). Function of polyphenol oxidases in higher plants, Physiol. Plant., 60:106–112.

Vaughn, K. C., and Duke, S. O. (1984b). Tentoxin stops the processing of polyphenoloxidase into an active protein, Physiol. Plant., 60:257–261.

Vidhyasekaran, P. (1988a). Physiology of Disease Resistance in Plants, Vol. I, CRC Press, Boca Raton, FL, pp. 149.

Vidhyasekaran, P. (1988b). Physiology of Disease Resistance in Plants, Vol. II, CRC Press, Boca Raton, FL, pp. 128.

Vidhyasekaran, P., Borromeo, E. S., and Mew, T. W. (1986). Host-specific toxin production by *Helminthosporium oryzae*, Phytopathology, 76:261–265.

Vidhyasekaran, P., Borromeo, E. S., and Mew, T. W. (1992). *Helminthosporium oryzae* toxin suppresses phenol metabolism in rice plants and aids pathogen colonization, Physiol. Mol. Plant Pathol., 41:307–315.

Vidhyasekaran, P., Chinnadurai, G., and Govindaswamy, C. V. (1971). HCN content of sorghum leaves in relation to rust disease incidence, Indian Phytopathol., 24:332–358.

Walker, J. C., Morell, C. S., and Foster, H. H. (1937). Toxicity of mustard oils and related compounds to certain fungi, Am. J. Bot., 24:536–541.

Walker, J. C., and Stahmann, M. A. (1955). Chemical nature of disease resistance in plants, Annu. Rev. Plant Physiol., 6:351–366.

Wang, P., Matthews, D. E., and VanEtten, H. D. (1992). Purification and characterization of cyanide hydratase from the phytopathogenic fungus *Gloeocercospora sorghi*, Arch. Biochem. Biophys., 298:569–575.

Wang, P., and VanEtten, H. D. (1992). Cloning and properties of a cyanide hydratase gene from the phytopathogenic fungus *Gloeocercospora sorghi*, Biochem. Biophys. Res. Commu., 187:1048–1054.

Ward, E. W. B., Cahill, D. M., and Bhattacharyya, M. K. (1989). Abscisic acid suppression of phenylalanine ammonia-lyase activity and mRNA and resistance of soybeans to *Phytophthora megasperma* f.sp. *glycinea*, Plant Physiol., 91:23–27.

Weidenborner, W., Hindorf, H., Hem, C. J., Tsotsonos, P., and Egge, H. (1990). Antifungal activity of isoflavonoids in different reduced stages on *Rhizoctonia solani* and *Sclerotium rolfsii*, Phytochemistry, 29:801–803.

Weigand, F., Koster, J., Weltzein, H. C., and Barz, W. (1986). Accumulation of phytoalexins and isoflavone glucosides in a resistant and a susceptible cultivar of *Cicer arietinum* during infection with *Ascochyta rabiei*, J. Phytopathol., 115:214–221.

Willeke, U., and Barz, W. (1982a). Degradation of the isoflavone biochanin A by *Fusarium javanicum*, Z. Naturforsch., 37:861–869.

Willeke, U., and Barz, W. (1982b). Catabolism of 5-hydroxy-isoflavones by fungi of the genus *Fusarium*, Arch. Microbiol., 132:266–269.

Willeke, U., Weltring, K. M., Barz, W., and VanEtten, H. D. (1983). Degradation of the isoflavone biochanin A by isolates of *Nectria haematococca* (*Fusarium solani*), Phytochemistry, 22:1539–1541.

Woodward, M. D. (1980). Phaseollin formation and metabolism in *Phaseolus vulgaris*, Phytochemistry, 19:921–927.

Yalpani, N., and Raskin, I. (1993). Salicylic acid: a systemic signal in plant disease resistance, Trends Microbiol., 1:88–92.

Yalpani, N., Silverman, P., Wilson, T. M. A., Kleier, S. A., and Raskin, I. (1991). Salicylic acid is a systemic signal and an inducer of pathogenesis-related proteins in virus-infected tobacco, Plant Cell, 3:809–818.

Part III

Molecular Events During Disease Symptom Development

7

Role of Toxins in Cell Membrane Dysfunction and Cell Death

I. CELL MEMBRANE IS AN ACTIVE SITE FOR INDUCTION OF DEFENSE MECHANISMS

As discussed in earlier chapters, host cell membrane appears to be an active site for induction of various defense mechanisms of the host. It seems to be the binding site for elicitors, the signal molecules from both fungal cell wall and host cell wall. Intracellular signal transduction occurs across cell membrane. Calcium ion, the second messenger, passively crosses plasma membranes via a channel-forming protein embedded in the lipid bilayer of the membrane. Cell membrane-bound enzymes such as phospholipases, protein kinases, and Ca^{2+} and H^+ transport ATPases take part in the signal transduction system. Phosphorylation of cell membrane-bound proteins may be crucial in transmembrane signaling. H^+-ATPase, a proton pump, is embedded within the plasma membrane. The fungal elicitors induce hyperpolarization, which is involved in signal transmission.

Systemic signal molecules also may have receptor sites in plasma membrane. Systemin serves as a systemic signal after binding to a plasma membrane receptor. Jasmonic acid and methyl jasmonate are the other important systemic signaling molecules in plants, and their release is controlled by the release of membrane fatty acids by membrane-bound phospholipases.

The elicitors, by activating the signal transduction system, induce synthesis of various defense chemicals such as phenolics, phytoalexins, lignin, callose, hydroxyproline-rich glycoproteins, and pathogenesis-related proteins. Active oxygen species are induced by membrane deterioration and they are involved in activation of various defense mechanisms of the host. This chapter discusses how successful pathogens suppress the defense mechanisms of the plasma membrane.

II. PATHOGENS PRODUCE PHYTOTOXINS THAT ACT ON CELL MEMBRANES AND SUPPRESS DEFENSE MECHANISMS OF THE HOST

A. Phytotoxin

Most of the plant necrotrophic pathogens produce metabolites toxic to plants. These toxins induce characteristic symptoms of diseases (Yoder, 1980; Vidhyasekaran, 1993a). Some pathogens produce host-specific toxins that are toxic only to the host of those pathogens when used at physiological concentrations. Others produce nonspecific toxins that are toxic both to hosts and nonhosts of the pathogen (Graniti, 1991; Vidhyasekaran, 1993b).

Alternaria alternata f.sp. *lycopersici*, the tomato pathogen, produces a host-specific toxin, named AAL toxin (Bottini and Gilchrist, 1981; Bottini et al., 1981; Siler and Gilchrist, 1982, 1983; Clouse et al., 1985; Wisenboer et al., 1988; Kodama et al., 1991). AAL-toxin consists of two compounds: T_A and T_B (Bottini et al., 1981; Bottini and Gilchrist, 1981). The toxin causes interveinal necrosis of tomato leaflets (Grogan et al., 1975; Gilchrist and Grogan, 1976). Two host-specific toxins, AL-toxin I and II, have been purified from Japanese isolates of *A. alternata* tomato pathotype (Kohmoto et al., 1982).

Alternaria alternata strawberry (*Fragaria glandiflora*) pathotype produces a host-specific toxin designated AF-toxin (Maekawa et al., 1984; Yamamoto et al., 1984, 1985; Namiki et al., 1986a,b). AF-toxin was characterized to consist of three related molecular species: AF-toxin I, II, and III (Maekawa et al., 1984; Yamamoto et al., 1984, 1985; Nakatsuka et al., 1986). *A. kikuchiana*, the pathogen of black spot disease of Japanese pear, produces a host-specific toxin, AK-toxin (Otani et al., 1985). The toxin contains three closely related toxins, two major and a minor one. AK-toxin I is highly toxic (Nishimura and Kohmoto, 1983). AK-toxin is released from germinating spores at the site of initial contact of the pathogen with host plant surface (Otani et al., 1975).

Alternaria alternata apple pathotype produces multiple host-specific toxins, AM-toxins I, II, and III, which selectively affect apple cultivars (Kohmoto et al., 1976). *A. alternata (citri)* rough lemon (*Citrus limon*) pathotype produces ACR-toxin, while *A. alternata* tangerine (*Citrus reticulata*) pathotype produces ACT-toxin and *A. alternata* tobacco pathotype produces AT-toxin (Kohmoto and Otani, 1991). *A. alternata* pathotype infecting knapweed (*Centaurea maculosa*) produces a host-specific toxin, maculosin, which causes chlorotic spots that develop into black necrotic lesions on the leaves of knapweed (Park et al., 1994). *A. macrospora*, the cotton pathogen, produces a nonspecific toxin (Krishnamohan and Vidhyasekaran, 1988). *A. helianthi*, the sunflower pathogen, produces several phytotoxins, radicinin, radianthin, deoxyradicinol, and 3-epideoxyradicinol (Tal et al., 1985). *A. chrysanthemi*, the pathogen of chrysanthemum (*Chrysanthemum morifolium*), produces radicinin (Tal et al., 1985). *A. alternata* produces a nonspecific toxin, tentoxin (Durbin et al., 1973; Durbin and Uchytil, 1977; Duke et al., 1982). *A. zinniae* produces a nonspecific toxin, zinniol (Robeson and Strobel, 1984), and this toxin is produced by four other *Alternaria* spp. (Vidhyasekaran, 1993a). *A. carthami*, the safflower (*Carthamus tinctorius*) pathogen produces three toxins: brefeldin A, zinniol, and 7-dehydrobrefeldin A (Tietjen et al., 1983; Kneusel et al., 1994).

Helminthosporium victoriae, the oat blight pathogen, produces a host specific toxin, HV-toxin (or victorin) (Wolpert and Macko, 1989). *H. maydis* T strain, the maize pathogen produces HM-T toxin, a host-specific one (Flavell, 1975; Turgeon et al., 1995). HM-T toxin is required for the high virulence of race T to maize carrying T cytoplasm (Turgeon et al., 1995). *H. sacchari*, the sugarcane pathogen, produces a host-specific toxin, HS-toxin (Strobel et al., 1975; Livingston and Scheffer, 1984). HS-toxin is a mixture of three isomers (A, B and C) that differ only in the position of a double bond in the sesquiterpene moiety (Duvick et al., 1984; Livingston and Scheffer, 1983, 1984). *H. carbonum* race 1, maize leaf spot pathogen, produces HC-toxin (Ciuffetti et al., 1983). The toxin is a cyclic tetrapeptide of structure cyclo (D-prolyl-L-alanyl-D-alanyl-L-2-Aeo). Aeo stands for amino-9, 10-epoxy-8-oxodecanoic acid (Walton et al., 1982; Kawai et al., 1983). The epoxide and the 8-carbonyl groups in Aeo are required for biological activity of HC-toxin (Walton and Earle, 1983; Kim et al., 1987). D-alanine is recognized and activated by one of the biosynthetic enzymes, HTS-2, in vitro, and radiolabeled alanine is incorporated into HC-toxin in vitro (Walton, 1987, 1991; Walton and Holden, 1988). Loss of HTS results in loss of toxin-producing ability and loss of pathogenesis (Panaccione et al., 1992). *H. carbonum* produces three more toxins (Rasmussen and Scheffer, 1988).

Ophiobolin A is produced by *Helminthosporium oryzae*, the rice pathogen (Nakamura and Ishibashi, 1958); *H. maydis*, the maize pathogen (Tipton et al., 1977); and *H. setariae*, the foxtail millet (*Setaria italica*) pathogen (Nukina and Marumo, 1976). *H. oryzae* produces several toxins besides ophiobolin A. They are ophibolin B, 6-epiophibolin A, anhydroophiobolin A, 6-epianhydroophiobolin A, and ophiobolin 1 (Xiao et al., 1991). All of them are nonspecific toxins. It produces a host-specific toxin (Vidhyasekaran et al., 1986). *H. carbonum* race 3 (syn. *Bipolaris zeicola*), the pathogen of maize and rice produces BZR-toxin (Xiao et al., 1992).

Pyricularia oryzae, the rice blast pathogen produces several toxins, piricularin, picolinic acid, pyriculol and tenuazonic acid (Tamari et al., 1966; Iwasaki et al., 1969; Umetsu et al., 1972). *Periconia circinata*, the sorghum pathogen produces a host-specific toxin, PC-toxin (Gardner et al., 1972). Two toxins peritoxins A and B have been purified from culture filtrates of *P. circinata* (Macko et al., 1992) *Phyllosticta maydis*, the maize pathogen produces a host specific toxin, Pm toxin (Danko et al., 1984; Kono et al., 1985).

The toxin ascochitine has been isolated from culture filtrates of *Ascochyta fabae* (Oku and Nakanishi, 1963), *A. pisi* (Bertini, 1956) and *Phoma clematidina* (Smith et al., 1994). Ascochitine has a carboxylic acid nature (Smith et al., 1994). Phytotoxins have been isolated from *Ascochyta imperfecta*, *Phoma herbarum*, and *P. medicaginis* also (Stoessl, 1981).

Large number of *Cercospora* spp. produce a toxin, designated cercosporin (Kuyama and Tamura, 1957; Balis and Payne, 1971; Lousberg et al., 1971; Yamazaki and Ogawa, 1972; Mumma et al., 1973; Assante et al., 1977; Lynch and Geoghegan, 1977; Fajola, 1978; Melouk and Schuh, 1987; Jenns et al., 1989). *Cercospora beticola* produces a toxin of unknown structure and the toxin is named CBT (Schlosser, 1971). Various *Fusarium* spp. produce a nonspecific toxin, fusaric acid (Paquin and Waygood, 1957; Shahin and Spivey, 1986). *Fusarium amygdali*, the causal agent of peach and almond wilts produces a toxin, designated fusicoccin (Marre, 1979). *Ceratocystis ulmi*, the Dutch elm (*Ulmus americana*) disease pathogen produces a toxin called ceratoulmin (Pijut et al., 1990).

Rhynchosporium secalis, the leaf scald pathogen of barley, produces three phytotoxic peptides (Wevelsiep et al., 1991). *Drechslera gigantea*, the pathogen of zonate eyespot disease in turfgrass (*Agrostis* sp.), produces several toxins belonging to eremophilane family (Kenfield et al., 1989).

Oxalic acid produced by *Sclerotium rolfsii* (Bateman and Beer, 1965), *Sclerotinia sclerotiorum* (Maxwell and Lumsden, 1970; Noyes and Hancock, 1981; Marciano et al., 1983; Magro et al., 1984; Godoy et al., 1990), *Endothia parasitica* (Havir and Anagnostakis, 1985), *Pythium aphanidermatum* (Koleosho et al., 1987), *Mycena citricolor* (Rao and Tewari, 1987), and *Sclerotium cepivorum* (Stone and Armentrout, 1985) is considered as toxin. *Gibberella zeae* produces a toxin trichothecene (Proctor et al., 1995). Trichothecene-deficient mutants of *G. zeae* were generated by gene disruption. The distrupted gene, Tri5, encodes the enzyme trichodiene synthase, which catalyzes the first step in trichothecene biosynthesis. The mutants did not secrete trichothecene and exhibited reduced virulence (Proctor et al., 1995).

Verticillium dahliae produces a low molecular weight peptide toxin (Nachmias et al., 1985). *Phytophthora infestans*, the potato late blight pathogen produces a toxin which causes necrotic symptoms in leaves (Seidel, 1961). The toxin has been partially purified (Mollers et al., 1992). *Pyrenophora tritici-repens* causes tan spot of wheat leaves. The pathogen produces a toxin which induces a typical tan spot symptom (Tomas and Bockus, 1987; Lamari and Bernier, 1989; Tomas et al., 1990; Orolazaa et al., 1995). The toxin has a high molecular weight of 14,700, unlike other low-molecular-weight toxins (Tomas et al., 1990). *Pyrenophora teres*, the net-blotch pathogen of barley, produces a toxin called aspergillomarasmine (Friis et al., 1991).

B. Toxins Suppress Defense Mechanisms of the Host

Toxins may modify the host's defense mechanisms. *Alternaria alternata* is a nonpathogen of rice. *Helminthosporium oryzae* the brown spot pathogen of rice produces ophiobolin A. When spore-germination fluids of *H. oryzae* were applied along with spores of *A. alternata*, the fungus caused a large number of disease lesions on rice leaves (Xiao et al., 1989). When spores of *A. alternata* were suspended in ophiobolin A solution, the spores germinated well and formed appressoria and infection pegs on the rice leaf sheath surface. Extensive inter- and intracellular hyphal growth was observed only in the sheath tissues that had been inoculated in the presence of ophiobolin A and/or spore germination fluids. The spores without ophiobolin A or spore germination fluids could not grow beyond the infection peg and did not grow inside the cell (Xiao et al., 1991). Ophiobolin A at 3 μg/ml induced susceptibility of rice leaf tissues to *A. alternata* (Xiao et al., 1991). The toxin produced by *Pyricularia oryzae* induced susceptibility to infection by nonpathogenic *A. alternata* of rice leaves (Fujita et al., 1994).

H. oryzae produces a host-specific toxin. When avirulent isolate of *H. oryzae* was inoculated along with the toxin, the avirulent isolate could penetrate the rice leaves and grow profusely. Typical brown spot lesions were also developed (Vidhyasekaran et al., 1992). Inoculation of leaves of susceptible maize genotypes with the weakly virulent race of *Helminthosporium carbonum* induces resistance to a challenge inoculation with the highly virulent race 1 (Cantone and Dunkle, 1990a). The host-specific toxin (HC-toxin) produced only by race 1 isolates abolished the induced resistance when applied with the challenge inoculum (Cantone and Dunkle, 1990a).

The limited infection by *H. carbonum* race 2, a weakly virulent, non-toxin-producing race induces the release of an inhibitory compound(s) from affected maize cells that then diffuses into water droplets on the leaf surface. The virulent race 1 conidia, when applied together with these inhibitory compounds are unable to cause susceptible type lesions (Cantone and Dunkle, 1990b). The synthesis of the inhibitory compound is suppressed by HC toxin (Cantone and Dunkle, 1991).

When spores of *Alternaria alternata* a nonpathogen of rice plants, were inoculated together with BZR-toxin produced by *Bipolaris zeicola* (syn. *H. carbonum*) race 3, the pathogen of rice and maize, the spores successfully penetrated rice tissues. Typical invasion and fungus development also occurred in maize tissues when BZR-toxin was added to spore drops of *H. victoriae*, the oat pathogen, which alone could not penetrate into maize tissues (Xiao et al., 1992).

When avirulent spores of *A. alternata* or *Cochliobolus heterostrophus* were inoculated on leaves of Tms-cytoplasm maize, along with a trace of HM-toxin produced by *H. maydis* T strain, the rate of formation of infection hyphae greatly increased and necrosis developed (Kohmoto et al., 1989).

Ceratocystis ulmi, the Dutch elm pathogen, produces a toxin: ceratoulmin.

The cytoplasm of the resistant variety responds to the fungal invasion by depositing phenolic material. In the susceptible variety, however, the toxin suppressed the phenolic deposition. (Pijut et al., 1990). HV-toxin, produced by *Helminthosporium victoriae*, reduces the accumulation of avenalumins, the phytoalexins, in oats (Mayama et al., 1986). Japanese pear tissues possess a potential resistance mechanism to *Alternaria kikuchiana* invasion. Dysfunction of plasma membranes by AK-toxin, produced by the fungus, results in abolishing the induction of resistance mechanism in susceptible pear tissues (Kohmoto et al., 1987). Papilla formation at the penetration sites of *A. kikuchiana* functions as a resistance mechanism in Japanese pear. When pear leaves were inoculated with avirulent spores plus AK-toxin, the rate of papilla formation reduced and the avirulent spores could invade susceptible leaves (Otani et al., 1995). When highly virulent isolate of *H. oryzae*, the rice brown spot pathogen, was inoculated on rice leaves, the spores germinated, formed appressoria, and penetrated the epidermal layers at 12 h after inoculation (Vidhyasekaran et al., 1986). Extensive inter- and intracellular mycelial growth was observed in mesophyll tissue of leaves at 48 h after inoculation. When a nonpathogenic isolate of *H. oryzae* was inoculated on rice leaves, the isolate did not grow beyond the infection peg. The pathogenic isolate induced typical brown spot symptoms with yellow halo at 48 h, while the nonpathogenic isolate did not produce any visible symptoms. When spore suspensions of the nonpathogenic isolate were prepared in *H. oryzae* toxin and inoculated on rice leaves, extensive inter- and intracellular mycelial growth was observed at 36 h. The growth rate of the mycelium of the nonpathogenic isolate was similar to that of the virulent isolate (Table 1; Vidhyasekaran et al., 1986). Typical brown spot symptoms appeared on leaves exposed to the toxin (Vidhyasekaran et al., 1986).

The results suggest that the toxin induces susceptibility in the host to the pathogen (Vidhyasekaran et al., 1986). In a similar study, it was observed that when a small amount of HV-toxin produced by *H. victoriae* was added, the

Table 1 Effect of *Helminthosporium oryzae* toxin on development of nonpathogenic *H. oryzae* within rice leaf tissues

Treatment	Extent of fungal growth inside leaf tissue p73[a]
Virulent isolate	+++
Nonpathogenic isolate	–
Toxin alone	–
Toxin + nonpathogenic isolate	+++

[a]–, No mycelial growth; +++, abundant inter- and intracellular mycelial growth
Source: Vidhyasekaran et al., 1986.

nonpathogenic mutant of *H. victoriae*, which did not produce toxin, developed throughout susceptible leaves just as rapidly as did the pathogenic *H. victoriae* (Yoder, 1972). Essentially the same results were obtained in a study of colonization of corn leaves by *H. carbonum* (Comstock and Scheffer, 1973). Topical application of toxin was necessary for successful infection of *Populus tremuloides* by spores of *Hypoxylon mammatum* (Schipper, 1975).

Thus the pathogens appear to suppress the defense mechanisms of the host by producing toxins. The primary sites of action of the toxins are associated with three target organelles: plasma membrane, mitochondrion, and chloroplast (Kohmoto and Otani, 1991). Cell membrane appears to be affected by most of the toxins.

III. PATHOGENS CAUSE MEMBRANE DYSFUNCTION

A. Structural Changes

Pathogens are known to modify the structure of cell membranes (Hirata and Kojima, 1962; Littlefield and Bracker, 1972; Aist, 1974). Early ultrastructural changes seen in *Cercospora*-infected sugar beets include the disruption of cell membranes, including thickening and swelling of ER and mitochondrial membranes and disruption of the plasmalemma, tonoplast, and chloroplast membranes (Steinkamp et al., 1979, 1981). The plasmalemma around the haustoria is thickened and strengthened due to *Erysiphe graminis* (Bracker, 1968) and *E. pisi* (Gil and Gay, 1977) infections in barley and pea, respectively. Barley tissues inoculated with *E. graminis* f.sp. *hordei* showed an increased percentage of cells showing a concave form of plasmolysis (Lee-Stadelmann et al., 1991). Electron microscopic observation showed that many invaginations in the plasma membranes of the susceptible strawberry cells occurred within 1–3 h after *Alternaria alternata* strawberry pathotype toxin treatment (Park et al., 1992).

B. Permeability Changes

1. Increases in Permeability

Plasma membrane in plant cells governs all permeability processes from simple diffusion to active transport (Otani et al., 1977). Modification of plasma membrane in the infected tissues may result in efflux of nutrients to the pathogen (Sol, 1965, 1967). Changes in membrane permeability are the first detectable events in the onset of diseases caused by different pathogens (Thatcher, 1942, 1943; Wheeler and Hanchey, 1968).

Changes in membrane are mostly induced by toxins produced by pathogens. The toxins induce electrolyte leakage from cells (Vidhyasekaran et al., 1986; Kohmoto and Otani, 1991). HV-toxin produced by *H. victoriae*, the oat blight pathogen, causes rapid loss of electrolytes (Wheeler and Black, 1963; Samaddar and Scheffer, 1971; Bronson and Scheffer, 1977).

Alternaria alternata apple pathotype toxin, AM-toxin, causes invagination of plasma membranes and an increase in electrolyte leakage (Kohmoto et al., 1976; Park et al., 1981a,b). The increase in electrolyte loss was observed at 1 h after the toxin treatment (Shimomura et al., 1991). Susceptible tissues of sorghum exposed to toxin from *Periconia circinata* had increased loss of electrolytes within 20 min (Gardner et al., 1972).

Fusaric acid (FA), the toxin produced by *Fusarium* spp., induced a sudden, large increase in electrolyte leakage (Marre et al., 1993). FA induced K^+ efflux from the cells of *Egeria densa* leaves. The addition of FA to the incubation medium induced a progressive increase in the concentration of K^+ in the medium reflecting a nett efflux of this ion from the leaves (Marre et al., 1993). Much of the PC-toxin induced efflux was potassium (Gardner et al., 1972).

The toxin produced by *Alternaria kikuchiana*, the pathogen of black spot disease of Japanese pear, caused rapid increases in efflux of K^+ from susceptible leaves of Japanese pear. On the contrary, the toxin caused a decrease in effluxes of Na^+, Mg^{2+}, and Ca^{2+}. The toxin-induced loss of K^+ was much greater in the presence of Na^+, Mg^{2+}, or Ca^{2+} in the ambient solution. Stimulation of toxin-induced leakage was evident immediately after adding these cations to toxin-treated tissues. When these cations were removed, loss of K^+ reverted to the lower level. When toxin-induced K^+ efflux was stimulated by other cations, the tissues took up such cations. The results suggest that AK-toxin affects K^+ transport across plasma membranes of susceptible cells; the transport is closely associated with movement of other inorganic cations in the cells (Otani et al., 1977). The toxin induced Na^+ and phosphate leakage also (Park et al., 1987).

Ophiobolin A is a nonspecific phytotoxin produced by *Helminthosporium oryzae*, *H. maydis*, and *H. setariae*. The toxin increases ion leakage (Tipton et al., 1977) and membrane permeability for potassium ions (Coccuci et al., 1983). Stomatal movement is driven by potassium fluxes in and out of the guard cells (Outlaw, 1983). Ophiobolin affects stomatal movement. The effect of ophiobolin on stomatal opening of *Commelina communis* was concentration-dependent. Ophiobolin at 0.01 μM and 0.1 μM significantly increased stomatal opening. At 10 μM, stomata were tightly closed (Table 2; Nejidat, 1987) and the tissue was damaged (Nejidat, 1987).

The inhibitory effect of the toxin at high concentration may be due to its effect on membrane permeability and potassium leakage from guard cells. The stimulating effect of ophiobolin at low concentration on stomatal opening may be due to inhibition of calcium transport (Leung et al., 1984, 1985).

Tipton et al. (1977) provided evidence to show that ophibolin A produced by *H. maydis* does not cause general membrane disruption but inhibits specific membrane transport processes. The toxin stimulates net leakage of electrolytes

Table 2 Effect of ophiobolin on stomatal opening in *Commelina communis*

Ophiobolin (μM)	Stomatal aperture (μM)	% of control
0	3.5	100
0.01	5.8	166
0.1	4.0	115
1.0	2.0	58
10.0	0.1	4

Source: Nejidat, 1987.

and glucose from maize seedling roots. Treatment of the roots with ophibolin A inhibits uptake of 2-deoxyglucose by 85%, but it does not show a similar effect on efflux of hexose (Tipton et al., 1977).

The fluidity of cell membranes of oat protoplasts was measured by electron-spin resonance with 5-doxylstearic acid as a spin label (Briggs et al., 1984). No changes in fluidity of the protoplast membranes were detected following exposure to HV toxin (Briggs et al., 1984). It suggests that the toxin-induced leakage of electrolytes may occur through the membrane matrix, possibly via a shuttle-type carrier.

2. Decreases in Permeability

A decrease in permeability to methyl and ethyl urea in susceptible barley cells, 48 h after inoculation with *Erysiphe graminis* f.sp. *hordei*, has been observed (Lee-Stadelmann et al., 1991). In the compatible host–parasite combination, structural alterations were induced in the plasmalemma that reduced passive movement of nonelectrolytes through the membrane and increased adhesion of the membrane to host cell walls (Lee-Stadelmann et al., 1991). Reduced permeability was observed in barley leaves inoculated with *E. graminis* f.sp. *hordei* during symptom development (Lee-Stadelmann et al., 1992). Reduction in permeability is known to indicate specific structural changes in the phospholipid bilayer of the plasmalemma, which results in reduced membrane fluidity (Margin et al., 1990).

C. Changes in Membrane-Bound ATPases

1. H^+-ATPase is Stimulated

The plasma membrane plays a major role in the control of cell processes. Using ATP as the energy source, it pumps protons from the cytoplasm to the cell exterior, thus creating an electrochemical gradient across the plasma membrane that constitutes the driving force for nutrient uptake (Briskin, 1990). Cell wall loosening, a prerequisite for cell growth, is induced by extensive acidification of the apoplast (Rayle and Cleland, 1992). It results in alkalinization of the cytoplasm, which triggers cell division. Thus H^+-ATPase is involved in nutrient uptake as well as cell growth and division (Serrano, 1990).

Fusicoccin, the toxin produced by *Fusarium amygdali*, the pathogen of peach and almond, stimulates the activity of the plasmalemma-localized H^+-ATPase (Marre, 1979). Fusicoccin leads to a twofold activation of ATP hydrolytic activity in the microsomal membrane fraction obtained from cell-suspension cultures of *Corydalis sempervirens* (Schulz et al., 1990). There was a stimulation of H^+ pumping by 50–130% and stimulation of ATP hydrolysis by 20–45% after incubation of microsomal membrane fraction from radish with fusicoccin (Rasi-Caldogno and Pugliarello, 1985).

Although fusicoccin stimulates H^+-ATPase, there does not seem to be direct interaction between the toxin and the H^+-ATPase activity, and a fusicoccin-binding protein could be separated (Stout and Cleland, 1980). Furthermore, stimulation by fusicoccin of the ATPase from *Vicia faba* was lost after solubilization of the enzyme (Blum et al., 1988). The results suggest that fusicoccin may have sites of action in addition to the H^+-ATPase (De Michelis et al., 1989). A high-affinity fusicoccin receptor 30–34 kDa, distinct from the 100 kDa H^+-ATPase, has been identified in plasma membrane fractions (De Boer et al., 1989; Meyer et al., 1989; De Michelis et al., 1989; Feyerabend and Weiler, 1988, 1989). Proteolytic removal of a 7–10 kDa fragment from the C-terminal end of the 100-kDa H^+-ATPase strongly activates ATP hydrolytic activity and H^+ pumping by the enzyme (Palmgren et al., 1990a, 1991). It was suggested that a part of the C-terminal region may constitute an autoinhibitory domain of the H^+-ATPase and this domain may be the ultimate target for toxins that function as regulators of H^+ pumping across the plasma membrane (Palmgren et al., 1991).

When spinach leaves were infiltrated with 5 μM fusicoccin, plasma membranes from the leaves showed a twofold increase in ATP hydrolytic activity (Johansson et al., 1993) and a threefold increase in H^+ pumping (Johansson et al., 1993) compared to controls.

Lysophosphatidylcholine stimulates the H^+-ATPase activity in isolated plasma membrane vesicles (Palmgren et al., 1988, 1990b; Palmgren and Sommarin, 1989). H^+-ATPase activated by lysophosphatidylcholine was not further activated by proteolytic removal of the C-terminal region, and it was suggested that activation by lysophosphatidylcholine involved a displacement of the C-terminal inhibitory domain (Palmgren et al., 1991). Activation by lysophosphatidylcholine was not additive to activation by fusicoccin in plasma membrane preparations from spinach leaves. It suggests that the latter process may also involve a displacement or removal of the C-terminal inhibitory domain (Johansson et al., 1993). Trypsin treatment of plasma membrane vesicles from fusicoccin-incubated leaves did not increase the ATPase or the H^+ pumping activity of these vesicles. Trypsin treatment resulted in the appearance of a 90-kDa band in addition to the native 100 kDa H^+-ATPase band in both the control and the fusicoccin-activated material (Johansson et al., 1993). The results suggest that activation of the H^+-ATPase by fusicoccin proceeds by a mechanism involving a displacement of the C-terminal inhibitory domain.

The fusicoccin-mediated increase in H^+-ATPase was rapid and largely completed within the first 4 min of incubation (Johansson et al., 1993). This relatively rapid activation suggests that de novo synthesis of H^+-ATPase did not contribute to the observed increase in activity. The intensity of the 100 kDa H^+-ATPase band is not increased but rather decreased in samples from fusicoccin-incubated leaves compared to controls (Johansson et al., 1993). No increase in H^+-ATPase polypeptide was observed after incubation with fusicoccin. These observations strengthen the conclusion that in vivo activation of the H^+-ATPase by fusicoccin proceeds by a mechanism involving a displacement of the C-terminal inhibitory domain (Johansson et al., 1993).

Fusicoccin opens stomata (Turner and Graniti, 1969; Brown and Outlaw, 1982). Fusicoccin enhances K^+ influx (Clint and Blatt, 1989; Blatt and Clint, 1989). The K^+ influx may be due to activation of H^+ pump in plasma membrane (Assmann and Zieger, 1987). Fusicoccin-promoted stomatal opening may be primarily due to the ability of fusicoccin to stimulate the plasma membrane H^+-ATPase of plant cells (Marre, 1979). In guard cells, activation of H^+ pump in plasma membrane created an electrochemical gradient that is presumed to drive the K^+ influx necessary for guard cell swelling and stomatal opening (Assmann and Zieger, 1987). K^+-selective ion channels present in the guard cell membrane may provide a pathway for K^+ movement across the cell membrane in response to an electrochemical driving force (Schroeder et al., 1988).

Fusicoccin may activate a carrier that mediates K^+ uptake against its electrochemical gradient and may inactivate ion channels that mediate K^+ efflux (Blatt, 1988; Blatt and Clint, 1989; Clint and Blatt, 1989). In 1 mM KCl, low concentrations of fusicoccin combined with exposure to white light synergistically stimulated stomatal opening (Assmann and Schwartz, 1992). With increasing fusicoccin concentrations, stomatal opening occurred even in the darkness and the synergistic effect diminished (Assmann and Schwartz, 1992). The synergistic effect of light and 0.1 μM fusicoccin also diminished as KCl concentrations were increased (Assmann and Schwartz, 1992). Stomata that were either opened by light on the intact leaf or opened in isolated peels by light and 0.1 μM fusicoccin could close to a significant extent upon imposition of darkness, indicating that 0.1 μM fusicoccin did not prevent ion efflux, and then reopen upon exposure to white light. These results show that the stomata remained functional after treatment with this low fusicoccin concentration. In the absence of fusicoccin, stomata in 1mM KCl closed in darkness and remained closed even in the light. The K^+ channel blocker, tetraethyl ammonium (TEA), was equally effective in inhibiting stomatal opening by either 0.1 μM fusicoccin + 1 mM KCl, or 120 mM KCl (Assmann and Schwartz, 1992). Fusicoccin (1 μM) prevented stomatal closure and this effect could be completely eliminated by inclusion of 1 mM vanadate, a H^+-ATPase inhibitor, in the incubation solution, despite little direct effect of vanadate on stomatal closure. In whole-cell patch-clamp experiments with guard cell protoplasts of *Vicia fabia*, fusicoccin (1 or 10 μM) stimulates an increase in outward current that is essentially voltage independent between –100 and +60 millivolts, and occurs even when the membrane potential is held at voltage (–60 millivolts) at which K^+ channels are inactivated and therefore could not be attributed to altered flux through K^+ channels. The fusicoccin-stimulated current was essentially voltage-independent over the range from –100 to +60 mv. This voltage independence contrasts sharply with the voltage dependence of K^+ and anion channels over the same voltage range. Hence it was concluded that the observed increase in outward current results from fusicoccin activation of H^+ ATPase at the guard-cell plasma membrane (Assmann and Schwartz, 1992). Fusicoccin may activate H^+ pumps and it may not function merely by blocking K^+ efflux channels. The plasma membrane H^+ ATPase of guard cells may be a primary locus for the fusicoccin effect on stomatal apertures.

Fusicoccin has been shown to open stomata in bean, tobacco, sorghum, cucumber, lucerne, and pokeweed (*Phytolacca americana*) (Turner and Graniti, 1969). By this process the toxin increases stomatal transpiration (Procacci and Graniti, 1966; Turner and Graniti, 1969; Lado et al., 1972). The uncontrolled water loss due to irreversible stomatal opening may result in a rapid reduction of available water in the intercellular spaces and in a subsequent loss of turgor of the mesophyll cells (Turner and Graniti, 1969). Large areas of the leaf blade wilt and wither (Turner and Graniti, 1969; Graniti and Turner, 1970).

Rhynchosporium secalis, the leaf scald pathogen of barley, grows subcuticularly, primarily above anticlinal epidermal walls during early stages of pathogenesis (Lehnackers and Knogge, 1990). Initial symptoms in barley cultivars include the swelling of epidermal cells and a subsequent loss of rigidity of anticlinal epidermal walls. This is accompanied by a separation of the plasmalemma from the cell wall (Lechnackers and Knogge, 1990). An increased opening of stomata is also observed (Ayers, 1972; Branchard and Laffray, 1987). It leads to increases in transpiration and permeability of cells (Jones and Ayers, 1972; Ayers and Jones, 1975). Cell necrosis occurs (Wevelsiep et al., 1991). A small family of necrosis-inducing peptides has been identified in culture filtrates of *R. secalis* (Wevelsiep et al., 1991). Two of these peptides, NIP1 and NIP3 correlated with the development of necrotic lesions. Both the peptides stimulated the activity of the plasmalemma-localized Mg^{2+}-dependent, K^+-stimulated H^+-ATPase (Wevelsiep et al., 1993). When ATPase from barley plasmamembrane vesicles was partially purified by centrifugation in a glycerol gradient, no stimulation of enzyme activity by peptides was detectable. Hence the peptides do not seem to interact directly with the ATPase. An NIP3-binding protein of about 65 kDa has been detected in barley (Wevelsiep et al., 1993).

2. H^+-ATPase is Inhibited

In some host–pathogen interactions, ATPase is inhibited. In pea infected by *Erysiphe pisi* and french bean infected with *Uromyces appendiculatus*, the plasma membrane lacked normal ATPase activity. ATPase activity was absent from extrahaustorial domains of the plasma membranes of cells infected with haustoria (Spencer-Philipps and Gay, 1981). AF-toxin produced by *Alternaria alternata* strawberry pathotype affected the electrogenic proton pump (H^+ pump) in the plasma membrane of susceptible strawberry cells (Namiki et al., 1986a), but no significant inhibition of the ATPase activity was observed by in vitro toxin treatment (Lee et al., 1990, 1992). A binding site for the toxin may be needed, as shown in case of action of *Rhynchosporium secalis* toxin (Wevelsiep et al., 1993). Similar observations have been reported in HS-toxin-treated sugarcane cells (Kohmoto and Otani, 1991). The toxin produced by *Cercospora beticola*, CBT, induces leakage of betacyanin and amino acids from red beet root tissue (Schlosser, 1971). CBT inhibits H^+ extrusion and depolarizes *trans* membrane potential (Macri and Vianello, 1979), but does not affect ATP and pyruvate levels (Macri et al., 1980). The toxin inhibits ATP-dependent proton translocation in microsomal vesicles and it has a primary effect on ATPases associated with plasmalemma or tonoplast membranes (Macri et al., 1983).

CBT was inhibitory to the ATP-induced intravesicular proton accumulation and this effect increased with the time of incubation. CBT inhibited ATPase activity and this inhibition also increased with the time of preincubation. However, CBT was more inhibitory to proton transport than to ATP hydrolysis. The results suggest a direct effect of CBT on ATPase associated with the plasma membrane (Blein et al., 1988).

D. Inhibition of Calmodulin Activity

Calmodulin is a Ca^{2+}-binding protein in plasma membrane. Calmodulin is a mediator of calcium transport (Hepler and Wayne, 1985). Ophiobolin A, the toxin produced by *Helminthosporium oryzae*, *H. maydis*, and *H. setariae*, reacts in vitro with bovine brain and spinach calmodulins so that they are unable to activate calmodulin-dependent cAMP phosphodiesterase (Leung et al., 1984). Activation of phosphodiesterase (PDE) is used to demonstrate the activity of calmodulin. When Ca^{2+} is removed from the medium by AGTA, the activation of PDE is abolished, thus showing the necessity of Ca^{2+}-binding the calmodulin for its stimulatory activity (Leung et al., 1985). Ophiobolin A inhibited maize calmodulin in the PDE assay. The inhibition of maize calmodulin by ophiobolin was strongly dependent of Ca^{2+} (Table 3; Leung et al., 1985).

Table 3 Ca^{2+} dependence of the inhibition of maize calmodulin by ophiobolin A

	PDE activity	
Additions	(n mol Pi/30 min)	%
None (Control)	10.3	100
$CaCl_2$	22.8	221
$CaCl_2$ + Ophiobolin A	10.8	105
EGTA + Ophiobolin A	20.6	200

Source: Leung et al., 1985.

The extract of ophiobolin A-treated maize roots had less active calmodulin than that of the untreated roots. It suggests that ophiobolin A interacts with calmodulin in situ (Leung et al., 1985). Calmodulin appears to be a target molecule for the toxicity of ophiobolin A in root cells. Transport enzymes such as calmodulin-dependent Ca-ATPase may be affected, and in turn may affect Ca transport in the membrane (Leung et al., 1985). Low concentrations of calcium inhibit stomatal opening (Fischer, 1972). Calcium chloride at a concentration of 0.1 mM inhibited stomatal opening of *Commelina communis* (Nejidat, 1987). Addition of 0.1 μM ophiobolin significantly abolished the Ca^{2+} effect. Chloropromazine, a potent inhibitor of calmodulin, enhanced stomatal opening and significantly reversed the effect of 0.1 mM $CaCl_2$ (Nejidat, 1987). Calmodulin inhibitors stimulate stomatal opening (Donovan et al., 1985). The results suggest that ophiobolin may affect calmodulin activity (Nejidat, 1987). Ophiobolin appears to be a natural calmodulin inhibitor (Leung et al., 1984, 1985).

E. Alteration in Membrane Potential

The electrical potential difference (PD) across biological membranes is due to a diffusion potential resulting from unequal distribution of ions (Dainty, 1962). There is also evidence that energy-requiring active transport processes (electrogenic pumps) contribute to potential difference in plant cells (Higginbotham, 1970; Higginbotham et al., 1970). Due to fungal infection depolarization of the membranes takes place. The toxins produced by *Helminthosporium maydis* (Mertz and Arntzen, 1978), *H. victoriae* (Novacky and Hanchey, 1974), and *H. sacchari* (Schroter and Novacky, 1985) depolarized the membranes of their respective susceptible hosts such as maize, oat, and sugarcane.

Treatment of susceptible oat roots with victorin, the toxin produced by *H. victoriae*, caused a measurable depolarization within 2–5 min and averaged approximately 30% deplorization within 10 min (Novacky and Hanchey, 1974).

Membrane depolarization was detected at concentrations as low as 50 nm *H. sacchari* toxin. The onset of membrane depolarization was observed after a lag period of 4–10 min. The energy-dependent component of membrane potential was inhibited by the toxin (Schroter and Novacky, 1985). HS-toxin may be causing a loss of H^+ gradient across the membrane (Schroter and Novacky, 1985).

Fusaric acid (FA) is the toxin produced by *Fusarium* spp. (Gaumann, 1957). Fusaric acid has both direct and indirect effects on the plasma membrane. The direct effect is indicated by a rapid depolarization, possibly due to change in membrane permeability. The indirect effects are indicated by transient hyperpolarization, possibly reflecting an fusaric acid-mediated influx of protons followed by their electrogenic extrusion; and a slow depolarization, possibly caused by reduced ATP levels inhibiting electrogenic extrusion of H^+ (D'Alton and Etherton, 1984).

Fusaric acid induces reductions in ATP concentrations (D'Alton and Etherton, 1984) that are necessary for electrogenic extrusion of H^+ and hence maintenance of the membrane potential difference (PD) between the inside of a plant cell and the external solution.

Fusaric acid induced an early transient hyperpolarization followed by a depolarization of transmembrane electrical potential difference (Em) in cells of *Egeria densa* leaves. Both effects were enhanced by higher concentration of fusaric acid. The effects of 1 mM fusaric acid were progressively reduced as the pH of the incubating medium increased, as the ratio of the uncharged to the charged form of the acid (pK fusaric acid = 5.59) was reduced (Marre et al., 1993).

The initial hyperpolarization may be due to an early stimulation of the proton pump by acidification of the cytosol consequent on the entry into the cell of the undissociated form of the acid. The depolarization observed after about 5 min of treatment with fusaric acid may be due to inhibition of oxidative phosphorylatin (Marre et al., 1993). FA inhibited malate dependent O_2 uptake. Succinate oxidation coupled to phosphorylation was also inhibited by FA (Marre et al., 1993). Inhibition of oxidative phosphorylation by FA has been reported in tomato roots (D'Alton and Etherton, 1984), maize roots (Arias, 1985), and sorghum seedlings (Dunkle and Wolpert, 1981). Inhibition of respiratory activity (Kuo and Scheffer, 1964; Arias, 1985) and decrease in ATP levels (Koehler and Bentrup, 1983) due to FA treatment in different plant species have been reported.

F. Modification of Membrane Phospholipids

The plasma membrane consists mainly of proteins and phospholipids. Changes in phospholipid content in cell organelles have been observed in several plants due to infection (Hoppe and Heitefuss, 1974; Sednina et al., 1981). Phospholipases may be of pathogen as well as of host origin. The crude extract of *Erysiphe pisi* caused pea mesophyll protoplasts to lose their contents and the crude extract contained phospholipases (Faull and Gay, 1983). It released palmitic and oleic acids from phosphatidyl choline 1-oley-2-palmitoyl (Faull and Gay, 1983). Phospholipase B or a mixture of phospholipase A_1 and A_2 has been detected in extracts of *Thielaviopsis basicola* (Lumsden and Bateman, 1968), *Sclerotium rolfsii* (Tseng and Bateman, 1969), and *Sclerotium sclerotiorum* (Lumsden, 1970).

When the cells are treated with toxins produced by pathogen, phospholipases of the host are activated. AF-toxin, produced by *Alternaria alternata* strawberry pathotype, induced a considerable decrease in phospholipid content in microsomes of strawberry protoplasts in parallel with a decline in cell viability (Table 4; Lee et al., 1992).

Phospholipase A_2 causes dysfunction of membranes by hydrolyzing phospholipids and releasing unsaturated fatty acids (Waite, 1985). When suspension-cultured cells of the susceptible cultivar were treated with AF-toxin I, an enhanced level of phospholipase A_2 activity in the microsomes was detected at relatively low concentrations of the toxin that are known to reduce cell viability (Lee et al., 1992). Furthermore, in vitro toxin treatment of microsomes prepared from susceptible cells significantly stimulated the phospholipase A_2 activity at much lower concentrations. Phospholipase A_2 activity was detected in strawberry microsomes in a Ca^{2+}-dependent manner. A significant increase in the enzyme activity was observed when 5 mM or more $CaCl_2$ was added to the reaction mixture. It suggests that AF-toxin I affects cytosolic Ca^{2+} concentrations, which, in turn affect activity of phospholipase A_2 (Lee et al., 1992).

The phospholipid-degarding enzymes, especially phospholipase A_2 and phospholipase C, are involved in signal transduction in plant cells in response to various stimuli (Martiny-Baron and Scherer, 1989; Palmgren and Sommarin, 1989; Scherer and Andre, 1989; Scherer et al., 1988, 1990). The toxins may serve as external stimuli and phospholipase A_2 activation in response to AF-toxin may be a consequence of signal transduction in strawberry cells (Lee et al., 1992). The removal of certain phospholipids from microsomes alters the activity of membrane-bound enzyme systems. ATPase activity is inhibited by the action of

Table 4 Decrease in phospholipid content in microsomes of susceptible strawberry protoplasts treated with AF-toxin I

		Phospholipid content (μmol/2.5 × 10^7 protoplasts)	
Strawberry cultivar	Cell viability	Control	Toxin-treated
Susceptible	18.9	3.05	1.87
Resistant	106.1	2.71	2.75

Source: Lee et al., 1992.

phospholipase (Faull and Gay, 1983) and it may lead to depolarization of active membrane potential (Lee et al., 1992). The removal of phospholipids from membranes causes damage to membrane structure, sometimes followed by cell death (Hoppe and Heitefuss, 1975).

G. Accumulation of Active Oxygen Species

When potato leaves were inoculated with compatible race of *Phytophthora infestans*, increase in hydroxyl radicals production was observed from 18 to 48 h after inoculation. Similar increase was not observed in leaf tissues inoculated with incompatible race (Jordan and Devay, 1990). Superoxide anions were also produced more in the compatible interactions. Production of the hydroxyl free radicals and superoxide anions was mainly a function of the aggressiveness of *P. infestans* (Jordan and Devay, 1990). Both hydroxyl radicals and superoxide anions cause peroxidation of membrane lipids and may induce disruption of lysosome membranes (Jordan and DeVay, 1990).

Cercospora spp. cause severe diseases in many crops. A large number of *Cercospora* cpp. produce a toxin, cercosporin. Cercosporin is structurally related to several photosensitizing compounds (i.e., compounds that sensitize cells to visible light) (Foote, 1976). Cercosporin causes peroxidation of plant lipids (Daub, 1982b), resulting in major changes in the structure and composition of plant cell membranes (Daub and Briggs, 1983).

Cercosporin absorbs light to form a long-lived electronically excited state (triplet state) that can then react with molecular oxygen to produce compounds toxic to living cells. The triple sensitizer (^3S) may react with O_2 in several ways. It may react with a reducing substrate (R or RH) by the transfer of a hydrogen atom or electron. The resulting sensitizer radical may then react with O_2 to produce superoxide ions O_2^-. As an alternative, the triplet sensitizer may react directly with O_2 by an energy transfer process yielding the electronically excited singlet state of O_2 ($^1O_2$) (Foote, 1976).

Cercosporin, when illuminated in the presence of O_2, reacted with cholesterol to form the 5α-hydroperoxide of cholesterol, which is only produced by reaction with singlet oxygen. Cercosporin, in the presence of light, O_2, and a reducing substrate, was able to reduce p-nitroblue tetrazolium chloride, a compound readily reduced by superoxide. Superoxide dismutase, a scavenger of superoxide, inhibited this reaction (Daub and Hangarter, 1983). The results suggest that cercosporin can induce the production of both $^1O_2$ and O_2^- (Daub and Payne, 1989).

Both $^1O_2$ and O_2^- are extremely toxic to cells, causing the oxidation of fatty acids, sugars, cellulosic materials, guanine, and several amino acids (cysteine, methionine, histidine, tryptophan and tyrosine), which in turn result in damage to DNA, inactivation of enzymes, and destruction of cell membranes (Foote, 1976;

Spikes, 1977). Cercosporin induces ion leakage from potato, carrot, beet, and corn tissue when these tissues are irradiated with an incandescent bulb and oxygen is involved in the reaction (Macri and Vianello, 1979).

Tobacco leaf discs treated with cercosporin showed a large increase in electrolyte leakage 1–2 min after irradiation with light. All tobacco protoplasts exposed to cercosporin in the light were damaged within 45 min. Tobacco cell culture mutants that show resistance to generation of O_2^- show decreased sensitivity to cercosporin (Daub and Hangarter, 1983). Compounds capable of quenching $^1O_2$ prevent the killing of tobacco suspension cultured cells by cercosporin (Daub, 1982a). Suspension-cultured tobacco cells were treated with bixin which is known to quench $^1O_2$. Bixin prevented killing response by cercosporin (Daub, 1982a).

H. Mitochondrial Membrane Dysfunction

Mitochondria control the respiratory activity of plant cells. They produce ATP by oxidative phosphorylation. Regulation of the molecular traffic through membranes is particularly important in mitochondrion, where electron transfer across the inner membrane creates a proton motive force that drives phosphorylation producing cellular energy (Braun et al., 1989b). Inhibition of ATP synthesis will result in suppression of defense mechanisms of the host and will facilitate extensive colonization of the pathogens (Miller and Koeppe, 1971; Koeppe et al., 1978; Berville et al., 1984; Holden and Sze, 1984). *Helminthosporium maydis* race T produces a host-specific toxin, HMT toxin. HMT toxin affects mitochondrial activities (Holden and Sze, 1989) and induces rapid swelling of mitochondria of the susceptible Texas line of cytoplasmic male-sterile corn (Koeppe et al., 1978). It increases permeability of the inner mitochondrial membrane to calcium (Holden and Sze, 1984) and protons (Klein and Koeppe, 1985). It dissipates mitochondrial membrane potential (Berville et al., 1984) and induces leakage of small molecules and ions (NAD^+ and Ca^{2+}) from the matrix space (Holden and Sze, 1984; Matthews et al., 1979). It induces uncoupling of oxidative phosphorylation (Bednarski et al., 1977; Holden and Sze, 1987). It stimulates NADH- or succinate-mediated respiration (Peterson et al., 1975; Bednarski et al., 1977) and mitochondrial ATPase activity (Bednarski et al., 1977; Holden and Sze, 1987). HMT toxin inhibits malate-driven respiration (Berville et al., 1984; Peterson et al., 1975).

A unique mitochondrial gene from cms-T plants, designated *µrf* 13-T, that encodes a 13-kDa protein URF 13-T, associated with the cms trait in maize has been isolated (Dewey et al., 1986, 1987; Wise et al., 1987a; Rottmann et al., 1987).

To determine whether this cms-associated protein is involved with susceptibility to the disease the *µrf* 13-T gene was cloned into the plasmid vector pATH3 and expressed in *Escherichia coli*. The pATH3-μ*rf* 13-T, construct, pATH 13-T was transcriptionally regulated by the *trp* promoter of *E. coli* and it produced a 13-kDa protein that cross-reacted with an antiserum to μrf 13-T protein and co-migrated electrophoretically with the mitochondrial μ*rf* 13-T protein product of cms-T maize as determined by protein blot analysis. Nucleotide sequence analysis confirmed that μ*rf* 13-T reading frame in pATH13-T was identical with the maize reading frame (Dewey et al., 1988). A truncated version of the μ*rf* 13-T gene, pJG 13-T, that produces a form of the 13 kDa protein missing amino acids 2 through 11 was also constructed and expressed in *E. coli* by means of the plasmid vector pJG 200. Oxygen consumption was completely inhibited by the addition of HM-T toxin in cell producing the complete 13-kDa protein. In contrast, respiration was not altered by toxin in control cells transformed with the pATH3 vector containing no insert, or in the cells producing the truncated version of the 13-kDa protein. HmT toxin induced dramatic swelling in spheroplasts from *E. coli* cells that produced 13-kDa polypeptide. No effect was seen with spheroplasts containing the pATH3 vector containing no insert or in the spheroplasts producing the truncated 13-kDa protein from plasmid pJG 13-T (Dewey et al., 1988).

Incubation of *cms*-T maize mitochondria with dicyclohexylcarbodiimide (DCCD) prevented toxin-induced inhibition of malate-dependent oxidation, dissipation of the membrane potential, and leakage of accumulated calcium (Bouthyette et al., 1985). Two major DCCD-binding proteins approximately 6 and 14 kDa in size, were observed when normal and *cms*-T mitochondria were incubated with ^{14}C-labeled DCCD. Immunoprecipitation of [^{14}C] DCCD-labeled maize mitochondrial proteins with an antiserum directed against the μrf 13-T protein product revealed that DCCD binds to the 13-kDa protein. Similar to its characteristics in *cms*-T mitochondria, the 13-kDa polypeptide produced in *E. coli* by pATH 13-T was also localized in the membrane and was shown to bind DCCD (Dewey et al., 1988). The results suggest that the amino-terminal region of the 13 kDa polypeptide is involved in conferring HmT toxin susceptibility to cms-T mitochondria (Dewey et al., 1988). The amino-terminal portion of the protein is the region most likely associated with the membrane, the site where the toxin action is believed to occur.

The observed inhibition of whole-cell respiration by HmT toxin could be explained by pore formation that leads to a loss of intermediate species associated with glycolysis and the TCA cycle as well as any unbound intracellular electrolytes. Addition of HmT toxin to *E. coli* cells expressing URF 13 induces a rapid increase in the conductivity of the external medium. A massive and immediate efflux of radiolabeled ^{86}Rb accumulated in *E. coli* cells expressing URF13 after the addition of HmT toxin (Braun et al., 1989a, 1989b, 1990). The rate at which ^{86}Rb is lost from cells after toxin addition (less than 1 min) is consistent with formation of hydrophilic pores within the plasma membrane; added uncouplers that break down the proton motive gradient driving Rb uptake but do not permeabilize the membrane to Rb, also promote a cellular Rb, but do so with a half-life of tens of minutes. The amount of accumulated ^{86}Rb in control cells lacking URF13 is not affected by addition of HmT toxin. These results provide evidence that effects of T-toxin in *E. coli* cells expressing URF 13 are analogous to those seen in *cms*-T mitochondria and result from the formation of hydrophilic pores within URF 13 containing membranes (Levings and Siedow, 1992).

A chimeric gene composed of T-μ*rf* 13 fused at its amino terminus to the mitochondrial targeting sequence of the ATPase subunit from *Neurospora crassa* has been constructed (Huang et al., 1990). When this construct was transformed into yeast, it produced a protein that was localized to the mitochondrial membrane fraction and was processed to a size similar to that of standard URF 13. Growth of yeast cells expressing the chimeric URF 13 construct was inhibited by added HmT toxin (Huang et al., 1990). Yeast transformed with full-length T-μrf 13 lacking the mitochondrial transit peptide was insensitive to HmT toxin (Huang et al., 1990). The results suggest that μ*rf* 13 needs to be localized in mitochondria to produce a toxin-sensitive phenotype. Glab et al. (1990) have also shown that T-μ*rf* 13 gene confers Hm toxin sensitivity to yeast.

Removal of carboxyl-terminal 32 amino acids producing a truncated URF 13 of 83 amino acids instead of the 115 found in the standard URF 13 had no effect on sensitivity to HmT toxin (Braun et al., 1989b). Deletion of the next amino acid (Leu-83) to give an 82 amino acid product, however resulted in a complete loss of toxin sensitivity (Braun et al., 1989b). A toxin insensitive mutant (T-4) was detected in *cms*-T maize cell tissue culture. In the T-4 mutant the T-μ*rf*13 reading frame was shortened to 74 codons by a frameshift mutation. The URF 13 protein in the mutant contained only 74 amino acids (Wise et al., 1987a,b).

Membrane-bound proteins had stretches of hydrophobic amino acid residues that may associate with the lipid bilayer. URF 13 contains several hydrophobic domains. Site-directed mutations demonstrated that DCCD binds to URF 13 at two separate residues, positions 12 and 39 (Braun et al., 1989b). DCCD binding to aspartate at position 12 is expected because it is located between the two largest hydrophobic domains and the residues adjacent to the aspartate at position 12 are similar to those found in other DCCD binding sites. DCCD binding to the aspartate at position 39 is unexpected because this residue is located in an area predicted to be hydrophilic; DCCD stably binds only in nonaqueous environments. Taken together, the results suggest that amino acids 12 and 19 are membrane bound. In addition, aspartate at position 39 is required for functional HmT-toxin–URF 13 interaction since substitutional mutations at this aspartate residue eliminate sensitivity to toxin. Although DCCD binds at two sites on URF 13, only binding at position 39 confers protection against T-toxin (Braun et al., 1989b). The binding of T-toxin to URF13 causes rapid permeabilization of the inner mitochondrial membrane, which results in leakage of NAD^+ and other ions from the matrix. A pore consisting of at least six *trans*-membrane α-helices is required for NAD^+ leakage (Levings et al., 1995). URF 13 oligomers are involved in hydrophilic pore formation (Levings et al., 1995).

The results suggest that HmT toxin inhibits *Cms*-T maize mitochondrial function by their capacity to permeabilize the inner mitochondrial membrane after interaction with the T-μ*rf*13 gene product, URF 13 which is able to form membrane pores. This, in turn, promotes the loss of metabolic integrity that results in the large-scale fungal colonization and subsequent necrotic lesions that are symptomatic of leaf blight in *cms*-T maize (Levings and Siedow, 1992).

Phyllosticta maydis also affects maize plants carrying *cms*-T cytoplasm. This fungus produces Pm toxin. Pm toxin is also capable of affecting *cms*-T mitochondria (Levings and Siedow, 1992). Fusaric acid produced by *Fusarium* spp. reduced the respiratory activity of isolated mitochondria of tomato (Paquin and Waygood, 1957). Mitochondria isolated from fusaric acid-treated tomato plants have reduced respiratory activity (Naef-Roth and Reusser, 1954; Kuo and Scheffer, 1964). Mitochondria in susceptible host cells are affected by toxins produced by *Alternaria alternata (citri)* rough lemon pathotype (ACR-toxin), *A. alternata* f.sp. *lycopersici* (AAL-toxin), and *A. alternata* tobacco pathotype (AT-toxin). AT-toxin- and AL-toxin induced ultrastructural damage to mitochondria. ACR-toxin increased the permeability of the inner mitochondrial membrane (Kohmoto and Otani, 1991). The toxin produced by *Ceratocystis ulmi*, the pathogen of Dutch elm disease, induced the swelling of mitochondria and disruption of the outer membrane (Pijut et al., 1990).

IV. HOW DO PATHOGENS INDUCE MEMBRANE DYSFUNCTION ONLY IN SUSCEPTIBLE HOSTS?

A. Detoxification of Phytotoxins that Occurs in Resistant Hosts Does Not Occur in Susceptible Hosts

Phytotoxins are produced by pathogens and induce membrane dysfunction in host cells. Phytotoxins cause membrane dysfunction only in compatible, susceptible hosts. These toxins are detoxified in many incompatible resistant hosts. *Cercospora oryzae* produced a red toxin, cercosporin (Batchvarova et al., 1992). Louisiana red rice is resistant to the toxin and the rice cells grew in the presence of cercosporin concentrations completely toxic to Labelle, the most susceptible variety. When the rice cells were treated with the toxin, red rice cells contained about one-tenth as much cercosporin as cells of Labelle. It suggests that resistant cells have a mechanism for excluding, exporting, or destroying the toxin. Treating cell suspensions of red rice with norflurazon abolished the cercosporin resistance in the cultivar (Batchvarova et al., 1992). Norflurazone is an inhibitor of carotenoid biosynthesis and reduced the carotenoid content of cells derived from red rice (Batchvarova et al., 1992). The results suggest that carotenoids are involved in resistance to the toxin. Illuminated cercosporin produces active oxygen species (Daub and Payne, 1989). Carotenoids quench such species (Goodwin, 1980), particularly singlet oxygen (Daub and Hangarter, 1983). The resistant varieties may show increased levels of superoxide dismutase, catalase, and peroxidase, which are all scavengers of oxygen species (Daub and Hangarter, 1983).

Alternaria alternata, a pathogen of spotted knapweed, produces a host-specific toxin, maculosin and when the toxin was applied on leaves of knapweed, it was converted into maculosin-β-o-glucoside which is not phytotoxic towards spotted knapweed (Park et al., 1994). Ferulic acid induced resistance to *Pyricularia oryzae* in rice seedlings and it showed ability to detoxify piricularin, the toxin produced by *P. oryzae* (Tamari et al., 1966). Phenylalanine and tyrosine detoxified *Alternaria macrospora* toxin (Krishnamohan and Vidhyasekaran, 1988). Petasol, a toxin produced by *Drechslera gigantea*, is conjugated with amino acids to yield an inactive conversion product (Bunkers and Strobel, 1991).

Helminthosporium carbonum, the maize leaf spot pathogen, produces a toxin named HC-toxin. The toxin is a cyclic tetrapeptide of structure cyclo (D-prolyl-L-alanyl-D-alanyl-L-2 Aeo and the Aeo stands for amino-9,10-epoxy-8-oxodecanoic acid (Kawai et al., 1983). The epoxide and the 8-carboxyl groups in Aeo are required for biological activity of HC toxin (Walton and Earle, 1983; Kim et al., 1987).

When HC-toxin is fed to maize leaves through the transpiration stream, it is metabolized to a single, nontoxic derivative in which 8-carboxyl group of Aeo is reduced to the corresponding alcohol (Meeley and Walton, 1991). Cell-free extracts of maize seedlings metabolize HC-toxin to the same compound, 8-hydroxy-HC-toxin, as whole green leaves. HC-toxin reductase has been shown to be involved in the metabolization of the toxin and this enzyme has been purified and characterized (Meeley and Walton, 1991). This enzyme inactivates HC-toxin by pyridine nucleotide-dependent reaction of an essential carbonyl group. This enzyme activity was detectable only in extracts of maize resistant to *H. carborum* race 1. The results indicate that detoxification of the toxin occurs in resistant varieties while such detoxification does not occur in susceptible varieties (Meeley et al., 1992).

Detoxification of the toxins induce resistance even in the susceptible varieties. Treating tomato roots with soil bacteria which are able to degrade and detoxify fusaric acid, as well as with toxin-degarding mutants of *Pseudomonas solanacearum* and clones of *Escherichia coli* transformed with engineered plasmids from *Cladosporium werneckii*, protected plants from wilting caused by *Fusarium oxysporum* f.sp. *lycopersici* (Utsumi et al., 1989). Bacterial strains capable of degrading oxalic acid, the toxin produced by *Sclerotinia sclerotiorum*, have been identified and these bacterial strains prevented *S. sclerotiorum* infection in *Arabidopsis thaliana* (Dickman and Mitra, 1992). These results also suggest the possibility of existence of detoxification mechanism in resistant plants and similar mechanisms may not exist in susceptible ones.

B. Susceptible Tissues May Have Toxin-Receptors that May Be Absent in Resistant Tissues

Toxin uptake by susceptible tissue is considered to involve adsorption to a receptor that may be lacking in resistant tissues (Scheffer and Pringle, 1964). Considerable circumstantial evidence supports the presence of toxin receptors in susceptible host tissues. In culture filtrates of pathogens, toxoids have also been detected along with toxins. Since toxoids are structurally related to the toxin, they may compete for the same toxin-receptor sites. Pretreatment of susceptible tissues with toxoids prevented toxicity of the corresponding native toxin in case of interactions between HS-toxin and sugarcane (Livingston and Scheffer, 1984), HV-toxin and oats (Wolpert and Macko, 1989) and AF-toxin and strawberry (Namiki et al., 1986b).

However, toxin receptor sites have been identified only in few cases. Only oat genotypes carrying the dominant Vb allele are both susceptible to *Helminthosporium victoriae* and sensitive to the toxin produced by the fungus, victorin (Wolpert et al., 1986; Wolpert and Macko, 1989). When oat leaf slices were incubated with [^{125}I]Bolton-Hunter derivative of victorin C(BHC), a 100-kDa protein was labeled only in the susceptible (s) oat genotype. The genotype-specific toxin-induced labeling of the 100 kDa protein appeared to be covalent as it persisted after phenol extraction, precipitation, and polyacrylamide gel electrophoresis (PAGE). Reduced form of victorin C(RC) is not toxic but it behaves as a protectant, presumably by competitive displacement of victorin. When oat leaf slices were incubated with vicotrin C, the toxic effects of victorin C were decreased or eliminated, depending on the concentration of RC (Wolpert and Macko, 1989).

In vitro binding analysis revealed a requirement for an exogenous reducing agent (2-mercaptoethanol or dithiothereitol). When leaf tissue homogenates were prepared in the presence or absence of reducing agent and tested for covalent binding, labeling was observed only in the presence of a reducing agent. In vitro binding experiments demonstrated the presence of a 100 kDa victorin-binding protein in both susceptible and resistant genotypes. Reduced form of victorin and methyl ester victorin C prevented [^{125}I]BHC-mediated labeling of the 100-kDa protein (Wolpert and Macko, 1989). The reduction of the aldehyde to a primary alcohol, as in RC, eliminates toxicity of victorin. RC may nevertheless associate with the active site, because treatment of oat leaf slices with RC prevents the toxic effect of victorin C. Many aldehydes are known to bind covalently to proteins. Hence the aldehyde may mediate a covalent association of victorin with its site of action and the covalent binding may be essential for toxicity (Wolpert and Macko, 1988).

Thus a 100-kDa protein binds [^{125}I]BHC in a genotype-specific manner only in vivo but has no genotype specificity in vitro. Binding in vitro required a reducing agent that was not required for in vivo binding. It is possible that the difference between S and R genotypes is due to the presence of a reducing group either in the 100 kDa protein or in associated molecules. The 100 kDa protein may be the victorin receptor (Wolpert and Macko, 1989).

Polyclonal antibodies against victorin were raised in rabbits immunized with a victorin bovine serum albumin conjugate (Akimitsu et al., 1992). The antibodies were purified from serum by protein A column chromatography. In vivo and in vitro covalent binding of victorin to proteins in susceptible and resistant oat tissue was examined by Western blot bioassays using anti-victorin antibody. In vivo binding of victorin to proteins of 100 and 45 kDa was observed in both susceptible and resistant cultivars of oats. Victorin also bound in vitro to proteins of 100, 65, and 45 kDa in both susceptible and resistant oats (Akimitsu et al., 1992). These results are in contrast to findings of Wolpert and Macko (1989). According to Akimitsu et al., (1992) there was no evidence for host-specific binding of victorin in vivo also. Binding in vivo occurred in both S and R oats, and not in just S oats. Hence host-specificity may be due to some other components in the victorin-transduction pathway after vicotrin binds to oat tissue proteins (Akimitsu et al., 1992). However Wolpert et al., (1994) argued that the genotype-specific binding could not be detected in vivo with the antivictorin antibody by Akimitsu et al. (1992), because the binding observed was as a result of in vitro binding that occurred as a consequence of homogenizing the tissue in a nondenaturing buffer. Wolpert and Macko (1994) developed antisera to the in vitro-labeled 100 kDa protein from the susceptible oat genotype and these preparations reacted with the in vivo-labeled 100 kDa protein from the susceptible genotype. The antibodies directed against the 100 kDa protein isolated from the susceptible genotype reacted with the 100 kDa protein from the resistant genotype. The results suggest that the proteins from the susceptible and resistant genotypes may differ by only a single amino acid residue, or may be structurally identical (Wolpert and Macko, 1991).

The 100 kDa victorin-binding protein (VBP) may be the site of action (Wolpert et al., 1994). A 3.4 kb cDNA clone was isolated that, when subjected to a 400 bp 5′ deletion, was capable of directing the synthesis of a protein in *Escherichia coli*, which reacted to an antibody specific for the 100 kDa protein. Peptide mapping, by limited proteolysis, indicated that the protein directed by the cDNA is the 100 kDa protein (Wolpert et al., 1994).

The 100 kDa VBP could be detected in many plant species such as *Arabidopsis*, pea, tomato, wheat, sorghum, maize, and oats (Wolpert et al., 1994). It could be detected in various organisms tested such as *Chlorella*, bovine lever, and *E. coli* (Wolpert et al., 1994). Thus the 100 kDa protein is some what universal (Loschke et al., 1994).

DNA sequence analysis of the 3427 bp cDNA for the VBP revealed a 3096 bp open reading frame coding a protein containing 1032 amino acid residues (Wolpert et al., 1994). A comparison of the deduced amino acid sequence of the 100 kDa VBP cDNA and the pea cDNA indicated 83% amino acid identity. The pea cDNA encodes for the P protein component of the glycine decarboxylase multienzyme complex. The mature form of the 100 kDa VBP also displays 71% amino acid similarity to the P protein of *E. coli*, humans, and chicken (Wolpert et al., 1994). The glycine decarboxylase is found in animals, plants, and prokaryotes and is located in the mitochondria of eukaryotic organisms (Kume et al., 1991). Hence Wolpert et al. (1994) concluded that the 100 kDa VBP is the P protein component of glycine decarboxylase based on the widespread species distribution of the 100-kDa VBP, suggesting a common and basic metabolic function of the 100 kDa VBP, and nucleotide sequence homology of the cDNA for the 100-kDA VBP with the P protein component of glycine decarboxylase. In plant tissues, glycine decarboxylase has a major role in the recovery of carbon and energy due to photorespiration, and mutations in the photorespiration are lethal to plants. The effect of victorin interaction with the P protein may be the inhibition of glycine decarboxylase activity, and this would cause cell death in green tissue (Wolpert et al., 1994).

Another victorin-binding protein, a 15 kDa protein has also been isolated (Wolpert et al., 1995). This protein was also a component of the multienzyme complex, glycine decarboxylase (Wolpert et al., 1995).

A putative receptor protein for dothistromin, the toxin produced by *Dothistroma pini*, the pathogen of *Pinus radiata*, has been identified (Jones et al., 1995). Results of experiments with treatments that protect sorghum seedlings against the effects of peritoxin produced by *Periconia circinata* (e.g., proteinase, heat shock, inhibitors of protein synthesis, and biotinylation of membrane proteins) suggest that disease symptoms result from an interaction of peritoxin with a proteinaceous receptor (Dunkle and Macko, 1995).

The plasmalemma of susceptible sugarcane contains a protein of 48 kDa that specifically binds to the toxin produced by *Helminthosporium sacchari* (HS-toxin) (Kenfield and Strobel, 1981). Plasmalemma of resistant sugarcane has a similar protein that differs by four amino acid residues and does not bind toxin (Strobel, 1973a,b; Strobel and Hess, 1974; Strobel et al., 1975). However some nonhosts of *H. sacchari* (tobacco and mink) bind as much or more ^{14}C-labeled toxin preparation as does susceptible sugarcane (Kenfield and Strobel, 1977). Conclusive evidence for the role of toxin-receptors in susceptibility of host plants to pathogens is still lacking (Yoder, 1980). In fact, toxin receptors or affinity sites may be existing in most of the plant tissues and host specificity may be a event after the toxin binding (Akimitsu et al., 1992).

C. Susceptible Tissues May be More Sensitive to the Toxins

Susceptible tissues may be highly sensitive to certain (host-specific) toxins. Tomato cultivars resistant to *Alternaria alternata* f.sp. *lycopersici* are at least 1000-fold less sensitive to the AAL-toxin than susceptible cultivars (Gilchrist and Grogan, 1976). The toxins at lower concentrations induce disease symptoms in the susceptible plants while at higher concentrations they induce disease symptoms in the resistant varieties also (Vidhyasekaran et al., 1986; Lalitha et al., 1989; Kohmoto et al., 1991). However, it should be noted that nonhost-specific toxins at low concentrations affect both susceptible and resistant plants almost equally and many pathogens are known to produce nonspecific toxins. Hence sensitivity of host tissues to the toxins may not be an important factor in their susceptibility to the pathogens.

D. Specific Protein Synthesized after Toxin Exposure may Confer Host Specificity

Treatment of sorghum seedlings with *Periconia circinata* toxin (PC-toxin) induces visible symptoms of the disease only in susceptible genotypes. These symptoms were associated with a loss of electrolytes from roots into the surrounding medium (Dunkle, 1978; Dunkle and Wolpert, 1981) and the enhanced synthesis of a family of proteins with Mr of 16 kDa and their respective mRNAs only in susceptible genotypes of the host (Wolpert and Dunkle, 1983; Traylor et al., 1988). Treatments that prevent the synthesis of 16 kDa proteins protect seedlings against the effects of PC-toxin (Wolpert and Dunkle, 1983). The treatments that induce the synthesis of 16 kDa proteins elicit the same disease symptoms as the toxin (Traylor et al., 1987). The results suggest that the 16 kDa proteins may confer host-specificity.

However, treatment with *Colletotrichum graminicola* elicitor also enhanced the synthesis of the 16 kDa proteins in sorghum seedlings root tips. Synthesis of the proteins was enhanced in seedlings pretreated with elicitor and then treated with toxin and in seedlings pretreated with toxin and subsequently treated with elicitor. The enhanced synthesis of the 16 kDa proteins was not observed in seedlings pretreated with toxin and transferred with water for 12 hr prior to pulse-labeling with [^{3}H]leucine. The results suggest that the seedlings recover from the effects of the toxin or that the effects are temporary and subside. The effects of the elicitor were genotype nonspecific; the enhanced synthesis of the 16 kDa proteins was induced in root tips of both susceptible and resistant seedlings (Ransom et al., 1992). The induction of enhanced synthesis of 16 kDa proteins may be a general stress response or a response to a generally toxic preparation. The elicitor is also phytotoxic like the toxin (Ransom et al., 1992).

Furthermore, these 16 kDa proteins have been detected in corn, barley, oat, rice, water hyacinth (*Eichhornia crassipes*), and spider plant (*Chlorophytum* sp.) (Ransom et al., 1994). The results suggest that the 16 kDa proteins are conserved proteins in several plant species. These studies do not confirm the importance of 16 kDa proteins in conferring susceptibility in plants to pathogens.

E. Proteins of Susceptible Hosts May Enhance Potential of Pathogens to Produce Toxins

Presence of susceptible host proteins(s) appears to enhance virulence of pathogens. *Cercospora kikuchii*, the soybean pathogen produces the important pathogenicity factor cercosporin in infected soybeans. The pathogen responds to soybean protein(s) with increased cercosporin production and changes in transcript accumulation. It suggests an alteration in fungal gene expression in the presence of host plant substances (Ehrenshaft and Upchurch, 1993).

F. Sucrose Influx May Have Correlation with Sensitivity to Toxin

Sucrose transport system may also be involved in conferring sensitivity to toxin in susceptible tissues. Sucrose transport system in tomato isolines resistant or susceptible to *Alternaria alternata* f.sp. *lycopersici* varied significantly. The young leaflets of the susceptible variety possessed an endogenous sucrose uptake rate three times higher than the young leaflets of the resistant variety. Thus a physiological difference in sugar transport associated with disease resistance mechanism has been identified in tomato (Moussatos et al., 1993). The pathogen produced a host-specific toxin, AAL-toxin. AAL-toxin significantly affected sucrose transport. AAL toxin treatment reduced sucrose influx to discs of the young tomato leaves, but the resistant isoline exhibited a recovery after 4 h while the susceptible isoline continued to decline. The toxin moves in the phloem and one of the disease symptoms associated with AAL toxin and *A. alternaria* f.sp. *lycopersici* in whole plants is a browning of phloem tissue. High initial or stimulated sucrose transport rates would function to increase uptake of AAL-toxin. The susceptible isolines showed higher initial sucrose transport (Moussatos et al., 1993). In general, sensitive to AAL-toxin involved a dramatic perturbation of sucrose influx while resistance had a reduced perturbation from a lower initial sucrose influx rate. The toxin causes reduction in sucrose transport and a higher initial uptake of the toxin along with the sucrose transport system in the susceptible isoline would have caused more perturbation in the sucrose transport (Moussatos et al., 1993).

PCMBS (P-chloro-mercuribenzene sulphonic acid) is a sucrose transport inhibitor. PCMBS treatment resulted in more interveinal necrosis in the susceptible isoline than in the resistant isoline in the absence of AAL-toxin. The results suggest that sucrose transport, and/or factors that regulate sucrose retrieval, are directly involved in AAL-toxin-induced metabolic changes that culminate in cell death and the sensitivity to the toxin depends upon the sucrose transport system (Moussatos et al., 1993). Thus a physiological process may be involved in the host-specificity. Sucrose transport system may be linked to ethylene biosynthesis (Moussatos et al., 1994). AAL-toxin induced ethylene biosynthesis in susceptible tomato cultivar and not in resistant isoline. Dihydroorotic acid, an intermediate in pyrimidine biosynthesis, abolished the AAL-toxin induced increase in 1-aminocyclopropane-1-carboxylic acid, the precursor of ethylene. Addition of N-(phosphonacetyl)-L-aspartate, a specific inhibitor of pyrimidine biosynthesis, elicited interveinal necrosis resembling AAL-toxin treatment (Moussatos et al., 1994). These results suggest a metabolic linkage between the action of sucrose transport, ethylene biosynthesis, pyrimidine metabolism, and cell death and the toxin had an important role in the metabolic linkage. This cascade of reactions induced by the toxin may be involved in determination of host-specificity.

V. CONCLUSION

Cell membrane is an active site for induction of several defense mechanisms. Elicitors act at the cell membrane and signals generated are transducted across the membrane. Both transmembrane signal system and systemic signal system exist in the plasma membrane. Successful pathogens should nullify the signals generated in the active membrane. Hence the pathogens cause membrane dysfunction. Membrane dysfunction is caused by both biotrophic and necrotrophic pathogens, but more studies have been conducted with necrotrophic pathogens that produce toxins. The primary site of action of toxins is cell membrane. The toxins increase membrane permeability. Leakage of cell membrane is commonly observed. The toxin may not cause general membrane disruption but may inhibit specific membrane transport system. The leakage may occur through the membrane matrix possibly via shuttle type carrier. Biotrophic pathogens may decrease permeability of cell membrane and it may be due to structural changes in the phospholipid bilayer of the plasmalemma which results in reduced membrane fluidity.

H^+-ATPase pumps protons across the plasma membrane. Some toxins stimulate the activity of H^+-ATPase. This stimulation may not be due to de novo synthesis of this but due to proteolytic removal of a part of the C-terminal end of the 100-kDa H^+-ATPase. The toxins may inhibit calmodulin activity affecting calcium transport. Depolarization of membrane potential occurs due to toxin application. Depolarization may be due to reduced ATP levels.

The pathogens may induce changes in phospholipid contents of the host cell membrane. The toxins activate phospholipases resulting in reduction in phospholipid content. The removal of phospholipids from membranes results in cell death. Hydroxyl radicals and superoxide anions are induced to accumulate due to fungal infection. They may cause disruption of lysosome membrane. Some toxins cause peroxidation of plant lipids. They release singlet oxygen and superoxide anions which are toxic to plant cells.

Mitochondrial membranes are also affected by toxins. The electron transfer occurs across the inner membrane of mitochondrion that drives oxidative phosphorylation producing cellular energy. The toxins inhibit ATP synthesis, which results in suppression of defense mechanisms of the host.

Membrane dysfunction generally leads to suppression of defense mechanisms of the host. It also induces efflux of nutrients to the pathogen. In resistant and incompatible host–pathogen interactions the action of the toxins is suppressed. Detoxification of the toxins is an important host defense mechanism. Some host enzymes detoxify the toxins. Absence of toxin-binding sites in resistant varieties has been suggested, but convincing evidence is lacking. The host specificity to the toxins may be due to a event after the toxin binding.

Membrane dysfunction followed by cell death may nullify all defense mechanisms of the host favoring development of the pathogen in the host tissue. Cell death is observed in both biotrophic and necrotrophic diseases. Cell death may eliminate the biotrophic pathogens, while it may favor the necrotrophic pathogens by releasing nutrients from dying cells. Every pathogen may have both biotrophic and necrotrophic phases in their life cycle. Biotrophic phase is much longer in case of biotrophic pathogens than the necrotrophic phase while the necrotrophic pathogens have longer necrotrophic phase with shorter biotrophic phase.

REFERENCES

Aist, J. R. (1974). A freeze etch study of membranes in *Plasmodiophora*-infected and non-infected root hairs, Can. J. Bot., 52:1441–1449.

Akimitsu, K., Hart, L. P., Walton, J. D., and Hollingsworth, R. (1992). Covalent binding sites of victorin in oat leaf tissues detected by anti-victorin polyclonal antibodies, Plant Physiol., 98:121–126.

Arias, J. A. (1985). Secretary organelle and mitochondrial alterations induced by fusaric acid in root cells of *Zea mays*, Physiol. Plant Pathol., 27:149–158.

Assante, G., Locci, L., Camarda, L., Merlini, L., and Nasini, G. (1977). Screening of the genus *Cercospora* for secondary metabolites, Phytochemistry, 16:243–247.

Assmann, S. M., and Schwartz, A. (1992). Synergistic effect of light and fusicoccin on stomatal opening. Epidermal peel and patch clamp experiments, Plant Physiol., 98:1349–1355.

Assmann, S. M., and Zeiger, E. (1987). Guard cell bioenergetics, Stomatal Function (E. Zeiger, F. Farquhar and C. Cowan, eds.), Stanford University Press, Stanford, pp. 163–194.

Ayers, P. G. (1972). Abnormal behaviour of stomata in barley leaves infected by *Rhynchosporium secalis* (Oudem.) J. J. Davis, J. Exp. Bot., 23:683–692.

Ayers, P. G., and Jones, P. (1975). Increased transpiration and the accumulation of root absorbed ^{86}Rb in barley leaves infected by *Rhynchosporium secalis* (leaf blotch), Physiol. Plant Pathol., 7:49–58.

Balis, C., and Payne, M. G. (1971). Triglycerides and cercosporin from *Cercospora beticola*: fungal growth and cercosporin production, Phytopathology, 61:1477–1484.

Batchvarova, R. B., Reddy, V. S., and Bennet, J. (1992). Cellular resistance in rice to cercosporin, a toxin of *Cercospora*, Phytopathology, 82:642–646.

Bateman, D. F., and Beer, S. V. (1965). Simultaneous production and synergistic action of oxalic acid and polygalacturonase during pathogenesis by *Sclerotium rolfsii*, Phytopathology, 55:204–211.

Bednarski, M. A., Izawa, S., and Scheffer, R. P. (1977). Reversible effects of toxin from *Helminthosporium maydis* race T on oxidative phosphorylation by mitochondria from maize, Plant Physiol., 59:540–545.

Bertini, S. (1956). Su di un composto ad antibiotica prodotto da *Ascochyta pisi* Lib., Ann. *Speriment. Agr.* 11:545–556.

Berville, A., Ghazi, A., Charbonnier, M., and Bonavent, J. F. (1984). Effects of methomyl and *Helminthosporium maydis* toxin on matrix volume, proton motive force, and NAD accumulation in maize (*Zea mays* L.) mitochondria, Plant Physiol., 76:508–517.

Blatt, M. R. (1988). Mechanisms of fusicoccin action: a dominant role for secondary transport in a higher plant cell, Planta, 174:187–200.

Blatt, M. R., and Clint, G. M. (1989). Mechanisms of fusicoccin action: kinetic modification and inactivation of K^+ channels in guard cells, Planta, 178:509–523.

Blein, J. P., Bourdil, J., Rossignol, M., and Scalla, R. (1988). *Cercospora beticola* toxin inhibits vanadate-sensitive H^+ transport in corn root membrane vesicles, Plant Physiol., 88:429–434.

Blum, W., Key, G., and Weiler, E. W. (1988). ATPase activity in plasmalemma-rich vesicles isolated by aqueous two-phase partitioning from *Vicia faba* mesophyll and epidermis-characterization and influence of abscisic acid and fusicoccin, Physiol. Plant., 72:279–287.

Bottini, A. T., Bowen, J. R., and Gilchrist, D. G. (1981). Phytotoxins II. Characterization of a phytotoxic fraction from *Alternaria alternata* f.sp. *lycopersici*, Tetrahedron Lett., 22:2723–2726.

Bottini, A. T., and Gilchrist, D. G. (1981). Phytotoxins I. A 1-aminodimethyl-heptadecapentol from *Alternaria alternata* f.sp. *lycopersici*, Tetrahedron Lett., 22:2719–2722.

Bouthyette, P. Y., Spitsberg, V., and Gregory, P. (1985). Mitochondrial interaction with *Helminthosporium maydis* race T toxin: blocking by dicyclohexylcarbodiimide, J. Exp. Bot., 36:511–528.

Bracker, C. E. (1968). Ultrastructure of the haustorial apparatus of *Erysiphe graminis* and its relationship to the epidermal cell of barley, Phytopathology, 58:12–30.

Branchard, M., and Laffray, D. (1987). A porometric comparison between the stomatal behaviour of in vitro cultivated barley plants infected with *Rhynchosporium secalis* and treated with rhynchosporoside-like toxins, Physiol. Mol. Plant Pathol., 31:251–260.

Braun, C. J., Siedow, J. N., and Levings, C. S. III. (1989a). The T-urf 13 gene is responsible for toxin sensitivity in maize and *E. coli*, The Molecular Basis of Plant Development (R. Goldberg, ed.), UCLA Symposia on Molecular and Cellular Biology, New Series, Vol. 92, Alan R. Liss, New York, pp. 79–85.

Braun, C. J., Siedow, J. N., and Levings, C. S. III. (1990). Fungal toxins bind to the URF 13 protein in maize mitochondria and *Escherichia coli*, Plant Cell 2:153–161.

Braun, C. J., Siedow, J. N., Williams, M. E., and Levings, C. S. III. (1989b). Mutations in the maize mitochondrial T-urf 13 gene eliminate sensitivity to a fungal pathotoxin, Proc. Natl. Acad. Sci. USA, 86:4435–4439.

Briggs, S. P., Scheffer, R. P., and Haug, A. R. (1984). Selective toxin of *Helminthosporium victoriae*: thermal relationships in effects on oat tissues and protoplasts, Phytopathology, 74:768–773.

Briskin, D. P. (1990). The plasma membrane H^+-ATPase of higher plant cells: biochemistry and transport function, Biochim. Biophys. Acta, 1019:95–109.

Bronson, C. R., and Scheffer, R. P. (1977). Heat- and aging-induced tolerance of sorghum and oat tissues to host-selective toxins, Phytopathology, 67:1232–1238.

Brown, P. H., and Outlaw, W. H. Jr. (1982). Effect of fusicoccin on the dark $^{14}CO_2$ fixation, by *Vicia faba* guard cell protoplasts, Plant Physiol., 70:1700–1703.

Bunkers, G. J., and Strobel, G. A. (1991). A proposed mode of action for green island induction by the eremophilane phytotoxins produced by *Drechslera gigantea*, Physiol. Mol. Plant Pathol., 38:313–323.

Cantone, F. A., and Dunkle, L. D. (1990a). Resistance in susceptible maize to *Helminthosporium carbonum* race 1 induced by prior inoculation with race 2, Phytopathology, 80:1221–1224.

Cantone, F. A., and Dunkle, L. D. (1990b). Involvement of an inhibitory compound in induced resistance in maize to *Helminthosporium carbonum*, Phytopathology, 80:1225–1230.

Cantone, F. A., and Dunkle, L. D. (1991). Reversible effects of inhibitory diffusates from maize inoculated with *Cochliobolus carbonum*, Physiol. Mol. Plant Pathol., 39:111–122.

Ciuffetti, L. M., Pope, M. R., Dunkle, L. D., Daly, J. M., and Knoche, H. W. (1983). Isolation and structure of an inactive product derived from the host-specific toxin produced by *Helminthosporium carbonum*, Biochemistry, 22:3507–3510.

Clint, G. M., and Blatt, M. R. (1989). Mechanisms of fusicoccin action: evidence of concerted modulations of secondary K^+ transport in a higher plant cell, Planta, 178:495–508.

Clouse, S. D., Martensen, A. N., and Gilchrist, D. G. (1985). Rapid purification of host-specific pathotoxins from *Alternaria alternata* f.sp. *lycopersici* by solid-phase adsorption on octadecylsilane, J. Chromatogr., 350:355–363.

Cocucci, S. H., Morgutti, S., Cocucci, M., and Ginani, L. (1983). Effects of ophiobolin A on potassium permeability, transmembrane electrical potential and proton extrusion in maize roots, Plant Sci. Lett., 32:9–16.

Comstock, J. C., and Scheffer, R. P. (1973). Role of host-selective toxin in colonization of corn leaves by *Helminthosporium carbonum*, Phytopathology, 63:24–29.

Dainty, J. (1962). Ion transport and electrical potentials in plant cells, Annu. Rev. Plant Physiol., 13:397–402.

D'Alton, A., and Etherton, B. (1984). Effects of fusaric acid on tomato root hair membrane potentials and ATP levels, Plant Physiol., 74:39–42.

Danko, S. J., Kono, Y., Daly, J. M., Suzuki, Y., Takeuchi, S., and McCrery, D. A. (1984). Structure and biological activity of a host-specific toxin produced by the fungal corn pathogen *Phyllosticta maydis*, Biochemistry, 23:759–766.

Daub, M. E. (1982a). Cercosporin, a photosensitizing toxin from *Cercospora* species, Phytopathology, 72:370–374.

Daub, M. E. (1982b). Peroxidation of tobacco membrane lipids by the photosensitizing toxin, cercosporin, Plant Physiol., 69:1361–1364.

Daub, M. E., and Briggs, S. P. (1983). Changes in tobacco cell membrane composition and structure caused by cercosporin, Plant Physiol., 71:763–766.

Daub, M. E., and Hangarter, R. P. (1983). Light-induced production of singlet oxygen and superoxide by the fungal toxin, cercosporin, Plant Physiol., 73:855–857.

Daub, M. E., and Payne, G. E. (1989). The role of carotenoids in resistance of fungi to cercosporin, Phytopathology, 79:180–185.

De Boer, A. H., Watson, B. A., and Cleland, R. E. (1989). Purification and identification of the fusicoccin binding protein from oat root plasma membrane, Plant Physiol., 89:250–259.

De Michelis, M. I., Pugliarello, M. C., and Rasi-Caldogno, F. (1989). Fusicoccin binding to its plasma membrane receptor and the activation of the plasma membrane H^+-ATPase, Plant Physiol., 90:133–139.

Dewey, R. E., and Levings, C. S. III., and Timothy, D. H. (1986). Novel recombinations in the maize mitochondrial genome produce a unique transcriptional unit in the Texas male-sterile cytoplasm, Cell, 44:439–444.

Dewey, R. E., Siedow, J. N., Timothy, D. H., and Levings, C. S. III. (1988). A 13-kilodalton maize mitochondrial protein in *E. coli* confers sensitivity to *Bipolaris maydis* toxin, Science, 239:293–295.

Dewey, R. E., Timothy, D. H., and Levings, C. S. III. (1987). A mitochondrial protein associated with cytoplasmic male sterility in the T cytoplasm of maize, Proc. Natl. Acad. Sci. USA, 84:5374–5378.

Dickman, M. B., and Mitra, A. (1992). *Arabidopsis thaliana* as a model for studying *Sclerotinia sclerotiorum* pathogenesis, Physiol. Mol. Plant Pathol., 41:255–263.

Donovan, N., Martin, S., and Donkin, M. E. (1985). Calmodulin binding drugs trifluoperazine and compound 48/80 modify stomatal responses of *Commelina communis* L., J. Plant Physiol., 118:177–187.

Duke, S. O., Wickliff, J. L., Vaughn, K. C., and Paul, K. N. (1982). Tentoxin does not cause chlorosis in greening mung bean leaves by inhibiting photophosphorylation, Physiol. Plant., 56:387–398.

Dunkle, L. D. (1978). Heterogenous reaction of shattercane to *Periconia circinata* and its host-specific toxin, Phytopathology, 69:260–262.

Dunkle, L. D., and Macko, V. (1995). Peritoxins and their effects on sorghum, Can. J. Bot., 73(Suppl.1):S444–S452.

Dunkle, L. D., and Wolpert, T. J. (1981). Independence of milo disease symptoms and electrolyte leakage induced by the host-specific toxin from *Periconia circinata*, Physiol. Plant. Pathol., 18:315–323.

Durbin, R. D., and Uchytil, T. F. (1977). A survey of plant insensitivity of tentoxin, Phytopathology, 67:602–603.

Durbin, R. D., Uchytil, T. F., and Sparapano, L. (1973). The effect of tentoxin on stomatal aperture and potassium content of guard cells, Phytopathology, 63:1077–1078.

Duvick, J. P., Daly, J. M., Kratky, Z., Macko, V., Acklin, W., and Arigoni, D. (1984). Biological activity of the isomeric forms of *Helminthosporium sacchari* toxin and of homologues produced in culture, Plant Physiol., 74:117–122.

Ehrenshaft, M., and Upchurch, R. G. (1993). Host protein(s) induces accumulation of the toxin cercosporin and mRNA in a phytopathogenic strain of *Cercospora kikuchii*, Physiol. Mol. Plant Pathol., 43:95–107.

Fajola, A. O. (1978). Cercosporin, a phytotoxin from *Cercospora* species, Physiol. Plant Pathol., 13:157–164.

Faull, J. L., and Gay, J. L. (1983). Phospholipase activity in *Erysiphe pisi*, Physiol. Plant Pathol., 22:55–63.

Feyerabend, M., and Weiler, E. W. (1988). Characterization and localization of fusicoccin-binding sites in leaf tissues of *Vicia faba* L. probed with a novel radioligand, Planta, 174:115–122.

Feyerabend, M., and Weiler, E. W. (1989). Photoaffinity labeling and partial purification of the putative plant receptor for the fungal wilt-inducing toxin, fusicoccin, Planta, 178:282–290.

Fischer, R. A. (1972). Aspects of potassium accumulation by stomata of *Vicia faba*, Aust. J. Biol. Sci., 25:1107–1123.

Flavell, R. (1975). Inhibition of electron transport in maize mitochondria by *Helminthosporium maydis* race T pathotoxin, Physiol. Plant Pathol., 6:107–116.

Foote, C. S. (1976). Photosensitized oxidation and singlet oxygen: consequences in biological systems, Free Radicals in Biology, Vol. II (W. A. Pryor, ed.), Academic Press, New York, pp. 85–113.

Friis, P., Olsen, C. E., and Moller, B. L. (1991). Toxin production in *Pyrenophora teres*, the Ascomycete causing net-spot blotch disease of barley (*Hordeum vulgare* L.), J. Biol. Chem., 266:13329–13335.

Fujita, K., Arase, S., Hiratsuka, H., Honda, Y., and Nozu, M. (1994). The role of toxin(s) produced by germinating spores of *Pyricularia oryzae* in pathogenesis, J. Phytopathol., 142:245–252.

Gardner, J. M., Mansour, I. S., and Scheffer, R. P. (1972). Effects of the host-specific toxin of *Periconia circinata* on some properties of sorghum plasma membranes, Physiol. Plant Pathol., 2:197–206.

Gaumann, E. (1957). Fusaric acid as a wilt toxin, Phytopathology, 47:342–357.

Gil, F., and Gay, J. L. (1977). Ultrastructure and physiological properties of the host interfacial components of haustoria of *Erysiphe pisi* in vivo and in vitro, Physiol. Plant Pathol., 10:1–12.

Gilchrist, D. G., and Grogan, R. G. (1976). Production and nature of host-specific toxin from *Alternaria alternata* f.sp. *lycopersici*, Phytopathology, 66:165–171.

Glab, N., Wise, R. P., Pring, D. R., and Jacq C. Slonimski, P. (1990). Expression in *Saccharomyces cerevisiae* of a gene associated with cytoplasmic male sterility from maize: respiratory dysfunction and uncoupling of yeast mitochondria, Mol. Gel. Genet., 223:24–32.

Godoy, C., Steadman, J. R., Dickman, M. B., and Dam, R. (1990). Use of mutants to demonstrate the role of oxalic acid in pathogenicity of *Sclerotinia sclerotiorum* on *Phaseolus vulgaris*, Physiol. Mol. Plant Pathol., 37:179–191.

Goodwin, T. W. (1980). The Biochemistry of the Carotenoids, Chapman and Hall, London, pp. 377.

Graniti, A. (1991). Phytotoxins and their involvement in plant diseases. Introduction, Experientia, 47:751–755.

Graniti, A., and Turner, N. C. (1970). Effect of fusicoccin on stomatal transpiration in plants, Phytopath. Medit., 9:75:86.

Grogan, R., Kimble, K., and Misaghi, I. (1975). A stem canker disease of tomato caused by *Alternaria alternata* f.sp. *lycopersici*, Phytopathology, 65:880–886.

Havir, E. A., and Anagnostakis, S. L. (1985). Oxalacetate acetylhydrolase activity in virulent and hypovirulent strains of *Endothia (Cryphonectria) parasitica*, Physiol. Plant Pathol., 26:1–9.

Hepler, P. K., and Wayne, R. O. (1985). Calcium and plant development, Annu. Rev. Plant Physiol., 36:397–439.

Higginbotham, N. (1970). Movements of ions and electrogenesis in higher plant cells, Am. Zool., 10:393–403.

Higginbotham, N., Graves, J. S., and Davis, R. F. (1970). Evidence for an electrogenic ion transport pump in cells of higher plants, J. Membr. Biol., 3:210–222.

Hirata, K., and Kojima, M. (1962). On the structure and sack of the haustorium of some powdery mildews with some considerations of the significance of the sack, Trans. Mycol. Soc. Jpn., 3:43–46.

Holden, M. J., and Sze, H. (1984). *Helminthosporium maydis* T toxin increased membrane permeability to Ca^{2+} in susceptible corn mitochondria, Plant Physiol., 75:235–237.

Holden, M. J., and Sze, H. (1987). Dissipation of the membrane-potential in susceptible corn mitochondria by the toxin of *Helminthosporium maydis* race T and toxin analogs, Plant Physiol., 84:670–676.

Holden, M. J., and Sze, H. (1989). Effects of *Helminthosporium maydis* race T toxin on electron transport in susceptible corn mitochondria and prevention of toxin actions by dicyclohexylcarbodiimide, Plant Physiol., 91:1296–1302.

Hoppe, H. H., and Heitefuss, R. (1974). Permeability and membrane lipid metabolism of *Phaseolus vulgaris* infected with *Uromyces phaseoli*. II. Changes in lipid concentration and ^{32}P incorporation into phospholipids, Physiol. Plant Pathol., 4:11–23.

Hoppe, H. H., and Heitefuss, R. (1975). Permeability and membrane lipid metabolism of *Phaseolus vulgaris* infected with *Uromyces phaseoli*. IV. Phospholipids and phospholipid fatty acids in healthy and rust infected bean leaves resistant and susceptible to *Uromyces phaseoli*, Physiol. Plant Pathol., 5:263–270.

Huang, J., Lee, S. H., Lin, C., Medici, R., Hack, E., and Myers, A. M. (1990). Expression in yeast of the T-URF 13 protein from Texas male-sterile maize mitochondria confers sensitivity to methomyl and to Texas-cytoplasm-specific fungal toxins, EMBO J., 9:339–347.

Iwasaki, S., Nozoe, R., Okuda, S., Sato, Z., and Kozaka, T. (1969). Isolation and structural elucidation of a phytoxic substance produced by *Pyricularia oryzae* Cavara, Tetrahedron Lett., 45:3977–3980.

Jenns, A. E., Daub, M. E., and Upchurch, R. G. (1989). Regulation of cercosporin accumulation in culture by medium and temperature manipulation, Phytopathology, 79:213–219.

Johansson, F., Sommarin, M., and Larsson, C. (1993). Fusicoccin activates the plasma membrane H^+-ATPase by a mechanism involving the C-terminal inhibitory domain, Plant Cell, 5:321–327.

Jones, P., and Ayers, P. G. (1972). The nutrition of the subcuticular mycelium of *Rhynchosporium secalis* (barley leaf blotch): permeability changes induced in the host, Physiol. Plant Pathol., 2:383–392.

Jones, W. T., Harvey, D., Jones, S. D., Sutherland, P. W., Nicol, M. J., Sergejew, N., Debnam, P. M., Cranshaw, N., and Reynold, P. H. S. (1995). Interaction between the phytotoxin dothistromin and *Pinus radiata* embryos, Phytopathology, 85:1099–1104.

Jordan, C. M., and DeVay, J. E. (1990). Lysosome disruption associated with hypersensitive reaction in the potato–*Phytophthora infestans* host–parasite interaction, Physiol. Mol. Plant Pathol., 36:221–236.

Kawai, M., Rich, D. H., and Walton, J. D. (1983). The structure and conformation of HC-toxin, Biochem. Biophys. Res. Commun., 111:398–403.

Kenfield, D. S., and Strobel, G. A. (1977). Biochemical aspects of plant disease resistance and susceptibility, Host Plant Resistance to Pests (P. A. Hedin, ed.), Am. Chem. Soc. Symp. Ser. No. 62, pp. 35–46.

Kenfield, D. S., and Strobel, G. A. (1981). Selective modulation of helminthosporoside binding by alpha-galactoside-binding proteins from sugarcane clones susceptible and resistant to *Helminthosporium sacchari*, Physiol. Plant Pathol., 19:145–152.

Kenfield, D. S., Sugawara, F., Bunkers, G., Wu, Y. H., Strobel, G., Fu, Y., and Clardy, J. (1989). Gigantenone, a novel sesquiterpene phytohormone mimic, Experientia, 45:900–902.

Kim, S. D., Knoche, H. W., and Dunkle, L. D. (1987). Essentiality of the ketone function for toxicity of the host-selective toxin produced by *Helminthosporium carbonum*, Physiol. Mol. Plant Pathol., 30:433–440.

Klein, R. R., and Koeppe, D. E. (1985). Mode of methomyl and *Bipolaris maydis* (race T) toxin in uncoupling Texas male-sterile cytoplasm corn mitochondria, Plant Physiol., 77:912–916.

Kneusel, R. E., Schiltz, E., and Matern, U. (1994). Molecular characterization and cloning of an esterase which inactivates the macrolide toxin brefeldin A, J. Biol. Chem., 269:3449–3456.

Kodama, M., Yoshida, T., Otani, H., Kohmoto, K., and Nishimura, S. (1991). Effect of AL-toxin produced by *Alternaria alternata* tomato pathotype on viability of cultured tomato cells determined by MTT-colorimetric assay, Ann. Phytopathol. Soc. Jpn., 57:663–670.

Koehler, K., and Bentrup, F. W. (1983). The effect of fusaric acid upon electrical membrane properties and ATP level in photoautotrophic cell suspension cultures of *Chenopodium rubrum* L., Z. fur Pflanzenphysiologie, 199 S5:355–361.

Koeppe, D. E., Cox, J. K., and Malone, C. P. (1978). Mitochondrial heredity: a determinant in the toxic response of maize to the insecticide methomyl, Science, 201:1227–1229.

Kohmoto, K., Akimitsu, K., and Otani, H. (1991). Correlation of resistance and susceptibility of citrus to *Alternaria alternata* with sensitivity to host-specific toxins, Phytopathology, 81:719–722.

Kohmoto, K., Khan, I. D., Renbutsu, Y., Taniguchi, T., and Nishimura, S. (1976). Multiple host-specific toxins of *Alternaria mali* and their effect on the permeability of host cells, Physiol. Plant Pathol., 8:141–153.

Kohmoto, K., and Otani, H. (1991). Host recognition by toxigenic plant pathogens, Experientia, 47:755–764.

Kohmoto, K., Otani, H., Kodama, M., and Nishimura, S. (1989). Host recognition: Can accessibility to fungal invasion be induced by host-specific toxins without necessitating necrotic cell death?, Phytotoxins and Plant Pathogenesis, (A. Graniti, R. D. Durbin, and A. Ballio, eds.), NATO ASI Series, Vol. H27, Springer-Verlag, Berlin, pp. 249–265.

Kohmoto, K., Otani, H., and Nishimura, S. (1987). Primary action sites for host-specific toxins produced by *Alternaria* species, Molecular Determinants of Plant Diseases (S. Nishimura, ed.), Japan Scientific Societies Press, Tokyo and Springer-Verlag, Berlin, pp. 127–143.

Kohmoto, K., Verma, V. S., Nishimura, S., Tagami, M., and Scheffer, R. P. (1982). New outbreak of *Alternaria* stem canker of tomato in Japan and production of host-selective toxins by the causal fungus, J. Fac. Agric. Tottori Univ., 17:1–8.

Koleosho, B., Ikotum, T., and Faboya, O. (1987). The role of oxalic acid and polygalacturonase in the pathogenicity of *Pythium aphanidermatum* on different cowpea varieties, Phytoparasitica, 15:317–323.

Kono, Y., Suzuki, Y., Takeuchi, S., Knoche, H. W., and Daly, J. M. (1985). Studies on the host-specific pathotoxins produced by *H. maydis* and *P. maydis*: absolute configuration of PM-toxins and HMT-toxins, Agric. Biol. Chem., 49:559–562.

Krishnamohan, G., and Vidhyasekaran, P. (1988). Detoxification of *Alternaria macrospora* toxin, J. Plant Dis. Protect., 95:630–634.

Kume, A., Koyata, H., Sakakibara, T., Ishiguro, Y., Kure, S., and Hiraga, K. (1991). The glycine cleavage system, J. Biol. Chem., 266:3323–3329.

Kuo, M. S., and Scheffer, R. P. (1964). Evaluation of fusaric acid as a factor in the development of *Fusarium* wilt, Phytopathology, 54:1041–1044.

Kuyama, S., and Tamura, T. (1957). Cercosporin. A pigment of *Cercospora kikuchi* Matsumoto et Tomoyasu. I. Cultivation of fungus, isolation and purification of pigment. J. Am. Chem. Soc., 79:5725–5726.

Lado, P., Pennachioni, A., Rasi Caldogno, F., Russi, S., and Silano, V. (1972). Comparison between some effects of fusicoccin and indole-3-acetic acid on cell enlargement in various plant materials, Physiol. Plant Pathol., 2:75–85.

Lalitha, B., Snow, J. P., and Berggren, G. T. (1989). Phytotoxin production by *Diaporthe phaseolorum* var. *caulivora*, the causal organism of stem canker of soybean, Phytopathology, 79:499–504.

Lamari, L., and Bernier, C. C. (1989). Toxin of *Pyrenophora tritici-repentis*. Host-specificity, significance in disease, and inheritance of host reaction, Phytopathology, 79:740–744.

Lee, S. S., Kawakita, K., Tsuge, T., and Doke, N. (1992). Stimulation of phospholipase A_2 activity in strawberry cells treated with AF-toxin I produced by *Alternaria alternata* strawberry pathotype, Physiol. Mol. Plant Pathol., 41:283–294.

Lee, S. S., Park, P., Tsuge, T., Nakatsuka, S., Doke, N., and Nishimura, S. (1990). Effect of AF-toxin I produced by *Alternaria alternata* strawberry pathotype on ATPase activity in plasma membrane of susceptible strawberry, Ann. Phytopathol. Soc. Jpn., 56:367.

Lee-Stadelmann, O. Y., Bushnell, W. R., Curran, C. M., and Stadelmann, E. J. (1991). Permeability to non-electrolytes as an indicator of change in membrane lipid structure in *Hordeum vulgare* cells infected by *Erysiphe graminis* f.sp. *hordei*, Physiol. Mol. Plant Pathol., 38:163–178.

Lee-Stadelmann, O. Y., Curran, C. M., and Bushnell, W. R. (1992). Incompatibility conditioned by the M1a gene in powdery mildew of barley: change in permeability to non-electrolytes, Physiol. Mol. Plant Pathol., 41:165–177.

Lehnackers, H., and Knogge, W. (1990). Cytological studies on the infection of barley cultivars with known resistance genotypes by *Rhynchosporium secalis*, Can. J. Bot., 68:1953–1961.

Leung, P. C., Taylor, W. R., Wang, J. H., and Tipton, C. L. (1984). Ophiobolin A. A natural product inhibitor of calmodulin, J. Biol. Chem., 259:2742–2747.

Leung, P. C., Taylor, W. R., Wang, J. H., and Tipton, C. L. (1985). Role of calmodulin inhibition in the mode of action of Ophiobolin A., Plant Physiol., 77:303–308.

Levings, C. S., Rhoads, D. M., and Siedow, J. N. (1995). Molecular interactions of *Bipolaris maydis* T-toxin and maize, Can. J. Bot., 73(Suppl. 1):S483–S489.

Levings, C. S., and Sedow, J. N. (1992). Molecular basis of disease susceptibility in the Texas cytoplasm of maize, Plant Mol. Biol., 19:135–147.

Littlefield, L. J., and Bracker, C. E. (1972). Ultrastructural specialisation at the host-parasite interface in rust infected flax, Protoplasma, 74:271–305.

Livingston, R. S., and Scheffer, R. P. (1983). Conversion of *Helminthosporium sacchari* toxin to toxoids by β-galactofuranoside from *Helminthosporium*, Plant Physiol., 72:530–534.

Livingston, R. S., and Scheffer, R. P. (1984). Toxic and protective effects of analogues of *Helminthosporium sacchari* toxin on sugarcane tissues, Physiol. Plant Pathol., 24:133–142.

Loschke, D. C., Tomaska, L. D., Chen, H., Gabriel, D. W., and Rolfe, B. G. (1994). The plant toxin victorin binds to a discrete number of proteins in all organisms tested, Aust. J. Plant Physiol., 21:243–253.

Lousberg, R. J. J. Ch., Weiss, U., Salemink, C. A., Arnone, A., Merlini, L., and Nasini, G. (1971). The structure of cercosporin, a naturally occurring quinone, Chem. Commun., 1971:1463–1464.

Lumsden, R. D. (1970). Phosphatidase of *Sclerotium sclerotiorum* produced in culture and in infected bean, Phytopathology, 60:1106–1110.

Lumsden, R. D., and Bateman, D. F. (1968). Phosphatide degrading enzymes associated with pathogenesis in *Phaseolus vulgaris* infected with *Thielaviopsis basicola*, Phytopathology, 58:219–227.

Lynch, F. J., and Geoghegan, M. J. (1977). Production of cercosporin by *Cercospora species*, Trans. Br. Mycol. Soc., 69:496–498.

Maekawa, N., Yamamoto, M., Nishimura, S., Kohmoto, K., Kuwata, M., and Watanabe, Y. (1984). Studies on host-specific AF-toxins produced by *Alternaria alternata* strawberry pathotype causing alternaria black spot of strawberry. I. Production of host-specific toxins and their biological activities, Ann. Phytopathol. Soc. Japan., 50:600–609.

Macko, V., Stimmel, M. B., Wolpert, T. J., Dunkle, L. D., Acklin, W., Banteli, R., Jaun, B., and Arigoni, D. (1992). Structure of the host-specific toxins produced by the fungal pathogen *Periconia circinata*, Proc. Natl. Acad. Sci. USA, 89:9574–9578.

Macri, F., Dell'antone, P., and Vianello, A. (1983). ATP-dependent proton uptake inhibited by *Cercospora beticola* toxin in pea stem microsomal vesicles, Plant Cell Environ., 6:555–558.

Macri, F., and Vianello, A. (1979). Inhibition of K^+ uptake, H^+ extrusion and K^+-activated ATPase, and depolarization of transmembrane potential in plant tissues treated with *Cercospora beticola* toxin, Physiol. Plant Pathol., 15:161–170.

Macri, F., Vianello, A., Cerana, R., and Rasi-Caldogno, F. (1980). Effects of *Cercospora beticola* toxin on ATP level of maize roots and on phosphorylating activity of isolated pea mitochondria, Plant Sci. Lett., 18:207–214.

Magro, P., Marciano, P., and Di Lenna, P. (1984). Oxalic acid production and its role in pathogenesis of *Sclerotinia sclerotiorum* (sunflower), FEMS. Microbiological Lett., 24:9–12.

Marciano, P., Di Lenna, P., and Magro, P. (1983). Oxalic acid, cell wall degrading enzymes and pH in pathogenesis and their significance in the virulence of two *Sclerotinia sclerotiorum* isolates on sunflower. Physiol. Plant Pathol., 22:339–345.

Margin, R. L., Niesman, M. R., and Bacic, G. (1990). Influence of fluidity on membrane permeability: correspondence between studies of membrane models and simple biological systems, Membrane Transport and Information Storage, Advances in Membrane Fluidity, Vol. 4 (R. C. Aloia, C. C. Curtain and L. M. Gordon, eds.), Alan R. Liss, New York, pp. 221–237.

Marre, E. (1979). Fusicoccin: a tool in plant physiology, Annu. Rev. Plant Physiol., 30:273–288.

Marre, M. T., Vergani, P., and Albergoni, F. G. (1993). Relationship between fusaric acid uptake and its binding to cell structures by leaves of *Egeria densa* and its toxic effects on membrane permeability and respiration, Physiol. Mol. Plant Pathol., 42:141–157.

Martiny-Baron, G., Scherer, G. F. E. (1989). Phospholipid-stimulated protein kinase in plants, J. Biol. Chem., 264:18052–18059.

Matthews, D. E., Gregory, P., and Gracen, V. E. (1979). *Helminthosporium maydis* race T toxin induces leakage of NAD^+ from T cytoplasm corn mitochondria, Plant Physiol., 63:1149–1153.

Maxwell, D. P., and Lumsden, R. D. (1970). Oxalic acid production by *Sclerotinia sclerotiorum* in infected bean and in culture, Phytopathology, 60:1395–1398.

Mayama, S., Tani, T., Ueno, T., Midland, S. L., Sims, J. J., and Keen, N. T. (1986). The purification of victorin and its phytoalexin elicitor activity in oat leaves, Physiol. Mol. Plant Pathol., 29:1–18.

Meeley, R. B., Johal, G. S., Briggs, S. P., and Walton, J. D. (1992). A biochemical phenotype for a disease resistance gene of maize, Plant Cell, 4:71–77.

Meeley, R. B., and Walton, J. D. (1991). Enzymatic detoxification of HC-toxin, the host selective cyclic peptide from *Cochliobolus carbonum*, Plant Physiol., 97:1080–1086.

Melouk, H. A., and Schuh, W. (1987). Cercosporin production and pathogenicity of *Cercospora arachidicola* isolates, Phytopathology, 77:642.

Mertz, S. M. Jr., and Arntzen, C. J. (1978). Depolarization of electrogenic transmembrane electropotential of *Zea mays* L. by *Bipolaris* (*Helminthosporium*) *maydis* race T toxin, azide, cyanide, dodecyl, succinic acid, or cold temperature, Plant Physiol., 62:781–783.

Meyer, C., Feyerabend, M., and Weiler, E. W. (1989). Fusicoccin-binding proteins in *Arabidopsis thaliana* (L.) Heynh, Plant Physiol., 89:692–699.

Miller, R. J., and Koeppe, D. E. (1971). Southern corn leaf blight: susceptible and resistant mitochondria, Science, 173:67–69.

Mollers, C., Zitzlsperger, J., and Wenzel, G. (1992). The effects of a toxin preparation from *Phytophthora infestans* on potato protoplasts and microspores, Physiol. Mol. Plant Pathol., 41:427–435.

Moussatos, V. V., Lucas, W. J., and Gilchrist, D. G. (1993). AAL toxin-induced physiological changes in *Lycopersicon esculentum* Mill: differential sucrose transport in tomato lines isogenic for the *Asc* locus, Physiol. Mol. Plant Pathol., 42:359–371.

Moussatos, V. V., Yang, S. F., Ward, B., and Gilchrist, D. G. (1994). AAL-toxin induced physiological changes in *Lycopersicon esculentum* Mill: roles for ethylene and pyrimidine intermediates in necrosis, Physiol. Mol. Plant Pathol., 44:455–468.

Mumma, R. O., Lukezic, F. L., and Kelly, M. G. (1973). Cercosporin from *Cercospora hayii*, Phytochemistry, 12:917–922.

Nachmias, A., Buchner, V., and Burstein, Y. (1985). Biological and immunochemical characterization of a low molecular weight phytotoxin isolated from a protein–lipopolysaccharide complex produced by a potato isolate of *Verticillium dahliae* Kleb., Physiol. Plant Pathol., 26:43–55.

Naef-Roth, S., and Reusser, P. (1954). Uber die Wirkung der Fusarinsaure auf den Gaswechsel von Tomatenblattgewebe, Phytopathol. Z., 22:281–287.

Nakamura, M., and Ishibashi, K. (1958). On new antibiotics "Ophiobolin" produced by *Ophiobolus miyabeanus*, J. Agric. Chem. Soc. Jpn., 32:739–744.

Nakatsuka, S., Ueda, K., Goto, T., Yamamoto, M., Nishimura, S., and Kohmoto, K. (1986). Structure of AF-toxin II, one of the host-specific toxins produced by *Alternaria alternata* strawberry pathotype, Tetrahedron Lett., 27:2753–2756.

Namiki, F., Okamoto, M., Katou, K., Yamamoto, M., Nishimura, S., Nakatsuka, S., Goto, T., Kohmoto, K., and Otani, H. (1986a). Studies on host-specific AF-toxins produced by *Alternaria alternata* strawberry pathotype causing alternaria black spot of strawberry (5). Effect of toxins on membrane potential of susceptible plants as assayed by electrophysiological method, Ann. Phytopathol. Soc. Jpn., 52:610–619.

Namiki, F., Yamamoto, M., Nishimura, S., Nakatsuka, S., Goto, T., Kohmoto, K., and Otani, H. (1986b). Studies on host-specific AF-toxins produced by *Alternaria alternata* strawberry pathotype causing Alternaria black spot of strawberry. 4. Protective effect of AF-toxin Ii on AF-toxin I-induced toxic action and fungal infection, Ann. Phytopathol. Soc. Jpn., 52:428–436.

Nejidat, A. (1987). Effect of Ophiobolin A on stomatal movement: role of calmodulin, Plant Cell Physiol., 28:455–460.

Nishimura, S., and Kohmoto, K. (1983). Host-specific toxins, and chemical structures from *Alternaria* species, Annu. Rev. Phytopathol., 21:87–116.

Novacky, A., and Hanchey, P. (1974). Depolarization of membrane potentials in oat roots treated with victorin, Physiol. Plant Pathol., 4:161–165.

Noyes, R. D., and Hancock, J. G. (1981). Role of oxalic acid in the sclerotinia wilt of sunflower, Physiol. Plant Pathol., 18:123–132.

Nukina, M., and Marumo, S. (1976). Aversion factor, antibiotics among different strains of a fungal species. Aversion factors of *Cochliobolus setariae*, Agric. Biol. Chem., 40:2121–2123.

Oku, H., and Nakanishi, T. (1963). A toxic metabolite from *Ascochyta fabae* having antibiotic activity, Phytopathology, 53:1321–1325.

Orolaza, N. P., Lamari, L., and Ballance, G. M. (1995). Evidence of a host-specific chlorosis toxin from *Pyrenophora tritici-repentis*, the causal agent of tan spot of wheat, Phytopathology, 85:1282–1287.

Otani, H., Haramoto, M., Kohmoto, K., and Nishimura, S. (1985). Two different phases in host cell damages induced by AK-toxin of *Alternaria alternata* Japanese pear pathotype, J. Fac. Agric. Tottori Univ., 20:8–17.

Otani, H., Kohmoto, K., and Kodama, M. (1995). *Alternaria* toxins and their effects on host plants, Can. J. Bot., 73(Suppl. 1):S453–S458.

Otani, H., Kohmoto, K., and Nishimura, S. (1989). Action sites for AK-toxin produced by the Japanese pear pathotype of *Alternaria alternata*, Host-Specific Toxins: Recognition and Specificity Factors in Plant Disease (K. Kohmoto and R. D. Durbin, eds.), Tottori University, Japan, pp. 107–120.

Otani, H., Morikawa, M., Nishimura, S., and Kohmoto, K. (1977). Nature of specific susceptibility to *Alternaria kikuchiana* in Nijisseiki cultivar among Japanese pears VIII. Effects of other cations on losses of K^+ after exposure to AK-toxin, Ann. Phytopath. Soc. Jpn., 43:562–568.

Otani, H., Nishimura, S., Kohmoto, K., Yano, K., and Seno, T. (1975). Nature of specific susceptibility to *Alternaria kikuchiana* in Nijisseiki cultivar among Japanese pears (V). Role of host-specific toxin in early step of infection, Ann. Phytopath. Soc. Jpn., 41:467–476.

Outlaw, W. H. (1983). Current concepts on the role of potassium in stomatal movements, Physiol. Plant., 59:302–311.

Palmgren, M. G., Larsson, C., and Sommarin, M. (1990a). Proteolytic activation of the plant plasma membrane H^+-ATPase by removal of a terminal segment, J. Biol. Chem., 265:13423–13426.

Palmgren, M. G., and Sommarin, M. (1989). Lysophosphatidylcholine stimulates ATP-dependent protein accumulation in isolated oat root plasma membrane vesicles, Plant Physiol., 90:1009–1014.

Palmgren, M. G., Sommarin, M., Serrano, R., and Larsson, C. (1991). Identification of an autoinhibitory domain in the C-terminal region of the plant plasma membrane H^+-ATPase, J. Biol. Chem., 266:20470–20475.

Palmgren, M. G., Sommarin, M., Ulvskov, P., and Jorgensen, P. L. (1988). Modulation of plasma membrane H^+-ATPase from oat roots by lysophosphatidylcholine, free fatty acids and phospholipase A_2, Physiol. Plant., 74:11–19.

Palmgren, M. G., Sommarin, M., Ulvskov, P., and Larsson, C. (1990b). Effect of detergents on the H^+-ATPase activity of inside-out and right side-out plant plasma membrane vesicles, Biochim. Biophys. Acta, 1021:133–140.

Panaccione, D. G., Scot-Craig, J. S., Pocard, J. A., and Walton, J. D. (1992). A cyclic peptide synthetase gene required for pathogenicity of the fungus *Cochliobolus carbonum* on maize, Proc. Natl. Acad. Sci. USA, 89:6590–6594.

Paquin, R., and Waygood, E. R. (1957). The effect of *Fusarium* toxins on the enzymatic activity of tomato hypocotyl mitochondria, Can. J. Bot., 35:207–218.

Park, P., Lee, S. S., Ohno, T., Tsuge, T., and Nishimura, S. (1992). Slightly damaged type of plasma membrane modification in strawberry leaves treated with AF-toxin I produced by *Alternaria alternata* strawberry pathotype, Ann. Phytopath. Soc. Japan, 58:431–445.

Park, P., Nishimura, S., Kohmoto, K., Otani, H., and Scheffer, R. P. (1981a). Comparative effects of host-specific toxins from four pathotypes of *Alternaria alternata* on the ultrastructure of host cells, Ann. Phytopathol. Soc. Jpn., 47:488–500.

Park, P., Nishimura, S., Kohmoto, K., Otani, H., and Tsujimoto, K. (1981b). Two action sites of AM-toxin I produced by apple pathotype of *Alternaria alternata* in host cells: an ultrastructural study, Can. J. Bot., 59:301–310.

Park, P., Ohno, T., Nishimura, S., Kohmoto, K., and Otani, H. (1987). Leakage of sodium ions from plasma membrane modification, associated with permeability changes, in host cells treated with a host-specific toxin from a Japanese pear pathotype of *Alternaria alternata*, Can. J. Bot., 65:330–339.

Park, S. H., Stierle, A., and Strobel, G. A. (1994). Metabolism of maculosin, a host-specific phytotoxin produced by *Alternaria alternata* on spotted knapweed (*Centaurea macuolosa*), Phytochemistry, 35:101–106.

Peterson, P. A., Flavell, R. B., and Barratt, D. H. P. (1975). Altered mitochondrial membrane activities associated with cytoplasmically-inherited disease sensitivity in maize, Theor. Appl. Genet., 45:309–314.

Pijut, P. M., Lineberger, R. D., Domir, S. C., Ichida, J. M., and Krause, C. R. (1990). Ultrastructure of cells of *Ulmus americana* cultured in vitro and exposed to the culture filtrate of *Ceratocystis ulmi*, Phytopathology, 80:764–767.

Procacci, R., and Graniti, A. (1966). Azione della fusioccina, tossina prodotta da *Fusicoccum amygdali* Del., silla transpirazione di foglie di pomodoro, Phytopathol. Medit., 5:149–150.

Proctor, R. H., Hohn, T. M., and McCormick, S. P. (1995). Reduced virulence of *Gibberella zeae* caused by disruption of a trichothecene toxin biosynthetic gene, Mol. Plant–Microbe Interact., 8:593–601.

Ransom, R. F., Hipskind, J., Leite, B., Nicholson, R. L., and Dunkle, L. R. (1992). Effects of elicitor from *Colletotrichum graminicola* on the response of sorghum to *Periconia circinata* and its pathotoxin, Physiol. Mol. Plant Pathol., 41:75–84.

Ransom, R. F., Wilder, J., and Dunkle, L. D. (1994). Purification and distribution of pathotoxin-enhanced proteins in sorghum, Physiol. Mol. Plant Pathol., 45:385–395.

Rao, D. V., and Tewari, J. P. (1987). Production of oxalic acid by *Mycena citricolor*, causal agent of the American leaf spot of coffee, Phytopathology, 77:780–785.

Rasi-Caldogno, F., and Pugliarello, M. C. (1985). Fusicoccin stimulates the H^+-ATPase of plasmalemma in isolated membrane vesicles from radish, Biochem. Biophys. Res. Commun., 133:280–285.

Rasmussen, J. B., and Scheffer, R. P. (1988). Isolation and biological activities of four selective toxins from *Helminthosporium carbonum*, Plant Physiol., 86:187–191.

Rayle, D. L., and Cleland, R. E. (1992). The acid growth theory of auxin-induced cell elongation is alive and well, Plant Physiol., 99:1271–1274.

Robeson, D. J., and Strobel, G. A. (1984). Zinniol induces chlorophyll retention in barley leaves: the selection action of a non host-specific phytotoxin, Phytochemistry, 23:1597–1599.

Rottmann, W. H., Brears, T., Hodge, T. P., and Lonsdale, D. M. (1987). A mitochondrial gene is lost via homologous recombination during reversion of CMS T maize to fertility, EMBO J., 6:1541–1546.

Samaddar, K. R., and Scheffer, R. P. (1971). Early effects of *Helminthosporium victoriae* toxin on plasma membranes and counteraction by chemical treatments, Physiol. Plant Pathol., 1:319–328.

Scheffer, R. P., and Pringle, R. B. (1964). Uptake of *Helminthosporium victoriae* toxins by oat tissue, Phytopathology, 54:832–835.

Scherer, G. F. E., and Andre, B. (1989). A rapid response to a plant hormone: auxin stimulates phospholipase A_2 in vivo and in vitro, Biochem. Biophys. Res. Commun., 163:111–117.

Scherer, G. F. E., Andre, B., and Martiny-Baron, G. (1990). Hormone-activated phospholipase A2 and lysophospholipid-activated protein kinase: a new signal transduction chain and a new second messenger system in plants? Curr. Top. Plant Biochem. Physiol., 9:190–218.

Scherer, G. F. E., Martiny-Baron, G., and Stoffel, B. (1988). A new set of regulatory molecules in plants: a plant phospholipid similar to platelet-activating factor stimulates protein kinase and protontranslocating ATPase in membrane vesicles, Planta, 175:241–253.

Schipper, A. L. Jr. (1975). *Hypoxylon mammatum* pathotoxin necessary for the infection of aspen by *Hypoxylon mammatum*, Proc. Am. Phytopathol. Soc., 2:46–47.

Schlosser, E. (1971). The *Cercospora beticola* toxin, Phytopathol. Medit., 10:154–158.

Schroeder, J. L., Raschke, K., and Neher, E. (1988). Voltage dependence of K^+ channels in guard cell protoplasts, Proc. Natl. Acad. Sci. USA, 85:621–624.

Schroter, H., and Novacky, A. (1985). Effect of *Helminthosporium sacchari*-toxin on cell membrane potential of susceptible sugarcane, Physiol. Plant Pathol., 26:165–174.

Schulz, S., Oelgemoller, E., and Weiler, E. W. (1990). Fusicoccin action in cell-suspension cultures of *Corydalis sempervirens* Pers., Planta, 183:83–91.

Sednina, G. V., Aksenova, V. A., and Khartina, G. A. (1981). Changes in lipid composition of mitochondrial membranes in kidney bean infected with *Rhizoctonia solani*, Fiziol. Rasten., 28:1030–1036.

Seidel, H. (1961). Untersuchunger uber den Nahrstoffbedarf und die Toxinbildung des Pilzes *Phytophthora infestans* (Mont.) de Bary in vollsynthetischen Nahrlosungen, Phytopathol. Z., 41:1–26.

Serrano, R. (1990). Plasma membrane ATPase, The Plant Plasma Membrane—Structure, Function and Molecular Biology (C. Larsson, and I. M. Moller, eds.), Springer-Verlag, Heidelberg, pp. 127–153.

Shahin, E. A., and Spivey, R. (1986). A single dominant gene for *Fusarium* wilt resistance in protoplast-derived tomato plants, Theor. Appl. Genet., 73:164–169.

Shimomura, N., Otani, H., Tabira, H., Kodama, M., and Kohmoto, K. (1991). Two primary action sites for AM-toxin produced by *Alternaria alternata* apple pathotype and their pathological significance, Ann. Phytopathol. Soc. Jpn., 57:247–255.

Siler, D. J., and Gilchrist, D. G. (1982). Determination of host-selective phytotoxins from *Alternaria alternata* f.sp. *lycopersici* as their maleyl derivatives by high-performance liquid chromatography, J. Chromatogr., 238:167–173.

Siler, D. J., and Gilchrist, D. G. (1983). Properties of host specific toxins produced by *Alternaria alternata* f.sp. *lycopersici* in culture and in tomato plants, Physiol. Plant Pathol., 23:265–274.

Smith, G. R., Munro, M. H. G., Fineran, B. A., and Cole, A. L. J. (1994). Evidence for the involvement of ascochitine in *Phoma* leafspot-wilt disease of *Clematis*, Physiol. Mol. Plant Pathol., 45:333–348.

Sol, H. H. (1965). Alteration in the susceptibility of *Vicia faba* to *Botrytis fabae* by various pretreatments of the leaves, Neth. J. Plant Pathol., 71:196–202.

Sol, H. H. (1967). The influence of different nitrogen sources on (1) the sugars and amino acids leached from leaves, and (2) the susceptibility of *Vicia faba* to attack by *Botrytis fabae*, Meded. Fac. Landbouwwet. Rijksuniv. Gent., 32:768–775.

Spencer-Philipps, P. T. N., and Gay, J. L. (1981). Domains of ATPase in plasma membranes and transport through infected plant cells, New Phytol., 89:393–400.

Spikes, J. D. (1977). Photosensitization, The Science of Photobiology, (K. C. Smith, ed.), Plenum Press, New York, pp. 87–112.

Steinkamp, M. P., Martin, S. S., Hoefert, L. L., and Ruppel, E. G. (1979). Ultrastructure of lesions produced by *Cercospora beticola* in leaves of *Beta vulgaris*, Physiol. Plant Pathol., 15:13–26.

Steinkamp, M. P., Martin, S. S., Hoefert, L. L., and Ruppel, E. G. (1981). Ultrastructure of lesions produced in leaves of *Beta vulgaris* by cercosporin, a toxin from *Cercospora beticola*, Phytopathology, 71:1278–1281.

Stoessl, A. (1981). Structure and biogenic relations: fungal nonhost specific, Toxins in Plant Disease (R. D. Durbin, ed.), Academic Press, New York, pp. 109–219.

Stone, H. E., and Armentrout, V. N. (1985). Production of oxalic acid by *Sclerotium cepivorum* during infection of onion, Mycologia, 77:526–530.

Stout, R. G., and Cleland, R. E. (1980). Partial characterization of fusicoccin binding to receptor sites on oat root membranes, Plant Physiol., 66:353–359.

Strobel, G. A. (1973a). The helminthosporoside-binding protein of sugarcane. Its properties and relationship to susceptibility to the eyespot disease, J. Biol. Chem., 248:1321–1328.

Strobel, G. A. (1973b). Biochemical basis of the resistance of sugarcane to eyespot disease, Proc. Natl. Acad. Sci. USA, 70:1693–1696.

Strobel, G. A., and Hess, W. M. (1974). Evidence for the presence of the toxin-binding protein on the plasma membrane or sugarcane cells, Proc. Natl. Acad. Sci. USA, 71:1413–1417.

Strobel, G. A., Steiner, G. W., and Byther, R. (1975). Deficiency of toxin-binding protein in mutants of sugarcane clone H54-775 as it relates to disease resistance, Biochem. Genet., 13:557–565.

Tal, B., Robeson, D. J., Burke, B. A., and Aasen, A. J. (1985). Phytotoxins from *Alternaria helianthi*: radicinin, and the structures of deoxyradicinol and radianthin, Phytochemistry, 24:729–731.

Tamari, K., Ogasawara, N., Kaji, J., and Togashi. (1966). On the effect of a piricularin-detoxyfying substance, ferulic acid, on the tissue resistance of rice plants to the blast-fungal infection, Ann. Phytopathol. Soc. Jpn., 32:186–193.

Thatcher, F. S. (1942). Further studies of osmotic and permeability relations in parasitism, Can. J. Res., C20:283–311.

Thatcher, F. S. (1943). Cellular changes in relation to rust resistance, Can. J. Res., C21:151–172.

Tietjen, K. G., Schaller, E., and Matern, U. (1983). Phytotoxins from *Alternaria carthami* Chowdhury: structural identification and physiological significance, Physiol. Plant Pathol., 23:387–400.

Tipton, C. L., Paulsen, P. V., and Betts, R. E. (1977). Effects of ophiobolin A on ion leakage and hexose uptake by maize roots, Plant Physiol., 59:907–910.

Tomas, A., and Bockus, W. W. (1987). Cultivar-specific toxicity of culture filtrates of *Pyrenophora tritici-repentis*, Phytopathology, 77:1337–1340.

Tomas, A., Feng, G. H., Reeck, G. R., and Bockus, W. W. (1990). Purification of a cultivar-specific toxin from *Pyrenophora tritici-repentis*, causal agent of tan spot of wheat, Mol. Plant–Microbe Interact., 3:221–224.

Traylor, E. A., Dunkle, L. D., and Schertz, R. F. (1988). Pathotoxin-induced alteration in protein synthesis associated with susceptibility of sorghum to milo disease, Crop Sci., 28:615–617.

Traylor, E. A., Shore, S. H., Ransom, R. F., and Dunkle, L. D. (1987). Pathotoxin effects in sorghum are also produced by mercuric chloride treatment, Plant Physiol., 84:975–978.

Tseng, T. C., and Bateman, D. F. (1969). A phosphatidase produced by *Sclerotium rolfsii*, Phytopathology, 59:359–363.

Turgeon, B. G., Kodama, M., Yang, G., Rose, M. S., Lu, S. W., and Yoder, O. C. (1995). Function and chromosomal location of the *Cochliobolus heterostrophus Tox* 1 locus, Can. J. Bot., 73(Suppl. 1):S1071–S1076.

Turner, N. C., and Graniti, A. (1969). Fusicoccin: a fungal toxin that opens stomata, Nature, 223:1070–1071.

Umetsu, N., Kaji, J., and Tamari, K. (1972). Investigation on the toxin production by several blast fungus strains and isolation of tenuazonic acid as a novel toxin, Agr. Biol. Chem., 36:859–866.

Utsumi, R., Hadama, T., Noda, M., Toyoda, H., Hashimoto, H., and Ouchi, S. (1989). Cloning of fusaric acid-detoxifying gene from *Cladosporium werneckii*: a new strategy for the prevention of plant diseases, J. Biotechnol., 8:311–316.

Vidhyasekaran, P. (1993a). Molecular sieve to select cells in tissue culture to develop disease resistant plants, Genetic Engineering, Molecular Biology and Tissue Culture for Crop Pest and Disease Management (P. Vidhyasekaran, ed.), Daya Publishing House, Delhi, India, pp. 295–309.

Vidhyasekaran, P. (1993b). Principles of Plant Pathology, CBS Publishers, Delhi, India, pp. 166.

Vidhyasekaran, P., Borromeo, E. S., and Mew, T. W. (1986). Host-specific toxin production by *Helminthosporium oryzae*, Phytopathology, 76:261–266.

Waite, M. (1985). Phospholipases, Biochemistry of Lipids and Membranes (D. E. Vance and J. E. Vance, eds.), Benjamin/Cummings, California, pp. 299–324.

Walton, J. D. (1987). Two enzymes involved in biosynthesis of the host-selective phytotoxin HC-toxin, Proc. Natl. Acad. Sci. USA, 84:8444–8447.

Walton, J. D. (1991). Genetics and biochemistry of toxin synthesis in *Cochliobolus* (*Helminthosporium*), Molecular Industrial Mycology (S. Leong and R. M. Berka, eds.), Marcel Dekker, New York, pp. 225–249.

Walton, J. D., and Earle, E. D. (1983). The epoxide in HC-toxin is required for activity against susceptible maize, Physiol. Plant Pathol., 22:371–376.

Walton, J. D., Earle, E. D., and Gibson, B. W. (1982). Purification and structure of the host-specific toxin from *Helminthosporium carbonum*, Biochem. Biophys. Res. Commun., 107:785–794.

Walton, J. D. and Holden, F. R. (1988). Properties of two enzymes involved in the biosynthesis of the fungal pathogenicity factor HC-toxin, Mol. Plant–Microbe Interact., 1:128–134.

Wevelsiep, L., Kogel, K. H., and Knogge, W. (1991). Purification and characterization of peptides from *Rhynchosporium secalis* inducing necrosis in barley, Physiol. Mol. Plant Pathol., 39:417–482.

Wevelsiep, L., Rupping, E., and Knogge, W. (1993). Stimulation of barley plasmalemma H^+-ATPase by phytotoxic peptides from the fungal pathogen *Rhynchosporium secalis*, Plant Physiol., 101:297–301.

Wheeler, H., and Black, H. S. (1963). Effects of *Helminthosporium victoriae* and victorin upon permeability, Am. J. Bot., 50:686–693.

Wheeler, H., and Hanchey, P. (1968). Permeability phenomena in plant disease, Annu. Rev. Phytopathol., 16:331–350.

Wise, R. P., Fliss, A. E., Pring, D. R., and Gengenbach, B. G. (1987a). Urf 13-T of cytoplasm maize mitochondria encodes a 13 kD polypeptide, Plant Mol. Biol., 9:121–126.

Wise, R. P., Pring, D. R., and Gengenbach, B. G. (1987b). Mutation to male fertility and toxin insensitivity in Texas (T)-cytoplasm maize as associated with a frameshift in a mitochondrial open reading frame, Proc. Natl. Acad. Sci. USA, 84:2858–2862.

Witsenboer, H. M. A., Schaik, C. F., van Bino, R. J., Loffler, H. J. M., Nijkamp, H. J. J., and Hille, J. (1988). Effects of *Alternaria alternata* f.sp. *lycopersici* toxins at different levels of tomato plant cell development, Plant Sci., 56:253–260.

Wolpert, T. J., and Dunkle, L. D. (1983). Alterations in gene expression in sorghum induced by the host-specific toxin from *Periconia circinata*, Phytopathology, 70:872–876.

Wolpert, T. J., and Macko, V. (1989). Specific binding of victorin to a 100-kDa protein from oats, Proc. Natl. Acad. Sci. USA, 86:4092–4096.

Wolpert, T. J., and Macko, V. (1991). Immunological comparison of the in vitro and in vivo labeled victorin binding protein from susceptible oats, Plant Physiol., 95:917–920.

Wolpert, T. J., Macko, V., Acklin, W., and Arigoni, D. (1988). Molecular features affecting the biological activity of the host-specific toxin from *Cochliobolus victoriae*, Plant Physiol., 88:37–41.

Wolpert, T. J., Macko, V., Acklin, W., Jaun, B., and Arigoni, D. (1986). Structure of the minor host-selective toxins from *Cochliobolus victoriae*, Experientia, 42:1296–1299.

Wolpert, T. J., Navarre, D. A., Lorang, J. M., and Moore, D. L. (1995). Molecular interactions of victorin and oats, Can. J. Bot., 73(Suppl. 1):S475–S482.

Wolpert, T. J., Navarre, D. A., Moore, D. L., and Macko, V. (1994). Identification of the 100-kD victorin binding protein from oats, Plant Cell, 6:1145–1155.

Xiao, J. Z., Nishimura, S., and Tsuda, M. (1989). Ophiobolins produced by *Helminthosporium oryzae* and their pathological role, Ann. Phytopathol. Soc. Jpn., 55:477.

Xiao, J. Z., Tsuda, M., Doke, N., and Nishimura, S. (1991). Phytotoxins produced by germinating spores of *Bipolaris oryzae*, Phytopathology, 81:58–64.

Xiao, J. Z., Tsuge, T., and Doke, N. (1992). Further evaluation of the significance of BZR-toxin produced by *Bipolaris zeicola* race 32 in pathogenesis on rice and maize plants, Physiol. Mol. Plant Pathol., 40:359–370.

Yamamoto, M., Namiki, F., Nishimura, S., and Kohmoto, K. (1985). Studies on host-specific AF toxins produced by *Alternaria alternata* strawberry pathotype causing alternaria black spot of strawberry (3): use of toxin for determining inheritance of disease reaction in strawberry cultivar Morioka-16, Ann. Phytopathol. Soc. Jpn., 51:530–535.

Yamamoto, M., Nishimura, S., and Kohmoto, K., Otani, H. (1984). Studies on host-specific AF-toxins produced by *Alternaria alternata* strawberry (2). Role of toxins in pathogenesis, Ann. Phytopathol. Soc. Jpn., 50:610–619.

Yamazaki, S., and Ogawa, T. (1972). The chemistry and stereochemistry of cercosporin, Agric. Biol. Chem., 36:1707–1718.

Yoder, O. C. (1972). Host-specific toxins as determinants of successful colonization by fungi, Phytotoxins in Plant Diseases (R. K. S. Wood, A. Ballio, and A. Graniti, eds.), Academic Press, London, pp. 457–463.

Yoder, O. C. (1980). Toxins in pathogenesis, Annu. Rev. Phytopathol., 18:103–129.

Index